# ANIMAL PARASITES

# ANIMAL PARASITES
## Their Life Cycles and Ecology

Third Edition

## O. Wilford Olsen, B.A., M.A., Ph.D.

Emeritus Head and Professor,
Department of Zoology and Entomology,
Colorado State University,
Fort Collins, Colorado

Parasitologist,
Agricultural Research Service,
U. S. Department of Agriculture

**Illustrations by the author**

**UNIVERSITY PARK PRESS**
Baltimore • London • Tokyo

To my students of the past who have inspired me to learn how to teach parasitology in a more interesting and informative manner, as well as to those of the future who may rightly expect similar treatment of this fascinating subject, and to my wife, Ione Palfreyman Olsen, who encouraged me to do it for them.

**UNIVERSITY PARK PRESS**
International Publishers in Science and Medicine
Chamber of Commerce Building
Baltimore, Maryland 21202

Printed in the United States of America

**Library of Congress Cataloging in Publication Data**

Olsen, Oliver Wilford, 1901–
   Animal parasites.

   Includes bibliographies.
   1. Parasitology.   I. Title.
[DNLM:  1. Parasites.  QX4 052a 1973]
QL 757.04   1973     591.5'24    73-17411
ISBN 0-8391-0643-2

# Contents

# List of Plates

# Preface

The third edition of *Animal Parasites* follows the format of the preceding ones. Much new information has been added, together with a large number of text figures. Most of the chapters have been rewritten to include recent advances. All references are given as complete citations. This edition is intended as a text and reference book for all persons interested in parasitology or concerned with it. It is a text for students of biology and parasitology, as well as a reference for veterinarians, extension agents, wildlife personnel, and investigators in advanced aspects of parasitology, including the protozoans and helminths.

The book presents what I consider the basic principles of parasitology essential to an understanding of the subject. They include (1) basic classification for understanding relationships of common parasites, (2) general morphology of adults and larval stages for recognition, (3) patterns of life cycles of the different groups for comprehension of how they live, (4) general ecological requirements of all stages of the parasites in the physical and biological environments, and (5) means of transmission of all stages.

An understanding of these fundamentals provides sound background for a career in applied parasitology or for investigating additional important and practical aspects of the subject. These include ecology of the parasites in the physical and biological habitats occupied by them; physiology of all stages of the parasites in their various microhabitats; investigation of ultrastructure of the different stages to recognize the organelles and the significance of their roles; nutrition of the various stages and how it is achieved; development of immune processes and their roles in limiting parasites in the hosts; host-specificity and the nature of its operation; geographic distribution of the parasites and their intermediate hosts; seasonal distribution of parasites; and, finally, the application of the great body of biological information to the development of methods of controlling parasites to prevent economic loss, morbidity, disease, suffering, and death.

The life cycles presented here have been chosen because they are representative types, are well known for the most part, and generally represent parasites that the students find in animals that they know and see about them. In selecting parasites from common hosts whose habits and ecological environments are familiar, students are able to understand and appreciate better the biological problems associated with the phenomenon of parasitism.

Parasites of humans are not excluded, but neither are they emphasized. Common species that are likely to occur, such as pinworms, entamoebas, taenias, and ascarids, are included. Other important species, such as the filarioids, medina worm, hookworms, and others known from man and treated extensively in texts on clinical parasitology, are represented here by closely related species found in domestic and wild animals of this country.

There is no extensive treatment of physiology, immunology, pathology, and medication. In cases where reference is made to these subjects, it pertains to the species under consideration and represents basic aspects of special interest.

Grateful acknowledgment is made to Dr. John L. Olsen from the Department of Pathology, College of Veterinary Medicine and Biomedical Sciences, Colorado State

University, and the Division of Veterinary Research, Food and Drug Administration, Agricultural Research Center, for his skillful and critical review of the manuscript, with timely comments.

Appreciation is expressed to the personnel of the Colorado State University Library for their willing and able assistance in obtaining literature and to the publishers of books and journals for permitting use of their printed material.

Special acknowledgment is made to Dr. Charles G. Wilber, Chairman, Department of Zoology and Entomology, Colorado State University, and to the University for providing facilities, space, and encouragement for conducting the work.

Finally, the author extends his gratitude to members of the staff of University Park Press, especially to Mrs. Frannie Levy, for their assistance, skill, and taste in producing a book that is pleasing and artistic in appearance and usable in its design.

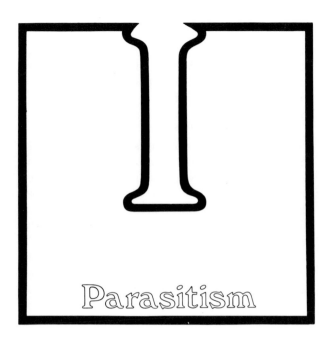

# Parasitism

Animal parasitism is a way of life in which one species, the *parasite*, living in or on another species, the *host*, gains its livelihood at the expense of the latter. The host furnishes both the habitat and the food for the parasites which are physiologically dependent on it for life. Moreover, the parasite always does damage in some degree to its host.

Parasitology is the study of parasitism. It includes the morphology, classification, biology, and physiology of the parasites. In addition to these, it involves the relationships between the parasites and their hosts, as well as the reactions toward each other. Its purpose is to lead to a fuller understanding of these relationships and the results of them on both the host and the parasite.

## Development of Parasitism

As phylogenetic groups of animals appeared in the beginning of our planet, they spread throughout the world, occupying all the available ecological niches of the physical environment. Their bodies, both inside and outside, constituted new ecological biohabitats ready for occupancy by those species that possessed the potential and capability of adapting to them. Many different phylogenetic groups of animals invaded this new living habitat but few were capable of adapting to it in a large measure of success. The protozoa, the helminths, and some of the arthropods were the most successful. They constitute the important groups of parasites known today. This biotic association is one of *symbiosis* in which animals live together in varying degrees of dependency between the *host* and the *symbiont*.

**Symbiotic Relationships.** Three degrees of symbiosis are generally recognized. They are *mutualism*, *commensalism*, and *parasitism*.

*Mutualism* constitutes one type of relationship in which the host and the symbionts are physiologically dependent upon each other and the relationship is mutually beneficial. Termites and their intestinal protozoa are an example. The termites provide the habitat and the food in the form of wood (cellulose) which they cannot digest. The protozoa in the intestinal habitat, however, are capable of hydrolyzing the wood for their own and the termites' use. Ruminants and other herbivores with their rich flora of bacteria and fauna of protozoa are additional examples of mutualism wherein both the host and the symbionts are physiologically dependent on each other.

*Commensalism* is the condition in which the host provides the habitat and food for its symbionts which live without benefit or harm to it. The symbionts, however, are physiologically dependent on the host for their existence. The host, on the other hand, is not dependent on them. Certain of the protozoa living in the alimentary canal of man or on the bodies of hydra are examples of commensals. Other examples of commensalism appear among the marine animals, particularly sea anemones and crabs.

*Parasitism* is that relationship in which the symbiont is physiologically dependent on the host for its habitat and sustenance and at the same time may be harmful to it. All of the trematodes, cestodes, acanthocephalans, and many of the protozoa and nematodes are examples of true animal parasites.

## Kinds of Parasites

The degree to which parasites are dependent on their hosts ranges from intermittent visits for food, as in the case of mosquitoes, to one of complete dependence inside the body with no free-living stage, as occurs in the plasmodia and trichinella. Several groups of parasites are recognized, depending on their relationship to the host.

Location *on* or *in* the body of the host serves as one basis for dividing parasitic animals into two groups. *Ectoparasites* live on the external surface of the body of the host or in cavities that open directly onto the surface. They include monogenetic trematodes, lice, mites, and ticks. *Endoparasites* live in the bodies of the hosts, occurring in the alimentary canal, lungs, liver, and other organs, tissues, cells, and body cavity. Examples are tapeworms, digenetic trematodes, nematodes, and protozoa.

The amount of time spent on or in the host serves as temporal basis for dividing parasites into two major groups. *Temporary parasites* visit the host for food. Having satisfied their hunger, they leave. Bloodsucking arthropods and leeches are examples. *Stationary parasites* spend a definite period of development on or in the body of the host. They may be divided into groups, according to the amount of time spent with the host. Those which remain with the host for only a part of their development and then leave to complete it and continue a nonparasitic life are known as *periodic parasites*. Botflies and mermithid nematodes represent this group. Parasites that spend their entire existence in hosts except for the times they occur free while transferring from one host to another are designated as *permanent parasites*. The trematodes, cestodes, acanthocephalans, nematodes, and protozoa are examples.

Parasites found in unusual hosts, or unusual places in normal hosts, are designated by terms indicating the nature of their abnormalities. *Incidental parasites* are those which occasionally appear in unusual hosts under natural conditions. The double-pored tapeworm of dogs is found sometimes in children, or the common liver fluke of sheep occurs in dogs or cats. *Erratic* or *aberrant parasites* are individuals of a species that wander into unusual places in the normal host. Ascarids of swine and man may wander from the intestine into the liver, body cavity, or nostrils.

All of the parasites listed in these categories are *obligate parasites*. They are unable to exist without some degree of development on or in the host. A few normally free-living animals, on the other hand, are able to exist for short periods in the bodies of other animals when accidentally introduced into them. They are spoken of as *facultative parasites*. Representatives of them are certain free-living nematodes of the genera

*Rhabditis* and *Turbotrix*, and fly maggots such as those of the cheese skippers (*Piophila casei*) and some of the blowflies.

### Properties of Parasitism

The basic properties of parasitism appear in the many adaptations necessary for the parasite to live in its biotic environment, to reproduce, and to infect new hosts. These interactions between the two living organisms give rise to both consequences and specializations for each participant, which become fixed through inheritance.

**Adaptations for Parasitic Life.** The basic physiological requirements of a parasite are similar to those of free-living animals. They are habitat, food, and reproduction. The problems in achieving these requirements under the conditions of parasitism are complex, and special adaptations have evolved to meet them. Probably the one basic underlying characteristic of parasitism is that of adaptation. One writer has called it the hallmark of parasitism.

In order to live on or in a host, the parasite must evolve structures for adhering to it. Such adaptations appear in the tarsi of Anoplura for holding on to hairs. Monogenetic trematodes and acanthocephalans have rigid hooks for attaching to the host. Suckers for the same purpose are highly developed by trematodes and cestodes.

Living in a host necessitates means of leaving it in order to reach new ones. Parasites of the alimentary canal, lungs, liver, and reproductive system utilize the natural outlets of these organ systems as avenues of exit for cysts or eggs. Those living in the bloodstream and tissues generally utilize other animals or means to leave their hosts. Bloodsucking arthropods serve as the route by which malaria and related protozoans and the microfilariae of filarioid nematodes escape from the body and in which further development occurs. Dracunculid worms in the subcutaneous tissues vesiculate the epidermis, forming openings through which the female releases active larvae into the water. Others (trichinella, cestodes, and trematodes) depend on the digestive processes of predators or scavengers which serve as definitive hosts to release them from the tissues of their intermediate hosts. *Capillaria hepatica* eggs in the liver of rodents may be freed by postmortem decomposition of the host or digestion of it by another animal, after which development of them proceeds under favorable conditions in the physical environment.

Means of survival and development are essential during the interval of transfer from one definitive host to the next. This transfer involves a period of development in the soil or water, in the case of parasites having a *direct* life cycle, or development in the body of one or more intermediate hosts with those having an *indirect* cycle. In parasites with a direct life cycle, protective cysts, thick egg shells, or the retained cuticle of larvae are adaptations to protect the stages free in the soil against the hazards of desiccation and freezing. Species having an indirect life cycle may depend on some of the features listed above for protection during the time of transfer between their several hosts. But since practically all of their development takes place in the body of one or more intermediate hosts, their adaptational requirements must be adjusted to the biotic environment of the bodies of several species, often from different phyla. An example is the lancet fluke, *Dicrocoelium dendriticum*, which lives and develops successively in sheep, snails, and ants.

Transmission of the infective stage of the parasite to the next host in the developmental cycle is accomplished by one of three methods. They are *passive, active,* or *inoculative.* Passive transmission occurs when the infective stages of the parasites contaminate or infect the food or water of the host and are swallowed with them. Examples are the eggs of ascarids, cysts of *Entamoeba*, and larvae of trichostrongyles. Infection of food occurs in the larval stages of trematodes, cestodes, acanthocephalans, and many nematodes. Active transfer occurs in the hookworms, and the miracidia and cercariae of trematodes. These parasites actively penetrate the bodies of their hosts upon coming in contact with them. Often responses such as thermotropic, geotropic, as in the case of hookworm larvae, or chemotropic, as occurs in miracidia toward their snail hosts, aid in bringing the host and parasite together. Inoculative transmission occurs when the infective stage of the parasite has developed in the body of a bloodsucking arthropod, as with *Plasmodium* in mosquitoes. Transfer back to the vertebrate host is accomplished when the arthropod inoculates the parasites into the host while feeding upon it.

Survival within the host is dependent upon the ability of the parasites to withstand the destructive action of the digestive juices and the immunological reactions of the host against them, or to reach microhabitats within the host where the required nutrients for growth and reproduction are available in adequate amounts.

The most successful parasites have evolved a biotic potential of great capacity in order to compensate for the tremendous losses of eggs or larvae, or both, incurred in the completion of their complicated life cycles. This is accomplished by an increased production of eggs, as in the ascarids and cestodes, the duplication of sex organs in segments, as in cestodes, or vegetative reproduction extending over long periods of time such as occurs in the sporocysts or rediae of trematodes and scolices in cestodes (*Echinococcus* and *Multiceps*).

**Specific Host-Parasite Relationships.**   Parasites normally do not infect different species of animals at random under natural conditions but show varying degrees of preference for hosts and for habitats within them. Thus parasites of horses, cattle, dogs, or humans are most likely to be found in their respective hosts. Moreover, parasites of the intestine, the liver, or tissues occur with marked regularity in these sites. This condition is designated as *host specificity, organ specificity,* or *tissue specificity,* as the case may be.

**Specificity of Host-Parasite Relationships.** Specificity of host-parasite relationships is determined by the success of the parasite to invade, occupy, and reproduce in certain microhabitats inside or on the outside of the bodies of hosts. The factors involved in the development of host specificity among parasites are opportunities for contact between them and their hosts, followed by entrance, adaptation, establishment, nourishment, and reproduction in them.

Opportunities for infection of the host are present only when it and the parasite come in contact with each other under favorable conditions. This occurs during periods when developmental, kinetic, or behavioral activities of the parasite and the host bring them together.

Ability to invade the host and survive in it involves the capabilities, requirements, and susceptibilities of the parasite. Structural properties of the integument may be beyond the capability of the parasite to penetrate, thereby serving as a barrier. The physiological properties of the tissues or organs may not meet the requirements of the parasite, making growth and development impossible. Susceptibility to antibodies produced by the host in response to the presence of the parasite may prove lethal to it or produce an environment unfavorable to growth and reproduction.

Chemotropic response by parasites to substances elaborated by hosts leads to host, organ, or tissue specificity. The miracidia of trematodes are attracted to their specific snail hosts by certain substances in the mucus secretions. Parasites in the body of hosts are directed to specific organs or tissues through their chemotropic responses to them. Young liver flukes, *Fasciola hepatica*, *Dicrocoelium dendriticum*, and *Clonorchis sinensis*, liberated from their cysts in the small intestine of the definitive host, find their way to the bile ducts, each by its own route, doubtless being directed by different stimuli originating in the liver. In cases where distribution is directed by chemotropic responses, parasites accumulate in specific hosts or tissues within them. In the absence of such responses, they occur at random in species or tissues. The cercariae of echinostome flukes encyst in a number of species of snails, and trichinella larvae may be found in the skeletal muscles of many species of mammals.

A single species of parasite may be limited to one species of host, as in the case of the beef tapeworm *Taeniarhynchus saginatum* in man. On the other hand, a species occurs in hosts as widely separated as orders, such as the trematode *Plagiorchis muris* in birds, rats, and humans.

Substances in certain organs or tissues become inimical to parasites and interfere with their development in them. After chicks become a few weeks old, the goblet cells of the mucosal lining of the intestine develop and secrete mucus which hinders the establishment and development of the larvae of *Ascaridia galli*.

The specific nutritional requirements of parasites must be satisfied if they are to succeed. These are developed through the evolutionary processes of host and parasite, resulting in host specificity.

## Effect of Parasitism on the Evolution of the Host Species

Parasitism may be considered as an evolutionary pressure in which the host and parasite adapt to each other through a selective process. Since parasitism has a deleterious effect on the host, it may be manifested in lowered vitality, reduced

rate of reproduction, slower growth, or death of infected individuals. This can result in extinction of a species or it can lead to a change in the population through the selection of resistant species or strains. Host adaptation to parasitism develops from selective pressure resulting in strains better adapted to resist or tolerate parasitism. This is accomplished through the evolution of antibody responses on the part of the host.

## Effect of Parasitism on Individual Host

The effect of parasitism on the individual host may be *injurious* or *defensive*. Injurious effects show a wide range of severity, leading to manifested disease often resulting in the death of the host. When parasites, such as hookworms in man or trichostrongyles in cattle and sheep, affect the entire population, the effect of the disease is masked and often not recognized. The less severe effects of parasitism are kept in repair by the host, and therefore are not readily detectable. The mechanisms of injury are mechanical, chemical, inflammatory, and the introduction of pathogens.

Mechanical injury involves destructive action such as perforation of an organ (ascarids, acanthocephalans), destruction of cells (coccidia, plasmodia), piercing tissues (whipworms, mosquitoes), chewing (Mallophaga), obstruction of a lumen (ascarids, cestodes), or the interference in transfer of foods across cell membranes (*Giardia*).

Chemical injury results from secretions by the parasites. Hookworms secrete substances from the cephalic glands that interfere with the blood. One of them is an anticoagulant that causes the blood to flow freely, even after the worms detach and move to new sites, and another depresses hematogenesis. *Cysticercus fasciolaris* of the cat tapeworm is carcinogenic in the liver of rats.

Parasites rob the host of essential nutrients. Trichostrongyles and hookworms feed on blood. Moreover, the injection of the anticoagulant factor by hookworms while feeding results in a great loss of blood through bleeding into the intestine. The broad fish tapeworm, *Diphyllobothrium latum*, of man absorbs vitamin $B_{12}$ from the tissues of the intestinal mucosa and stores it in its own body.

The introduction of pathogenic organisms such as bacteria (*Clostridium novyi*) by the common liver fluke into the liver of sheep, rickettsiae (*Neorickettsia helminthoeca*) into the tissues of the intestine by the salmon poisoning fluke of dogs, or protozoa (*Histomonas meleagridis*) through the eggs of the cecal nematode into the alimentary canal of turkeys results in a high mortality among these hosts.

Defensive reactions to invasion or attack by parasites are aspects of physiological or conscious responses by the host against them. Inflammatory responses of a general or local nature are often the first reaction to the presence of parasites. They are primarily cellular and are divided into relatively distinct but overlapping stages of 1) temporary localization and destruction of parasites, as in the case of the cercariae of blood flukes of birds that infect humans, causing papules; 2) walling off or encapsulation of the invaded area, as seen in infection by the large American liver fluke, *Fascioloides magna*, of Cervidae in cattle; and 3) repair of damaged tissue, as occurs in light infections of amebiasis. Antibodies may destroy, localize, neutralize, or interfere with the reproduction of parasites and thereby serve as secondary defenses. Conscious efforts are made by the host to avoid parasitism by fleeing, as cattle do from botflies, or bunching, in the case of sheep as a protective measure against nose bots.

## Evolutionary Effects of Parasitism

Once parasitism has been achieved, it proceeds by its very nature to exert a definite directive influence on the evolution of the host and the parasite. In the biotic environment, the parasite is isolated in the body of the host and confronted with the dynamic defensive mechanisms of it. The host, which constitutes the environment, is attacked and injured by the parasite. Improved relations on the part of each are indicated. In order to be successful, the evolution of parasitism must be toward better adaptation between the host and parasite. Any other course would lead to the ultimate destruction of one or both.

The direct force in achieving better adaptation between parasites and hosts is natural selection. The conditions for it are excellent. Through isolation of the parasites in the host, a population of genotypes of parasites is evolved that is adapted to meet more successfully the resistive efforts of the host. By the same selective forces, strains of genotypes of host animals appear whose antibody systems have evolved to cope more adequately with the destruction caused by the parasites in them. Thus, through the processes of isolation and natural selection, populations of parasites and hosts with specific tolerances for each other have developed.

## Life Cycles of Parasites

The life cycles of parasites are complex and the physical and biological requirements for completing them are manifold and exacting. In order to surmount the many adverse conditions encountered in the course of maintaining themselves, parasites have developed a great reproductive potential together with the means of protection against physical hazards. In spite of the great odds, sufficient numbers of them succeed to assure continuity of the species.

The life cycles fall into two basic types. They are 1) the direct one with only the definitive host, and 2) the indirect one with a definitive host and one or more intermediate hosts. Parasites with a direct life cycle have a free-living phase, except for a few species, during which they develop to the infective stage. Those with an indirect life cycle usually, though not always, have a free-living stage between some of the hosts.

The Protozoa have both types of life cycles among their members. In the Sporozoa, which have both types of life cycles, there are intermediate stages such as trophozoites, merozoites, sporozoites, and gametocytes which may occur in a single host in the case of the direct type of cycle or in two hosts in the indirect type (Plate 33).

Among the Trematoda, the Monogenea and most of the Aspidogastrea have the direct type of life cycle. The stage hatching from the egg is a larva that develops directly into the adult. Others of the Aspidogastrea and all of the Digenea have an indirect life cycle involving mollusks. In the Digenea, the egg produces a miracidium whose development in a mollusk produces the sporocyst, rediae, and cercariae. The cercariae, with the exception of several families of blood flukes, encyst on objects or in other animals and develop into metacercariae which are infective to the definitive host upon entering the alimentary canal (Plate 83).

In nearly all cases, the Cestoda have one or two intermediate hosts in the life cycle. The eggs hatch into a six- or ten-hooked larva. In the Proteocephala and Pseudophyllidea, there is a procercoid and plerocercoid larval stage, usually each in a different intermediate host (Plate 100). The Cyclophyllidea have three basic types of larvae in the intermediate hosts (Plates 92, 100). They are the cysticercoid, the dithyridium, and the vesicular types. The vesicular type has such variations as the cysticercus, strobilocercus, coenurus, and hydatid.

All species of Acanthocephala have an intermediate host. An acanthor hatches from the egg in the intestine of the intermediate host and develops into an acanthella and a cysticanth in succession. Species infecting aquatic vertebrates usually have aquatic invertebrates as the intermediate host; those in terrestrial vertebrates usually have terrestrial invertebrates as the intermediaries.

The Nemathelminthes have both a direct (Plate 141) and an indirect (Plate 142) type of life cycle. The larvae pass through a series of four molts to become adults. Some of the molts take place in the egg, in the free-living phases, and others in the intermediate and definitive hosts.

While these are basic patterns on which broad principles of the parasite life cycles can be based, the individual cycles in each of them often vary greatly within the fundamental plan. Some parasites, such as the sporozoans, trypanosomes, trichinella, and filarioids, with intermediate hosts, have no free-living stages. In others, such as *Probstmyria viviparus* and *Hymenolepis nana* (Plate 98), the eggs may hatch and the larvae develop to sexual maturity without ever leaving the definitive host.

The life cycle of each parasite presents its own peculiarities. Solution of it requires a broad knowledge of the physiology, biology, and ecology of the parasite and all of the hosts involved. The elucidation of the life cycle of a parasite, or, indeed, a part of it, is a distinct challenge and a gratifying achievement.

### SELECTED REFERENCES

Baer, J. G. (1951). Ecology of Animal Parasites. University of Illinois Press, Urbana, 224 pp.

Caullery, M. (1952). Parasitism and Symbiosis. Sidgewick and Jackson, Ltd., London, 340 pp.

Huff, C. G., L. O. Nolf, R. J. Porter, C. P. Read, A. G. Richards, A. J. Riker, and L. A. Stauber. (1958). An approach toward a course in the principles of parasitism. J. Parasitol. 44:28–45.

Kudo, R. R. (1966). Protozoology. 5th ed. Charles C Thomas, Springfield, Ill., 1174 pp.

Read, C. P. (1970). Parasitism and Symbiology: An introductory text. Ronald Press, New York, 316 pp.

Self, J. T. (1961). The biological significance of parasitism and its evolutionary accomplishments. Bios 32:51–61.

Whitlock, J. H. (1958). The inheritance of resistance to trichostrongyloidosis in sheep. I. Demonstration of the validity of the phenomena. Cornell Vet. 48:127–133.

### Ecology of Parasitism

Ecology refers to the relationship of an organism to its environment for essentials necessary for the life processes. In the case of parasitic animals, the requirements for development and survival in the physical environment outside their hosts include favorable temperature, adequate moisture, sufficient oxygen, and, in some parasitic stages, nutriment. In addition, there must be protection from the lethal effects of freezing, heat from direct sunlight, and desiccation. For some species, the physical environment provides stimuli for 1) hatching of the eggs of many trematodes and nematodes, 2) direction of movement, such as cercariae swimming to or away from light, 3) periodic activity, shown by trichostrongyles that migrate up and down grass blades in response to light intensity, diurnal periodicity of microfilariae, and time of day when cercariae leave their snail hosts, and 4) molting of first and second stage strongyles and trichostrongyles.

The internal, or biological, environment in or on the bodies of host animals provides parasitic animals with stimuli for hatching of eggs and growth of larval stages, nutriment, site-finding, and a habitat favorable for reproduction. The biological environment, like the physical, may be inhospitable in some respects. The host may react toward the presence of the parasite by producing antigens that prevent it from becoming established and reproducing.

Only when conditions in the external and internal environments are met at an adequate level can parasitic animals survive, develop to the reproductive stage, and be transmitted to other hosts.

The ecological adaptations of parasites of terrestrial hosts have evolved to meet the exigencies of the external environment. They include thick-shelled eggs, ensheathed larval stages, or intermediate hosts that provide protection during the period of growth and transmission. Those parasitizing aquatic hosts often lack such protective features, since the eggs or larval stages are in the water and have no need for them. Thus the evolution of the free-living stages of parasites parallels the requirements for protection according to the environment in which transmission takes place.

**Transmission.** Transmission is involved with ecology and linked in some manner with the food chain. The infective stage may be a contaminant of the food or water and accidentally swallowed by the host (Plate 15) or it may be encysted on forage (Plate 64) or in the body of an intermediate host that serves as food for the definitive host (Plate 90).

Other parasites such as plasmodia (Plate 29) and some filarioids (Plate 133) are injected into or deposited on the skin, respectively, of the host by bloodsucking intermediate hosts such as mosquitoes. A few actively penetrate the skin, as do cercariae of blood flukes from the water (Plate 58) and larval hookworms (Plate 110) and strongyloidids (Plate 109) from the soil.

**Escape from Hosts.** Escape of cysts, eggs, or larval parasites from the digestive, urinary, reproductive, and respiratory tracts is through the natural openings and in the discharges of the respective systems. Certain flagellates [*Tritrichomonas foetus* (Plate 13) and *Trypanosoma equiperdum*] of the reproductive tract are transmitted by contact during coitus. Stages of parasites in the blood (plasmodia) and tissues (filarioids) are withdrawn by bloodsucking arthropods that serve as hosts.

**Life Cycles.** The requirements for successful development of the parasite and its transmission to a suitable host are somewhat different in those with a direct life cycle and those with an indirect one.

In the case of parasites with a direct life cycle, cysts, oocysts, and eggs are passed in the feces of the host, being dropped at random by all except humans under most conditions. Following development of the parasite to the infective stage, it has a limited span of longevity to the adverse action of the various facets of the environment and the finite store of nutriment in its body. Transmission is dependent upon the frequency with which the host appears in the particular areas where the infective stages were deposited. Domestic ruminants confined to permanent pastures have a high degree of exposure, frequently with heavy infection. Range animals and wild ruminants, on the other hand, may not return to the same spot for some time, during which the infective stages may have developed and died. Behavioral patterns and territorial boundaries of the hosts influence the rate of infection. Feeding habits also play an important role in infection. As an example, sheep crop grass close to the ground, acquiring extremely heavy infections of trichostrongyles (Plate 116), whereas cattle feeding on the higher grass are less heavily parasitized by these worms. Swine and humans, particularly children, spread ascarids by contaminating their immediate and limited habitats with feces containing eggs that are subsequently swallowed with soil (Plate 121). Contamination of the water supply of communities or families with feces containing the cysts of *Giardia lamblia* and *Entamoeba histolytica* (Plate 15) is a means of infection.

Parasites with a direct life cycle often are widespread in their geographic distribution. Their presence is dependent primarily on the occurrence of the definitive host, although some species such as those of *Ostertagia* of sheep and cattle favor cold climates, whereas species of *Haemonchus* from the same hosts prefer warm ones.

Transmission of monogenetic trematodes is dependent upon the host and oncomiracidia coming together (Plates 43–45). Contact occurs in the territorial range of the host while performing its behavioral patterns of feeding, patrolling, and resting.

Parasites with an indirect life cycle have more restrictive requirements for its completion. In the first place, the geographic distribution of the intermediate host automatically determines that of the parasite. Digenetic trematodes usually require aquatic mollusks, mostly snails, as the first intermediate host, often a single or at most a few closely related species (Plate 83). Eggs of these flukes are able to develop to the infective stage only if they are in a suitable aquatic habitat with the appropriate molluscan intermediary. Complications are introduced with the interpolation of intermediate hosts into the life cycle. The hosts must be present at the right time, in the right place, must be readily infected, must survive sufficiently long for the parasite to develop, and must release the cercariae in a location where they can encyst on vegetation or penetrate the next host, be it a second intermediate or a definitive host.

The common liver fluke *Fasciola hepatica* (Plate 64) is a classical example of a trematode with a single intermediate host. The cercariae can act only in water where they encyst on objects to transform into infective metacercariae. Those on emergent forage are available to grazing herbivorous mammals. However, the longevity of the metacercariae is influenced by physical factors such as moisture, temperature, and sunlight in the microenvironment surrounding them. Blood flukes, on the other hand, do not encyst. Their cercariae are stimulated to activity when the water is agitated and they attach to the host by means of the suckers. With the aid of muscular action and histolytic substances from the penetration glands, they burrow through the epidermis and enter the circulatory system (Plates 57–60).

Trematodes with a second intermediate host in the life cycle have a more complicated situation. In such cases, the geographic distribution is limited to the presence and simultaneous occurrence of two different species of intermediate hosts. They include snails together with other aquatic organisms such as arthropods (Plate 71), fish (Plate 56), and amphibians (Plate 67). The second intermediate host contains the infective metacercariae and is often food of the definitive host (Plates 74, 79, 91).

When a terrestrial snail and arthropod serve as the intermediate hosts of a digenetic trematode, a whole new set of ecological expediencies is required. Fully embryonated eggs enclosed in thick, impervious shells are passed in the feces of the definitive host. Hatching occurs through stimuli supplied in the gut of the land snails, the first intermediary, that swallow the eggs. Cercariae released into the air chambers of the snail provoke excessive secretion of mucus which surrounds them in a liquid environment in the form of slimeballs that are discharged onto the vegetation or soil. Discharge of the slimeballs usually follows rain and a lowering of the temperature. When the slimeballs and contained cercariae are eaten by a terrestrial arthropod, the cercariae migrate through the gut wall to the body cavity and transform into infective metacercariae (Plate 69). Those encysting around the central nervous system of ants affect the normal behavior of these insects, causing

them to rest sluggishly on grass with which they are readily ingested by the definitive hosts.

In some instances, the same species of terrestrial snail may serve as both first and second intermediate hosts (Plate 61). Cercariae escape from crawling snails and are left in their slime trails. The behavioral characteristic of snails following the slime trails of other individuals provides a means of contact with the cercariae. They are stimulated to activity, enter the openings of the ureter, migrate to the kidney and pericardial cavity, and transform to unencysted metacercariae.

Cestode life cycles are indirect, involving one or two intermediate hosts, except for certain primitive, neotenic caryophyllaeids (Plate 85) which are direct. Undeveloped eggs of pseudophyllidean tapeworms deposited in water embryonate and hatch into motile, ciliated hexacanths which, when eaten by cyclops, develop into procercoids. In the digestive tract of fish, they escape from the cyclops and migrate to the muscles and transform into plerocercoids that are infective to mammals and birds (Plate 87). Other species of cestodes utilize mammals (Plates 89–92) or arthropods (Plates 93, 95–99) as intermediate hosts. The ubiquity of mammalian intermediaries of taenioid cestodes of carnivores (Plate 92) and soil-dwelling acarine intermediaries of anoplocephalid tapeworms of ruminants (Plate 99) together with eggs highly resistant to desiccation and freezing accounts for the cosmopolitan distribution of these particular parasites. The soil mites live in the turf and on the grass, where they are well protected, extremely abundant, frequently infected, and readily available to grazing hosts.

The ecological adaptation of Acanthocephala to the type of habitat of the definitive hosts is marked. Species of aquatic definitive hosts utilize aquatic crustacean intermediate hosts (Plate 104), whereas those in terrestrial vertebrates generally have land insect intermediaries (Plate 105). The egg shells of species occurring in terrestrial hosts are thicker and more resistant to the vagaries of their microenvironment than those deposited in water by aquatic definitive hosts. The eggs hatch inside the intermediate hosts where the larvae are well protected during and after development to the infective stage.

The life cycles of nematodes are both direct and indirect. In either case, there is a common pattern consisting of an egg and five developmental stages. These include the first, second, third, and fourth larval stages followed by the adult stage. Each stage except the first is preceded by a molt (p. 402). In general, third stage larvae are infective to the definitive host, whether it is in the soil or in an intermediate host.

Eggs of the strongyles, trichostrongyles, oxyurids, ascarids, and some trichurids are undeveloped when laid. They depend on favorable environmental conditions for embryonation. Ascarid larvae develop to the second stage in the egg, where they remain without further change until swallowed by a host in whose intestine hatching occurs, followed by migration and growth (Plate 121). These eggs, both in the unembryonated and embryonated stages, are extremely resistant to normal environmental adversities. Embryonated eggs of oxyurids (Plate 119) and certain trichurids (Plate 135) hatch when swallowed by the definitive host.

Eggs of strongyles and trichostrongyles hatch in the soil, where the larvae of most species undergo two molts. The cuticle of the first molt is usually cast off by the second stage larvae but that of the second molt is retained as a protective sheath for the third stage infective larvae. Stages one and two feed on bacteria in the soil and fecal masses in which they hatch, but stage three is unable to feed because of the enclosing sheath. Its energy must be derived from the finite amount stored in the intestinal cells.

Infective hookworm larvae respond negatively to gravity by migrating to the highest particles of soil and react positively to temperature by becoming active under warm conditions. Their response to body heat and possibly carbon dioxide of the host stimulates them to penetrate the skin. Third stage trichostrongyle larvae respond positively to dim light and migrate up grass blades in a film of moisture in twilight and on cloudy days, retreating as the light intensity increases. This crepuscular migratory response places large numbers of them on edible herbage at times favored by ruminants for grazing (Plate 116).

A few species such as the hookworm of fur seals (Plate 111) and *Nematodirus* spp. of sheep complete the first two molts in the egg and hatch as ensheathed third stage, infective larvae.

Metastrongylid, dracunculid, spirurid, filarioid, some capillarid, and dioctophymatid nematodes require intermediate hosts for development to the infective stage.

Some metastrongylids (Plate 118), dracunculids (Plate 124), spirurids (Plate 128), and filarioids of the lungs and air sacs (Plate 131) lay embryonated eggs. Of the metastrongylids, the eggs of *Metastrongylus* hatch only when eaten by oligochaete earthworms. The unhatched eggs are extremely resistant and third stage larvae in earthworms are in a most favorable microenvironment for long survival (Plate 118). Eggs of other species metastrongylids, which include *Protostrongylus* and *Muellerius*, hatch while passing from the lungs to the intestine and appear in the feces as first stage larvae infective to certain land snails which they penetrate. They are exposed to the external environment only as first stage larvae (Plate 117).

Eggs of spirurid and air sac-dwelling filarioid nematodes are thick shelled and resistant. They hatch when eaten by terrestrial insects (Plates 125, 131) or aquatic crustaceans (Plate 127). These are cosmopolitan parasites because of the prevalence of suitable hosts.

Dracunculids, tissue-dwelling filarioids, and trichinellids give birth to hatched larvae. Each has different ecological requirements for reaching the intermediate and definitive hosts. Female dracunculids migrate to the subcutaneous tissues of the definitive host, mature, extend the anterior end through an ulcer in the skin when stimulated by water, and release first stage larvae through a rent in the prolapsed uterus. The active larvae in the water attract various species of cyclops which eat them. Infected cyclops become sluggish and are easily captured or swallowed (Plate 124). The microfilariae of the tissue-inhabiting filarioids circulate in the blood, often showing marked periodicity. Development proceeds to the infective stage only after microfilariae are drawn into the digestive tract of bloodsucking arthropods. Development is in specific parts of the body. The infective larvae migrate to the mouth parts and escape on the skin of the definitive host as the insect feeds. Adequate moisture is essential for survival of the larvae as they burrow through the skin (Plate 133).

Larval trichinellids born in the lymph spaces of the intestinal wall are carried via the circulation to the voluntary muscles, where they encyst and develop to the infective stage. Infection normally occurs when the encysted larvae are eaten. There is no free-living stage (Plate 139).

In some species of helminths that have an indirect cycle, that intermediary is not normally in the food chain or associated closely enough with it to assure favorable opportunity for regular transmission. In such cases, an additional host that normally feeds on the intermediary and is in turn eaten by the definitive host serves as a transfer or paratenic host. Upon entering the transfer host, the infective stage migrates from the gut to the body cavity or tissues, where it remains without further development. When the infective stage from the transfer host is liberated in the intestine of the definitive host, it develops in the same manner that it would have in case the intermediary had been swallowed (Plates 55, 88, 130).

**Factors of the Environment.**   Requirements from the external environment are adequate moisture in the soil to prevent desiccation and to allow development, temperature to promote optimal development, and sufficient oxygen for metabolic processes. Shade and cover protect against drying and freezing conditions. The combination of these factors promotes development. In cases where hatching of eggs occurs in the soil, it is triggered by chitinase and proteinase enzymes secreted by the fully developed first stage larvae that break down the egg membranes and shell, allowing the larvae to escape. In nematodes where molting is part of the development in the soil or water, the fully developed larva of each stage becomes quiescent and secretes molting fluids that cause the cuticle to loosen and be shed.

The internal environment provided by the body of the intermediate or definitive host supplies protection against unfavorable physical factors and provides food for growth and stimuli for initiating the developmental processes such as production of hatching, excysting, and exsheathing fluids.

While protected against unfavorable facets of the external environment, parasites in the bodies of hosts are confronted with the antagonistic reactions resulting from antigens produced in response to the presence of the organisms and their metabolites.

**Intermediate Hosts.**   The type of intermediate host utilized by a parasite in the completion of the life cycle involves various aspects of ecology. Aquatic organisms which act as intermediate hosts are predominantly mollusks and aquatic annelids and arthropods (crustaceans and insects) among the invertebrates, and fish and amphibians among the vertebrates. Terrestrial animals commonly acting as intermediaries are polychaete annelids, land snails, crustaceans (pill bugs and sow bugs), insects, reptiles, and mammals.

Protection provided by the aquatic environment to the larval stages, cysts, and eggs results in the absence of highly developed protective structures. For development of the infective stage, its contact with and development in the intermediate host, the water must persist in a condition favorable to both the parasite and the host.

Eggs and larvae infective to terrestrial intermediate hosts require protective covering such as impervious cuticular sheaths or thick egg shells and oocysts that retard desiccation and freezing. Moreover, the habitats, presence, and behavior of the intermediate hosts must coincide to permit infection and transmission.

**Entrance of Parasites into Hosts.**   Entrance of infective stages of parasites into their hosts is attained either passively or actively.

Cysts (Plate 15), oocysts (Plate 22), embryonated eggs (Plate 70), metacercariae (Plate 66), cysticerci (Plate 90), cysticercoids (Plate 99), plerocercoids (Plate 87), acanthella (Plate 106), and many hatched larvae such as strongyles (Plate 114), trichostrongyles (Plate 116), metastrongyles (Plate 117), dracunculids (Plate 124), and filarioids (Plate 134) enter the host passively. They are dependent upon the host swallowing them while feeding or drinking.

Miracidia (Plate 83), cercariae (Plate 83), and certain larval nematodes such as hookworms (Plate 110), strongyloidids (Plate 109), and filarioids (Plate 133) actively penetrate the epithelial covering of their hosts as a result of stimuli provided by them. Miracidia are attracted by chemical substances in the mucus of snail hosts. Hookworm and filarioid larvae are excited in part to burrowing activity by the body temperature and perhaps carbon dioxide provided by the host.

**Role of Infective Stage.**   Upon reaching the infective stage, the parasite is unable to develop further until it enters an appropriate definitive host and receives specific signals in the form of stimuli to begin the next stage of development. Thus upon attaining the infective stage, the young parasite enters a resting period from which it is unable to emerge without the stimuli from the host. Upon entering the proper host, stimuli from it trigger secretions of hatching and exsheathing fluids that restart the developmental processes. These stimuli may be the high carbon dioxide content of the alimentary canal, bile salts, body temperature, certain digestive enzymes, and others.

**Localization of Parasites in Hosts.** It is well known that species of parasites are found primarily in rather restricted sites in their hosts, be they definitive or intermediate hosts. Moreover, these sites which represent ecological niches are reached and adjusted to by rather definitive behavioral patterns on the part of the parasites. The sites may vary within the host, whether intermediate or definitive, according to the stage of development.

Miracidia entering snails transform into mother sporocysts often in locations near the point of penetration. It may be in the epidermis if entry is from the outside or on the gut wall if hatching occurs in the intestine. Daughter sporocysts and rediae migrate via the circulatory system to the liver and other visceral organs presumably for a more nutritious source of food required for production of the numerous cercariae that follow. They leave the snails by various routes, such as intestine, female genital ducts, ureter, via bloodstream to the mantle and emerge, or simply burrow through the tissues.

Cercariae appear to seek definite sites in which to encyst, judging from the regularity with which metacercariae occur in the intermediate hosts. Common locations include the branchial basket of dragonflies by *Haematoloechus* (Plate 71), lens of the eyes and in the brain of fish by Diplostomum (Plate 54), the tentacles of snails by *Leucochloridium*, and the brain of ants by *Dicrocoelium dendriticum* (Plate 69).

Adult flukes occur with such regularity in specific sites that a high degree of preference is indicated. Well-known routes of migration are followed. How the sites are found is unknown. Perhaps the flukes are attracted to hormones produced by the organs and they seek the source. Liver flukes (Plate 64), kidney flukes, and lung flukes (Plate 78) are examples of migrating forms. A single bisexual lung fluke (Plate 78) will be sought and found by another of its kind introduced into the host at a later date. Blood flukes of separate sexes seek and locate each other in the host (Plate 57). Pheromones as are found in insects have been suggested as attractants.

In cestodes, the oncospheres enter the hemocoel of insects (Plate 93) and crustaceans (Plate 97) or the gonads of aquatic annelids (Plate 85) for development. In vertebrates, they go to the viscera (Plate 91), subcutaneous tissue (Plate 92, b), striated muscles (Plate 90), and nervous tissue (Plate 92, b) where the developed larval stages occur. Adults often arrange themselves in portions of the small intestine to gain the greatest nutritional advantage.

The acanthor of Acanthocephala migrates to the hemocoel of the insect (Plate 105) or crustacean (Plate 104) intermediary. In the intestine of the definitive host, they, like cestodes, move to sites in the small intestine that provide most abundantly the nutritional needs at a specific stage in their development and physiological activity.

Nematodes, like digenetic trematodes, occur in many parts of the definitive host. These sites are reached by rather specific routes of migration (Plates 110, 117, 121, 133, 140). While the means by which these sites are found is unknown, it seems reasonable to postulate that chemical attractants to which the chemoreceptors of the worms are physiologically attuned direct their course. Common sites of adult worms include various parts of the alimentary canal by strongyles, trichostrongyles, oxyurids, ascarids, spirurids, trichurids, and trichinellids with their different pH, *Skrjabingylus* in the frontal sinuses of mustelids, *Tetrameres* in the proventricular crypts of ducks (Plate 127), filarioids in the air sacs of birds and reptiles (Plate 131), tissues and pulmonary artery of birds and mammals (Plate 133) and pleural cavity of mammals (Plate 132), metastrongylids in the respiratory tract of mammals (Plates 117, 118), strongyles (Plate 113) and dioctophymatids (Plate 140) in the kidneys of mammals, and dracunculids in the subcutaneous tissues of mammals (Plate 124), birds, and fish.

Larval nematodes in intermediate hosts occur in specific sites. Many are in the hemocoel (Plate 128). Others are in the Malpighian tubules (Plates 126, 133), tissue and lymph spaces (Plate 117), and intestinal wall (Plate 130).

Such distribution in the bodies of both intermediate and definitive hosts points to the highly diverse ecological conditions to which a single species must adapt in the course of its developmental and reproductive periods. These constitute complex physiological and physical conditions as well as behavioral patterns involving two or more animal hosts about which much remains to be learned.

**Host-Parasite Specificity as an Ecological Factor.**    Just as unfavorable conditions in the external environment exercise adverse effects on the development and survival of free-living stages of the parasite, so do incompatible circumstances in the internal habitat where the requirements of the parasite are equally if not more exacting.

It is considered that a state of compatibility between a host and its parasites is the result of a long association that has evolved into a mutually tolerable association. In the case of most parasites, this relationship results in specialization on their part to live in a specific host or type of host. In other words, they are attuned to the specific aspects of the physiological environment provided by that host. This type of relationship is referred to as host-parasite specificity.

The requirements of some species of parasites are so specific that they are able to live and reproduce only in a single species of vertebrate host. The malarial organisms (Plate 29) and pork and beef tapeworms (Plates 89, 90) of man are examples. Avian cestodes of given species are restricted to certain orders of birds. Other parasites such as the common sheep liver fluke (Plate 64), *Plagiorchis muris* (Plate 70), and *Echinostoma revolutum* (Plate 63) have a wide spectrum of definitive hosts. Obviously their requirements of the bioenvironment are not so restrictive.

**Immunity as an Ecological Factor.**    The presence of parasites and their metabolites stimulates the host to react by producing antigens. These alter the biochemical characteristics of the bioenvironment. The extent to which these changes react on the parasites ranges from slight interference with development and reproduction to complete elimination of them from the host. The phenomenon of self-cure of stomach worms (*Haemonchus contortus*) in sheep (Plate 116) is an example. When sheep already harboring a population of adult worms receive a new infection of larvae which embed in the gastric mucosa, the immune reaction to them alters the environment by raising the pH of the stomach to a level

that is beyond the tolerance of the adult worms. They detach and are expelled. Further change induced in the gastric environment by their disappearance enables the larvae to emerge and develop.

**Ecological Niche.**    The concept of the ecological niche is not violated by the presence of different species of parasites in the intestine of a single animal. *Giardia*, *Entamoeba*, *Ascaris*, *Ancylostoma*, *Hymenolepis*, and *Taenia* may all occur simultaneously in the small intestine of man. Each has its own specific requirements and, therefore, may be regarded as occupying its own niche which supplies each species with its general and specific requirements for all the life activities.

**Age of Host.**    The age of the host alters the bioenvironment to the extent that infection is difficult, if not impossible. In many species of snails, only young individuals are susceptible to infection by miracidia. Young chickens which lack mucus cells in the intestine are readily susceptible to infection by *Ascarida galli* (Plate 120), whereas older birds in which the mucus cells have appeared are refractory. *Toxocara canis* develops readily in the intestine of puppies but not in adult dogs (Plate 122). Adult hookworms of fur seals (Plate 111) are lost when the pups are 3 to 6 months of age. They never occur in older seals.

**Diet.**    Diet of the host may alter the environment and essential nutrients in the intestine necessary for development of parasites residing in it. A milk diet appears to produce an unfavorable environment for ascarids. The type of carbohydrate in the diet of the host affects the growth of *Hymenolepis diminuta*. Tapeworms in rats eating starch are larger than those in animals receiving only dextrose or sucrose.

Animals on a low level of nutrition often are more heavily parasitized than adequately nourished individuals on a well-balanced diet. With insufficient intake of proteins, the physiological processes required for an antigenic response to the presence of the parasites is lacking. Thus, the bioenvironment has been altered because of malnutrition.

## Conclusions

The points discussed above point to the relationship of parasites to the physical and biological environments in which they must develop. Each constitutes a set of highly variable microenvironments in which the responses of the parasite are very complex on the biochemical and physiological levels. These are only partially understood at the present time. Discussion of these responses is beyond the scope of this chapter. Reference to the list of selected readings will provide information for understanding some of the complexities evolved and practiced by animals of free-living origin that have adapted the complex parasitic way of life with its advantages, disadvantages, and hazards.

## SELECTED REFERENCES

Baer, J. G. (1951). Ecology of Animal Parasites. University of Illinois Press, Urbana, 224 pp.

Croll, N. A. (1966). Ecology of Parasites. Harvard University Press, Cambridge, 136 pp.

Duke, B. O. L. (1971). The ecology of onchocerciasis in man and animals. In: A. M. Fallis, ed. Ecology and Physiology of Parasites. University of Toronto Press, pp. 213–222.

Hammond, D. M. (1971). The development and ecology of coccidia and related intracellular parasites. In: A. M. Fallis, ed. Ecology and Physiology of Parasites. University of Toronto Press, pp. 4–20.

Kates, K. C. (1965). Ecological aspects of helminth

transmission in domesticated animals. Amer. Zool. 5:95–130.

Lee, D. L. (1971). Helminths as vectors of microorganisms. In: A. M. Fallis, ed. Ecology and Physiology of Parasites. University of Toronto Press, pp. 104–122.

Noble, E. R., and G. A. Noble. (1971). Parasitology. The Biology of Animal Parasites. 3rd ed. Lea & Febiger, Philadelphia, pp. 459–542.

Read, C. P., Jr. (1950). The vertebrate small intestine as an environment for parasitic helminths. Rice Inst. Pamphlet Monogr. in Biol. 37(2):1–94.

Rogers, W. P. (1962). The Nature of Parasitism. Academic Press, New York, 287 pp.

Smyth, J. D. (1962). Introduction to Animal Parasitism. Charles C Thomas, Springfield, Ill., pp. 9–24.

Sprent, J. F. A. (1963). Parasitism. Bailliére, Tindall and Cox, London, 145 pp.

Ulmer, M. J. (1971). Site-finding behaviour in helminths in intermediate and definitive hosts. In: A. M. Fallis, ed. Ecology and Physiology of Parasites. University of Toronto Press, pp. 121–160.

Vávra, J. (1971). Physiological, morphological, and ecological consideration of some microsporidia and gregarines. In: A. M. Fallis, ed. Ecology and Physiology of Parasites. University of Toronto Press, pp. 92–103.

# Phylum Protozoa

Among the Protozoa are many symbiotic species living in various degrees of relationship with other animals, both invertebrates and vertebrates. Numerous species of Protozoa are parasitic, living in cells (cytozoic), tissues (histozoic), or cavities (celozoic). Some of them are notorious as causes of serious diseases in humans, as well as in domestic and wild animals from honeybees to cattle.

The phylum Protozoa consists of an assembly of unicellular animals of varying degrees of complexity in both structure and biology. It is only reasonable, therefore, to expect that various schemes of classification have evolved. They continually trend toward greater sophistication as information becomes more precise through the use of new tools of investigation such as phase contrast and electron miscroscopy and through more detailed studies on life cycles.

The basic classification set forth by Honigberg's committee and generally accepted by protozoologists divides the phylum Protozoa into four subphyla. They are the Sarcomastigophora, Sporozoa, Cnidospora, and Ciliophora.

Subphylum Sarcomastigophora has flagella or pseudopods, or both. It is divided into the super-classes Mastigophora with flagella, Opalinata with cilia-like organelles, and Sarcodina with pseudopods.

The subphylum Sporozoa has spores without polar filaments. It contains the classes Telosporea with the subclasses Gregarinia and Coccidia having spores and sexes, Toxoplasmea without spores or sexes, and Haplosporea with spores but without sexes.

Subphylum Cnidospora possesses spores, each with one or more polar filaments. It includes the classes Myxosporidea with spores which originate from several cells and Microsporidea whose spores originate from a single cell.

The final subphylum is Ciliophora, whose bodies are covered with cilia, at least in the early stage of development, and which have two types of nuclei. There is one class, the Ciliata, which includes the subclasses Holotrichia with simple, uniform somatic cilia; Peritrichia, essentially without somatic cilia in the adult stages but with conspicuous oral cilia; Suctoria, in which cilia are replaced in adults by tentacles; and Spirotrichia, which have sparse somatic cilia.

While this system is still in skeletal form and the early stages of development, it promises to provide a scheme of classification that should show basic relationships between the various groups of Protozoa. However, in its present form, it does not provide information at the species level needed by students in parasitology. Besides, marked differences of opinion exist regarding the classification of the Sporozoa. Since Kudo has included the species in "Protozoology," his classification is followed in general until the newer system is more fully developed.

The phylum Protozoa, as given by Kudo (1966) in "Protozoology," consists of the subphyla Plasmodroma and Ciliophora. The Plasmodroma contains the class Mastigophora whose members bear flagella, the class Sarcodina whose members have pseudopodia at some stage in their development, and the class Sporozoa that are without apparent organelles of locomotion, except in the microgametes. The subphylum Ciliophora consists of the class Ciliata with cilia throughout the trophic life and the class Suctoria in which cilia in the early stages are replaced by tentacles.

Each class of Protozoa has one or more genera whose species parasitize humans, causing dreaded disease that results in much suffering, morbidity, and mortality. There are *Trypanosoma gambiense* and *Leishmania donovani* of the Mastigophora, causing African sleeping sickness and kala azar, respectively; *Entamoeba histo-*

## Explanation of Plate 1

A, Profile of invasive trophozoite. B, Section of crypt of cecum with invasive phase of amebiform parasites. C, Trophozoite with food granules and extranuclear granules. D, Trophozoite with light areas indicating digestion of food particles. E, Large vegetative forms prior to division, packed in liver. F, Four cells in so-called "resistant" stage. G, Four "resistant" cells in giant cell. H, Groups of "resistant" cells in mucosa of cecum. I, Chicken host. J, Turkey host. K, Liver with characteristic saucer-shaped lesions caused by *Histomonas meleagridis*. L, M, Anterior and posterior ends, respectively, of male cecal worm *(Heterakis gallinae)*.

1, Nucleus; 2, nucleolus; 3, extranuclear bodies; 4, flagella; 5, food particles; 6, food particles being digested; 7, epithelial cells of cecum; 8, trophozoite in cecal crypt; 9, invasive trophozoites in cecal epithelium; 10, liver parenchyma; 11, large vegetative protozoa preparing for division; 12, "resistant phase" of protozoa; 13, thicker outer tissue covering laid down by host; 14, giant cell of host; 15, cecal mucosa.

a, "Resistant stage" in ceca; b, "resistant stage" passed in feces; c, d, "resistant phase" in feces; e, "resistant stage" swallowed; f, trophozoite developing; g, flagellated trophozoite in mucosa; h, adult cecal worm; i, eggs of cecal worm presumably containing histomonads; j, k, egg passed in feces; l-n, development of egg in feces; o, infective egg swallowed; p, egg hatches in alimentary tract; q, larva containing protozoan; r, larva in cecum where it is believed to liberate pathogenic histomonads.

a', (over head of turkey) Histomonad from feces of chickens; b', histomonad swallowed; c', trophozoite active in intestine; d', trophozoite enters cecum; e', trophozoite enters interstices between cells; f', vegetative cells multiply; g', mass of vegetative cells causes destruction of epithelial cells; h', adhesions resulting from destruction of masses of epithelial cells; i', caseous core in lumen of ceca due to reaction to protozoa; j', "resistant" cells of protozoa in mucosa; k', "resistant" cell passed in feces; l', "resistant" cell in feces; m', "resistant" cell in feces a source of infection to turkeys; n', trophozoites enter hepatic portal vein from intestine and cecum; o', trophozoites enter liver parenchyma from capillaries; p', "resistant stages" in saucer-shaped lesions in liver.

a'', *Heterakis* egg passed in feces of chicken; b'', developing egg; c'', embryonated egg containing larva and histomonads; d'', infective egg swallowed by turkey; e'', egg hatches; f'', larva free in intestine goes to ceca where histomonads are liberated; g'', adult cecal worm; h'', eggs of cecal worm deposited in ceca; i'', eggs passed in feces; j'', eggs in feces; k'' –m'', eggs embryonate and larvae contain infective histomonads (see Plate 120).

Figures A–H adapted from Tyzzer, 1919, J. Med. Res. 40:1; K, from photograph (Fig. 11) in Richard and Kendall, 1957, Veterinary Protozoology, Oliver and Boyd, London; L, M, from Clapham, 1933, J. Helminthol. 11:67.

---

*lytica* of the Sarcodina, producing amebic dysentery; *Plasmodium vivax* of the Sporozoa as an etiological agent for malaria; and *Balantidium coli* of the Ciliophora, which is least serious but still responsible for damage to the intestine. Each of these species has its counterpart in the domestic and wild animals. The species discussed below represent some of the major ones from all of the classes except Suctoria.

## SELECTED REFERENCES

Hall, R. P. (1953). Protozoology. Prentice-Hall, New York, 682 pp.

Honigberg, B. M., W. Balamuth, E. C. Bovee, J. O. Corliss, M. Gojdics, R. P. Hall, R. R. Kudo, N. D. Levine, A. R. Loeblich, Jr., J. Weiser, and D. H. Wenrich. (1964). A revised classification of the Phylum Protozoa. J. Protozool. 11:7–20.

Jahn, T. L., and F. F. Jahn. (1949). How to Know the Protozoa. W. C. Brown, Dubuque, Iowa, 234 pp.

Kudo, R. R. (1966). Protozoology. 5th ed. Charles C Thomas, Springfield, Ill., 1174 pp.

Levine, N. D. (1961). Protozoan Parasites of Domestic Animals and of Man. Burgess Publishing Co., Minneapolis, 412 pp.

Levine, N. D. (1969). Problems in systematics of parasitic Protozoa. In: G. D. Schmidt, ed. Problems in Systematics of Parasites. University Park Press, Baltimore, pp. 109–121.

Levine, N. D. (1970). Taxonomy of the Sporozoa. J. Parasitol. 56 (No. 4, Sect. II, Pt. 1): 208–209.

Scholtyseck, E., and H. Mehlhorn. (1970). Recent problems of taxonomy and morphology of Coccidia. J. Parasitol. 56 (No. 4, Sect. II, Pt. 1): 306–307.

### SUBPHYLUM PLASMODROMA

Of the classes of Plasmodroma, the Mastigophora bear flagella for locomotion and the Sarcodina

**Plate 1**   *Histomonas meleagridis*                                                                17

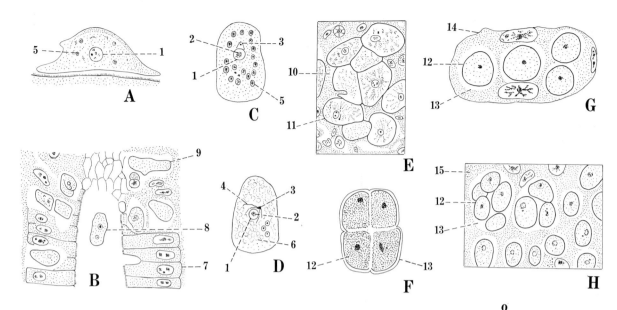

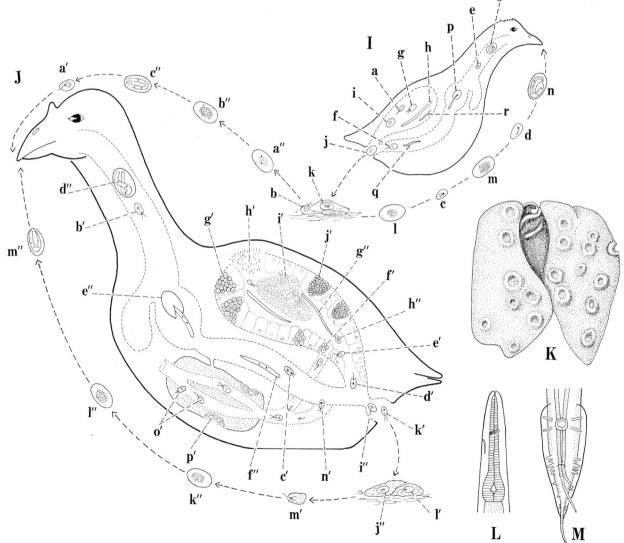

bear pseudopods for traveling and capturing food. The Sporozoa lack organelles of locomotion as adults but are capable of gliding and ameboid movement; microgametes bear flagella.

## CLASS MASTIGOPHORA

The members of this class are both plants (subclass Phytomastiga) and animals (subclass Zoomastiga) and are characterized by the presence of one to eight flagella located at one end of the body. Generally there is a single nucleus.

Zoomastiga lack chromatophores characteristic of their plant counterparts. Organelles, in addition to flagella, such as parabasal body, axostyle, pelta, cytostome, and undulating membrane are present. Parasitic Zoomastiga are parasites of the alimentary canal of both invertebrates and vertebrates, and of the circulatory, reproductive, and muscle systems of vertebrates.

Their life cycles are both direct and indirect. In the latter case, an invertebrate and a vertebrate are involved. Some cause serious diseases in their hosts and are of major medical and veterinary significance. These forms have been studied most intensively and form the basis for much of the knowledge pertaining to the biology of this group of parasites.

## ORDER RHIZOMASTIGIDA

Members of this order have ameboid characteristics such as pseudopodia, in addition to flagella.

## FAMILY MASTIGAMOEBIDAE

The species belonging to this family have one to three flagella, sometimes four. One species, *Histomonas meleagridis*, is parasitic in gallinaceous birds.

### *Histomonas meleagridis*
### (Smith, 1910) (Plate 1)

*Histomonas meleagridis* is a flagellate parasite of the cecum of turkeys, chickens, ruffed grouse, quail, pheasants, Hungarian partridges, and pea fowls. It produces a serious disease in turkeys, but not in chickens, known as blackhead, infectious enterohepatitis, or histomoniasis.

**Description.**    The parasite includes both ameboid and flagellated forms, measuring 8 to 12 $\mu$ in diameter. The vesicular nucleus is round or pyriform in shape and is accompanied by an extranuclear body or endosome. When flagellated, usually a single flagellum arises from the endosome, but as many as four may be present.

Flagellated forms whose cytoplasm contains particles of food occur in the lumen of the ceca. Aflagellated stages crowded in the liver or mucosa of the ceca assume variable shapes due to the pressure exerted on them. Small oval bodies without flagella surrounded by a thick membrane laid down in the liver have been regarded as resistant stages, but existence of true resistant forms is doubtful. Multiplication is by binary fission.

**Life Cycle.**    The life cycle is unique in the manner by which the parasite has adapted for successful transmission from host to host. Organisms in the cecal contents constitute the natural infective stage but require special adaptations to successfully transfer from host to host. They infect the host by the oral route.

When the naked, nonencysting trophozoites are voided with the feces, they survive only a short time. During this brief period, they are capable of infecting birds when ingested if they survive passage through the acidic environment in the crop, proventriculus, and gizzard. The erratic success of experimental infections by feeding susceptible birds feces containing the naked trophozoites suggests that this is not the normal mode of infection. The consistency with which birds can be infected by injecting contents from infected ceca directly into these organs of susceptible birds, however, is evidence that this stage is regularly the infective form. In order to achieve the high incidence of infection that normally occurs in chickens and turkeys, there must be a means of protecting the parasites as they pass through the acidic part of the digestive tract.

This protection is provided by the ubiquitous cecal worm *Heterakis gallinae*. It provides places for multiplication of the parasite and a means of safe passage through the acidic parts of the alimentary canal. Worms feeding on the cecal contents of the birds ingest histomonads. Through their ameboid action, they invade the cells of the intestinal epithelium, where extensive multiplication takes place. The infected cells rupture, allowing the protozoa to enter the body cavity and penetrate the reproductive system. In the females, they invade the ovary and are incorporated in the eggs, where feeding and further multiplication occur. There is gradual decrease in size of the protozoans from about 10 $\mu$ in the intestine down to 4 to 5 $\mu$ in the eggs. Apparently there is no damage to the eggs as they produce infective larvae. Upon hatching in the intestine and ceca, the protozoa are released from the larval nematodes and infect the cecal epithelium of the bird host. They occur as small individuals in compact masses between the cells. Some enter the venules of the hepatic portal system and are carried to the liver, where multiplication continues with the formation of characteristic lesions in turkeys.

The small quiescent vegetative stage apparently feeds by secreting proteolytic enzymes which digest the cell cytoplasm surrounding them and then taking in small bits by pinocytosis or fluid parts by diffusion.

**Ecology.**    While *H. meleagridis* is infective as it occurs in the ceca, it does not encyst and the naked trophozoites are not adapted to surviving passage through the highly unfavorable acidic conditions in the stomach complex of the birds.

By their ability to survive in the intestine of the cecal worms and eventually to be incorporated in the eggs of the nematodes, the parasites have developed a means of evading the hazards of both the physical and biological environments through which they must pass.

*Histomonas meleagridis* is able to survive only a few hours when passed in the feces of infected birds. However, when incorporated in the eggs of cecal worms, they remain alive and infective for as long as 3 years. Since the eggs of cecal worms hatch in the intestine, the protozoa are protected while traversing the highly acidic portion of the alimentary canal.

Coprophagous invertebrates such as earthworms, houseflies, grasshoppers, and sow bugs ingest eggs of cecal worms which are unharmed in the digestive tract. Being part of the food chain of birds, these invertebrates serve as a means of collecting and conveying the eggs of the nematode to the birds. The nematodes serve as a transport host.

Earthworms carry eggs of cecal worms deep into the soil where the temperature and humidity are favorable for prolonged survival. When eggs are returned to the surface at a later time by earthworms or tilling of the soil, infections of birds with nematodes and histomonads result. For this reason, turkeys should not be put in runs formerly occupied by poultry.

Chickens, which are almost universally infected, serve as natural reservoirs without disease, whereas turkeys, especially poults, are very

susceptible and suffer high mortality, sometimes up to 100 percent, from it.

**Diagnosis and Pathology.**    Finding the characteristic histomonads in the feces of birds is evidence of infection. In turkeys, extensive lesions in the cecal epithelium and liver are evidence of the disease. In the liver, the large saucer-shaped lesions on the surface are first dark red in color, later yellowish, and finally yellowish green. Often a large caseous core forms in the ceca. In advanced cases, the enlarged and inflamed ceca are attached through adhesions to the body wall.

**Resistance and Immunity.**    Chickens, while susceptible at an early age to mild cecal lesions, generally exhibit a natural resistance that protects them from the disease. Turkeys of all ages are highly susceptible and suffer high mortality.

Turkeys that recover from the disease generally are immune to further attacks. Histomonads penetrating the cecal epithelium or introduced by migrating heterakid larvae burrowing into it probably stimulate the formation of antibodies found in the sera of immune birds. It is not known what factor is responsible for the immunity.

**Prevention and Control.**    Proper management of flocks of turkeys is essential to preventing the disease. In addition, medication helps keep it in check.

Chickens, which serve as reservoir hosts, and turkeys should not be kept together. Neither must turkeys be allowed on areas previously occupied by infected birds because infected heterakid eggs that are brought to the surface of the soil are capable of producing new outbreaks of the disease.

Medication may be used along with good husbandry practices to help control the histomonads. Low concentrations of proprietary compounds of Entheptin administered continuously in food or water keep the histomonads in check and provide useful prophylaxis. Self-medication of the birds with phenothiazine in the feed is highly efficacious in destroying the cecal worms and thereby removing the essential transport host required by *H. meleagridis*.


## EXERCISE ON LIFE CYCLE

Infection may be transmitted to healthy young turkey poults by 1) feeding cecal worms (*Heterakis gallinae*) or feces from turkeys infected with *Histomonas gallinae*, 2) placing poults on turkey yards, or 3) quartering poults with grown chickens. Feeding the eggs of cecal worms from chickens or turkeys is a fairly certain means of producing an infection. To demonstrate that the worm eggs convey the infection, they should be fully incubated, sterilized thoroughly to destroy any histomonads that might possibly be present on the outside, and fed to young poults. When the eggs hatch in the intestine, the causitive organism produces the lesions in the ceca and liver characteristic of the disease in about 3 weeks.

## SELECTED REFERENCES

Becker, E. R. (1959). Protozoa. In: H. E. Biester and L. W. Schwarte, eds. Diseases of Poultry. 4th Ed. Iowa State University Press, Ames, p. 828.

Farr, M. M. (1956). Survival of the protozoan parasite *Histomonas meleagridis,* in feces of infected birds. Cornell Vet. 46:178–187.

Lapage, G. (1956). Veterinary Parasitology. Oliver and Boyd, London, p. 751.

Lee, D. L. (1971). Helminths as vectors of microorganisms. In: A. M. Fallis, ed. Ecology and Physiology of Parasites. University of Toronto Press, pp. 104–122.

Lee, D. L., P. L. Long, B. J. Millard, and J. Bradley. (1969). The fine structure and method of feeding of the tissue parasitizing stages of *Histomonas meleagridis.* Parasitology 59:171–184.

Levine, N. D. (1961). Protozoan Parasites of Domestic Animals and Man. Burgess Publishing Co., Minneapolis, p. 74.

Lund, E. E. (1956). Oral transmission of *Histomonas* in turkeys. Poultry Sci. 35:900–904.

Lund, E. E., and A. M. Chute. (1970). Infectivity of *Histomonas meleagridis* of cecal and liver origins compared. J. Protozool. 17:284–287.

Lund, E. E., E. E. Wehr, and D. J. Ellis. (1966). Earthworm transmission of *Heterakis* and *Histomonas* to turkeys and chickens. J. Parasitol. 52:899–902.

Reid, W. M. (1967). Etiology and dissemination of the blackhead disease syndrome in turkeys and chickens. Exp. Parasitol. 21:249–275.

Ruff, M. D., L. R. McDougald, and M. F. Hansen. (1970). Isolation of *Histomonas meleagridis* from embryonated eggs of *Heterakis gallinarum.* J. Protozool. 17:10–11.

Spindler, L. A. (1967). Experimental transmission of *Histomonas meleagridis* and *Heterakis gallinarum* by the sow-bug, *Poricellio scaber,* and its im-plications for further research. Proc. Helminthol. Soc. Wash. 34:26–29.

Tyzzer, E. E. (1919). Developmental phases of the protozoon of "blackhead" in turkeys. J. Med. Res. 40:1–30.

Tyzzer, E. E. (1926). *Heterakis vesicularis* Frölich 1791: A vector of an infectious disease. Proc. Soc. Exp. Biol. Med. 23:708–709.

## ORDER PROTOMONADA

This group of protozoa is characterized by the presence of one or two free flagella. Nutrition is holozoic or saprozoic. Reproduction is by binary, longitudinal fission but multiple fission and budding occur. The order includes the biflagellate forms of Honigberg's order Kinetoplastida.

Several families of the Protomonada are parasitic, including some parasites of great medical and veterinary importance. The Trypanosomatidae, Cryptobiidae, and Bodonidae are discussed.

### FAMILY TRYPANOSOMATIDAE

This family consists of seven common genera of monoflagellates parasitic in invertebrates (leeches and arthropods) and vertebrates.

**Description.**    The body may be oval and without a flagellum, or elongate and slender with a single flagellum. The flagellum may extend freely from the anterior end or run along the free margin of a delicate undulating membrane attached to the side of the body. In some cases, the flagellum terminates at the anterior end of the undulating membrane and in others it extends beyond as a free structure. The basal portion of the flagellar complex inside the body is the axoneme, a nine-fibered cylinder which terminates as the centriole. Near the base of the axoneme is the distinct, darkly staining kinetoplast containing DNA. A reservoir consisting of an invagination of the cuticle surrounds the axoneme along its full length; the flagellum emerges through the opening. A contractile vacuole empties into the reservoir from the side.

The following six distinct body types, each representing a genus, are recognized. 1) The amastigote (Plate 2, A), formerly the leishmanial form, is a small rounded or oval body without a flagellum. It occurs as one stage in all of the life cycles and is representative of the genus *Leishmania.* 2) The promastigote (Plate 2, C) is the former leptomonad type and is characterized by the slender, spindle-shaped body with a free flagellum. The kinetoplast is near the anterior end of the body. There is no undulating membrane. This form represents the genus *Leptomonas*. 3) Choanomastigotes (Plate 2, B) are small, flagellated bodies with broad ends, a truncated anterior and a rounded posterior. The flagellum emerges from the funnel-shaped reservoir. There is no undulating membrane. The kinetoplast is anterior and lateral to the nucleus. This somewhat aberrant form is characteristic of the genus *Crithidia* and does not appear in the life cycle of any other genus. 4) The opisthomastigote (Plate 2, D) represents the genus *Herpetomonas*. The kinetoplast is posterior to the nucleus and the axoneme runs through almost the entire length of the body. There is no undulating membrane. 5) The epimastigote (Plate 2, E) has the kinetoplast near the anterior margin of the nucleus. A short undulating membrane is present. This form represents the genus *Blastocrithidia*. 6) The trypomastigote (Plate 2, F) includes the genus *Trypanosoma*, in which the kinetoplast is near the posterior end of the body and an undulating membrane runs the full length of the parasite. There may or may not be a free flagellum.

**Classification.**    The taxonomy of the genera is based on a combination of the morphological features and the life cycles. There are six types of bodies, as described above (Plate 2, A–F), and two basic types of life cycles.

The monogenetic cycle is completed in a single invertebrate host. While an amastigote is common to all of the cycles, the final form in each is different and characteristic of the genus (Plate 2, G–J). In the digenetic cycle, there are two hosts, one an invertebrate and the other either a plant (milkweed) or a vertebrate (Plate 2, K–M). Only in the genus *Trypanosoma* does a trypomastigote—the most advanced body form —appear. It is characteristic of this genus (Plate 2, M).

### Monogenetic Trypanosomatids

The genera in this group include *Leptomonas*, *Herpetomonas*, *Crithidia*, and *Blastocrithidia*. *Blastocrithidia* is the most advanced of the monogenetic trypanosomatids, having an amastigote, promastigote, and epimastigote in its life cycle.

### Leptomonas

This genus is the simplest and most generalized of the Trypanosomatidae. The developmental forms consist of an amastigote and a promastigote (Plate 2, G).

The promastigote is pointed posteriorly and rounded anteriorly. A short reservoir opens narrowly at the anterior end, through which the flagellum passes, beginning near the kinetoplast and lying in the anterior fourth or fifth of the body. There are many species that infect invertebrates, including nematodes, mollusks, and arthropods.

*Leptomonas oncopelti* in the spotted milkweed bug (*Oncopeltus fasciatus*) is representative of the species. Bugs become infected when the cyst-like amastigotes voided with the feces are ingested. In the crop, the flagellates are freed. A stimulus supplied by the host initiates growth and multiplication by binary fission. Numerous promastigotes are formed. In the pylorus and midgut, unequal fission, or budding, is the principal mode of reproduction. The small amastigotes arising from the budding process persist in the rectum, where multiplication continues. They are the infective stages that appear in the feces.

*Leptomonas ctenocephali* in the gut of dog fleas (*Ctenocephalides canis*) has a similar life cycle. The mandibulate larvae probably acquire the infections by eating the amastigotes in the feces of one another and of adult fleas (Plate 2, G).

### Crithidia

Species of *Crithidia* are common parasites of the alimentary canal of many different kinds of insects. As a group, the genus is easily recognized.

The flagellates have small, short, wide bodies that are truncated anteriorly and broadly rounded posteriorly. A funnel-shaped reservoir opens on the anterior end from which the flagellum emerges. There is no undulating membrane. The nucleus is in the posterior part of the body and the kinetoplast slightly anterior, being displaced laterally to the base of the reservoir. The flagellated form is a choanomastigote and characteristic of this genus (Plate 2, B).

*Crithidia fasciculata* (Plate 2, H) occurs in the intestines of mosquitoes. The swimming choanomastigotes are 6 to 8 $\mu$ long by 2 to 3 $\mu$ wide. Smaller rounded amastigotes attached to the intestinal mucosa are 3 to 4 $\mu$ long by 2 to 4 $\mu$ wide.

Infection results from ingestion of amasti-

## Explanation of Plate 2

A, *Leishmania* (amastigote form). B, *Crithidia* (choanomastigote). C, *Leptomonas* (promastigote). D, *Herpetomonas* (opisthomastigote). E, *Blastocrithidia* (epimastigote). F, *Trypanosoma* (trypomastigote). G, Life cycle of *Leptomonas ctenocephali* with amastigote and promastigote in dog flea (*Ctencephalides canis*). H, Life cycle of *Crithidia fasciculata* with amastigote and choanomastigote in mosquitoes. I, Life cycle of *Herpetomonas muscarum* with amastigote, promastigote, and opisthomastigote in houseflies. J, Life cycle of *Blastocrithidia gerridis* with amastigote, promastigote, and epimastigote in water striders. K, Life cycle of *Phytomonas elmassiani* with amastigote and promastigote from milkweed bug (*Oncopeltus fasciatus*) and milkweeds (*Asclepias servus*). L, Life cycle of *Leishmania donovani* with promastigotes in sand flies (*Phlebotomus*) and amastigotes in reticuloendothelial cells of humans, dogs, and other mammals. M, Life cycle of *Trypanosoma lewisi* with trypomastigotes in blood of rats and epimastigotes, promastigotes, and amastigotes in the gut of rat fleas (*Nosopsyllus fasciatus*).

1, Nucleus; 2, blepharoplast; 3, kinetoplast; 4, axoneme; 5, flagellum; 6, undulating membrane; 7, 8, contractile vacuole.

a, Larval flea; b, promastigote; c, amastigote; d, adult mosquito; e, larval mosquito (wriggler); f, choanomastigote; g, amastigote; h, housefly; i, fully developed opisthomastigote; j, developing opisthomastigote; k, promastigote; l, amastigote; m, water strider; n, epimastigote; o, promastigote; p, amastigote; q, common milkweed bug; r, milkweed; s, promastigote; t, amastigote; u, v, mammalian hosts (dogs, humans); w, sand fly invertebrate host; x, promastigote; y, amastigote; a', flea invertebrate host; b', rat vertebrate host; c', trypomastigote; d', epimastigote; e', promastigote; f', amastigote.

Figures A–F adapted from Clark, 1959, J. Protozool. 6:227 and Wallace, 1963, Proc. 1st Int. Conf. Protozool., Prague, p. 70.

---

gotes voided with feces that contaminate the habitat. Adult mosquitoes feeding on nectar contaminate the flowers with infective amastigotes which are ingested by other mosquitoes subsequently visiting the blossoms in search of food. Mosquitoes flying or blown over marshes, ponds, pools, or other aquatic habitats contaminate them with their feces and dead bodies, thus assuring continuous and widespread distribution of the parasites.

Larval mosquitoes acquire their infections by ingesting amastigotes which are constantly being passed into the aquatic habitat. While infections are lost with each molt of the larval instars when the peritrophic membrane with its load of parasites is shed, new infections are quickly acquired with the food of foraging larvae. Natural infections in larval mosquitoes consist mostly of multiplying amastigotes in the rectum.

When fourth instar larvae transform to pupae, remnants of the larval gut with its contents, including the parasites, are retained and become the source of infection. Multiplication of the amastigotes resumes as soon as the pupae are formed. Parasites in molting pupae are passed on to the newly formed adults through the metamorphosing gut. Every stage in the life cycle of the mosquito host is involved in the multiplication and transmission of the parasite.

Isolates of clones of *Crithidia* indicate there are many species which have a rather loose host specificity. Serological tests group them into dipteran, hemipteran, and hymenopteran classes.

### Herpetomonas

Species of this genus are commonly found in the intestine of houseflies and blowflies. The final stage is an opisthomastigote.

The body is long, slender, and pointed or truncated posteriorly. The kinetoplast may be located anywhere between the anterior and posterior ends, depending on the stage of development between the promastigote and opisthomastigote. A narrow reservoir containing the axoneme extends from near the kinetoplast to a small opening at the anterior tip of the body through which the flagellum emerges. There is no undulating membrane.

*Herpetomonas muscarum* (Plate 2, I) is the common species in houseflies. The body of specimens from flies is 12 to 18 $\mu$ long by 2 $\mu$ wide. Those grown in cultures are larger. In addition to the opisthomastigotes characteristic of *Herpetomonas*, old cultures yield amastigote and promastigote forms. Thus, three forms occur in the life cycle of species of this genus. Biflagellate forms are seen as dividing individuals in which the new flagellum and kinetoplast appear before division of the body has been completed.

Infection of new hosts occurs under natural conditions when the encysted amastigotes are ingested. In response to the stimulus provided

**Plate 2**    *Genera of Trypanosomatidae and Their Basic Life Cycles*    23

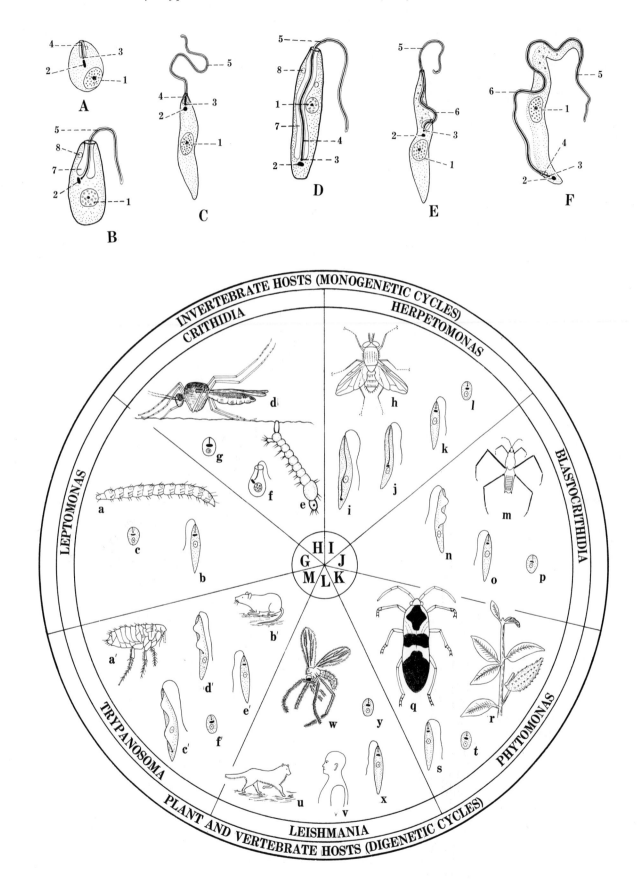

in the gut of the host, the amastigotes develop into promastigotes which multiply by binary fission. As these mature, the kinetoplast migrates posteriorly to a position caudad from the nucleus. Rounded amastigotes are formed and voided with the feces.

### Blastocrithidia

This genus is characterized by the epimastigote stage with its pointed ends and kinetoplast usually near the anterior or lateral margin of the nucleus (Plate 2, J). The flagellum passes laterally through a short reservoir and emerges on the side of the body at about the level of the nucleus. It continues anteriorly along the margin of a short undulating membrane, finally extending free beyond the end of the body.

*Blastocrithidia gerridis* occurs commonly in water striders (*Gerris remigis*) and *B. euschisti* in the milkweed bug *Euschistus servus*. In the case of *B. gerridis*, epimastigotes are abundant in the first stomach, with promastigotes and amastigotes in the rectum of water striders.

In axenic cultures of *Blastocrithidia euschisti*, only promastigotes and epimastigotes appear. When injected into the rectum of assassin bugs (*Triatoma infestans*), all the stages seen in *B. gerridis* are present.

### EXERCISE ON LIFE CYCLE

Houseflies, blowflies, water striders, and mosquito larvae provide a ready source of living material of the monogenetic trypanosomatids.

In order to find the parasites in insects, the alimentary canal should be removed to a microscope slide, teased apart in a drop of physiological saline, covered with a cover glass, and examined, using the lowest illumination feasible. A phase contrast microscope is preferable for seeing the reservoir and contractile vacuole. Smears of the contents of infected intestines stained with Giemsa blood stain are necessary for detailed study of the internal features.

The flagellates may be cultivated by placing a piece of infected mid- or hindgut in tubes of Neopeptone blood agar prepared as follows:

> Neopeptone (Difco), 18 gm
> NaCl, 4.95 gm
> Water, 900 ml
> Agar, 18 gm
> Put 5 ml of the medium in tubes, add 0.5 or 1 ml of fresh rabbit bloodslant tubes, and allow

mixture to harden. Two to 400 units each of streptomycin and penicillin added to each tube will prevent growth of bacteria. Incubate the tubes at 20°C.

### SELECTED REFERENCES

Clark, T. B. (1959). Comparative morphology of four genera of Trypanosomatide. J. Protozool. 6: 227–232.

Clark, T. B., W. B. Kellen, J. E. Lindgren, and T. A. Smith. (1969). The transmission of *Crithidia fasciculata* Leger 1902 in *Culiseta incidens* (Thomson). J. Protozool. 11:400–402.

Drbohlav, S. T. (1926). The cultivation of *Herpetomonas muscarum* (Leidy 1856) Kent 1881 from *Lucilia sericata*. J. Parasitol. 12:183–190.

Hanson, W. L., and R. B. McGhee. (1961). The biology and morphology of *Crithidia acanthocephali* n. sp., and *Leptomonas leptoglossi* n. sp., and *Blastocrithidia euschisti* n. sp. J. Protozool. 8:200–204.

Hanson, W. L., R. B. McGhee, and J. H. DeBoe. (1968). Experimental infection of *Triatoma infestans* and *Rhodnius prolixus* with Trypanosomatidae of the genera *Crithidia* and *Blastocrithidia*. J. Protozool. 15:346–349.

Hoare, C., and F. G. Wallace. (1966). Developmental stages of trypanosomatid flagellates: A new terminology. Nature 212:1385–1386.

Kusel, J. P., K. E. Moore, and M. M. Weber. (1967). The ultrastructure of *Crithidia fasciculata* and morphological changes induced by growth in acriflavin. J. Protozool. 14:283–296.

Laird, M. (1959). *Blastocrithidia* n. g. (Mastigophora: Protomonadina) for *Crithidia* (in part) with a subarctic record for *B. gerridis* (Patton). Can. J. Zool. 37:749–752.

McGhee, R. B., and W. L. Hanson. (1964). Comparison of life cycle of *Leptomonas oncopelti* and *Phytomonas elmassiani*. J. Protozool. 11:555–562.

McGhee, R. B., W. L. Hanson, and S. M. Schmittner. (1969). Isolation, cloning, and determination of biologic characteristics of five new species of *Crithidia*. J. Protozool. 16:514–520.

Wallace, F. G. (1943). Flagellate parasites of mosquitoes with special reference to *Crithidia fasciculata* Léger, 1902. J. Parasitol. 29:196–205.

Wallace, F. G. (1963). Criteria for the differentiation of genera among the trypanasomid parasites of insects. Progr. Protozool. Proc. 1st Int. Conf. Protozool. (Prague, Aug. 22–31, 1961), pp. 70–74.

Wallace, F. G. (1966). The trypanosomatid parasites of insects and arachnids. Exp. Parasitol. 18: 124–193.

Wallace, F. G., and T. B. Clark. (1959). Flagellate

parasites of the fly *Phaenicia sericata* (Meigen). J. Protozool. 6:58–61.

### Digenetic Trypanosomatids

Trypanosomatids with a digenetic type of life cycle include an invertebrate as one host and a plant or a vertebrate as the other. The genus *Phytomonas* completes its cycle in milkweed bugs and milkweed plants. Genera occurring in vertebrates include *Leishmania*, with triatome bugs and mammals as hosts, and *Trypanosoma* requiring leeches or bloodsucking insects as the invertebrate hosts and vertebrates of various kinds as the alternate host.

### Trypanosomatids of bugs and plants

Trypanosomatids occurring in the latex of milkweeds belong to the genus *Phytomonas*. They are transmitted by a number of species of sap-sucking hemipteran bugs whose piercing mouth parts serve in transferring the parasites from plant to plant.

*Phytomonas* resembles the monogenetic genus *Leptomonas* of insects and the digenetic *Leishmania* of mammals in having an amastigote and a promastigote in its life cycle. *Phytomonas* might be considered as a transitional form between the monogenetic species in invertebrates and the digenetic ones having an alternation of hosts between insects and vertebrates.

**Morphology.**    The pronuclear portion of the body of the promastigotes of *Phytomonas* is cylindrical and the postnuclear section is flattened and twisted screw-like.

**Life Cycle.**    *Phytomonas elmassiani* is parasitic in the milkweed *Asclepias syrica* and the spotted milkweed bug *Oncopeltus fasciatus*, both of which are widespread throughout this country (Plate 2, K).

Amastigotes of relatively few numbers and numerous small promastigotes occur in the latex of the milkweeds. When ingested by the bugs with the latex, multiplication of the flagellates ceases temporarily and no growth takes place in the esophagus and crop. Upon arriving in the pylorus and midgut, rapid growth begins, producing giant promastigotes. These large forms leave the digestive tract for the hemocoel, where growth continues. From here, they go into the salivary glands. In these organs, reproduction is resumed with the formation of a few amastigotes and numerous small promastigotes similar in size and form to those appearing in the milkweeds when ingested by the insects.

Bugs sucking latex from the milkweeds inject the parasites with the salivary secretions. Inside the plants, the parasites remain small, being 13.5 $\mu$ long by 2.5 $\mu$ wide, but they proliferate rapidly by binary fission, forming both promastigotes and amastigotes. At first, they are localized in the area of the bites but the infection soon spreads, becoming generalized, with the flagellates distributed throughout all parts of the plant.

### Trypanosomatids of Mammals and Insects

The trypanosomatid flagellates of mammals include a number of species notorious for the decimating diseases they cause in man and his domestic mammals. The pathogenic species normally occur in tropical and semitropical regions, being limited by the distribution of their insect vectors such as the triatome bugs and tsetses.* A few species that have dropped the insect vector from their life cycles are transmitted mechanically or by venereal contact.

Because of their medical and veterinary importance, much effort has been devoted to an understanding of the taxonomic and biological relationships between the species infecting mammals. A practical and workable classification based on a combination of morphological and biological characteristics has been proposed by Hoare. He divided them into two groups according to 1) morphology of trypanosomes in the blood, 2) stage of multiplication in the vertebrate host, 3) place of development of the metacyclic trypanosomes in the invertebrate host, and 4) the mode of infection of the vertebrate host by the vector. Each major category is based primarily on the site where the metacyclic, or infective, trypanosomes develop in the vector and the means by which they are transmitted to the vertebrate host. Metacyclic trypanosomes developing in the rectum, or posterior station, and voided with the feces are allocated to the section Stercoraria (*sterus* (Latin) = dung). The other group developing in the proboscis and salivary glands, or anterior station, and injected with the saliva by feeding tsetses are the section Salivaria. The outline of this classification appears in Table 1.

*According to Glasgow, the word tsetse comes from the Sechuana language spoken in Bechuanaland. Since it refers to these bloodsucking flies, the English usage "tsetse flies" is tautological, in spite of the impressive precedents in the literature.

## Table 1. Classification of Trypanosomes from Mammals

A. SECTION STERCORARIA

Description: Posterior end of body pointed; a free flagellum present; kinetoplast large and in posterior part of body but not near end; multiplication in mammal discontinuous and typically in amastigote or epimastigote stages; generally nonpathogenic (except *Trypanosoma cruzi*); development in posterior station of vector; transmission contaminative through feces. The species in this section were known formerly as the *lewisi* group.

Subgenus *Megatrypanum*
Large body; kinetoplast far from posterior end of body. Representative species: *T. (M.) theileri* (Fig. 1, a) of cattle, *T. (M.) melophagium* of sheep.

Subgenus *Herpetosoma*
Medium-sized body; kinetoplast subterminal. Representative species: *T. (H.) lewisi* (Plate 3, A; Fig. 1, b) of rats, *T. (H.) duttoni* of house mice.

Subgenus *Schizotrypanum*
Small body, C-shaped; kinetoplast very large and near posterior end of body. Representative species: *T. (S.) cruzi* (Plate 4, A; Fig. 1, C) of humans, raccoons, opossums, rodents, *T. (S.) vespertilionis* of bats.

B. SECTION SALIVARIA

Description: Posterior end of body usually blunt; a free flagellum present or absent; multiplication continuous as trypomastigotes in mammals; typically pathogenic; development of metacyclic trypanosomes in anterior station of vector. Some species lack developmental stages in arthropod vectors and are transmitted mechanically or by contact (coitus) of mammalian hosts.

Subgenus *Duttonella* (formerly *vivax* group)
Monomorphic forms with free flagellum; kinetoplast large and terminal; development only in proboscis of tsetses. Representative: *T. (D.) vivax* (Fig. 1, d) of cattle, sheep, goats, antelopes.

Subgenus *Nannomonas* (formerly *congolense* group)
Small, mono- or polymorphic forms; kinetoplast medium-sized and typically marginal near posterior end of body; development in midgut and proboscis of tsetses. Representative species: *T. (N.) congolense* (Fig. 1, e, f), a monomorphic form with a long or short body and with or without a short free flagellum, from cattle, horses, sheep, swine; and *T. (N.) simiae*, a polymorphic form with short body without a free flagellum or long and sometimes stout forms with conspicuous undulating membrane with or without a flagellum, from cattle, horses, camels, swine.

Subgenus *Pyconomonas* (formerly *suis* group)
Small, stout (thick) monomorphic species with pointed posterior end and a free flagellum; kinetoplast small and near tip of body; development in midgut and salivary glands of tsetses. Representative species: *T. (P.) suis* (Fig. 1, g) from pigs.

Subgenus *Trypanozoon* (formerly *bruceievansi* group)
Generally polymorphic forms with free flagellum; develops in midgut and salivary glands of tsetses. Representative species of *brucei* group: *T. (T.) brucei* (Fig. 1, h–j) of domestic animals together with *T. (T.) rhodesiense* and *T. (T.) gambiense* of man; and the *evansi* group which includes species with aberrant life cycles without development in an invertebrate vector, such as *T. (T.) evansi* and *T. (T.) equinum* which are transmitted mechanically to horses, cattle, camels, and others, and *T. (T.) equiperdum* of horses which is transmitted by venereal contact.

*Trypanosoma*

The *Trypanosoma* are parasites of the blood, lymph, tissues, or cavities of all classes of vertebrates. Because of their importance as the causative agents of decimating diseases of man and his domestic mammals, species infecting these hosts have been studied more intensively than those occurring in other vertebrates.

Transmission of the infective stage from one vertebrate host to another may take place in several ways. These include 1) inoculative and contaminative means during feeding of the in-vertebrate vectors, 2) mechanical transfer by vectors, and 3) transmission by venereal contact of two vertebrate hosts.

While taking blood meals, vectors, such as leeches feeding on fish and amphibians and tsetses on mammals and crocodiles, inject infective trypomastigotes directly into the host. In the case of bugs, keds, and fleas, which defecate while feeding, the trypomastigotes are voided with the feces, contaminating the area around the wound which they enter. Mechanical transmission occurs with *T. equinum* and *T. evansi*

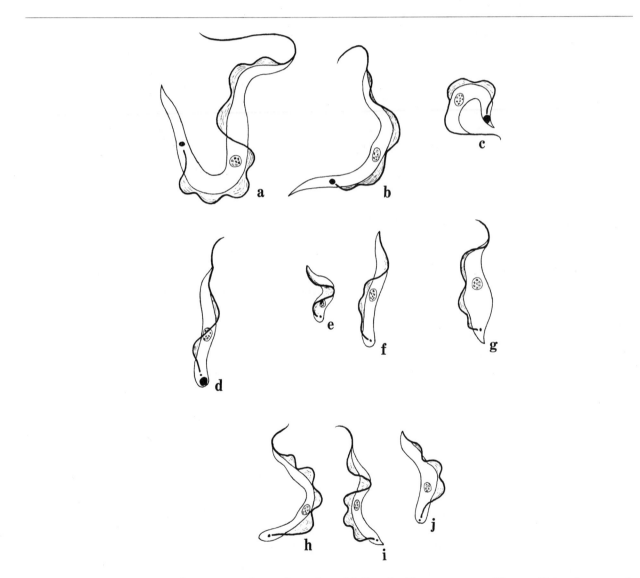

**Fig. 1.** Representative subgenera and species of the genus *Trypanosoma* from mammals. **I**, Stercoraria (a–c): **a**, *Trypanosoma (Megatrypanum) theileri;* **b**, *Trypanosoma (Herpetosoma) lewisi;* **c**, *Trypanosoma (Schizotrypanum) cruzi.* **II.** Salivaria (d–j): **d**, *Trypanosoma (Duttonella) vivax;* **e**, **f**, *Trypanosoma (Nannomonas) congolense;* **g**, *Trypanosoma (Pycnomonas) suis;* and **h**, **i**, **j**, *Trypanosoma (Trypanozoon) brucei.* Adapted from Hoare, 1964.

when infective trypomastigotes are transferred directly from one host to another in blood on the piercing mouth parts of bloodsucking flies, especially tabanids. Transfer of *T. equiperdum* is by venereal contact.

**Morphology.** The stages occurring in the life cycle include four forms. They are the amastigote (leishmanial), promastigote (leptomonad), epimastigote (blastocrithidal) and trypomastigote (trypanosomal) (Plate 2, M).

**Basic Life Cycle.** There are two hosts, a vertebrate and a bloodsucking invertebrate, in the life cycle with specific development of the parasite in each. Exceptions occur in the species *T. evansi*, *T. equinum*, and *T. equiperdum* in which the insect vector has been dropped from the life cycle. Multiplication is by binary fission and budding. When taken into the alimentary canal of the invertebrate vector, the parasite undergoes cyclic transformation through one or more stages such as an amastigote, promastigote, epimastigote, and finally the trypomastigote. The first three forms are not infective to the vertebrate host. Only the trypomastigote is the infective stage and is known as a metatrypanosome or a metacyclic trypanosome. They are small trypomastigotes that develop in the anterior part of the alimentary canal and salivary glands in leeches and tsetses and in the posterior part of the digestive tract of fleas, hippoboscid flies, and bloodsucking bugs.

In rat fleas, keds, and reduviid bugs, the metacyclic trypanosomes of *T. lewisi*, *T. melophagium*, and *T. cruzi*, respectively, develop in the rectum, or posterior station as it is called, and are expelled in the feces. These belong to the section Stercoraria. Infection of the vertebrate hosts in this group is by the contaminative method in which the feces containing the metacyclic trypanosomes are rubbed into the wounds made by the bites or abrasions on the skin.

In tsetses (*Glossina* spp.), development of the metacyclic trypanosomes takes place in the proboscis and salivary glands. This is designated as the anterior station and is where trypanosomes of the *vivax* and *congolense* groups develop. Infection of the vertebrate is through injection of the metacyclic trypanosomes with the saliva of the feeding flies. This is the inoculative means of infection. Species of trypanosomes injected into the vertebrate host with the saliva of the vector belong to the section Salivaria. A similar means of infection occurs between the leech vectors and aquatic vertebrates such as fish and amphibians.

As mentioned above, species of the *evansi-brucei* group, which includes *T. (T.) brucei*, *T. (T.) evansi*, and *T. (T.) equiperdum*, no longer require an invertebrate host for development and transfer. They are passed from one vertebrate to another mechanically by biting flies or through venereal contact. Multiplication takes place in the bloodstream.

### Trypanosomes of mammals

The trypanosomes of mammals will be considered under the two major groups based on whether the metacyclic trypanosomes develop in the posterior or anterior stations of the vector.

### Section Stercoraria

Metacyclic trypanosomes of this group develop in the rectum of the vectors and are discharged from the body in the feces. Morphologically, the posterior end of the trypomastigotes from the blood is pointed and the kinetoplast is subterminal from the posterior end of the body. Biologically, the species of this group are characterized by the site of development in the rectum of the invertebrate vector, manner of infection of the vertebrate host, and place and type of multiplication in the vertebrate host.

The metacyclic trypanosomes develop in the rectum or posterior station, of fleas, keds, hippoboscid flies, and reduviid bugs and are voided in the feces while the insects are feeding. Transmission from vector to vertebrate host through the feces is by the contaminative method. The metacyclic trypanosomes invade the vertebrate host through wounds made by the biting vectors, through abrasions on the skin, or via the intestinal mucosa when infected vectors are swallowed, entering the bloodstream from these various places of contact. Multiplication of the parasites in the mammalian host occurs in different forms and locations. In *T. lewisi* and related species, multiplication is by binary or multiple fission of epimastigotes in the fluid portion of the circulating blood. In *T. cruzi* and similar species, multiplication is through binary fission of amastigotes inside reticuloendothelial cells such as macrophages in the liver, spleen, and bone marrow, and in myocardial fibers.

### I. The *lewisi* group

All the species of Stercoraria in mammals belong to the *lewisi* group, typified by *Trypanosoma lewisi* of rats.

They are characterized morphologically by a

body with both ends pointed and a large kinetoplast separated from the posterior tip. Multiplication of the parasite in the mammalian host is discontinuous and by division of amastigotes, promastigotes, or epimastigotes but not by trypomastigotes. The different locations in the vertebrate host in which reproduction occurs makes it convenient to discuss *T. lewisi* and related species and *T. cruzi* with its relatives separately for a clearer understanding.

### A. The *lewisi* complex

Multiplication in the fluid portion of the blood is by binary or multiple fission of the epimastigotes.

### *Trypanosoma lewisi* (Kent, 1880) (Plate 3)

This species is a cosmopolitan parasite of rats.

**Description.** The slender body of trypomastigotes in stained smears of blood is about 25 $\mu$ long, curved, and with the attenuated posterior end pointed. The nucleus lies anterior to the middle of the body; the kinetoplast is large and located some distance from the posterior tip of the body. A weakly developed undulating membrane with the attached marginal flagellum extends along the full length of the body. The free part of the flagellum is well developed.

**Life Cycle.** Upon entering the bloodstream of rats, the metacyclic trypanosomes from the feces of fleas transform to epimastigotes and appear in 4 to 6 days. Reproduction begins by multiple fission in which division is incomplete so that the newly formed individuals remain attached by the posterior tip of the body, forming rosettes of epimastigotes. When division is completed, they detach from each other and repeat the process. Eventually, division ceases and the rosettes disappear by day 8 to 9. The small epimastigotes transform into trypomastigotes that are infective to rat fleas (*Nosopsyllus fasciatus*) when ingested with the blood. The trypanosomes in the blood gradually decline in number, generally disappearing completely in about 3 to 4 weeks. However, the time may be as short as 3 weeks or as long as 8 weeks.

In order to continue development, the trypanosomes must undergo a period of cyclic development in rat fleas to produce metacyclic forms infective to rats. Within 6 hours after entering a flea's stomach, the trypomastigotes penetrate the epithelial cells and fold to form a U-shaped structure whose arms fuse, producing a pear-shaped body. As these bodies increase in size, the kinetoplast and nucleus divide. New flagella are formed from axonemes arising from daughter kinetoplasts while the original axoneme and its flagellum persist. The flagella are scattered over the surface of the pyriform bodies which produce eight to 10 nuclei and as many kinetoplasts. Division in the epithelial cells terminates within 18 hours or it may continue 4 or 5 days.

When the infected cells rupture, the newly formed trypomastigotes enter the lumen of the stomach. They may penetrate other cells and divide again. Eventually, the small trypomastigotes migrate to the hindgut and rectum, where they transform into epimastigotes which undergo another period of multiplication, producing various forms. The result of this period of multiplication is the production of great numbers of small metacyclic trypanosomes in the rectum. These are the infective stages voided in the feces of feeding fleas.

Rats attempting to remove the biting fleas lick the area and in the process swallow the feces containing the metacyclic trypanosomes which initiate infections in susceptible animals.

**Immunity.** Following infection of rats, the trypanosomes undergo a reproductive period of about 1 week, during which time great numbers of dividing forms appear in the blood. After day 8 or 9, dividing forms disappear but numerous trypomastigotes are present for 1 to 4 months. During this time, the number of trypomastigotes declines and they eventually disappear from the blood. Rats having gone through this period remain immune to subsequent infections.

Disappearance of the trypanosomes and the development of an immunity in the rats to subsequent infection is due to the appearance of two antibodies in the infected animals. One is a reproduction-inhibiting antibody known as ablastin (from Greek, without sprouting) that appears only during infection and subdues reproduction by the end of the first week. The other antibody is trypanocidal, gradually killing the adult trypanosomes sometime between 1 and 4 months after infection. The immunity resulting from the infection continues for the life of the rat without benefit of a low-level residual infection to maintain continuous premunition.

**Pathology and Resistance.** While *T. lewisi* is nonpathogenic to adult rats, young ones die as the result of heavy infections. This lack of resistance in the young may be due to the

## Explanation of Plate 3

**A–H,** Trypanosomes from blood of rat. **A,** Typical mature trypomastigote. **B,** Epimastigote preparing to divide. **C,** Early stage of multiple division forming epimastigotes. **D,** Multiple division of original trypomastigote into rosette of incompletely separated epimastigotes. **E, F,** Free epimastigote from blood. **G,** Young trypomastigote. **H,** Older, fully developed trypomastigote. **I,** Metacyclic trypanosome from rectum of rat flea. **J,** Rat vertebrate host. **K,** Rat flea *(Nosopsyllus fasciatus)* invertebrate host.

1, Characteristic pointed posterior end of mature trypomastigote; 2, flagellum; 3, basal body; 4, kinetoplast; 5, axoneme; 6, undulating membrane; 7, nucleus; 8, basal body and kinetoplast combined; 9, individual epimastigote of rosette.

a, Metacyclic trypanosome from feces of flea being swallowed by rat; b, metacyclic trypanosome migrating through wall of esophagus into blood vessels; c, transforming into an epimastigote and dividing; d, a step further in the process of division; e, beginning of multiple fission forming rosette of epimastigotes in heart; f, further division of epimastigotes in pulmonary artery; g, new generation of epimasti-gotes in pulmonary vein; h, further multiplication of epimastigotes in left ventricle; i, recently transformed trypomastigotes with long, slender posterior ends; j, newly formed trypomastigotes infective to fleas when sucked up by them; k, trypomastigotes in foregut of flea and entering stomach; l, trypomastigotes in stomach of flea often adhere to each other; m, shortened trypomastigotes attached to epithelium by flagella; n, trypomastigote entering epithelial cell; o, trypomastigote assuming pear shape inside cell; p–r, pear-shaped cell enlarging and dividing to form trypomastigotes; s, trypomastigotes produced by intracellular multiplication; t, cell ruptures and trypomastigotes enter lumen of stomach, some re-entering cells **(n)** to begin a new generation; u, trypomastigotes enter rectum; v, trypomastigotes attach to rectal wall and undergo transformation leading to epimastigotes; w, multiplying epimastigotes; x, transformation of epimastigotes to metacyclic trypanosomes which fill rectum; y, metacyclic trypanosomes voided with feces while fleas are feeding; z, infective metacyclic trypanosomes in feces.

Protozoa adapted from Minchin and Thomson, 1915, Quart J. Microsc. Sci. 60:463.

---

inability of the reticuloendothelial system of the host at this stage of development to produce sufficient antibodies to overcome and destroy the parasites. Age resistance to parasitic infections is a common phenomenon.

**Other Species.** Related species of *Trypanosoma* occur in different mammals. *Trypanosoma duttoni* is parasitic in house mice. Its life cycle and activities are similar to those of *T. lewisi*. *Trypanosoma melophagium* of sheep is extremely rare in blood smears but can be demonstrated readily by cultures prepared from the blood. They are abundant in the intestine of keds (*Melophagus ovinus*), the invertebrate vector, in whose intestine cyclic development similar to that of *T. lewisi* in fleas takes place.

## EXERCISE ON LIFE CYCLE

A ready source of mammalian trypanosomes for experimental studies is wild rats infected with *T. lewisi*. A drop of blood from an infected rat examined under the 4-mm lens of a microscope with reduced light will reveal the presence of the active parasites by the movement of the erythrocytes. A phase contrast microscope makes the living trypanosomes visible. Blood (3 to 5 ml) from an infected rat injected into the body cavity of young white rats will initiate an infection. The trypanosomes are able to penetrate the peritoneal lining and enter the bloodstream where they appear in 4 to 5 days. Rosettes of dividing epimastigotes appear by day 6 and are gone by days 8 or 9. After this time, only trypomastigotes are present. They decline in number, usually disappearing by the end of week 4, but it may be as early as week 3 or as late as week 8. Recovered rats are immune to reinfection. Challenge the immunity of a recovered rat by injecting blood from an infected animal.

Dividing trypomastigotes occur in the mucosal cells of the stomach of rat fleas and epimastigotes in the cells of the midgut. Metacyclic trypanosomes are present in the lumen of the rectum. Examine live and stained specimens for the different stages of the parasite.

Experimentally infect young white rats from infected rat fleas. Feed some rats entire fleas and others only feces obtained from fleas. Rub feces of fleas into an abraded area of skin of another rat. Examine the blood of each rat for parasites 4 to 6 days after these means of exposure to infection.

If desired, blood of sheep may be cultured, using the method described by Hoare (1923), Noble (1955), or Simpson and Green (1959), to demonstrate the presence of *T. melophagium*.

**Plate 3**  *Trypanosoma lewisi*  31

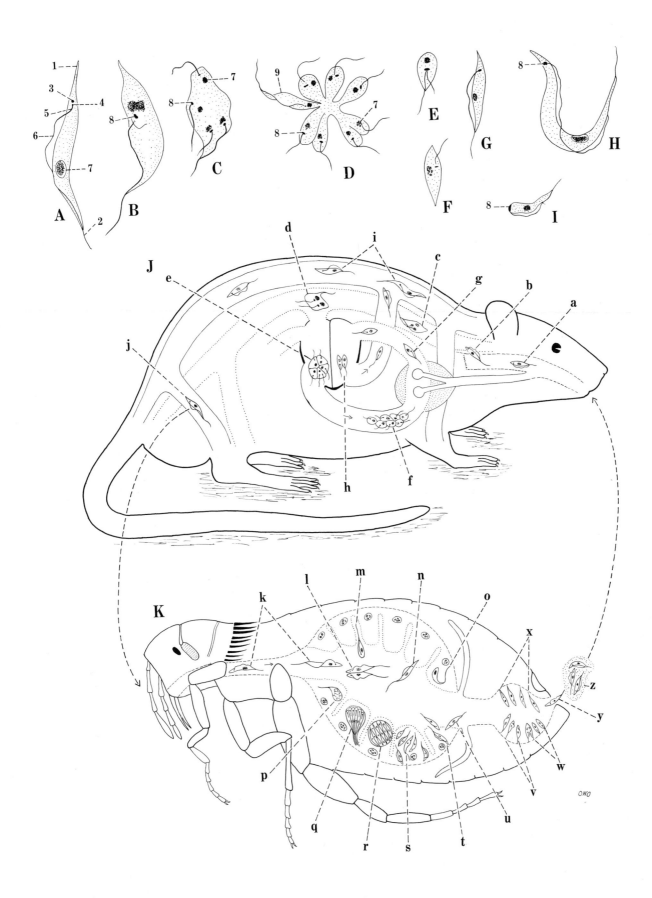

## SELECTED REFERENCES

Doflein, F., and E. Reichenow. (1953). Lehrbuch der Protozoenkunde. 6th ed. Gustav Fischer, Jena, p. 528.

Glasgow, J. P. (1963). The distribution and abundance of tsetse. Pergamon Press, Oxford, p. 3.

Hoare, C. A. (1923). An experimental study of the sheep-trypanosome (*T. melophagium* Flu, 1908), and its transmission by the sheep-ked (*Melophagus ovinus* L.). Parasitology 15:365–424.

Hoare, C. A. (1936). Morphological and taxonomic studies on mammalian trypanosomes. I. The method of reproduction in its bearing upon classification, with special reference to the *lewisi*-group. Parasitology 28:98–109.

Kartman, L. (1954). Observations on *Trypanosoma lewisi* and *Grahamella* sp. in the blood of rats from the Hamakua District, Island of Hawaii. J. Parasitol. 4:571–579.

Levine, N. D. (1961). Protozoan Parasites of Domestic Animals and of Man. Burgess Publishing Co., Minneapolis, p. 43.

Minchin, E. A., and J. D. Thomson. (1915). The rat trypanosome, *Trypanosoma lewisi,* in its relation to the rat-flea, *Ceratophyllus fasciatus.* Quart. J. Microsc. Sci. n. s. 60:463–692.

Noble, E. R. (1955). The morphology and life cycles of trypanosomes. Quart. Rev. Biol. 30:1–28.

Simpson, C. F., and J. H. Green (1959). Cultivation of *Trypanosoma theileri* in liquid medium at 37°C. Cornell Vet. 49:192–193.

Taliaferro, W. H. (1932). Trypanocidal and reproduction-inhibiting antibodies to *Trypanosoma lewisi* in rats and rabbits. Amer. J. Hyg. 16:32–84.

Wenyon, C. M. (1926). Protozoology. William Wood and Co., New York (Reprinted by Hafner Publishing Co., New York, 1965), Vol. 1, p. 463.

### B. The *cruzi* complex

*Trypanosoma cruzi* and *T. rangeli* are parasites of humans and a number of animal reservoir hosts in the tropical regions of the western hemisphere. The geographic range of *T. cruzi* extends northward into the southeastern, southern, and southwestern parts of the United States, where it is parasitic normally in reservoir hosts and bug vectors. At least one autochthonous case of infection of a child has been reported from North America in southern Texas. *Trypanosoma cruzi* is pathogenic to humans, whereas *T. rangeli* is not.

### *Trypanosoma cruzi*
(Chagas, 1909) (Plate 4)

In the vertebrate hosts, the principal sites of *T. cruzi* are blood, reticuloendothelial cells of the blood vessels, liver, spleen, lymph glands, bone marrow, the glial cells, and cardiac and skeletal muscles. Other tissues are also invaded.

In addition to man, common natural reservoir hosts include bats, opossums, raccoons, armadillos, dogs, cats, and pigs in South and Central America. In the United States, wood rats (*Neotoma* spp.) are the most common reservoir host, but raccoons and opossums are infected.

The invertebrate hosts include a number of species of reduviid bugs. The common ones in the tropics are *Panstrongylus megistus, Triatoma infestans,* and *Rhodnius prolixus.* In North America, more than a dozen species of reduviid bugs are naturally infected with *T. cruzi.* Trypanosomes isolated from *Triatoma heidemani* from Texas produced infections of *T. cruzi* with the characteristic eye lesion in humans.

**Description.**    Both slender and broad trypanosomes averaging 20 $\mu$ in length occur in the blood. The body of stained specimens in blood smears is U- or C-shaped. The posterior end of the body is pointed, the nucleus is located near the middle of the body, and the dark kinetoplast, near but not at the posterior end of the body, is so large as to produce bulges on each side of the body. The undulating membrane is weakly developed, having two or three convolutions, and the flagellum is always free. Intracellular amastigotes are oval in shape and measure 1.5 to 4 $\mu$ in diameter.

**Life Cycle.**    The life history of *Trypanosoma cruzi* consists of two cyclic parts, one in the vertebrate host and the other in a number of species of hemipteran vectors.

Upon entering the skin of the vertebrate host through punctures or abrasions of the skin, the metacyclic trypanosomes do not undergo multiplication in the blood. After entering the reticuloendothelial cells of the liver, spleen, glial cells, and myocardial and skeletal muscles, the trypomastigotes transform to amastigotes by first going through the epimastigote stage. Inside the various cells, the amastigotes multiply by binary fission, producing numerous individuals. When the destroyed cells rupture, the parasites are liberated in the blood, where they transform successively to promastigotes, epimastigotes, and finally trypomastigotes. No further multiplication occurs. Trypomastigotes in the peripheral blood are infective to the arthropod vectors.

When sucked up with blood by reduviid bugs, the trypomastigotes begin the cycle of development in the intestine. They first transform to epimastigotes in the midgut, where the initial process of multiplication by binary fission occurs. By the end of 8 to 10 days of multiplication, the epimastigotes change to small metacyclic trypanosomes in the rectum, known as the posterior station, of the bugs.

While feeding, preferably on the eyelids or lips, the bugs habitually defecate, voiding feces that contain numerous metacyclic trypanosomes. Infection of the vertebrate host is by the contaminative process in which fecal matter enters the skin through punctures made by the biting bugs or through abrasions of the skin.

**Reservoir Hosts and Zoonoses.** *Trypanosoma cruzi* occurs in numerous wild animals, including bats, opossums, raccoons, armadillos, dogs, and cats in South and Central America, and wood rats, raccoons, and opossums in North America. These and other mammals serve as natural reservoirs of trypanosomes from which humans are infected. The bloodsucking reduviid bugs depend on these mammals, among others, for their food. Thus like most parasites, *T. cruzi* is tied in with the food chain of one of its hosts for completion of its life cycle and transmission to new hosts.

The occurrence of *T. cruzi* in humans is a zoonotic relationship in that the parasites occur naturally in animal reservoirs from which they are transmitted to man.

**Epidemiology.** The occurrence of *T. cruzi* in humans is dependent on the bites of infected bug vectors which acquire their infections from the wild reservoir mammalian hosts. In the tropics, the nocturnal bugs utilize the mud and thatched houses as habitats for resting during the days and for hunting at nights when they feed on the sleeping people. Infection occurs most commonly in rural and ghetto areas where poor housing provides a suitable habitat for the bugs.

In North America, the conenosed bugs (*Triatoma protracta* and others) live in the houses of wood rats and burrows of other hosts where they are in close contact with their principal sources of food, from which they acquire infections of *T. cruzi*. These bugs frequently migrate to human habitations, possibly attracted by lights at night, when they are most active.

The common occurrence of trypanosomes in the urine of infected wild mammals suggests that it might be the source of transfer between these hosts in areas such as the southeastern United States, where triatome bugs are rare.

**Pathology.** Trypanosomiasis cruzi appears in an acute form especially in children and in a chronic stage mainly in adult humans. It apparently does not affect the reservoir hosts.

Early symptoms of the acute phase appear as swelling of one, usually, or both eyelids where the metacyclic trypanosomes have entered the tissues at the site of the bug bites. The inflammatory swellings at the location of the bites are known as chagomas, named after Chagas, who discovered the organism and the disease it causes. Later, enlargement of the cervical and other lymph glands, liver, and spleen occurs. Anemia, headache, and fever are prevalent.

Upon reaching the tissues, the trypanosomes enter or are engulfed by different cells, where they transform to amastigotes and multiply. From here, they eventually enter and multiply in the major lymph nodes, liver, spleen, glial cells, macrophages, and myocardium, as well as other tissues and organs. Destruction of the cells is the end result of infection.

The chronic form of the disease usually appears in adults and may vary in the different geographic areas. In general, the results stem from damage initiated in the acute stage. These include cardiac damage resulting in heart dysfunction, and enlargement of the esophagus and colon, causing impaired peristalsis. Additional manifestations involve parts that are affected such as nerve tissues, reticuloendothelial system, and others.

*Trypanosoma rangeli* Téjera, 1920

This is a South American species that occurs in the blood of man and reservoir hosts such as dogs, cats, opossums, and monkeys. Possibly other hosts occur. It is nonpathogenic to the vertebrate hosts, including man, but may be injurious to the bug vector *Rhodnius prolixus*.

The trypomastigotes leave the intestine of the vector, enter the hemocoel, and transform to metacyclic trypanosomes that are transmitted through the bite by the inoculative method rather than in the feces by the contaminative route as occurs in the other species of the *lewisi* group.

## Explanation of Plate 4

**A,** Typical C-shaped trypomastigote from blood of vertebrate host. **B,** Reduviidae bug *(Panstrongylus megistus)* invertebrate host. **C,** Side view of head of bug. **D–I,** Some common vertebrate hosts. **D,** humans. **E,** bats. **F,** opossums. **G,** armadillos. **H,** dogs. **I,** cats (raccoons also serve as natural reservoirs).

**1,** Nucleus; **2,** basal body; **3,** kinetoplast; **4,** axoneme; **5,** undulating membrane; **6,** free flagellum; **7,** proboscis in resting position.

**a,** Metacyclic trypanosome that has entered blood from feces through wound made by feeding bug; **b,** trypanosome enters macrophage cell in blood; **c,** trypanosome transforms to amastigote stage; **d,** multiplication of amastigote; **e,** amastigotes transform to epimastigotes that multiply; **f,** epimastigotes transform to trypomastigotes which rupture cells and escape into blood; **g,** trypomastigotes free in blood; **h–n,** repetition of cycle depicted in a–g with apearance of trypomastigotes in blood for second time (trypomastigotes appear in blood periodically by repetition of reproductive cycle); **o,** trypomastigotes enter skeletal muscles, as well as cardiac muscles, also reticuloendothelial cells of liver, spleen, and bone marrow; **p,** trypomastigotes transform to amastigotes in cyst-like bodies; **q,** amastigotes multiply, forming new amastigotes in cyst-like bodies; **r,** amastigotes transform to epimastigotes; **s,** epimastigotes transform to trypomastigotes and cyst ruptures; **t,** trypomastigotes are freed; **u–x,** repetition of o–t; **y,** nests of amastigotes in heart muscle; **z,** nest of amastigotes in brain.

**a′,** Bug invertebrate host sucking up trypomastigotes from blood; **b′,** trypomastigote passing through esophagus; **c′,** epimastigotes in stomach transform from trypomastigotes; **d′,** dividing epimastigote in stomach; **e′,** multiplication of epimastigotes in midgut; **f′,** epimastigotes in rectum transform to metacyclic trypanosomes; **g′,** metacyclic trypanosomes in feces infect host through wound made by bug while feeding or by contamination of mucous membranes of lips (**h′**) or eyes (**i′**).

Figures A, B, part of C, and those of protozoa adapted from Medical Protozoology and Helminthology, 1955, Naval Medical School, Bethesda, Maryland, p. 92.

## SELECTED REFERENCES

Doflein, F., and E. Reichenow. (1953). Lehrbuch der Protozoenkunde. 6th ed. Gustav Fischer, Jena, p. 522.

Elkeles, G. (1951). On the life cycle of *Trypanosoma cruzi*. J. Parasitol. 37:379–386.

Faust, E. C. (1949). The etiological agent of Chagas' disease in the United States. Bol. Ofic. Sanit. Panamer. 28:455–461.

Faust, E. C., P. F. Russell, and R. C. Jung. (1970). Craig and Faust's Clinical Parasitology. 8th ed. Lea & Febiger, Philadelphia, p. 113.

McKeever, S., G. W. Gorman, and L. Norman. (1958). Occurrence of a *Trypanosoma cruzi*-like organism in some mammals from southwestern Georgia and northwestern Florida. J. Parasitol. 44: 583–587.

Mehringer, P. J., and S. F. Wood. (1958). A resampling of wood rat houses and human habitations in Griffith Park, Los Angeles, for *Triatoma protracta* and *Trypanosoma cruzi*. Bull. S. Calif. Acad. Sci. 57:39–46.

Packchanian, A. (1943). Infectivity of the Texas strain of *Trypanosoma cruzi* to man. Amer. J. Trop. Med. 23:309–314.

WHO. (1969). Comparative studies of American and African trypanosomiasis. World Health Organ. Tech. Rep. Ser. 411.

Woody, N. C., and H. B. Woody. (1955). American trypanosomiasis (Chagas' disease). First indigenous case in the United States. J. Am. Med. Ass. 159: 676–677.

### Section Salivaria

The species of Salivaria occur primarily in tropical Africa. The distribution is predetermined for most species by the geographic distribution of the tsetses (*Glossina* spp.), which serve as the invertebrate vectors in whose bodies cyclic development takes place.

Generally, the African species of trypanosomes are pathogenic to man and his domestic mammals, causing severe disease and much mortality in both. They are nonpathogenic to the wild reservoir hosts. The salivarian species form a compact group that is closely related morphologically and biologically.

Morphologically, these trypanosomes are characterized by a body that is blunt posteriorly and with the kinetoplast terminal or subterminal. The trypomastigotes of a single species in the blood may appear as different morphological forms. Some, like *T. vivax*, are monomorphic, being uniform in size and structure. Others, such as *T. gambiense* and *T. congolense*, are polymorphic, in which case the body of some specimens is stumpy and aflagellate while in others it is slender and flagellate. Forms of intermediate size occur (Fig. 1, d–j).

**Plate 4** *Trypanosoma cruzi* 35

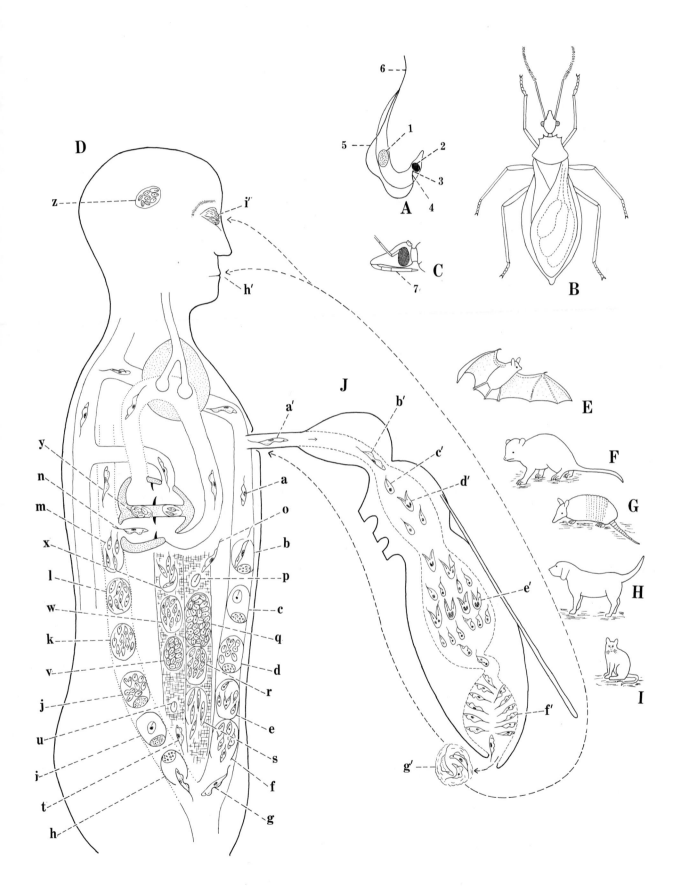

Studies on the ultrastructure of *T. brucei* provide significant insights to some structural and physiological properties of the different morphological forms. The form in the blood of the mammalian host and that developing in the midgut of the tsetse have some distinct morphological differences. These are thought to be associated with the physiological requirements imposed by the habitats of the trypanosomes in the different hosts.

The basic morphology of the trypomastigote in the blood of the mammal and the epimastigote in the midgut of the tsetse is similar. In the forms from the bloodstream (Fig. 2, A), the flagellum arises from the reservoir formed by an inpocketing of the pellicle at a point near the posterior end of the body. The flagellum extends along the side of the body, being attached by an accessory membrane of lattice-like structure. In cross section, the flagellum consists of a pellicle-like membrane surrounding a core formed of an outer ring of nine double and two central fibrils (Fig. 2, B). The internal end of the axoneme consists of the basal body. Close to it, but not obviously connected, is the disc-shaped kinetoplast, containing DNA. Extending anteriorly from the kinetoplast is a large, simple, tubular mitochondrion. Longitudinal fibers on the underside of the pellicle extend the length of the body and are thought to be contractile, thereby providing means for body movement. Other internal structures in the cytoplasm include granules believed to be volutin and lysosomes. A Golgi apparatus and a tubular reticulum are present.

Forms in the midgut (Fig. 2, C), as well as in cultures, differ in some important and significant ways after transformation first to the epimastigotes and finally to the metacyclic trypanosomes. The kinetoplast is located far anterior

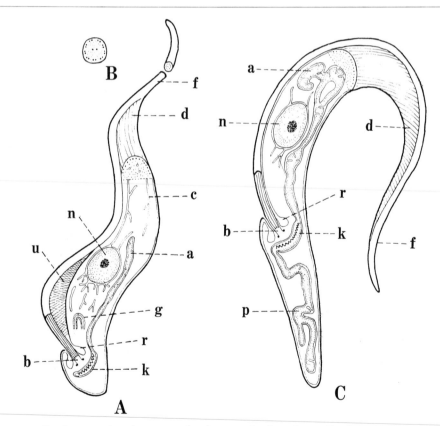

**Fig. 2.** Diagrammatic figures showing some basic ultrastructure of *Trypanosoma*. A, Trypomastigote. B, Cross section of flagellum showing nine double peripheral fibrils and two central fibrils. C, Epimastigote in transitional stage. **a,** Anterior mitochondrium; **b,** basal body of flagellum; **c,** tubule of cytoplasmic reticulum; **d,** pellicular fibrils; **f,** flagellum; **g,** Golgi apparatus; **k,** kinetoplast; **n,** nucleus; **p,** posterior mitochondrion; **r,** reservoir; **u,** reticulated undulating membrane. Adapted from Vickerman, 1962, Trans. Roy. Soc. Trop. Med. Hyg. 56:487.

from the end of the body. In addition to the large, highly convoluted anterior mitochondrion, an equally voluminous posterior one has been formed that extends to the hind end of the body. The presence and size of the posterior mitochondrion is responsible for the displacement of the kinetoplast, axoneme, Golgi apparatus, and reservoir. The presence of the posterior mitochondrion and dislocation of the kinetoplast may be associated with physiological requirements of the epimastigotes in the gut of tsetses. Upon transformation of the epimastigotes to the metacyclic trypanosomes, the posterior mitochondrion disappears and the kinetoplast together with axoneme and reservoir move posteriorly to the end of the body. The nucleus moves back from the anterior position occupied during this stage.

The role of the mitochondria may be associated with the physiological requirements of the trypanosomes in their different hosts and are produced by the kinetoplast to meet these needs at specific times. It is postulated that the enlarged, double mitochondrion of the epimastigote might be associated with the greater demands of respiration in the tsetse gut, where the oxygen tension is low, and with utilization of glucose present in limited concentration. Trypomastigotes, on the other hand, in the bloodstream with its high oxygen tension and rich glucose content, do not require the same biochemical action and are able to function with the aid of the single, simple mitochondrion.

If the presence or absence of the posterior mitochrondrion and associated changes in morphology are influenced by the metabolic needs of the parasite, it would appear that these anatomical differences are related to and a function of the physiological requirements of the parasite at specific times in its different hosts.

Biologically, cyclic development takes place only in tsetses. The metacyclic trypanosomes develop in the proboscis, salivary glands, or intestine. Infection of the vertebrate occurs when the metacyclic trypanosomes are injected with the saliva from the anterior station by the flies while sucking blood. Multiplication in the blood of the vertebrate host is by simple binary fission of the trypomastigotes, the only stage present.

All species of trypanosomes can be transferred directly on the mouth parts of bloodsucking insects such as tabanid and stomoxys flies.

In three species, *T. evansi*, *T. equinum*, and *T. equiperdum*, the invertebrate host has been dropped from the life cycle. This condition has made it possible for these species to extend far beyond the geographic range of the tsetses.

The section Salivaria is divided on a biological basis into three parts. They are the *vivax* group, the *congolense* group, and the *brucei-evansi* group. Each group is based on the location in the tsetses where cyclic development takes place.

### I. The *vivax* group

The cyclic development in the tsetses of members of this small group is limited to the upper part of the alimentary tract, including the esophagus and epipharynx. The species belong to the subgenus *Duttonella*. They are parasitic in and pathogenic to cattle, sheep, goats, horses, and camels. Humans are not infected.

While indigenous to Africa, where antelope and giraffes are the natural reservoir hosts and two tsetses, *Glossina morsitans* and *G. tachinoides*, are the principal vectors, these trypanosomes have been distributed through transportation of livestock to other parts of the world beyond the range of the tsetses, including the West Indies and Central and South America, where deer serve as reservoir hosts. In the absence of the tsetses, the trypanosomes have adapted to mechanical transfer by bloodsucking flies, particularly tabanids, stomoxys, and their relatives, as the only means of infection between mammalian hosts.

*Trypanosoma vivax*
Zieman, 1905 (Plate 5, I)

This species is widespread in tropical Africa, where it infects antelope and giraffes as the principal reservoir hosts, in which it is nonpathogenic. Its importance is due to the disease known as nagana that it produces in domestic animals, especially cattle and sheep.

**Description.** The body of the trypomastigotes, the only stage in the blood, is 20 to 26 $\mu$ long, typically with a decidedly blunt posterior end (Fig. 1, d). A large terminal kinetoplast, a free flagellum, and an inconspicuous undulating membrane are present. They are monomorphic. The trypanosomes travel rapidly across the field of the microscope in fresh blood smears.

**Life Cycle.** Trypomastigotes circulating in the blood of the vertebrate host multiply by binary fission, producing only similar forms. When sucked up by tsetses, particularly *Glossina* *palpalis*, *G. morsitans*, and *G. tachinoides*, they undergo a process of cyclic reproduction in the anterior portion of the digestive tract, where stages of the parasite appear. In the esophagus, the trypomastigotes transform to epimastigotes that multiply and migrate to the pharyngeal region. Upon arrival there, they attach and continue multiplication. The mass of newly formed epimastigotes move into the pharynx and hypopharynx, where they transform into small metacyclic trypanosomes, the final and infective stage. All of the cyclic development in *T. vivax* takes place in the pharynx and hypopharynx of the tsetse vectors. The infective trypanosomes are injected into the blood of the vertebrate host with the saliva of the feeding flies. In the vertebrate, the metacyclic trypanosomes mature and commence multiplication.

In regions outside the geographic range of the tsetse vectors where *T. vivax* is established, transmission is mechanical. Tabanids and stomoxys flies interrupted while feeding on an infected animal may transmit viable trypanosomes on or in the proboscis to susceptible hosts. Upon being introduced into another mammalian host, the trypanosomes multiply in the same manner as when injected by tsetses.

*Trypanosoma uniforme* Bruce *et al.*, 1911 is similar to *T. vivax* except that it is longer. It infects the same mammalian hosts and tsetse vectors, with similar development in each.

## II. The *congolense* group

The cyclic development of this group begins in the stomach and is completed in the proboscis of the tsetses. The trypomastigotes are monomorphic or polymorphic, and may possess a free flagellum. *Trypanosoma congolense* belongs in the subgenus *Nannomonas*, a group of small species. They are parasitic in most kinds of domestic mammals. Principal reservoir hosts are antelope, zebra, and warthogs. Being unable to develop in the absence of tsetses, this group of trypanosomes is limited to the range of the flies.

*Trypanosoma congolense*
Broden, 1904 (Plate 5, II)

This trypanosome was found first in the Congo. It is indigenous to Africa and occurs widely in the tropical belt of this continent.

**Description.**    This is one of the smallest trypanosomes, measuring 9 to 18 $\mu$ (average 14) in length and less than 3 $\mu$ in width. It is mono-

morphic and without a free flagellum. The posterior end of the body is rounded with the kinetoplast nearby and slightly displaced toward the margin. A weakly developed undulating membrane is present. The nucleus is near the middle of the body (Fig. 2, e, f). Although active in blood smears, the trypomastigotes do not travel.

**Life Cycle.**    Upon arriving in the midgut of the tsetses, particularly *Glossina morsitans*, *G. palpalis*, *G. longipalpis*, *G. pallidipes*, and *G. austeni*, the trypomastigotes change in shape from the typical short, thick forms into long, slender ones as the first step in the course of cyclic development. The newly formed trypomastigotes migrate forward from the midgut to the hypopharynx, bypassing the salivary glands. In this new location, they undergo the second morphological change, transforming into epimastigotes that are attached to the walls. A period of multiplication by binary fission follows, ending the second phase of development. The newly formed epimastigotes transform into aflagellate infective metacyclic trypanosomes, terminating the third and final stage of development. Infection of the mammalian host takes place when the metacyclic trypanosomes are inoculated from the anterior station by the feeding flies. The only form present in the blood of the vertebrate host is the aflagellate, stumpy, trypomastigote. Multiplication in the blood is continuous.

## III. The *brucei-evansi* group

The species of this heterogeneous group, belonging to the subgenus *Trypanozoon*, are divided into two subgroups of related trypanosomes whose basic life cycles in the tsetses differ markedly.

The *brucei* subgroup includes *T. brucei*, *T. gambiense*, and *T. rhodesiense*. Being morphologically similar, they are considered by some workers as representatives of a biological race. Each is characterized by the differences in the cyclic development in the tsetses and the clinical manifestations in their respective hosts. *Trypanosoma brucei* is pathogenic to domestic livestock only, *T. rhodesiense* to domestic animals and humans, and *T. gambiense* to man only.

Members of the *evansi* subgroup include the aberrant species of *T. evansi*, *T. equinum*, and *T. equiperdum* that have forsaken the tsetse vector together with the associated cyclic development that normally takes place in them. They are transmitted mechanically by bloodsucking flies or by venereal contact. Because of their liberation from the tsetses, these trypanosomes have a wide geographic distribution beyond Africa. They are pathogenic to cattle, camels, equines, dogs, and other domestic animals.

## A. The *brucei* subgroup (Plate 5, III)

The species are polymorphic, having slender and intermediate flagellate forms, together with stumpy, aflagellate ones. Morphologically, the species in this subgroup are indistinguishable from each other. Their identity is based largely on epidemiological and physiological differences.

### Trypanosoma gambiense
### Dutton, 1902 (Plate 5, III)

*Trypanosoma gambiense* is used as the model representing the life cycle of the species of this subgroup. It, along with *T. rhodesiense*, causes a severe and often fatal disease in humans known commonly as African sleeping sickness. These species are confined to the tropical regions of Africa, between 15° North and 18° South latitude. The principal tsetse vectors are *Glossina palpalis* and *G. tachinoides*, both riverine species, living in wooded areas along streams and lake shores.

**Description.**    These trypanosomes are designated as polymorphic because of the variation in size and shape in the blood. There are long, slender flagellate forms, intermediate flagellate specimens, and short, stumpy, aflagellate representatives present simultaneously.

Stained specimens of the slender forms are 29 to 33 $\mu$ in length. The free flagellum is long, the posterior end of the body somewhat pointed, and the kinetoplast subterminal. Intermediate forms average 23 $\mu$ long and the flagellum is of medium length. The posterior end of the body is blunt and the kinetoplast subterminal. Stumpy forms are stout, 12 to 26 $\mu$ long, average 18 $\mu$. Usually there is no free flagellum. The posterior end of the body is broadly rounded and the kinetoplast is terminal. The undulating membrane of all the forms is broad and makes several folds. A nucleus is located centrally in the body (Fig. 1, h–j).

**Life Cycle.**    In the human host, the trypanosomes are in the blood and other places such as the lymph nodes, spleen, liver, and cerebrospinal fluid. Multiplication is by continuous fission of the trypomastigotes. Division begins

with the kinetoplast and is followed by that of the nucleus and undulating membrane. Since the flagellum and axoneme do not divide, a new set develops for the daughter trypomastigote.

The cyclic development in the tsetse is characteristic, differing from that in the *vivax* and *congolense* groups. Upon being sucked up with the blood by the tsetses, development begins in the midgut. The trypomastigotes become long, slender individuals that multiply and migrate into the hindgut, dividing continuously. They pass through the opening of the free end of the peritrophic membrane, a long chitinous, sleeve-like tube attached to the anterior end of the midgut and hanging free in it. Once outside the peritrophic membrane, the trypomastigotes migrate anteriorly between it and the intestinal epithelium without attaching to either. After a period of multiplication lasting about 10 days, large numbers of the slender trypomastigotes and sometimes epimastigotes are present. By day 15, the slender trypomastigotes are migrating through the fluid portion of the anterior end of the peritrophic membrane, going into the proventriculus, and finally into the labial cavity, thence into the salivary glands via their ducts. Inside the salivary glands, the trypomastigotes attach and rapidly transform into broad epimastigotes, filling the lumen of each gland. Within 2 to 5 days, the epimastigotes transform into small metacyclic trypanosomes that detach, later becoming free in the saliva. Completion of the cycle in the tsetses requires about 20 days. Infection of the mammalian host takes place when feeding tsetses inject the metacyclic trypanosomes contained in the saliva.

The life cycles of *T. rhodesiense* and *T. brucei* are essentially the same as that of *T. gambiense*.

## B. The *evansi* subgroup

The species of this subgroup have dropped the tsetse vectors and cyclic phases of development from their life cycles. Transmission is mechanical, whereby the trypomastigotes are transferred directly from host to host on the mouth parts of bloodsucking flies, as occurs in the case of *T. evansi* and *T. equinum*, or by venereal contact, as takes place in *T. equiperdum*.

Having adapted themselves to survival without cyclic development in a tsetse, these species have been transported in infected horses to all parts of the world. Individual species have be-

## Explanation of Plate 5

**I.** The *vivax* group *(Trypanosoma vivax)*.

A, Big game serve as reservoir hosts from which the parasite is transmitted to domestic mammals; man is not infected. B, Tstese (*Glossina* spp.).

1, Labrum-epipharynx of proboscis; 2, hypopharynx of proboscis; 3, proventriculus; 4, midgut; 5, peritrophic membrane; 6, salivary gland.

a, Trypomastigote sucked up with blood by tsetse; b, trypomastigotes have transformed to epimastigotes that divide in esophagus and pharynx; c, epimastigotes enter pharynx; d, they attach to wall of pharynx and divide; e, metacyclic trypanosomes transformed from epimastigotes, pass down hypopharynx, and are injected into vertebrate host; f, trypomastigotes multiply as such in blood of mammal.

**II.** The *congolense* group *(Trypanosoma congolense)*.

A and B, As in *vivax* group above.

1–6, As in *vivax* group.

a, Trypomastigote sucked up with blood by tsetse; b, aflagellated trypomastigote enters midgut; c, begins first phase of development by elongation; d, e, elongated aflagellated trypomastigotes; f, long epimastigotes, beginning second phase of development; g, epimastigotes enter salivary glands; h, epimastigotes shorten, attach to wall of pharynx, and divide; i, metacyclic trypanosomes derived from epimastigotes initiate third and final phase which is infective when injected into mammal; j, trypomastigotes in blood divide and multiply.

**III.** The *brucei-evansi* group *(Trypanosoma brucei, T. gambiense, T. rhodesiense, T. evansi, T. equinum, T. equiperdum)*.

A, Humans and domestic mammals with African game as reservoir host, except as given in text. B, Tstese.

1–6, As in *vivax* group; 7, spinal cord; 8, brain.

a, Trypomastigotes sucked up with blood by tsetses; b, trypomastigotes reach midgut; c, trypomastigotes elongate, becoming slender; d, multiplication of trypomastigotes; e, they enter space between peritrophic membrane and midgut wall, migrating forward; f, they migrate through soft proventricular portion of peritrophic membrane into proventriculus; g, trypomastigotes continue forward, entering esophagus and pharynx; h, they migrate into salivary glands; i, trypomastigotes change to epimastigotes which multiply; j, epimastigotes transform into metacyclic trypanosomes which are inoculated into mammal; k, medium-sized trypomastigotes; l, stumpy trypomastigotes; m, long, slender trypomastigotes dividing; n, trypomastigotes in central nervous system.

Figures A and protozoa, from Medical Protozoology and Helminthology, 1955, Naval Medical School, Bethesda, Maryland, p. 89.

---

come established in different major regions of the globe.

**Morphology.** *Trypanosoma evansi* and *T. equiperdum* are morphologically indistinguishable, being similar to *T. brucei*. They measure 15 to 34 $\mu$ long, averaging 24 $\mu$, and are typically monomorphic. The well-defined kinetoplast is subterminal from the somewhat rounded end of the body, and both the undulating membrane and free flagellum are well developed. *Trypanosoma equinum* averages smaller in size than the other two species, being 22 to 24$\mu$ long. It differs distinctly from them in consistently lacking a kinetoplast.

*Trypanosoma evansi* is in the blood of horses, cattle, camels, and other mammals, both wild and domestic. The parasite occurs throughout the Far East, Near East, northern Africa, and in South and Central America. Transmission is mechanical by the major bloodsucking flies such as tabanids, stomoxys, and related forms. It is the cause of a highly fatal disease known as surra in horses. Bovines are an important reservoir of infection.

*Trypanosoma equinum* is parasitic in the blood of a number of domestic mammals, and is particularly pathogenic to horses. It occurs in the western hemisphere, being prevalent in South and Central America. Transmission is mechanical by bloodsucking flies. The disease is prevalent in swampy areas where tabanids breed in abundance.

*Trypanosoma equiperdum* is mainly a parasite of horses and is transmitted during coitus. It causes a debilitating and fatal disease known as dourine in horses. The trypomastigotes occur in mucus secretions of the reproductive organs. Donkeys are not affected by dourine but they harbor the trypanosomes and serve as an important reservoir of infection.

Dourine occurs in Asia and southeastern Europe, as well as outside the tsetse belt in North and South Africa. The disease has disappeared recently from the United States and western Europe, where it was once prevalent.

**Pathology.** Trypanosomiasis produces fever, anemia, and progressive emaciation. In domestic mammals, *T. vivax, T. brucei,* and *T.*

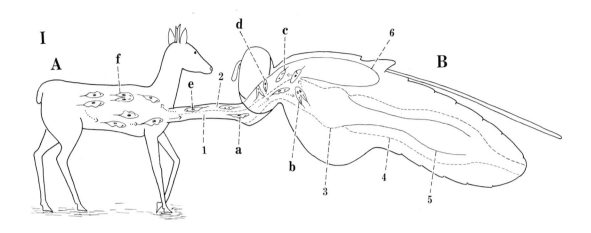

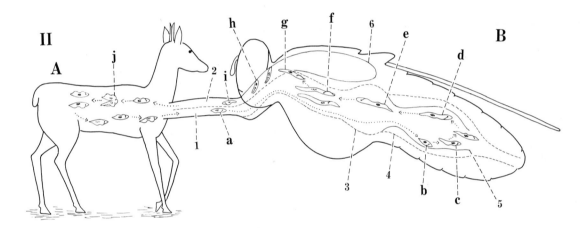

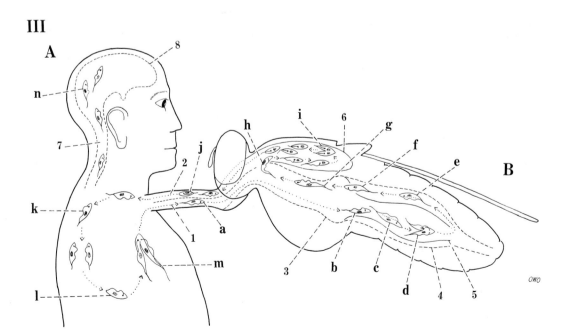

*rhodesiense* are the etiological agents of nagana, a Zulu word indicating emaciation and weakness. *Trypanosoma evansi* produces a disease known as surra, a Hindi word meaning rotten, in a wide range of domestic mammals, particularly horses. Emaciation and edema are the common symptoms. Death usually follows if treatment is not administered. *Trypanosoma equinum* infects horses primarily, producing *mal de caderas*, referring to a sickness of the hindquarters. Early and rapid emaciation leads to progressive weakness of the hindquarters until the victim is down and unable to rise. *Trypanosoma equiperdum* produces a venereal disease in horses known as dourine, Arabic for unclean. The course of the disease goes through three phases. Initial manifestations are mild fever and edema of the genitalia, accompanied by discharges. These symptoms are followed by the appearance of circular, itching plaques 1 to 4 inches in diameter in the flanks and later over the body. The final and fatal phase begins with paralysis of the facial muscles and nostrils that becomes generalized, involving the general muscular system.

In humans, *T. gambiense* and *T. rhodesiense* are pathogenic, producing fever and emaciation that terminates in the disease known generally as African sleeping sickness, or more specifically as Gambian and East African sleeping sickness, respectively. *Trypanosoma gambiense* is a parasite of man and is transmitted by tsetses, whereas *T. rhodesiense* is from domestic and wild mammals and is transmitted to man by tsetses, thereby being zoonotic in nature.

Trypanosomiasis gambiense is a chronic disease, developing through three principal stages. Upon injection, the metacyclic trypanosomes undergo a period of extensive multiplication. Following the parasitemia, they leave the blood and accumulate in the lymph nodes, which along with the liver and spleen become enlarged. The patient develops a fever. In the final stage of the disease, the parasites migrate to and accumulate in the central nervous system. They invade the brain, stimulating proliferation and infiltration of plasma cells and lymphocytes. The disease terminates with headache, mental deterioration, muscular incoordination, and finally immobility accompanied by apathy and emaciation. Trypanosomiasis rhodesiense appears in an acute form with rapid development. Symptoms are somewhat similar to those produced by trypanosomiasis gambiense but are more intense and the

course of the disease progresses rapidly. Death results in both cases.

**Life Cycle and Ecology of Tsetse Vectors.** Tsetses which are members of the Muscidae are indigenous to Africa. They inhabit wooded areas, being dependent on the shade to protect them from the lethal effects of the sun. Some species live in rain forests and others occur in woods along streams, lake shores, and in grasslands. The latter group is important in the transmission of trypanosomes because of their relationship to game, domestic mammals, and man.

Female flies appear to mate but once. They give birth to five to seven third instar larvae during an average longevity of about 75 days. Adult flies may travel up to 1450 yards from the place of pupation.

Temperature is a controlling factor in the geographic distribution of the tsetses. They do not occur above 6000 feet at the equator. Favorable temperatures for normal activity are between 21° and 28°C. Females do not feed when the temperature is below 20° nor mate when it is above 29°C. Puparia develop most favorably at 25°C. They have no diapause enabling them to survive at low temperatures.

Humidity is important to all stages of the flies. The newly born larvae must enter moist soil soon after birth. Larvae, puparia of most species, and tenerals require about 80 percent relative humidity to maintain a satisfactory water balance for survival. The puparia of a few species can survive in dry situations. Adults often seek microclimates for resting where the relative humidity and temperature are most favorable to survival.

Light controls the feeding of females as they do not venture from the shade into intense sunlight and dangerous heat. Feeding generally occurs in dim light, usually about 1 to 2 footcandles.

Tsetses feed exclusively on the blood of vertebrates (reptiles, birds, and mammals). While food may be taken daily or at periods of 2 weeks apart, usually feeding occurs at intervals of 3 to 5 days. They prefer feeding in the shade on resting animals, warthogs being a favored host.

Distribution of tsetses within the broad limits of their range is limited. They are absent from the short grasslands. Their presence is correlated with that of wooded areas which provide shade necessary for protection against excessive heat, suitable relative humidity, a resting place for game and domestic mammals, and desirable sites for human habitations.

Control of trypanosomiasis in man and his domestic mammals must be based on sound knowledge of the life cycles and ecology of the tsetses. In addition, the management of livestock and location of human habitations need be planned so as to gain the advantage of the present information on tsetses, their distribution, and habits. Medication of infected individuals is an adjunct to but not a substitute for management based on broad biological information.

## SELECTED REFERENCES

Doflein, F., and E. Reichenow. (1953). Lehrbuch der Protozoenkunde. 6th ed. Gustav Fischer, Jena, p. 519.

Fairbairn, H., and H. C. J. Watson. (1953). The transmission of *Trypanosoma vivax* by *Glossina palpalis*. Ann. Trop. Med. Parasitol. 51:464–470.

Faust, E. C., P. F. Russell, and R. C. Jung. (1970). Craig and Faust's Clinical Parasitology. 8th ed. Lea & Febiger, Philadelphia, p. 98.

Glasgow, J. P. (1963). The Distribution and Abundance of Tsetse. Pergamon Press, Oxford, 241 pp.

Hoare, C. A. (1967). Evolutionary trends in mammalian trypanosomes. In: B. Dawes, ed. Advances in Parasitology. Academic Press, New York, Vol. 5, pp. 47–91.

Langride, W. P., R. J. Kernaghan, and P. E. Glover. (1963). A review of recent knowledge of the ecology of the main vectors of trypanosomiasis. Bull. WHO 28:671–701.

Levine, N. D. (1961). Protozoan Parasites of Domestic Animals and Man. Burgess Publishing Co., Minneapolis, p. 42.

Lumsden, W. H. R. (1970). Biological aspects of trypanosomiasis research, 1965; A retrospect, 1969. In: B. Dawes, ed. Advances in Parasitology. Academic Press, New York, Vol. 8, pp. 227–249.

Nevu-Lemaire, M. (1943). Traité de Protozoologie Medical et Vétérinaire. Vigot Freres, Paris, p. 174.

Richardson, U. F., and S. B. Kendall. (1957). Veterinary Protozoology. 2nd ed. Oliver and Boyd, London, p. 44.

Soulsby, E. J. L. (1968). Helminths, Arthropods and Protozoa of Domestic Animals (Mönnig). 6th ed. Williams & Wilkins Co., Baltimore, p. 546.

Vickerman, K. (1962). The mechanism of cyclical development of trypanosomes of the *Trypanosoma brucei* subgroup: An hypothesis based on ultrastructural observations. Trans. Roy. Soc. Trop. Med. Hyg. 56:487–495.

Wenyon, C. M. (1926). Protozoology. William Wood and Co., New York (Reprinted by Hafner Publishing Co., New York, 1965), Vol 1, p. 442.

Willet, K. C., and R. M. Gordon. (1957). Studies on the deposition, migration, and development to blood forms of trypanosomes belonging to the *Trypanosoma brucei* group. II. An account of the migration of the trypanosomes from the site of their deposition in the rodent host to their appearance in the general circulation, with some observations on their probable routes of migration in the human host. Ann. Trop. Med. Parasitol. 51:471–492.

### Trypanosomes of Birds, Reptiles, Amphibians, and Fish

The classification and biology of the trypanosomes of these hosts are not as well understood as those of mammals. They belong to the section Stercoraria by virtue of all cyclic development occurring in the intestine of the vector.

Often, naturally infected animals of these groups have such small numbers of trypomastigotes in the blood that it is difficult to find them in stained smears. Cultures of blood are more reliable than smears for detecting infections.

#### Birds

Trypanosomes occur commonly in birds. *Trypanosoma avium* Danilewsky, 1885 was the first species described from them. Eight species are recognized at present. While trypomastigotes appear in blood smears, usually in small numbers, they can be found more frequently in smears of bone marrow, either fresh or stained.

**Morphology.** In general, trypomastigotes from birds are large and spindle-shaped, measuring 35 to 60 $\mu$ in length. Longitudinal parallel lines formed by myonemes appear on the body. The nucleus is variable in shape, being oval, circular, or triangular. A prominent kinetoplast is near the posterior end of the body. The undulating membrane and flagellum are well developed. Epimastigotes in cultures occur in a variety of transitional stages.

**Life Cycle.** *Trypanosoma avium* occurs as trypomastigotes in the blood and bone marrow and as epimastigotes in the midgut and metacyclic trypanosomes in the hindgut of louse flies (Hippoboscidae) and in cultures. In Europe, *Ornithomyia avicularia* is the known vector, whereas in North America, *Stilbometopa impressa* and *Pseudolynchia canariensis* serve in this capacity. Natural infection of birds occurs when the vectors are swallowed instead of by contamination with fecal material as in other stercorarians.

After infected louse flies are swallowed by birds, a prepatent period of 17 to 24 hours follows before trypomastigotes appear in the peripheral blood. During the first 6 hours, the flagellates are in the intestinal and cecal tissue, kidneys, lungs, mesenteries, peritoneal fluid, and the blood. By 12 hours, some have entered the liver, spleen, and bone marrow, and at 16 hours the brain and heart muscles have been invaded. By 24 hours, many trypomastigotes are in the peripheral blood, reaching a peak by the second day. Multiplication does not take place in the blood. The number of trypomastigotes steadily declines until they disappear after 1 or 2 months. During the winter, the flagellates are more or less restricted to the bone marrow and lungs. Relapses may occur with parasitemia in the peripheral blood.

Development in the louse flies begins in the midgut. The trypomastigotes quickly transform into large, blunt epimastigotes. The first three divisions, beginning within 2 hours after ingestion, produce small epimastigotes and the fourth large ones after 24 hours. These free-swimming flagellates migrate to the hindgut, attach by the flagellum to the peritrophic membrane, and divide twice, producing more small epimastigotes. The kinetoplast migrates posterior to the nucleus to form the infective metacyclic trypanosomes. Development in the hindgut begins about 48 hours after ingestion of the blood forms and is completed by 60 to 72 hours.

### Reptiles

All of the major groups of reptiles are hosts to trypanosomes. Of nearly 40 species described, they occur predominantly in hosts from warm climates. *Trypanosoma chrysemydis* from the painted turtle (*Chrysemys belli*) and the snapping turtle (*Chelydra serpentina*) is found in the central part of the United States.

In general, the trypomastigotes are rare in blood smears, indicating a low parasitemia in the peripheral circulation. Because of this condition, cultures of blood are more likely to reveal infections than blood smears.

**Description.** The reptilian species of trypanosomes are generally large and the body is covered with longitudinal lines formed by parallel myonemes. Like other stercorarians, the pos-

terior end of the body is pointed. There is a well-developed undulating membrane and a free flagellum.

**Life Cycle.** The trypomastigotes are transmitted when infected vectors are ingested by the vertebrate hosts. Leeches serve as vectors for species in aquatic reptiles and bloodsucking arthropods for terrestrial ones.

*Trypanosoma grayi* of African crocodiles is the best known species infecting reptilian hosts. In the crocodile, trypomastigotes occur primarily in the epidermal capillaries. Infections persist 2½ years or longer. The prepatent, or incubation, period is 4 days, during which time the metacyclic trypanosomes mature, growing to gigantic size. There is no multiplication in the blood, resulting in a low parasitemia. The crocodiles appear to have a partial immunity or tolerance to the flagellates.

The tsetse vector *Glossina morsitans* acquires its infection by sucking blood from crocodiles. The entire cyclic development takes place in the alimentary canal of the flies, which take up one to six trypomastigotes per feeding. Development begins in the midgut about 6 hours after ingestion. The body shortens and the kinetoplast moves forward, forming epimastigotes. Multiplication is intensive during the first 3 days, with diminishing size of the flagellates. By day 14, they move to the hindgut, attach, and transform to metacyclic trypanosomes in 6 days. Crocodiles become infected only by swallowing parasitized tsetses.

The sand fly *Lutzomyia* (*Phlebotomus*) *vexatrix* is probably the vector of *Trypanosoma scelopori* of the fence lizard (*Sceloporus occidentis*) in California and *L.* (*P.*) *trinidadensis* for trypanosomes in forest geckos (*Thecadactylus rapicaudus*) in Panama. Cyclic development of the flagellates occurs in the alimentary canal of the sand flies in such great numbers as to occlude it in 3 to 14 days. Infection of the lizards is by ingestion of the sand flies, especially those filled with blood.

Trypanosomes of turtles and snakes frequenting an aquatic environment are transmitted by leeches such as *Glossiphonia* sp. for *Trypanosoma vittatae* from the soft tortoise (*Emyda vittata*) and *Placobdella* spp. for *Trypanosoma primeti* from the water snake *Tropidonotus piscator* and others. Cyclic development of the flagellates occurs in the crop. Infection of the vertebrate hosts is most likely to be through the bites of leeches. Some workers suggest that it may result from ingestion of leeches.

### Amphibians

These animals harbor a number of large species of trypanosomes whose bodies appear striated due to subpellicular myonemes.

### Morphology

The mature body is comparatively large and slender with the size variable, ranging from 54 to 78 $\mu$ (average 64 to 68 $\mu$) long. The posterior end of the body is pointed, the kinetoplast is subterminal, the undulating membrane is well developed, and a free flagellum is ⅓ to ½ the length of the body. The nucleus is near the middle of the body.

**Life Cycle.**    The life cycle of *Trypanosoma diemyctyli* of the American newt or eft *Diemyctylus v. viridescens* is used as an example.

Only adult newts are infected, larval forms being free of the trypanosomes. Upon injection by the leech vectors, the metacyclic trypanosomes are carried from the point of inoculation by the blood to the liver, spleen, kidneys, heart, and brain where growth occurs. By days 8 to 10, the flagellates have attained an average length of 60 $\mu$ and are present in the bloodstream. Multiplication does not occur in the vertebrate host.

The glossiphoniid leech (*Batrachobdella picta*), which feeds naturally on newts, serves as the vector. When ingested, there is no change in the trypomastigotes for 2 to 3 hours. By hour 6, the flagellates in the stomach have transformed to epimastigotes. A rapid succession of multiplication by binary fission produces numerous flagellates. They leave the stomach by day 6 for the gastric ceca, attach to the mucosa, and become infective metacyclic trypanosomes at the end of week 2. Infection of newts takes place when feeding leeches inject metacyclic trypanosomes.

There are several additional species of trypanosomes from amphibians whose life cycles have been studied. *Trypanosoma ambystomae* of the Pacific Coast newt *Taricha* (*Triturus*) *torosus* develops in an erpobdellid leech (*Erpobdella* sp.) and *Trypanosoma granulosae* from the same host develops in *Actinobdella* sp., a glossiphoniid leech. *Trypanosoma rotatorum* from frogs probably consists of a complexity of species of which *Trypanosoma ranarum* from *Rana pipiens* is one. Species of *Hemiclepsis*, another

glossiphoniid leech, probably are the natural vectors.

### Fish

Trypanosomes occur commonly in fish, both freshwater and saltwater species. Leeches of the genera *Hemiclepsis* and *Piscicola* serve as vectors for trypanosomes from freshwater fishes and *Pontobdella* and *Trachobdella* for those from marine fishes.

**Morphology.**    The body of trypanosomes of fish normally is long and slender, ranging from 15 to 130 $\mu$, depending on the species. The usual size is about 50 $\mu$ long by 2 to 5 $\mu$ in diameter. The flagellum varies in length and there is a well-developed undulating membrane. Some species are dimorphic, being of two sizes, while in the blood. Certain of the large trypanosomes show longitudinal myonemes.

**Life Cycle.**    Leeches act as the vectors. Cyclic development in the leeches falls into three basic patterns. They are 1) trypomastigotes from the blood of the host transform into epimastigotes and later into metacyclic trypanosomes in the stomach that are injected directly into the blood by feeding leeches, as occurs with *Trypanosoma percae* of the yellow perch (Plate 6); 2) trypomastigotes from the blood a) transform to epimastigotes in the stomach and multiply, b) go to the intestine without further change, and c) return to the stomach where they change to metacyclic trypanosomes which d) migrate to the proboscis sheath and are injected into the fish, as in the case of *Trypanosoma granulosum* of eels; and 3) trypomastigotes that transform first into epimastigotes, multiply, and then change to metacyclic trypanosomes in the stomach, following which they go directly to the proboscis sheath to be injected into the host, as occurs with *Trypanosoma danilewskyi* of carp.

The life cycle of *Trypanosoma danilewskyi* is typical of trypanosomes in fish. All development in the leech *Hemiclepsis marginata* takes place in the stomach, requiring about 10 days for completion. During day 1 in the leech, the trypomastigotes transform to epimastigotes that begin multiplication by binary fission. At the end of day 3, normal stumpy and slender epimastigotes are present. Slender forms predominate and are dividing from days 4 to 6. Development ceases after day 7 and the flagellates increase in size. Metacyclic trypanosomes have migrated into the proboscis sheath by day 10. After injec-

## Explanation of Plate 6

A, *Trypanosoma percae* from blood of perch. B, Trypomastigote in process of transforming into epimastigote in stomach of leech. C, Epimastigote stage dividing. D, Developed epimastigote. E, Metacyclic trypanosome. F, Fish vertebrate host. G, Leech (*Piscicola* spp.) invertebrate host.

1, Nucleus; 2, myonemes; 3, basal body; 4, kinetoplast; 5, axoneme; 6, undulating membrane; 7, flagellum.

a, Trypomastigotes in blood; b, trypomastigote being sucked into pharynx of leech; c, trypomastigote in stomach; d, trypomastigote in process of changing to epimastigote; e, dividing trypomastigote and formation of epimastigote; f, epimastigote multiplying; g, h, metacyclic trypanosomes in stomach; i, metacyclic trypanosome entering esophagus; j, injection of metacyclic trypanosomes into bloodstream; k, metacyclic trypanosome entering bloodstream; l, metacyclic trypanosome growing into mature trypomastigote; m, proboscis sheath; n, intestine.

Figure A adapted from Minchin, 1909, Proc. Zool. Soc. London (1), Jan.–Feb. p. 2; B–E, G, from Wenyon, 1926, Protozoology, Vol. 1, Fig. 244.

---

tion into the blood, they mature, growing to 51 $\mu$ long by 1 $\mu$ wide.

## SELECTED REFERENCES

### General

Doflein, F., and E. Reichenow. (1953). Lehrbuch der Protozoenkunde. Gustav Fischer, 6th ed. Jena, pp. 553–561.

Wenyon, C. M. (1926). Protozoology. William Wood and Co., New York (Reprinted by Hafner Publishing Co., New York, 1965), pp. 577–607.

### Birds

Baker, J. R. (1956). Studies on *Trypanosoma avium* Danilewsky, 1885. I. Incidence in some birds of Hertfordshire. Parasitology 46:308–320.

Baker, J. R. (1956). Studies on *Trypanosoma avium* Danilewsky, 1885. II. Transmission by *Ornithomyia avicularia* L. Parasitology 46:321–334.

Baker, J. R. (1956). Studies on *Trypanosoma avium* Danilewsky, 1885. III. Life cycle in vertebrate and invertebrate hosts. Parasitology 46:335–352.

Baker, J. R. (1966). Studies on *Trypanosoma avium*. IV. The development of infective metacyclic trypanosomes in cultures grown *in vitro*. Parasitology 56:15–19.

Baker, J. R., and R. G. Bird. (1968). *Trypanosomum avium*: Fine structure of all developmental stages. J. Protozool. 15:298–308.

Diamond, L. S., and C. M. Herman. (1954). Incidence of trypanosomes in the Canada goose as revealed by bone marrow culture. J. Parasitol. 40:195–202.

Herman, C. M. (1945). Hippoboscid flies as parasites of game animals in California. Calif. Fish Game 31:16–25.

Stabler, R. M. (1961). Studies on the age and seasonal variations in the blood and bone marrow parasites of a series of black-billed magpies. J. Parasitol. 47:413–416.

Zajicek, D. (1968). *Trypanosoma dafilae* sp. n., and *Leucocytozoon anatis* Wickware, 1915 in the northern pintail *(Dafila acuta acuta)*. Folia Parasitol. 15:169–171.

### Reptiles

Anderson, J. R., and S. C. Ayala. (1968). Trypanosome transmitted by Phlebotomus. First report from the Americas. Science 161:1023–1025.

Ayala, S. C. (1970). Two new trypanosomes from California toads and lizards. J. Protozool. 17:370–373.

Chaniotis, B., and J. R. Anderson. (1968). Age structure, population dynamics and vector potential of *Phlebotomus* in northern California. Part II. Field populations dynamics and natural flagellate infections in parous females. J. Med. Entomol. 5:273–292.

Christensen, H. A., and S. R. Telford. (1971). *Lutzomyia trinidadensis*, an intermediate host for trypanosomes infecting the forest gecko *Thecadactylus rapicaudus*. J. Parasitol., 46th Annu. Meet., Aug. 23–27, p. 31.

Hoare, C. A. (1929). Studies on *Trypanosoma grayi*. II. Experimental transmission to the crocodile. Trans. Roy. Soc. Trop. Med. Hyg. 23:39–56.

Hoare, C. A. (1931). Studies on *Trypanosoma grayi*. III. Life cycle in tsetse-fly and in crocodile. Parasitology 23:449–484.

Roudabush, R. L., and G. R. Coatney. (1937). Some blood protozoans of reptiles and amphibians. Trans. Amer. Microsc. Soc. 56:291–297.

Shortt, H., and C. Swaminath. (1931). Life-history and morphology of *Trypanosoma phlebotomi* (Mackie, 1914). Indian J. Med. Res. 19:541–564.

Walliker, D. (1965). *Trypanosoma superciliosae* sp. nov. from the lizard *Uranoscodon superciliosa*. Parasitology 55:601–606.

**Plate 6**   *Trypanosoma percae*                                                                47

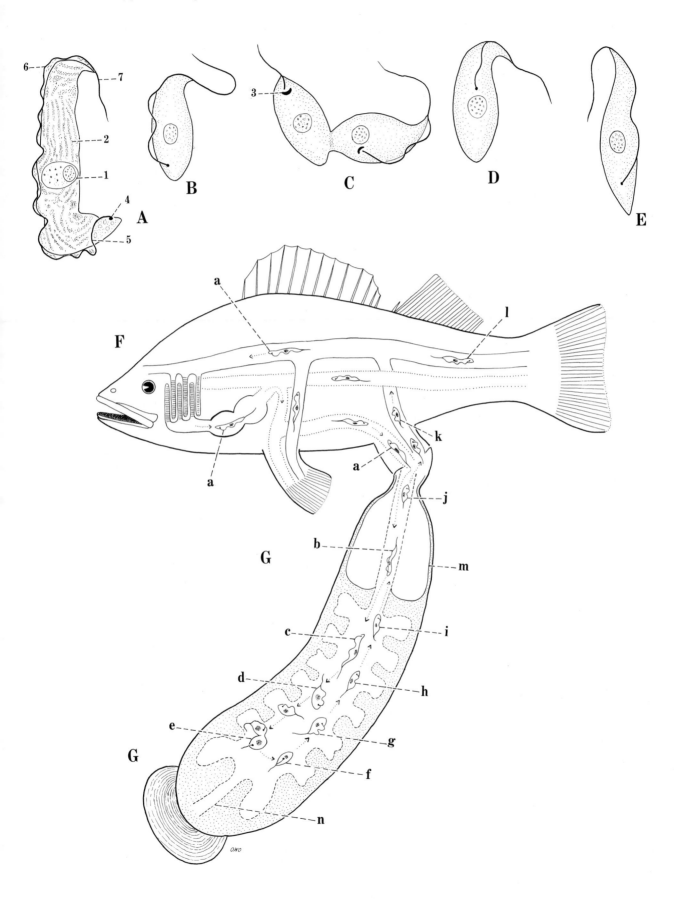

*Amphibians*

Barrow, J. H. (1953). The biology of *Trypanosoma diemyctyli* Tobey. I. *Trypanosoma diemyctyli* in the leech *Batrachobdella picta* Verril. Trans. Amer. Microsc. Soc. 72:197–216.

Barrow, J. H. (1954). The biology of *Trypanosoma diemyctyli* Tobey. II. Cytology and morphology of *Trypanosoma diemyctyli* in the vertebrate host. Trans. Amer. Microsc. Soc. 73:242–257.

Barrow, J. H. (1958). The biology of *Trypanosoma diemyctyli* Tobey. III. Factors influencing the cycle of *Trypanosoma diemyctyli* in the vertebrate host *Triturus v. viridescens*. J. Protozool. 5:161–170.

Lehmann, D. L. (1952). Notes on the life cycle and infectivity of *Trypanosoma barbari*. J. Parasitol. 38:550–553.

Lehmann, D. L. (1958). Notes on the biology of *Trypanosoma ambystomae* Lehmann, 1954. II. The life cycle in the invertebrate host. J. Protozool. 5:96–98.

Lehmann, D. L. (1959a). The invertebrate cycle of *Trypanosoma granulosae* from *Taricha granulosa*. J. Protozool. 6(Suppl.):26–27.

Lehmann, D. L. (1959b). The cultivation of some trypanosomes from urodeles. J. Protozool. 6:340–343.

Wallace, F. G. (1956). Cultivation of *Trypanosoma ranarum* on a liquid medium. J. Protozool. 3:47–49.

*Fish*

Qadri, S. S. (1962). An experimental study of the life cycle of *Trypanosoma danilewskyi* in the leech, *Hemiclepsis marginata*. J. Protozool. 9:254–258.

## EXERCISE ON LIFE CYCLE

The stages in the life cycle include trypomastigotes in vertebrate hosts and epimastigotes in invertebrate vectors and cultures.

Since trypomastigotes often are scarce in the circulating blood, depending on the season and age of the infection, it is often difficult to find them in fresh or stained smears. Being more abundant in the bone marrow, especially in birds during the winter, they can be demonstrated more readily in smears, both fresh and stained from this source.

Epimastigotes similar to those in the vectors may be obtained by culturing blood or bone marrow on a variety of media such as NNN, Diamond's SNB-9, or proteose-peptone blood agar. For preparation of the media and culturing of the flagellates from various hosts, consult Baker (1966), Diamond and Herman (1954), Lehmann (1959b), or Wallace (1956).

Observe live specimens under a phase contrast microscope and smears stained with Giemsa or Wright's blood stain under a light microscope for morphological and behavioral characteristics. Note the position of the kinetoplast in mature trypomastigotes in the blood and developing epimastigotes in cultures to see the varying locations of it in the transitional stages of the flagellates from one form to the other.

Infection of readily available parasite-free vertebrate hosts is accomplished experimentally by subcutaneous or intraperitoneal injections of cultured epimastigotes into newly hatched chicks for avian trypanosomes and goldfish from pet shops for piscine forms. Appropriate vectors, preferably laboratory-reared specimens, may be infected by permitting them to feed on parasitized vertebrate hosts. Follow the course of development in the alimentary canal.

### Leishmania

The members of this genus appear as minute, oval amastigotes in the cells of the reticuloendothelial system of the skin, bone marrow, liver, spleen, and lymph nodes of mammals and reptiles, and as spindle-shaped promastigotes in the alimentary canal of sand flies (*Phlebotomus* spp.), the invertebrate vector, and in cultures.

*Leishmania* occurs throughout the tropical and subtropical regions of the world, and is the cause of serious disease in humans. Its geographic distribution corresponds with that of the sand flies responsible for transmission among the reservoir hosts and between them and man.

The species infecting man are generally recognized as *Leishmania donovani* producing visceral leishmaniasis or kala azar, *L. tropica* the cause of cutaneous leishmaniasis or oriental sore of two types, and *L. braziliensis* the etiological agent of cutaneous-mucocutaneous leishmaniasis. While the organisms of the respective species are indistinguishable morphologically, their identity is claimed on the clinical manifestations of the diseases they produce and differences in serological tests.

### Leishmania donovani
### (Laveran and Mesnil, 1903) (Plate 7)

*Leishmania donovani* is the etiological agent of visceral leishmaniasis, also known as kala azar. While present in parts of South America, Africa, and southern Europe, it is prevalent in Asia, occurring over a wide expanse of the warm areas of the subcontinent.

**Description.** The amastigotes are round or oval in shape and measure 1.5 to 4 μ in diameter. Stained specimens in the cytoplasm of macrophages viewed under the light microscope show a nucleus, kinetoplast, and axoneme. A light halo surrounds each parasite. Electron micrographs reveal much more. The pellicle, or periplast, consists of three layers, of which the two outer ones form thin membranes and the innermost consists of slender fibrils. A short flagellum barely emerges from a flagellar pocket, or reservoir, lined by two membranes. There are nine pairs of peripheral and two central fibrils in the flagellum (Fig. 2, B), the internal end of which forms the basal body. The sausage-shaped kinetoplast, transverse to but not in contact with the base of the flagellum, is surrounded by a double membrane. Cristae extend medially from the inner layer. An elongate skein of densely coiled fibrils consisting of DNA lies lengthwise inside the kinetoplast (Fig. 3).

Promastigotes in the midgut of the sand flies or in cultures are spindle-shaped bodies 15 to 25 μ long by 1 to 3.5 μ in diameter and with a free flagellum 15 to 28 μ in length. Other differences from the amastigotes include pronounced changes of internal structure. The pellicle is a single membrane over the body, flagellum, and in the reservoir. The two nucleoli have fused to form a single, stellate one. The kinetoplast has enlarged and sent out several branches of new mitochondria, and the Golgi apparatus has enlarged.

**Life Cycle.** The life cycle of all three species of *Leishmania* is basically the same, each consisting of two different types of organisms: amastigotes in mammals and promastigotes in sand flies and cultures. The parasites are constantly changing the external form and internal structure as they go from one host to the other. The manner in which this is done and the influences of the respective hosts constitute the life cycle.

Amastigotes ingested with blood by female flies pass to the midgut, where they are liberated by digestion of the macrophages. In the midgut, the amastigotes transform to promastigotes, multiply, and move forward, where they attach by

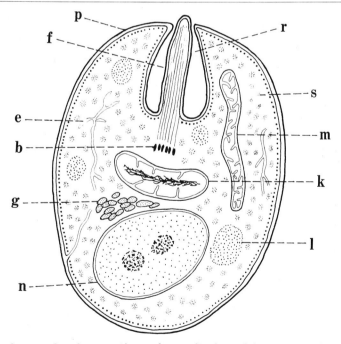

Fig. 3. Diagrammatic figure showing section of *Leishmania donovani* reconstructed from electron micrographs. b, Basal body; e, endoplasmic reticulum; f, flagellum containing paired peripheral fibrils; g, Golgi apparatus; k, kinetoplast with crista and fibrillar material of DNA; l, lipid bodies; m, mitochrondrion; n, nucleus with two nucleoli; p, pellicle, or periplast, consisting of two separate membranous outer and one inner fibrillar layers; r, flagellar pocket, or reservoir; s, particles of RNA (Palade's particles).

## Explanation of Plate 7

A, Amastigote stage. B, Promastigote stage. C, Sand fly *(Phlebotomus)* invertebrate host. D, Vertebrate host.

1, Nucleus; 2, parabasal body; 3, kinetoplast; 4, axoneme; 5, flagellum.

a, Macrophage containing amastigotes being swallowed by sand fly; b, macrophage being digested and amastigotes released; c, binary division of amastigotes; d, formation of promastigotes; e, binary fission of promastigotes; f, multiplication of promastigotes in gut; g, migration of promastigotes toward proboscis; h, promastigotes in proboscis of feeding sand fly; i, promastigote recently injected by sand fly into bloodstream of vertebrate; j, macrophage ingesting promastigote; k, amastigote from transformed promastigote; l, amastigotes divide by binary fission; m, two amastigotes; n, multiplication of amastigotes; o, macrophage completely filled with amastigotes; p, macrophage ruptures, releasing amastigotes in blood; q, macrophages ingest amastigotes; r, infected macrophages pass through heart and into general circulation, (r–v); s–u, multiplication of amastigotes begins upon ingestion of them by macrophages; v, infected macrophages in peripheral circulation available to biting flies; w, amastigotes in reticuloendothelial cells of liver (also in spleen and bone marrow); x, reticuloendothelial cells rupture, releasing amastigotes which are ingested by phagocytic cells of liver, spleen, and bone marrow, where multiplication occurs, with eventual destruction of the host cells.

Adapted from Medical Protozoology and Helminthology, 1955, Naval Medical School, Bethesda, Maryland, p. 89.

---

the tip of the flagellum to the epithelium of the midgut and esophageal valve, often occluding the lumen with their great number. The flagellates move anteriorly into the pharynx and buccal cavity whence they are injected into the skin by feeding flies.

Promastigotes grow well in cultures at temperatures of 25° to 28°C, duplicating the cycle as it occurs in the alimentary canal. When the promastigotes are conditioned gradually to higher temperatures from 28° to 34°C, they transform to typical amastigotes that grow in the absence of cells, appearing similar to those in the mammalian host. Such cultured amastigotes are infective.

Upon being injected into the skin by a biting sand fly, the promastigotes exist as foreign bodies that are attacked by the various cells of the reticuloendothelial system. The cells surround the flagellates with a pseudopod that is withdrawn, depositing the parasite inside the body. In the macrophages, the promastigotes round up and undergo transformation to amastigotes. They develop a second pellicular layer, the nucleoleus divides to form two, the mitochondria and Golgi apparatus become less complex, and the free flagellum disappears. Inside the phagocytes, the newly formed amastigotes commence a process of multiplication. Division is initiated by the kinetoplast, which becomes active and elongates. It and the basal body divide, followed by mitotic division of the nucleus. A new axoneme is formed and the body completes the division by binary fission. Infected reticuloendothelial cells released from the skin enter the visceral organs, where the parasites multiply rapidly. Heavily parasitized cells rupture, liberating amastigotes that are engulfed by other macrophages in which reproduction is repeated.

In the midgut of sand flies, the amastigotes gradually transform to promastigotes. The body changes to the characteristic spindle shape with a long flagellum. Internal changes include loss of one pellicular layer, fusion of the nucleoli to form one, and the kinetoplast changes shape and produces new mitochondria in the form of large branches. The process may be completed in about 24 hours under favorable temperature of 25° to 28°C.

**Types of Leishmaniasis and Their General Geographic Distribution.** Three distinct types of leishmaniasis are recognized, each being caused by what is generally believed to be a separate species of flagellate. However, some workers think only one species may be involved and the different clinical symptoms are manifestations of separate biological races of the parasite.

*Leishmania donovani* produces the visceral type of disease in which the number of macrophages increases greatly, accumulating in the liver, spleen, lymph nodes, and bone marrow, causing great enlargement of the visceral organs. Since the reticuloendothelial cells are parasitized, they are no longer able to protect the host against other infectious agents. The disease is often fatal. *Leishmania donovani* occurs in

Plate 7    *Leishmania donovani*                                                51

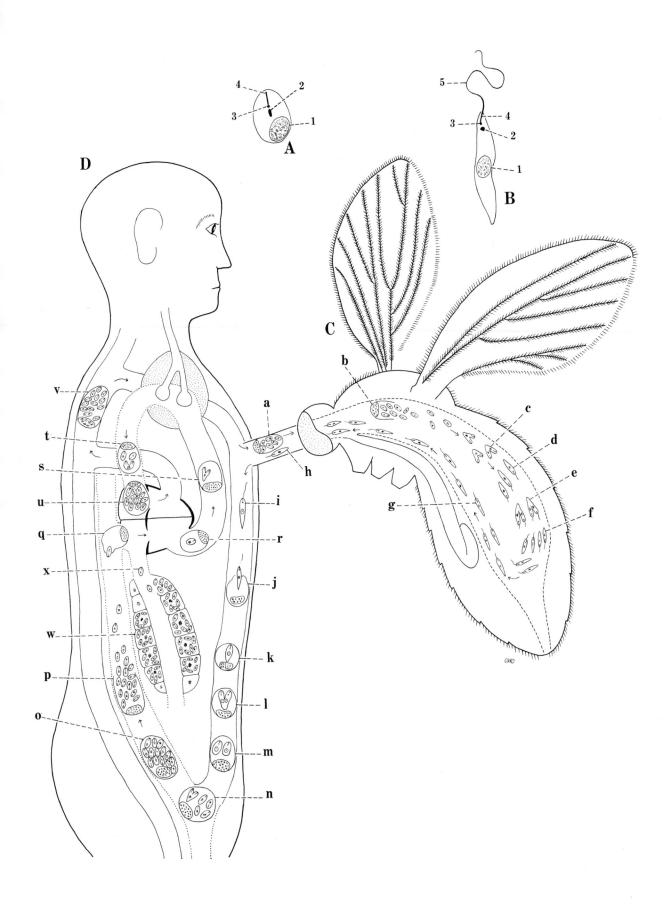

Asia, the Middle East, southern Europe, northern Africa, parts of South and Central America, and Mexico. In India, transmission is from man to man but in most other regions dogs serve as reservoirs.

*Leishmania tropica* is the cause of a skin lesion known as oriental sore, occurring in two forms. The dry, or urban, type is prevalent in towns and has a chronic course of late ulceration. The other type is the moist, or rural, which is present in open country. It runs an acute course with early ulceration and exudation. Recovery is usual with resultant resistance to further infection. Oriental sore occurs predominantly in warm dry areas such as northern Africa, the Sudan, the Middle East, southern USSR, and part of India. Rodents serve as the reservoir host.

*Leishmania braziliensis* is the etiological agent of American leishmaniasis known as the cutaneomucocutaneous type. Local names for the cutaneous lesions are uta in the Andes of Peru and chiclero's disease in Venezuela, Guatamala, and southern Mexico, and espunda for the mucocutaneous type in Brazil. Uta and chiclero's disease appear as weeping sores on the skin of exposed parts of the body, whereas in espunda the parasites invade the mucocutaneous junctions and deep tissues, causing ulcerative lesions. Cutaneomucocutaneous leishmaniasis is contracted by people who work or live in forested areas. Rodents are the principal reservoirs.

Sand flies of many species serve as vectors of *Leishmania*. Some common ones are *Phlebotomus papatasii* and *P. caucasius* for *L. tropica*; *P. intermedius* and *P. whitmani* for *L. braziliensis* in Brazil, *P. verrucarum* and *P. peruensis* in Peru, and *P. panamensis* and *P. cruciatus* from Venezuela to Mexico; and *P. argentipes*, *P. sergenti*, and *P. papatasii* for *L. donovani*.

**Ecology of Sand Flies.** Sand flies are widespread in the tropics and subtropics, occurring in various habitats ranging from deserts to rain forests and sea level to the highlands of Peru. Species in the Palearctic have a diapause and pass the winter as fourth instar larvae. Great numbers of them emerge early in the spring.

Females are hematophagous. Some species show a definite preference for a specific kind of host on which to feed, such as humans (anthrophilous types) or rodents, dogs, etc. (zoophilous kinds).

Humidity must be near saturation before females will oviposit. This requirement is more or less constant in tropical forests. Desert forms find suitable humidity in the burrows of rodents which in many cases are reservoirs of the parasite. Eggs and larvae require free or capillary-bound water to develop.

**Epidemiology.** Infection of the fly population with leishmania is maintained from feeding on reservoir hosts. Infection of humans varies somewhat, depending on the specific parasite involved. In the case of *L. donovani*, only people, mainly adults, are infected in India but in other areas dogs serve as reservoir hosts and children are commonly infected. The presence of dogs and specific sand flies provide the source of infection associated directly with the human populace. *Leishmania tropica* occurs in rodent reservoirs living primarily in open country. Their burrows provide natural habitats for the sand flies. People traveling through or living in or near such areas are bitten by the infected flies. *Leishmania braziliensis* is a parasite of forest rodents. People associated with the forests as inhabitants or workers harvesting timber or collecting chicle are exposed to the bites of the sand flies with subsequent infection.

**Physiology.** Morphological changes in the amastigotes and promastigotes are related to the physiological requirements for transformation and growth of the different stages of the parasites in the reticuloendothelial cells of the homoiothermic host and in the alimentary canal of the poikilothermic host.

Amastigotes in *in vitro* cultures change from the oval, aflagellate form with a double periplast to a spindle-shaped promastigote having a single periplast similar to that seen in the gut of sand flies. Internal changes appear to be initiated by the kinetoplast. It changes from the regular sausage-like form with the dense fibrillar band of DNA into an expanded body with diffuse contents that produces several large branch-like mitochondria with numerous cristae. Since mitochondria produce major enzymes and are the site of oxidative phosphorylation, these morphological changes are related to the physiological requirements of the parasite in the intestinal environment.

The change from two rounded nucleoli in the amastigote (Fig. 3, n) to one stellate type in the promastigote is probably associated with synthesis of enzymes. Changes in the nucleus, the kinetoplast-mitochondrial complex, and other structures such as the Golgi apparatus and endoplasmic reticulum are preparation for the new type of life associated with the change of form and related metabolic activities. The loss of the double periplast of the amastigote is expected since it probably functioned in an antigenic manner in the macrophages.

A reversal of these morphological changes with regression of the expanded mitochondria, restoration of the sausage shape of the kinetoplast, and formation of an outer pellicular membrane takes place when the promastigotes enter the mammal and are engulfed by the macrophages, where their physiological requirements are different.

## SELECTED REFERENCES

Adler, S., and O. Theodor. (1957). Transmission of disease agents by phlebotomine flies. Annu. Rev. Entomol. 2:203–226.

Faust, E. C., P. F. Russell, and R. C. Jung. (1970). Craig and Faust's Clinical Parasitology, 8th ed. Lea & Febiger, Philadelphia, p. 77.

Hoare, C. A. (1955). The epidemiological role of animal reservoirs in human leishmaniasis and trypanosomiasis. Vet. Rev. Annot. 1:62–68.

Lemma, A., and E. L. Schiller. (1964). Extracellular cultivation of the leishmanial bodies of species belonging to the protozoan genus Leishmania. Exp. Parasitol. 15:503–513.

Miller, H. C., and D. W. Twohy. (1967). Infection of macrophages in culture by leptomonads of Leishmania donovani. J. Protozool. 14:781–789.

Pyne, C. K., and J. Chakraborty. (1958). Electron microscopic studies on the basal apparatus of the flagellum in the protozoon Leishmania donovani. J. Protozool. 5:264–268.

Rudzink, M. A., P. A. D'Alesandro, and W. Trager. (1964). The fine structure of Leishmania donovani and the role of the kinetoplast in the leishmania-leptomonad transformation. J. Protozool. 11:166–191.

### FAMILY CRYPTOBIIDAE

The members of this family occur in the blood and alimentary canal of fish, and in the seminal vesicles and spermatophores of mollusks. These flagellates are similar in body shape to the trypanosomes but have two flagella. One extends anteriorly while the other is directed posteriorly and is adherent to the body, is with or without an undulating membrane, and has a free trailing portion. Members of the genus *Cryptobia* do not have an undulating membrane, whereas those of *Trypanoplasma* do.

The presence of glycogen granules in the cytoplasm of these parasites suggests that they derive energy from the metabolism of carbohydrates.

### *Cryptobia helicis* Leidy, 1846 (Plate 8)

This species occurs commonly in the seminal vesicle and intestine of pulmonate land snails (*Helix aspersa, Triodopsis albolabris, T. tridentata, Anguispira alternata,* and *Monadenia fidelis*).

**Description.** The organism is typically elongate and slender, measuring 6 to 20 $\mu$ in length, with the body flattened. Some forms are short and broad. A darkly staining mass near the anterior end of the body consists of an elongate structure and two kinetosomes. The axoneme of one kinetosome gives rise to the anterior free flagellum, and that of the other extends backward, giving rise to the posterior flagellum which is attached to the surface of the body except for the extermity which trails freely behind. No undulating membrane is present (Plate 8, D).

**Life Cycle.** Multiplication is by binary fission. The kinetosomes each divide, thus making two pairs. The original axoneme and flagellum are retained by one kinetosome and new ones are formed by the other. The kinetosome and nucleus divide and cleavage of the organism follows, forming two new individuals.

No encysted forms have been observed. It is believed that the parasites, being in the seminal vesicle, are transferred from one snail to another during copulation.

### EXERCISE ON LIFE CYCLE

To collect specimens, remove the shell from an anesthetized snail, open the animal, and excise the orange-colored spherical spermatheca. Express a drop of the contents onto a slide, add a drop of physiological saline, cover with a coverslip, and examine with the oil immersion lens, preferably on a phase contrast microscope. Note the characteristic movements.

While it is presumed that *Cryptobia helicis* is transmitted during copulation of the snails, it would be interesting to demonstrate whether this actually is the case and, if so, whether it is the only means of transfer of the parasite. It is the apparent absence of encysted forms that has

## Explanation of Plate 8

A–C, *Chilomastix mesnili*. A, Trophozoite containing food particles. B, Trophozoite. C, Cyst. D, *Cryptobia helicis*. E, *Helix* sp. F, *Trypanoplasma borreli*. G, Leech (*Piscicola* sp.) vector of *Trypanoplasma*.

1, Blepharoplast; 2, anterior flagella; 3, posterior flagellum hanging in cytostome; 4, lip of cytostome; 5, cytostome; 6, nucleus; 7, food particles; 8, cyst wall; 9, anterior flagellum; 10, trailing posterior flagellum; 11, basal body; 12, dividing nucleus; 13, undulating membrane.

Figures adapted from various sources.

led to this conclusion rather than actual observations.

Information on whether transmission may take place from the soil may be obtained by placing uninfected laboratory-reared snails in terrariums previously occupied by infected ones.

Uninfected sexually immature snails placed with infected sexually mature ones should suggest whether infection is via the soil. Uninfected and infected sexually mature snails placed together, in addition to the experiments mentioned above, would provide strong evidence of the mode of infection.

### *Trypanoplasma borreli*
Laveran and Mesnil, 1901 (Plate 8)

Except for the presence of an undulating membrane in species of *Trypanoplasma*, they are very similar morphologically to *Cryptobia*. Some authors consider them as being the same. Differences occur in that *Cryptobia* is in the seminal vesicles and digestive tract of snails, while *Trypanoplasma* is found only in the blood of marine and freshwater fishes. *Trypanoplasma borreli* occurs in such freshwater fish as *Catostomus* (suckers), *Cyprinus* (carp), and *Leuciscus* (rudd).

**Description.** The body is flattened, blunt anteriorly, and pointed posteriorly. It is curved, with the nucleus just behind the anterior third, and is around 20 $\mu$ long. The two blepharoblasts lie near the anterior end of the elongated basal body. The axoneme from one kinetosome emerges anteriorly to form the free anterior flagellum, and that of the other passes posteriorly along the margin of an undulating membrane on the convex side of the body to the caudal end to form a free flagellum (Plate 8, F).

**Life Cycle.** Multiplication in the blood of fish is by binary fission, much as described above for *Cryptobia helicis*. Transmission among the fish is by leeches, particularly *Hemiclepsis* and *Piscicola* (Plate 8, G). Trypanoplasms ingested

by the leeches divide by binary fission in the crop. After several days of division, slender forms migrate forward to the proboscis sheath, where they occur in great numbers either free or attached to the wall of the sheath by their flagella. By the time the leech is ready to feed at about 10-day intervals, all of the trypanoplasms appear to have left that part of the intestine posterior from the crop. Feeding apparently clears the leech of the trypanoplasms present in the proboscis. It becomes filled again, presumably by dividing forms from the crop. Trypanoplasms appear in the blood of fish on day 7 after they have been injected by the leeches.

The trypanoplasms are inoculable from one species of fish to another.

## EXERCISE ON LIFE CYCLE

Attempt to culture the trypanoplasms from blood of carp, using media and methods suited for trypomastigotes of amphibians (p. 48).

## SELECTED REFERENCES

Doflein, F., and E. Reichenow. (1953). Lehrbuch der Protozoenkunde. 6th ed. Gustav Fischer, Jena, p. 566.

Keysselitz, G. (1906). Generations- und Wirtswechsel von *Trypanoplasma borreli* Laveran et Mesnil. Arch. Protistenk. 7:1–74.

Kipp, H. (1967). Zur Kenntnis auffallend grosser Formen der im Blut von Schleie, Karpfen und Barbe lebenden Gattung *Cryptobia* (Protozoa, Flagellata). Naturwissenschaften. 54:576.

Kozloff, E. N. (1948). The morphology of *Cryptobia helicis*. J. Morphol. 83:253–279.

Kudo, R. R. (1966). Protozoology. 5th ed. Charles C Thomas, Springfield, Ill., p. 425.

Mavor, J. W. (1915). On the occurrence of a trypanosome, probably *Trypanoplasma borreli*. J. Parasitol. 2:1–6.

Mackinnon, D. L., and R. S. J. Hawes. (1961). An Introduction to Protozoa. Oxford, London, p. 105.

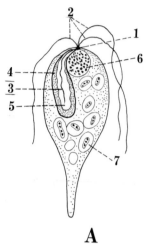

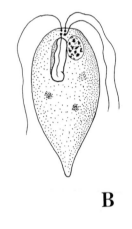

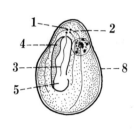

A

B

C

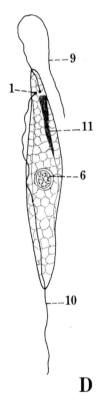

D

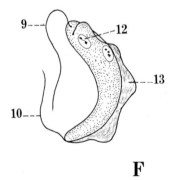

E

F

G

## Explanation of Plate 9

A, Mature flagellate, elongate form. B, Shortened form of mature flagellate. C, Dividing flagellate. D, Two new individuals resulting from longitudinal fission. E, Young flagellate from cyst. F, Coming together by two small flagellates. G, Union of two flagellates inside cyst. H, Complete fusion of two flagellates, union of two nuclei, and formation of vacuoles. I, Complete fusion of nuclei. J, Small uninucleate cyst with large central vacuole that crowds cytoplasm into a thin layer against cyst wall, producing a blastocystis-like appearance. K, Small binucleate cyst. L, Large multinucleate cyst. M, Formation of young flagellates. N, Fully developed and presumably infective flagellates in cyst which occurs in posterior stretch of hindgut of host. O, Gecko *(Tarentola mauritanica)* host. P, Infective cysts pass in feces.

1, Nucleus; 2, karyosome; 3, basal body; 4, kinetosome; 5, axoneme or rhizoplast; 6, chromatin ring surrounding axoneme; 7, posterior or trailing flagellum; 8, anterior flagellum; 9, cyst wall; 10, vacuole; 11, thin layer of cytoplasm pressed against cyst wall by expanding vacuole; 12, developing flagellates in large cyst; 13, fully developed flagellates in cyst.

a, Ingested cyst has ruptured and is releasing young flagellates; b, free flagellate just released from cyst; c, fully grown flagellate; d, dividing flagellate; e, two new flagellates resulting from longitudinal fission; f, union of two flagellates which have secreted a cyst about themselves; g, individuals have fused, forming a uninucleate individual with a long central vacuole; h, binucleate form; i, large multinucleated cyst; j, infective cyst in posterior region of hindgut which is voided with feces (P).

Figures redrawn from various sources.

---

Shindera, M. (1922). Beiträge zur Biologie, Agglomeration und züchtung von *Trypanoplasma helicis*. Arch. Protistenk. 45:200–240.

Wenyon, C. M. (1926). Protozoology. William Wood and Co., New York (Reprinted by Hafner Publishing Co., New York, 1965), p. 642.

### FAMILY BODONIDAE

The members of this family are characterized by two flagella at the anterior end of the body, one of which is directed anteriorly and the other posteriorly along the side of the body, often attached for a short distance, and extending beyond. Most of the species are free-living.

### *Proteromonas lacertae*
(Grassi, 1879) (Plate 9)

*Proteromonas lacertae* occurs in the colon of lizards (*Lacerta* and *Tarentola*) and *P. longifila* in urodeles (*Diemyctylus, Plethodon, Batrachoseps*) throughout their geographic range.

Transmission is direct, the infective stages presumably being ingested. The parasites do little if any harm.

**Description.** The fully developed parasites are elongated pyriform bodies 10 to 30 $\mu$ in length and flattened like a blade of grass. The posterior end is slender, sometimes twisted and somewhat screw-like. The blunt anterior end bears two flagella. One is directed anteriorly and is 3 to 5 times as long as the body; the other trails behind and is about twice as long as the body to which it may be attached for a short distance. The nucleus, located near the anterior end of the body, has a large central karyosome. One to several basal bodies lie close by the nucleus. An axoneme, or rhizoplast, extends from the anterior side of the nucleus to the margin of the body, where there forms a kinetosome whence the two flagella arise. Two rings of chromatin surround the axoneme.

Cysts range from 10 to 100 $\mu$ in size. The smallest ones have a large central vacuole and one or two nuclei located near the wall. As nuclear division progresses, the cyst enlarges and the nuclei are arranged over the inside surface of the cyst, being crowded by the vacuole. Bits of cytoplasm associate with the nuclei to form small flagellates.

**Life Cycle.** Asexual reproduction in the gut takes place by longitudinal fission, and sexual reproduction occurs in cysts following the fusion of two individuals. In dividing forms, the organisms become shortened and the nucleus divides so that each part has a kinetosome and an axoneme which continue as a flagellum. The basal body is divided between the two nuclei.

In encysting forms, the body shortens, becoming ovoid in shape, and the flagella are lost. Two forms come together and secrete a cyst about themselves which is about 10 $\mu$ in diameter. The individuals in the cyst lose their identity and the two nuclei fuse. A large central vacuole develops and the nuclear division begins. On the first division, the two nuclei move to opposite poles of the cyst, giving it the appearance of a blastocystis. As nuclear division con-

**Plate 9**     *Proteromonas lacertae*                                                                                                57

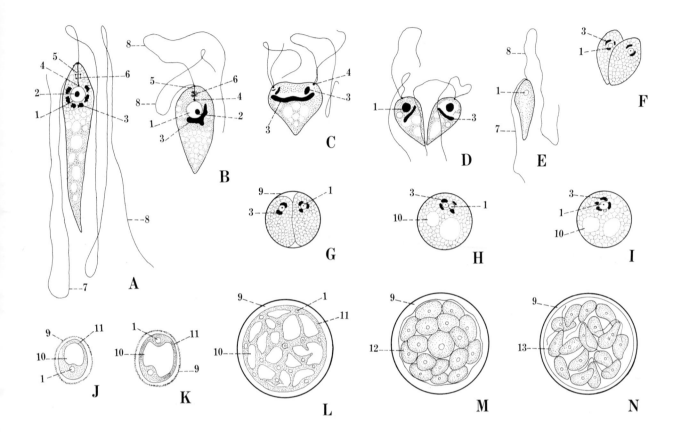

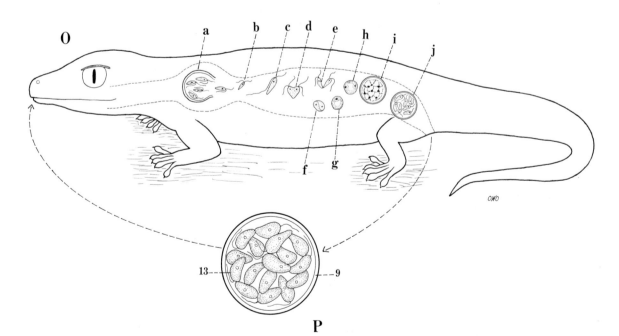

tinues, the cyst increases in size, up to 100 $\mu$, presumably by absorption of fluid. The cytoplasm and nuclei appear to be pressed against the cyst wall by the internal pressure of the fluid. When approximately 32 nuclei have developed, a scanty bit of cytoplasm surrounds each of the nuclei to form small flagellates which fill the cyst.

Since the cysts are in the hind part of the gut of lizards, it is presumed that they are the infective stage to other hosts which ingest them with the feces in which they are voided. This point, however, has not been demonstrated experimentally.

## EXERCISE ON LIFE CYCLE

The parasites are abundant in the mucus of the upper part of the rectum. They can be found by examining a bit of the rectal fluid on a microscope slide. Specimens stained with neutral red and unstained should be observed.

Inasmuch as the complete life cycle has not been demonstrated experimentally, it offers an excellent opportunity to conduct investigations on the manner of infection of the hosts. Infected hosts, as determined by recognition of the cysts in the feces, are necessary to provide a continuous source of infective material. The other, and more difficult part, is a colony of uninfected hosts. These would of necessity be acquired by hatching them from eggs and maintaining them under conditions in which natural infection cannot occur.

When cysts are fed to uninfected animals, the development of the infection should be followed daily by postmortem examination of specimens. Other animals should be kept to determine when cysts appear in the feces. Uninfected control animals should be maintained.

## SELECTED REFERENCES

Doflein, F., and E. Reichenow. (1953). Lehrbuch der Protzoenkunde. 6th ed. Gustav Fischer, Jena, p. 564.

Grassé, P. P. (1926). Contribution à l'étude des flagellés parasites. Arch. Zool. Exp. Gén. 65:345–602.

Grassé, P. P. (1952). Traité de Zoologie, Masson et Cie, Paris, Vol. 1, Fasc. 1, p. 694.

Mackinnon, D. L., and R. S. J. Hawes. (1961). An Introduction to the Study of Protozoa. Oxford, London, p. 107.

Wenyon, C. M. (1921). Observations on the intestinal Protozoa of three Egyptian lizards, with a note on a cell-invading fungus. Parasitology 12: 350–365.

Wenyon, C. M. (1926). Protozoology. William Wood and Co., New York (Reprinted by Hafner Publishing Co., N.Y., 1965), p. 611.

## ORDER POLYMASTIGIDA

The Polymastigida possess three to eight flagella; usually there is a stiffened rod-like axostyle extending through the body and beyond the posterior end. In the Trichomonadidae, a second rod-like structure, the costa, lies along the base of an undulating membrane. Most of the Polymastigida are inhabitants of the alimentary canal.

The order consists of three suborders, two of which will be considered. The suborder Monomonadina is characterized by having a single nucleus, and the suborder Diplomonadina, by two nuclei.

### Suborder Monomonadina
### FAMILY TETRAMITIDAE

The Tetramitidae bear four equal or unequal flagella on one end of the body, one or two of which may trail behind. The species of the family include both free-living and parasitic forms. Members of the genus *Costia* are ectoparasites on freshwater fish in Europe, North America, and possibly throughout the world.

### *Costia necatrix*
(Henneguy, 1883) (Plate 10)

This and a related species occur on the gills and skin of trout, carp, goldfish, and tench. In addition to parasitizing the fish, *C. necatrix* attacks and destroys roe of carp. It produces a dermatitis in which a bluish or grayish film of mucus spreads over the body, fins, and tail. Upon examination of the discolored mucus film, the parasites may be seen.

**Description.**    These small flagellates vary from 5 to 18 $\mu$ in length by 2.5 to 7.7 $\mu$ in width. The shape, likewise, is highly variable. Attached individuals are pyriform in shape while freeswimming ones are ovoid with a convex dorsal and concave ventral side. A groove, or flagellar pocket, extends obliquely along the ventral side of the body. Two to four flagella emerge from the flagellar pocket, or reservoir. In cases where four flagella occur, one pair is shorter than the body and the other pair longer. The latter are prominent in swimming forms. The pellicle consists of two membranes between which is a fibrillar layer whose fibrils run lengthwise in the body and converge in a stalk at the end of the body. The pellicle lines the flagellar pocket and forms the outer covering of the flagella.

Internal structures include the usual nine peripheral and central pairs of flagellar fibrils. The central pair ends in an axial granule while the nine outer ones extend beyond to form a kinetoplast-mitochondrion complex. Other structures in the cytoplasm include a double-membraned nucleus with outward bulges and a single nucleolus, contractile vacuoles that empty every 5 to 10 minutes, and food vacuoles.

**Life Cycle.**    Attached parasites are active, moving rather vigorously. They detach readily from the host and swim about, trailing the two long flagella. When preparing to attach to the host, the flagellates swim backwards with the long flagella extended. According to Tavolga and Nigrelli, the parasites attach by inserting the two long flagella into the epithelial cell, as seen in both living and iodine-stained cells. Schubert, on the other hand, reported that electron microscope micrographs showed finger-like extensions from the stalk of the parasite entering the cells. Attached parasites are pyriform in shape and free-swimming ones ovoid.

Division is by binary fission. In the case of attached forms, dividing individuals number about 1 in 500. Numerous specimens appear in the debris on the bottom of the pond, some swimming and some attached to fish scales. As many as 10 to 100 may be adhering to a single loose scale. Among these saprophytic forms, as many as 3 to 5 percent of the population may be in a stage of binary fission. They are active in temperatures between 8° and 38°C.

Small resistant cysts have been reported, but in laboratory observations none has been found.

In mild infections, the parasites are attached exclusively to the cells lining the pockets between the overlapping scales. Up to 200 and more parasites may be attached to a single scale. Heavy infections are manifested by a bluish or grayish film, especially around the base of the fins and tail. The gills are attacked. Heavy losses occur among severely infected fish, especially young ones.

There is some question whether *Costia pyriformis* Davis, 1943, described from trout might not be only a small form of *C. necatrix.*

Schubert is of the opinion that *Costia* belongs to the family Bodonidae because of the nature and structure of the kinetoplast-mitochondrion complex.

## EXERCISE ON LIFE CYCLE

Infection is common in aquarium fish and hatchery trout. Presence of the parasite in heavily infected fish may be detected by the light bluish or grayish mucus layer at the base of the fins or caudal peduncle. Examination of a bit of the mucus on a slide by means of a high dry objective of a microscope will reveal the flagellates.

Occasionally, dividing individuals may be seen attached to scales that have been removed carefully from the fish. They are far more common, however, on detached scales on the bottom of the aquarium. Infected scales stained with iodine in 70 percent alcohol show the ends of the posterior flagella anchored within the epidermis. Sections of skin fixed in 10 percent formalin and stained in hematoxylin and eosin are useful for showing the parasites in the scale pockets.

Once infected fish are found, uninfected ones may be introduced into the aquarium and the course of the infection followed. Observe the gills as well as the scales for infection.

Infected fish may be freed of the parasites by immersing them in a solution of 1 part acetic acid by volume to 500 parts of water or a single treatment for 1 hour in a 1 to 4000 solution of formalin.

## SELECTED REFERENCES

Davis, H. S. (1943). A new polymastigine flagellate, *Costia pyriformis,* parasitic on trout. J. Parasitol. 29:385–386.

Davis, H. S. (1946). Care and diseases of trout. Fish Wildl. Serv. U.S. Res. Rep., No. 12, 98 pp.

Hłond, S. (1963). Occurrence of *Costia necatrix* Henneguy on the roe of carp. [In Polish; English summary.] Wiad. Parazytol. 9:249–251.

Schubert, G. (1966). Zur ultracytologie von *Costia necatrix* Leclerq, unter Besonderer Berucksichtigung des kinetoplast-mitochondrions. Z. Parasitenk. 27:271–286.

Tavolga, W. N., and R. F. Nigrelli. (1947). Studies on *Costia necatrix* (Henneguy). Trans. Amer. Microsc. Soc. 66:366–378.

Wenyon, C. M. (1926). Protozoology. William Wood and Co., New York (Reprinted by Hafner Publishing Co., New York, 1965), p. 305.

## Explanation of Plate 10

A, Sinistral view of flagellate, showing flagella, nucleus, and contractile vacuole. B, Ultrathin section of flagellate attached to host cell. C, Dividing form, undergoing binary fission. D, Ventral view, showing origin of flagella. E, Similar view of D but from an angle. F, G, Shapes of swimming *Costia necatrix*. H, Section of epidermal cells with three *Costia necatrix* attached. I, Section through two scales, showing flagellates attached to epidermis in scale pockets and swimming in exuded mucus.

1, Anterior flagella; 2, posterior, or trailing, flagella; 3, ventral groove; 4, basal body; 5, nucleus with single nucleolus; 6, contractile vacuole; 7, cell inclusion; 8, flagellate attached to epidermal cell; 9, epidermal cell; 10, trailing flagella inserted into epidermal cell for attachment of flagellate to host; 11, flagellar pocket showing cross sections of two flagella each with nine peripheral and two central fibrils; 12, pellicular membrane consisting of outer and inner layers with intermediate fibrillar layer; 13, mitochondrion; 14, endoplasmic reticulum; 15, food vacuole; 16, stalk with adhesive disc; 17, cytoplasmic extension projecting into host cell.

a, Scale; b, epidermis; c, connective tissue; d, epidermal pocket between scales; e, free-swimming flagellates; f, mass of mucus; g, flagellates swimming in exuded mucus; h, macrophages; i, flagellates in epidermal pocket.

Figures A, C, F–I redrawn from Tavolga and Nigrelli, 1947, Trans. Amer. Microsc. Soc. 66:366; B composite from Schubert, 1966, Z. Parasitenk. 27:271; D, E redrawn from Davis, 1946, Fish Wildl. Serv. U.S. Res. Rep. (12):35.

### FAMILY CHILOMASTIGIDAE

The members of this family possess a cytostomal groove with a cytoplasmic fibril extending across the anterior end and posteriorly along each side. A posteriorly directed flagellum lies in the cytostomal groove.

### *Chilomastix mesnili*
(Wenyon, 1910) (Plate 8)

This is a common inhabitant of the large intestine of humans, especially young children, both as motile, flagellated trophozoites and cysts. It is cosmopolitan in distribution but more common in warm climates than in cool ones.

**Description.** The trophozoites are somewhat pear-shaped with a ventral groove, or cytostome, in the anterior half of the body. Three flagella extend anteriorly and a short delicate one lies in the obliquely arranged cystostomal groove, the margin of which is formed by a fibril extending around the anterior end and posteriorly along each side. Trophozoites measure 6 to 20 $\mu$ long by 3 to 10 $\mu$ wide. The nucleus is near the anterior end. The cystic forms are lemon- or pear-shaped. They are colorless, typically uninucleate, have a thick wall, and measure 7 to 10 $\mu$ long by 4 to 6 $\mu$ wide. The cytostome is visible inside the cyst (Plate 8, A–C).

**Life Cycle.** The life cycle is direct, being transmitted by cysts in water and food. Multiplication of trophozoites in the intestine is by binary fission following division of the nucleus. Cysts containing two individuals have been observed. Both cysts and trophozoites appear in the feces.

Other species have been reported from a variety of hosts. They include *C. caulleryi* from frogs, *C. bettencourti* from rats and mice, *C. cuniculi* from rabbits, *C. intestinalis* from guinea pigs, *C. caprae* from goats, and *C. gallinarum* from poultry. Monkeys of several species harbor *Chilomastix*. Whether these various forms are different from *C. mesnili* in humans is uncertain.

### EXERCISE ON LIFE CYCLE

Simple experiments on host specificity may be conducted by cross infections, using forms from frogs to infect mice, and vice versa. Care must be taken to use parasite-free experimental animals.

### SELECTED REFERENCES

Doflein, F., and E. Reichenow. (1953). Lehrbuch der Protozoenkunde 6th ed. Gustav Fischer, Jena, p. 569.

Levine, N. D. (1962). Protozoan Parasites of Domestic Animals and Man. Burgess Publishing Co., Minneapolis, p. 111.

Tumka, A. F. (1968). Geographical and age distribution of parasitic Flagellata from intestine of man. [In Russian; English summary.] Parasitologiya 2:477–482.

Wenyon, C. M. (1926). Protozoology. William Wood and Co., New York (Reprinted by Hafner Publishing Co., New York, 1965), p. 621.

Wishahy, A., G. Abdel-Hamid, M. A. Rifaat, and S. A. Salem. (1969). Parasitic infection in infants presenting gastroenteritis in U.A.R. J. Trop. Med. Hyg. 72:280–281.

Plate 10    *Costia necatrix*                                                        61

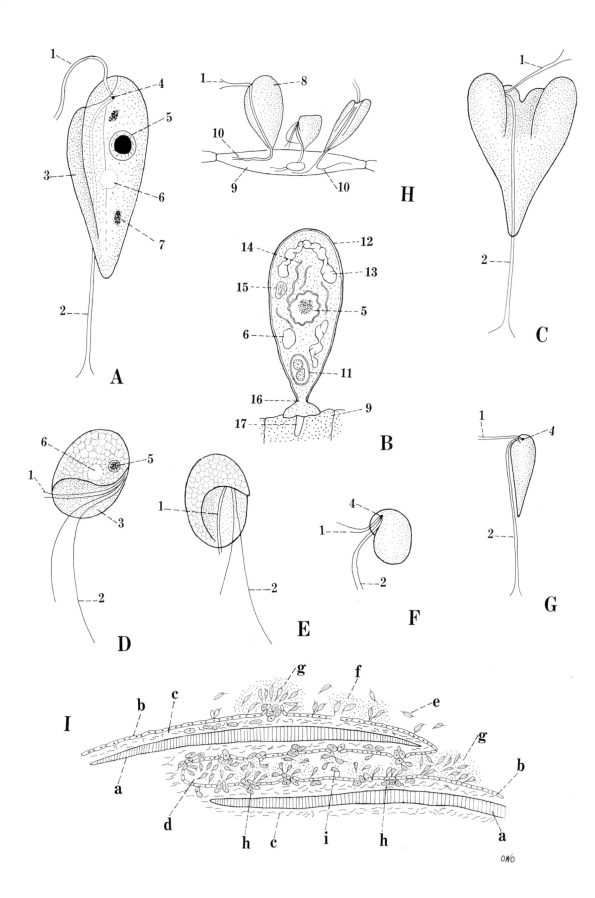

## Explanation of Plate 11

A, Trophozoite of *Hexamita ovatus* from American newt. B, Trophozoite of *H. ovatus* dividing by binary fission. C, Trophozoite of *H. ovatus* undergoing multiple fission. D, Trophozoite of *H. intestinalis* from American newt. E, Trophozoite of *H. muris* from house mouse. F, Cyst of *H. muris*. G, Trophozoite of *H. salmonis*. H, Trophozoite of *H. salmonis* from trout dividing by binary fission. I, Cyst of *H. salmonis*. J, Section of a pyloric cecum of trout infected with intracellular stages of *H. salmonis*. K, Trophozoite of *Octomitus pulcher* from marmot. L, Anterior end of trophozoite of *O. pulcher,* showing arrangement of flagella.

1, Anterior flagella; 2, posterior flagella; 3, basal body; 4, nucleus; 5, axoneme; 6, chromatin ring; 7, protruding portion of axoneme; 8, cyst wall; 9, encysted stage; 10, circular muscle of intestine with its nuclei; 11, epithelial cells of intestine with their nuclei; 12, intracellular forms of *H. salmonis*.

Figures A–D, redrawn from Swezy, 1915, Univ. Calif. Publ. Zool. 16:71; E, F, from Doflein and Reichenow, 1953, Lehrbuch der Protozoenkunde, p. 582; G–J, from Davis, 1947, Fish Wildl. Serv. U.S. Res. Rep. 12; K, L, from Gabel, 1954, J. Morphol. 94:473.

### Suborder Diplomonadina

The principal characteristic of this group of flagellates is the duplication of the nuclei, flagella, basal bodies and axostyles so that the body is bilaterally symmetrical. In this respect, they differ from all others. The bodies are pyriform with nuclei near the anterior end. They are parasites of the alimentary canal.

### FAMILY HEXAMITIDAE

The Hexamitidae are intestinal parasites of invertebrate and vertebrate hosts. The organisms are bilaterally symmetrical and somewhat oval in shape. There are two nuclei, six or eight flagella, two axostyles, and basal bodies. The genera *Hexamita*, *Octomitus*, and *Giardia* are important pathogens.

### *Hexamita* Dujardin, 1838 (Plate 11)

The *Hexamita* are pyriform, bilaterally symmetrical animals with bodies somewhat flattened dorsoventrally. There are eight flagella, three pairs extending anteriorly from the broadly rounded end and one posteriorly from the more pointed end. Three anterior and one posterior flagella, one nucleus, and one axostyle are in each half of the bilateral body, giving the animal the appearance of consisting of two individuals fused along their longitudinal axes. The posterior tips of the axostyles do not extend beyond the cytoplasm of the body. Cysts are formed by some species. They are common parasites of oysters, insects, and all classes of vertebrates.

Well-known species include *H. nelsoni* Schlicht and Mackin from oysters, *H. salmonis* (Moore), frequently referred to as *Octomitus*, from hatchery trout, *H. ovatus* Swezy from the American newt, *H. batrachorum* Swezy from the grassfrog, *H. intestinalis* Dujardin from tadpoles of frogs, *H. meleagridis* McNeal, Hinshaw, and Kofoid from turkeys, pheasants, chukar partridges, and quail, and *H. muris* (Grassi) from rats and house mice.

**Life Cycle.** The life cycles of these species are not well understood, as none of them has been completely worked out. It is recognized generally, however, that wherever life cycles are known, they are direct. Resistant cysts are known to occur in *H. salmonis*, *H. intestinalis*, *H. muris*, and *H. meleagridis*.

Two types of multiplication occur. Binary longitudinal fission, preceded by division of the nuclei, basal bodies, and axonemes, is common. Multiple division is seen frequently in smears of unencysted forms from the intestine of newts and turkeys. It consists of three divisions of the nuclei and organelles. The eight attached individuals with their 48 anterior flagella roll along somewhat like a *Volvox*. There is no evidence that multiple division takes place in encysted forms. Some workers claim that the multiple division of the nucleus seen in these flagellates is a process of schizogony. Rather, it is a process of multiple binary fission.

The life cycles of two species have been studied somewhat extensively because of their economic importance. *H. meleagridis* of turkeys occurs as both flagellated and encysted forms. The flagellated trophozoites are abundant in the lower part of the small intestine and rectum but not in the duodenum and ceca. In the early stage of the infection, the flagellates are imbedded in the intercellular spaces, reticuloendothelial cells of the intestine, and between the

**Plate 11**  *Hexamita ovatus, H. intestinalis, H. muris, H. salmonis, and Octomitis pulcher*

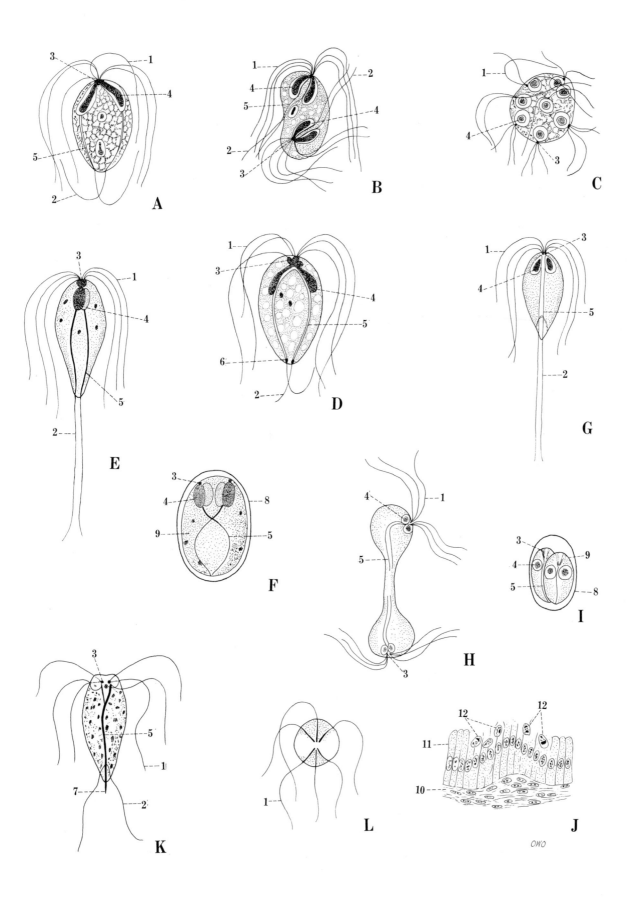

OWO

villi. As multiplication of the flagellates proceeds, they fill the crypts of Lieberkühn and myriads of them appear in the intestinal fluid. The masses voided with the feces die quickly and do not constitute a source of infection of new hosts.

Encysted flagellates begin to appear in flecks of intestinal mucus soon after infection of the host. They are present in enormous numbers in the mucus in fresh droppings. The cysts are highly resistant to unfavorable environmental conditions and survive long periods outside the host. Infection occurs when the cysts are ingested with contaminated food and water. Excystment takes place in the intestine. The flagellate laboriously emerges through a narrow slit in the cyst and enters the glandular crypts and intercellular spaces, where growth is completed and multiplication begins.

*Hexamita salmonis* is a common parasite of hatchery trout and salmon, especially under conditions of low sanitation. It has been reported as a serious pathogen of fingerlings in hatcheries but one series of experimental infections of fish failed to produce disease from the strain of cultured flagellates used. Evidence from observations on epizootics among hatchery fish indicate that the parasite is pathogenic, resulting in death of fish up to 1 year of age.

The life cycle of *H. salmonis* is complex, comprising two stages. In one stage consisting of the mature flagellates in the lumen of the intestine, the parasites divide rapidly by binary fission producing myriads of individuals. After a period of multiplication, they encyst and the enclosed flagellate divides once inside. These resistant cysts are voided with the feces and are the source of infection to fish that swallow them. The other stage begins with the excysted flagellates. They penetrate the epithelial cells of the intestine and undergo a process of multiple fission, forming minute individuals. Upon rupture of the parasitized epithelial cells, the tiny flagellates enter other cells to repeat the multiplication process. Eventually, the small forms no longer penetrate cells but mature in the lumen of the intestine and encyst.

### *Octomitus* von Prowazek, 1904 (Plate 11)

In general, this genus is very similar in body structure and conformation, differing in the nature of the axostyles. In *Octomitus*, the posterior tips of the axostyles fuse inside the body and extend beyond the aflagellar end as a single pointed structure, whereas in *Hexamita* they remain separate and do not protrude. *Octomitus* is considered by some workers to be a synonym of *Hexamita*, but the difference in the arrangement of the axostyles appears to be sufficient justification for retaining them as separate genera. Cysts have not been reported. They are parasites of vertebrates, and *O. pulcher* occurs in marmots in North America.

**Symptoms and Disease.**     Hexamitiasis of turkeys is a cosmopolitan disease, resulting in enormous losses, especially in young poults where entire flocks may die. Symptoms of ruffled feathers, drooping wings, listlessness, and discharge of liquid green, fetid feces appear as soon as 3, but usually 6 to 7 days after infection. Infected birds crowd together and are silent. They refuse food and water.

The entire intestine becomes filled with watery, frothy mucus, resulting in excessive loss of body fluid and with subsequent dehydration. Beginning in the rectum and hind part of the intestine, the parasites invade the intestinal epithelium and reticuloendothelial cells. The intestine is tightly contracted at numerous points.

In salmonid fish, *Hexamita salmonis* occurs in the blood, intestinal epithelium, brain, and reproductive organs. It causes whirling disease, destruction of intestinal epithelium, castration of both sexes, and loss of appetite resulting in the appearance of emaciated fingerling pinheads with large heads and attenuated bodies.

### EXERCISE ON LIFE CYCLE

Of the hosts listed above, newts, frogs, tadpoles, and mice are likely sources of various species of *Hexamita* for experimental studies. When available, sick and dying fingerling trout from hatcheries will provide an abundance of *H. salmonis*.

Flagellates removed from the intestine with its juices and placed on slides can be viewed under the microscope. By sealing the margins of the coverslip with Vaseline, such mounts may be kept for long periods, during which encystment may be observed. Staining of the flagellates and cysts is necessary to see the details of internal structure.

*Hexamita salmonis* may be cultured on a limited scale by introducing flagellates into a highly alkaline and dilute fish broth where the various growth stages occur (Moore), or a more sophisticated medium may be prepared for production of large numbers of them (Uzmann and Hayduk).

## SELECTED REFERENCES

Allison, L. N. (1963). An unusual case of *Hexamita (Octomitus)* among yearling rainbow trout. Progr. Fish-Cult. 25:220.

Davis, H. S. (1953). Culture and Diseases of Fishes. University of California Press, Berkeley, p. 240.

Doflein, F., and E. Reichenow. (1953). Lehrbuch der Protozoenkunde. 6th ed. Gustav Fischer, Jena, p. 582.

Feng, S. Y., and L. A. Stauber. (1968). Experimental hexamitiasis in the oyster *Crassostrea virginica*. J. Invert. Pathol. 10:94–110.

Gabel, J. R. (1954). The morphology and taxonomy of the intestinal protozoa of the American woodchuck, *Marmota monax* Linnaeus. J. Morphol. 94: 473–540.

Hoare, C. A. (1955). Life cycle of *Hexamita meleagridis*. Vet. Rec. 67:324.

Levine, N. D. (1962). Protozoan Parasites of Domestic Animals and Man. Burgess Publishing Co., Minneapolis, p. 15.

Medcof, J. C. (1961). Trial introduction of European oysters *(Ostrea edulis)* to Canadian east coast. Proc. Nat. Shellfish. Ass. 50: 113–124.

Meshorer, A. (1969). Hexamitiasis in laboratory mice. Lab. Anim. Care 19:33–37.

Moore, E. (1923). A report of progress on the study of trout diseases. Trans. Amer. Fish. Soc., 53rd Annu. Meet. St. Louis, pp. 74–94.

Moore, E. (1924). The transmission of *Octomitus salmonis* in the egg of trout. Trans. Amer. Fish. Soc., 54th Annu. Meet. Quebec, pp. 54–56.

Salvin, D., and J. E. Wilson. (1960). A fuller conception of the life cycle of *Hexamita meleagridis*. Poultry Sci. 39:1559–1576.

Schlicht, F. G., and J. G. Mackin. (1968). *Hexamita nelsoni* n. sp. (Polymastigina: Hexamitidae) parasitic in oysters. J. Invert. Pathol. 11:35–39.

Swezy, O. (1915). Binary and multiple fission in Hexamitus. Univ. Calif. Publ. Zool. 11:71–88.

Uzmann, J. R., and J. W. Jesse. (1963). The *Hexamita* (*=Octomitus*) problem: A preliminary report. Progr. Fish-Cult. 25:141–143.

Uzmann, J. R., and S. H. Hayduk. (1963). *In vitro* culture of the flagellate protozoan *Hexamita salmonis*. Science 140:290–291.

Uzmann, J. R., G. J. Paulik, and S. H. Hayduk. (1965). Experimental hexamitiasis in juvenile coho salmon *(Oncorhynchus kisutch)* and steelhead trout *(Salmo gairdneri)*. Trans. Amer. Fish. Soc. 94:53–61.

Wilson, J. E., and D. Salvin. (1953). Hexamitiasis of turkeys. Vet. Rec. 67:236–242.

## *Giardia lamblia* Stiles, 1915 (Plate 12)

This species is one of many of the genus and is the best known because of its common occurrence in the intestine of humans. It is cosmopolitan, being more common in children than in adults and in warm than cold climates. Related species occur in the intestine of many species of vertebrates.

**Description.** The anterior end is broadly rounded and the posterior drawn to a point. The dorsal side is convex and the ventral concave, forming a functional sucking disc-like depression in the anterior half of the body.

Trophozoites are constant in shape, 9 to 30 $\mu$ long by 5 to 10 $\mu$ wide. The sucking disc-like ventral surface serves as an organelle of attachment on the surface of the intestinal epithelium. Two axostyles extend the length of the body, remaining separate. Two nuclei are near the junction of the first and second thirds of the body. There are eight flagella in four pairs. Two originate near the anterior ends of the axostyles, cross each other and follow the anterolateral margin of the sucking "disc," becoming free; two toward the anterior part of the axostyles near the nuclei, leaving the body about two-thirds the distance from the anterior end; two others begin near the middle of the axostyles and two at the caudal end and extend backward as free filaments. A deeply staining median mass occurs over the axostyles. A basal body occurs at the end of each flagellum.

Cysts are ovoid, 8 to 14 $\mu$ long by 6 to 10 $\mu$ wide, and are thin-walled. Two nuclei appear in immature and four in mature cysts. Fibrils, flagella, axostyles, and darkly staining masses are visible.

**Life Cycle.** The trophozoites are most common in the duodenum, occurring in the crypts. Multiplication is by longitudinal binary fission, involving the nuclei, sucking disc, and other parts, followed by separation of the parasite into two daughter trophozoites. The number may be so great as to virtually "line" the surface of the duodenum and adjoining part of the intestine with their bodies. Individual trophozoites rest on the cells, adhering by the sucking disc,

## Explanation of Plate 12

A, Trophozoite of *Giardia agilis* from tadpole of frog, ventral view of stained specimen. B, Trophozoite of *G. muris*, ventral view of unstained individual. C, Ventral view of unstained *G. lamblia*. D, Ventral view of stained *G. lamblia*. E, Sinistral view of unstained *G. lamblia*. F, Recently formed unstained cyst of *G. lamblia*. G, Mature unstained cyst of *G. lamblia*. H, Stained mature cyst of *G. lamblia.*

1, Ventral "sucker"; 2, anterior pair of flagella; 3, lateral pair of flagella; 4, ventral pair of flagella; 5, caudal pair of flagella; 6, nucleus with endosome; 7, axostyle; 8, basal body; 9, chromatin mass; 10, "sucking-disc" fibril; 11, cyst wall.

Figures adapted from various sources.

and waving the flagella. They appear in diarrheic stools.

After a period of multiplication by binary fission in the intestine, the trophozoites prepare for encystment. They withdraw the flagella, shorten the body, and secrete a tough, haline cyst about themselves. Inside the cyst, each flagellate undergoes a process of maturation. When completed, each encysted individual has four nuclei and two sucking discs.

Mature cysts, which appear in formed stools, are infective as well as resistant to unfavorable environmental conditions. Infection of new hosts occurs when fully developed cysts are swallowed with contaminated water and food. Upon excystation in the duodenum, the mature tetranucleate flagellate quickly undergoes cytoplasmic division, forming two binucleate daughter trophozoites.

A number of species are known from domestic and wild animals. They include *G. agilis* Künstler from tadpoles of frogs, *G. bovis* Fantham from cattle, *G. cati* Descheins from cats, *G. canis* Hegner from dogs, *G. caviae* Hegner from guinea pigs, *G. duodenalis* (Davaine) from rabbits, *G. equi* Fantham from horses, *G. muris* (Grassi) from rats and house mice, *G. ondatrae* Travis from muskrats, and *G. simoni* Lavier from the Norway rat and various other wild rodents.

The life cycle of each of these is basically similar to that of *G. lamblia*.

### EXERCISE ON LIFE CYCLE

*Giardia muris* from mice, *G. duodenalis* from rats, and *G. agilis* of tadpoles of frogs provide suitable material for study, depending on the facilities and material available.

In *G. duodenalis*, encystment begins in the iliocecal region of the intestine. Cysts may be collected by breaking up fecal pellets, straining them through fine cloth, and placing the screened material in a tall cylinder of tap water where they settle to the bottom. Repeated decantation and sedimentation will remove coloring matter and fine debris. The cysts may be removed from the bottom by means of a fine pipette, placed on a slide, covered with a cover glass, and examined under high magnification. Staining with tincture of iodine will reveal internal structures.

The use of a centrifuge facilitates the collection of cysts. About 1 ml of feces is strained and placed in a cone-shaped 15-ml centrifuge tube with normal saline. The mixture is centrifuged for 1 to 2 minutes and the supernatant fluid is decanted. It is replaced by a solution of $ZnSO_4 \cdot 7H_2O$ (331 gm in 1 liter of water for a specific gravity of 1.18), thoroughly mixed with the sediment, and centrifuged for 1 to 2 minutes. The cysts rise to the surface of the fluid and may be transferred by means of a bacteriological loop to a slide for examination. For accumulating them, the supernatant fluid should be poured into 100 ml of water and this centrifuged to concentrate numerous cysts for experimental use.

Rats to be used for experimental infection must be free, or as nearly so as possible, of giardias before feeding them cysts. By keeping the cages scrupulously clean and by prompt removal of feces, the infection can be kept low, if not completely eliminated. Compare the number of cysts in the feces prior to and after administering a large number of them. A sharp increase in the number of cysts would indicate that a new infection had been established and also give some idea of the time necessary for them to excyst, multiply, and produce cysts. Trophozoites occur in diarrheic feces and cysts in formed ones. Trophozoites may be obtained most commonly in the small intestine of rats at a level about 10 cm anterior to the iliocecal valve, where they occur in great numbers. Determine where excystation takes place.

Recently formed cysts have two nuclei and mature ones have four near the anterior end of the body. Fibrils representing the flagella appear

**Plate 12** *Giardia agilis, G. muris, G. lamblia* 67

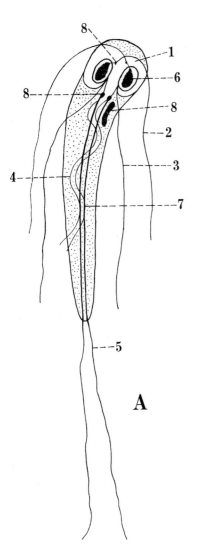

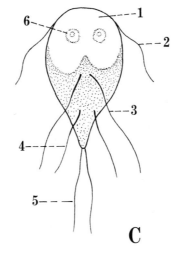

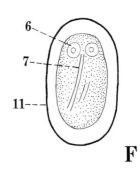

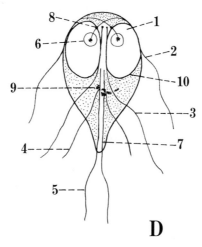

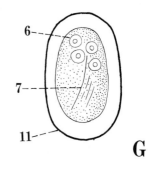

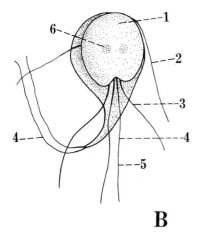

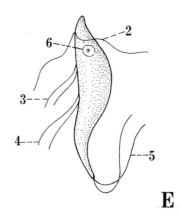

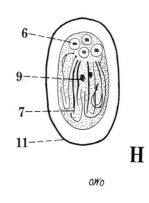

owo

in the cytoplasm. Infection is by ingestion of mature cysts contaminating food and water, or by contact between infected and uninfected hosts under unsanitary circumstances. There are no known methods for culturing species of *Giardia*.

## SELECTED REFERENCES

Doflein, F., and E. Reichenow. (1953). Lehrbuch der Protozoenkunde. 6th ed. Gustav Fischer, Jena, p. 586.

Filice, F. P. (1952). Studies on the cytology and life history of Giardia from the laboratory rat. Univ. Calif. Publ. Zool. 57:53–143.

Hegner, R. W. (1922). The systematic relationship of *Giardia lamblia* Stiles, 1915, from man and *Giardia agilis* Kunstler, 1882, from the tadpole. Amer. J. Hyg. 2:435–441.

Hegner, R. W. (1923). Giardias from wild rats and mice and *Giardia caviae* sp. n. from the guinea pig. Amer. J. Hyg. 3:345–349.

Hegner, R. W. (1927). Excystation and infection in the rat with *Giardia lamblia* from man. Amer. J. Hyg. 7:433–477.

Levine, N. D. (1961). Protozoan Parasites of Domestic Animals and of Man. Burgess Publishing Co., Minneapolis, p. 118.

Rothenbacher, H. (1970). Giardiosis in a wild mouse *(Peromyscus leucopus)* colony. J. Amer. Vet. Med. Ass. 157:685–688.

Wenyon, C. M. (1926). Protozoology. William Wood and Co., New York (Reprinted by Hafner Publishing Co., New York, 1965), p. 691.

### ORDER TRICHOMONADIDA

The typical body bears four to six flagella, one of which is recurrent, running along the margin of the undulating membrane when the latter is present. An axostyle, extending through the length of the body, and a nondividing basal body are present. Cysts are unknown.

#### FAMILY TRICHOMONADIDAE

The members of this family are pear-shaped parasites of the alimentary canal and genital tract. There are four to six flagella originating from a group of kinetosomes lying anterior to the single large nucleus; one flagellum passes posteriorly along the free margin of an undulating membrane and may continue as a free-trailing portion. The filamentous costa arising from a kinetosome extends posteriorly along the basal margin of the undulating membrane. The elongated basal body and its accompanying basal filament arise from a kinetosome and project posteriorly alongside the nucleus. A hyaline rod-like or tube-like axostyle extends posteriorly from the blepharoplast complex through the body, projecting beyond the hind end. Chromatin dots may be present inside it. A darkly staining chromatic ring may surround the axostyle at the point where it emerges from the body. A shield-shaped pelta appears over the anterior end of the body in specimens prepared with silver stains. A cytostome may be present near the anterior end of the side opposite the undulating membrane.

The genera are characterized in part by the number of free anterior flagella. *Ditrichomonas* Duboscq and Grassé, with two anterior flagella, is of doubtful status. Recent studies on specimens from sheep and cattle indicate that they have four flagella, hence belong to the genus *Trichomonas*. *Tritrichomonas* Kofoid from the genital tract of cattle and the alimentary canal of amphibians, *Tricercomitus* Kirby, a small form with a long trailing flagellum, and *Pseudotrypanosoma* Grassi from the gut of termites, have three flagella each. *Trichomonas* Donné, the type genus, from the genital tract of humans, the mouth and alimentary canals of humans and horses, the alimentary canal of dogs, cats, pigeons, chickens, ducks, geese, reptiles, and amphibians, possesses four flagella. *Pentatrichomonas* Mesnil from the intestine of primates and *Pentatrichomonoides* Kirby from the gut of termites have five free flagella.

### *Tritrichomonas foetus*
(Riedmüller, 1928) (Plate 13)

This species is parasitic in the genital tract of male and female cattle in many parts of the world, particularly in North America, Eurasia, Japan, and South Africa. Its prevalence is high in the United States and Europe and its economic importance great since it causes abortion and sterility in cows.

**Description.**    Normally, the flagellate is spindle- or pear-shaped and measures from 10 to 25 $\mu$ long with the width being about one-third to two-fifths the length. There are three anterior flagella as long as or longer than the body and a single recurrent one passing along the margin of the undulating membrane and extending beyond as a free-trailing portion about equal in length to the anterior ones. The kinetosomes lie at the anterior extremity of the body and are the point of origin of the flagella, costa, and axostyle. An undulating membrane equal in length to the body lies along the dorsal side; it has, in addition to the recurrent flagellum, an accessory filament near the free margin of the undulating membrane and lying parallel to the flagellum. A costa, or chromatic basal rod, extends along the base of the undulating membrane. The oval nucleus is relatively large and lies in the anterior third of the body; it contains a small karyosome. The axostyle is a stout hyaline rod with an enlarged end, the capitulum, which contains chromatin granules, and a sharp tip that extends beyond the posterior end of the body. At the point of its emergence, the axostyle is surrounded by a ring of chromatin. A cylindrical or club-shaped parabasal body lies between the nucleus and costa, terminating shortly before the posterior margin of the nucleus. A mouth, or cytostome, near the anterior end and on the ventral side, is visible in some specimens.

**Life Cycle.**    The flagellates occur first in the vagina of cattle, being introduced by infected bulls during coitus. From here, they move into the uterus, which appears to be the permanent location. Disappearance of them from the vagina is due probably to an immunity that develops in the epithelium of the organ. In bulls, the parasites occur in highest numbers on the glans penis and in the adjacent part of the prepuce, with fewer in the other areas.

Multiplication is by longitudinal binary fission, beginning with the kinetosomes and nucleus. The axostyle divides longitudinally, beginning at the anterior end. Multiple fission has been reported.

Transmission is almost entirely venereal. Infected bulls transmit the flagellates to uninfected cows. In infected females where the parasites are in the uterus, discharges from it containing the protozoa enter the vagina whence clean males servicing them become infected. Once infected, bulls remain permanently so and transmit the infection to all uninfected females with which they mate.

Infected cows commonly abort during weeks 1 to 16 of pregnancy. Two kinds of abortion occur. In one, the fetal membranes detach from the uterine walls and are expelled together with the fetus. In the other type, a cervical seal develops and the pregnancy is terminated. The embryonic membranes and dead fetus are retained, eventually being macerated. A retained fetus results in filling of the uterus with fluid, macerated tissues, and great numbers of flagellates, a condition known as pyrometra. Progressive enlargement of the pyrometric uterus suggests, from external appearance of the cow, that the pregnancy is progressing normally. In cases in which pyrometra occurs, cows generally become sterile. Infected cows that expel the fetus re-enter estrus and accept the bull. Such cows may conceive and bear a calf. Recurrence of estrus at short intervals suggests that the cow is infected and that abortion has occurred.

The abomasum of aborted calves contains enormous numbers of the flagellates in almost pure culture.

Although transmission of the parasites is normally venereal, infections have been observed in virgin heifers. Infection might result naturally from contact between infected animals or through transfer by insects of viable flagellates from infected to noninfected individuals. Houseflies have been shown to ingest the tritrichomonads and regurgitate living ones within 5 minutes. They may be passed in the feces of flies for 8 hours, and numerous active ones remain in the intestine for longer periods. It is believed that flies feeding on vaginal exudates from infected cows might transmit the flagellates to uninfected ones. Some workers believe, however, that nonvenereal transmission is insignificant in the epizootiology of tritrichomoniasis.

Control of the disease is dependent on proper management of the breeding program. Since the sire is largely responsible for its spread and remains permanently parasitized, infected bulls should be eliminated from all breeding programs. Aborting cows that have expelled the fetus and its membranes may be freed of the parasites by withholding them from breeding for 90 days.

Other species of *Tritrichomonas* occur in the intestine or respiratory tract of a number of different kinds of animals. Levine listed the principal species in domestic animals, together with an extensive bibliography of pertinent papers.

*Tritrichomonas muris* of house mice and rats was studied extensively by Wenrich. He described the structure of the flagellate and the method of division. According to him, the blepharoplasts and nucleus divide and separate from each other, moving toward opposite ends of the cell but remaining connected by a thread-like

## Explanation of Plate 13

A, Trophic stage of *Trichomonas vaginalis* of humans. B, Trophic stage of *Tritrichomonas foetus* of bovine. C, Trophic stage of *Tritrichomonas batrachorum* of amphibians and reptiles. D, Dividing *T. batrachorum*. E, Trophic stage of *Tritrichomonas augusta* of amphibians. F, Early telophase of *T. augusta*. G, Late telophase of *T. augusta*. H, Mitosis completed, cytoplasm dividing in *T. augusta*. I, Cow showing reproductive tract with *Tritrichomonas foetus* in vagina and uterus. J, Bull showing reproductive tract with *T. foetus* in preputial pouch and on the glans penis. K, Aborted fetus with placenta. L, Pyrometric uterus filled with fluid and tritrichomonads from a chronically infected cow.

1, Anterior flagella; 2, trailing flagellum; 3, undulating membrane; 4, costa; 5, kinetosomes; 6, axostyle; 7, endaxostylar chromidia; 8, capitulum of axostyle; 9, parabasal body; 10, filament of basal body; 11, nucleus with chromatin granules and karyosome; 12, paradesmose; 13, centrosome; 14, chromosomes.

a, Vulva; b, vagina; c, cervix; d, uterus; e, Fallopian tube; f, ovary; g, urinary bladder; h, trichomonads; i, anus; j, penis; k, glans penis; l, prepuce; m, scrotum; n, testis; o, vas deferens; p, prostate gland; q, retractor muscle; r, rumen; s, placenta; t, cotyledons; u, fetus outside placenta but still attached by umbilical cord.

Figures A, B redrawn from Wenrich and Emmerson, 1933, J. Morphol. 55:193; C, D, from Honigberg, 1953, J. Parasitol. 39:191; E–H, from Kofoid and Swezy, 1915, Proc. Nat. Acad. Sci. 1:315; I, J, from Hammond *et al.*, 1956, U.S. Dep. Agr. Yrbk., p. 277; K, from photograph by B. B. Morgan; L, from Boyd, 1937, Lederle Vet. Bull. 6:3.

---

fiber, the paradesmose. New flagella and undulating membranes develop while the nucleus is undergoing division with six chromosomes. He believed the old axostyle disappeared and new ones developed as outgrowths of the kinetosomes. Wenyon, however, stated that the original one divided longitudinally, beginning at the capitulum and extending caudad.

*Tritrichomonas augusta* from the digestive tract of salamanders, frogs, and toads is described as having longitudinal binary fission with the axostyle dividing longitudinally, beginning at the anterior end.

### EXERCISE ON LIFE CYCLE

Although life cycle studies on the trichomonads are difficult to perform, significant data on their morphology and biology may be obtained through observations on living and prepared specimens.

*Tritrichomonas augusta* and *T. batrachorum* from the recta of frogs are excellent species for study as they are common, large, and easily cultured, especially the latter species. Specimens stained by the Giemsa method, which is easy to use, are excellent for observing morphological details.

Cultures, particularly of *T. batrachorum* made by the simple technique of Kofoid and Swezy, and of Rosenberg, yield results satisfactory for observation of the flagellates. To prepare the culture, the rectal contents of a frog are mixed with an equal volume of 0.85 percent NaCl solution prepared from glass-distilled water, and the pH is adjusted to 7.64 by adding a few drops of N/20 NaOH. A few drops of the mixture are placed in a cell formed by ringing a slide with Vaseline or petroleum jelly, and a coverslip is pressed in position to form an airtight seal. The preparation should be kept in the dark at room temperature. Directions for the preparation and use of more sophisticated media were discussed by Morgan, Wenrich, and Hibler *et al.*

Division of living flagellates may be observed. They may be fixed and stained for critical observation of progressive division, as shown in Plate 13.

### SELECTED REFERENCES

Doflein, F., and E. Reichenow. (1953). Lehrbuch der Protozoenkunde. 6th ed. Gustav Fischer, Jena, p. 594.

Hammond, D. M., P. R. Fitzgerald, and J. L. Shupe. (1956). Trichomoniasis of the reproductive tract. U.S. Dep. Agr. Yrbk. Agr., pp. 277–288.

Hibler, C. P., D. M. Hammond, F. H. Caskey, A. E. Johnson, and P. R. Fitzgerald. (1960). The morphology and incidence of trichomonads of swine, *Tritrichomonas suis* (Gruby and Delafond), *Tritrichomonas rotunda*, n. sp. J. Protozool. 7:159–171.

Kirby, H. (1947). Flagellate and host relationships of trichomonad flagellates. J. Parasitol. 33:214–228.

Kofoid, C. A., and O. Swezy. (1915). Mitosis in trichomonas. Proc. Nat. Acad. Sci. 1:315–321.

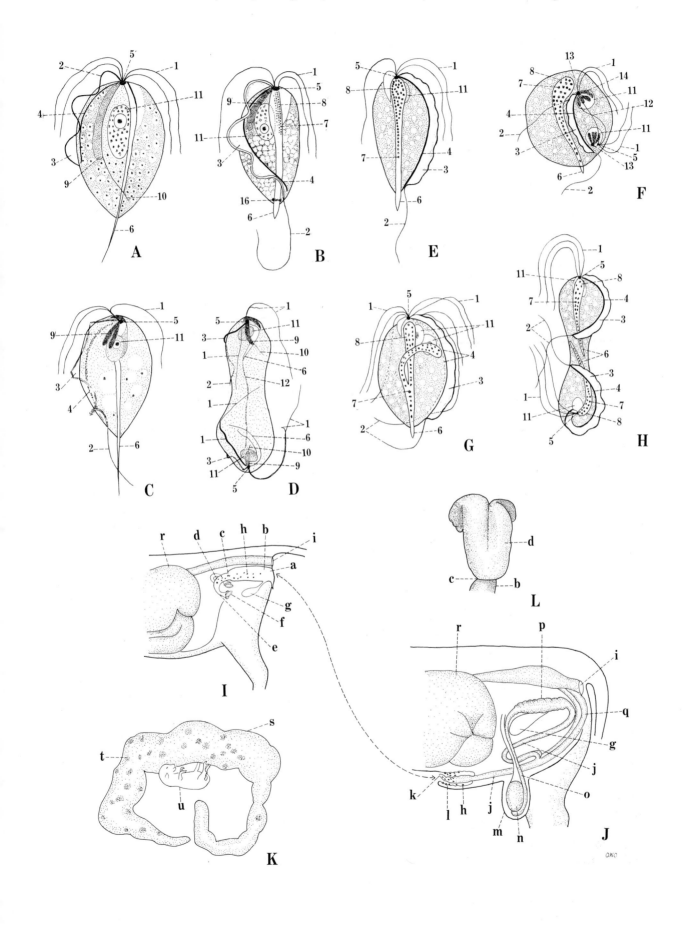

Levine, N. D. (1961). Protozoan Parasites of Domestic Animals and of Man. Burgess Publishing Co., Minneapolis, p. 82.

Levine, N. D., and F. L. Anderson. (1966). Frozen storage of *Tritrichomonas foetus* for 5.6 years. J. Protozool. 13:199–202.

Morgan, B. B. (1942). The viability of *Trichomonas foetus* (Protozoa) in the house fly *(Musca domestica)*. Proc. Helminthol. Soc. Wash. 9:17–20.

Morgan, B. B. (1944). Bovine Trichomoniasis. Burgess Publishing Co., Minneapolis, 150 pp.

Morgan, B. B., and B. A. Beach (1942). The geographical distribution of bovine trichomoniasis. Vet. Med. 37:459–462.

Rosenberg, L. E. (1936). On the viability of *Tritrichomonas augusta*. Trans. Amer. Microsc. Soc. 55:313–314.

Samuels, R. (1957). Studies on *Tritrichomonas batrachorum* (Perty) 2. Normal mitosis and morphogenesis. Trans. Amer. Microsc. Soc. 76:295–307.

Wenrich, D. H. (1921). The structure and division of *Trichomonas muris* (Hartman). J. Morphol. 36:118–155.

Wenrich, D. H. (1945). The cultivation of *Trichomonas augusta* (Protozoa) from frogs. J. Parasitol. 31:375–380.

Wenyon, C. M. (1926). Protozoology. William Wood and Co., New York (Reprinted by Hafner Publishing Co., New York, 1965), p. 122.

### *Trichomonas* Donné

The species of this genus have four anterior flagella and a recurrent one along the undulating membrane. There are numerous species in birds and mammals.

### *Trichomonas vaginalis* Donné, 1837

This species is cosmopolitan in the urinogenital tract of mature humans, being the most common of the trichomonads infecting them. It is transmitted venereally and is a serious pathogen of the reproductive tract of women.

**Description.**   The body is usually ellipsoidal or oval in shape, frequently with pseudopods for attachment or feeding (Fig. 4). They measure 7 to 23 (average 13) $\mu$ long by 5 to 12 (average 6.5 to 7.5) $\mu$ wide. Each of the four anterior flagella originates in a kinetosome arranged somewhat in a circle near the anterior end of the body with the fifth or recurrent flagellum coming from a kinetosome inside the circle. The tip of each anterior flagellum may be reflexed to form a hook-like structure or be straight and pointed. Each flagellum consists of a filament containing a circle of nine pairs of peripheral and one pair of central fibrils. The recurrent flagellum is attached to the margin of the full length of the undulating membrane, both of which arise from a kinetosome. The axostyle begins at the anterior end of the body as a thick capitulum with a cresent-shaped pelta, a thin median part, and a slender, pointed terminal end that extends beyond the caudal end of the body. The sheath consists of a layer of 50 to 55 parallel tubular fibrils. An oval nucleus with three membranes lies near the middle of the body. A Golgi complex lies dorsal to the nucleus. A long costa with flattened end near the kinetosome is marked by cross striation. Endoplasmic reticulum, ribosomes, and food vacuoles are present, but mitochondria are absent.

**Life Cycle.**   *Trichomonas vaginalis* occurs naturally only in the urinogenital tract of humans. The flagellates multiply by binary fission with mitotic nuclear division. After division of the nucleus and locomotor apparatus is completed, the cytoplasm divides to form two daughter cells. Multiple division is rare and sexual reproduction does not occur. Cysts are not formed. Normal transmission from host to host is through venereal contact, hence its high prevalence in mature individuals and rarity in children. Infection persists for years during the period of sexual activity but disappears in about 2 years in nonparticipating persons.

**Pathology.**   *Trichomonas vaginalis* is a recognized clinical entity in women and produces a true venereal disease with well-known symptoms. Upon introduction of the flagellates into the vagina, proliferation is rapid and vast numbers appear. Inflammation of the vaginal mucosa occurs, with desquamation of the epithelial cells. Trichomonads, epithelial cells, and leukocytes appear in copious vaginal secretions, producing intense irritation. The acute infection eventually passes into a chronic stage. Sterility may result from long standing infection.

While as prevalent in men as in women, the parasite remains latent without producing disease in the former. The flagellates are in the urethra and prostate glands.

Medication is not successful because of the difficulty of removing the flagellates from the urethra.

Other common species of trichomonads from humans include *T. hominis* from the intestine and *T. tenax* from the mouth.

*Trichomonas gallinae* is a frequent parasite of pigeons, occurring most commonly in the crop and that portion of the alimentary canal anterior to it. The disease is manifested by the presence of caseous masses in the pharyngeal region, as well as in the tissues. In the former area, they may be so massive as to prevent occlusion of the beak. In advanced cases, lesions appear in the liver. Carriers of the parasite appear quite normal in health.

The epizootiology of this parasite among pigeons is remarkable in its relationship to the manner in which the parents feed the young squabs. Birds surviving infection as squabs re-main carriers with the parasites localized in the mouth, esophagus, and crop. When the young birds are fed with the regurgitated secretions of the crop known as pigeon milk, the flagellates are transferred with it.

There are avirulent and virulent strains of the parasite. In the case of the latter, lesions appear in 7 days as small, yellowish areas on the buccal mucosa. They grow rapidly, forming large caseous masses in the pharyngeal region. Fluid accumulates in the crop and the infected birds die in about 10 days.

While pigeons and doves appear to be the natural hosts of *T. gallinae*, other birds may

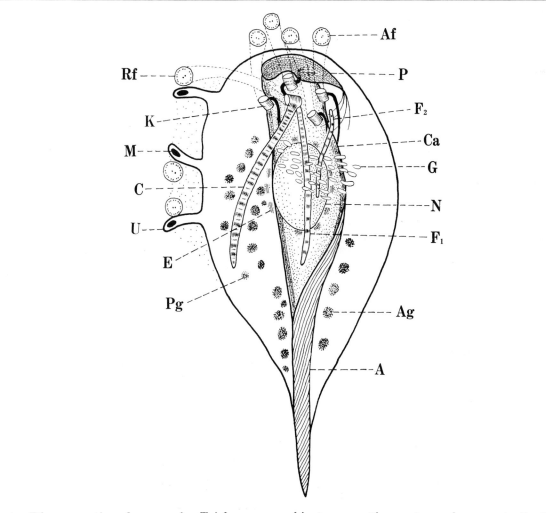

**Fig. 4.** Diagrammatic figure of *Trichomonas vaginalis* reconstructed from light and electron microscopy. **A**, axostyle; **Af**, anterior flagella (four); **Ag**, axostylar granules; **C**, costa; **Ca**, capitulum of axostyle; **E**, endoplasmic reticulum; **F**, accessory filaments; **G**, Golgi apparatus (parabasal body); **K**, kinetosomes (four outer and one central); **M**, marginal lamella of undulating membrane; **P**, pelta; **Pg**, paracostal granules; **Rf**, recurrent fibril; **U**, undulating membrane. Adapted from Neilson *et al.*, 1966, In: Adv. Parasitol., 1968, 6:123, and Mattern *et al.*, 1967, J. Protozool. 14:320.

harbor them. Hawks become infected by eating parasitized pigeons and doves, and chickens and turkeys through contamination of food and water. A number of different species of birds have been infected experimentally.

Inasmuch as the parasites are able to live for some time in water and have been recovered from contaminated drinking vessels, it is believed that these may be a source of infection among chickens and turkeys.

## SELECTED REFERENCES

Jirovec, O., and P. Miroslav. (1968). *Trichomonas vaginalis* and trichomoniasis. In: B. Dawes, ed. Advances in Parasitology. Academic Press, New York, Vol. 6, pp. 117–188.

Mattern, C. F. T., B. M. Honigberg, and W. A. Daniel. (1967). The mastigont system of *Trichomonas gallinae* (Rivolta) as revealed by electron microscopy. J. Protozool. 14:320–339.

Stabler, R. M. (1947). *Trichomonas gallinae*, pathogenic trichomonad of birds. J. Parasitol. 33: 207–213.

Trussell, R. E. (1947). *Trichomonas vaginalis* and Trichomoniasis. Charles C Thomas, Springfield, Ill., 277 pp.

## CLASS OPALINIDA

Although the body is clothed with rows of cilia-like organelles, the opalinids are no longer considered as belonging to the subphylum Ciliophora. Grassé in 1952 placed them in the Mastigophora. Corliss pointed out that the opalinids are more like flagellates than ciliates in that 1) the nuclei are of the same kind, 2) there is true sexual fusion of small microgametes with large macrogametes, and 3) the mode of binary fission produces two mirror image filial organisms, as in the case of the flagellates. Since none of these fundamental characteristics appear in the ciliates, he considered their occurrence in the opalinids as a justifiable reason for separating the two groups. Honigberg *et al.* raised them to the rank of superclass Opalinata, placing it between the superclasses Mastigophora and Sarcodina. Although indicating that the opalinids are related to the Mastigophora, Kudo left them unassigned to any group.

For uniformity in the scheme of classification followed here with respect to their relationships with other groups, the opalinids are assigned to the rank of class and placed between the classes Mastigophora and Sarcodina.

**Morphology.**   The body is oval, flat, and bi- or multinucleate or cylindrical, elongate and bi- or multinucleate. The organisms are densely clothed with longitudinal rows, or kineties, of short flagella. Two short parallel sickle-shaped rows of large flagella, the falx, are on the anterio-ventral side. Between the kineties, the pellicle forms ribbon-like bands of fine fibrils unique to the opalinids. The flagella consist of an outer membrane continuous with the pellicle and contain nine pairs of peripheral and one pair of central fibrils, typical of these organelles. The central fibrils terminate at the level of the cell membrane and the peripheral fibers continue a short distance into the ectoplasm, where they end abruptly to form the kinetosomes which are open-ended fibrous cylinders. The number and appearance of the protoplasmic bodies vary greatly.

### FAMILY OPALINIDAE

The family contains four genera. *Opalina* Purkinje and Valentin is highly flattened and multinucleate; *Zelleriella* Metcalf is similarly flattened but has two nuclei. *Cepedea* Metcalf is cylindrical, being circular in cross section, and multinucleate; while *Protoopalina* Metcalf is similar in shape but binucleate.

The opalinids are worldwide in distribution, occurring mostly in the Salientia. Insofar as known, they are nonpathogenic and are considered by some workers as commensals.

### *Opalina ranarum*
Purkinje and Valentin, 1835 (Plate 14)

This is probably the best known of the many species of opalinids and will serve as an example of the life cycle of a member of the family. It is the common European form, occurring regularly in the rectum of *Rana temporaria* and *Bufo bufo*, also in turtles and salamanders. *Opalina obtrigonoidea* Metcalf is a common form in various species of North American frogs and toads (*Rana, Hyla, Bufo, Gastrophryne*, and others) throughout the continent.

**Description.**   Mature *O. ranarum* are basically flat, oval organisms which vary considerably in general outline. In size, they range from 188 to 384 $\mu$ long by 163 to 200 $\mu$ wide. They appear to the naked eye as motile opalescent bodies. The flagella are arranged in parallel oblique rows. Each flagellum terminates in the ectoplasm as a fibrous cylindrical kinetosome.

There are numerous nuclei of the same kind located in the endoplasm, together with many cytoplasmic bodies. A mouth is lacking.

*Opalina obtrigonoidea* which is similar in appearance to *O. ranarum* is much larger, being 400 to 840 μ long, 175 to 180 μ wide, and 20 to 25 μ thick. A large variety may be even bigger, attaining a length of 1.3 mm and a width of 0.27 mm.

**Life Cycle.** During the greater part of the year, only trophic forms are present and dividing in the recta of frogs. With the approach of spring and activation of the sexual hormones of the frogs, the flagellates are stimulated to begin their own sexual reproductive cycle. Individual opalinids undergo repeated division, giving rise to small forms with one to several nuclei. These small individuals produce cysts 30 to 70 μ in diameter about themselves in the rectum which are expelled with the feces into the water where they sink to the bottom of the ponds.

Newly hatched tadpoles swallowing the cysts with food gleaned from the submerged vegetation and bottom mud become infected. Within 4 hours after reaching the rectum, the opalinids are released from the cysts as male or female gametocytes. They divide repeatedly, eventually giving rise to flagellated uninucleate gametes 28 to 30 μ long. The minute slender microgametes attach to the larger oval macrogametes and conjugate with fusion of the nuclei to form a zygote which resembles a small opalinid. They grow in size and multiply actively to form daughter opalinids which attain maturity at the same time the frogs metamorphose to the adult form.

Brumpt pointed out that some of the excysting zygotes reencyst in the tadpoles and are passed in the feces. When swallowed by uninfected tadpoles, they act similarly to individuals in adult frogs by forming gametes which fuse and produce by division many new opalinids. Thus infection of adult frogs may occur in two ways. Tadpoles ingesting original cysts from adult frogs (Plate 14, I) retain the infection during and after metamorphosis or they may in turn produce additional cysts that infect other tadpoles (Plate 14, J).

## EXERCISE ON LIFE CYCLE

For study, adult ciliates should be mounted in rectal fluid on slides and the coverslip sealed with Vaseline by running a hot needle around the edge. Observe them for movement, division, and encystment. Encysted forms should be placed in gastric and intestinal fluid of tadpoles or frogs on slides in order to observe excystation and release of the small ciliates.

## SELECTED REFERENCES

Brumpt, E. (1915). Cycle evolutif des opalines. Bull. Soc. Pathol. Exot. 8:397–404.

Corliss, J. O. (1955). The opalinid infusorians: Flagellates or ciliates? J. Protozool. 2:107–114.

Doflein, F., and E. Reichenow. (1953). Lehrbuch der Protozoenkunde. 6th ed. Gustav Fischer, Jena, p. 1081.

Grassé, P. P. (1952–53). Traité de Zoologie, Anatomie, Systematic, Biologie. Masson et Cie, Paris, Vol. 1, Fasc. 1, p. 983.

Grosgrove, W. B. (1947). Fibrillar structures in *Opalina obtrigonoidea* Metcalf. J. Parasitol. 33: 351–357.

Kudo, R. R. (1966). Protozoology, 5th ed. Charles C Thomas, Springfield, Ill., pp. 828, 1029.

Metcalf, M. M. (1923). The opalinid ciliate infusorians. Bull. U.S. Nat. Mus. 120, 484 pp.

Metcalf, M. M. (1940). Further studies on the opalinid infusorians and their hosts. Proc. U.S. Nat. Mus. (3077). 87:465–635.

Neresheimer, E. R. (1907). Die Fortpflanzung der Opalinen. Arch. Protistenk. (Suppl. 1):1–42.

Pitelka, D. R. (1956). An electron microscope study of cortical structures of *Opalina obtrigonoidea*. J. Biophys. Biochem. Cytol. 2:423–432.

Wenyon, C. M. (1926). Protozoology. William Wood and Co., New York (Reprinted by Hafner Publishing Co., New York, 1965), p. 1153.

### CLASS SARCODINA

The class Sarcodina contains a variety of orders of freshwater and marine protozoa with thin hyaline and granular pseudopods for locomotion and capturing food. They may be naked, or bear shells.

The order Amoebida, consisting of naked forms, has representatives in fresh, brackish, and saltwater, in moist soil and the alimentary tract of animals. Members of the family Endamoebidae are symbionts living in the gut of invertebrate and vertebrate animals. Some species are parasites and others commensals. One appears to be free-living in sewage plants.

### FAMILY ENDAMOEBIDAE

The trophozoites of this family are small, pseudopod-forming individuals that inhabit the

## Explanation of Plate 14

A–F, Trophic stages of four genera of Opalinidae from Amphibia. A, *Opalina ranarum* (multinucleate flat form). B, *Zelleriella elliptica* (binucleate flat form). C, Diagrammatic cross section of *Opalina* and *Zelleriella*, showing flattened nature of body characteristic of these two genera. D, *Cepedea cantabrigensis* (multinucleate cylindrical form). E, *Protoopalina mitotica* (binucleate cylindrical form). F, Diagrammatic cross section of *Cepedea* and *Protoopalina*, showing cylindrical shape of body characteristic of these two genera. G, Adult frog with vegetative stages of *Opalina ranarum* (a–d) in rectum, multiplying by binary fission during summer, fall, and winter. H, Formation of gametocytes (e–l) in rectum of frogs during spring concurrently with breeding activities of the amphibians. I, Formation of gametes and zygote (m–s) in the intestine of tadpoles. J, Development of trophic stages, beginning in tadpoles and continuing in frogs following metamorphosis (t–w).

1, Cilia; 2, nucleus.

a–d, Vegetative development by binary fission in gut of frogs during summer, fall, and winter; a, fully developed trophic stage which is the beginning of both the vegetative and sexual cycles; b–d, cells undergoing vegetative binary fission and in various stages of completion; e–k, in frogs during spring when they are engaged in breeding activities; e–g, division of trophic stage into gametocytes; h–j, formation and maturation of sexual nuclei; k, cyst formed around gametocyte; l, encysted gametocytes voided in feces of frogs into water where they are ingested by feeding tadpoles; m–s, in alimentary canal of tadpole; m, excystation; n, division of cytoplasm with reduction of nuclei in each daughter cell; o, formation of uninucleate gametes; p, two gametes; q, fusion of gametes in fertilization; r, zygote; s, encysted zygote expelled from intestine with feces; s–w, in intestine of tadpole during growth and metamorphosis; s, encysted zygote swallowed by tadpole; t, excystation of zygote in tadpole; u, young trophic stage; v, w, growth of opalinid in tadpole, followed by binary fission in rectum of metamorphosed frog.

Figure A adapted from various sources; B, from Chen, 1948, J. Morphol. 88:281; D, E, from Metcalf, 1923, U.S. Nat. Mus. Bull. 120.

---

intestine of invertebrate and vertebrate animals. Food vacuoles are present. Contractile vacuoles are lacking in trophozoites in the isotonic contents of the intestine but they appear in specimens reared in hypotonic media. Multiplication involves mitotic division of the nucleus followed by binary fission. Cysts are formed by most species but are unknown for a few. Generic differences are based primarily on the arrangement of the nuclear chromatin as seen in stained specimens.

Well-known genera of the family include *Entamoeba*, *Iodamoeba*, *Endolimax*, and *Dientamoeba* from humans, *Martiniezia* from iguanas, *Endamoeba*, the type genus, and *Dobellina* from insects, and *Hydramoeba* from hydras.

### Entamoeba

This genus contains a number of species distributed among different groups of animals. It forms four natural morphological groups based on whether encystment occurs and on the number of nuclei in mature cysts.

Encysting species
  Eight-nucleated group (lumen dwellers)
    *E. coli* from man
    *E. muris* from mice
  Four-nucleated group (tissue invaders)
    *E. histolytica* from man
    *E. invadens* from snakes
    *E. ranarum* from frogs
  One-nucleated group (lumen dwellers)
    *E. bovis* from cattle
    *E. debliecki* from goats
    *E. polecki* from man and pigs
Nonencysting species (lumen dweller)
    *E. gingivalis* from man

The tetranucleate group contains the pathogenic, tissue-invading species while those of the other groups are nonpathogenic commensals.

### Entamoeba histolytica
Schaudinn, 1903 (Plate 15)

Following the discovery of ameba by Losch in 1875 in bloody, diarrheic stools of a soldier in Russia, much confusion existed regarding the identity of quadrinucleate amebic cysts of large and small size in the intestine of man. Even the naming of the protozoan was fraught with controversy until finally *E. histolytica* prevailed.

Eventually, it was established that the *E. histolytica* complex actually comprises two forms consisting of large, pathogenic individuals and small, nonpathogenic ones. Some authorities con-

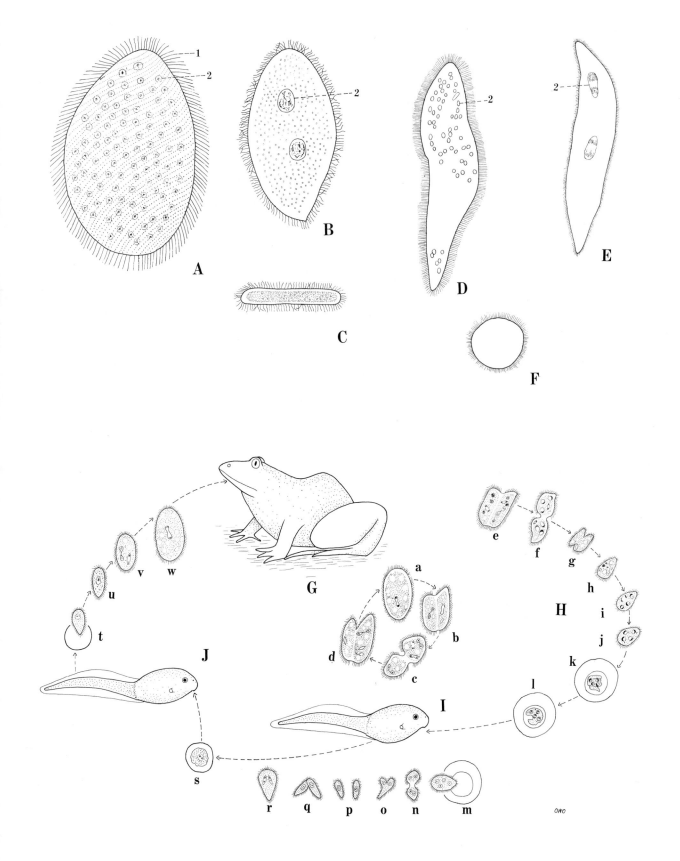

## Explanation of Plate 15

A, Active trophozoite. B, Trophozoite containing erythrocytes. C, Precystic stage. D, Young mononucleate cyst with chromatoidal bars. E, Binucleate cyst. F, Trinucleate cyst. G, Mature cyst with four nuclei and chromatoidal bar. H, Mature cyst without chromatoidal bars. I, Human host. J, K, Sections of intestine showing progression of lesions in mucosa and wall.

1, Nucleus; 2, nucleolus; 3, linin network; 4, chromatin granules on nuclear membrane; 5, pseudopods; 6, ectoplasm; 7, endoplasm; 8, erythrocyte; 9, cyst wall; 10, chromatoidal bodies; 11, glycogen vacuole.

a, Quadrinucleate cyst being swallowed; b, excystation of ameba in small intestine; b, four metacystic trophozoites; d, metacystic trophozoite invading intestinal epithelium; e, formation of small ulcer in epithelium by ameba; f, enlargement of ulcer; g, adjacent ulcers; h, fusion of adjacent ulcers, forming a large one with intervening portion being sloughed; i, consolidation of three ulcers form a single large one with sloughing portions; j, amebae passing through intestinal wall into hepatic portal vein; k, ameba entering hepatic parenchyma; l, hepatic ulcer; m, ameba entering lung parenchyma; n, lung ulcer; o, ameba in general circulation; p, ameba entering brain; q, brain ulcer; r, cyst and trophozoite passing from intestine (trophozoite dies); s–u, development of cyst outside host; s, uninucleate cyst; t, binucleate cyst; u, infective quadrinucleate cyst.

a′, Lumen of intestine; b′–g′, layers of intestinal wall; b′, mucosa with villi and crypts; c′, muscularis mucosae; d′, submucosa; e′, circular muscle layer; f′, longitudinal muscle layer; g′, serosa; h′, body cavity; i′, excysting quadrinucleate ameba in lumen of intestine; j′, four metacystic amebae; k′, ameba entering mucosa at tip of villus; l′, amebae entering mucosa in depths of crypts; m′, multiplication of ameba forming small ulcer in crypt; n′, ulcer at tip of villus where numerous amebae develop and are liberated to attack other epithelial cells; o′, formation of numerous ulcers in depth of crypts; p′, individual ulcers coalesce, forming a single large one; q′, villi sloughing because of undermining by coalescing ulcers; r′, ulcer extended into muscularis mucosae and submucosal tissue; s′, amebae enter blood vessels through damaged sections in deep ulcers and are carried to other parts of the body.

Adapted from Faust and Russell, 1957, Craig and Faust's Clinical Parasitology, 6th ed. Lea & Febiger, Philadelphia, p. 189.

---

sider them as large and small races of *E. histolytica*, others as distinct species with *E. histolytica* as the large tissue-invaders and *E. hartmanni* as the small lumen dwellers.

**Description.** Trophozoites are actively motile, with pseudopods that form quickly and somewhat explosively. The ectoplasm is thin and hyaline and the endoplasm is a grayish and granular substance. Trophozoites range in size from 10 to 60 $\mu$ in diameter. Those appearing in dysenteric feces generally are large and usually contain red blood cells in the food vacuoles; those in formed stools usually are small and do not contain erythrocytes. The two sizes represent the large and small races.

The nucleus of stained specimens is spherical, having a thickened membrane lined on its inner surface with a single layer of fine dots of dark chromatin. The small central endosome consists of a group of small discrete dots from which fine lines radiate spoke-like to the nuclear membrane. Glycogen masses and food vacuoles are present. Electron micrographs show a double nuclear membrane perforated by a system of pores. The endoplasmic reticulum is rudimentary; a Golgi apparatus and mitochrondria are lacking. Food vacuoles have a lining similar to the plasma membrane.

The sluggish precysts are round to oval, smaller than the trophozoites, but larger than the cysts that follow. Cysts are more or less spherical, with a smooth, refractile, unstained wall. Immature cysts have one nucleus, developing stages two, and mature ones four similar to but smaller than those in the trophozoites. Densely stained, elongate, chromatoidal bars with rounded ends are present in the cytoplasm of young cysts but are lacking in older ones, having been absorbed. The cysts fall into two size groups. Those of the small race are 6 to 8 $\mu$ in diameter and those of the large race 10 to 14 $\mu$.

**Life Cycle.** The large and small races occur simultaneously in the large intestine where they live and feed for a time in the crypts and folds of the mucosa. In multiplication of the trophozoites, mitotic division of the nucleus is followed by separation of the cytoplasm to form two daughter cells.

Members of the small race feed on bacteria

**Plate 15**    *Entamoeba histolytica*

79

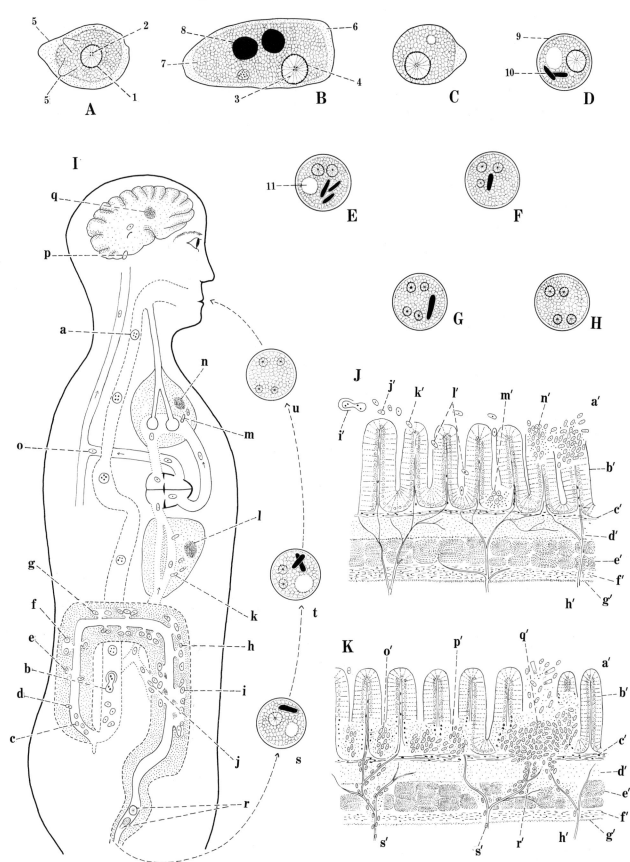

and intestinal contents in the lumen of the large intestine. As multiplication proceeds, some of the trophozoites round up as precysts, expel the contents of the food vacuoles, and finally complete encystment. As the cysts mature in the intestine, the nucleus undergoes two mitotic divisions resulting in a tetranucleate cystic ameba with small nuclei. Being nonpathogenic, dysentery does not occur and both the small trophozoites and cysts appear in formed stools.

Trophozoites of the large race attack the mucosal epithelium, beginning to lyse the cells. As destruction progresses, the amebae penetrate deeper into the mucosa. Once inside the tissue, they feed on it and erythrocytes. Destruction of the tissue and blood vessels results in bloody dysentery. Large trophozoites containing red blood corpuscles appearing in diarrheic stools soon die. As the intestinal contents become dehydrated in the lower part of the colon, the trophozoites expel food contained in the vacuoles, round up, form precysts, and secrete a cyst about themselves. Newly formed cysts contain a single nucleus and two kinds of food substances, glycogen masses and chromatoidal rods. As maturity proceeds, the food material is metabolized and the nucleus undergoes mitotic divisions, forming four small ones. The resistant, mature cystic amebae of both the small and large races are infective and constitute the transfer stage from host to host.

Food contaminated with feces containing mature cysts or drinking water polluted with them are the common sources of infection. Upon being swallowed, the cysts pass unchanged through the stomach and upper part of the intestine, where acid conditions prevail. Upon entering a neutral or alkaline environment, they excyst in about 2 to 4 hours. The small multinucleate trophozoite emerging from the cyst undergoes nuclear and cytoplasmic division that produces eight minute, uninucleate metacystic daughter trophozoites in about 12 hours.

Upon reaching the large intestine, where the movement of the intestinal contents is less vigorous, the metacyclic amebae establish themselves in the mucosal and glandular crypts, producing colonies of large or small trophozoites, depending on which race produced them.

**Epidemiology and Pathology.**    The wide distribution and pathogenic properties of *E. histolytica* rank it as an important parasite of humans. Cysts are able to survive up to 12 days under favorable conditions of temperature and moisture. Infected nonclinical persons are passers of cysts. If engaged in handling food or food products, these people serve as carriers from which new infections originate. Drinking and culinary water contaminated with fecal material from improperly designed or leaky sewage systems or polluted by drainage from privies or cesspools is an important source of infection. Other avenues of contamination resulting in infection are insects whose vomitus or feces containing viable cysts are deposited on food, and from vegetables normally eaten raw that have been grown on soil fertilized with human feces.

The highest incidence of infection generally occurs among people living in warm, moist climates where low standards of sanitation prevail. The rate of infection is lower in children under 5 years of age than in older ones and adults.

The basic pathology is destruction of the colonic epithelium and invasion of the gut wall by the parasites. When destruction of the epithelial cells by the trophozoites is faster than their repair by the host, pathology results. The amebae enter the damaged mucosa and multiply, forming colonies. Soon craterous, flask-like ulcers are formed which expand laterally and extend deeper into the intestinal wall. When the bases of adjacent ulcers coalesce, the overlying tissue sloughs, forming enlarged ulcers. They are most abundant in the cecum, ascending colon, and sigmoid rectum, where movement of the intestinal contents is slowest, thus giving the amebae a better opportunity to establish themselves. Destruction of tissue on a massive basis results in severe dysentery, accompanied by blood coming from eroded vessels. The appearance of hematophagus trophozoites in dysenteric feces is evidence of damage to the intestine by the large race of amebae.

With erosion of the blood vessels, the amebae enter them and are transported via the hepatic portal system to the liver. The lytic action of their secretions form openings in the small blood vessels through which the amebae escape into the liver parenchyma, forming colonies. As they develop, large nonpurulent abscesses filled with necrotic fluid develop. Multitudes of amebae aggregate next to the living cells around the inner periphery of the abscesses where they feed and continue to multiply. Occasionally some are carried by the blood to the lungs or brain, where they colonize and produce abscesses. These are rare.

### Other Amebae from Man

Man is host to a number of species of commensal lumen-dwelling amebae whose importance lies in the fact that they must be recognized and differentiated from the pathogenic, tissue-invading *E. histolytica*.

Three commensal species of *Entamoeba* occur in man. *Entamoeba coli* from the intestine has octonucleate cysts when mature. The chromatin plaques on the inner side of the nuclear membrane are fewer and larger than in *E. histolytica* (Plate 16, B). Ripe cysts vary greatly in size, ranging from 10 to 30 μ in diameter. *Entamoeba polecki* of swine occurs occasionally in the intestine of man. The ripe cysts, 5 to 11 μ in diameter, have a single nucleus with a large particulate endosome and a small number of large plaques of chromatin on the inner surface of the nuclear membrane. *Entamoeba gingivalis* occurs only as a trophozoite between the gums and teeth of humans. It measures 5 to 35 μ, averaging 10 to 20 μ. The nuclear membrane is lined with fine chromatin granules. Fragments of tissue debris, including nuclei, appear in the food vacuoles. *Iodamoeba bütschlii* is in the intestine of man and pigs. The nucleus contains a large, compact endosome surrounded by a ring of clear vesicles (Plate 16, C). A large iodinophilic glycogen mass may be present. The ripe cyst is more oval than spherical and normally contains one nucleus. *Endolimax nana* is a small form 6 to 15 μ in diameter in the intestine. The endosome is a large, irregular, centrally located mass. A few large chromatin plaques are on the inner surface of the nuclear membrane (Plate 16, D). Cysts are distinctly oval, tetranucleate, and 5 to 14 μ in length. *Dientamoeba fragilis* is a small, 3 to 22 μ in diameter, nonencysting ameba of the intestine. The endosome consists of several distinct bodies centrally located (Plate 16, E). Inasmuch as cysts do not appear, the method of transfer from host to host is unknown. Some workers postulate that it might be in the eggs of pinworms, similar to what occurs in *Histomonas meleagridis* in cecal worms (p. 18).

### Amebae from Other Animals

Many species of *Entamoeba* are known from various animal hosts. *Entamoeba ranarum* from frogs and *E. invadens* from snakes are of the *histolytica* type. They are tissue invaders with tetranucleate cysts. *Entamoeba muris* of mice and *E. gallinarum* of fowl are of the *coli* type, being nonpathogenic lumen dwellers with octonucleate cysts.

Species with uninucleate cysts prevalent in some domestic mammals include *E. bovis* (4 to 15, average 8.8 μ in diameter) from cattle, *E. debliecki* (4 to 12, average 6.4, μ) from goats and pigs, *E. polecki* (4 to 17, average 8, μ) from pigs, and *E. ovis* (4 to 13, average 7.2, μ) from sheep.

*Endamoeba*, the type genus and once considered to be synonymous with *Entamoeba*, is a parasite of cockroaches and termites. The chromatin granules of the nucleus are distributed concentrically near the inner periphery of the nuclear membrane (Plate 16, A). The cysts are multinucleate.

*Hydramoeba hydroxena* is parasitic on and in hydra, attacking and destroying the tentacular and celenteric epithelium. The parasitized hydra die within a week at which time the amebae form spherical cysts 27 to 29 μ in diameter and with one or more nuclei. A variously shaped ring of chromatin lies about midway between the endosome and the nuclear membrane (Plate 16, F).

*Schizamoeba* Davis is a parasite of the stomach of salmonid fishes. There is no endosome; the chromatin material is arranged in a few large plaques on the inner surface of the nuclear membrane (Plate 16, G).

## EXERCISE ON LIFE CYCLE

Species of *Entamoeba* of the *histolytica* and *coli* groups from naturally infected lower animals are available for life history studies.

*Entamoeba ranarum* occurs in the rectum of frogs and their tadpoles. *Entamoeba invadens* is parasitic in captive snakes, particularly in zoos. Both species are of the *histolytica* type, having tetranucleate cysts and being tissue invaders which form abscesses in the liver. They are virtually indistinguishable at any stage from *E. histolytica*.

*Entamoeba muris* with octonucleate cysts is a representative of the *coli* group. It is a harmless commensal in the lumen of the cecum of roughly 10 to 20 percent of rats and up to 50

## Plate 16

percent of mice and voles. *Entamoeba gallinarum* of the ceca of fowl is a member of this group.

Standardized procedures for finding both trophozoites and cysts in the intestinal contents and feces of hosts give best results. A few proven methods are given.

Microscopic examination of smears prepared in physiological solution provides a simple means of detecting trophozoites and cysts, especially in cases of heavy infection. Certain stains are helpful in revealing the entamoebae. A drop of Lugol's iodine or D'Antoni's standardized iodine solution added to a fecal smear makes the chromatin material of the amebae appear as light colored objects; the cytoplasm is yellowish brown, and the glycogen masses are dark, mahogany brown. Supravital brilliant cresyl blue stains the living parasites. Background staining with eosin colors everything pink except living amebae which remain translucent, unstained spherical objects. The Merthiolate-iodine-formaldehyde fixative stain, designated as MIF, is one of the most useful techniques for preserving and staining amebae in fecal material for microscopic examination.

Concentration of cysts increases the probability of finding them. The most efficient method for this is the zinc-centrifugal-flotation technique.

The procedures for finding trophozoites and cysts of entamoebae in feces are outlined by Faust *et al.* (1970, p. 785).

Live specimens of *E. muris* may be obtained from mice. Open the abdominal cavity of a freshly killed mouse and remove the cecum and large intestine to a warm dish and keep moist with warm physiological solution. Prepare stained smears of the cecal contents and examine for trophozoites which are 25 to 30 $\mu$ in diameter. Cysts are more likely to occur in the large intestine. They represent two races with mean diameters of 14 and 17 $\mu$. The chromatoidal bodies have splintered ends and the glycogen masses diffuse margins. Eight nuclei are present in ripe cysts.

*Entamoeba ranarum* which are not always prevalent should be sought in the cloaca and liver abscesses of frogs and their tadpoles. Ripe cysts are tetranucleate and contain chromotoidal bars with smooth, rounded ends. Methods for culturing amebae from frogs and snakes appear in the list of references.

## SELECTED REFERENCES

Barret, H. P., and N. M. Smith. (1926). The cultivation of *Entamoeba ranarum*. Ann. Trop. Med. Parasitol. 20:85–88.

Diamond, L. S. (1960). The axenic cultivation of two reptilian parasites, *Entamoeba terrapinae* Sanders and Cleveland, 1930, and *Entamoeba invadens* Rodhain, 1934. J. Parasitol. 46:484.

Dobell, C. C. (1928). Researches on the intestinal Protozoa of monkeys and man. I. General introduction. II. Description of the whole life-history of *Entamoeba histolytica* in cultures. Parasitology 20:357–412.

Dobell, C. C. (1938). Researches on the intestinal Protozoa of monkeys and man. IX. The life-history of *Entamoeba coli*, with special reference to metacystic development. Parasitology 30:195–238.

Elsdon-Dew, R. (1968). The epidemiology of amoebiasis. In: B. Dawes, ed. Advances in Parasitology. Academic Press, New York, Vol. 6, pp. 1–62.

Entner, N. (1961). Genetics of *Entamoeba histolytica*: Initial experiments. J. Protozool. 8:131–134.

Faust, E. C., P. F. Russell, and R. C. Jung. (1970). Craig and Faust's Clinical Parasitology. 8th ed. Lea & Febiger, Philadelphia, pp. 129–176, 783.

Geiman, Q. M., and H. L. Ratcliffe. (1936). Morphology and life-cycle of an amoeba producing amoebiasis in reptiles. Parasitology 28:208–228.

Goldman, M., and V. Davis. (1965). Isolation of different size strains from three stock cultures of *Entamoeba histolytica* with observations on spontaneous size changes affecting whole populations. J. Protozool. 12:509–523.

Levine, N. D. (1961). Protozoan Parasites of Domestic Animals and of Man. Burgess Publishing Co., Minneapolis, p. 133.

Mackinnon, D. L., and R. S. J. Hawes. (1961). An Introduction to the Study of Protozoa. Clarendon Press, Oxford, p. 32.

Meerovitch, E. (1961). Infectivity and pathogenicity of polyaxenic and monoxenic *Entamoeba invadens* to snakes kept at normal and high temperatures and the natural history of reptile amoebiasis. J. Parasitol. 47:791–795.

Miller, J. H., J. C. Swartzwelder, and J. E. Deas. (1961). An electron microscopic study of *Entamoeba histolytica*. J. Parasitol. 47:571–595.

Neal, R. A. (1950). An experimental study of *Entamoeba muris* (Grassi, 1879); its morphology, affinities, and host-parasite relationship. Parasitology 40:343–365.

Noble, G. A., and E. R. Noble. (1952). Entamoebae in farm animals. J. Parasitol. 38:571–595.

**Plate 16**  *Endamoebidae (nuclei)*                                                              83

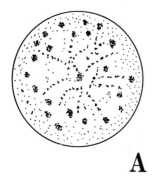

**A**

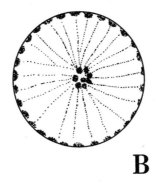

**B**

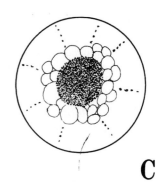

**C**

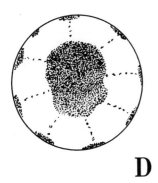

**D**

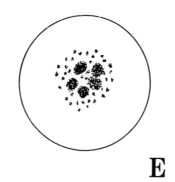

**E**

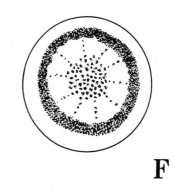

**F**

**G**

Reynolds, B. D., and J. B. Looper. (1928). Infection experiment with *Hydramoeba hydroxena*. J. Parasitol. 15:23–30.

Sanders, E. P. (1931). The life cycle of *Entamoeba ranarum* Grassi (1879). Arch. Protistenk. 74:365–371.

Sapero, J. J., E. G. Hakansson, and C. M. Louttit. (1942). The occurrence of two significantly distinct races of *Entamoeba histolytica*. Amer. J. Trop. Med. 22:191–208.

WHO Expert Committee on Amoebiasis. (1969). Amoebiasis. Report of a WHO expert committee. World Health Organ. Tech. Rep. Ser. No. 421, 52 pp.

### CLASS SPOROZOA

All members of this class are parasitic and produce spores, as indicated by the name. Reproduction is both sexual and asexual in all species. Individual spores when fully developed contain one to many infective sporozoites, depending on the species. In species that develop in a single host, the spore (oocyst) membrane is thick and resistant to protect the sporozoites while developing outside and passing from host to host. In species with two hosts, the spore membrane is always thin and delicate as all development is inside the invertebrate host which transmits the infective sporozoites without a free-living phase in the life cycle.

The life cycles of the Sporozoa appear complicated at the outset. However, they follow a pattern of development that makes comprehension of them rather easy. Basically, the three phases of development of a typical sporozoan life cycle are 1) schizogony, 2) gametogony, and 3) sporogony. Each phase is initiated by a specific stage of the parasite which through a process of development culminates in the succeeding one in the cycle (Fig. 5). The morphological type resulting from each phase of development includes merozoites from schizogony, zygotes from gametogony, and sporozoites from sporogony.

The development of each phase follows a consistent pattern. The process of schizogony begins with the sporozoites when they enter the host. Each sporozoite enters a cell and transforms into a trophozoite. Multiplication is asexual by a process of schizogony in which the nucleus divides repeatedly to form a schizont. As each new nucleus is surrounded by a bit of cytoplasm the whole becomes a segmenter. The individual parts separate to form merozoites, the end product of schizogony. The merozoites enter other cells to produce either trophozoites which repeat schizogony or to develop into gametocytes, the beginning of the sexual phase known as gametogony. It produces the male and female sex cells designated collectively as gametocytes. The male gametocyte is known as a microgametocyte, and by multifission of the nucleus, it produces numerous, minute microgametes. Each female gametocyte, designated as a macrogametocyte, develops into a large, uninucleate macrogamete. Entrance of a microgamete into a macrogamete results in fertilization and formation of a zygote. If motile, the zygote is termed an ookinete. This completes the sexual or gametogenous phase.

The mature zygote, designated as an oocyst, initiates the sporogenous phase which is one of asexual multiplication similar to that occurring in schizogony. As the nuclear material of the oocyst divides, each part associates with a bit of cytoplasm to form a definite number of sporocysts. The nucleus of each sporocyst divides to produce a constant number of enclosed sporonts that develop into sporozoites.

The class Sporozoa is divided into the orders Gregarinida, Coccidida, Haemosporida, and Haplosporida.

### ORDER GREGARINIDA

Members of the Gregarinida are known commonly as gregarines. They are frequent parasites

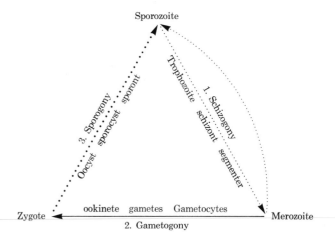

**Fig. 5.** Diagram of the three phases of a typical sporozoan life cycle. Adapted from Baer, 1951, *Ecology of Parasites*, University of Illinois Press, Urbana, p. 16.

of the intestinal tract of arthropods and reproductive organs of annelids. Most of them do not undergo schizogony but a few have a primitive type of it which takes place in the intestinal lumen. Sporogony also occurs in the lumen of the intestine with the formation of spores.

### Suborder Eugregarinina

The members of this suborder are chiefly celozoic parasites of arthropods. The young trophozoites enter epithelial cells of the intestine for a short time but soon return to the lumen, remaining attached to the cells until maturity, at which time they become detached and are free in the lumen of the intestine. Fully developed trophozoites, known as sporadins, unite in pairs and secrete a cyst about themselves to form a gametocyst. Following repeated nuclear division, male and female isogametes, that is, gametes of equal size, are formed. Union of the isogametes produces numerous zygotes, each of which forms an oocyst. Through division of the nucleus and cytoplasm, sporozoites are formed.

Two superfamilies are recognized. They are the Acephalinoidea which are with neither transverse septa dividing the body into compartments nor an organelle of attachment, and the Cephalinoidea which have both septa and a well-developed organelle for attachment within cells.

### Superfamily Acephalinoidea

The trophozoites are not divided into distinct linear parts by transverse septa.

### FAMILY MONOCYSTIDAE

The members of this family are common parasites of earthworms throughout the world. They live in the seminal vesicles, where the sporozoites enter the sperm morulae and develop into trophozoites, absorbing the germinal cells in the process. Reproduction is completed within the seminal vesicle. Of the many species, *Monocystis lumbrici* from *Lumbricus terrestris* is one of the commonest and best known.

### *Monocystis lumbrici*
### (Henle, 1845) (Plate 17)

Trophozoites live in the sperm morulae floating about in the seminal vesicles. As the trophozoites grow, they consume all but the tails of the sperms which adhere to them, giving the appearance of a covering of cilia.

**Description.**    Mature trophozoites are active and measure about 200 $\mu$ long by 60 to 70 $\mu$ wide. Each has a single nucleus containing one or more karyosomes. Electron micrographs of the cortical regions show a layer of ectoplasm differentiated from the endoplasm. The superficial portion of the ectoplasm consists of three layers, outer, middle, and inner. It is formed into thin folds that cover the entire body. The outer margin of the folds is thickened and the base consists of complex tunnel-like cavities. There are ectoplasmic vacuoles that open to the outside. The ectoplasm has no basement membrane but it contains many mitochondria and dense granules of paraglycogen. Fibrils are rare. Conjugating adults come in contact with each other along the long axis of the body, round up, and secrete a two-layered cyst approximately 162 $\mu$ in diameter around themselves. The outer cyst wall is roughened externally and the inner one is smooth and thin. The navicular spores measure 11 to 12 $\mu$ long by 5 to 6 $\mu$ wide. When ripe, they contain eight sporozoites.

**Life Cycle.**    The minute sporozoites in the seminal vesicles of the earthworms penetrate the sperm morulae, or mother sperm cells, and transform into trophozoites which grow at the expense of the germ cells. By the time the trophozoites have matured, the sperm morulae have been absorbed, leaving only the tails of sperms adhering to the trophozoites.

The mature trophozoites mark the end of the trophic or vegetative stage and the beginning of the reproductive phase. They are now known as sporonts or sporadins. In reality, they are gametocytes. Two of them come together, adhering along the long axis of the body, shorten, and secrete a double-layered cyst about themselves, still remaining separate individuals. The cyst may be considered as a gametocyst.

Upon completion of the cyst, nuclear division begins in each gametocyte and continues until a great number of small nuclei is present throughout the cytoplasm. These migrate to the surface of each gametocyte. Bits of cytoplasm detach from the main body and surround each nucleus, forming the gametes. A residual mass of cytoplasm remains, to which the gametes are attached. At this time, the cell wall of each gametocyte disappears and the individuals become indistinguishable. The gametes show movement in the cyst. Two gametes, presumably from different gametocysts, conjugate to form a zygote in which the two nuclei fuse. Each zygote

## Explanation of Plate 17

A, Sperm morula, or sperm mother cell, from seminal vesicle. B, Sperm mother cell containing a young trophozoite of *Monocystis*. C, Mature trophozoite "clothed" with the tails of spermatozoa remaining from the mature sperm cluster. D, Association of two mature gametocytes covered with tails of sperm cells preparatory for encystment and sporulation. E, Encysted mononuclear gametocytes. F, Segment of gametocyst, showing gametes that have formed from division of nuclei of the gametocytes and remain attached to the residual protoplasm of the gametocytes. G, Segment of gametocyst showing fertilization and formation of zygotes. H, Zygotes develop into sporocysts in gametocyst. I, Fully developed spore. J, Binucleate spore. K, Tetranucleate spore. L, Octonucleate spore. M, Sporulated spore containing eight sporozoites. N, Ruptured sporocyst liberating sporozoites. O, Cross section of spore, showing arrangement of sporozoites. P, Earthworm (*Lumbricus terrestris*), host of *Monocystis*, showing endogenous phase of life cycle of parasite.

1, Spermatogonium of mother sperm cell; 2, cytoplasm of mother sperm cell; 3, young trophozoite within mother sperm cell; 4, nucleus; 5, tails of sperms from destroyed mother sperm cell; 6, conjugating gametocytes; 7, epicyst or external layer of gametocyst; 8, endocyst or internal layer of gametocyst; 9, gametocyte; 10, gamete; 11, residual protoplasm; 12, nucleus of residual protoplasm; 13, gamete in residual protoplasm; 14, fertilization and formation of zygote; 15, fusion of nuclei and further development of zygote; 16, complete fusion of nuclei and full development of zygote; 17, sporoblast resulting from development of zygote; 18, boat-shaped spore or sporocyst; 19, sporozoite; 20, wall of spore.

a, Mature spore leaving seminal vesicle via the seminal funnel; b, spore leaving male genital pore to be free in soil or retained in egg cocoon; c, mature spore in soil; d, binucleate spore; e, tetranucleate spore; f, octonucleate spore; g, sporulated spore containing eight sporozoites; h, mature spore in pharynx, having been swallowed with contaminated soil by earthworm; i, spore in crop; j, k, spores in gizzard and intestine, respectively, opening and releasing sporozoites; l, sporozoite free in lumen of alimentary canal; m, sporozoite penetrating intestinal epithelium and entering capillaries of blood vessels in gut wall; n, sporozoites in dorsal vessel, being carried forward by blood; o, sporozoite in one of the hearts; p, testes; q, young sperm morula or sperm mother cell; r, seminal vesicle; s, sporozoite entering developing sperm mother cell with well-formed spermatogonia; t, young trophozoite in sperm mother cell; u, associated elongate gametocytes attached to seminal funnel; v, gametocytes in gametocyst; w, formation of gametes and fertilization with resultant zygotes; x, formation of mature spores which will escape via sperm funnel (a).

Figures adapted from various sources, except P, which is original.

---

may now be considered as a sporoblast. It secretes a tough cyst about itself and assumes the typical fusiform shape.

Final development includes three successive amitotic nuclear divisions resulting in eight nuclei which take up an equatorial position in the spore. A bit of cytoplasm surrounds each to form the minute sporozoites, leaving a residuum. Numerous spores are present in each gametocyst.

Infection of earthworms apparently occurs only when the ripe spores are expelled from the seminal vesicles and subsequently ingested.

They may be passed through the male genital pores into the soil or into the egg cocoons. Doubtless oligochaetophagus birds and mammals play an important role in releasing them by the digestive processes and disseminating them in their feces. Probably disintegration of dead worms in the soil is a means of freeing the spores.

Earthworms become infected by swallowing ripe spores. The route of migration of the sporo-zoites from the alimentary canal to the seminal vesicles is unknown. It is probable, however, that they are released from the spores in the alimentary canal, where they burrow into the intestinal wall and enter the capillaries of the circulatory system. Once inside the vascular system, they are transported via the dorsal vessel and hearts to the seminal vesicles, where they leave the capillaries. Inside the seminal vesicles, they seek out the sperm mother cells, penetrate them, and transform to trophozoites.

### EXERCISE ON LIFE CYCLE

The various stages of the parasite, such as the trophozoites, gametocytes in the gametocysts, gametes, and spores containing the sporozoites, can be demonstrated in stained sections of infected seminal vesicles. Fresh contents of the seminal vesicles examined with the aid of a microscope will reveal the various stages. Con-

**Plate 17**   *Monocystis lumbrici*                                    87

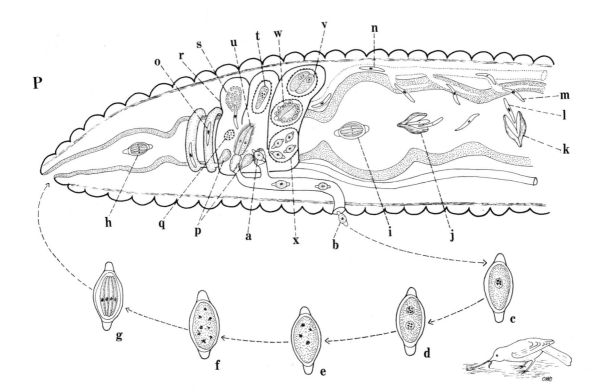
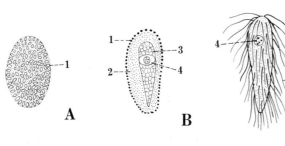
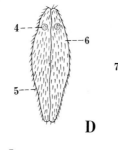
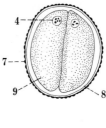
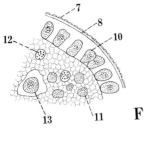
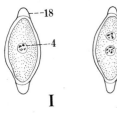
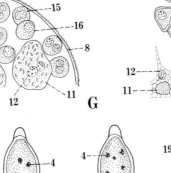
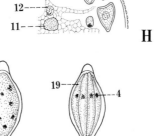
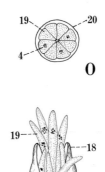

tents of the seminal vesicles spread on slides, dried, and stained with Wright's blood stain show various stages.

For experimental purposes, infective spores and uninfected earthworms are necessary. Spores may be obtained from the seminal vesicle of infected *Lumbricus terrestris*. Young, uninfected earthworms must be hatched from eggs placed in soil previously sterilized by heat to assure that natural infections may not result from it. Captive *L. terrestris* produce small numbers of lemon-shaped cocoons 5 to 7 mm long, each with several eggs, only one of which hatches. Other earthworms such as the fecal earthworm, *Eisenia foetida*, a hardy species that grows well under cultural conditions (Galtsoff *et al.*), may be used. Also, it is parasitized by gregarines of both Monocystidae and Zygocystidae.

## SELECTED REFERENCES

Bhatia, B. L. (1929). On the distribution of gregarines in oligochaetes. Parasitology 21:120–131.

Berlin, H. (1924). Untersuchungen über Monocystiden in den Vesiculae Seminales der schwedischen Oligochaeten. Arch. Protistenk. 48:1–124.

Cuenot, L. C. M. J. (1901). Recherches sur l'evolution et la conjugaison des gregarines. Arch. Biol. 17:581–652.

Galtsoff, P. S., F. E. Lutz, P. S. Welch, and J. G. Needham. (1959). Culture Methods for Invertebrate Animals. Dover Publications, New York, p. 195.

Hesse, E. (1909). Contribution à l'etude des monocystidees des oligochètes. Arch. Zool. Exp. Gén. 43:27–301.

Lankester, E. R. (1903). A Treatise on Zoology, Part I: Introduction and Protozoa. Adam and Charles Black, London, Fasc. 2, p. 154.

Mackinnon, D. L., and R. S. J. Hawes. (1961). An Introduction to the Study of Protozoa. Clarendon Press, Oxford, p. 169.

Mickel, C. E. (1925). Notes on *Zygocystis cometa* Stein, a gregarine parasite of earthworms. J. Parasitol. 11:135–139.

Miles, H. B. (1962). The mode of transmission of the acepaline gregarine parasites of earthworms. J. Protozool. 9:303–306.

Troisi, R. A. (1933). Studies on the gregarines of the Oligochaeta annelids. Trans. Amer. Microsc. Soc. 52:326–352.

Vinckier, D. (1969). Organization ultrastructurale corticale de quelques monocystidées parasites du ver oligochète *Lumbricus terrestris* L. Protistologica 5:505–517.

## FAMILY DIPLOCYSTIDAE

The members of this family are acephaline gregarines from the intestine of flatworms, insects, and tunicates. They produce numerous spores (oocysts) in a single gametocyst. Each mature spore contains eight spindle-shaped sporozoites.

The morphology and biology of species of the genus *Lankesteria* from mosquitoes have been studied extensively and are used as representatives of the family.

*Lankesteria culicis* (Ross, 1898) (Plate 18)
This gregarine is parasitic in the intestine and Malpighian tubules of the yellow fever mosquito *Aedes aegypti* in many parts of the world, including the southern part of the United States. In areas where it occurs, the incidence of infection may include almost the entire population, with large numbers of parasites per individual mosquito.

**Description.** Stages in mosquitoes include 1) trophozoites that develop in the epithelial cells of the midgut, break out, and remain attached by the epimerite, 2) sporonts free in the lumen of the larval midgut, 3) gametocysts in the lumen of the Malpighian tubules of pupae, and 4) gametocysts and free oocysts in the lumen of the hindgut of adult mosquitoes.

Mature intracellular trophozoites are 150 to 200 $\mu$ long. The body consists of three parts: the anterior umbrella- or mushroom-shaped epimerite attached by a slender stalk to the broad anterior part of the body called the protomerite, and the remainder comprising the transversely wrinkled deutomerite. The nucleus is equatorial in position.

Detachment of the epimerite from the body frees the trophozoite that becomes a sporadin which is somewhat spatulate in shape and 150 to 200 $\mu$ long by 30 to 41 $\mu$ wide. Gametocysts are spheres 60 to 100 $\mu$, occasionally larger, in diameter, oocysts are oval, 10 by 6 $\mu$ in size, and contain eight spindle-shaped sporozoites when mature.

In electron micrographs of trophozoites, the outer layer of the body is the epicyte. It shows a dark outer and inner membrane separated by a clear middle one (Fig. 6, A). The epicyte forms longitudinal folds of uniform size extending the length of the body (Fig. 6, B). The apex of the folds contains 4 to 8 longitudinal fibrils on the inner membrane. Pellicular pores between the folds open into subpellicular vacuoles. A layer of ectoplasm immediately under the epicyte contains numerous paraglycogen granules and vesicular mitochondria with internal cristae (Fig. 6, C). The ectoplasm of the protomerite is separated from the endoplasm by a layer of mitochondria (Plate 18, D). Endoplasmic organelles include the nucleus with nucleolus, paraglycogen granules, cytoplasmic vacuoles, lipid bodies, numerous free ribosomes, and rarely parallel arrays of endoplasmic reticulum.

**Life Cycle.** Life cycles of the Eugregarinina, of which *Lankesteria culicis* is a typical example, are characterized by the absence of a schizogenous phase in the process of development.

Infection of the host occurs when ripe oocysts in the water are ingested by larval mosquitoes. Sporozoites released from oocysts in the midgut enter the epithelial cells and transform to trophozoites. As growth proceeds, the trophozoites elongate, show transverse wrinkles, and fold upon themselves in the limited confines of the enlarging cells. Mature trophozoites with fully developed epimerite (mucron), protomerite, and deutomerite burst from the cells. They remain attached for a short time by the epimerite anchored in the cytoplasm of the epithelial cell with the remainder of the body hanging free between the epithelium and peritrophic membrane. After a short period in this position, the stalk of the epimerite is pinched off and the freed trophozoite becomes a sporadin.

As the larval mosquito transforms to a pupae, the sporadins migrate into the Malpighian tubules to begin the gametogenous phase. They unite in pairs, first by the anterior ends, later maneuvering to a side-by-side position to become gametocytes enclosed in a gametocyst. The nucleus of each gametocyte divides many times with each part moving to the surface of the reticulated cytoplasm. Each one buds off with a small amount of cytoplasm to form gametes having either large or small nuclei. Gametes with a small nucleus unite with those having a large one to produce many zygotes inside each gametocyst, thus ending the gametogenous phase of the life cycle.

With transformation of the pupae to adult mosquitoes, the gametocysts filled with zygotes are passed from the damaged Malpighian tubules

to the hindgut. In this location, gametogony begins and is completed, resulting in the transformation of each zygote into an oocyst with eight spindle-shaped sporozoites. Rupture of the gametocyst releases the oocysts, which are voided with the feces, contaminating the habitat of the mosquitoes with myriads of them.

*Lankesteria culicis* appears to be host-specific for the yellow fever mosquito *Aedes aegypti*. Larvae of other species of mosquitoes collected from the same pools with infected *A. aegypti* are always free of *L. culicis*.

*Lankesteria barretti* is parasitic in *Aedes triseriatus* from Texas. Morphologically, it differs from *L. culicis* in having primary and secondary folds of the epicyte and the nucleus in the anterior third of the body. The life cycle is basi-

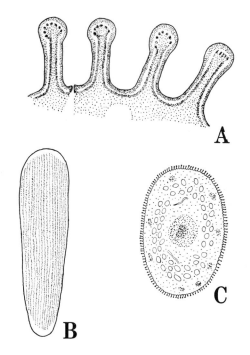

Fig. 6. Diagrammatic representations of the epicyte or cortical layer and cross section of *Lankesteria*. A, Enlarged section of outer wall of trophozoite showing longitudinal folds consisting of dark outer, clear middle, and dark inner membranes, fibrils in enlarged apex of folds, and cytopore between two folds. B, Trophozoite showing longitudinal folds of epicyte. C, Cross section of trophozoite showing nature of folds of epicyte and cytoplasmic inclusions, including nucleus with nucleolus, mitochondria (stippled), paraglycogen granules (open), and endoplasmic reticulum. Adapted from electron micrographs by Vávra, 1969, J. Protozool. 16:546.

## Explanation of Plate 18

A, Mature oocyst with eight sporozoites. B, C, Intracellular phases of trophozoite. D, Extracellular trophozoite attached and free trophozoite known as a sporadin. E, Mature gregarine. F, Larva of *Aedes* mosquito. F′, Midgut of larval mosquito. G, Pupa of mosquito. G′, Malpighian tubules of pupal mosquito. H, Adult mosquito. H′, Hindgut of adult mosquito.

1, Wall of oocyst; 2, sporozoites; 3, residual mass of cytoplasm; 4, young intracellular trophozoite; 5, nucleus with karyosome; 6, epithelial cell of midgut of larval mosquito; 7, nucleus of epithelial cell; 8, epimerite; 9, enlarged epimerite embedded in cytoplasm of host epithelial cell; 10, mature trophozoite in lumen of midgut but still attached to host cell by epimerite; 11, epimerite remains embedded in host cell but detached from trophozoite; 12, mature trophozoite, known as a sporadin, free in midgut of larva.

a, Ruptured oocyst releasing sporozoites in midgut of larval mosquito; b, sporozoite entering epithelial cell; c, d, intracellular trophozoites; e, young extracellular trophozoite attached to epithelial cell; f, mature extracellular trophozoite attached to epithelial cell; g, detached epimerite remains embedded in epithelial cell; h, mature trophozoite, known as a sporadin, free in midgut goes to Malpighian tubule of pupa and becomes a gametocyte; i, association of two gametocytes enclosed in a gametocyst; j–l, division of nucleus and formation of numerous nuclei; m, nuclear multiplication completed; n, gametes formed by budding from cytoplasm; o, gametes formed, leaving a large residual mass of cytoplasm; p, conjugation of gametes which are of two kinds, based on difference in size of nuclei; q, zygotes (one nucleus), division of nucleus (two nuclei), division of two nuclei, and finally the formation of sporozoites in oocyst; r, gametocyst containing oocysts passes into hindgut and ruptures when pupa transforms into adult mosquito, releasing oocysts; s, ripe oocysts voided in feces of mosquito.

Adapted from Ray, 1933, Parasitology 25:392.

---

cally similar to that of *L. culicis* but with minor differences.

**Pathology.** Trophozoites of *L. culicis* developing in the epithelium of the midgut destroy the cells, thus interfering with the nutrition of the developing larval mosquitoes. Larvae with 250 or more trophozoites are stunted and thin, with many dying. Individuals with 100 or fewer trophozoites appear normal in size and development.

The Malpighian tubules of infected pupae are damaged, presumably through pressure necrosis by the gametocysts. One or two gametocysts destroy cells at the tips of the tubules, 8 to 25 destroy the apical third of the tubules, and masses of oocysts destroy the distal two-thirds or more. Destruction of the Malpighian tubules interferes with or disrupts their excretory activities.

Adult mosquitoes produced from stunted larvae and pupae with damaged Malpighian tubules are themselves dwarfed and unable to function normally. Many die.

During 2 years of mosquito collections in east Texas and the coastal region of the state, *Aedes aegypti* was absent where normally present. The reason for their absence is uncertain but the presence of *L. culicis* is suspected. Following this lead, medical entomologists are investigating the feasibility of using the gregarine to control yellow fever mosquitoes.

## EXERCISE ON LIFE CYCLE

Two species of *Lankesteria* are available, one from *Aedes aegypti* and one from *A. triseriatus*, for study. In order to find the various stages of the parasite, it is necessary to examine 1) larvae for trophozoites and sporadins, 2) pupae for gametocysts containing gametocytes, and 3) adults for gametocysts and free oocysts.

Remove the intestine and Malpighian tubules from each stage of the mosquitoes in a dish of physiological saline. Examine the midgut of larvae for trophozoites developing inside the epithelial cells and mature ones attached by means of the epimerite. Note their location between the epithelium and peritrophic membrane.

Check the Malpighian tubules of pupae for paired sporadins and gametocysts containing gametes with different sized nuclei fusing to form zygotes. Note the reticulate nature of the cytoplasm that provides greater surface area for the gametes to develop on.

Gametocysts containing oocysts with eight sporozoites each and free oocysts are in the hindgut of adult mosquitoes.

## SELECTED REFERENCES

Barrett, Jr., W. L. (1968). Damage caused by *Lankesteria culicis* (Ross) to *Aedes aegypti*. Mosquito News 28:441–444.

Ganapati, P. N., and P. Tate. (1949). On the grega-

**Plate 18**　*Lankesteria culicis*　91

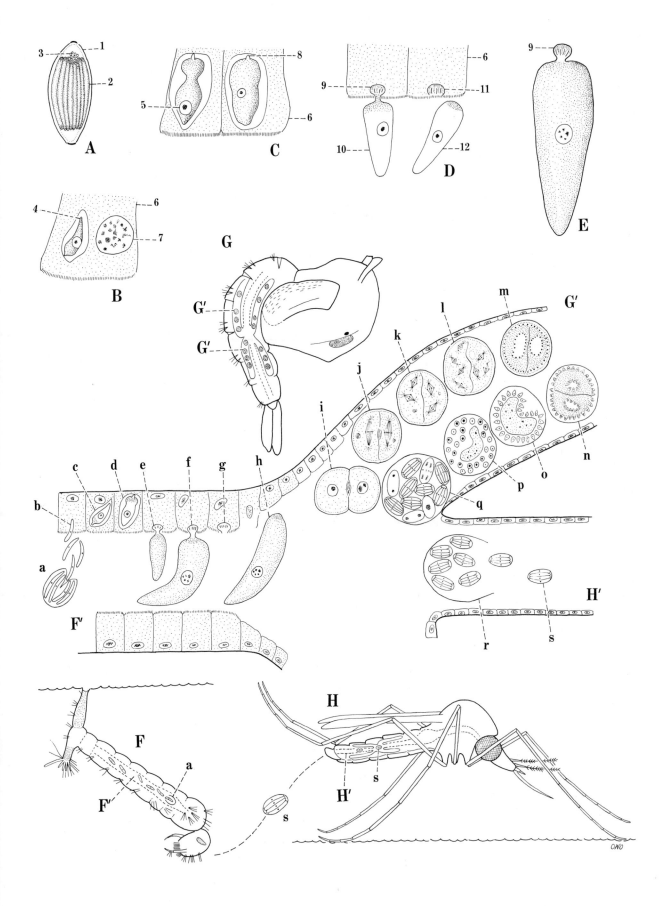

## Explanation of Plate 19

A, Two sporadins in syzygy, i.e., attached end to end. B, Sporonts in conjugation. C, Conjugation of sporonts in newly secreted cyst. D, Cyst with gametes. E, Ripe cyst with sporoducts and filled with spores. F, Partial chains of spores. G, Developed spore inside mucoid covering. H, Sporulated spore. I, Sporozoite attached to epithelial cell. J, Growing trophozoite partially embedded in epithelial cell of gut. K, Older trophozoite partially embedded in gut cell, showing developing protomerite and deutomerite. L, *Blatta orientalis* host.

1, Protomerite; 2, deutomerite; 3, nucleus; 4, sporocyst wall; 5, gametes; 6, gelatinous wall of cyst; 7, invaginated spore duct; 8, evaginated spore duct; 9, chain of spores being released; 10, mass of spores; 11, three spores from chain; 12, mucoid covering; 13, immature spore; 14, sporozoite; 15, epithelial cell of midgut of cockroach; 16, striated surface of cell; 17, sporozoite attached to epithelial cell; 18, partially embedded trophozoite; 19, protomerite of developing trophozoite; 20, deutomerite.

a, Sporulated spore swallowed by cockroach; b, spore ruptured, releasing sporozoites; c, free sporozoites; d, sporozoite attached to intestinal epithelium; e, developing trophozoite; f, fully developed trophozoite attached to epithelium by epimerite; g, sporadin (a detached trophozoite) free in gut; h, two sporadins in syzygy; i, two sporadins in conjugation, preparing to encyst; j, formation of gametes, which will be followed by fertilization and formation of spores; k, ripe cyst with sporoducts and containing numerous spores; l, ripe cyst passed in feces of cockroach; m, free unsporulated spore; n, infective sporulated spore.

Figure adapted from various sources, except L, which is original.

---

rine *Lankesteria culicis* (Ross), 1898, from the mosquito *Aëdes (Finlayia) geniculatus* (Olivier). Parasitology 39:291–294.

Scheffield, H. G., P. C. C. Garnham, and T. Shiroishi. (1971). The fine structure of the sporozoite of *Lankesteria culicis*. J. Protozool. 18:98–105.

Vávra, J. (1969). *Lankesteria barretti* n. sp. (Eugregarinida, Diplocystidae), a parasite of the mosquito *Aedes triseriatus* (Say) and a review of the genus *Lankesteria* Mingazzini. J. Protozool. 16:546–570.

Walsh, R. D., Jr., and C. S. Callaway. (1969). The fine structure of the gregarine *Lankesteria culicis* parasitic in the yellow fever mosquito *Aedes aegypti*. J. Protozool. 16:536–545.

Wenyon, C. M. (1926). Protozoology. William Wood and Co., New York (Reprinted by Hafner Publishing Co., New York, 1965), p. 1121.

### Superfamily Cephalinoidea

The body of a complete trophozoite of the Cephalinoidea consists of two major parts. They are the smaller anterior protomerite with a slender forward extension, the epimerite, bearing a little terminal knob and the longer posterior deutomerite. The two parts are divided by a transverse, membraneous septum.

#### FAMILY GREGARINIDAE

The Gregarinidae are parasites of the alimentary canal of arthropods, especially insects. Kamm and Watson gave extensive lists of them, including figures.

Members of the family are characterized by a simple symmetrical epimerite attached by a slender stalk to the protomerite which is followed by the much larger deutomerite. The sporadins unite in pairs and form sporocysts with or without tubular ducts through which symmetrical spores are released in chains.

### *Gregarina blattarum*
von Siebold, 1839 (Plate 19)

This species is common in cockroaches (*Periplaneta americana*, *Blatta orientalis*, and *Blatella germanica*), having been reported from many parts of the world. Up to 30 percent of the cockroaches may be infected with 30 or more sporocysts in the winter, and a lower incidence during the summer in some areas.

**Description.** A complete trophozoite consists of two parts, as stated above. In intact trophozoites, the epimerite is embedded in the intestinal epithelium forming a direct contact with it and the organism. Since the stalk of the epimerite breaks easily, the epimerite usually remains within the cell. The freed trophozoites are sporadins or sporonts. They are large, measuring 500 to 1100 $\mu$ long by 160 to 400 $\mu$ in diameter. Sporocysts are thick-walled, spherical or ovoidal, with eight to 10 or more spore ducts. The spores within the sporocyst are cylindrical to barrel-shaped with truncate ends and measure 8 to 8.5 $\mu$ long by 3.7 to 4 $\mu$ in diameter. They are extruded in chains from the spore ducts.

The epicyte of both parts of the body is formed into numerous folds. Through undulatory movement of the folds, fluid and mucus be-

**Plate 19**    *Gregarina blattarum*                                                                 93

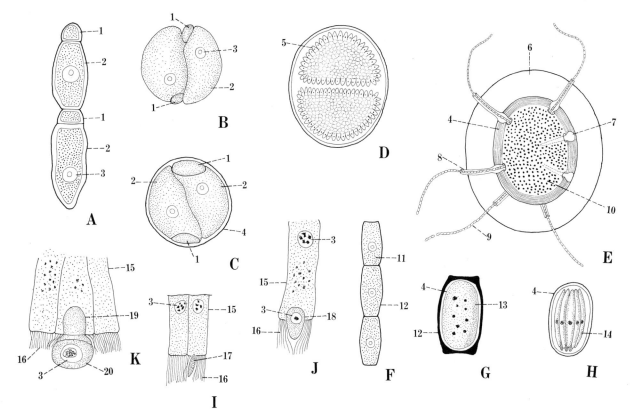

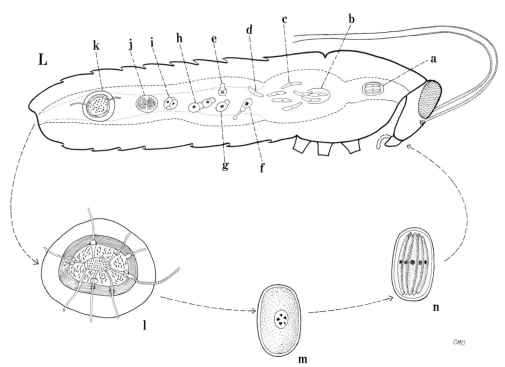

tween them is forced backward, providing pressure against the liquid medium of the gut to push the gregarines forward in the gliding movement characteristic of them.

**Life Cycle.** The life cycle is typically gregarine, being direct and without the succor of a vector. Ripe sporocysts evaginate their spore ducts and extrude chains of undeveloped spores in the intestine. These are voided with the feces. Development of the spore proceeds through three divisions of the nucleus with the formation of eight nuclei. A fragment of cytoplasm surrounds each nucleus and forms a sporozoite. Ripe spores containing the eight sporozoites are infective to cockroaches eating them. Upon being freed from the spore in the gut of the cockroach, the sporozoites attach to the epithelium of the midgut and begin development as a trophozoite.

The embedded portion outgrows the posterior extracellular part to form the epimerite. When it is fully developed, growth takes place in the posterior region. When fully formed, an annulus appears diametrically across posterior to the epimerite dividing the body into the protomerite and deutomerite. At about this time, the pendent portion usually detaches from the epimerite and is free in the intestine as the sporadins. While embedded in an epithelial cell, the epimerite served both as an anchor and direct contact with the cytoplasm of the host cell for absorption of nutritive substances. When the trophozoites are grown and have separated from the epimerite, the epicyte grows rapidly, causing it to form longitudinal parallel folds over the entire body. These mature individuals are in reality the sexual forms known as gametocytes. Two gametocytes of opposite sex attract each other and attach end to end in a complicated process known as syzygy. The anterior member of the pair is the primite and the posterior one the satellite, being female and male, respectively. As soon as syzygy occurs, the male satellite responds to the stimulus provided by the female primite by rapid modification of the protomeritic epicyte. In *Gregarina polymorpha*, it undergoes rapid growth and forms a cup-like structure in which the folds on the posterior end of the primite interlock with diminished ones on the anterior end of the satellite (Fig. 7).

Eventually the gametocytes conjugate by attaching to each other on the long axis of the body and secrete a cyst about themselves to form the sporocyst. The nucleus of each gametocyte undergoes repeated division, forming many small nuclei scattered throughout the cytoplasm. These migrate to the surface of the cytoplasmic masses where each becomes clothed with a bit of cytoplasm to form gametes. Union of the gametes produces zygotes which develop into spores within the sporocyst. Continued development of the sporocyst results in the formation of a thick cyst wall with a number of sporoducts through which the spores will be released when mature.

**Pathology.** The effect of *G. blattarum* on cockroaches is unknown. *Gregarina polymorpha* of the mealworm *Tenebrio molitor*, however, produces adverse effects in larvae living under suboptimal conditions of diet, temperature, and humidity. Both larvae and pupae are underweight and experience difficulty in completing metamorphosis.

## EXERCISE ON LIFE CYCLE

Widespread distribution of cockroaches, high incidence of infection, and ease of handling the

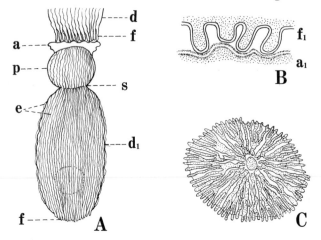

Fig. 7. Schematic protrayal of syzygy of two sporadins of *Gregarina polymorpha* and others. **A,** Entire satellite attached to posterior end of primite. **B,** Details of epicytal structure of posterior end of primite and anterior end of satellite, showing nature of attachment. **C,** Details of posterior end of deutomerite showing epicytal folds converging on papilla. **a,** Anterior portion of protomerite; **a₁,** high magnification of (a), showing three membranes of epicyte and its modification for syzygy; **d,** deutomerite of primite; **d₁,** deutomerite of satellite; **e,** epicytal folds; **f,** extension of epicytal folds of deutomerite; **f₁,** detailed digitation of (f), showing three membranes and details of fusion of sporadins in syzygy; **p,** protomerite of satellite; **s,** septum. Adapted from Devauchelle, 1968, J. Protozool. 16:629, and Vávra and Small, 1969, J. Protozool. 16:745.

hosts makes this an ideal gregarine parasite for experimental studies.

Place the gut of a cockroach in a dish of physiological solution and open it carefully. Mount segments of the midgut on a slide and examine it for trophozoites anchored by means of the epimerite to the epithelial cells. These may be difficult to find. For the free stages, mount some of the intestinal contents on a slide, place a coverslip on it, and seal with Vaseline. Identify the large, nearly spherical trophozoites with two parts (protomerite and deutomerite). Find partners in syzygy and recognize the primite and satellite.

For observing encystation, mount associated gametocytes in egg albumen. Watch their movements as they maneuver into a head-tail position along the lateral margins of the bodies. Following proper orientation and union, a thick-walled sporocyst may appear *in vitro*. If it does, note the development of the gametes followed by their union to form zygotes. Sporoducts appear, extend from the sporocyst, and evert within 48 hours after its formation. The barrel-shaped spores are extruded in strings. Watch them for development of the sporozoites.

## SELECTED REFERENCES

Cuénot, L. C. M. J. (1901). Recherches sur l'évolution et la conjugaison des grégarines. Arch. Biol. 17:581–652.

Devauchelle, G. (1968). Etude de l'ultrastructure de *Gregarina polymorpha* (Hamm.) en syzygy. J. Protozool. 15:629–636.

Harry, O. G. (1965). Studies on early development of the eugregarine *Gregarina garnhami*. J. Protozool. 12:296–305.

Kamm, M. W. (1922). Studies on gregarines. II. Synopsis of the polycystid gregarines of the world, excluding those from the Myriapoda, Orthoptera, and Coleoptera. Ill. Biol. Monogr. 7:1–104.

Mackinnon, D. L., and R. S. J. Hawes. (1961). An Introduction to the Study of Protozoa. Clarendon Press, Oxford, p. 190.

Mano-Sabaratnam. (1971). The biology of *Gregarina fernandoi* n. sp. in the cockroach *Pycnoscellus surinamensis* with observations on its development and cytochemical nature. J. Protozool. 18:141–146.

Sprague, V. (1941). Studies on *Gregarina blattarum* with particular reference to the chromosome cycle. Ill. Biol. Monogr. 18:1–57.

Vavra, J., and E. B. Small. (1969). Scanning electron microscopy of gregarines (Protozoa, Sporozoa) and its contribution to the theory of gregarine movement. J. Protozool. 16:745–757.

Watson, M. E. (1916). Studies on gregarines including descriptions of twenty-one new species and a synopsis of the eugregarine records from the Myriapoda, Coleoptera, and Orthoptera of the world. Ill. Biol. Monogr. 11:1–258.

### Suborder Schizogregarinina

Schizogregarines undergo schizogony, as indicated by the name, in the intestine of arthropods, annelids, and tunicates. The trophozoites develop intra- or extracellularly. Those developing outside the cells are attached by means of a small, pseudopod-like epimerite. When mature, the schizonts developing outside the cells have the appearance of clusters of grapes.

The suborder consists of two families. Species of Ophrycystidae undergo two types of schizogony. In the first, the schizont produces merozoites with small nuclei which in turn develop into trophozoites. The second schizogony resulting from development of these trophozoites forms merozoites with large nuclei. These merozoites develop into gametocytes. Species of the second family Caulleryellidae (also listed as Schizocystidae) have but one type of schizogony in which the nuclei of the merozoites and gametocytes are of one size only. The genus *Schizocystis* of the Caulleryellidae will be considered as an example of the suborder because the nature of its life cycle has been studied extensively.

### *Schizocystis gregarinoides*
(Léger, 1908) (Plate 20)

*Schizocystis gregarinoides* is a parasite of larval midges (*Ceratopogon solstitialis* and possibly other species of the genus) that develop in water and moist humus.

This species of gregarine is of interest because it is one of a group that shows schizogony in its simplest form. Schizonts are extracellular, hanging to the midgut epithelium or lying free in the lumen, and therefore, the merozoites develop in the lumen of the gut. Gametogony and sporogony are typically gregarine, as in *Lankesteria culicis*.

**Description.**    Mature vermicular trophozoites are up to 400 $\mu$ long by 15 $\mu$ in diameter with the anterior end bearing a pseudopod-like organelle (epimerite) of a sucking type for adhering to the gut. When fully mature, each con-

## Explanation of Plate 20

A, Adult vermicular schizont. B, Schizont in the process of segmentation and formation of merozoites. C, Merozoites formed. D, Sporulated oocyst. E, Larval midge, showing enlarged section of intestine.

1, Holdfast organelle at anterior end; 2, nuclei; 3, merozoites forming from schizont; 4, residual cytoplasm; 5, merozoites; 6, wall of oocyst; 7, sporozoites.

a, Ripe sporocyst in intestine; b, ruptured sporocyst liberating sporozoites; c, attachment of sporozoite to intestinal epithelium; d, growing schizont with two nuclei; e, older schizont with more nuclei; f, mature vermicular schizont; g, segmentation of schizont to form merozoites; h, merozoite initiating a succeeding generation of schizonts; i, merozoites initiating gametogenous phase; j, merozoites become gametocytes; k, union of gametocytes; l, enclosure of gametocytes in gametocyst; m, nuclear division in gametocytes to form gametes; n, formation of pointed microgametes and round macrogametes; o, union of micro- and macrogametes to form zygotes within gametocyst; p, formation of young oocysts inside gametocyst and beginning of sporogenous phase; q, sporulated oocyst free in intestine; r, sporulated oocyst passing through intestine in feces; s, infective oocyst free in water.

Adapted from Léger, 1909, Arch. Protistenk. 18:83.

---

tains numerous nuclei, up to 200, and is now a schizont. Each fully developed schizont forms into grape-like clusters of merozoites. Gametocysts contain two gametocytes, one of which forms round macrogametes and the other pyriform microgametes. Oocysts inside the gametocysts are oval, each with eight sporozoites.

**Life Cycle.** Sporulated occysts passed in the feces of midge larvae are the source of infection of other midges when swallowed with their food. Fusiform sporozoites 8 $\mu$ long are liberated in the intestine and attach to the epithelium. They gradually develop into elongate trophozoites adhering to the mucosa. When they attain a length of about 30 $\mu$, the epicyte shows fine striations and the nucleus starts to divide. This is the beginning of the schizogenous phase of the life cycle. Growth continues until the schizont acquires a vermiform shape about 400 $\mu$ long by 15 $\mu$ in diameter and contains up to 200 nuclei. At this point in its development, the schizont loses its striations and divides into as many merozoites as there are nuclei, forming a grape-like cluster. As the individual merozoites detach from the cluster, they produce a second generation of large, oval trophozoites, instead of the vermiform type, that lack an organelle of attachment, hence live free in the intestinal lumen. When mature, the nucleus divides repeatedly, as before, and the schizont produces a rosette of merozoites which separate from the group to become gametocytes. The products of both generations of schizonts have nuclei of the same size and the process is considered as a single schizogony characteristic of this family.

The gametocytes grow in size and have epicytal striations. There is no division of the nucleus at this time. When mature, the gametocytes unite in syzygy and secrete a gametocyst about themselves, thus beginning the gametogenous phase.

The nucleus of each gametocyte now divides a number of times with each part migrating to the surface of the cytoplasmic mass, where a bit of cytoplasm surrounds it to form gametes of equal size (isogametes) but of different form. The round ones from one gametocyte are macrogametes (female) and the pyriform ones from the other gametocyte are microgametes (male). When fully developed, a micro- and a macrogamete fuse to form a zygote. Each zygote secretes a double wall about itself, becoming an oval oocyst about 8 $\mu$ long by 4 $\mu$ in diameter. Each pole of the oocyst is slightly thickened. Sporogony takes place inside the still intact gametocyst in the intestine of the midge larva.

The nucleus of each zygote divides three times, resulting in a total of eight. Each new nucleus is surrounded by a bit of cytoplasm and develops into a minute, spindle-shaped sporozoite about 8 $\mu$ long. The gametocysts, containing the fully developed oocysts, are voided from the intestine with the feces into the water. Larval midges become infected by swallowing the infective oocysts with their food.

*Lymphotropha triboli* Ashford, 1965, is a schizocystid gregarine from the hemocoel of the rust-red flour beetle, *Tribolium castaneum*. Sporozoites pass from the gut to the hemocoel, where they transform and develop into large, ovoid trophozoites. Schizogony is of the single type and produces up to eight merozoites. Game-

**Plate 20**    *Schizocystis gregarinoides*                                                   97

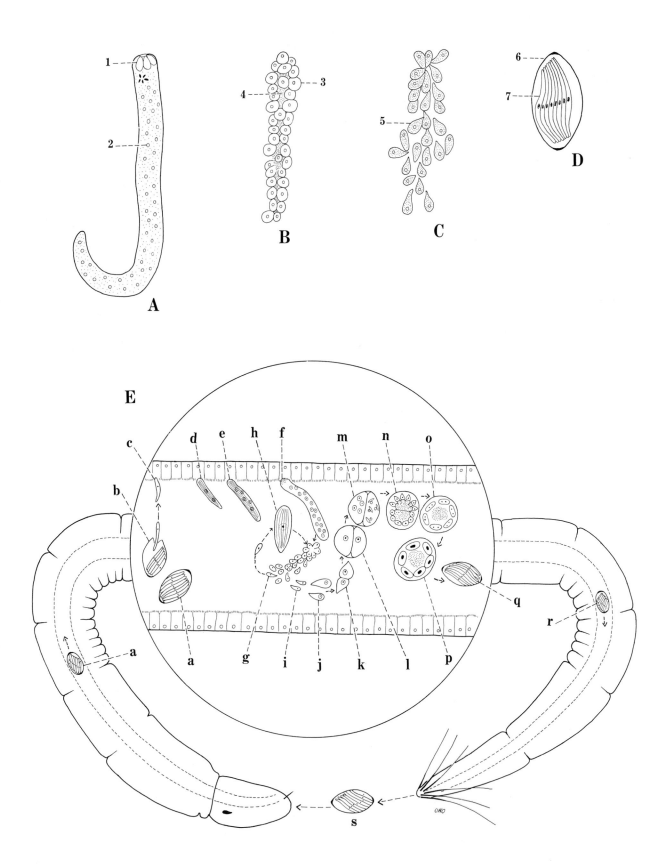

tocytes unite and produce 16 nuclei which form gametes. Oocysts contain eight sporozoites. These parasites cause high mortality among young larvae.

## EXERCISE ON LIFE CYCLE

Remove the entire intestine of ceratopogon midge larvae and examine the wall and contents for the characteristic stages of *Schizocystis*. The mature trophozoites, gametocysts with gametocytes, and spores can be seen. Vital strains used on the intestinal contents and sections of the gut aid in finding and identifying the various stages.

Contents of the hemocoel of the rust-red flour beetle, and possibly other species of flour beetles, should be examined in physiological solution on slides for the various stages of development of schizocystid gregarines.

## SELECTED REFERENCES

Ashford, R. W. (1965). *Lymphotropha triboli* gen. nov., sp. nov., Neogregarinida, Schizocystidae, from the haemocoele of *Tribolium castaneum*. J. Protozool. 12:609–615.

Doflein, F., and E. Reichenow. (1953). Lehrbuch der Protozoenkunde. 6th ed. Gustav Fischer, Jena, p. 786.

Grassé, P. P. (1953). Traité de Zoologie. Masson et Cie, Paris, Vol. 1, Fasc. 2, p. 668.

Léger, L. (1909). Les schizogrégarines des trachéates. 2. Le genre *Schizocystis*. Arch. Protistenk. 18:83–110.

Weiser, J. (1955). A new classification of the Schizogregarina. J. Protozool. 2:6–12.

Wenyon, C. M. (1926). Protozoology. William Wood and Co., New York (Reprinted by Hafner Publishing Co., New York, 1965), Vol. 2, p. 1128.

## ORDER COCCIDIDA

This order, according to Kudo, consists of the suborders Eimeriina and Adeleina. They are intracellular parasites mainly of the intestinal epithelium but also of the liver, kidneys, and blood cells. In this respect, they differ from the Gregarinida which are chiefly celozoic. Reproduction is both asexual by schizogony and sporogony and sexual by gametogony. Micro- and macrogametes differ greatly in size. Sporozoites of variable number are enclosed in a sporocyst inside an oocyst. Development may be in one host or it may require two.

The Coccidida show a transition in the types of life cycles from the Gregarinida which are chiefly celozic, through the coccidian which is mainly in the intestinal epithelium, to the hemogregarine which is primarily in blood cells. The progression of this development is shown by the genera *Selenococcidium*, *Eimeria*, *Aggregata*, and *Schellackia* of the Suborder Eimeriina, and by *Hepatozoon* and *Haemogregarina* of the suborder Adeleina (Plate 33).

### Suborder Eimeriina

The members of this suborder are chiefly intracellular parasites of the intestinal epithelium, although some occur in the liver and kidney cells. Both schizogenic and sporogenic development may occur in one host, but two hosts appear in the life cycle of some genera. The gametocytes develop separately. Microgametes are numerous, minute, biflagellated, comma-shaped bodies and macrogametes are large oval ones.

*Selenococcidium* has a life cycle reminiscent of the gregarine *Schizocystis* in that the schizont begins development in the lumen of the intestine, but differs in that it is completed in the intestinal epithelium. Gametogony is coccidian in character, taking place in the intestinal epithelium.

*Eimeria* is typically coccidian in its life cycle, having an intracellular alternation of generations in the intestinal epithelium of a single host. Oocysts are thick-walled and sporulation occurs outside the host with two sporozoites contained in each of four sporocysts, making a total of eight. Infection is by swallowing sporulated oocysts.

*Aggregata* introduces an alternation of hosts, but it is a primitive relationship. Schizogony occurs in the intestinal epithelium of a crab and sporogony in the same cells of a cuttlefish. Infection of crabs takes place when they eat cuttlefish and of the latter when they devour crabs. The bivalved oocyst contains three free sporozoites. The appearance of an alternation of hosts points toward the hemosporidian type of life cycle.

*Schellackia* has a more advanced life cycle in that reproduction occurs in two organs in the vertebrate host and in having a bloodsucking arthropod involved as a mechanical transport of the sporozoites. Schizogony is completed in the intestinal epithelium of lizards. Development of the microgametocytes likewise takes place in the intestinal epithelium, a coccidian trait, but that of the macrogametocytes is in the intestinal wall where the zygote is formed and the sporozoites develop in thin-walled oocysts. Upon release from the oocyst, the sporozoites enter the bloodstream and penetrate erythrocytes, a hemosporidian characteristic. Infected blood cells swallowed by lizard mites are engulfed by epithelial cells of the stomach and digested, liberating the sporozoites in them. No development takes place in the mite. It serves only to collect sporozoites and as a means of transporting them to the intestine of the lizards when eaten by them. This cycle, which foreshadows that of the Haemosporida, is advanced further by the members of the Suborder Adeleina, which is discussed in a later section.

### FAMILY SELENOCOCCIDIIDAE

The schizonts are elongated vermiform bodies which together with the gametocytes are extracellular at first, but later enter the intestinal epithelium to complete their development.

*Selenococcidium intermedium*
Léger and Duboscq, 1909 (Plate 21)

*Selenococcidium* is a parasite of the intestine of European lobsters. It is remarkable in that the early stages of both the schizogenous and gametogenous cycles occur in the lumen of the intestine as vermicular trophozoites, a characteristic of the gregarines. Schizogony and gametogony are completed in the epithelial cells of the intestine, similar to that which occurs in the Coccidida. Thus, this life cycle appears to be a connecting link between that of the gregarines and the coccidia. Sporogony presumably occurs outside the host, but is unknown at present.

**Description.** Trophozoites are vermiculate bodies in the intestinal lumen where schizogony begins. The vermiculate octonucleate schizont enters the epithelial cells and produces eight merozoites; large and small vermiform merozoites in the intestinal lumen give rise to quadrinucleate and octonucleate macro- and microgametocytes, respectively. The gametes, being of distinctly different sizes, are known as anisogametes. The microgametes are minute, elongate bodies and the macrogametes relatively large and oval or spherical in shape. Formation of gametes takes place in the intestinal epithelium. Sporogony is unknown.

**Life Cycle.** Presumably infection of lobsters takes place when ripe oocysts thought to be free in the water are swallowed. This part of the life cycle, however, remains unknown.

The young vermiculate trophozoite in the intestine undergoes three nuclear divisions, producing an octonucleate, mature schizont. It enters the intestinal epithelium, rounds up, becomes a segmenter, and produces eight vermiform merozoites. Several cycles of schizogony of this type may take place. Finally, a generation of merozoites of two sizes but smaller than those forming the schizonts appear. The small ones develop into oval octonucleate bodies in the lumen of the intestine and produce eight vermicular microgametocytes. Upon entering the epithelial cells, they round up and the nucleus divides many times, producing as many microgametes. The large merozoites transform into quadrinucleate vermicules that produce four similar macrogametocytes in the lumen of the intestine. These enter the epithelial cells, where they round up and each one becomes a macrogamete. Fertilization occurs and a zygote is formed.

It is seen that the trophozoites, schizonts, and gametocytes develop in the lumen of the intestine, much the same as the gregarines. Completion of schizogony with the formation of merozoites, and of gametogony with the formation of gametes, takes place in the epithelial cells of the intestine, as occurs in the Coccidida. In these respects, *Selenococcidium intermedium* appears to occupy a position intermediate between the gregarines and the Coccidida, thereby bridging the gap between the extracellular development of the gregarine on the one hand, and the intracellular development of the Coccidida and Haemosporida on the other.

### EXERCISE ON LIFE CYCLE

In regions where living lobsters are available for examination, a search should be made in species off the coast of North America for *Selencoccidium intermedium* or related species. The presence of these parasites would provide an opportunity to find the oocysts and thereby complete the life cycle.

### SELECTED REFERENCES

Kudo, R. R. (1966). Protozoology. 5th ed. Charles C Thomas, Springfield, Ill., p. 678.

## Explanation of Plate 21

A, Vermicule with eight nuclei of schizogenous phase. B, Vermicule forming schizont inside epithelial cell of intestine and whose merozoites will reinitiate a second schizogenous cycle. C, Five merozoites which will become macrogametocytes. D, Macrogamete. E, Schizont preparing to divide to form eight merozoites that will become microgametocytes. F, Microgametocyte with numerous microgametes. G, Lobster host. H, Section of intestine of lobster showing schizogony. I, Section of intestine of lobster showing gametogony (sporogony unknown).

1, Nucleus; 2, merozoites; 3, microgametes.

a–g, Schizogony. a, large uninucleate trophozoite; b, quadrinucleate schizont; c, octonucleate schizont; d, octonucleate schizont enters epithelial cell and assumes spherical form; e, vermiculate merozoites forming; f, fully formed vermiculate merozoites; g, vermiculate merozoites escaping into lumen of gut enter other cells and repeat schizogenous cycle (a–g) or initiate gametogenous cycle (h–x); h, small merozoite destined to form microgametes; i, schizont with eight nuclei that will produce eight merozoites; j, eight merozoites that will form microgametocytes; k, merozoite entering cell; l, young microgametocyte formed from merozoite; m, developing microgametocyte with many nuclei; n, fully developed microgametocyte with numerous microgametes; o, large merozoite destined to produce macrogametes; p, binucleate merozoite; q, quadrinucleate merozoite; r, formation of four merozoites that will develop into four macrogametocytes; s, merozoite entering cell; t, u, merozoite transforming into macrogametocyte; v, macrogamete; w, fertilization (small microgamete attached on upper side); x, zygote (sporogony unknown).

Adapted from Léger and Duboscq, 1910, Arch. Zool. Exp. Gén. 40, 5 s. 5 (4) : 187.

Léger, L., and O. Duboscq. (1910). *Selenococcidium intermedium* Lég. et Dub. et la systematique des sporozoaires. Arch. Zool. Exp. Gén. 40:187–238.

Grassé, P. P. (1953). Traité de Zoologie. Masson et Cie, Paris, Vol. 1, Fasc. 2, p. 744.

Wenyon, C. M. (1926). Protozoology. William Wood and Co., New York (Reprinted by Hafner Publishing Co., New York, 1965), Vol. 2, p. 801.

### FAMILY EIMERIIDAE

Species of this family are widespread among the higher invertebrates and all classes of vertebrates. They develop inside and completely destroy infected cells of the intestine, as well as those of the liver of rabbits and kidneys of geese. Among the domestic animals, species of *Eimeria* are of special importance because of the losses, estimated at over 23 million pounds per year, among poultry, rabbits, sheep, and cattle. Added to the loss in meat, inefficient utilization of feed by infected animals and a lower return for labor required to care for them must be included in the cost to the livestock industry caused by this group of parasites.

The family consists of about 25 genera, two of which are considered here. *Eimeria* has oocysts with four sporocysts, each containing two sporozoites, and *Isospora* has two sporocysts, each with four sporozoites (Fig. 8). There is a total of eight sporozoites per oocyst in each genus. They are best known from the oval, thick-walled oocysts voided with the feces of infected hosts.

Schizonts and micro- and macrogametocytes develop separately inside host cells, usually of a specific type in a definite location. Each microgametocyte produces numerous, minute microgametes that disperse from it. Each macrogametocyte transforms into a large, single macrogamete that remains inside the host cell. Upon fertilization, the characteristic immobile oocyst is formed within and finally released from the host cell.

### *Eimeria tenella*
Railliet and Lucet, 1891 (Plate 22)

This is the common cecal coccidian of chickens. It is the cause of a fatal disease recognized as cecal coccidiosis manifested by severe bloody diarrhea, especially in young birds 4 to 8 weeks of age. This species is used as an example of the group.

**Description.**    Oocysts are broadly ovoid, measuring 29 to 19.5 $\mu$ long by 22.8 to 16.5 $\mu$ in diameter with an average size of 26.6 by 19 $\mu$. The oocyst wall is smooth and there is no micropyle. The wall consists of two layers of about equal thickness. The outer layer is of uniform structure but the inner one shows a light middle zone. Unsporulated oocysts contain an undifferentiated mass of cytoplasm with nucleus, whereas fully sporulated ones contain four sporocysts each with two sporozoites. A large residual mass of cytoplasm remains inside the oocyst and a small one within each sporocyst. Sporozoites are fusiform in shape, each with a single nucleus

**Plate 21** *Selenococcidium intermedium* 101

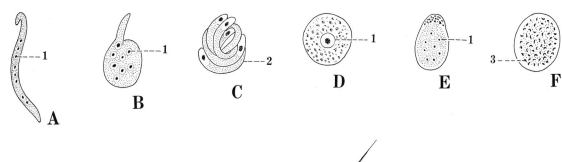

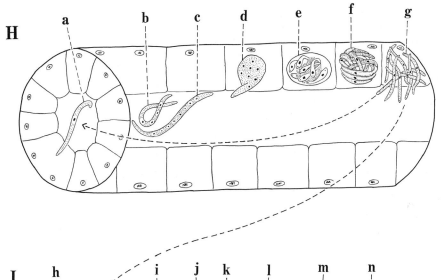

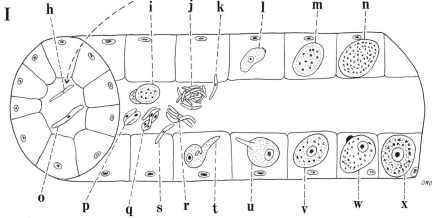

## Explanation of Plate 22

A, Unsporulated oocyst. B, Early stage of sporulation. C, Oocyst with four sporoblasts. D, Fully developed oocyst with eight sporozoites. E–I, Stages of life cycle in cecal epithelium of chick. E, First generation of merozoites. F, Second generation of merozoites. G, Development of micro- and macrogametocytes. H, Development of micro- and macrogametes, formation of zygotes, and escape of oocysts from epithelium. I, Chicken host.

1, Oocyst wall; 2, cytoplasm with central nucleus; 3, dividing cytoplasm; 4, sporoblasts; 5, sporocyst; 6, sporozoites; 7, polar granule.

a–h, First generation of merozoites in cecum; a, epithelium of cecum; b, crypt of Lieberkühn; c, sporozoite from lumen of cecum entering crypt; d, sporozoite entering epithelial cell; e, trophozoite; f, schizonts; g, segmenter with formed merozoites; h, merozoites escape from segmenters into crypts and lumen; i–o, second generation of merozoites; i, mass of first generation merozoites entering crypt; j, mass of young trophozoites; k, mass of fully developed subepithelial trophozoites; l, schizonts; m–o, second generation of merozoites causing extensive destruction of tissue; p–s, formation of gametocytes; p, q, second generation of merozoites enter crypts and penetrate cells; r, young developing gametocytes; s, fully developed gametocytes; t–x, formation of gametes, fertilization of macrogamete, and its release from cells as an oocyst; t, developing microgametocyte; u, microgametes escaping from epithelial cell; v, developing macrogametocyte; w, fertilization of macrogamete and formation of zygote; x, mass of oocysts breaking out of tissue and later voided in feces.

a′–d′, Sporogony of oocyst; a′, unsporulated oocysts passed in feces; b′, developing oocysts with sporoblasts; c′, sporocysts developed; d′, fully sporulated and infective oocyst; e′–i′, infection of host; e′, sporulated oocyst in crop; f′, sporocysts escape from oocyst; g′, sporozoites escape from sporocysts in small intestine; h′, sporozoites enter ceca; i′, sporozoites enter epithelial cells of ceca; j′–m′, schizogony; j′, trophozoites; k′, schizont; l′, segmenter; m′, merozoites which are the final stage of first generation that is repeated (n′, j′–m′) for second generation; o′–t′, gametogony of male gamete; o′, merozoite of second generation enters epithelial cell and initiates microgametogony; p′, developing microgametocyte; q′, early division of nucleus; r′, nuclear material arranged around periphery of microgametocyte preparatory to forming microgametes; s′, developing microgametes; t′, fully developed microgametes escaping from epithelial cell; u′–y′, gametogony of female gamete; u′, second generation merozoite entering epithelial cell to develop into macrogametocyte; v′–x′, developing macrogametocyte (w′, x′ shows extruded nuclear material); y′, macrogamete ready for fertilization; z′, fertilization. a″, Union of nuclei and formation of zygote; b″, oocyst in cell; c″, oocyst escapes from cell, enters lumen of cecum, and is voided with feces.

Figures A–H adapted from Tyzzer, 1929, Amer. J. Hyg. 10:269.

---

and a refractile body near each end (Fig. 8, A).

The body of the sporozoite is surrounded by two bilamellar membranes separated by a clear intermediate zone. The anterior end is complex and the most impressive part of the cell. It consists basically of a hollow cone-shaped body, the conoid, from which a number of structures origi-

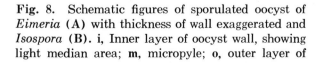
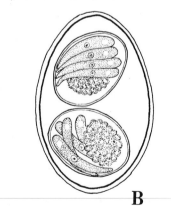

Fig. 8. Schematic figures of sporulated oocyst of *Eimeria* (A) with thickness of wall exaggerated and *Isospora* (B). i, Inner layer of oocyst wall, showing light median area; m, micropyle; o, outer layer of oocyst wall, showing uniform consistency; p, polar granule; r, residual cytoplasm of sporocyst; R, residual cytoplasm of oocyst; s, sporozoite; S, sporocyst with two sporozoites and residual cytoplasm.

**Plate 22**   *Eimeria tenella*                                                                 103

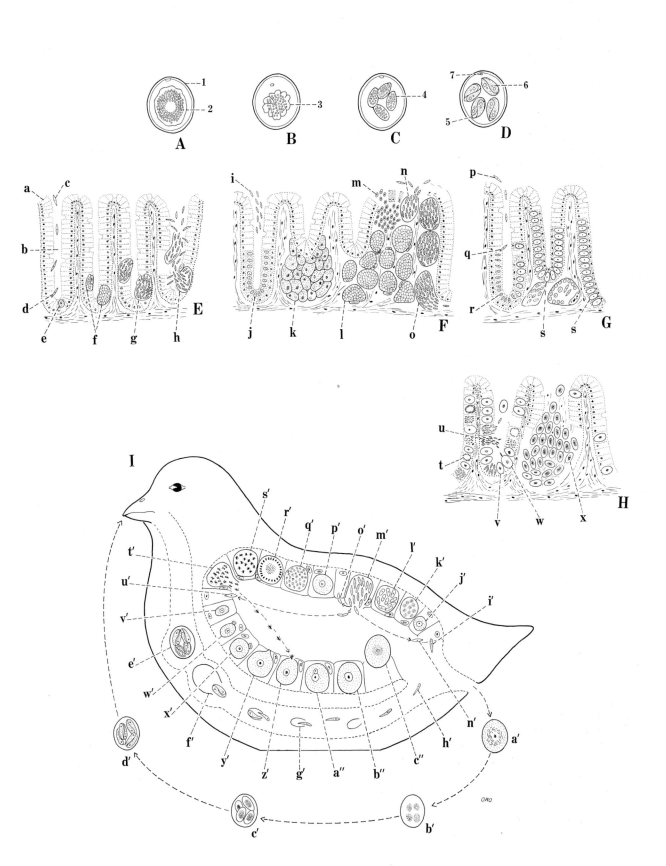

nate. These consist of subpellicular fibrils that extend to the posterior end of the body, together with shorter, thicker toxonemes, and a pair of large club-shaped organelles that equal one-fourth to one-third the length of the body. Invaginations, or micropyles, along the side of the body open into subpellicular vesicles that break away and occur free in the cytoplasm as food vacuoles. Scattered in the cytoplasm are mitochondria, lipid bodies, endoplasmic reticulum, Golgi apparatus, and other structures (Fig. 11).

Sporozoites resemble merozoites in fine structure (Fig. 9, B).

Trophozoites are intracellular, uninucleate bodies of variable shape. First generation schizonts have the cytoplasm divided into many lobes to increase surface areas that are covered with nuclei, resulting from repeated division of the original one, that are destined to form merozoites. These are uninucleate, somewhat fusiform bodies (Fig. 9, B). First generation merozoites which are 2 to 4 μ long by 1 to 1.5 μ in

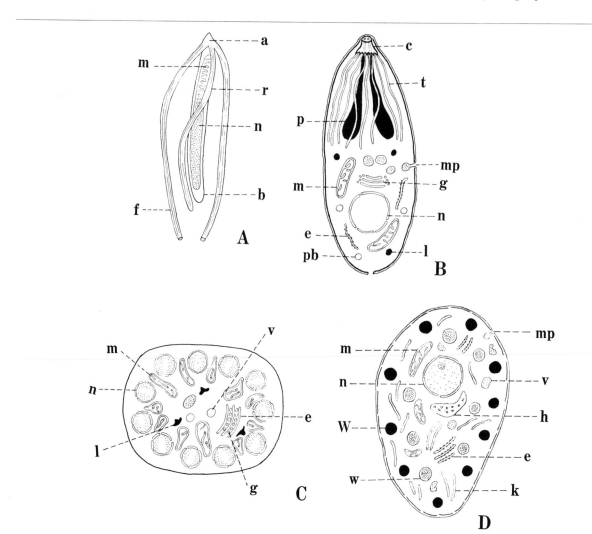

Fig. 9. Schematic figures of different stages of *Eimeria* reconstructed from electron micrographs. A, Microgamete. B, Merozoite. C, Young microgametocyte. D, Young macrogametocyte. a, Basal apparatus; b, body; c, conoid; d, glycogen; e, endoplasmic reticulum; f, trailing flagellum; g, Golgi apparatus; h, half-moon–shaped body; k, canal system; l, lipid mass; m, mitochondrion; mp, micropyle with vesicle (food vacuole) forming; n, nucleus with pores in double membrane; p, paired organelles; pb, protein body; r, recurrent flagellum; t, toxonemes; v, vesicle; W, outer wall-forming body; w, inner wall-forming body. A adapted from Scholtyseck, 1965, Z. Parasitenk. 66:635; B adapted from various sources; C, from Hammond *et al.*, 1967, J. Parasitol. 53:236; D, from Scholtyseck *et al.*, 1969, Z. Parasitenk. 33:31.

diameter transform into large second generation schizonts, measuring up to 49 $\mu$ in diameter. They produce large merozoites 16 $\mu$ long by 2 $\mu$ in diameter.

Merozoites of various species of *Eimeria* have basic morphological similarities. In addition, they resemble those of *Lankesterella* and *Toxoplasma*, suggesting some relationship. Merozoites are surrounded by a double-layered pellicle separated by a clear zone. Inside the anterior end is a hollow cone-shaped conoid with central pores. A pair of club-shaped organelles and numerous thread-like fibrils, or toxonemes, extend posteriorly from inside the conoid into the cytoplasm for about one-fourth to one-third the length of the body. Subpellicular fibrils extend from the conoid to the posterior end of the body. The nuclear membrane is double and has a number of pores in it. A Golgi apparatus lies near the nucleus. Endoplasmic reticulum, lipid bodies, and mitochondria appear in the cytoplasm (Fig. 9, B).

Mature microgametocytes fill the host cell. Minute nuclei with a porous double membrane are arranged in a peripheral layer over the surface of the cytoplasm, giving it a warty appearance. Each one develops into a microgamete. Many mitochondria are intermingled with the nuclei. An endoplasmic reticulum and Golgi apparatus are located in the cytoplasm (Fig. 9, C).

Microgametes are slender, triflagellate bodies, rather than biflagellate as formerly believed, about 6 $\mu$ long, consisting mainly of an elongate nucleus almost equaling the body in size. A groove in the anterior third of the nucleus contains an elongate mitochondrion. Three kinetosomes each with two concentric cylinders are at the anterior end of the microgamete. Each kinetosome gives rise to a flagellum. One is recurrent and attached except for its tip to the surface of the body. The other two are longer and free. Both kinetosomes and flagella have an inner structure consisting of nine pairs of peripheral and one pair of central fibrils (Fig. 9, A).

Fully developed macrogametes containing numerous eosinophilic granules fill the host cells. The large, vesicular, centrally located nucleus contains coarse, deeply staining chromatin material. Ultrastructural studies show characteristics common to most species of the genus. The macrogamete is surrounded by a single membrane which has osmophilic masses beneath it. Wall-forming bodies and other structures are in the cytoplasm similar to those in the microgametocyte (Fig. 9, D).

**Life Cycle.** Infection occurs when susceptible chickens swallow sporulated oocysts. In the anerobic environment of the gizzard, excystation of the sporozoites begins. First, the sporocysts with their sporozoites are liberated from the oocyst while passing through the gizzard. Upon reaching the small intestine, the sporozoites become active, extend the anterior tip of the body through the sporocyst wall, decrease in diameter, and slide through the tiny opening. Excystation in the small intestine may be under way as soon as 1.5 hours after swallowing oocysts. If the oocysts do not open in the gizzard, they remain intact and are passed in the feces without the sporozoites becoming freed.

Within 2 hours after infection, sporozoites are in the cecal contents and by 4 hours some are already in the cecal cells. Many have reached the lamina propria by hour 6. In this location, they are phagocytized by macrophages and transported to the fundi of the glands of Lieberkühn. By the end of day 2, many sporozoites have left the macrophages and are in the glandular cells below the nuclei. Some have transformed into trophozoites and a few have already gone through schizogony to produce the first generation of small merozoites 2 to 4 $\mu$ long by 1.5 $\mu$ in diameter. They are liberated *en masse* by 2.5 to 3 days after infection.

These merozoites, of which there are as many as 80 per schizont, penetrate the epithelium of the glands without the aid of macrophages, as in the case of sporozoites, and initiate the second generation of large schizonts. Large groups of them are in the submucosa. The merozoites are much larger than those of the first generation, being 16 $\mu$ long by 2 $\mu$ in diameter. Great numbers of them are packed in the submucosa. They mature during day 4 after infection and are released into the lumen of the ceca during day 5. Large areas of cecal epithelium and capillary beds are destroyed by the developing schizonts. Hemorrhage begins on day 4 but becomes copious by day 5.

Second generation merozoites penetrate epithelial cells, where they develop into gametocytes. In the microgametocytes, the nucleus undergoes many divisions (Fig. 9, C). These small nuclei migrate to the surface of the cytoplasmic mass and bud off with a bit of cytoplasm to form minute, slender, triflagellate microgametes 6 $\mu$

long, whose elongate nucleus stains dark blue in hematoxylin (Fig. 9, A).

Macrogametocytes grow in size but the nucleus does not multiply. As they mature and differentiate into single macrogametes, prominent eosinophilic granules appear in them.

Nutrition of the various intracellular stages is by means of micropores through which food is taken by pinocytosis into vesicles that detach and become free in the cytoplasm as food vacuoles (Figs. 9, B; 11).

Fertilization is accomplished when a microgamete enters a macrogamete and the two nuclei fuse to form a zygote. A thick, two-layered covering produced by wall-forming bodies en-

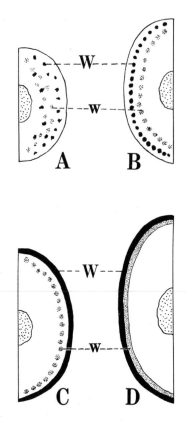

**Fig. 10.** Schematic representation of development of oocyst wall of *Eimeria* from wall-forming bodies. **A,** Young macrogamete with scattered wall-forming bodies. **B,** Outer wall-forming bodies (**W**) lining up next to unit membrane of macrogamete and inner wall-forming bodies (**w**) forming a close inner layer. **C,** Outer wall-forming bodies consolidate to form outside layer of oocyst wall. **D,** Inner wall-forming bodies coalesce to form compact inner wall, thus completing the double layered wall of the oocyst. Adapted from Scholtyseck *et al.,* 1969, Z. Parasitenk. 31:289.

closes the oocyst with its mass of cytoplasm and single nucleus (Fig. 10). Oocysts escape from the host cells shortly after fertilization and appear in the feces in great numbers, beginning about day 7 after infection and continuing for about 17 days, after which they disappear. Since the final generation of merozoites produces only gametocytes, the development and elimination of them terminates the infection. Hence, *Eimeria* is a self-limiting parasite by virtue of the nature of its life cycle.

Sporogony, the next step in the development of the life cycle, occurs outside the host under favorable environmental conditions of moisture and temperature. In the process of sporulation, the cytoplasm soon divides into four masses called sporoblasts, each with its individual nucleus. Each sporoblast produces an enclosing membrane to become a sporocyst. The cytoplasm and nucleus of each sporocyst develops into two sporozoites, which completes sporogony and constitutes the infective stage of the parasite (Fig. 8, A).

Other Species of Coccidia

Species of *Eimeria* and *Isospora* occur in many wild and domestic animals. In the latter group of hosts, some of the species are pathogenic. The severity of their pathogenicity is aggravated by the intensity of infections that develops from the crowded conditions in which domestic animals are kept.

Some pathogenic species of *Eimeria* in domestic animals include *E. acervulina, E. brunetti, E. mitis, E. mivati,* and *E. necatrix* of chickens; *E. adenoeides* and *E. meleagrimitis* of turkeys; *E. stiedae* of rabbits; *E. arlongi, E. parva* and *E. ninakohlyakimovae* of sheep; and *E. bovis* of cattle. The life cycles of these and other species of *Eimeria* are basically similar to that of *E. tenella.* Variations occur in the time required for sporulation of oocysts, specific location in the intestine, number of generations of schizonts, and length of prepatent period.

Species of *Isospora* include *I. suis* of pigs; *I. canis* and *I. rivolta* of dogs and cats; and *I. lacazii* of passerine birds, especially caged ones such as canaries. Life cycles of *Isospora* are essentially the same as those of *Eimeria,* except that a few species such as *I. rivolta* and *I. bigemina* of dogs and cats sporulate in the intestine.

The different species of *Eimeria* show preference for specific locations in the intestine of their respective hosts where development normally occurs. When sporulated oocysts of *Eimeria* of poultry are injected subcutaneously, intravenously, or intraperitoneally, the sporozoites are liberated and find their way, presumably via the circulatory system, to the preferred site in the intestine or ceca where development proceeds normally as in the case of natural infections.

Merozoites have been found in various parts of the body of hosts other than the natural site of the intestinal epithelium. Such unusual distribution is to be expected in cases where severe damage to the intestine results in opening of the vascular network of the hepatic portal vein and lacteals, allowing merozoites to enter and be transported to the liver and lymph nodes, or beyond.

**Ecology.**    Movement is confined to sporozoites, merozoites, and microgametes. These are the stages responsible for penetrating cells and eventually producing oocysts for the purpose of infecting new hosts and perpetuating the species.

Each species shows strong host specificity. There is usually strict tissue preference and site limitations within a host. In some hosts such as fish, there is indifference to host species.

The duration of infection initiated by a single sporozoite is limited by the nature of coccidial life cycles. The eventual progeny of a single sporozoite consists of enormous numbers of oocysts which are expelled from the tissues and finally voided in the feces. No stages capable of continuing the infection remain in the host.

Chronic infections undoubtedly result from reinfections. Prolonged or recurrent infections beyond the time limit characteristic of a pathogenic species such as *E. tenella* might be from merozoites that have wandered from the ceca into other body tissues and eventually returned to the preferred site of development by means of the circulation.

When oocysts leave the host, specific environmental conditions are necessary to initiate and complete sporulation. They include sufficient oxygen, moisture, and temperature. Oocysts in putrifying feces or decaying litter are unable to develop because of the anerobic conditions produced by bacterial action. Although provided with a thick wall, oocysts do not tolerate desiccation and direct sunlight. However, they may survive up to 1 year in moist soil. Under experimental conditions, they withstand freezing at $-70°C$ and when thawed are able to infect hosts.

Under moderate winter conditions, oocysts survive but development is delayed. With the onset of spring and warm weather, development is completed. The accumulated, ripe oocysts are sources of outbreaks of coccidiosis in young birds. Infection is seasonal, being greatest from May through August.

**Pathology.**    The species vary in pathogenicity. Some are highly pathogenic, others only slightly, and a few produce no symptoms of disease. All destroy cells in the course of development of schizonts and gametocysts. Clinical manifestations of pathogenicity in chickens are drooping wings, huddling, loss of appetite, decline of weight, loss of feathers, inactivity, weakness, and diarrhea with feces sometimes streaked or filled with blood. Young animals generally are severely affected, whereas mature ones usually suffer no ill effects and act as carriers.

Basic pathology begins with the entrance of sporozoites and merozoites into the intestinal mucosa and underlying cells. When either of them enters a host cell, a vacuole, called a parasitophorous vacuole, forms around the parasite. Next a cytoplasmic membrane surrounds the vacuole. Thus the host cell has taken the first step to isolate the parasite (Fig. 11). But when the vacuolar membrane is destroyed, pathogenic effects proceed, resulting in destruction of the infected cells. As the different stages of the parasite develop, the infected cells hypertrophy and then rupture. Pressure from the expansion of the large schizonts and growth of the gametocytes destroys intestinal mucosa, subepithelial tissues, and associated capillaries and lymphatics. The result is hemorrhage, anemia, reduced efficiency in secreting enzymes and absorbing nutrients, and frequently high mortality.

**Control.**    Because of the severity of coccidia as pathogenic agents of poultry, much research has been devoted to controlling the disease. These efforts have been directed along two lines. One includes the development of plans for managing flocks to minimize exposure of the birds to infection. Such programs are based on a thorough knowledge of the life cycle and ecology of the species of parasite involved and the immunological response of the host. The other, used as a supplement to a program of management, is medication of flocks with coccidiocidal drugs at appropriate ages of the birds and stages of development of the parasites.

Sanitation is an important aspect of management for controlling coccidiosis. Clean poultry houses with deep, dry litter prevent excessive exposure of birds to infection and denies the oocysts favorable conditions of moisture for development. Care must be taken to avoid carrying infections on soiled utensils or boots from flock to flock.

Preventative, or prophylactic, medication with sulfa drugs supplied continuously at a low level in water or mash suppresses development of schizonts. It protects birds against severe damage while natural immunity is being produced by the light infections that result. Curative treatment with heavier, interrupted—rather than light, continuous—doses of sulfa drugs is necessary when coccidiosis appears in flocks. Caution in the use of drugs is necessary to avoid severe toxic effects.

Immunity produced by intracellular stages of the parasites is proportional to the size of the infective dose. Continuous light infections such as often acquired under field conditions maintain a low level of immunity sufficient to protect birds against clinical coccidiosis. The immune response by the host toward intracellular sporozoites greatly interferes with their ability to develop into schizonts.

## EXERCISE ON LIFE CYCLE

*Eimeria tenella*, when available, provides an interesting exercise on life history studies. Oocysts usually can be obtained from the feces of young chickens 4 to 8 weeks old that are kept under reasonably unsanitary conditions. Oocysts may be separated from the fecal material by macerating the feces with water and allowing them to sediment for 1 hour in a tall container, after which the supernatant fluid containing the coloring matter and fine debris is decanted. Repeat the process until the water is clear.

Oocysts will sporulate in 24 hours at room temperature in a shallow solution of 2 percent dichromate, which prevents an excessive growth of bacteria and fungi, or in shallow water, or if the oocysts are poured on filter paper that is kept moist.

Fully sporulated oocysts fed with mash to young chicks or placed in the crop by means of a tube produce an infection. Since 2.5 to 3 days are required for the first generation of merozoites to develop, experimentally infected chicks should be sacrificed at intervals of 24 hours or less, and a portion of the fresh cecum fixed in one of the standard fixing agents for sectioning. These sections, stained with hematoxylin and eosin, will provide progressive stages of the development of first generation schizonts. Smears of the contents of the ceca made on day 3 after infection and stained with Wright's or Giemsa's blood stains should show the first generation of minute merozoites.

Chicks sacrificed during days 4 and 5 after infection will show the second generation of schizonts and the extensive damage done by them. Smears of cecal contents taken on day 5 will show the large second stage merozoites mixed with erythrocytes, resulting from extensive hemorrhaging. Sections taken between days 5 and 7 will show the stages of development of the gametocytes and the formation of the micro- and macrogametes and oocysts. Oocysts will appear in the feces by day 8.

## SELECTED REFERENCES

Becker, E. R. (1934). Coccidia and Coccidiosis. Monogr. No. 2, Div. Ind. Sci., Iowa State College Press, Ames, p. 29.

Challey, J. R., and W. C. Burns. (1959). The invasion of the cecal mucosa by *Eimeria tenella* sporozoites and their transport by macrophages. J. Protozool. 6:238–241.

Doran, D. J. (1969). Cultivation and freezing of poultry coccidia. Acta Vet. 38:25–30.

Farr, M. M., and D. J. Doran. (1962). Comparative excystation of four species of poultry coccidia. J. Protozool. 9:403–407.

Fayer, R., and J. L. Mahrt. (1972). Development of *Isospora canis* (Protozoa; Sporozoa) in cell culture. Z. Parasitenk. 38:313–318.

Hammond, D. M., B. Chobotar, and J. V. Ernst. (1968). Cytological observations on sporozoites of *Eimeria bovis* and *E. auburnensis*, and *Eimeria* species from the Ord kangaroo rat. J. Parasitol. 54:550–558.

Hammond, D. M., J. V. Ernst, and M. L. Miner. (1966). The development of first generation schizonts of *Eimeria bovis*. J. Protozool. 13:559–564.

Hammond, D. M., E. Scholtyseck, and B. Chobotar. (1967). Fine structures associated with nutrition of the intracellular parasite *Eimeria auburnensis*. J. Protozool. 14:678–683.

Hammond, D. M., E. Scholtyseck, and M. L. Miner.

(1967). The fine structure of microgametocytes of *Eimeria perforans, E. stiedae, E. bovis,* and *E. auburnensis.* J. Parasitol. 53:235–247.

Horton-Smith, C., and P. L. Long. (1963). Coccidia and coccidiosis in the domestic fowl and turkey. In: B. Dawes, ed. Advances in Parasitology. Academic Press, New York, Vol. 1, p. 67–107.

Kheysin, E. M. (1971). Life Cycles of Coccidia of Domestic Animals. K. S. Todd, Jr., ed. University Park Press, Baltimore, 264 pp.

Lund, E. E., and M. M. Farr. (1965). Protozoa. In: H. E. Biester and L. H. Schwarte, eds. Diseases of Poultry. 5th ed. Iowa State College Press, Ames, pp. 1056–1148.

McLaren, D. J., and G. E. Paget. (1968). A fine structural study on the merozoite of *Eimeria tenella* with special reference to the conoid apparatus. Parasitology 58:561–571.

Nyberg, P. A., D. H. Bauer, and S. E. Knapp. (1968). Carbon dioxide as the initial stimulus for excystation of *Eimeria tenella* oocysts. J. Protozool. 15:144–148.

Pellérday, L. (1965). Coccidia and Coccidiosis. Akademiai Kiado, Budapest, 657 pp.

Reid, W. M. (1964). A diagnostic chart for nine species of fowl coccidia. Georgia Agr. Exp. Sta. Tech. Bull. N. S. 39, 18 pp.

Ryley, J. F. (1969). Ultrastructural studies on the sporozoite of *Eimeria tenella.* Parasitology 59:67–72.

Scheffield, H. G., and D. M. Hammond. (1967). Electron microscope observations on the development of first-generation merozoites of *Eimeria bovis.* J. Parasitol. 53:831–840.

Scholtyseck, E. (1953). Beitrag zur Kenntnis des Entwicklungsganges des Hühnercoccids *Eimeria tenella.* Arch. Protistenk. 98:415–456.

Scholtyseck, E. (1965). Die Mikrogametenentwicklung von *Eimeria perforans.* Z. Zellforsch. 66:625–643.

Scholtyseck, E. (1969). Electron microscope studies on the effect upon host cells of various developmental stages of *Eimeria tenella* in natural chicken host and in cultures. Acta Vet. 38:153–156.

Scholtyseck, E., G. Gönnert, and A. Haberkorn. (1969). Die Feinstruktur der Makrogameten des Hühnercoccids *Eimeria tenella.* Z. Parasitenk. 33:31–43.

Scholtyseck, E., A. Rommel, and G. Heller. (1969). Licht- und elektronenmikroskopische Untersuchungen zur Bildung der Oocystenhülle bei Eimerien (*Eimeria perforans, E. stiedae* und *E. tenella.*) Z. Parasitenk. 31:289–298.

Scholtyseck, E., R. G. Strout, and A. Haberkorn.

(1969). Schizonten und Merozoiten von *Eimeria tenella* in Macrophagen. Z. Parasitenk. 32:284–296.

Senaud, J., and Z. Černá. (1969). Etude ultrastructurale des merozoites et de la schizogonie des coccidies (Eimeriina): *Eimeria magna* (Perard, 1925) de l'intestin des lapins et *E. tenella* (Railliet et Lucet, 1891) des coecums des poulets. J. Protozool. 16:155–166.

Shah, H. L. (1971). The life cycle of *Isospora felis* Wenyon, 1923, a coccidium of the cat. J. Protozool. 18:3–17.

Sharma, N. N. (1964). Response of the fowl *(Gallus domesticus)* to parenteral administration of seven coccidial species. J. Parasitol. 50:509–517.

Shaw, B. T., ed. (1954). Losses in Agriculture. A.R.S., U.S. Dep. Agr., ARS–20–1.

Soulsby, E. J. L. (1968). Helminths, Arthropods, and Protozoa of Domestic Animals. 6th ed. Williams and Wilkins Co., Baltimore, p. 615.

Tyzzer, E. E. (1929). Coccidiosis in gallinaceous birds. Amer. J. Hyg. 10:269–383.

## FAMILY AGGREGATIDAE

The Aggregatidae are parasites of marine annelids, mollusks, and crustaceans. Two hosts are required to complete the life cycle with schizogony occurring in one, and both gametogony and sporogony in the other. Schizogony is characterized by preliminary division of the cytoplasm into large bodies or cytomeres followed by the appearance of merozoites over the greatly increased surface area of them. Thin-walled oocysts contain numerous sporocysts, each with 2 to 30 sporozoites.

*Aggregata eberthi* (Labbé, 1895) (Plate 23)

*Aggregata eberthi* is a coccidian parasite of crabs and cuttlefish. Like other coccidians considered up to this point, there is an alternation of generations, but unlike them an alternation of hosts appears. All stages of the parasite develop in the wall or epithelium of the intestine of these two hosts. The appearance of two hosts and the development of the parasite in cells of the intestine of both hosts makes the life cycle in this group more complicated than in the Eimeriidae (Plate 22).

**Description.** Sporocysts are hemispheric or somewhat flattened capsules consisting of two parts, or valves. Mature ones are 8 to 9 $\mu$ in diameter and normally contain three free sporozoites 10 $\mu$ long by 2 $\mu$ in diameter and fusiform in shape. The structure of the sporozoites is basically similar to that of *Eimeria* (Fig. 11)

## Explanation of Plate 23

A, Mature sporocyst with full complement of three sporozoites. B, Sporocyst open and liberating sporozoites. C, Cross section of gut of crab *(Portunus depurator)* with four schizonts containing merozoites. D, Mature sporocyst showing two valves and three sporozoites. E, Crab intermediate host becomes infected by eating sporocysts and harbors schizogenous phases of life cycle. F, Segment of gut of crab, showing schizogony. G, Infected crab harboring extraintestinal merozoites (as in C, F) infective to definitive host. H, Cuttlefish *(Sepia officinalis)* definitive host. I, Segment of gut of cuttlefish in which gametogony and sporogony take place.

1, Sporocyst; 2, sporozoite; 3, intestinal wall of crab; 4, extraintestinal cysts with merozoites; 5, periintestinal tissue.

a–i, Schizogony in gut of crab; a, segment of intestine of crab; b, sporocyst with two valves open and releasing the three sporozoites; c, sporozoites entering epithelial cell; d, early trophozoites in epithelial cell; e, young trophozoite in periintestinal tissue; f, large trophozoite with characteristic nucleus; g, early schizont with multiplying nuclei arranged over surface of cytoplasm (the cytoplasm may become folded or separated into individual clumps); h, cytomere; i, merozoites clustered over surface of cytomeres (this ends the schizogenous phase); j–s, gametogony in cuttlefish; j, segment of gut of cuttlefish; k, cytomere with merozoites which enter submucous connective tissue of gut and develop into gametocytes; l–p, development of microgametes; l, young microgametocyte in epithelial cell; m, older microgametocyte in submucous connective tissue; n, multiplication of nuclei which are precursors of microgametes; o, filamentous microgametes adhering to ball of residual cytoplasm; p, biflagellate microgamete in lumen of intestine; q–s, development of macrogametocyte; q, young macrogametocyte in epithelial cell; r, older macrogametocyte in submucous connective tissue; s, macrogamete being fertilized by microgamete; t–v, sporogenous phase; t, sporont with multiplying nuclei arranged over surface of cytoplasm; u, cluster of sporoblasts; v, individual sporocysts developed from sporoblasts, each with three sporozoites which are released individually from epithelial cells into lumen or carried out in strips of necrotic intestinal epithelium discharged from the intestine.

Figures of protozoa adapted from Dobell, 1925, *Parasitology* 17:1.

---

and other coccidians. A cylindrical polar ring at the anterior end contains the conoid. Subpellicular fibrils, toxonemes, and paired organelles originate from the conoid and extend posteriorly into the cytoplasm. A single nucleus is present.

Schizonts in the periintestinal tissue of crabs begin as uninucleate masses of cytoplasm that soon separate into individual uninucleate clumps called cytomeres. As maturity proceeds, the nucleus of each cytomere divides many times, forming daughter nuclei that migrate to the surface of the cytomeres and develop into merozoites. They are fusiform in shape, 7 to 9 $\mu$ long, and similar in structure to those of other coccidians (Fig. 9, B). They develop into gametocytes.

Microgametocytes have a smooth surface at the outset. As development proceeds with multiple division of the nucleus, the cytoplasm divides into separate spheres. The daughter nuclei and mitochondria wander to these surfaces, forming numerous minute bumps. These are the developing microgametes. As development proceeds, they form finger-like extensions from the surface of the cytoplasm. Mature microgametes are 25 to 30 $\mu$ long and have three flagella. In general, they resemble those of other coccidians in (Fig. 9, A) except that an undulating membrane occurs which is lacking in other forms.

Macrogametocytes are up to 200 $\mu$ in diameter. The enclosing membrane consists of an outer and inner layer separated by a middle zone. The entire surface is thrown into deep folds. Numerous minute pores connect with subpellicular vesicles. The conoid and its associated organelles seen in the merozoites disappear but the nucleus and other cytoplasmic structures remain. Fertilization of the macrogamete produces a zygote.

Zygotes are large, up to 200 $\mu$ in diameter, and similar in the beginning to macrogametes. As they mature, an outer wall 8 $\mu$ and an inner one 20 $\mu$ thick are formed. The outer one bears numerous microvilli that extend into the cytoplasm of host cells. As the zygote matures and transforms into an oocyst, numerous sporocysts appear within it, each with three sporozoites.

**Life Cycle.**    As already pointed out, there is an alternation of both hosts and generations in the life cycle of *Aggregata eberthi*. The crab is the intermediate host in which schizogony, the asexual phase with production of merozoites, occurs and the cuttlefish is the definite host where

Plate 23    *Aggregata eberthi*                                                                111

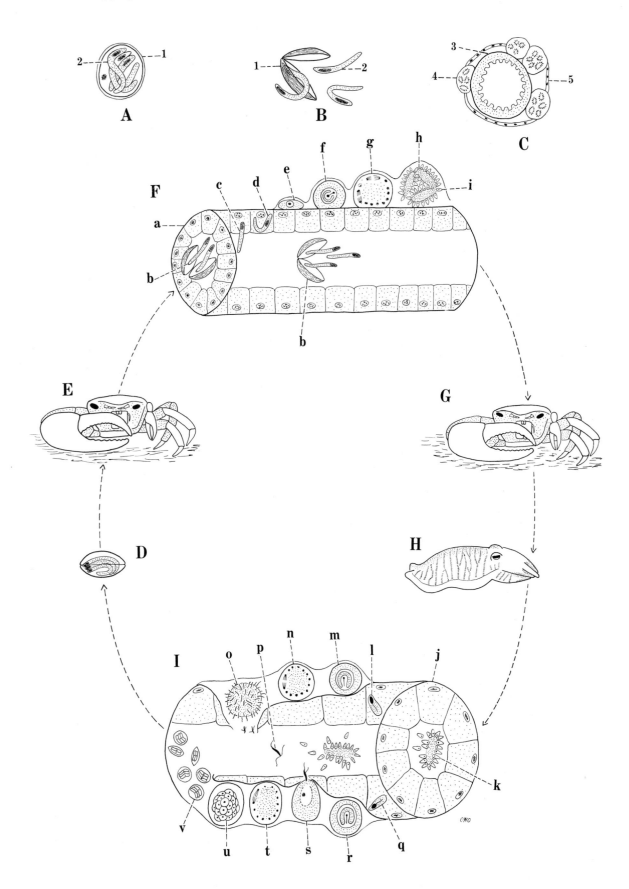

sexual development takes place, with formation of gametocytes and sporocysts containing sporozoites.

Masses of sporocysts developing in the intestinal epithelium of cuttlefish destroy large areas of the mucosa. Ripe sporocysts passed in the feces or contained in necrotic sloughs of epithelium or in the intact intestine are infective to crabs when eaten by them.

Sporozoites released in the intestine of crabs enter the epithelial cells and transform to trophozoites which migrate to the periintestinal tissue to undergo schizogony. First the cytoplasm of each trophozoite divides into large uninucleate masses called cytomeres, thereby increasing the surface area greatly. The nucleus of each cytomere undergoes multiple division, forming many daughter nuclei that migrate to the surface of the cytomeres. Each nucleus with a bit of cytoplasm develops into a merozoite with its complex inner structure. No further development of the merozoites takes place in the crabs.

Infection of cuttlefish occurs when crabs harboring merozoites are eaten. Merozoites liberated in the gut from the periintestinal cysts penetrate the epithelial cells and go to the submucosal tissue, where they transform into gametocytes that produce anisogametes, i.e., tiny microgametes and massive macrogametes. Each microgametocyte produces numerous triflagellated microgametes by the usual process of multiple division of the nucleus. A single macrogamete is formed from each macrogametocyte. Fertilization results in the formation of a zygote. Multiple division of the nucleus produces many sporocysts, each of which forms three sporozoites. The epithelium becomes necrotic because of the vast number of sporocysts in it. Large pieces containing ripe sporocysts are sloughed and passed from the intestine into the water. Crabs eating these sloughs, carcasses of cuttlefish, or fragments of them left by porpoises become infected.

## SELECTED REFERENCES

Dobell, C. (1925). The life-history and chromosome cycle of *Aggregata eberthi* (Protozoa: Sporozoa: Coccidia). Parasitology 17:1–136.

Grassé, P. P. (1953). Traité de Zoologie. Masson et Cie, Paris, Vol. 1, Fasc. 2, p. 795.

Heller, G. (1969). Elektronenmikroskopische Unter-

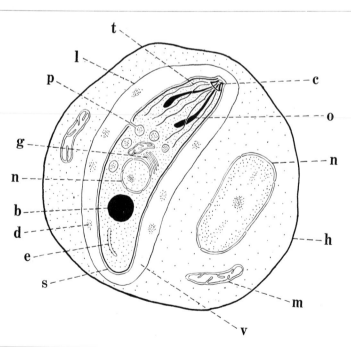

Fig. 11. Schematic sketch of sporozoite in host cell, showing organelles and parasitophorous vacuole. **b,** Paranuclear body; **c,** conoid; **d,** detritus of host cell; **e,** endoplasmic reticulum; **g,** Golgi apparatus; **h,** host cell; **l,** membrane of parasitophorous vacuole; **m,** mitochondrion; **n,** nucleus; **o,** organelle; **p,** micropore with attached vesicle (several are in cytoplasm); **s,** sporozoite; **t,** toxoneme; **v,** parasitophorous vacuole. Adapted from Scholtyseck, 1968, Z. Parasitenk. 31:67.

suchungen an *Aggregata eberthi* aus dem Spiral-
darm von *Sepia officinalis* (Sporozoa: Coccidia) I.
Die Feinstrukturen der Merozoiten, Makrogameten
und Sporen. Z. Parasitenk. 33:44–64.

Heller, G. (1970). Elektronenmikroskopische Untersuchungen an *Aggregata eberthi* aus dem Spiral-
darm *Sepia officinalis* (Sporozoa, Coccidia) II. Die
Entwicklung der Mikrogameten. Z. Parasitenk. 33:
183–193.

Wenyon, C. M. (1926). Protozoology. William
Wood and Co., New York (Reprinted by Hafner
Publishing Co., New York, 1965), Vol. 2, p. 869.

## FAMILY LANKESTERELLIDAE

*Lankesterella* and *Schellackia* of this family are
placed in the Eimeriidae by some authors. Because two hosts, an invertebrate and a vertebrate, are involved to complete the life cycle
under natural conditions and all development is
in the intestinal tissues, certain blood cells, or
endothelial cells of blood vessels of the vertebrate host, they are put in a different family.
Oocysts are without sporocysts but contain eight
or more free sporozoites, depending on the genus.

Species of these genera show an interesting
evolutionary trend in their life cycles toward
that of the Haemosporida by the adoption of a
mode of life in blood cells and the introduction
of a hematophagous invertebrate, albeit in a
passive though ecologically essential role, in the
life cycle. In these respects, they stand between
the basic coccidian type of life cycle in which
sporogony takes place outside the host and the
hemosporidian line in which it occurs inside
the host through which the parasite is actively
transmitted during the feeding process.

### *Schellackia* Reichenow, 1919

Two species of *Schellackia* are considered. They
are *S. bolivari* from the fringed-toed lacertid
lizard (*Acanthodactylus vulgaris*) and the plated
lacertid (*Psammodromus hispanicus*) in the
Mediterranean region of Europe and Africa,
and *S. occidentalis* from rough-scaled lizards
(*Sceloporus occidentalis*) and ground utas (*Uta
stansburiana*) in southern California. An un-
identified species occurs in *Sceloporus undulatus*
and *Anolis carolinensis* in Georgia.

### *Schellackia bolivari*
Reichenow, 1919 (Plate 24)

This Old World representative of the genus
will be used as an example.

**Description.**    Schizonts and microgametocytes develop in the intestinal epithelium and
macrogametocytes in the subepithelial connective tissue of the intestine. Merozoites are of two
sizes. Microgametes are slender and said to have
two flagella, however, it is likely that they have
three as in the case of *Eimeria* and other coccidians that have been studied with the electron
microscope. Spherical oocysts, 15 to 18 $\mu$ in
diameter, located in the intestinal wall contain
eight free sporozoites 8 to 10 $\mu$ long without
sporocysts. Ripe oocysts rupture in the gut wall,
releasing the sporozoites which penetrate erythrocytes that eventually appear in the general
circulation.

**Life Cycle.**    Lizard mites, *Liponyssus
saurarum*, feeding on infected lizards ingest
parasitized blood cells. Upon release from them,
the sporozoites are engulfed by the epithelial
cells of the midgut. Since no development takes
place in the mite, the sporozoites tend to accumulate in them. The mites serve merely as col-
lectors of sporozoites and a vehicle for transporting them back to the lizards when the latter
eat them.

Upon digestion of the infected mites in the
gut of the lizards, the sporozoites are freed.
They enter the epithelial cells of the small intestine and transform to trophozoites that undergo schizogony. Two sizes of merozoites are
formed. The small ones produce another generation of schizonts and the large ones develop into
gametocytes. Those destined to form microgametocytes develop in the intestinal epithelium
along with the schizonts, eventually producing
microgametes by multiple division of the nucleus,
the method characteristic of the coccidians.
Merozoites that form macrogametocytes migrate
to the subepithelial connective tissue, where they
develop into macrogametes, are fertilized by the
microgametes, and become zygotes. This ends
the gametogenous phase of the life cycle.

Sporogony begins and is completed in the
subepithelial tissues where the zygote is formed.
When completed, the thin-walled oocyst contains
eight free sporozoites. The ripe oocysts rupture,
releasing sporozoites that enter the circulation
and penetrate blood cells in a manner similar to
the hemogregarines. In the fringed-toed lizard
(*Acanthodactylus*), the sporozoites enter erythrocytes, while in the plated lizard (*Psammodromus*), they penetrate lymphocytes. These cells
are ingested by lizard mites. Since no development of the sporozoites occurs in the mites, they

## Explanation of Plate 24

A, Schizont from intestinal epithelium. B, Segmenter with fully developed merozoites. C, Oocyst containing sporozoites from submucosal tissue of intestine of lizard. D, Microgametocyte with microgametes from intestinal epithelium of lizard. E, Macrogamete with polar body from submucosa of intestine of lizard. F, Lizard (*Acanthodactylus vulgaris* and *Psammodromus hispanicus*). G, Lizard mite (*Liponyssus saurarum*).

1, Nuclei of schizont; 2, merozoites in schizont; 3, sporozoites in oocyst; 4, microgametes; 5, residual protoplasm of microgametocyte; 6, macrogametocyte; 7, polar body.

a, Infective mite in process of being digested; b, sporozoites escaping from mite in stomach of lizard; c, sporozoite in lumen of small intestine; d, sporozoite entering cell of intestinal epithelium; e, trophozoite; f, schizont; g, mature schizont with fully developed merozoites; h, merozoites enter other intestinal epithelial cells to form gametocytes; i, j, young microgametocytes; k, mature microgametocyte with microgametes; l, merozoite enters submucosa of intestine, where it develops into macrogametocyte; m, young macrogametocyte; n, macrogamete with polar body; o, zygote showing nucleus and microgamete inside; p, young oocyst; q, oocyst with sporozoites (no sporocyst); r, sporozoites enter hepatic portal system (and lymphatics); s, sporozoites enter macrophages or erythrocytes (this one is in blood vessels of liver); t, infected blood cell in pulmonary artery, having gone through right side of heart and entering lungs; u, infected blood cell in left side of heart; v, w, infected blood cells in dorsal aorta and general circulation; x, infected blood cell sucked up by mite; y, z, infected blood cells being phagocytized by intestinal cells.

a', Sporozoites freed in intestinal cells by digestion of blood cells accumulate in great numbers and are source of infection of lizards when eaten by them.

Protozoa adapted from Reichenow, 1921, Arch. Protistenk. 42:179.

---

serve only as accumulators, coming to contain an indeterminate number of parasites.

*Schellackia occidentalis* Bonorris and Ball, 1955, from scaled and ground lizards in southern California, is ingested by the lizard mite *Geckobiella texana*. As in the case of *S. bolivaria*, the mites act as a passive vector. After swallowing infected mites by the lizards, sporozoites 5.6 to 9.8 $\mu$ long by 2.8 to 5.6 $\mu$ in diameter appear in the circulating erythrocytes in 30 to 45 days. Schizonts 8 to 11.2 $\mu$ in diameter with 13 to 30 merozoites occur in clusters in the intestinal epithelium. Thin-walled spherical oocysts, 10 $\mu$ in diameter, appear only in the lamina propria.

### Lankesterella Labbé, 1899

Representatives of this genus occur in birds and amphibians. *Atoxoplasma* Garnham, 1950, is in reality *Lankesterella garnhami* Lainson, 1959, of English sparrows and canaries; possibly other birds as well are common hosts.

The basic morphology and life cycle are similar to those found in *Schellackia bolivari*. Additional investigations on the ultrastructure are instructive in interpreting the relationships among the Sporozoa.

### Lankesterella garnhami Lainson, 1959

This species is very common in nestling and fledgling English sparrows. Being crowded and in close contact with the adult birds during warm weather provides ideal conditions for development of the mites and transmission of the parasites.

**Description.**    Gross morphology and ultrastructure of the sporozoite are similar to that of *Eimeria* (Fig. 11). The pellicle consists of an outer and inner membrane separated by an electron-light middle layer. The anterior end consists of a longitudinally striated apical ring continuing as the conoid. There are about 30 subpellicular fibrils extending from the apical ring to the posterior end of the body. Paired clavate organelles and numerous toxonemes extend caudad from the conoid. Typical double-walled mitochondria and nucleus, Golgi apparatus, and the usual other cytoplasmic inclusions are present.

Similarity of the sporozoites of *Lankesterella*, *Eimeria*, *Sarcocystis*, and *Toxoplasma* indicate a close evolutionary relationship between these groups.

Infected monocytes and lymphocytes show a clear zone around the sporozoites. This is the parasitophorous vacuole which is soon surrounded by a thin membrane.

**Life Cycle.**    The life cycle is similar to that of *Schellackia* with all development taking place in the vertebrate host and the mites serving only as passive vectors. All stages of schizogony are in lymphoid and macrophage cells of the spleen, liver, and bone marrow. Gametocytes

**Plate 24**  *Schellackia bolivari*  115

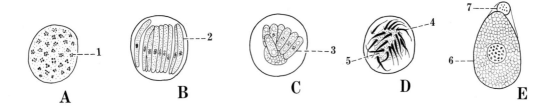

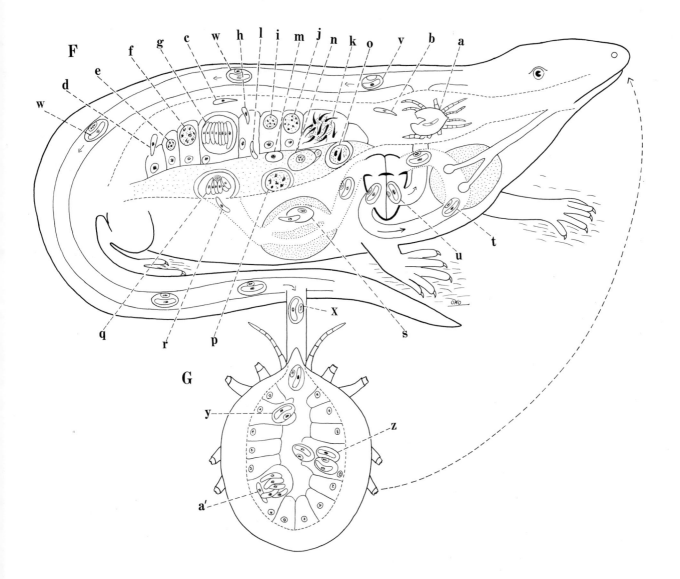

develop and mature in similar cells in the liver, lungs, and kidneys where fertilization occurs. The zygote produces a thin wall to form an oocyst containing many naked, sausage-shaped sporozoites averaging 5 $\mu$ long by 3 $\mu$ in diameter.

Sporozoites liberated from ruptured oocysts enter the circulation and burrow into lymphocytes and monocytes. When the infected cells are ingested by the bloodsucking red poultry mites (*Dermanyssus gallinae*), the blood corpuscles along with the parasites are engulfed by the cells of the midgut, where the sporozoites are retained without damage to them. No development takes place in the mites.

Birds become infected by eating mites containing sporozoites. These are released in the intestine, make their way into the bloodstream and go to the liver, spleen, and bone marrow to enter the lymphoid and macrophage cells, where development ensues.

*Lankesterella hylae* of the European green tree frog (*Hyla caerula*) and *L. minima* from the frog *Rana esculenta* are similar to *L. garnhami* in structure. In *L. hylae*, all development is in the lamina propria of the intestine. Sporozoites enter the erythrocytes which are passive carriers. *Lankesterella minima* develops in endothelial cells of the blood vessels of the visceral organs, especially liver and spleen. Sporozoites penetrate erythrocytes. Transmission is passive by leeches (*Hemiclepsis marginata*).

## EXERCISE ON LIFE CYCLE

Examine blood smears stained with Wright's or Giemsa's blood stains from lizards, especially species of *Sceloporus* and *Uta*, for sporozoites in erythrocytes. Make sections of the small intestine and examine the epithelium for schizonts together with nests of merozoites and for microgametocytes. Look for macrogametocytes, macrogametes, and oocysts containing sporozoites in the subepithelial layers of the gut wall.

Remove the midgut of lizard mites, prepare smears of it, stain, and examine for sporozoites in the epithelial cells. Note that there are no stages of division nor evidence of development in the mites.

To infect parasite-free lizards, draw a drop of blood from a clipped toe of an infected individual into a tuberculin syringe coated inside with an aqueous solution of 0.5 percent NaCl containing 0.068 percent heparin and inject it subcutaneously. Another method of infecting lizards is to feed them mites removed from infected animals. Note when sporozoites appear in the erythrocytes, which should be 30 to 45 days later, depending on the temperature at which the lizards are kept.

Follow similar procedures for studying the life cycle and morphology of *Lankesterella garnhami* in sparrows and their mites, as well as *L. hylae* or *H. minima* in frogs and leeches.

## SELECTED REFERENCES

Bonorris, J. S., and G. H. Ball. (1955). *Schellackia occidentalis* n. sp., a blood-inhabiting coccidian found in lizards in southern California. J. Protozool. 2:31–34.

Doflein, F., and E. Reichenow. (1953). Lehrbuch der Protozoenkunde. 6th ed. Gustav Fischer, Jena, pp. 340, 882.

Garnham, P. C. C., J. R. Baker, and R. G. Bird. (1962). The fine structure of *Lankesterella garnhami*. J. Protozool. 9:107–114.

Jordan, H. B., and M. B. Friend. (1971). The occurrence of *Schellackia* and *Plasmodium* in two Georgia lizards. J. Protozool. 18:485–487.

Lainson, R. (1959). *Atoxoplasma* Garnham, 1950, as a synonym for *Lankesterella* Labbé, 1899. Its life cycle in the English sparrow (*Passer domesticus domesticus* (Linn.)). J. Protozool. 6:360–371.

Lainson, R. (1960). The transmission of *Lankesterella* (*Atoxoplasma*) in birds by the mite *Dermanyssus gallinae*. J. Protozool. 7:321–322.

Reichenow, E. (1920). Der Entwicklungsgang der Hämococcidien *Karyolysus* and *Schellackia* nov. gen. Sitzungsber. Ges. Naturforsch. Fr. Berlin (10), Dez. 1919, pp. 440–447.

Stehbens, W. E. (1966). Obsevations on *Lankesterella hylae*. J. Protozool. 13:59–62.

Stehbens, W. E. (1966). The ultrastructure of *Lankesterella hylae*. J. Protozool. 13:63–73.

Wenyon, C. M. (1926). Protozoology. William Wood and Co., New York (Reprinted by Hafner Publishing Co., New York, 1965), Vol. 2, p. 878.

### Suborder Adeleina

This suborder includes the families Adeleidae and Haemogregarinidae. It is characterized by early fusion of a large macro- and a small microgametocyte within a cyst produced by them in which gametogony and sporogony occurs. The microgametocyte produces few microgametes, one of which enters the macrogamete to form a zygote. The zygote transforms into an oocyst with many sporocysts, each containing two to four sporozoites.

The members of this suborder show affinities to the suborder Eimeriina through the family Adeleidae and to the order Haemosporida through the family Haemogregarinidae.

### FAMILY ADELEIDAE

By developing in the epithelium of the gut and its glands or in the fat bodies of invertebrates and in the kidneys of mammals, the Adeleidae show relationship to the suborder Eimeriina. There is no intermediate host.

### Adelina cryptocerci
Yarwood, 1937 (Plate 25)

This species occurs in the fat bodies as an intracellular parasite of the wood-eating roach *Cryptocercus punctulatus* that lives in rotting logs in northern California, Oregon, and the southern United States.

**Description.**    Oocysts in the fat bodies are spherical, thick-walled, and measure 46 to 51 $\mu$. They contain 5 to 21 sporocysts 10 to 12 $\mu$ in diameter, each with two sporozoites 15 to 17 $\mu$ long by 2 to 3 $\mu$ in diameter and with rounded anterior and narrow posterior ends. The sporozoites lie in close contact with the walls of the sporocyst, being separated by a central residual mass of cytoplasm. Schizonts are of two types. One contains elongate merozoites and the other short, broad ones arranged in two groups, one at each pole and separated by a residual mass of cytoplasm. Mature gametocytes differ greatly in size. The small microgametocyte attaches prematurely along the side of the large macrogametocyte. Later in development, the mature microgametocyte migrates to the anterior end of the fully grown macrogametocyte. The two are surrounded by a membrane and the nucleus of the former divides into four parts each of which becomes a microgamete. One of these enters the macrogamete to produce a zygote. An oocyst with many sporocysts, each with two sporozoites, forms.

The ultrastructure of the stages, such as merozoites and gametocytes, is basically the same as that seen in the *Eimeria*.

**Life Cycle.**    Being in the hemocoel of roaches, the ripe oocysts containing the infective sporozoites are not voided from the body as in the case of intestinal or renal forms. Infection of new hosts takes place when infected dead or incapacitated roaches are eaten by their companions.

In the alimentary canal, the sporozoites are freed, enter the epithelium of the midgut, and migrate through it into the hemocoel, where they enter the fat bodies for further development. In the fat bodies, the sporozoites transform into schizonts which produce up to 40 first generation elongate merozoites, arranged when mature to give the superficial appearance of barrel staves. The merozoites upon release enter other fat bodies and produce a second generation of schizonts whose merozoites are similar in shape to their predecessors but somewhat longer, being 18 to 20 $\mu$ in length instead of 11 to 13 $\mu$.

Merozoites of the second generation produce bipolar schizonts in fat bodies with small blunt merozoites arranged in a group at each end and separated by a mass of residual cytoplasm. Each group of merozoites in the schizont represents a different sex. After escaping into the celom, they develop into gametocytes. In the young macrogametocytes, the nucleus with its large nucleolus is surrounded by a nuclear membrane and the cytoplasm is vacuolated, whereas in the microgametocytes, the nucleus consists of few granules and the cytoplasm is not vacuolated.

Macrogametocytes grow rapidly, quickly exceeding the microgametocytes in size. The two join laterally and the microgametocyte moves to the anterior end of the macrogametocyte; a single membrane encloses both of them. Gametogony begins with nuclear division of the microgametocyte which undergoes two divisions, producing four microgametes and the formation of a macrogamete. One minute microgamete enters a massive macrogamete to produce the zygote. The remaining three microgametes retain their position on the outer surface of the zygote. A new membrane appears over the first, enclosing the free microgametes and cytoplasmic residuum of the microgametocyte. Thus a double wall is formed. Later, a third membrane forms around the zygote within the two-layered cyst wall.

Sporogony begins by repeated division of the nucleus of the zygote and terminates with the appearance of sporozoites. Each newly formed nucleus of the zygote is surrounded by a portion of cytoplasm to produce a uninucleate sporont enclosed by a thin membrane. Development of the sporont into a sporocyst is completed when the nucleus has divided once and each part is

## Explanation of Plate 25

A, Ripe oocysts filled with sporocysts, each with two sporozoites. B, Free sporozoite. C, Trophozoite developed from sporozoite, showing nucleus with ring of chromatin granules and central nucleolus. D, Schizont with four nuclei. E, Multinucleate schizont, showing nuclei in telophase, each with eight chromosomes. F, Trophozoite developed from merozoite, showing large central karyosome. G, Merozoites arranged like staves of a barrel. H, Free merozoite of first generation. I, Bipolar division of schizont, with short merozoites arranged at each pole (these become gametocytes). J, Growing macrogametocyte. K, Macro- and microgametocytes in early stage of union. L, Large macrogametocyte with two associated microgametocytes. M, Gametocytes enclosed in cyst; nucleus of microgametocyte undergoing division to form microgametes. N, Young oocyst with uninucleate sporoblasts. O, Older but still immature oocyst with binucleate sporoblasts which will produce two sporozoites each (see A). P, Wood-eating roach (*Cryptocercus punctulatus*).

1, Oocyst wall; 2, sporocyst wall; 3, sporozoite; 4, nucleus; 5, dividing nucleus in telophase, showing eight chromosomes; 6, long slender merozoites; 7, short merozoites that will develop into gametocytes; 8, residual cytoplasm; 9, macrogametocyte; 10, microgametocyte; 11, division of nucleus of microgametocyte in formation of microgametes; 12, macrogamete; 13, uninucleate sporoblasts; 14, binucleate sporoblasts that produce sporozoites (see 3); 15, fat body of roach.

a, Ripe oocyst in stomach, acquired by eating an infected wood roach; b, ruptured oocyst allowing sporozoites to escape in intestine; c, sporocysts free in intestine; d, ruptured sporocyst allowing sporozoites to escape in intestine whence they penetrate the intestinal epithelium; e, sporozoite entering hemocoel, having migrated through the intestinal wall; f, sporozoite entering fat body; g–i, first schizogony; g, young trophozoite with small anterior end; h, schizont with dividing nuclei; i, bundles of merozoites arranged in barrel-stave fashion; j, merozoite from first generation of schizonts entering fat body; k, trophozoite with narrowed anterior end and nucleus with halo of chromatin dots around nucleolus; l, schizont with dividing nuclei; m, bundles of merozoites arranged in barrel-stave fashion; n, merozoites escaping from schizont; o, merozoite entering fat body; p–r, third schizogony; p, trophozoite with narrow anterior end and with nucleolus surrounded by ring of chromatin dots; q, schizont with dividing nuclei; r, bipolar schizont with a group of small merozoites at each end, separated by a residual mass of protoplasm; these will become gametocytes; s–u, formation of gametocytes and their union; s, undifferentiated gametocytes in fat body; t, male and female gametocytes; u, union of male and female gametocytes; v, formation of four microgametes from male gametocyte at end of female gametocyte and appearance of a second membrane to form the gametocyst; w, migration of one microgamete into macrogamete; x, union of two nuclei with appearance of eight pairs of chromosomes to form the zygote with additional membranes; y, reduction of number of chromosomes; z, a′, b′, sporogony; z, division of nuclei to form sporoblasts.

a′, Formation of sporoblasts; b′, mature oocyst with sporocysts each containing two sporozoites.

Figures A–O redrawn from Yarwood, 1937, Parasitology 29:37.

---

clothed in a bit of cytoplasm to form two sporozoites.

Other representatives of the Adeleidae whose life cycles are known to resemble those of *Adelina* are 1) *Adelea ovata* Schneider from the centipede *Lithobius forficatus*, *A. mesnili* Pérez from the webbing clothes moth *Tineola bieseliella* and the Mediterranean flour moth *Ephestia kuehniella*; 2) *Klossia helicana* Schneider in the kidneys of land snails of the genera *Helix* and *Vitrina* and slugs of the genus *Limax*; and 3) *Orcheobius herpobdella* Schuberg and Kunze from the testes of the freshwater leech *Erpobdella atomaria*. Additional examples are given by Doflein and Reichenow and Kudo.

An undetermined species of *Adelina* occurs in cultures of dermestid beetles (*Trogoderma parabile*), causing high mortality among the larvae, pupae, and adults. The parasite is not incorporated in the eggs of the beetles. The occurrence of such a pathogenic protozoan leads to the question of whether it might be used as a means of biological control for insect pests.

## EXERCISE ON LIFE CYCLE

*Adelina tribolii* of the cosmopolitan red flour beetle *Tribolium ferrugineum* is excellent for study because of the universal availability of the host and high incidence of infection.

First and second instar larvae are unlikely to be infected, but all subsequent stages harbor the oocysts. Live material may be obtained by teasing out the fat bodies in Ringer's solution

**Plate 25** *Adelina cryptocerci* 119

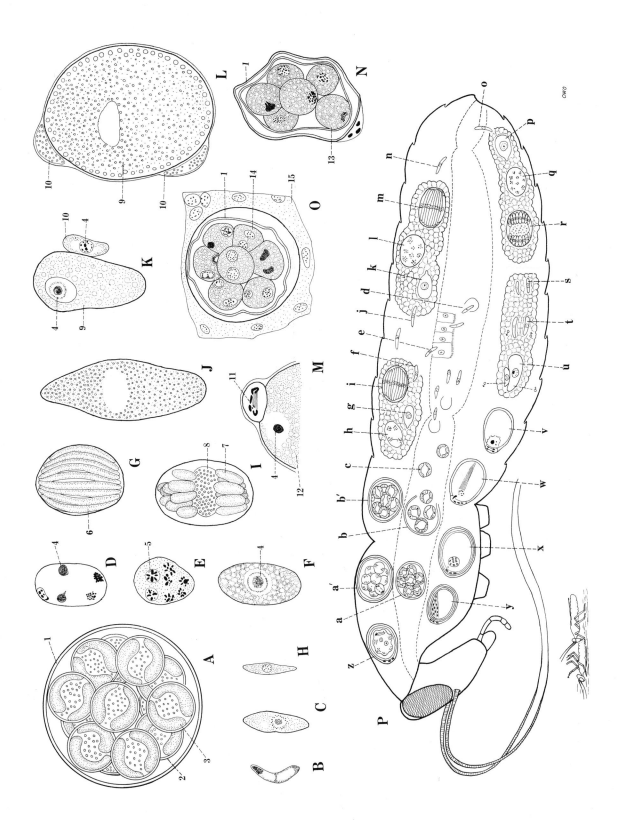

and placing them on slides with the coverslips sealed by means of petroleum jelly placed along the edges and spread evenly with a hot needle. The parasites will survive for up to 24 hours in such preparations, during which time the various stages may be studied.

In order to get sporozoites to escape on slides, mature oocysts should be placed in the digestive juices of intestines dissected from beetles. Digestive enzymes from different kinds of insects should be tried, as well as artificial ones such as pepsin and trypsin. Both sporocysts and sporozoites may be liberated through mechanical pressure on the coverslip.

Means of infection have not been observed in this species but it is most likely to occur through cannibalism among the beetles. Experiments should be designed to demonstrate the means by which infection occurs. Sections of the intestine should be made to follow liberation of the sporozoites and the route taken by them in passing to the hemocoel. See Bhatia, Riley and Krogh, and Loschiavo for details.

## SELECTED REFERENCES

Bhatia, M. L. (1937). On *Adelina tribolii*, a coccidian parasite of *Tribolium ferrugineum* F. Parasitology 29:239–246.

Doflein, F., and E. Reichenow. (1953). Lehrbuch der Protozoenkunde. 6th ed. Gustav Fischer, Jena, p. 825.

Kudo, R. R. (1966). Protozoology. 5th ed. Charles C Thomas, Springfield, Ill., p. 701.

Loschiavo, S. R. (1969). A coccidian pathogen of the dermestid *Trogoderma parabile*. J. Invert. Pathol. 14:89–92.

Riley, W. A., and L. Krogh. (1927). A coccidian parasite of the flour beetle, *Tribolium confusum* and *T. ferrugineum*. J. Parasitol. 13:224.

Tuzet, O. (1970). Recherches ultrastructurales sur les mérozoïtes et les gamontes de la coccidie *Adelina dimidiata* Schneider parasite du myripode Chilopoda *Scolopendra cingulata* Latrielle. Natur. Can. 97:369–386.

Yarwood, E. A. (1937). The life cycle of *Adelina cryptocerci* sp. nov., a coccidian parasite of the roach *Cryptocercus punctulatus*. Parasitology 29:370–390.

Žižka, Z. (1969). The fine struture of the macrogametocytes of *Adelina tribolii* Bhatia, 1937 (Eucoccidia, Telosporea) from the fat body of the beetle *Tribolium castaneum* Hbst. J. Protozool. 16:111–120.

## FAMILY HAEMOGREGARINIDAE

The family consists of four genera: *Haemogregarina*, *Hepatozoon*, *Karyolysis*, and *Klossiella*. Representatives of these genera occur in the blood of all classes of vertebrates. All of the genera except *Klossiella* require two hosts for development and completion of the life cycle.

The Haemogregarinidae show kinship to the order Haemosporida in that both have schizogony in blood cells or visceral organs, such as the liver, lungs, spleen, kidneys, bone marrow, or wall of the intestine, and both gametogony and sporogony in the gut or hemocoel of bloodsucking invertebrates (Plate 29).

Stages in the circulating erythrocytes include sporozoites, gametocytes, or schizonts which may be uninucleate, multinucleate, or even more developed and breaking up into merozoites. Development of schizonts usually occurs in the internal organs where the hypertrophied erythrocytes are trapped in the capillaries.

### Haemogregarina stepanowi
Danilewsky, 1885 (Plate 26)

This species is representative of the group of coccidians whose schizogenous phase of the life cycle occurs in the erythrocytes of reptiles, amphibians, and fishes, and the gametogenous and sporogenous phases in leeches. This type of life cycle is more advanced than that of *Schellackia bolivari* where all development is in the vertebrate host, and the mites serve only as a passive receptacle for collecting and transmitting sporozoites when eaten by the lizards.

*Haemogregarina stepanowi* occurs worldwide in freshwater turtles. Its life cycle was described from the European water tortoise *Emys orbicularis* and the leech *Placobdella catenigra*. Geographic distribution of the parasite does not necessarily coincide with that of the vertebrate host but is dependent on that of the invertebrate host, which varies considerably.

**Description.**    Identification on the basis of the erythrocytic stages alone is both difficult and hazardous. It must be supported by the species of hosts infected, the location in the vertebrate host where schizogony occurs, and in the invertebrate host where gametogony and sporogony take place.

Vermicular sporozoites enter circulating erythrocytes and transform into uninucleate trophozoites. In the course of development, each one becomes U-shaped with arms of unequal size which gradually fuse to form a large oval body. With the first nuclear division, they become schizonts with two generations. The hypertrophied erythrocytes containing the mature schizonts lodge in the bone marrow, where schizogony takes place. Individuals of the first generation are large in size and designated as macroschizonts. Each macroschizont produces 13 to 24 large merozoites. Upon liberation, these merozoites enter new erythrocytes and develop into small schizonts, known as microschizonts, of the second generation. Each microschizont produces six small merozoites which become gametocytes. They penetrate erythrocytes which appear in the circulating blood. Some of the gametocytes transform into macrogametocytes with a small nucleus, while others become microgametocytes with a large nucleus and darkly staining bands near one end. Gametogony and sporogony are completed with production of sporozoites in the intestinal lumen of leeches.

The ultrastructure of the various stages that have been studied is similar to that of the *Eimeria.*

**Life Cycle.** Sporozoites injected into the blood of tortoises by feeding leeches initiate the schizogenous phase of the life cycle. They penetrate erythrocytes and begin growth, eventually forming U-shaped vermicular trophozoites. As growth proceeds, the arms of the trophozoite fuse, forming large ovoid schizonts known as macroschizonts.

Hypertrophied erythrocytes containing these schizonts become lodged in the capillaries of the bone marrow, where schizogony is completed with the production of 13 to 14 large merozoites. These escape into the bloodstream and enter uninfected erythrocytes to produce the second generation of small schizonts known as microschizonts. Each one produces about six little merozoites which are in reality gametocytes, thus terminating the schizogenous phase. Upon penetrating red blood cells, these small merozoites differentiate into micro- and macrogametocytes. This is the beginning of gametogony. No further development takes place in the tortoises.

When blood cells containing the gametocytes are sucked up by leeches and enter the intestine, they survive and undergo gametogony and sporogony with the ultimate production of sporozoites. Upon release from the erythrocytes in the intestine of the leeches, the freed microgametocytes and macrogametocytes fuse in pairs and produce a thin membrane that encloses both of them. The macrogametocyte becomes a large oval body and the microgametocyte a small oval one.

The nucleus of the microgametocyte divides twice, forming four minute microgametes, while that of the macrogametocyte undergoes meiotic division, discarding one nucleus, to become a single large macrogamete. Fertilization of the macrogamete by one of the microgametes results in a zygote which upon development produces an oocyst containing eight naked sporozoites. When mature, the oocyst ruptures, releasing the sporozoites in the intestinal lumen. They migrate through the epithelium and into the blood vessels, eventually appearing in the dorsal artery, and are distributed in the body. Vigorous muscular activity of the proboscis by feeding leeches ruptures some of the surrounding blood vessels, allowing the sporozoites to escape into the proboscis. While feeding on turtles, these leeches inject sporozoites into the bloodstream. They penetrate erythrocytes and transform into trophozoites to begin the cycle anew.

The life cycle of *H. nicoriae* from the Ceylon Lake tortoise *Nicoria trijuga* and the leech *Ozobranchus shipleyi* is similar to that of *H. stepanowi*. Recent studies on other species of *Haemogregarina* indicate similar life cycles.

## EXERCISE ON LIFE CYCLE

Hemogregarines commonly parasitize fish, amphibians, reptiles, and birds. Trophozoites and gametocytes are in the erythrocytes of circulating blood. Schizonts with merozoites are in erythrocytes of circulating blood and the internal organs such as the liver and bone marrow. Zygotes and oocysts with sporozoites are in the gut of leeches.

Smears from circulating blood and impressions from the liver and bone marrow prepared and stained by standard procedures are satisfactory for detecting and recognizing infection of hemogregarines. Examine the intestine of leeches that have fed on infected turtles for sexual stages together with oocysts containing naked sporozoites.

## SELECTED REFERENCES

Baker, J. R., and R. Lainson. (1967). The fine structure of the gametocytes of an adeleine haemogregarine. J. Protozool. 14:233–238.

## Explanation of Plate 26

A, U-shaped developing trophozoite. B, Trophozoite with arms of U fused to form ovoid body. C, Nucleus breaks into fragments to form schizont. D, Macroschizont with large merozoites of first generation. E, Microschizont with small merozoites of second generation. F, Macrogametocyte. G, Microgametocyte. H, Water tortoise (*Emys orbicularis*), vertebrate host. I, Leech (*Placobdella catenigera*), invertebrate host.

1, Arms of growing trophozoite; 2, nucleus of trophozoite; 3, erythrocyte; 4, nucleus of erythrocyte; 5, small multiple nuclei; 6, large merozoites of first generation; 7, small merozoites of second generation; 8, macrogametocyte; 9, microgametocyte.

a, Leech feeding on hind leg of tortoise; b, sporozoite injected into bloodstream; c, sporozoite entering erythrocyte; d, sporozoite within erythrocyte circulated through bloodstream; e, young trophozoite; f, older, U-shaped trophozoite; g, bone marrow (represented by stippled area); h, mature trophozoite in erythrocyte entering bone marrow; i, macroschizont with several nuclei; j, segmenter with large first generation merozoites; k, large merozoites escape from erythrocyte and enter bloodstream (represented by unstippled area); l, merozoite entering erythrocyte in blood to initiate second generation of schizonts; m, n, U-shaped trophozoites; o, mature trophozoite in bone marrow (stippled area); p, microschizont; q, smaller but fewer second generation merozoites; r, rupture of erythrocyte and escape of merozoites into bloodstream to become gametocytes; s, second generation merozoites free in blood initiate gametogenous cycle; t, merozoite entering erythrocyte to form gametocytes; u, v, growing gametocytes in circulation; w, mature microgametocyte; x, mature macrogametocyte in tortoise; y, gametocytes sucked up and swallowed by leech definitive host.

a', b', macro- and microgametocytes in erythrocytes in stomach of leech; c', macro- and microgametocytes free in stomach; d', association of micro- and macrogametocytes; e', microgametocyte produces four microgametes; f', fertilization of macrogamete; g', zygote with fragment of microgametocyte attached; h'–j', growth and division of nucleus of zygote; k', oocyst with eight sporozoites; l', oocyst ruptures and sporozoites are released in stomach; m', sporozoites enter dorsal blood vessel; n', sporozoites in dorsal blood vessel; o', sporozoite enters proboscis of leech; p', sporozoites injected into tortoise by feeding leech, and cycle begins anew (a) in vertebrate host.

Figures of protozoa adapted from Reichenow, 1910, Arch. Protistenk. 20:251.

Ball, G. H. (1958). A haemogregarine from a water snake, *Natrix piscator* taken in the vicinity of Bombay, India. J. Protozool. 5:274–281.

Ball, G. H. (1967). Some blood sporozoans from East African reptiles. J. Protozool. 14:198–210.

Clark, G. W., and J. Bradford. (1969). Blood parasites of some reptiles of the Pacific Northwest. J. Protozool. 16:578–581.

Grassé, P. P. (1953). Traité de Zoologie. Masson et Cie, Paris, Vol. 1, Fasc. 2, pp. 749, 754.

Helmy Mohammed, A. H., and N. S. Mansour. (1966). Development of *Haemogregarina boueti* in the toad *Bufo regularis*. J. Protozool. 13:259–264.

Hull, R. H., and J. H. Camin. (1960). Haemogregarines in snakes: The incidence and identity of erythrocytic stages. J. Parasitol. 46:515–523.

Lehmann, D. L. (1959). The description of *Haemogregarina boyli* n. sp. from the yellow-legged frog, *Rana boyli boyli*. J. Parasitol. 45:198–203.

Mansour, N. S., and A. S. Helmy Mohammed. (1966). Development of *Haemogregarina pestanae* in the toad *Bufo regularis*. J. Protozool. 13:265–267.

Marquardt, W. C. (1966). *Haemogregarines* and *Haemoproteus* in some reptiles of southern Illinois. J. Parasitol. 52:823–824.

Reichenow, E. (1910). *Haemogregarina stepanowi*. Die Entwicklungsgeschicte einer Hämogregarine. Arch. Protistenk. 20:251–350.

Robertson, M. (1910). Studies on haematozoa. No. 2. Notes on the life-cycle of *Haemogregarina nicoriae* Cast. and Willey. Quart. J. Microsc. Sci. n. s. (220), 55:741–762.

Saunders, D. C. (1955). The occurrence of *Haemogregarina bigemina* Laveran and Mesnil and *H. achiri* n. sp. in marine fish from Florida. J. Parasitol. 41:171–176.

Stehbens, W. E., and M. R. L. Johnston. (1968). Cystic bodies and schizonts associated with a haemogregarine (Sporozoa) parasitic in *Gehyra variegata* (Reptilia: Gekkonidae). J. Parasitol. 54:1151–1165.

Wang, C. C., and S. H. Hopkins. (1965). *Haemogregarina* and *Haemoproteus* (Protozoa, Sporozoa) in blood of Texas turtles. J. Parasitol. 51:682–683.

Wenyon, C. M. (1926). Protozoology. William Wood and Co., New York (Reprinted by Hafner Publishing Co., New York, 1965), Vol. 2, p. 1095.

**Plate 26**   *Haemogregarina stepanowi*                                           123

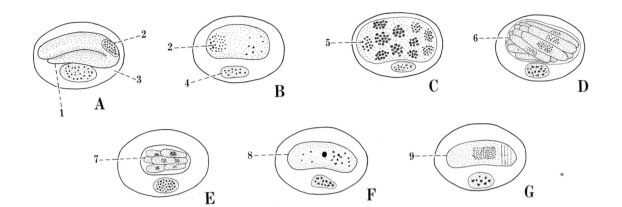

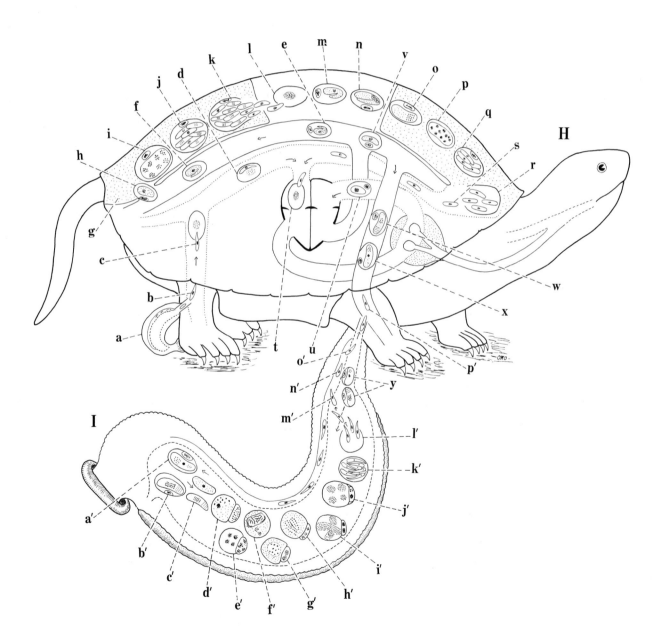

124    ANIMAL PARASITES

## Explanation of Plate 27

A, Section of liver of rat showing developing schizonts. B, Section of intestine of mite showing zygotes. C, Section of mite showing oocyst containing sporocysts and sporozoites. D, Rat vertebrate host. E, Rat mite (*Echinolaelaps echidninus*) invertebrate host.

1–4, Developing schizonts in numerical order of progression; 5, zygotes; 6, oocyst; 7, sporocyst filled with sporozoites; 8, sporozoites free in oocyst.

a, Infected mite being digested in stomach and intestine of rat with release of sporocysts; b, ruptured sporocyst freeing sporozoites; c, sporozoite passing through intestinal epithelium into blood vessels; d, sporozoite in hepatic portal vein; e, sporozoite passing from blood vessel into hepatic cell; f, early schizont; g, nuclei gathering near ends of schizont; h, mature schizont with merozoites assembled near ends; i, schizont ruptures, freeing merozoites; j, merozoites attack other hepatic cells, initiating several succeeding generations of schizonts and merozoites; k, after about third generation of schizonts, gametocytes are produced that are engulfed by mononuclear leukocytes, usually in the liver; l, infected leukocytes entering general circulation from heart after having passed through lungs; m, infected leukocyte being sucked up by mite; n, infected leukocytes in stomach of mite being digested with release of gametocytes; o, gametocytes unite; p, macrogamete enlarges and partially encircles microgamete; q, zygote; r, ookinete leaves stomach and enters body tissues; s, ookinete grows; t, u, developing sporont; v, nucleus multiplies and nuclei assemble over surface of sporont; w, budding on surface of sporont, forming sporoblasts; x, oocyst containing developing sporoblasts; y, oocyst with sporoblasts whose nuclei are dividing and arranging themselves at ends; z, oocyst with sporocysts each containing a number of sporozoites.

Adapted from Miller, 1908, U.S. Pub. Health Serv. Hyg. Lab. Bull. 46.

## *Hepatozoon* Miller, 1908

Members of the genus *Hepatozoon* parasitize reptiles, birds, and mammals. Gametocytes usually appear in mononuclear leukocytes, but *H. fusifex* is reported from erythrocytes in boas where the cells are swollen and attenuated at both ends similar to the condition seen in *Leucocytozoon*. Gametocytes of other genera of hemogregarines and hemosporidians are in erythrocytes.

Two hosts are required for development. Schizogony occurs in various organs such as the intestinal wall, liver, lungs, spleen, bone marrow, and endothelial cells of vertebrates, according to the requirements of the individual species of parasite.

There are two or more generations of schizonts. Merozoites of the first generation develop into schizonts of the succeeding one. The last generation of schizonts produces merozoites that develop into gametocytes in mononuclear leukocytes in most species. Both gametogony and sporogony take place in the hemocoel of bloodsucking invertebrates such as insects, mites, ticks, and probably leeches. Microgametes are said to be aflagellate but some authors report the presence of two flagella.

### *Hepatozoon muris*
(Balfour, 1905) (Plate 27)

This species is a parasite of rats and mice. Later, it was erroneously named *H. perniciosum*, indicating the severe pathology produced in rats. Schizogony occurs in liver cells and possibly in endothelial cells of the blood vessels or wandering Kupffer cells of endothelial origin. Gametocysts appear in mononuclear leukocytes. Gametogony and sporogony take place in the hemocoel of the rat mite *Echinolaelaps echidninus*.

The life cycle of this hemogregarine retains certain definite coccidian characteristics on the one hand and portrays hemosporidian features on the other. It is coccidian in having schizogony in a visceral organ, the liver, in the vertebrate host. It is hemosporidian in that the gametocytes appear in blood cells and both gametogony and sporogony take place in mites. Few microgametes are produced. Transmission is primitive in nature. It occurs only when infected mites are swallowed by the vertebrate host instead of being inoculated by feeding arthropods, as happens with the more advanced Haemosporida.

**Description.** Schizonts in hepatic cells are ovoid bodies enclosed in a cyst, the oocyst, measuring up to 28 by 35 $\mu$ in size. Mature oocysts contain many sporocysts, each with 12 to 20 vermiform merozoites usually arranged in two groups, one at each pole. Gametocytes in mononuclear leukocytes are somewhat quadrilateral in outline with rounded ends. The nucleus consists of skeins or bands of chromatin. A delicate parasitophorous cyst surrounds each

**Plate 27**   *Hepatozoon muris*                                                                    125

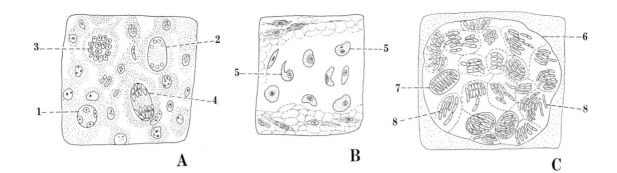

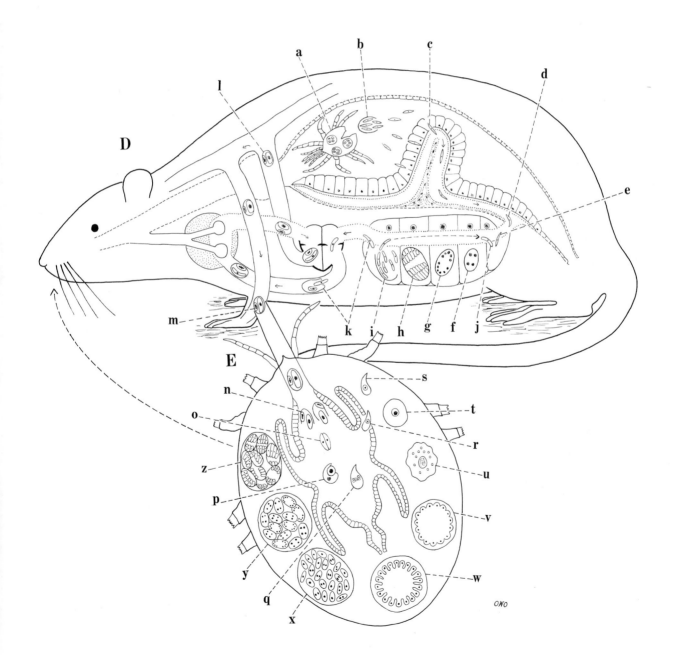

parasite, separating it from the cytoplasm of the leukocyte.

Mature oocysts 200 to 250 $\mu$ in diameter, containing a number of sporocysts 25 by 30 $\mu$ in size, fill the hemocoel of the mites. Each sporocyst contains 12 to 24 vermicular sporozoites averaging 14 by 5 $\mu$ in size.

**Life Cycle.** Infection of rats is initiated when parasitized mites are swallowed. The multitude of sporozoites released from the mites by action of the duodenal digestive juices becomes very active. They migrate through the intestinal epithelium and into the blood vessels of the hepatic portal vein. Upon arriving in the liver, most of them escape from the capillaries and enter hepatic and Kupffer's cells. Some remain in the capillaries and appear in the endothelial cells. Schizogony takes place in the liver cells.

Inside a liver cell, each sporozoite transforms into a spherical, uninucleated trophozoite about 10 $\mu$ in diameter surrounded by a cavity. After a short period of growth, the nucleus divides into two. The first nuclear division initiates the schizogenous stage and the parasite becomes a schizont.

Schizonts are oval bodies of two sizes, each producing merozoites of different dimensions. The first generation of schizonts is large and known as macroschizonts. When mature, they are 28 by 35 $\mu$ in size and are surrounded by a thin membranous cyst. Reproduction begins by repeated division of the nucleus, resulting in 12 to 20 daughter nuclei located in buds of cytoplasm over the surface of the ends of the schizont. After a period of development, each bud pinches off to form a large, motile merozoite beginning day 4 and ending by day 5. These large merozoites are fully mature when released. They disperse in the liver where the majority enter other hepatic cells to develop into the second generation of schizonts. These are small and known as microschizonts. Other merozoites enter capillaries and simply circulate in the body without further growth.

Development of the microschizonts is similar to that of the macroschizonts, requiring about the same time. Small merozoites are produced and are fewer in number than the large ones. These small merozoites are in reality gametocytes of uniform shape and size. Upon being released from the sporocysts, they do not penetrate liver cells but intermingle with leukocytes in the liver sinusoids.

The gametocytes are soon engulfed by the mononuclear leukocytes, which enter the bloodstream and appear throughout the circulatory system. More than two generations of schizonts may appear. The final one in the series produces the gametocytes. The appearance of them marks the end of schizogony, the asexual phase of reproduction.

Development of the gametocytes proceeds only in the stomach and hemocoel of the rat mites. When parasitized mononuclear leukocytes are sucked up by mites, they are digested in 6 to 24 hours, liberating the vermicular gametocytes in the stomach. Two similar gametocytes pair within 24 hours, adhering along one side. The nuclei become vesicular. One of the gametocytes grows rapidly, becoming longer and larger. It is the macrogamete which encircles the smaller microgamete and the two finally fuse to form an elongate, motile zygote, or ookinete, 10 by 25 $\mu$ in size. This completes the gametogenous phase of the life cycle. The ookinete matures in 48 hours and has doubled in size, being 25 to 50 $\mu$. After 3 to 4 days, only a few ookinetes remain in the stomach. Some have been voided with the feces but most of them have migrated through the stomach wall into the hemocoel. In the new location, sporogony begins. They grow in size and become oval or spherical sporonts, each with a giant nucleus 10 $\mu$ in diameter.

When mature, the sporonts are 60 to 75 $\mu$ in diameter and by day 4 or 5 of age each one is enclosed in a large, loose membrane, the oocyst, which is 100 to 150 $\mu$ in diameter. Nuclear division begins and by day 6 or 7 the surface of the sporont within the oocyst is covered with 50 to 100 or more bud-like projections, each enclosing one of the daughter nuclei. The buds break away from the cytoplasmic residue to form uninucleate sporoblasts 10 by 15 $\mu$ in size. Nuclear division in each sporoblast results in 10 to 20 daughter nuclei that migrate to the surface of the cytoplasmic mass, which by this time is enclosed in a thin membrane to form the sporocyst. Each daughter nucleus is enclosed in a rod-shaped bit of cytoplasm and becomes a fully developed crescentic sporozoite 14 by 5 $\mu$ in size with the posterior end the narrower. This terminates the sexual phase of development. Ripe oocysts containing sporozoites appear in mites in 10 to 14 days after feeding on infected rats. The oocysts may fill the hemocoel of the mites, imparting a whitish and swollen appearance to them. No further development takes place in the mites. When infected individuals are swallowed by rats, sporozoites reaching the liver begin the cycle anew.

The life cycles where known of other species of *Hepatozoon* are basically similar to that of *H. muris. Hepatozoon canis* of dogs develops in the tick *Rhipicephalus sanguineus; H. criceti* of hamsters in *Liponyssus arcuatus; H. gerbilli* of gerbils in the louse *Haematopinus stephensi; H. balfouri* of the jumping mouse *Jaculus jaculus* in the flea *Pulex cleopatra* and the mite *Haemolaelaps aegyptius; H. pettiti* of crocodiles in the tsetse *Glossina palpalis; H. mesnili* of the gecko in the mosquito *Culex fatigans; H. mauritanicum* of the turtle *Testudo mauritanica* in the tick *Hyalomma aegyptium;* and *H. triatomae* of the lizard *Tupinambus teguizin* in the bug *Triatoma rubrovaria* (experimental); *H. rarefaciens* of snakes in the mosquitoes *Culex tarsalis, Anopheles albimanus,* and *Aedes sierrensis; H. griseiscuri* of squirrels in the mites *Echinolaelaps echidninus* and *E. ambulans;* and *H. breinli* of varanid lizards in *Culex fagitans.* Common vertebrate hosts of species of *Hepatozoon* may be found in the references.

Another aspect of the life cycle of species of *Hepatozoon* appears in *H. domerguei* and possibly all species. Schizonts, after several successive generations, give rise to gametocytes in the usual manner and to pseudocysts by endogenesis. The role of the pseudocysts is similar to that of oocysts, namely, the production of sporozoites. Predatory animals become infected by eating mature pseudocysts. This type of endogenetic reproduction and transmission already known for *Toxoplasma* is now recognized in other coccidians.

**Pathology.**    Rats dying from hepatozoonosis muris show multiple manifestations from infection. Grossly, the liver is enlarged with the thin edges rounded. It is dull yellow or mottled in color. Sections show fatty changes and large areas of destroyed liver cells. In chronic infections, the rats are emaciated and anemic. Muscles and mucous membranes are blanched. The spleen is enlarged five or more times the normal size. Mesenteric glands, the small intestine, and lungs show evidence of hemorrhage.

In jerboas, a single mite infected with *H. balfouri* produces a fulminating, fatal infection. Acute infections in them result in tremendous destruction of hepatic cells and extensive hyperplasia of the spleen. Heavy infection in mites often terminates in their death.

### EXERCISE ON LIFE CYCLE

Infected rats, either wild or white ones, harboring mites provide excellent material for observing the developmental stages of the parasite and for conducting experiments on its life cycle.

The gametocytes appear only in the leukocytes and may be seen in blood smears stained with Wright's or Giemsa's blood stain. The various stages of the schizonts may be found in smears or sections made from the liver.

Mites from infected rats should be dissected in saline to obtain the stages of sporogony occurring in the body cavity. Isogametes and ookinetes may be found in the stomach by direct examination of contents or by making smears and staining them with blood stains.

For life history studies, one must have rats free of *Hepatozoon,* as proved by parasite-free blood smears taken daily over a period of a week, and infected mites as shown by dissections. Feed infected mites to an uninfected rat and examine blood smears from it daily until parasites appear.

Mites may be infected experimentally by allowing them to feed on heparinized blood from parasitized rats. Ten clean mites are placed in a small vial with the end closed with bolting silk. The silk is pressed into one end of the vial to a depth of about 4 mm and held in place by a rubber or plastic ring. Blood is placed on the bolting silk and the whole is kept at about 95°F. Mites of various species feed readily on blood that is in contact with the silk.

### SELECTED REFERENCES

Ball, G. H., J. Chao, and S. R. Telford, Jr. (1967). The life history of *Hepatozoon rarefaciens* (Sambon and Seligmann, 1907) from *Drymarchon corais* (Colubridae) and its experimental transfer to *Constrictor constrictor* (Boidae). J. Parasitol. 53:897–909.

Ball, G. H., J. Chao, and S. R. Telford, Jr. (1969). *Hepatozoon fusifex* sp. n., a hemogregarine from *Boa constrictor* producing marked morphological changes in infected erythrocytes. J. Parasitol. 55:800–813.

Booden, T., J. Chao, and G. H. Ball. (1970). Transfer of *Hepatozoon* sp. from boa constrictor to a lizard, *Anolis carolinensis,* by mosquito vectors. J. Parasitol. 56:832–833.

Clark, G. M. (1958). *Hepatozoon griseiscuri* n. sp.; a new species of *Hepatozoon* from the grey squirrel (*Sciurus carolinensis* Gmelin, 1788), with studies on the life cycle. J. Parasitol. 44:52–63.

Clark, G. M., and B. Swinehart. (1966). Blood parasitism in cliff swallows from the Sacramento Valley. J. Protozool. 13:395–397.

Cross, H. F. (1954). Feeding tests with blood sucking mites on heparinized blood. J. Econ. Entomol. 47:1154–1155.

Doflein, F., and E. Reichenow. (1953). Lehrbuch der Protozoenkunde. 6th ed. Gustav Fischer, Jena, p. 832.

Furman, D. P. (1966). *Hepatozoon balfouri* (Laveran, 1905): sporogonic cycle, pathogenesis, and transmission by mites to jerboa host. J. Parasitol. 52:373–382.

Hoogstraal, H. (1961). The life cycle and incidence of *Hepatozoon balfouri* (Laveran, 1905) in Egyptian jerboas (*Jaculus* spp.) and mites (*Haemolaelaps aegyptius* Keegan, 1956). J. Protozool. 8: 231–248.

Landau, I., A. G. Chabaud, J.-C. Michel, and E. R. Brygoo. (1970). Données nouvelles sur le cycle évolutif d'*Hepatozoon domerguei*: Importance de l'endogenèse: analogies avec d'autres cycles de coccidies. C. R. Acad. Sci. Ser. D. 271:1679–1681.

Mackerras, M. J. (1962). The life history of a *Hepatozoon* (Sporozoa: Adeleida) of varanid lizards in Australia. Austr. J. Zool. 10:35–44.

Miller, W. W. (1908). *Hepatozoon perniciosum* (N. G., N. Sp.); a haemogregarine pathogenic for white rats; with a description of sexual cycle in the intermediate host, a mite (*Lelaps echidninus*). U.S. Pub. Health Serv. Hyg. Lab. Bull. No. 46, 48 pp.

Ohbayashi, M. (1971). *Hepatozoon* sp. in northern voles, *Microtus economus* on St. Lawrence Island, Alaska. J. Wildl. Dis. 7:49–51.

Pessôa, S. B., L. Sacchetta, and J. Cavalheiro. (1970). Notas sôbre hemogregarinas de serpentes Brasileiras. X. Hemogregarinas da *Hydrodynastes gigas* (Doméril et Bibron) e sua evolucão. Rev. Latinoamer. Biol. 12:197–200.

Wenyon, C. M. (1926). Protozoology. William Wood and Co., New York (Reprinted by Hafner Publishing Co., New York, 1965), Vol. 2, p. 1085.

*Karyolysus lacertae*
(Danilewsky, 1866) (Plate 28)

This blood parasite of the European wall lizard *Lacerta muralis* is used as an example of the life cycle of an advanced hemococcidian. It infects the endothelial cells of the capillaries of the liver of lizards and the gut epithelium and eggs of lizard mites *Liponyssus saurarum* of the family Gamasidae.

**Description.**    The genus was established by Labbé in 1894 for intraerythrocytic parasites of reptiles that were said to lyse and fragment the nucleus of the host cell. Subsequently, it was realized that such lytic or fragmentary action by the parasite cannot be regarded as of generic value and accordingly was dropped as a characteristic.

*Karyolysus* is difficult to separate from *Haemogregarina* on the basis of gametocytes since both occur in the erythrocytes and are similar. Identification is based on the host cells in which schizogony occurs. In *Karyolysus*, it takes place in endothelial cells of visceral organs, and in *Haemogregarina* in the bone marrow. Final confirmation of *Karyolysus* must be on the basis of finding schizogenous stages in impressions or sections of the liver, lungs, spleen, or heart.

Gametocytes in the erythrocytes range from young vermicular forms resembling merozoites to mature sausage-shaped individuals. As the parasite grows in size, the nucleus of the erythrocyte may be indented and pushed to one side. A membrane surrounds each parasite. A karyosome, or nucleolus, appears at one end of the nucleus of the macrogametocyte, but not in the microgametocyte. Free merozoites may appear in the blood and show up in stained smears.

**Life Cycle.**    The life cycle of *Karolysus lacertae* completes a sequence of development among the hemogregarines, or blood coccidia, leading to the type that occurs in the more advanced hemosporidians, as exemplified by *Plasmodium*. Schizogony with formation of gametocytes occurs in a vertebrate host and sporogony with production of sporozoites takes place in an invertebrate host. The life cycle of *K. lacertae* may be divided into three parts: 1) two or more schizogonies in endothelial cells of lizards with production of gametocytes by the last generation of schizonts, 2) gametogony in adult female mites and the first phase of sporogony to produce sporoblasts, and 3) the second, or final, phase of sporogony to form sporozoites in daughter nymphal mites from infected mothers.

Beginning with lizards that have swallowed infected nymphal mites feeding on them, the vermiform sporozoites are liberated from the sporocysts by the action of the digestive juices and become extremely active. They burrow into the intestinal wall and enter the lymph and blood vessels. In the viscera, particularly the liver, they penetrate endothelial cells of the capillaries and become ovoid schizonts, each enclosed in a membranous cyst. They grow and accumulate food reserves. Nuclear division, which begins in advance of completion of growth, initiates schizogony. Eventually 8 to 30 daughter nuclei are produced per schizont of the first generation. As many merozoites as nuclei are formed. They remain in the cyst for a considerable time attached to and feeding on the residual mass of cytoplasm. Finally, the cyst ruptures and the merozoites enter the bloodstream and penetrate endothelial cells, where schizogony is repeated.

Schizonts of the second generation, like those of the first, are enclosed in a thin, membranous cyst. Each schizont produces a large number of small merozoites generally similar to the large ones except that a nucleolus cannot be distinguished. This ends the schizogenous cycle.

These small merozoites are in reality young gametocytes. Upon rupture of the enclosing membrane, they soon enter red blood cells, often pushing the nucleus to one side. They appear in the erythrocytes about 42 days after the lizards have swallowed infected mites. The microgametocytes remain slender and do not acquire a distinct nucleolus. The macrogametocytes, on the contrary, grow in size and have a large nucleolus at one end of the nucleus. Each gametocyte is surrounded by a thin, clear membrane. No further development of the gametocytes takes place in the erythrocytes.

When infected blood cells are ingested by female mites, (males do not feed on blood), the gametocytes are released in the gut by the digestive processes. They associate in pairs, forming elongate, spindle-shaped bodies. In this condition, the pairs are engulfed by the gut cells of the mites. Once inside, each pair becomes enclosed in a membrane, where development proceeds. The macrogametocyte increases in size and the nucleus becomes larger while the microgametocyte decreases in size. Its nucleus divides once to produce two flagellated microgametes. One enters the macrogamete to complete fertilization and form a motile zygote, or ookinete. Reduction division occurs following union of the nuclei. After a period of intensive growth, the zygote is mature, thus completing the gametogenous stage.

Sporogony begins by the production of a

membrane around the zygote to form the oocyst. Nuclear division commences and is repeated until a number of daughter nuclei appear. They migrate to the surface of the cytoplasmic mass where, in connection with bits of cytoplasm, uninucleate vermicular sporoblasts 40 to 50 $\mu$ long are formed. They are very active and, except for being larger, are similar to ookinetes. Being active precursors of sporozoites they are called sporokinetes to distinguish them from ookinetes.

Upon release from the oocyst, the sporokinetes leave the gut epithelium for the body cavity, wandering into all parts. Of those entering the ovaries of the mites, some go into the yolk of the eggs. Here they grow into an ovoid shape that becomes enclosed in a membrane to form the sporocyst. The next and final step in sporogony is repeated nuclear division, resulting in 20 to 30 daughter nuclei. Each one in association with a bit of cytoplasm forms a sporozoite. During development of the embryonic mite, the sporocysts enter the endodermal cells of the larval gut. By the time the nymphal mite has developed, numerous mature sporocysts filled with sporozoites are already present in the epithelial cells of the gut. Following each meal of blood by the nymphs, epithelial cells, including those containing sporocysts, are shed into the gut. Some are retained and others are voided with the feces. Lizards become infected upon swallowing the sporocysts or mites containing them.

The life cycle of the other species of *Karyolysus* are either the same or basically similar to that of *K. lacertae*. In *K. lacazei*, slight differences occur in development in the mother mite.

In addition to lizards, *Karyolysus sonomae* occurs in the yellow-legged frog (*Rana boyli*) from California. While its life cycle is unknown, the chigger mite, *Hannemania dunni*, which is parasitic on amphibians during its larval stage could possibly serve as the intermediate host.

**EXERCISE ON LIFE CYCLE**

In areas where lizards infected with *Karolysus* occur, the animals should be collected together with their mite parasites for observations on the life cycle. By keeping them together in cages designed to prevent the mites from escaping, a large population of infected animals of both species are available for study.

Various developmental stages of the parasites in the mites can be found by sectioning

## Explanation of Plate 28

**A,** Erythrocyte of lizard infected with macrogametocyte. **B,** Erythrocyte infected with microgametocyte. **C,** Microgametocyte freed from erythrocyte in gut of mite. **D,** Macrogametocyte freed from erythrocyte in gut of mite. **E,** Developing schizont. **F,** Schizont with large merozoites developed from sporozoite. **G,** Schizont with residual body and small merozoites which become gametocytes upon entering erythrocytes. **H,** Large sporokinete that develops in adult female mite, enters eggs, and gives rise to sporocysts in larvae. **I,** Sporocyst with sporozoites in larval and nymphal mites. **J,** Wall lizard showing development of parasites and final infection of erythrocytes. **K,** Infection of adult female mite, development of sporokinetes, and infection of eggs. **L,** Infection of epithelial cells of gut of larval mite and development of sporozoites. **M,** Nymphal mite containing sporocytes with sporozoites infective to lizards.

1, Erythrocyte; 2, displaced nucleus of erythrocyte, a characteristic of the action of the parasite on the red blood cells; 3, macrogametocyte; 4, microgametocyte; 5, nucleus; 6, nucleolus; 7, nucleus of schizont; 8, nucleolus; 9, schizont originating from either a merozoite or a sporozoite; 10, merozoites; 11, schizont from merozoite; 12, gametocyte; 13, residual cytoplasmic body; 14, reserve body of sporokinete; 15, sporocyst; 16, sporozoite.

a–v, Infection of lizards and development of parasites in them; **a,** infected nymph being digested and sporocysts liberated; **b,** freed sporocyst; **c,** ruptured sporocyst liberating sporozoites; **d,** free sporozoites in gut lumen; **e,** sporozoite entering gut epithelial cell; **f,** sporozoite passing through gut wall and entering hepatic portal vein; **g,** sporozoite entering endothelial cells of capillaries of liver (other organs also infected); **h,** young schizont in endothelial cell; **i,** young schizont with first division nucleus; **j,** schizont with dividing nuclei; **k,** mature schizont with merozoites; **l,** ruptured schizont freeing merozoites in bloodstream; **m,** merozoites entering endothelial cells to initiate new generation of merozoic schizonts; **n,** merozoite entering endothelial cell to initiate generation of gametic schizonts; **o,** young schizont; **p,** young schizont in first nuclear division; **q,** multinucleate schizont; **r,** mature schizont with gametocytes; **s,** ruptured sporocyst with gametocytes being released into bloodstream; **t,** gametocyte entering red blood cell; **u,** gametocyte in circulating red blood cell; **v,** infected red blood cell being ingested by mature female mite.

a′–n′, Development in adult female mite; **a′,** infected erythrocytes; **b′,** gametocytes freed from erythrocytes; **c′,** gametocytes in lateral union; **d′,** gametocytes in intestinal epithelial cell; **e′,** growth of micro- and macrogametocytes; **f′,** large macrogametocyte, nucleus of microgametocyte has divided once to form two microgametes; **g′,** one microgamete has entered macrogamete, fertilization spindle has formed; **h′,** zygote enlarges, one microgamete remains outside; **i′,** first nuclear division of zygote (sporont); **j′,** multinucleate sporont about to produce sporoblasts; **k′,** sporocyst with fully developed large sporokinetes; **l′,** large sporokinete free in gut lumen; **m′,** sporokinete has migrated from lumen of gut to ovary of mite and has entered egg; **n′,** egg of mite.

a″–f″, Development in larval and nymphal mites; **a″,** sporokinete in tissues and gut of newly hatched larva; **b″,** sporokinete entering newly formed intestinal endoderm; **c″,** sporokinete in intestinal epithelial cell; **d″,** growth of sporokinete into young schizont; **e″,** multinucleate sporont forming sporoblasts; **f″,** sporocyst with a residual cytoplasmic body and numerous sporozoites, sporulation completed simultaneously with molting of larva to nymph.

Figures **A–I** redrawn from Reichenow, 1921, Arch. Protistenk. 42:180.

---

engorged gravid adult females, larvae, and nymphs obtained from infected lizards. Gametocytes in erythrocytes and merozoites in the plasma may be found in thin smears stained with Wright's or Giemsa's stains. Smear impressions and sections of the liver reveal all the stages of schizogony in the endothelial cells of the capillaries.

For infecting unparasitized mites, allow them to feed on heparinized blood from infected lizards, as described for *Hepatozoon muris*.

### SELECTED REFERENCES

Grassé, P. P. (1953). Traité de Zoologie. Masson et Cie, Paris, p. 757.

Hyland, K. E. (1950). The life cycle and parasitic habit of the chigger mite *Hannemania dunni* Sambon, 1928, a parasite of amphibians. J. Parasitol. 36(6/2):32–33.

Lehmann, D. L. (1959). *Karyolysus sonomae* n. sp., a blood parasite from the California yellow-legged frog, *Rana boyli boyli*. Proc. Amer. Phil. Soc. 103: 545–553.

Reichenow, E. (1921). Die Hämococcidien der Eidechsen. Vorbemerkungen und I. Teil: Die Entwicklungsgeschichte von *Karyolysus*. Arch. Protistenk. 43:179–291.

Wenyon, C. M. (1926). Protozoology. William Wood and Co., New York (Reprinted by Hafner Publishing Co., New York, 1965), Vol. 2, p. 1095.

**Plate 28**   *Karyolysus lacertae*                                                                131

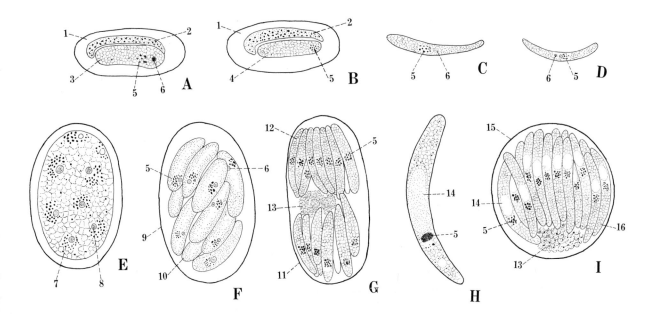

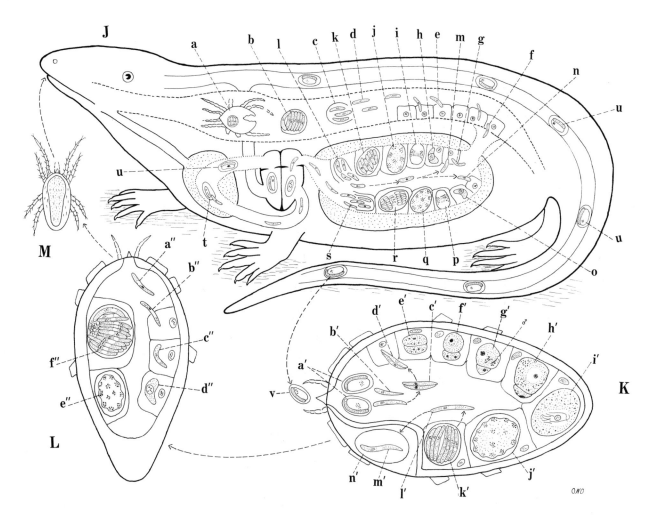

## ORDER HAEMOSPORIDA

Haemosporida are parasites of blood cells and tissues of reptiles, birds, and mammals. Asexual reproduction, or schizogony, takes place in the internal organs, blood cells, reticuloendothelial cells, or combinations of them in the vertebrate host. Sexual reproduction begins in a vertebrate host and is completed in a bloodsucking dipteran host. Gametogony commences in the vertebrate where the gametocytes are formed and enter blood cells. It is completed in the intestine of the arthropod host with fertilization and formation of the zygote. Sporogony is in the body tissues of arthropods, where development of the oocysts and sporozoites takes place. Since there is no free-living stage required for transmission, the sporozoites develop in thin-walled oocysts instead of the thick-walled protective ones that must pass from host to host as in *Eimeria*. Transmission is through bites of infected arthropod hosts which inject the sporozoites.

The order, as presented here and with some deviation from Kudo's classification, contains three families. They are Plasmodiidae, Haemoproteidae, and Leucocytoidae.

## FAMILY PLASMODIIDAE

The family was erected by Mesnil in 1903 for the malarial parasites of the genus *Plasmodium*. It is characterized by three phases in the life cycle. The first consists of exoerythrocytic schizogony without pigmentation in reticuloendothelial cells and monocytes of viscera in birds and hepatic cells in mammals. The second schizogony with pigmentation is in circulating erythrocytes of the vertebrate host where periodic febrile paroxysms in mammals coincide with liberation of the merozoites. The third and final phase is sexual, in which pigment-producing gametocytes appear in the erythrocytes of the vertebrate host. Fertilization and sporogony are in the mosquito host. Infection is initiated when sporozoites are injected by parasitized mosquitoes.

Species of malarial parasites infecting humans are worldwide in distribution, occurring in both the tropical and temperate zones. Simian species are common and those infecting birds are cosmopolitan.

The family is presently divided into eight subgenera, four each from mammalian and avian hosts. Subgenera from the two groups of vertebrate hosts are readily distinguishable. Those from mammals are in anucleated erythrocytes and those from birds in nucleated ones.

### Mammalian Species of *Plasmodium*

*Plasmodium* (*Plasmodium*) contains those primate species of malaria with spherical gametocytes. Representative species in humans are *P.* (*P.*) *vivax*, *P.* (*P.*) *malariae*, and *P.* (*P.*) *ovale*. A common simian species is *P.* (*P.*) *knowlesi*. *Plasmodium* (*Laverana*) *falciparum* is the human species with crescentic gametocytes. *Plasmodium* (*Vinckeia*) *berghei* of rodents has spherical gametocytes and is representative of the species from nonhuman primates. *Plasmodium* (*P.*) *vivax* is used to represent the life cycle of a malarial parasite.

#### *Plasmodium* (*P.*) *vivax*
#### (Grassi and Feletti, 1890) (Plate 29)

*Plasmodium vivax* is a parasite of the liver cells and erythrocytes of humans, and the intestinal lumen and wall, hemocoel, and salivary glands of many species of anopheline mosquitoes. This parasite is the cause of benign tertian malaria, one of the serious diseases of mankind. It is worldwide but flourishes in temperate climates within isotherms of 16° to 20°C in the northern hemisphere and 20°C in the southern. Indigenous cases occur as far north as England, Siberia, and Manchuria, and south into Argentina and South Africa. Susceptible mosquito hosts are prevalent in temperate regions. If malaria is introduced, the mosquitoes become infected and the disease appears in the human population of the area.

With the introduction of potent insecticides and antimalarial drugs, hope was raised that malaria could be controlled, indeed, eliminated from vast areas. While much was accomplished in this direction, the appearance of new strains of mosquitoes and plasmodia resistant to the present drugs has raised questions as to how far control and eradication on this basis can be carried by drugs and chemicals.

**Description.** The life cycle consists of three separate phases, each involved with multiplication and the production of a specific morphological and functional type necessary for completion of the life cycle.

In the first phase, development is outside the blood cells and is exo- or preerythrocytic. Sporozoites enter liver cells of the human host to form uninucleate trophozoites. At first, they are round or oval. Each is surrounded by a double unit membrane with a light-colored layer between. An opening called the cytostome at one end engulfs cytoplasm of the host cell for nutritive purposes. Inside the trophozoite are food vacuoles, a large nucleus with nucleolus and enclosed in a double membrane, mitochondria, and other structures (Fig. 12, A). As growth proceeds, the nucleus divides and subdivides many times to become a large, irregularly shaped schizont. Giemsa stain colors the cytoplasm light blue and the nuclei red. Minute protrusions of cytoplasm each containing a nucleus pinch off to form oval or round hepatic merozoites. These enter blood cells to initiate the second phase.

The erythrocytic phase includes a schizogenous process similar to that in the liver. Upon entering the red blood cells, the hepatic merozoites undergo the following sequence of developmental changes: 1) ring-like trophozoites with a thin rim of cytoplasm and a small nucleus perched on one side; 2) large uninucleate trophozoites with enlarged, irregularly shaped cytoplasm containing brownish pigment inside swollen erythrocytes with numerous pink Schüffner's dots dispersed throughout the cytoplasm; 3) mature schizonts with 15 to 20 nuclei; 4) segmenters in which a bit of cytoplasm surrounds each of the nuclei to form individual sausage-shaped merozoites attached to a residual mass of cytoplasm (Fig. 12, B); and 5) large pigmented, oval gametocytes derived from merozoites. The microgametocytes stain light blue and contain a large, diffuse reddish nucleus; the macrogametocytes are larger, darker blue and have a small, compact, deeply staining nucleus (Plate 29, I, J).

The final stages, including formation of the gametes and fertilization, occur in the stomach of mosquitoes. Upon release from the red blood corpuscles, the nucleus of the microgametocyte divides three times to produce eight daughter nuclei. Each one migrates into a slender, elongate extension of cytoplasm from the microgametocyte to form eight uninucleate microgametes. Each microgamete contains a basal body at the attached end, a recurrent (enclosed) flagellum with one pair of central and nine pairs of peripheral fibrils (Fig. 12, C). Mitochondria have not been seen in microgametes. Upon fertilization of the macrogamete, the zygote elongates to become a motile, sausage-shaped ookinete (Plate 29, K, i) that migrates to a position

between the midgut epithelium and the thin outer membranous covering of the intestine. It becomes an oocyst. Development results in multiplication of the nucleus and division of the cytoplasm into uninucleate masses called sporoblasts without a cyst covering. Each sporoblastic nucleus divides and subdivides many times. The resultant daughter nuclei migrate to the surface where in conjunction with a bit of cytoplasm they form sporozoites (Fig. 12, D). Sporozoites are slender, elongate bodies with blunt anterior and pointed posterior ends. The thick pellicle provides rigidity and is entire except at two points. The anterior end is surrounded by several thickened, concentric rings in the center of which is an opening called the anterior cup. A pair of club-shaped structures, the organelles, extends from the anterior cup backward to near the anterior margin of the nucleus. Another opening, the micropyle, is situated laterally near the level of the nucleus. Hollow subpellicular fibrils extend from the anterior end of the body posteriorly to a point behind the nucleus. Mitochondria, convoluted tubules, and other structures are present (Fig. 12, E).

**Life Cycle.** About 40 species of female mosquitoes of the genus *Anopheles* are susceptible to infection with *P. vivax* and are capable of transmitting it. Male mosquitoes, which do not feed on blood, play no direct role in the transmission of malaria. *Anopheles labranchiae* in Europe, *A. quadrimaculatus* and *A. freeborni* in the United States, and *A. stephensi* of Asia are highly susceptible to infection.

Tissue stages of malaria in the vertebrate host begin with injection of sporozoites. When introduced into a human by the bite of a mosquito, or by experimental injection, the sporozoites quickly accumulate in the liver via the bloodstream, leave the capillaries, and enter the hepatic cells to begin the exoerythrocytic phase of development. The sporozoites round up to form the feeding and growing trophozoites. Food consists of the cytoplasm of the host cell engulfed through the cytosome by a process of pinocytosis. By day 7, trophozoites have attained maturity and a diameter of up to 40 $\mu$. At the moment multiple nuclear division begins, the trophozoite becomes a schizont. Numerous minute daughter nuclei numbering up to 10,000 migrate to the surface of the schizont. A bit of cytoplasm condenses around each nucleus to form a spherical exoerythrocytic merozoite about

## Explanation of Plate 29

A, Ring stage in erythrocyte. B, C, Growing trophozoites. D, E, Young schizonts with few nuclei. F, G, Older schizonts with more nuclei. H, Segmenter with merozoites. I, Macrogametocyte. J. Microgametocyte. K, Anopheline mosquito host in characteristic feeding position. L, Human host.

1, Nucleus of parasite in erythrocyte; 2, cytoplasm; 3, Schüffner's dots; 4, pigment granules; 5, merozoites.

a, Gametocyte being sucked up by mosquito and entering stomach; b, microgametocyte; c, exflagellation of microgametocyte; d, microgamete; e, macrogametocyte; f, macrogamete; g, microgamete fertilizes macrogamete; h, zygote; i, ookinete; j, ookinete passing through intestinal epithelium; k, ookinete rounds up between epithelium and basement membrane to form young oocyst; l, growing oocyst; m, formation of sporoblasts; n, ripe oocyst with sporozoites escaping into hemocoel; o, sporozoites migrating through hemocoel toward salivary glands; p, sporozoites in cells of salivary glands; q, sporozoites being injected into blood by biting mosquito.

a', Sporozoites injected into blood stream; b', sporozoites distributed throughout blood; c', sporozoite entering hepatic cell to initiate exoerythrocytic schizogony; d', segmenter (preceding stages of trophozoites omitted); e', first generation of exoerythrocytic merozoites; f', merozoites enter other hepatic cells to produce second generation; g', segmenter (trophozoite omitted); h', second generation of merozoites which enter bloodstream; i', merozoite entering red blood cell to begin erythrocytic schizogony; j', ring stage; k', trophozoite stage; l', m', developing schizonts; n', segmenter; o', erythrocytic merozoites liberated 48 hours after entrance of hepatic merozoites (chills and fever at the time merozoites are liberated into bloodstream); p', merozoites entering red blood cells to initiate second generation; q', ring stage; r', trophozoite; s', t', schizonts; u', segmenter; v', liberation of second generation of erythrocytic merozoites (chills and fever); w', merozoite entering erythrocyte; x'–z', a'', formation of microgametocyte; b'', merozoite entering erythrocyte; c''–g'', development of macrogametocyte.

Figures redrawn from various sources.

---

1.2 $\mu$ in diameter. The schizont ruptures by day 8 and the merozoites enter other liver cells to produce a second generation with fewer merozoites which appear on day 15. Liver schizonts may persist for many years with intermittent production by secondary schizogony that results in recurrences of the disease.

Exoerythrocytic merozoites of the final schizogony quickly enter red blood cells by unknown processes to begin the erythrocytic phase of asexual development. At first, the merozoites develop into ring-like forms in which a large vacuole is surrounded by a thin layer of cytoplasm. Within a few hours, the rings have enlarged and the parasites show ameboid movement by sending out pseudopods in all directions. By hours 6 to 8 of growth, brown pigment granules appear as a result of metabolism of the hemoglobin by the parasites. As growth of the trophozoites proceeds, the infected erythrocytes enlarge, lose their color, and show a stippling of lilaceous spots known as Schüffner's dots as early as hour 5 of development.

By hour 24 of development, the vacuole disappears, ameboid movement declines, and pigment granules accumulate in clumps. Between hours 36 and 42, the nucleus begins dividing until 12 to 24 daughter nuclei appear. This is the erythrocytic schizont. Cytoplasmic condensation

around each nucleus produces merozoites. A parasite consisting of fully formed merozoites arranged in a rosette is called a segmenter. The infected red blood cells are hypertrophied and have a diameter of 10 $\mu$ or more.

The infected erythrocyte and segmenter rupture by hour 48, liberating the merozoites, residual cytoplasmic body, and membrane of the schizont together with products of metabolism. The cycle of schizogonic development and liberation of merozoites every 48 hours is remarkably constant. Host response to the liberation of the cellular detritus into the blood is manifested in the periodic paroxysms of chills and fever which usually occur in the afternoon. Since this appears every third day, *P. vivax* is referred to as tertian malaria. The paroxysms continue with increasing severity to a maximum intensity attained in about 2 weeks, following which the parasites decline in number. After a month, recrudescences (appearance of erythrocytic stages from hepatic merozoites) appear at irregular intervals over long periods of time.

At some point during the exoerythrocytic or erythrocytic phases of development, certain merozoites differentiate into gametocytes which are the sexual forms. In the case of tissue origin, they appear in the blood cells as early as day 3 after infection, whereas those of erythrocytic

**Plate 29**   *Plasmodium vivax*                                                                    135

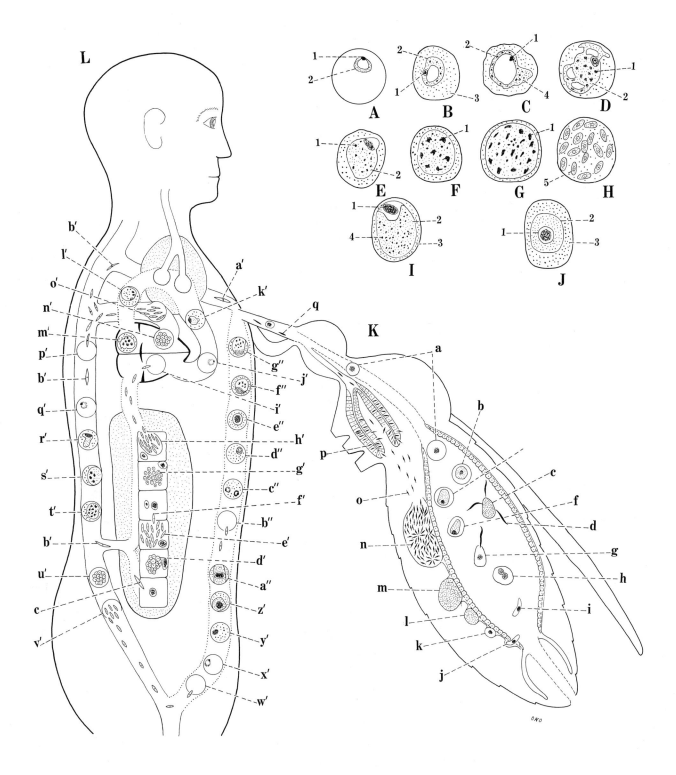

origin are present 4 to 6 days after the peak of parasitemia. Inside the erythrocytes, they mature to micro- and macrogametocytes but do not develop further. Through metabolism of the cytoplasm of the host cells, pigment is deposited in the gametocytes.

The phases of development in the mosquitoes include formation of the gametes and fertilization with production of a zygote, and subsequent multiplication, or sporogony, to form sporozoites, the stage infective to humans.

Erythrocytes containing gametocytes are digested in the stomach of the mosquitoes and the parasites are liberated. Maturation of the gametocytes takes place in the gut lumen. The cytoplasm of the microgametocytes becomes turbulent and the nucleus divides three times, producing eight daughter nuclei. Each one migrates

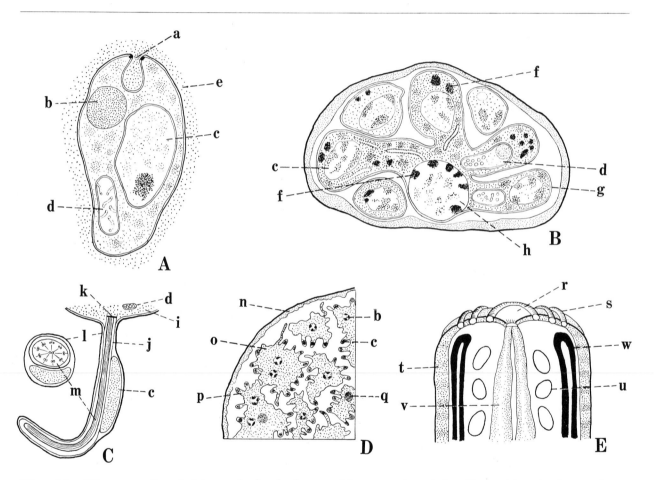

**Fig. 12.** Diagrammatic sketches of developing stages of malarial parasites. **A,** Young trophozoite of *Plasmodium gallinaceum* in host cell. **B,** Budding schizont of *P. knowlesi*, showing formation of merozoites. **C,** Developing microgamete from microgametocyte. **D,** Section of oocyst, showing sporoblasts with individual sporozoites forming. **E,** Anterior end of sporozoite, showing detailed structures. **a,** Cytosome engulfing cytoplasm of host cell (e); **b,** food vacuole; **c,** nucleus; **d,** mitochondrion; **e,** cytoplasm of host cell; **f,** malarial pigment; **g,** developing merozoite; **h,** residual body of schizont; **i,** microgametocyte; **j,** microgamete; **k,** basal body; **l,** outer membrane with middle and inner layers inside; **m,** recurrent flagellum with paired central and nine peripheral pairs of fibrils with spoke-like lines; **n,** oocyst wall; **o,** budding sporoblast; **p,** developing sporozoites; **q,** lipid body; **r,** anterior cup; **s,** concentric rings around anterior cup; **t,** pellicle; **u,** hollow subpellicular fibril; **v,** paired organelle; **w,** convoluted tubule. Figs. A and B adapted from electron micrographs from Aikawa *et al.*, 1966, Milit. Med. 131 (Suppl.):969; C adapted from Garnham, 1966, *Malaria Parasites and Other Haemosporidia*, Blackwell, Oxford, p. 23; D adapted from Terzakis *et al.*, 1966, Milit. Med. 131 (Suppl.):984; E adapted from Garnham *et al.*, 1961, Trans. Roy. Soc. Trop. Med. Hyg. 55:98.

to the surface of the microgametocyte and into a finger-like extension of the cytoplasm to form a microgamete whose free end lashes about. The process of microgamete formation is known as exflagellation and takes place in a few minutes after the microgametocytes are freed from the red blood cells in the stomach of the mosquito. After a brief period of vigorous activity, the microgametes detach and are free. In the meantime, a small prominence has appeared on the surface of each macrogametocyte to which a microgamete attaches and is drawn inside, accompanied by violent activity of the macrogametic cytoplasm. Fertilization is completed in about 10 minutes and the two nuclei fuse within a few hours.

The resultant oval to round zygote elongates to form a motile sausage-shaped organism called the ookinete that is 15 to 20 $\mu$ long by 3 $\mu$ in diameter. The pigment in it accumulates at one end. Formation of the ookinete is completed in about 18 hours.

By gregariniform movement of gliding and bending, ookinetes move to the brush border of the epithelial cells, penetrate them, and come to rest between the base of the cells and the thin membranous layer surrounding the gut within 24 to 48 hours after the blood meal of the mosquito.

In this location, sporogony begins and is completed. The ookinete rounds up and nuclear division begins within 48 hours. The first division is meiotic. The diploid chromosomes divide to form the haploid number of two in each nucleus. This number remains constant until fertilization in a new cycle occurs. The ookinete is now an oocyst in which multiple nuclear division takes place. The next step in development of the oocyst is breaking up of the cytoplasm into uninucleate masses called sporoblasts. The nucleus of each sporoblast divides and subdivides many times. Mature oocysts measure up to 50 $\mu$ or more in diameter and contain scattered granules of brown pigment. By day 8, tiny finger-like processes of cytoplasm project from the surface of the sporoblasts and one of the daughter nuclei passes into each filament to form a sporozoite. Up to 10,000 sporozoites are formed per oocyst in 9 days after infection of the mosquitoes. Numerous oocysts may be present in a single mosquito.

Mature oocysts rupture, releasing the slender sporozoites into the hemocoel. They scatter to all parts. Some enter the acinar cells of the salivary glands where they line up. When the mosquitoes feed and the cells of the salivary glands secrete saliva, the sporozoites are released with it and pass into the acinal ducts and are carried into the vertebrate host to begin the exoerythrocytic phase of the life cycle.

There are three other well-known species of malaria that infect man. The life cycles are basically similar to that of *P. vivax* but with differences that will be considered briefly. Detailed accounts of them are available in a number of advanced texts.

*Plasmodium falciparum* is the cause of malignant tertian malaria of the subtropical and tropical regions of the world. In southeast Asia, the species has evolved strains resistant to antimalarial drugs.

The exoerythrocytic phase is limited to a single generation of merozoites. Young schizonts appear as crescent-, oval-, or round-shaped bodies inside the erythrocytes (Fig. 13, B). Red blood cells containing developing schizonts retreat from the peripheral blood to the visceral

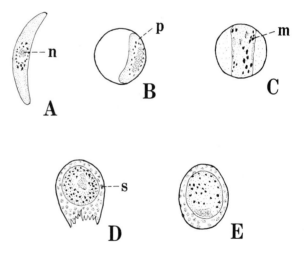

Fig. 13.   Three species of *Plasmodium* from humans. A, B, *P. falciparum;* A, mature crescentic macrogametocyte (microgametocyte is similar) free in blood; B, young schizont in erythrocyte without any cytoplasmic stippling before disappearance into visceral organs where schizogony takes place; C, *P. malariae* showing common band-like schizont in erythrocyte with fine cytoplasmic stippling; D, E, *P. ovale* gametocytes in red blood cells with large cytoplasmic Schüffner's dots; D, frayed erythrocyte characteristic of many infected cells; E, ovalized infected cell; n, nucleus; m, malarial pigment; p, parasite; s, Schüffner's dots.

organs where schizogony takes place. Dividing schizonts are rarely seen in circulating blood. Each schizogonic cycle is completed in 36 to 48 hours. Death of the human host usually occurs when 25 percent or more of the erythrocytes are infected. Gametocytes are crescent-shaped bodies free in the blood (Fig. 13, A).

*Plasmodium malariae* is quartan malaria with paroxysms occurring at 72-hour intervals. It is prevalent in tropical regions of the world except South America, where occurrences are uncommon. The exoerythrocytic phase is completed in 13 to 16 days. There are reports that hepatic schizonts may live up to 40 years, as indicated by relapses following the last exposure to infected mosquitoes. The erythrocytic schizonts are band-like (Fig. 13, C) and produce up to 12, usually 8, merozoites. Gametocytes are round to oval. Stippling in parasitized red blood cells is very fine.

*Plasmodium ovale* is a benign tertian form with paroxysms 49 to 50 hours apart, being slightly longer than in *P. vivax*. The exoerythrocytic stage requires 9 days for development and is repeated. Hepatic merozoites invade reticulocytes (young erythrocytes) and complete schizogony in 49 to 50 hours. Infected erythrocytes may have frayed or tagged margins and the cytoplasm contains large Schüffner's dots (Fig. 13, D, E). Schizonts and gametocytes are similar to those of *P. vivax* and difficult to separate.

**Pathology and Immunity.**    Pathological changes are produced in a variety of ways. While intracellular stages of the parasites destroy cells, clinical symptoms are not apparent. Rupture of erythrocytes by mature schizonts and release of merozoites, residual body, and metabolites into the blood are responsible for the febrile and chill paroxysms. It is thought that the parasites produce lecithinase which damages mitochondria, resulting in inhibition of respiration of host cells, especially of the liver. Other toxins are produced that cause constriction of small veins in liver lobules resulting in back pressure of the blood and extensive shock.

*Plasmodium falciparum*, the cause of malignant malaria, causes infected red blood cells to retreat to the capillaries of visceral organs where clumping occurs, blocking circulation with secondary effects such as hemorrhage and necrosis of cells.

Gross hyperplasia of lymphocytes and phagocytes as well as bone marrow, liver, and spleen, are common. Pigmentation from cells digested by the parasites darkens host organs.

Immunity begins at the time erythrocytes infected with schizonts rupture and release the metabolites together with by-products of schizogony into the bloodstream. The lymphoid-macrophage system responds to the presence of these by-products by producing humoral antibodies and substances that stimulate macrophages to greater activity. This reaction of the host enables it to reduce the bulk of the parasites. When only a few remain, they exert a continuous stimulation on the lymphoid-macrophage system to maintain the immunity already present, thereby preventing subsequent infection. Immunity maintained by a small population of parasites living almost as commensals is termed premunition.

In addition to immunity developed in the vertebrate host by the presence of plasmodial parasites, mosquito hosts vary in their susceptibility to infection. Susceptibility or insusceptibility of mosquitoes to infection are based on biological, biochemical, environmental, or genetic factors. Any or a combination of them affect the ability of mosquitoes to produce sporozoites. The genetic basis for mosquito resistance is believed to be multiple reacting genes. The factors responsible for susceptibility of mosquitoes are systematic rather than being confined to the environment of the stomach and its wall. When the stomach is bypassed by injecting gametocytes into the hemocoel, development proceeds with the production of sporozoites capable of infecting vertebrates bitten by the mosquitoes. Such development does not occur in refractory mosquitoes.

### Avian Species of *Plasmodium*

The species of *Plasmodium* infecting birds are worldwide in distribution and prevalent in many orders. Their wide geographic distribution is based on the mobility and migratory patterns of the avian hosts. In addition, a single species of malaria is able to infect species of mosquitoes from several genera. The combination of mobility of the bird hosts and ready availability of many species of susceptible mosquitoes results in widespread distribution.

Since the discovery of avian malaria parasites by Danilewsky in 1885, studies on them have contributed greatly to present day knowledge of human malaria. In 1898, Ross, the British army surgeon stationed in India, working with sparrows discovered and described the sexual life cycle in mosquitoes.

For many years after the discovery of the mosquito phase of the life cycle, it was believed that sporozoites upon injection into the vertebrate host penetrated directly into erythrocytes. By the 1930's, the belief prevailed among malariologists that a stage of development in the vertebrate host existed between injection of sporozoites into a vertebrate host and appearance of trophozoites in erythrocytes. Huff, in 1930, described *P. elongatum* in which segmenting schizonts were abundant in the hematopoietic tissue. This observation led to the discovery and clarification of the exoerythrocytic phase of the life cycle, first in birds and later in mammals.

A shortage of quinine during World War II led to an extensive search in Europe and America for antimalarial drugs. Bird malaria became important for screening drugs. A number of compounds highly effective against various developmental stages of malaria in humans were discovered in this gigantic effort.

**Classification and Identification.** Avian species of *Plasmodium* are divided into four subgenera. They are *P. (Haemamoeba)* with round gametocytes and *P. (Giovannolaia)*, *P. (Novyella)*, and *P. (Huffia)* with elongate ones. Identification of the subgenera and species is based on morphological characters evident in blood smears and biological characters in the vertebrate and invertebrate hosts.

Morphological characters seen in blood smears that are of value in identifying the parasites are 1) age of erythrocytes infected; 2) differences in shape of pigment granules in microgametocytes; 3) shape and position of young trophozoites in host cells and location of the nucleus of the latter; 4) shape of gametocytes and their relation to the nucleus of the host cell; and 5) location of segmenters in erythrocytes, number of merozoites, and location of the nucleus of the host cell in relation to the parasite. These features are summarized in Plate 30 for some of the common species.

Biological and morphological features indicative of the subgenera and species based on experimental studies include 1) tissues in which exoerythrocytic schizonts develop; 2) morphology of exoerythrocytic merozoites; 3) time required for exoerythrocytic schizogony; and 4)

sporogonic stages. The last involves a) length of cycle in the mosquito host; b) morphological characters of oocysts and sporozoites; c) generic group of mosquito vectors; and d) susceptibility as measured in the natural vertebrate host and compared with that in canaries and chickens.

**Life Cycle.** The life cycle of malaria in birds, both in the vertebrate and invertebrate hosts, is basically similar to that in mammals. Variations occur among the species in the different types of cells selected for exoerythrocytic schizogony and the time required for development of the separate phases.

Exoerythrocytic schizogony begins from sporozoites injected into the skin where they enter mesodermal cells and produce merozoites. It is continued when they invade reticuloendothelial cells of the bone marrow, liver, spleen, heart, kidneys, and endothelial cells of the capillaries. Merozoites produced in the fixed cells and liberated into the bloodstream are adapted to entrance into and development inside red blood cells. Erythrocytic schizogony begins when merozoites enter red blood cells. They undergo periodic schizogony at fixed intervals of time with production of merozoites and gametocytes. Merozoites from either the exo- or erythrocytic phases may begin a secondary invasion of reticuloendothelial cells to continue to the exoerythrocytic phase.

Relapses apparently do not occur in birds as happens in mammals.

Mosquito vectors are not confined to species of *Anopheles*, as in the case of human malaria, but include those of *Culex*, *Anopheles*, *Aedes*, *Culiseta*, *Psorophora*, *Armigeres*, and *Mansonia*. A single species of bird malaria is able to develop in species of mosquitoes from several genera. Sporogonic development in mosquito vectors is similar for both avian and mammalian malaria.

**EXERCISE ON LIFE CYCLE**

In order to gain a full appreciation of the life cycle of *Plasmodium*, it is necessary to observe exo- and erythrocytic stages of schizogony, exflagellation in blood or the stomach of mosquitoes, and sporogony in mosquitoes. Fortunately these stages of development can be seen in well-known species of avian malaria prevalent in many common birds and the ubiquitous mosquito vectors.

*Exoerythrocytic schizonts.*    Prepare smear

## Explanation of Plate 30

I. Ring stages
  A. In young erythrocytes
    1. *P. cathemerium*\*, *P. circumflexum*†, *P. elongatum*‡, *P. nucleophilum*¶, *P. relictum*† (Fig. 1)
  B. In mature erythrocytes
    1. *P. lophurae*†, unknown for other species (Fig. 2)
II. Difference in pigment in microgametocytes
  A. Round pigment granules
    1. *P. relictum* (Fig. 3)
  B. Elongate pigment granules
    1. *P. cathemerium* (Fig. 4)
III. Shape and location of young trophozoites and displacement of nucleus of host cell
  A. Shape of trophozoites
    1. Small and elongate
      a. *P. circumflexum*, *P. lophurae* (Fig. 5)
    2. Small, irregular but not round or elongate
      a. *P. cathemerium*, *P. gallinaceum*\*, *P. nucleophilum*, *P. polare*¶, *P. relictum* (Figs. 6, 7, 8)
  B. Location of trophozoites
    1. Nucleus of erythrocyte in normal position
      a. Free in cytoplasm
        (1) Lateral to nucleus of host cell
          (a) *P. circumflexum*, *P. lophurae* (Fig. 5)
        (2) Terminal to nucleus
          (a) *P. polare* (Fig. 7)
      b. Adherent to nucleus of host cell
        (1) *P. nucleophilum* (Fig. 6)
  C. Location of nucleus of host cell
    1. Nucleus in normal position
      a. *P. circumflexum*, *P. lophurae*, *P. nucleophilum*, *P. polare* (Figs. 5, 6, 7)
    2. Nucleus displaced
      a. *P. cathemerium*, *P. gallinaceum*, *P. relictum* (Fig. 8)
IV. Shape of gametocytes and location of nucleus of host cell

A. Shape
  1. Round
    a. *P. cathemerium*, *P. gallinaceum*, *P. relictum* (Fig. 9)
  2. Elongate
    a. Thin, does not surround nucleus
      (1) *P. elongatum*, *P. hexamerium*¶, *P. polare*, *P. rouxi* (Fig. 10)
    b. Thick, partially or completely encircles nucleus
      (1) *P. circumflexum*, *P. lophurae* (Figs. 11, 12)
  B. Location of nucleus of host cell
    1. Normal position
      a. *P. circumflexum*, *P. elongatum*, *P. hexamerium*, *P. lophurae*, *P. polare*, *P. rouxi*¶, *P. vaughani*¶ (Figs. 10, 11, 12)
    2. Displaced position
      a. *P. cathemerium*, *P. gallinaceum*, *P. relictum* (Figs. 3, 4, 9)
V. Segmenters
  A. Nucleus of host cell not displaced
    1. Small segmenter, characteristically with four merozoites located terminal to nucleus
      a. *P. rouxi* (Fig. 15)
    2. Large segmenter, partially or completely surrounding nucleus
      a. *P. circumflexum*, *P. lophurae* (Fig. 16)
  B. Nucleus of host cell displaced
    1. Segmenter small, attached to nucleus, few merozoites
      a. *P. nucleophilum* (Fig. 14)
    2. Segmenter large, not attached to nucleus, many merozoites
      a. *P. cathemerium*, *P. gallinaceum*, *P. relictum* (Fig. 13)

---

\* *Plasmodium* (*Haeamoeba*).
† *P.* (*Giovannolaia*).
‡ *P.* (*Huffia*).
¶ *P.* (*Novyella*).
Adapted from Hewitt, 1940, Amer. J. Hyg. Monog. Ser. No. 15.

---

impressions and sections of visceral organs from infected birds and stain by the Wright or Giemsa method to demonstrate schizonts. Following treatment with these stains, the cytoplasm of the parasites is blue and the nuclei are a deep reddish hue. Pigment granules are absent. Find and identify trophozoites, schizonts, and segmenters with their merozoites. Note the host cells that are parasitized.

*Erythrocytic schizonts.* Stained blood smears are necessary for observing erythrocytic schizonts and gametocytes. Smears are obtained by pricking a brachial vein or tarsal artery with a sharp needle and allowing several drops of blood to accumulate. Touch the edge of a microscope slide to the drop and quickly draw it over a clean slide, spreading the blood in a very thin layer. Wave the slide vigorously in the air to dry

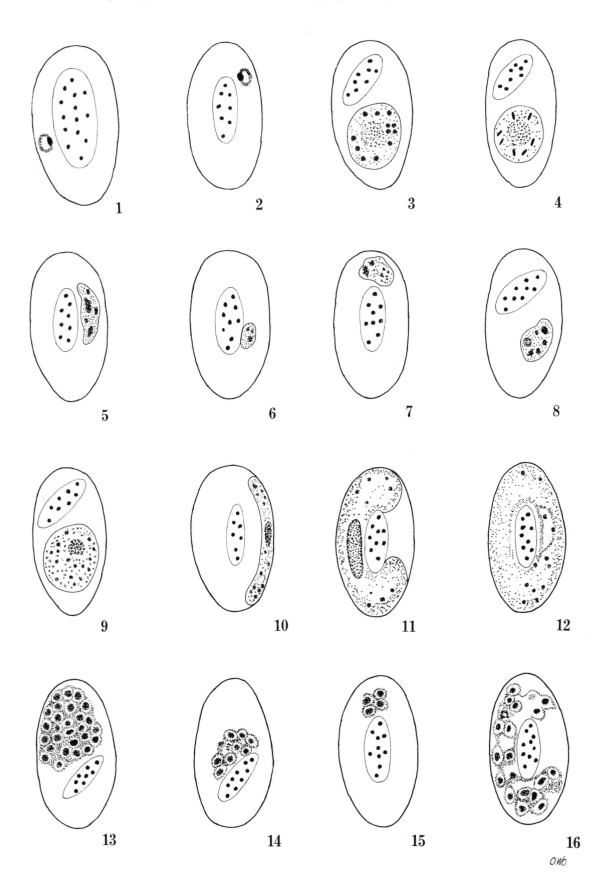

OWG

the blood quickly. Follow standard procedures in staining the smears with Wright's or Giemsa's stains.

Examine the smears under an oil immersion lens and attempt to identify the parasite either to species or closely related group of species. Note the 1) ring stage; 2) age of erythrocyte based on the large nucleus for young ones; 3) large uninucleate trophozoites; 4) multinucleate schizont; and 5) segmenter with merozoites, including number. Check 1) whether the nucleus of the host cells remains in the normal position or is dislocated by the parasites; 2) the shape of the young schizonts; 3) whether they are in contact with the nucleus; and 4) its location in the cell. Compare these features with those shown in Plate 30. Note that pigment granules appear in the erythrocytic schizonts.

*Gametocytes.* Both micro- and macrogametocytes appear in erythrocytes, the only place where they can develop. They may be either round and displace the nucleus of the blood cell or elongate. If elongated, they are slender and separate from the nucleus or they partially or completely surround it but do not displace it from the normal location. They are uninucleate, with the nucleus of the microgametocyte diffuse and that of the macrogametocyte more compact and somewhat fibrillar. The cytoplasm of the macrogametocytes is blue and that of the microgametocytes usually pinkish. Pigment granules are present and vary in shape characteristic of the species.

*Exflagellation.* Formation of microgametes may be seen in fresh blood smears or in the stomach of a mosquito. The simple procedure is to obtain a drop of blood from a heavily infected bird, cover it with a cover glass, and find microgametocytes under high power microscopy. Observe them under oil immersion. Four slender, finger-like projections appear on the surface of the microgametocyte, become very active, and finally detach to form individual microgametes (Fig. 12, C).

Exflagellation may be observed in the stomach of mosquitoes shortly after they have fed on a heavily infected bird. Remove the stomach in a dish of physiological saline, mount on a slide, dissect open, cover with a cover glass, and observe under oil immersion. Note the activity within the microgametocyte.

*Sporogonic stages.* These may be observed in susceptible mosquitoes that have fed on infected birds.

Ookinetes, which are sausage-shaped and active, are present in the intestine of mosquitoes 12 to 24 hours after the blood meal.

Oocysts develop as oval to spherical bodies from ookinetes that have migrated from the intestine to a position between the epithelial cells of the gut and the membrane surrounding it. Remove the gut of the mosquito in physiological saline and examine it for oocysts which appear as spherical bodies located on the outside.

Sporozoites are first contained in ripe oocysts. When liberated by rupture of the oocyst, many of them accumulate in the acinar cells of the salivary glands. Smears, whole mounts, or sections of infected glands colored with blood stains show the parasites lined up in the cells or free, depending on the manner of preparation.

*Infection of hosts.* In order to obtain infections, birds must be bitten by infected mosquitoes or injected with infected erythrocytes. Gametocytes will develop only in birds infected with sporozoites. Mosquitoes are infected by allowing them to feed on infected birds.

## SELECTED REFERENCES

*Mammalian malaria*

Aikawa, M., C. G. Huff, and H. Sprinz. (1966). Comparative feeding mechanisms of avian and primate malarial parasites. Mil. Med. (Suppl.) 131: 969–983.

Garnham, P. C. C. (1966). Malaria Parasites and Other Haemosporidia. Blackwell Scientific Publications, Oxford, pp. 3–513.

Garnham, P. C. C., R. G. Bird, and J. R. Baker. (1963). Electron microscope studies of motile stages of malaria parasites. IV. Fine structure of the sporozoites of four species of *Plasmodium*. Trans. Roy. Soc. Trop. Med. Hyg. 57:27–31.

Garnham, P. C. C., R. G. Bird, W. R. Baker, and R. S. Bray. (1961). Electron microscope studies of motile stages of malaria parasites. II. Fine structure of the sporozoites of *Laverania* (*Plasmodium*) *falcipara*. Trans. Roy. Soc. Trop. Med. Hyg. 55:98–102.

Ladda, R., J. Arnold, and D. Martin. (1966). Electron microscopy of *Plasmodium falciparum*. I. The structure of the trophozoites in erythrocytes of human volunteers. Trans. Roy. Soc. Trop. Med. Hyg. 60:369–375.

Sadun, E. H., and H. S. Osborne, eds. (1966). Research in malaria. Mil. Med. (Suppl.) 131:847–1272.

Short, H. E., and P. C. C. Garnham. (1948). The pre-erythrocytic development of *Plasmodium cynomolgi* and *Plasmodium vivax*. Trans. Roy. Soc. Trop. Med. Hyg. 41:785–795.

Short, H. E., N. H. Fairley, G. Covell, P. G. Shute, and P. C. C. Garnham. (1949). The pre-erythrocytic stage of *Plasmodium falciparum*. Trans. Roy. Soc. Trop. Med. Hyg. 44:405–419.

Trager, W., M. A. Rudzinska, and P. C. Bradbury. (1966). The fine structure of *Plasmodium falciparum* and its host erythrocytes in natural malarial infections in man. Bull. WHO 35:883–885.

Weathersby, A. B. (1966). Mosquito transmission of malaria. Mil. Med. (Suppl.) 131:929–931.

Yoeli, M., R. S. Upmanis, J. Vanderberg, and H. Most. (1966). Life cycle and patterns of development of *Plasmodium berghei* in normal and experimental hosts. Mil. Med. (Suppl.) 131:900–914.

*Avian malaria*

Garnham, P. C. C. (1966). Malarial Parasites and Other Haemosporidia. Blackwell Scientific Publications, Oxford, pp. 514–741; 1003–1073.

Hewitt, R. (1940). Bird Malaria. Amer. J. Hyg. Monog. Ser. No. 15. The Johns Hopkins Press, Baltimore, 228 pp.

Huff, C. G. (1930). *Plasmodium elongatum* n. sp., an avian malarial organism with an elongate gametocyte. Amer. J. Hyg. 11:385–391.

Huff, C. G. (1934). Comparative studies on susceptible and insusceptible *Culex pipiens* in relation to infections with *Plasmodium cathemerium* and *P. relictum*. Amer. J. Hyg. 19:123–147.

Huff, C. G. (1951). Observations on the pre-erythrocytic stages of *Plasmodium relictum, Plasmodium cathemerium,* and *Plasmodium gallinaceum* in various birds. J. Infec. Dis. 88:17–26.

Huff, C. G. (1957). Organ and tissue distribution of exoerythrocytic stages of various avian malarial parasites. Exp. Parasitol. 6:143–162.

Huff, C. G. (1966). Experimental research in avian malaria. In: B. Dawes, ed. Advances in Parasitology. Academic Press, New York, Vol. 1, pp. 1–65.

Huff, C. G. (1968). Recent experimental research on avian malaria. In: B. Dawes, ed. Advances in Parasitology. Academic Press, New York, Vol. 6, pp. 293–311.

Huff, C. G., and F. Coulston. (1944). The development of *Plasmodium gallinaceum* from sporozoites to erythrocytic trophozoite. J. Infec. Dis. 75:231–249.

Huff, C. G., and F. Coulston. (1946). The relation of natural and acquired immunity to the cryptozoites and metacryptozoites of *Plasmodium gallinaceum* and *Plasmodium relictum*. J. Infec. Dis. 78:99–117.

## FAMILY HAEMOPROTEIDAE

Hemoproteids are parasites of mammals, birds, and reptiles. The family is characterized by having all phases of schizogony in fixed tissue cells and only gametocytes appearing in the erythrocytes. Insect vectors, insofar as known, are pupiparous flies and ceratopogonid midges.

Three genera occur in mammals. *Hepatocystis* are parasites of simians, fruit bats, squirrels, and hippopotamuses. Species of *Culicoides* are the invertebrate vectors. *Nycteria* and *Polychromophilus* infect insectivorous bats. Bat flies of the family Nycteribiidae are known vectors of the latter genus and possibly of the former as well.

Two genera infect birds. *Haemoproteus columbae* of pigeons and *H. lophortyx* of quail are common and well-known species. Invertebrate vectors are pigeon or louse flies of the family Hippoboscidae. Exoerythrocytic schizogony is in endothelial cells of blood vessels, especially in the lungs. *Parahaemoproteus* is represented by *P. nettionis* of ducks, *P. canachites* of grouse, and *P. fringillae* and *P. garnhami* of various species of sparrows. Sporogony is in midges of the genus *Culicoides* and exoerythrocytic schizogony is in visceral organs.

Two genera are found in reptiles. Their life cycles are unknown. *Haemocystidium* is in lizards and *Simondia* in turtles.

*Haemoproteus columbae* is the best known species in the family and is used to illustrate the life cycle.

*Haemoproteus columbae*
Celli and Sanfelice, 1891 (Plate 31)

*Haemoproteus columbae* is a common parasite of columbriforme birds, especially in warm climates where the hippoboscid vectors flourish.

**Description.**    Gametocytes, the only form in the blood cells, are sausage-shaped and bent at each end so as to partially surround the nucleus of the host cell, as does *Plasmodium circumflexum* (Plate 30, 11). They contain pigment derived from erythrocytes. The host cell may become enlarged and its nucleus displaced laterally as the parasites attain full size. Micro-

## Explanation of Plate 31

A, Mononuclear leukocyte with parasite. B, Erythrocyte with ring stage of gametocyte. C, D, Growing gametocytes. E, Macrogametocyte. F, Microgametocyte. G, Pigeon host. H, Pigeon louse fly ( *Pseudolynchia canariensis*) invertebrate host.

1, Nucleus of blood cell; 2, young schizont; 3, ring stage of gametocyte in erythrocyte; 4, 5, growing gametocytes; 6, nucleus of mature macrogametocyte; 7, nucleus of mature microgametocyte.

a–u, In pigeon; a, sporozoite injected into bloodstream; b, sporozoites in blood go through heart and to lungs; c, sporozoite entering endothelial cell in lungs; d, growth of sporozoite in endothelial cell; e–k, schizogony in endothelial cells; e, formation of uninuclear cytomeres (host cell enlarges as parasite grows); f, g, cytomeres become multinucleate; h, cytomeres with large number of nuclei; i, great increase in number of small nuclei; j, formation of numerous minute merozoites in schizonts; k, escape of merozoites from schizonts into bloodstream (this process of schizogony may be repeated); l, merozoite entering red blood cell; m–u, gametogony in erythrocytes; m, ring stage in erythrocyte; n, o, formation of macrogametocyte; p, merozoite entering red blood cell; q, ring stage; r, s, formation of microgametocyte; t, mature macrogametocyte; u, mature microgametocyte in general circulation.

a′–l′, In pigeon louse fly; a′, microgametocyte and b′, macrogametocyte sucked up by pigeon louse fly; c′, exflagellation of microgametocyte with formation of microgametes; d′, macrogamete being fertilized by microgamete; e′, zygote; f′, ookinete in stomach; g′, ookinete migrating through epithelium of stomach; h′, young oocyst between epithelium and basement membrane; i′, oocyst with sporoblasts; j′, ripe oocyst filled with sporozoites; k′, rupture of ripe oocyst and liberation of sporozoites; l′, migration of sporozoites into salivary glands; m′, injection of sporozoites into blood by feeding pigeon louse fly.

Figure adapted from Aragao, 1908, Arch. Protistenk. 12:154.

---

gametocytes are 11.9 to 15.3 $\mu$ long by 2.5 to 4.3 $\mu$ wide with a large oval nucleus, 2.8 by 5.8 $\mu$, that stains light red; the cytoplasm is light blue. Microgametes are short, slender bodies enclosed in a single pellicular membrane. The elongate, somewhat irregular nucleus, enclosed in a double poreless membrane, occupies most of the length of the body. Electron micrographs reveal two large fibrils embedded in the cytoplasm, each with the characteristic pattern of one pair of central and nine pairs of peripheral fibrils. They are reflexed in the anterior end of the body so that in cross sections four of them appear; they extend singly to near the posterior end where two cross sections are seen (Fig. 14, A, B).

Macrogametocytes are 13.7 to 17.1 $\mu$ long by 2.3 to 2.8 $\mu$ in diameter with a small, almost spherical nucleus 2.1 by 2.3 $\mu$ in size. Both the nucleus and cytoplasm stain darker than in the microgametocyte. Electron micrographs show a double pellicle separated by a light layer. Cytostomes are present. The nuclear membrane with pores consists of two layers, the outer of which forms extensive evaginations into the cytoplasm. Food vacuoles, pigment, convoluted tubules, two fibrils, and double-membraned mitochondria are present (Fig. 14, C).

The unpigmented schizonts are primarily in the endothelial cells of capillaries of the lungs. Oocysts are attached to the stomach wall of the common pigeon fly *Pseudolynchia canariensis*

( = *Lynchia maura*). *Pseudolynchia brunea* and *Microlynchia pusilla* are vectors in Brazil.

**Life Cycle.**    Sporozoites, upon being injected into the skin of the vertebrate host by feeding pigeon flies, make their way into the circulation and through it to the various internal organs. Development is primarily in the lungs with their rich supply of oxygen and to a lesser extent in other organs such as the liver, spleen, and bone marrow.

Sporozoites entering the endothelial cells of lung capillaries round up to form small uninucleate trophozoites 3 to 4 $\mu$ in size. After 2 days of growth, the nucleus divides into 12 to 15 daughter nuclei followed by division of the cytoplasm into the same number of uninucleate masses called cytomeres. The mass constitutes the oocyst which eventually attains a size of 60 $\mu$. The nucleus of each cytomere divides repeatedly, forming many daughter nuclei. Following further growth, the multinucleate cytomeres become enclosed in a delicate membrane and continue to grow by extending along the branching capillaries. The cytomeres form branched masses that occlude the blood vessels. As development proceeds, the nuclei of each cytomere migrate to the surface where each one becomes clothed in a bit of cytoplasm to form individual merozoites. This process requires about 4 weeks following inoculation of sporozoites. Upon rupture of the membrane surrounding the ripe

Plate 31    *Haemoproteus columbae*                                                    145

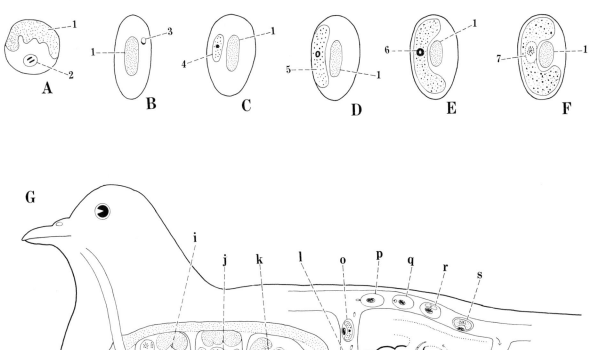

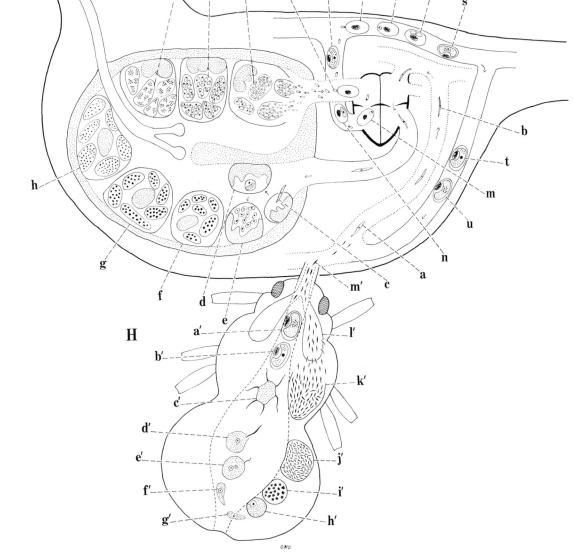

oocyst, a multitude of merozoites about 1 μ in diameter enters the bloodstream. Some reenter endothelial cells to initiate another but somewhat transitory exoerythocytic cycle. Most of them penetrate erythrocytes and develop into gametocytes.

Two kinds of schizonts occur in the lungs. The macroschizonts are much the larger, cytomeres are formed, and the cytoplasm stains bright blue, whereas microschizonts are about half the size of the former, cytomeric division may not occur, and the cytoplasm is lighter blue.

Duration of the preerythrocytic period is reported by workers to vary from 14 to 28 days.

The earliest stage of gametocytes in the red blood cells appears as minute rings similar to the ring stages of *Plasmodium*. When several appear in a single cell, only one or two survive. During the 5 or 6 days of growth necessary for full development, the gametocytes increase in size and change in shape. Growth continues until

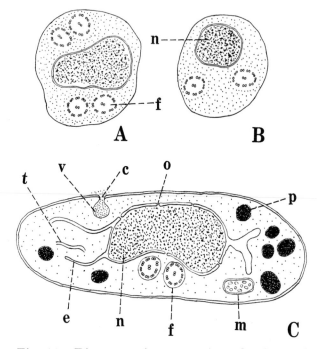

A    B

v    c    o

t    p

e    n    f    m    C

Fig. 14. Diagrammatic composite sketches of microgamete and macrogametocyte adapted from electron micrographs of *Haemoproteus columbae*. A, Cross section of anterior end of microgamete. B, Cross section of microgamete near posterior end. C, Longitudinal section of macrogametocyte. c, Cytostome; e, evagination of outer membrane of nucleus; f, fibril; m, mitochondrion; n, nucleus; o, opening or pore in nuclear membrane; p, malarial pigment; t, convoluted tubule; v, food vacuole. Adapted from Bradbury and Trager, 1968, J. Protozool. 15:89, 700.

they partially surround the nucleus, sometimes pushing it slightly to one side. Mature macrogametocytes are about 14 μ long and have become halter-shaped (named for the shape of stone weights used in broad jumping by the ancient Greeks). They contain about 14 small pigment granules. Microgametocytes are smaller (13 μ), less halter-shaped, and have about six to eight large pigment granules. No further growth or development takes place in the red blood cells.

Upon being ingested by pigeon flies, the gametocytes are freed by digestion of the blood corpuscles in the midgut. The nucleus of the microgametocytes undergoes three divisions, producing eight daughter nuclei. At the same time, short, finger-like extensions of the cytoplasm appear, into each of which a single nucleus is drawn to form eight flagellate, uninucleate microgametes by the process of exflagellation. Through lashing movements, the microgametes soon become detached and free.

During the period of exflagellation, the macrogametocyte has transformed by a less dramatic process into a macrogamete. A single microgamete enters to fertilize the macrogamete and form a zygote, thus completing the gametogenous phase of the cycle.

Sporogony begins with transformation of the rounded zygotes into motile, sausage-shaped ookinetes (meaning active eggs) up to 23 μ long. These glide to the brush border of the epithelial cells, penetrate them, and come to rest between the base of the cells and the thin membrane surrounding the gut. By day 4, they are about 4 μ in diameter and at maturity on day 9 about 40 μ. They protrude into the hemocoel as glistening somewhat spherical bodies still attached to the gut. Limited nuclear division provides a nucleus for each of the cytomeres, or sporoblasts, that forms. Multiple division of each cytomeric nucleus produces myriads of daughter nuclei, each of which in conjunction with a bit of cytoplasm forms a sporozoite somewhere between 6 and 11 μ long by about 1 μ in diameter. By days 10 to 12 after infection of the flies, the ripe oocysts rupture, releasing the mature sporozoites into the hemocoel. These disperse throughout the hemocoel. Some of them reach the salivary glands, where they enter and accumulate in the cells. Infection of pigeons results when flies feeding on them inject sporozoites into the skin with their saliva.

**Pathology.** About half of the erythrocytes are infected and ultimately destroyed, resulting in anemia and anorexia. The liver hypertrophies, becoming congested, and the spleen darkens with pigment and enlarges. Developing schizonts destroy endothelial cells in the lungs, as well as block capillaries, producing back pressure and accumulation of fluid which results in pneumonia. Young squabs die from complications produced by heavy infections.

## SELECTED REFERENCES

Adie, H. (1924). The sporogony of *Haemoproteus columbae*. Bull. Soc. Exot. 17:605–613.

Bennett, G. F., P. C. C. Garnham, and A. M. Fallis. (1965). On the status of the genera *Leucocytozoon* Ziemann, 1895 and *Haemoproteus* Kruse, 1890 (Haemosporoidiida: Leucocytozoidae and Haemoproteidae). Can. J. Zool. 43:927–932.

Bradbury, P. C., and W. Trager, (1968). The fine structure of the mature gametes of *Haemoproteus columbae*. J. Protozool. 15:89–102.

Bradbury, P. C., and W. Trager. (1968). The fine structure of microgametogensis in *Haemoproteus columbae* Kruse. J. Protozool. 15:700–712.

Coatney, G. R. (1936). A check-list and host-index of the genus *Haemoproteus*. J. Parasitol. 22:88–105.

Doflein, F., and E. Reichenow. (1953). Lehrbuch der Protozoenkunde. 6th ed. Gustav Fischer, Jena, p. 890.

Fallis, A. M., and G. F. Bennett. (1960). Description of *Haemoproteus canachites* n. sp. (Sporozoa: Haemoproteidae) and sporogony in *Culicoides* (Diptera: Ceratopogonidae). Can. J. Zool. 38:455–464.

Fallis, A. M., and G. F. Bennett. (1961). Ceratopogonidae as intermediate hosts for *Haemoproteus* and other parasites. Mosquito News 21:21–28.

Fallis, A. M., and D. M. Wood. (1957). Biting midges (Diptera: Ceratopogonidae) as intermediate hosts for *Haemoproteus* of ducks. Can. J. Zool. 35:425–435.

Garnham, P. C. C. (1966). Malaria Parasites and Other Haemosporidia. Blackwell Scientific Publications, Oxford, pp. 823–960.

Herman, C. M., and B. Gladding. (1942). The protozoan blood parasite *Haemoproteus lophortyx* O'Roke in quail at San Joaquin experimental range, California. Calif. Fish Game 28:150–153.

Mohammed, A. H. H. (1965). Studies on the schizogony of *Haemoproteus columbae*. Proc. Egypt. Acad. Sci. 19:37–46.

O'Roke, E. C. (1930). The morphology, transmission, and life history of *Haemoproteus lophortyx* O'Roke, a blood parasite of the California valley quail. Univ. Calif. Publ. Zool. 36:1–50.

Tarshis, I. B. (1955). Transmission of *Haemoproteus lophortyx* of the California quail by hippoboscid flies of the species *Stilbometopa impressa* (Bigot) and *Lynchia hirsuta* Ferris. Exp. Parasitol. 4:464–492.

Wenyon, C. M. (1926). Protozoology. William Wood and Co., New York (Reprinted by Hafner Publishing Co., New York, 1965), Vol. 2, p. 885.

## FAMILY LEUCOCYTOZOIDAE

Leucocytozoidae are blood sporozoans of birds, being both widespread in distribution and the most prevalent of the hemosporidians in these hosts. The family is characterized by sporogony in bloodsucking insects other than mosquitoes, and schizogony in hepatic and lymphoid-macrophage cells of birds. Unpigmented round gametocytes are in young red blood cells and elongated ones in leukocytes. Two genera, *Leucocytozoon* and *Akiba*, are recognized.

*Leucocytozoon* undergoes sporogony in ornithophilic simuliid flies. The oocysts are small, nonexpanding, and produce less than 100 sporozoites each with blunt anterior and pointed posterior ends. Schizonts in the form of megaloschizonts with cytomeres and centrally located host cell nucleus are in internal organs of birds where megaloschizogony takes place. Gametocytes, either round or elongate, are in blood cells. Common species are *L. danilewskyi* from owls, *L. simondi* from anseriformes, *L. sakharoffi* from corvine and other birds, *L. smithi* from turkeys, *L. bonasae* from grouse, and *L. fringillarum* from passeriformes.

*Akiba* has its sporogonic development in ceratopogonid midges of the genus *Culicoides*. Oocysts are small, nonexpanding, and produce sporozoites with both ends pointed. Schizonts in internal organs of birds are in the form of megaloschizonts with a laterally displaced host cell nucleus. Gametocytes are in blood cells. *Akiba caulleryi*, the only species known, is parasitic in chickens.

*Leucocytozoon simondi* serves as the example for the life cycle.

*Leucocytozoon simondi*
Mathis and Léger, 1910 (Plate 32)

This hemosporidian is parasitic in ducks and geese. It is widespread, occurring indigenously wherever the vertebrate hosts and black flies of

## Explanation of Plate 32

A, Young microgametocyte in blood cell (presumably a leukocyte). B, Larger microgametocyte in blood cell. C, Mature microgametocyte in distorted blood cell. D, Young macrogametocyte in blood cell. E, Older macrogametocyte. F, Mature macrogametocyte. G, Duck vertebrate host. H, Black fly (Simuliidae) invertebrate host.

1, Young microgametocyte; 2, host cell; 3, nucleus of host cell varies in shape, depending on the extent of crowding by developing gametocytes; 4, older microgametocyte; 5, mature microgametocyte; 6, nucleus; 7, young macrogametocyte; 8, older macrogametocyte; 9, mature macrogametocyte; 10, nucleus.

a–z, Schizogony (a–h, hepatic schizogony); a, sporozoites injected by black flies; b, sporozoites enter hepatic cells; c, trophozoite; d, young hepatic schizont; e, merozoites forming in schizont; f, mature hepatic schizont; g, ruptured schizont releasing hepatic merozoites and an "island" of cytoplasm containing merozoites (hepatic merozoites produce 1) another generation of hepatic schizonts, 2) megaloschizonts, and 3) round gametocytes); h, some hepatic merozoites enter hepatic cells to initiate second cycle of hepatic schizogony; i–z megaloschizogony; i, some hepatic merozoites enter macrophages of liver to initiate cycle of megaloschizonts; j, trophozoite; k, developing megaloschizont with grouping of nuclei to form cytomeres; l, megaloschizont with cytomeres; m, mature megaloschizont ruptures to release megaloschizogonic merozoites which produce only elongate gametocytes; n, immature megaloschizont in heart originating from hepatic merozoite; o, mature megaloschizont in heart releasing megaloschizogonic merozoites; p, hepatic merozoite leaving lung capillary; q, hepatic merozoite entering macrophage to form megaloschiz-

ont; r, young megaloschizont (preceding trophozoite omitted); s, megaloschizont with cytomeres; t, mature megaloschizont releasing merozoites; u, hepatic merozoite entering macrophages of intestine (also they enter macrophages of spleen, lymph nodes, and brain, which are omitted from drawing, where development is similar); v, young megaloschizont with cytomeres forming (trophozoite omitted); w, young megaloschizont with formed cytomeres; x, mature megaloschizont releasing megaloschizogonic merozoites; y, macrophage engulfing "island" from hepatic schizont (see g); z, hepatic and megaloschizogonic merozoites destined to become gametocytes.

a′–e′, Beginning of gametogony in duck; a′, megaloschizogonic merozoite of either male or female sex entering leukocytic cell to produce elongate gametocyte; b′, mature microgametocyte with open nucleus; c′, mature macrogametocyte with compact nucleus; d′, hepatic merozoite entering polychromatic cell to form round gametocyte; e′, mature round gametocyte.

a″–g″, Gametogony in black fly; a″, micro- and macrogametocytes (round gametocytes also occur) in stomach of black fly; b″, exflagellation of microgametocyte to form four microgametes; c″, macrogamete; d″, microgamete enters macrogamete to accomplish fertilization; e″, zygote; f″, motile ookinete; g″, ookinete migrates through intestinal epithelium; h″–l″, sporogony in black fly; h″, young oocyst between epithelium and membrane surrounding intestine; i″, developing oocyst with formation of sporoblasts; j″, ripe oocyst filled with sporozoites; k″, oocyst wall ruptures and sporozoites are freed in hemocoel; l″, sporozoites migrate to and enter salivary glands whence they are injected into vertebrate host.

Figures A–F adapted from O'Roke, 1934, Univ. Mich., School Forestry Conserv. Bull. 4.

---

the family Simuliidae come together. The parasite causes high mortality in ducklings.

**Description.**    Mature gametocytes, the only stage in the peripheral blood cells, are of two types, each appearing in a different kind of cell. Round gametocytes are in immature erythrocytes (Fig. 17, A, B). The blue-stained parasite with reddish nucleus fills almost the entire host cell. The more prevalent elongate gametocytes are in leukocytes which are greatly attenuated, attaining a length of 48 $\mu$, and with the nucleus abnormally long and slender. The parasites are 14 to 15 $\mu$ long by 4.5 to 5.5 $\mu$ in diameter. The cytoplasm of macrogametocytes stains dark blue and the nucleus is small, com-

pact, and deep red. Microgametocytes are similar in shape but slightly smaller. The cytoplasm is lighter blue and the nucleus is large, open, and lighter red. No pigment is present in the gametocytes.

Microgametes are filamentous bodies 23 $\mu$ long by about 1 $\mu$ in diameter, bounded by a single trilaminar membrane. The elongate nucleus, about equal in diameter to the microgamete, occupies the middle third of the body. Two internal axonemes, with the conventional single central and nine peripheral doublets, intertwine about each other for the entire length of the body (Fig. 15, A). Mitochondria, ribosomes, and other cytoplasmic organelles are lack-

**Plate 32**  *Leucocytozoon simondi*                                                                                     149

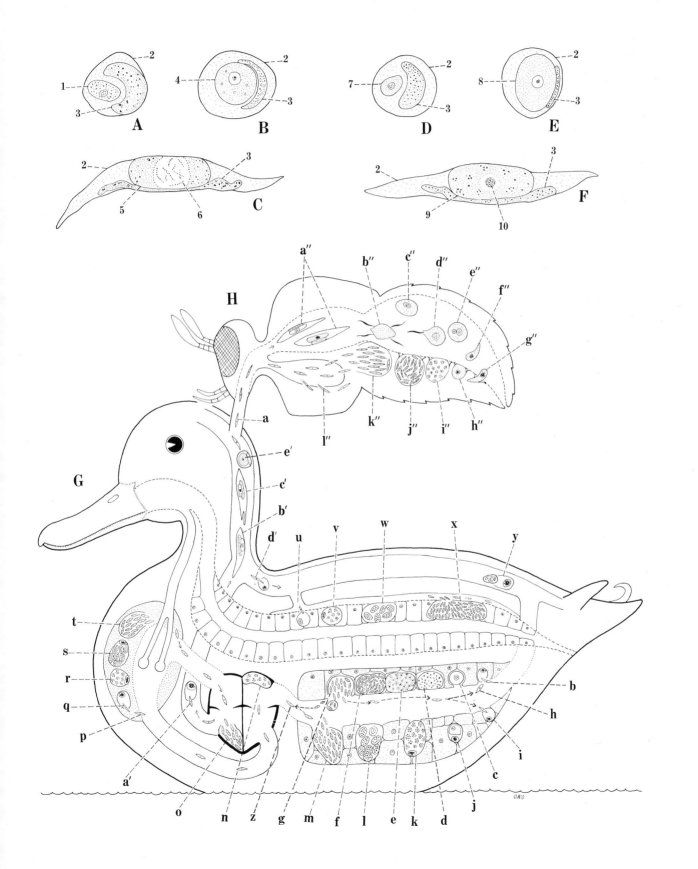

ing. The arrangement of the axonemes differs from that in the microgametes of *Plasmodium* and *Haemoproteus*.

Schizonts are of two types and occur in different kinds of cells. Hepatic schizonts in parenchymal cells of the liver are spherical and relatively small (Fig. 15, B). Megaloschizonts are large, consisting of many cytomeres. They are inside lymphoid-macrophage cells primarily in the spleen, liver, heart, kidneys, lungs, brain, and intestine. The parasite together with the enclosing host cell is the megaloschizont (Fig. 15, C).

Merozoites from hepatic schizonts and megaloschizonts are enclosed in a single membrane. Cytoplasmic structures include a single nucleus and mitochondrion, and paired organelles asso-

ciated with an anterior polar ring, or conoid. The cytoplasm of female merozoites is darker blue in Giemsa-stained specimens than that of the male merozoites.

Ookinetes in the stomach of simuliids are sausage-shaped and 24 to 30 $\mu$ long. The anterior end bears a conoid with an external opening. Micromeres, subpellicular microtubules, and other structures extend internally from the conoid (Fig. 17, C).

Oocysts are spherical bodies lying partially enclosed in the basal part of the epithelial cells of the midgut and the outer membrane surrounding the intestine. They protrude into the hemocoel. As development proceeds, the nucleus undergoes multiple division to form sporozoites (Fig. 16, A, B).

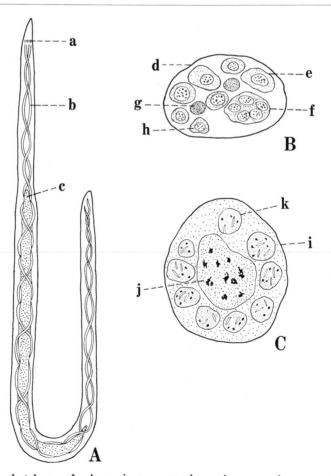

**Fig. 15.** Diagrammatic sketches of A, microgamete; B, hepatic schizont; and C, megaloschizont of *Leucocytozoon simondi*. **a,** Basal body; **b,** double axonemes; **c,** nucleus; **d,** hepatic schizont; **e,** developing merozoites; **f,** "island" of cytoplasm with developing merozoites; **g,** paired organelle; **h,** nucleus; **i,** macrophage containing megaloschizont; **j,** nucleus of macrophage; **k,** cytomere of parasite showing nuclei and clefts in cytoplasm. A adapted from Desser, 1970, Can. J. Zool. 48:647; B, from Desser *et al.,* 1970, Can. J. Zool. 48:331; C, from Desser, 1970, Can. J. Zool. 48:417.

Two types of sporozoites appear in the salivary glands of black flies. One is 6 to 7 $\mu$ long, stout, and with blunt ends; the other is 7 to 11 $\mu$ long, slender, and with a blunt anterior and a pointed posterior end. They are enclosed in a double-layered pellicle. Paired organelles extend internally from the conoid. The nuclear membrane is double, with the outer membrane evaginating and probably continuing with the endoplasmic reticulum (Fig. 16, C).

**Life Cycle.**    Sporozoites injected into the skin of ducks by infected simuliid flies work their way into the circulatory system and appear in the liver within 48 hours to begin the schizogenous phase of the life cycle. Here they leave the capillaries, penetrate the parenchymal cells of the liver, and round up to form uninucleate trophozoites 6 to 7 $\mu$ in diameter. Already the host cell has become distorted. Within 2½ days after infection and continued growth, multiple division of the nucleus initiates the schizogenous, or sexual stage. The daughter nuclei migrate to the surface of the cytoplasm and become incorporated in bits of cytoplasm to form the first generation of merozoites. Many schizonts have matured and ruptured by day 4 after infection. Merozoites appear in the liver sinusoids and circulating blood by day 5. They constitute the first generation of hepatic merozoites. Schizonts of sporozoite origin occur only in the liver, hence the designation hepatic schizonts. The first generation is completed by days 7 and 8 after infection. These merozoites may follow one of several courses. Some invade liver cells to produce another hepatic cycle, others penetrate young, or polychromatic, eryth-

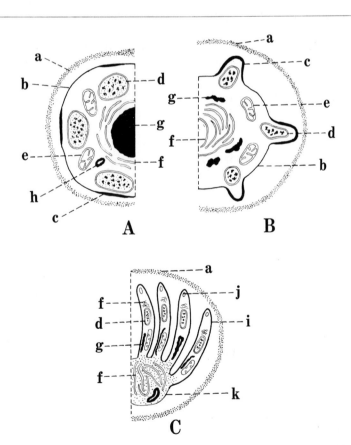

**Fig. 16.**    Schematic representation of development of sporozoites of *Leucocytozoon simondi*. A, Migration of daughter nuclei and other organelles toward surface of oocyst; B, formation of cytoplasmic buds with nuclei, mitochondria, and lipid material moving into them; and C, sporozoites developed and ready to separate from the residual cytoplasmic mass of the oocyst. a, Oocyst wall; b, oocyst; c, sporozoite bud; d, nucleus; e, mitochondrion; f, endoplasmic reticulum; g, electron dense material; h, lipid material; i, sporozoite; j, paired organelle; k, residual body of oocyst. Adapted from Desser and Wright, 1967, Can. J. Zool. 46:303.

rocytes to give rise to round gametocytes, and those phagocytized by macrophages grow into megaloschizonts throughout the body of the duck host, especially in the liver, spleen, lungs, kidneys, heart, intestine, and brain.

Some hepatic schizonts fail to divide completely into individual merozoites but separate off as clumps of cytoplasm known as "islands" containing several developing merozoites (Fig. 15, B, f). These "islands" circulate throughout the body in the blood. When phagocytized by macrophages in the various organs, they develop into megaloschizonts. The cytoplasm separates into nucleated masses called cytomeres (Fig. 15, C, k).

The megaloschizonts increase rapidly in size, forming large, irregularly shaped bodies up to 189 $\mu$ in size by days 8 to 9. They occur throughout the various organs but the greatest concentration is in the spleen and lymph nodes. During this time, the nuclei divide and subdivide many times, resulting in myriads of daughter nuclei. They migrate to the surface of the cytomeres, literally covering it. Each nucleus becomes enclosed in a bit of cytoplasm to form minute megaloschizogonic merozoites about 1 $\mu$ in size. By days 12 to 13, most of the megaloschizonts have reached maturity and ruptured, releasing up to a million merozoites each into the circulatory system. Remains of the megaloschizonts appear as hemorrhagic scars. Megaloschizonts in the brain are of smaller size and develop more slowly, requiring 14 to 15 days to mature. Megaloschizogonic merozoites of both male and female sexes penetrate white blood cells, primarily leukocytes, to develop into elongate gametocytes.

Gametogony begins in the vertebrate host when the merozoites enter blood cells and develop into gametocytes (Fig. 17 A, B). It is completed with the formation of zygotes in the gut of the black flies.

Hepatic merozoites entering either immature (polychromatic) or mature erythrocytes are fully developed round gametocytes in about 72 hours. The host cells become distorted and the nucleus compressed by the developing parasites.

At about the time round gametocytes appear in the erythrocytes, megaloschizogonic merozoites showing external dimorphism are released from mature megaloschizonts. They enter only leukocytes, mainly lymphocytes and monocytes,

to develop into elongate gametocytes. Elongation of the host cells begins while the parasites are still very small, indicating a cause other than pressure that produces distortion.

On days 10 to 12 after infection, round gametocytes outnumber elongated ones. By days 12 to 13, the two types are about equal in number. From day 14 on, the elongate gametocytes greatly exceed the round ones in number, the ratio being about 9 : 1.

Nutrition of developing gametocytes is by osmosis and of mature ones by ingestion of cytoplasm of host cells through cytostomes (Fig. 17, B).

Gametocytes ingested with the blood meals by black flies are released from the enclosing cells by the digestive processes. Microgametocytes promptly become active internally. The nucleus divides three times, producing eight daughter nuclei. As small cytoplasmic elevations appear on the microgametocyte, a daughter nucleus migrates into one of the protuberances and is drawn into the filamentous structure to form a microgamete (Fig. 15, A). Upon separation from the residual cytoplasmic mass, the microgamete moves to a macrogamete, enters it, and the two nuclei fuse to form a zygote. This act completes the gametogenous phase.

Sporogony begins soon after fertilization and formation of the zygote. Within 12 hours, the zygotes elongate into sausage-shaped motile ookinetes 24 to 30 $\mu$ long (Fig. 17, C). Upon contacting the epithelial cells of the midgut, the ookinetes penetrate them, coming to rest between the base of the cells and the outer membrane surrounding the gut where they develop into oocysts. Growth is rapid. Oocysts of various sizes project into the hemocoel as glistening, spherical bodies. By 36 to 72 hours after ingestion, daughter nuclei are present in the cytoplasm (Fig. 16, A). As growth proceeds, the nuclei migrate into cytoplasmic elevations (Fig. 16, B) covered with thickened pellicular plaques to form sporozoites in 2 days at optimal temperature (Fig. 16, C). Mature oocysts rupture, releasing active sporozoites into the hemocoel from where they scatter to all parts. Those entering the salivary glands are injected into the skin of ducks by feeding black flies.

**Epizootiology of Leucocytozoonosis.**    The coming of spring and its influence on the activities of the avian and fly hosts combine to produce favorable circumstances for intense and often fatal infection.

Gametocytes are scarce in infected adult ducks during the winter. With the onset of spring, the lengthening of daylight hours prompts migration and breeding activities of the birds, and bring on internal stresses. Infected carrier adult birds relapse under these conditions, with the production of merozoites and appearance of gametocytes in the peripheral blood. Their presence in the blood coincides with the spring emergence of the black flies, providing the latter

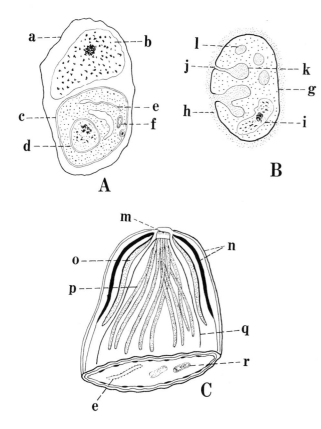

Fig. 17. Diagrammatic representations of round gametocytes (A, B) and anterior end of ookinete (C) of *Leucocytozoon simondi*. a, Polychromatic erythrocyte; b, nucleus of host cell; c, parasite with double membrane; d, nucleus with double membrane; e, endoplasmic reticulum; f, mitochondrion; g, mature round macrogametocyte; h, cytoplasm of host polychromatic erythrocyte; i, single-membraned nucleus; j, cytostome; k, food vacuole forming; l, food vacuole; m, conoid; n, pellicle; o, struts; p, micronemes; q, subpellicular microtubule; r, mitochondrion. A and B adapted from Desser *et al.*, 1970, Can. J. Zool. 48:331; C, from Desser, 1970, Can. J. Zool. 48:641.

with an abundant supply of gametocytes at a time when favorable temperature facilitates development of them, and when there is a large population of susceptible ducklings.

Various species of black flies serve as vectors, according to the region and season. *Simulium venustum* is an important host in Michigan. *Simulium croxtoni* and *S. euryadminiculum* transmit the infection during the spring in Ontario and *S. rugglesi* throughout the summer.

**Pathology.** Infection by *Leucocytozoon* may cause severe pathology, resulting in high mortality of ducklings. Host tissues respond variously to the presence of the parasite. Megaloschizonts in the brain and lungs, but not in other organs, are encapsulated by a double or single layer of fibers formed for lymphocytes and large monocytes. Ruptured mature megaloschizonts depleted of merozoites are invaded by eosinophils and large monocytes. They finally heal, leaving scars.

Pathological changes are most evident in blood cells, liver, spleen, lungs, heart, and bone marrow. Parasitism of liver cells by hepatic schizonts and macrophages by megaloschizonts, together with infection of erythrocytes and leukocytes by gametocytes results in the destruction of great numbers of these cells.

The liver is altered in several ways. Numerous necrotic areas containing macrophages, parasitized blood cells, and dead parenchymal cells expand and eventually coalesce until much of the organ is involved. Extensive periportal and general lymphatic infiltration, together with congestion of sinusoids and hepatic cords occur.

The spleen, whether megaloschizonts are present or absent, becomes enlarged due to congestion and proliferation of macrophages. Lungs show pulmonary congestion due to infiltration of the septa of air sacs by large histocyte-type cells. Round and elongate gametocytes concentrate in the lungs, liver, and spleen, blocking capillaries and impeding circulation.

Lymphocytic infiltration occurs around the blood vessels of the heart. Bone marrow undergoes excessive hyperplasia with replacement of fat by proliferating cellular elements.

Death results from severe anemia due to destruction of blood cells and bone marrow, pathological alterations of the liver originating from necrosis, anemia, and blockage of capillaries, enlargement of the spleen, and respiratory difficulty due to congestion of the lungs by lymphocytes and gametocytes.

## EXERCISE ON LIFE CYCLE

Many species of birds are parasitized by *Leucocytozoon*. The gametocytes, both round and elongate forms, may be found in blood smears stained with Wright's or Giemsa's stains. Young birds are more likely to show infections than older ones. Gametocytes are more prevalent in the blood of the latter during the spring than at other times.

Schizogenous stages are most likely to appear in sections or smear imprints of liver and lymph nodes, but also in lungs, spleen, kidneys, intestine, and brain. They include small hepatic schizonts in the liver and megaloschizonts containing cytomeres in other organs.

Black flies allowed to feed on infected birds show fully developed, spherical oocysts on the outer surface of the stomach and midgut in 1 to 2 days when dissected in physiological saline. To demonstrate living sporozoites, dissect salivary glands in saline solution and examine under high power of the microscope. Stained smears of infected salivary glands show the characteristic blue cytoplasm and red nucleus of the parasites.

Studies on transmission of parasites to uninfected birds may be done experimentally by injecting homogenates of liver or spleen, or whole blood containing merozoites from birds in the early stages of infection.

## SELECTED REFERENCES

Coatney, G. R. (1937). A catalog and host-index of the genus *Leucocytozoon*. J. Parasitol. 23:202–212.

Desser, S. S. (1967). Schizogony and gametogony of *Leucocytozoon simondi* and associated reactions in the avian host. J. Protozool. 14:244–254.

Desser, S. S. (1970). The fine structure of *Leucocytozoon simondi*. II. Megaloschizogony. Can. J. Zool. 48:417–421.

Desser, S. S. (1970). The fine structure of *Leucocytozoon simondi*. III. The ookinete and mature sporozoite. Can. J. Zool. 48:641–645.

Desser, S. S. (1970). The fine structure of *Leucocytozoon simondi*. IV. The microgamete. Can. J. Zool. 48:647–649.

Desser, S. S., J. R. Baker, and P. Lake. (1970). The fine structure of *Leucocytozoon simondi*. I. Gametocytogensis. Can. J. Zool. 48:331–336.

Desser, S. S., and A. M. Fallis. (1967). A description of the stages of sporogony of *Leucocytozoon simondi*. Can. J. Zool. 45:275–277.

Desser, S. S., and K. A. Wright. (1967). A preliminary study of the fine structure of the ookinete, oocyst, and sporozoite formation of *Leucocytozoon simondi* Mathis and Leger. Can. J. Zool. 46:303–307.

Fallis, A. M., R. C. Anderson, and G. F. Bennett. (1956). Further observations on the transmission and development of *L. simondi*. Can. J. Zool. 38:389–404.

Fallis, A. M., and G. F. Bennett. (1958). Transmission of *Leucocytozoon bonasae* Clarke to ruffed grouse (*Bonasa umbellus* L.) by the black flies *Simulium latipes* Mg. and *Simulium aureum* Fries. Can. J. Zool. 36:533–539.

Fallis, A. M., D. M. Davies, and M. A. Vickers. (1951). Life history of *L. simondi* Mathis and Leger, in natural and experimental infections and blood changes produced in the avian host. Can. J. Zool. 29:305–308.

Garnham, P. C. C. (1966). Malaria Parasites and other Haemosporidia. Blackwell Scientific Publications, Oxford, pp. 963–987.

Huff, C. G. (1942). Schizogony and gametocyte development in *Leucocytozoon simondi*, and comparisons with *Plasmodium* and *Haemoproteus*. J. Infect. Dis. 71:18–32.

Newberne, J. W. (1957). Studies on the histopathology of *Leucocytozoon* infections. Amer. J. Vet. Res. 18:191–199.

O'Roke, E. C. (1934). A malaria-like disease of ducks caused by *Leucocytozoon anatis* Wickware. Univ. Mich. School of Forestry Conserv. Bull. 4, pp. 1–44.

### Summary of Life Cycles of Some Sporozoan Species

As seen in the preceding discussions of some representative genera of sporozoans, the life cycles follow a basic pattern, but variations among them occur. The cycles are summarized diagrammatically on Plate 33, where these differences are pointed out. The life cycles fall into two groups: those with a single host and those with two hosts. Variations occur within each of these major groups.

In the genera with one host, *Lankesteria* and *Schizocystis* complete all development in the lumen of the gut. Schizogony is lacking in *Lankesteria* and only a simple form of it is present in *Schizocystis*. Infection of new hosts is by swallowing ripe oocysts (Plate 33, *1, 2*). In the case of *Eimeria*, schizogony and gametogony are typically in the intestinal mucosa. Exceptions are *E. stiedae* in the liver of rabbits and *E. truncata* in the kidneys of geese. Sporogony is in the soil. Infection is by ingesting ripe oocysts (Plate 33, *3*).

Members of the two-host group have no free-living phase in the life cycle. All stages of development take place in the bodies of the two hosts.

*Aggregata* (Plate 33, *4*) occurs in two invertebrates, crabs and cuttlefish. Schizogony is in the intestinal epithelium of the crabs, with sporogony in a similar location in the cuttlefish. Infection is dependent on one host being eaten by the other. The other genera of the two-host group involves a vertebrate and an invertebrate host for completion of each life cycle. There are no free-living stages. In *Schellackia*, all phases of development, i.e., schizogony, gametogony, and sporogony, occur in the intestinal epithelium of lizards. Sporozoites enter the erythrocytes. Lizard mites sucking blood serve as a reservoir for sporozoites that are transmitted without further development to the rat when mites containing them are eaten (Plate 33, *5*). In the remaining genera, development in the vertebrate hosts is in tissues other than the intestinal epithelium, typically blood cells. Gametogony begins in blood cells and is completed in the alimentary canal of the invertebrate host. Sporogony takes place between the peritoneal membrane and longitudinal muscles of the midgut. *Hepatozoon* is transmitted to rats when infected mites are eaten (Plate 33, *6*). In the remaining genera, *Haemogregarina*, *Plasmodium*, *Haemoproteus*, and *Leucocytozoon*, the infective stages for the bloodsucking invertebrate hosts are taken with blood meals and after appropriate development in the wall of the midgut the infective sporozoites are inoculated into the vertebrates (Plate 33, *7–10*).

### ORDER PIROPLASMIDA

This group includes blood parasites of the families Babesiidae and Theileriidae. The present organization is used as one of convenience because of the current state of uncertainty among protozoologists regarding the classification of the group.

The parasites appear as small, nonpigmented, round, oval, pyriform, or rod-shaped bodies in erythrocytes of mammals. Cattle, sheep, goats, horses, swine, dogs, and cats are commonly infected. In addition, a number of wild animals, including rodents, are infected.

Diseases caused by these parasites, especially in cattle, are of great economic importance in warm countries where the tick hosts occur. Under natural conditions, ticks serve as vectors. They acquire the parasites in blood sucked from infected animals. After a period of development in the ticks, infective stages appear in the salivary glands of the young and are transmitted to new hosts while feeding on them.

### FAMILY BABESIIDAE

The Babesiidae are minute nonpigmented parasites of the erythrocytes of mammals. Transmission is by ticks.

Traditionally, the Babesiidae have been placed in the Sporozoa, based on the concept that sexual phases occur in their developmental cycles. Some recent workers, however, believe that sexuality does not occur. They claim, therefore, that these forms belong in a separate group. Leaving the Babesiidae in the Sporozoa, as is done here, does not in any way constitute a challenge to this point of view.

### *Babesia* Starcovici, 1893

Species of *Babesia* occur in a variety of mammals in various parts of the world. These include rodents of many species, hares, carnivores, swine, sheep, goats, cattle, horses, and primates. Forms which may be *Babesia* are found in insectivores.

The parasites may be divided into two groups on the basis of size. Some common large species more than 3 $\mu$ long include *B. canis* (dogs), *B. caballi* (horses), *B. bigemina* (cattle), *B. motasi* (sheep), and *B. trautmani* (swine). Small species less than 3 $\mu$ long include *B. gibsoni* (dogs), *B. felis* (cats), *B. bovis* (cattle), and *B. equi* (horses).

### *Babesia bigemina* (Babes, 1881) (Plate 34)

*Babesia bigemina* is a parasite of the erythrocytes of cattle in warm regions of the world. Its normal distribution in the United States was limited by that of its tick host [*Boophilus* (*Margaropus*) *annulatus*] which occurs in tropical and subtropical climates.

Smith and Kilborne, studying babesiosis in the southern part of the United States in the early 1890's, discovered that development of the infective stage of the parasite takes place in

## Explanation of Plate 33

I. Order Gregarinida
  A. Single host.
   1. No schizogony; gametogony and sporogony in lumen of Malpighian tubules and hindgut of insect; infection by ingestion of sporocysts containing sporozoites (*Lankesteria*, Fig. 1.)
   2. Simple schizogony followed by gametogony in lumen of gut of invertebrate; infection by ingestion of sporocysts containing sporozoites. (*Schizocystis*, Fig. 2.)
II. Order Coccidida
  A. Single host.
   1. Schizogony and gametogony in intestinal epithelium; sporulation outside; infection by ingestion of sporocysts containing sporozoites. (*Eimeria*, Fig. 3.)
  B. Two hosts
   1. Schizogony submucosal in gut of crab and gametogony and sporogony submucosal in gut of cuttlefish; infection by ingestion of merozoites in crab by cuttlefish and sporozoites in cuttlefish by crabs. (*Aggregata*, Fig. 4.)
   2. Schizogony, gametogony, and sporogony submucosal in gut of lizards; sporozoites in erythrocytes of lizards and later accumulate in intestinal epithelium of lizard mites; infection by ingestion of infected mites. (*Schellackia*, Fig. 5.)

3. Schizogony in hepatic cells of rat; gametocytes in leukocytes; gametogony in gut and sporogony in hemocoel of rat mites; infection by ingestion of infected mites. (*Hepatozoon*, Fig. 6.)
4. Schizogony and gametocytes in erythrocytes of water tortoise; gametogony and sporogony in intestine of leeches; infection by injection of sporozoites. (*Haemogregarina*, Fig. 7.)
III. Order Haemosporida
  A. Two hosts
   1. First schizogony in liver cells, second schizogony and gametocytes in erythrocytes; gametogony in lumen of intestine and sporogony in wall of intestine of mosquitoes; infection by injection of sporozoites. (*Plasmodium*, Fig. 8.)
   2. Schizogony in endothelial cells of lungs; gametocytes in erythrocytes of birds; gametogony in lumen of intestine and sporogony in intestinal wall of hippoboscid flies; infection by injection of sporozoites. (*Haemoproteus*, Fig. 9.)
   3. Megaloschizonts in macrophages of intestine, spleen, heart; hepatic schizonts in macrophages of liver; gametocytes in mononuclear leukocytes; gametogony in lumen of intestine and sporogony in wall of intestine of black flies; infection by injection of sporozoites. (*Leucocytozoon*, Fig. 10.)

ticks. This was the first instance in which a protozoan parasite was demonstrated to develop in and be transmitted by an arthropod. Following clarification of the biology and ecology of the tick host, a program of its eradication by dipping cattle in a tickicide resulted in elimination of the parasite from the United States.

The organism causes an acute, often fatal, disease of adult cattle known as Texas cattle fever, red water fever, or hemoglobinuria fever. The two latter names refer to the red color of the urine due to the presence of hemoglobin resulting from the massive destruction of red cells by the parasites.

**Description.**    The parasites are nonpigmented organisms in the erythrocytes. Usually, they are pear-shaped but round, oval, or irregularly shaped forms appear. They are 4 $\mu$ long by 1.5 $\mu$ at the widest point. They commonly occur in pairs and are united at the pointed tips;

sometimes up to four individuals are present in a single blood cell. They extend across the cell. The nucleus is small; there is a fine line of chromatin granules extending from it toward the pointed end of the parasite.

Upon entering cells of the ovarian tissue of the tick, each parasite assumes a spherical shape within a space known as the parasitophorous vacuole. Following growth and development, the merozoites become somewhat elongated and typical in shape (Fig. 18). The pellicle and nuclear membrane are double. The outer membrane of the nucleus forms tubules that are continuous with the pre- and postnuclear areas of roughened endoplasmic reticulum. A Golgi complex, mitochondria, and paired organelles appear to be lacking, but vesicular organelles and numerous micronemes are present. Anteriorly, the merozoites are somewhat umbrella-shaped and have a prominent polar ring. About 32 ribs,

**Plate 33**   *Summary of Development of Some Sporozoan Parasites*   157

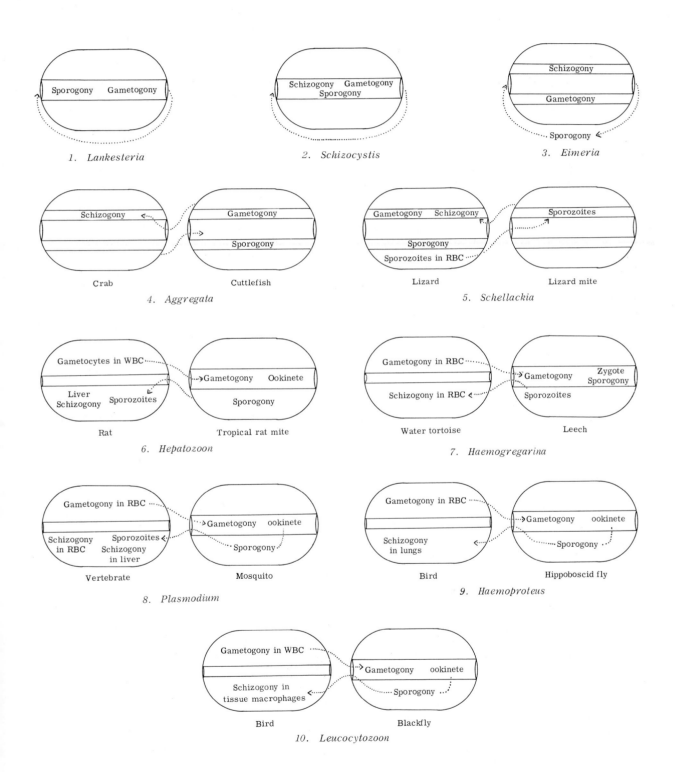

1. *Lankesteria*

2. *Schizocystis*

3. *Eimeria*

4. *Aggregata*

5. *Schellackia*

6. *Hepatozoon*

7. *Haemogregarina*

8. *Plasmodium*

9. *Haemoproteus*

10. *Leucocytozoon*

## Explanation of Plate 34

A, Bovine host. B, Tangential section of adult female tick attached to host. C, Unembryonated egg. D, Embryonated egg. E, Larval tick attached to host. F, Nymphal tick attached to host.

1, Chelicera; 2, infected erythrocytes being swallowed; 3, parasites escaping from erythrocyte in stomach; 4, spherical multinucleate parasite developed from pyriform stage in erythrocytes; 5, spindle-shaped stage presumably resulting from division of multinucleate sphere (4); 6, uninucleate cell probably originating from pyriform stage in red blood cells; 7, binucleate spherical cell believed to result from union of spindle-shaped cell (5) and uninucleate spherical cell (6); 8, vermiform stage thought to develop from binucleate sphere (7); 9, vermicule in intestinal epithelium; 10, multifission of nuclear material; 11, formation of vermicules from multifission body; 12, rupture of cell, releasing vermicules into hemocoel; 13, vermicule from intestinal generation in cells of Malpighian tubules; 14, multifission of nuclear material in Malpighian tubules; 15, rupturing Malpighian cell with release of vermicules into hemocoel; 16, egg in ovary; 17–19, vermicules accumulating in developing eggs in ovary, oviduct, and uterus; 20, unparasitized mature egg; 21, vagina; 22, rectum; 23, coxa; 24–28, details of development of parasites in intestinal epithelium of adult female tick; 24, vermicule in cell; 25, spherical stage with nuclear material increasing in amount; 26, multifission stage; 27, formation of vermicules; 28, individual vermicule free in hemocoel; 29–34, stages of development in cells of Malpighian tubules of adult female tick; 29, vermicule in cell; 30, 31, growth with increase in the amount of nuclear material; 32, multifission of nuclear material; 33, formation of vermicules in cells of Malpighian tubules; 34, vermicule free in hemolymph; 35, vermicules in yolk of undeveloped egg; 36, egg shell; 37, unhatched larva; 38, intestine; 39–42, growth of parasite in intestinal epithelium of unhatched larval tick; 39, vermicule; 40, vermicule inside cell; 41, growth of chromatin material; 42, multifission of chromatin; 43, formation of vermicules; 44–48, details of development in intestinal epithelium of unhatched larval tick; 44, vermicule in cell; 45, increase of chromatin material; 46, multifission of nuclear body; 47, fully developed stage with vermicules; 48, free vermicule in hemolymph; 49, vermicules in intestinal epithelium of feeding larva (E); 50, rupture of infected cell, freeing vermicules in hemolymph; 51–59, development in salivary glands of nymph (F); 51, vermicule in hemolymph; 52, vermicule entering salivary gland; 53, vermicule in cell of salivary gland; 54, 55, growth of chromatin; 56, multiple fission of nuclear material; 57, vermicules in cells; 58, rupture of cell with liberation of vermicules in lumen of salivary glands; 59, vermicules being injected into vertebrate host by feeding nymph; 60–65, details of developmental forms in salivary cells in nymphal stages; 60, vermicule in cell; 61, 62, growth of chromatin; 63, multifission of nuclear material; 64, formation of short oval vermicules; 65, details of vermicules in lumen of salivary glands which are infective to erythrocytes when injected into vertebrate host.

a, Parasite injected by tick, entering erythrocyte; b, "ring" stage; c, ameboid form; d, e, trophozoites in stages of binary fission; f, two trophozoites escaping into blood stream from destroyed erythrocyte and ready to enter other erythrocytes.

Figures adapted from various sources, including Riek, 1954, Austr. J. Agr. Res. 15:802; A, original.

---

each with an inner microtubule, radiate outward from the polar ring toward the pellicle. There is no opening in the pellicle at the polar ring.

**Life Cycle.**    *Babesia bigemina* is a parasite of ticks (*Boophilus, Rhipicephalus, Haemaphysalis*) and cattle. It is transmitted to cattle by the offspring of female ticks that have fed on infected bovines.

*Boophilus annulatus* has but one host in its life cycle and is known as one-host tick since all of the feeding and mating takes place on a single animal. After mating and engorging, the females drop to the ground, oviposit, and die. Larval ticks, commonly referred to as seed ticks, hatching from the eggs, climb on vegetation and attach to animals as they brush by the plants.

Parasites injected into the bloodstream by the young ticks enter erythrocytes. They form first into ring-shaped and then into ameboid trophozoites. Multiplication is by binary fission, producing two pyriform bodies which are at first attached by the pointed ends, but separate later. Upon destruction of the blood cells, they enter other erythrocytes and divide, thus continuing the cycle.

Development of the parasite in the ticks proceeds through a series of complicated stages of multiplication repeated in adult females and their young, resulting in tremendous numbers of the infective forms known as vermicules or merozoites. Transmission of the parasite from the female tick to her offspring is through the eggs and is designated as transovarial.

The first report on the life cycle of *Babesia*

**Plate 34** *Babesia bigemina* 159

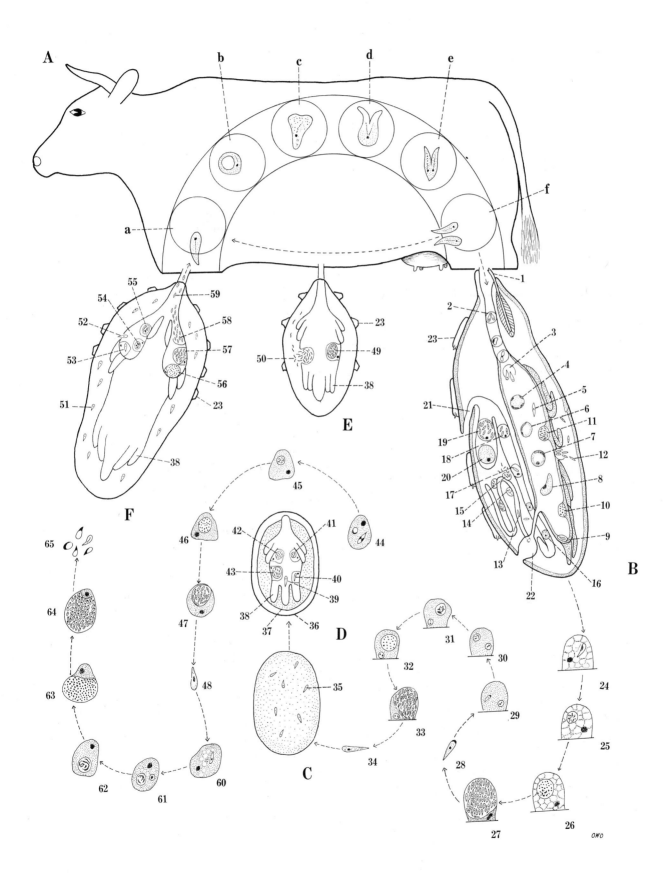

*bigemina* was by Dennis. He believed the parasite to be a sporozoan in which a sexual phase appears in its life history. According to him, the isogametes in the blood of the vertebrate host conjugate in the gut lumen of engorged female ticks (*Boophilus annulatus*) to form motile zygotes. These migrate through the wall of the gut into the uterine eggs. Through a process of sporogony, the zygotes form numerous sporozoites that enter the salivary glands of larval ticks whence they are injected into the vertebrate host when these feed. He called this transovarial transmission from one generation of ticks to the next.

Regendanz and Reichenow, working on the life cycle of *Babesia canis* of dogs, could find no evidence of sexual reproduction in the tick *Dermacentor reticulatus* as reported by Dennis. They stated that most of the forms freed from the erythrocytes in the gut of ticks died. The survivors penetrate the epithelial cells of the gut without change, grow, and divide by repeated binary fission, forming vermicules that pass into the hemolymph. These stages enter the eggs of the ticks where they divide, forming relatively few rounded forms. After hatching of the tick and its first molt to form the nymphal stage, the vermiculate bodies migrate to the cells of the salivary glands and continue development by repeated binary fission. Eventually, they enter the lumen of the salivary glands and are injected into the host by the feeding ticks. Regendanz reported that *B. bigemina* has a similar development in *Boophilus microplus*.

More recent studies by Riek on the life cycle of *Babesia bigemina* in the tick vector *Boophilus microplus* in Australia show differences from the cycles reported by Dennis and by Regendanz and Reichenow. This cycle is followed, even though all of the points are not entirely clear.

According to Riek, the trophozoites in the erythrocytes of cattle vary in shape, although typically they are pyriform and paired with the pointed ends approximating each other. Generally, they are uninucleate but at times during the infection some show two nuclei, one of which stains more deeply than the other and may be extruded into the cytoplasm of the erythrocyte.

Upon ingestion of the infected blood cells by ticks, the parasites are freed by the digestive processes. Most of them show pycnotic nuclei and spikey rayed cytoplasm. These forms are destroyed. The survivors are of three types. The most common type is 3 to 5 $\mu$ in diameter and has a large central vacuole surrounded by a thin layer of cytoplasm. The chromatin is usually arranged as a thin layer around the periphery of the parasite but it may also appear as a dot at each pole, or a single or double dot. The second and comparatively rare form has three to four separate chromatin bodies around the periphery. It divides into as many elongated spindle-shaped bodies as there are nuclei, each bearing a central nucleus and measuring 4 to 7 $\mu$ long by 1 to 2 $\mu$ wide. The third form, likewise, is spherical and has two nuclei, one of which is elongated and located at the periphery, and the other is round. The origin of this cell is uncertain but Riek suggested that it may be the result of union of the uninucleate round type mentioned first and the spindle-shaped cells arising from division of the second multinucleate type. He pointed out, however, that the true relationship of these cells in the life cycle is not entirely clear. It should be recalled that Regendanz and Reichenow did not observe this type of transformation of the trophozoites in the gut of the replete female ticks.

Toward the end of 24 hours, the developmental stages appear as blunt cigar-shaped bodies 8 to 10 $\mu$ long with a central nucleus. They penetrate the epithelial cells of the gut and develop into spherical bodies 9 to 16 $\mu$ in diameter. During day 2, they undergo multiple fission of the nucleus, with the chromatin being distributed as numerous small dots throughout the cytoplasm. A ring of cytoplasm surrounds each nuclear particle, forming small nucleated bodies 3 to 6 $\mu$ in diameter. On day 3, these small bodies develop into vermicules, measuring 9 to 13 $\mu$ long by 2 to 3 $\mu$ wide, which migrate from the epithelial cells of the gut into the hemolymph. After 4 days, the vermicules enter the cells of the Malpighian tubules, round up, and undergo multiple fission similar to that occurring in the gut epithelium. The resultant vermicules, which are similar to their predecessors, migrate to the eggs. At first they are scattered in the yolk, but as the larval ticks develop they enter epithelial cells of the gut where multiple fission of the nucleus takes place and more vermicules or merozoites (Fig. 18) are formed. Upon rupture of the infected epithelial cells, these vermicules enter the gut lumen and the hemolymph, presumably remaining there during the larval stage which is 5 to 7 days after attachment. They migrate to the salivary glands of the nymphs, round up, grow in size, and undergo multiple nuclear division. Enormous numbers of minute pyriform bodies about 2 to 3 $\mu$ long by 1 to 2 $\mu$ wide are produced. These are injected into the host by the feeding nymphs. The parasites appear in the erythrocytes 8 to 10 days after larval attachment which is 1 to 5 days after the onset of the nymphal stage.

This complicated life cycle may be summarized in four steps as follows: 1) binary fission of trophozoites occurs in the erythrocytes of the bovine host; 2) trophozoites ingested by adult female ticks undergo growth and a process of multifission in the epithelial cells of the gut, producing vermiculate bodies that enter the cells of the Malpighian tubules and repeat multifission; vermicules from the Malpighian tubules enter and accumulate in developing eggs; 3) vermicules in the yolk migrate to the gut epithelium of the larval ticks (seed ticks), grow, undergo multifission, and are released into the hemolymph; and 4) the vermicules invade the salivary glands

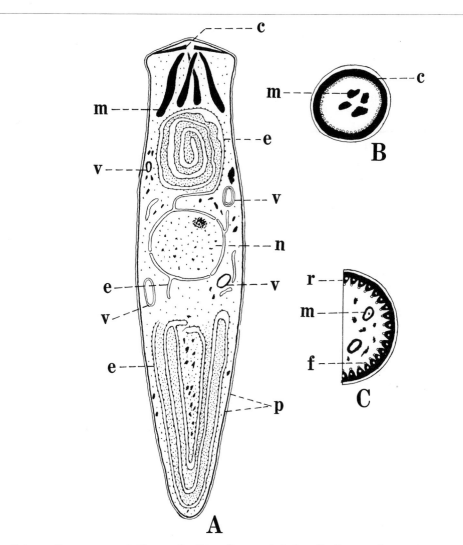

Fig. 18. Schematic representation of the fine structure of an extended merozoite of *Babesia bigemina*. **A,** Longitudinal section; **B,** cross section through conoid or polar ring; and **C,** cross section immediately posterior to polar ring. **c,** Conoid or polar ring lined on inner surface with tubular fibrils; **e,** endoplasmic reticulum which is continuous with space between nuclear membranes; **f,** subpellicular tubular fibril; **m,** microneme; **n,** nucleus surrounded by a double membrane with pores and containing a nucleolus; **p,** pellicular membrane (thick inner and thin outer) layers separated by a space; **r,** ribs; **v,** vacuoles (thick- and thin-walled). Adapted from Friedhoff and Scholtyseck, 1969, *Z. Parasitenk.* 32: 266.

of newly formed nymphal ticks, grow, and divide by multifission to form vermiculate progeny that are injected into the vertebrate host by feeding nymphs and subsequent developmental stages of the ticks.

While *Babesia bigemina* is transmitted by a one-host tick in America, one-, two-, and three-host ticks serve as vectors in other parts of the world. Transovarial transmission occurs only in the one-host ticks.

Inasmuch as all stages of the one-host ticks, including the adults, remain on one animal and do all of their feeding on it, direct transmission from one host to another by them does not occur. When the gravid females drop to the ground, they die following oviposition. In the process of their development, the babesias go to the eggs in the female and to the salivary glands of the larvae which transmit them to a new host.

In the two-host tick, the larvae and nymphs feed on a parasitized host, become infected, and drop to the ground to molt to the adult stage. The infected adults, upon reattaching to and feeding on another animal, transmit the disease.

In the three-host ticks, each stage feeds, drops to the ground, molts, and reattaches to feed again. Thus, infection acquired in any stage of the ticks may be transmitted by the next one.

Because of this change of animals by the two- and three-host ticks, transovarial transmission is unnecessary and, indeed, does not occur.

Other species of babesiids occurring in cattle, sheep, goats, horses, swine, primates, elephants, dogs, cats, hyenas, civets, mice, rats, and hares in various parts of the world are given by Doflein and Reichenow and Reik.

**Ecological Factors Affecting Infection of Ticks.** Several factors, such as degree of infection of bovine host, environmental temperature, and strains of ticks, are important in the development and production of infective stages of the parasite.

The extent of infection of female ticks, their survival after feeding, and the rate of egg production are adversely affected by the number of parasitized erythrocytes available in the blood. When 20 percent or more of the erythrocytes are infected, 90 percent of the engorged females die within 7 days after dropping off the host. Presumably death is the result of damage done by the developing parasites. Surviving ticks produce fewer eggs than uninfected ones. Four to 25 percent of ticks feeding on cattle with 20 to

50 parasites per ml of blood become infected. The number of vermicules in the hemolymph and eggs is low.

The temperature and humidity in the microenvironment exert an influence on the survival and development of ticks off the host. Low temperature retards development of both the parasite and vector. Too dry conditions result in the death of the developing ticks.

All ticks are not equally susceptible to infection even when exposed to a similar or higher rate of infection. This is probably the result of variation in susceptibility of different strains of ticks.

**Pathology.** Pathogenicity of babesiosis varies between animals native to an infested area and those from regions where the parasites do not occur. In endemic areas, the cattle develop a degree of immunity resulting from continuous infection and seldom show severe clinical symptoms.

Within 5 to 10 days after infection of nonimmune animals, the parasitemia has developed to a high level, with destruction of many infected erythrocytes, release of hemoglobin in the bloodstream, and accompanying anemia. Clinical symptoms generally are fever, malaise, loss of appetite, listlessness, anemia, and hemoglobinuria. Visceral organs are involved. The spleen enlarges, showing dark red pulp, the liver degenerates and becomes icteric, and the kidney tubules degenerate from blockage.

Death results from occlusion of capillaries of visceral organs by parasitized erythrocytes, free parasites, and cellular detritis. The combination of these conditions produces toxic metabolites and anoxia.

**Control.** Control of the disease is dependent on eliminating the vector and utilizing the phenomenon of premunition.

Eradication of the ticks by the systematic application of tickicides to cattle, horses, and mules has resulted in the elimination of the parasite. This method was used successfully in the southern part of the United States during the early part of the present century.

Since young calves are highly resistant to infection, they acquire the parasites slowly under natural conditions as their normal resistance declines. By the time they are yearlings, the slowly acquired infections have produced suffcient immunity by premunition to keep the parasite population at such a low level that clinical symptoms do not develop. Before shipping young cattle from parasite-free areas into regions where babesiosis is prevalent, they are immunized by light infections artificially induced. Immunized and recovered animals remain carriers of the piroplasms.

## EXERCISE ON LIFE CYCLE

Where babesiids can be found in dogs or rodents, observations may be made on the growth of trophozoites in the erythrocytes. The stages in the ticks may be detected in stained smears of the organs in which the developing forms occur.

## SELECTED REFERENCES

Dennis, E. W. (1930). The morphology and binary fission of *Babesia bigemina* of Texas cattle-fever. Univ. Calif. Publ. Zool. 33:179–192.

Doflein, F., and E. Reichenow. (1953). Lehrbuch der Protozoenkunde. 6th ed. Gustav Fischer, Jena, p. 957.

Friedhoff, K., and E. Scholtyseck. (1969). Feinstrukturen der Merozoiten von *Babesia bigemina* im Ovar von *Boophilus microplus* und *Boophilus decoloratus*. Z. Parasitenk. 32:266–283.

Holbrook, A. A., D. W. Anthony, and J. A. Johnson. (1968). Observations on the development of *Babesia caballi* (Nuttall) in the tropical horse tick *Dermacentor nitens* Neumann. J. Protozool. 15:391–396.

Levine, N. D. (1961). Protozoan Parasites of Domestic Animals and of Man. Burgess Publishing Co., Minneapolis, p. 285.

Levine, N. D. (1971). Taxonomy of piroplasms. Trans. Amer. Microsc. Soc. 90:2–33.

Regendanz, P. (1936). Ueber den Entwicklungsgang von *Babesia bigemina* in der Zecke, *Boophilus microplus*. Zentralbl. Bakteriol. 1. Abt. Orig. 137: 423–428.

Regendanz, P., and E. Reichenow. (1933). Die Entwicklung von *Babesia canis* in *Dermacentor reticulatus*. Arch. Protistenk. 79:50–71.

Reichenow, E. (1935). Uebertragenweise und Entwicklung der Piroplasmen. Zentralbl. Bakteriol. 1. Abt. Orig. 135:108–119.

Reik, R. F. (1964). The life cycle of *Babesia bigemina* (Smith and Kilborne, 1893) in the tick vector *Boophilus microplus* (Canastrini). Austr. J. Agr. Res. 15:802–821.

Reik, R. F. (1968). Babesiosis. In: D. Weinman and M. Ristic, eds. Infectious Blood Diseases of Man and Animals. Academic Press, New York, Vol. 2, pp. 219–268.

Smith, T., and F. L. Kilborne. (1893). Investigations into the nature, causation, and prevention of Texas or southern cattle fever. U.S. Dep. Agr. Bur. Ani. Ind. Bull. 1, pp. 1–301.

## FAMILY THEILERIIDAE

The species of this family occur commonly in cattle, sheep, and goats in Africa, southern Europe, Australia, and Asia. Wild ruminants, including water buffalo, antelope, and deer, including white-tailed deer in the southern part of the United States, are infected. Some species are the cause of a serious disease of cattle.

The parasites are minute, round, oval, or bacilliform uninucleate bodies in the erythrocytes and schizonts in lymphocytes and reticulocytes. The vectors are ticks of the family Ixodidae, including one-, two-, and three-host species.

The two genera in the family are separated in part on the basis of the kinds of blood cells the schizonts parasitized. Species of *Theileria* are in leukocytes and *Cytauxzoon* in erythrocytes, liver parenchyma cells, and capillaries. *Theileria parva*, the best known species, is chosen as an example of the life cycle.

### *Theileria parva* (Theiler, 1904) (Plate 35)

In Africa, *T. parva* is the cause of a serious and highly fatal disease known as East Coast Fever of cattle. It is especially dangerous to animals brought into endemic areas where infected cattle and ticks already occur.

**Description.** The forms occurring in erythrocytes appear under the light microscope as minute rings, ovals, comma-shaped bodies, or rods, the last form predominating. In Giemsa-stained smears, the rods have a small red nucleus at one end surrounded by a bit of blue cytoplasm extending as a short, blunt, bacilliform body. They measure 1.5 to 2 $\mu$ long by 0.5 to 1.0 $\mu$ in diameter.

Multiplying schizonts are of two types, the macro- and microschizonts, developing in lymphocytes and occasionally endothelial cells of lymphatic glands and the spleen. Macroschizonts, 3.6 by 1.5 $\mu$ in size, contain up to 90 large chromatin granules, or nuclei, 0.4 to 2 $\mu$ in diameter. The parasite is enclosed in a single membrane. Cytoplasmic organelles include a nucleus with a double membrane, the outer being connected with the endoplasmic reticulum, numerous ribosomes, and food vacuoles. The nuclei are budded

## Explanation of Plate 35

A, Bovine host. B, Tick egg. C, Empty egg shell. D, Newly hatched and unfed larval tick. E, Feeding larval tick with infected erythrocytes in stomach. F, Detached larval tick on ground that has just molted to nymphal stage, with dividing parasites in salivary glands. G, Feeding nymph injects all parasites into bovine host. H, Cleaned nymph in G ingesting infected red blood cells. I, Detached nymph on ground molting to adult, with dividing parasites in salivary glands. J, Feeding infected adult tick injects parasites into bovine and becomes free of them. K, Clean gravid female lays eggs which hatch into uninfected larvae.

1, Uninfected salivary glands; 2, salivary glands with dividing parasites; 3, salivary glands containing infective fusiform parasites; 4, exhausted uninfected salivary glands; 5, empty stomach; 6, extended stomach of engorged tick; 7, anus; 8, infected erythrocytes in stomach of recently fed tick; 9, exuvia; 10, developing uterus; 11, gravid uterus filled with eggs; 12, vulva.

a, Piroplasm in erythrocyte; b, piroplasm in stomach of larval tick; c, salivary glands with multiplying parasites; d, fully developed infective stage in salivary glands; e–j, development of macroschizont and macromerozoites; e, infective stage injected into bovine host by feeding nymph (G) and adult (J); f, parasite penetrating leukocyte; g, developing macroschizont originating from infective stage injected by ticks, (G, J); h, mature macroschizont containing macromerozoites; i, ruptured schizont releasing macromerozoites; j, macromerozoite entering leukocyte to produce another generation or cycle of macroschizonts; k–p, development of microschizont and micromerozoites; k, macromerozoite penetrating leukocyte to develop into a microschizont; l, leukocyte with developing micromerozoites; m, leukocyte containing mature microschizont; n, ruptured leukocyte releasing micromerozoites; o, micromerozoite entering erythrocyte; p, piroplasm in erythrocyte infective to larval tick (E) and nymphal tick (H) when ingested.

Figures B–K adapted from Richardson and Kendall, 1957, Veterinary Protozoology. 2nd ed. Oliver and Boyd, London, p. 162.

---

off in a bit of cytoplasm to produce macromerozoites 2 to 2.5 $\mu$ in length that reenter lymphocytes to continue the cycle. Microschizonts, also in leukocytes, have about 80 to 90 smaller chromatin granules 0.3 to 0.8 $\mu$ in diameter. The parasites inside the leukocyte are about 4.8 to 3.3 $\mu$ in size and enclosed in a membrane formed by the host cell. Each parasite consists of a residual body 1 by 2 $\mu$ in size and many micromerozoites, formed in the same manner as the macromerozoites. Micromerozoites are enclosed in a double membrane with the nucleus at one end. Upon entering the erythrocytes, they retain their same general appearance and are generally known as piroplasms (Fig. 19) which are infective to the tick hosts.

Schizonts, or Koch's Blue Bodies, in the blood or lymphoid tissue smears constitute the most reliable stage for identifying the Theileriidae.

**Life Cycle.**    The life cycle has two separate phases of multiplication. One occurs in the tick vector and the other in the bovine host.

The principal vector is the brown cattle tick *Rhipicephalus appendiculatus*. It is a three-host tick, i.e., each stage (larva, nymph, and adult) feeds, drops to the ground to molt, and reattaches to a bovine host for the next blood meal necessary for each molt, and finally for oviposi-

tion. All of this may take place on a single bovine or on several different individuals.

Ticks of each stage can acquire infection only when erythrocytes containing piroplasms are ingested with blood meals. Since they drop off the host after feeding and molt, the infection acquired in one stage is transmitted to the vertebrate host by the following one. Transmission is stage by stage. For these reasons, only nymphs and adults are transmitters. Larval ticks, uninfected when hatched, obtain piroplasms with their first blood meal and then drop off to molt and transform into nymphs. Infectivity of the parasites in the salivary glands takes place during transition from larva to nymph and from nymph to adult.

When ticks of each stage reach a host, they seek the preferred areas, attach, feed to repletion, and drop off. The ingested erythrocytes are digested and many of the piroplasms are destroyed by enzymes or phagocytic intestinal cells. Survivors quickly disappear from the intestine. They cannot be found in the tissues and hemolymph possibly because of their minute size. By 24 to 48 hours after the infective blood meal, uninucleate forms appear in the secretory droplets of the salivary glands. Growth is rapid and multiple division of the nucleus produces a large mass consisting of vast numbers of tiny

**Plate 35**   *Theileria parva*                                                                165

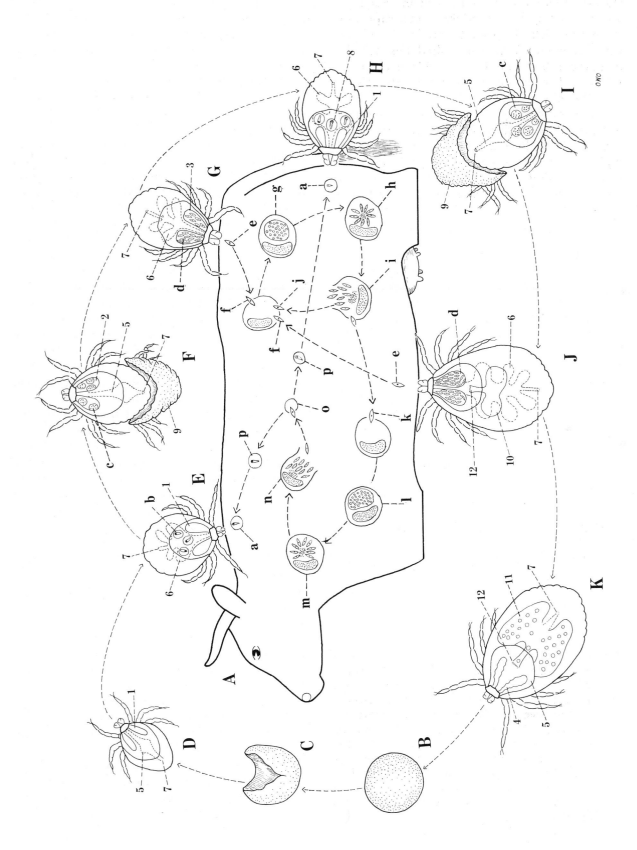

uninucleate infective bodies. Multiplication and maturation are completed in 24 hours in nymphs and up to 4½ days in adults. Upon rupture of the mature cell, the parasites are free in the salivary secretion. When the molted tick attaches to another host and takes its blood meal, the infective bodies are injected into the tissues with the saliva.

It is not clear by what process multiplication is achieved in the ticks. Some investigators have believed it to be sexual but others could not find satisfactory evidence to support this concept. Accordingly, binary fission or schizogony has been suggested as the manner of multiplica-

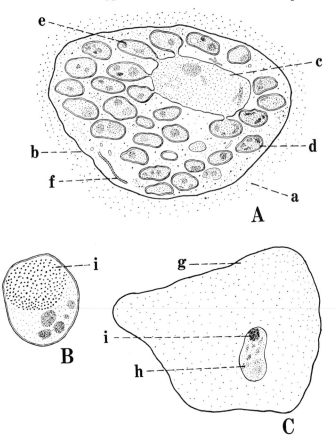

Fig. 19.  Diagrammatic representations from electron micrographs of *Theileria parva.* A, Microschizont in leukocyte showing micromerozoites free and still attached to residual body. B, Micromerozoite. C, Piroplasm (micromerozoite) in erythrocyte. **a,** Cytoplasm of host cell (leukocyte); **b,** membrane formed around microschizont by host cell; **c,** residual body of microschizont; **d,** free micromerozoite; **e,** attached micromerozoite; **f,** endoplasmic reticulum; **g,** erythrocyte; **h,** piroplasm (micromerozoite); **i,** nucleus. Adapted from Jarrett and Brocklesby, 1966, J. Protozool. 13:301.

tion. While a sexual stage has not been confirmed neither has it been disproved. If a sexual stage does occur in the tick, the developing bodies in the salivary glands would be oocysts, the process of multiplication sporogony, and the product sporozoites.

*Theileria* passes only from stage to stage of the tick and not in eggs as in *Babesia.* This requires multiplication of the parasite in the salivary glands in each instar after the blood meal. The act of molting and reorganization of the tissues is essential for migration of the parasites to the salivary glands and subsequent development. Infected ticks loose their infection at each feeding. To be infective in the next stage, they must acquire more piroplasms at each meal. When ticks feed, all the parasites in the salivary glands at that time are injected into the host.

The fate of the early stages of the parasite in the bovine is uncertain. Undoubtedly they are carried via the lymph vessels to the lymphoid tissues, especially the lymph nodes. When first recognizable, they are small uninucleate bodies in the lymphocytes of nodes nearest the site where the ticks attached and fed. Nuclear division is rapid and up to 90 daughter nuclei soon appear in each schizont. One to 3 days later, schizonts appear in lymphoid tissue throughout the body, especially in the liver, lungs, kidneys, bone marrow, lymph nodes, and lymphocytes of the blood. The presence of schizonts in the leukocytes apparently stimulate the latter to divide. At the same time, the schizont separates into two clumps so that each daughter leukocyte contains a multinucleate schizont.

Two types of schizonts appear. Those developing in leukocytes from the infective stages injected by the ticks are macroschizonts. They multiply in number for several days. Each produces many merozoites known as macromerozoites, a name indicating their origin rather than size. When mature, the macroschizonts and their host cells rupture releasing the macromerozoites that enter lymphocytes. Some of them continue the cycle while others grow into the second type of schizont known as microschizonts.

Because of the large blue colored cytoplasmic residual bodies in the two types of schizonts in leukocytes stained with Giemsa, they are called Koch's Blue Bodies after their discoverer. Their presence is a diagnostic character of infection with species of *Theileria.*

Nuclear division in the microschizonts produces 80 to 90 comma-shaped micromerozoites. If the microschizonts rupture in lymphoid tissue, the micromerozoites enter lymphocytes and repeat the cycle. If they rupture in the bloodstream, the micromerozoites that enter reticulocytes or erythrocytes are called piroplasms. Initially, the piroplasms are identical in shape and size with the micromerozoites but round, oval, and bacilliform types appear later. The number of infected erythrocytes continues to increase until the host dies or recovers. If the host recovers, the piroplasms of *T. parva* gradually disappear from the erythrocytes until ticks feeding on the recovered animals do not become infected. Piroplasms of *T. annulata* and *T. mutans*, on the other hand, continue at a low level in the erythrocytes of recovered animals. Some investigators believe that their presence is maintained by division of the piroplasms of these two species in the erythrocytes rather than by schizogony in leukocytes. Blood from recovered animals is infective to ticks.

Some other species of *Theileria* include *T. annulata* and *T. mutans* of cattle, *T. hirci* and *T. ovis* of sheep, and *T. cervi* of deer. *Theileria cervi* is not transmissible to cattle or sheep. *Amblyomma americanum*, the lone star tick, is a common vector of *T. cervi* in Texas.

Species of *Cytauxzoon* occur in African antelope such as the eland, kudu, and duiker deer.

**Etiology and Pathology.** While several species of *Hyalomma* and *Rhipicephalus* are able to transmit *T. parva*, *Rhipicephalus appendiculatus* is the principal vector. High susceptibility to infection by *T. parva* together with a strong preference in all stages for cattle as a host are factors that give this species such a prominent place in the etiology of theileriosis. It is the principal vector in East, Central, and South Africa when annual precipitation is about 25 inches year. The ticks are common in savannah, bush, and forest areas but not in tropical rain forests.

Seasonal activity of ticks due to unfavorable variations of temperature and moisture is reflected in the incidence of theileriosis in the cattle.

Under natural conditions on the pastures, ticks lose their infections in 12 to 15 months in the absence of cattle and other susceptible hosts. Increasing temperature results in progressive loss of infectivity among the ticks. Virulence declines in ticks with the passage of time.

The clinical syndrome is fairly constant.

After a period of incubation of about 12 days, fever appears with an abrupt rise in body temperature to 103° to 104°F the first day and up to 106° within 2 days. There is malaise followed by swelling of lymph nodes. Appetite falls off with accompanying loss of condition. Respiration accelerates followed by difficulty in breathing. Constipation at the onset of fever is followed by diarrhea, often with blood. In the final stages, the afflicted animals are emaciated, recumbent, and discharge froth from the nose. Anemia is not marked.

## SELECTED REFERENCES

Barnett, S. F. (1968). Theileriasis. In: D. Weinman and M. Ristic. eds. Infectious Blood Diseases of Man and Animals Caused by Protista. Academic Press, New York, Vol. 2, pp. 269–328.

Howe, D. L. (1971). Theileriosis. In: J. W. Davis and R. C. Anderson, eds. Parasitic Diseases of Wild Mammals. Iowa State University Press, Ames, pp. 343–353.

Hulliger, L. (1965). Cultivation of three species of *Theileria* in lymph cells *in vitro*. J. Protozool. 12: 649–655.

Jarrett, W. F. H., and D. W. Brocklesby. (1966). A preliminary electron microscopic study of East Coast Fever (*Theileria parva* infection). J. Protozool. 13:301–310.

Kuttler, K. L., R. M. Robinson, and R. R. Bell. (1967). Tick transmission of theileriasis in a white-tailed deer. Bull. Wildl. Dis. Ass. 3:182–183.

Purnell, R. E., and L. P. Joyner. (1968). The development of *Theileria parva* in the salivary glands of the tick *Rhipicephalus appendiculatus*. Parasitology 58:725–732.

Reichenow, E. (1940). Der Entwicklungsgang des Küstenfieberregers in Rinde und in der übertragenden Zecke. Arch. Protistenk. 94:1–56.

Riek, R. F. (1966). The development of *Babesia* spp. and *Theileria* spp. in ticks with special reference to those occurring in cattle. In: E. J. L. Soulsby, ed. The Biology of Parasites. Academic Press, New York, pp. 15–32.

Robinson, R. M., K. L. Kuttler, J. W. Thomas, and R. G. Marburger. (1967). Theileriasis in Texas white-tailed deer. J. Wildl. Manage. 31:455–459.

### ORDER TOXOPLASMIDA

This category in the scheme of classification is used as a provisional holding place for *Toxoplasma* and *Sarcocystis*. It was created originally for them when the concept prevailed that these genera could not be Sporozoa because of the absence of a sexual phase in the life cycle with production of oocysts. Recent studies have shown that *Toxoplasma* and *Sarcocystis* have a coccidian-like life cycle in which oocysts are produced in the small intestine of cats and dogs. Likewise, tissue cultures of *Sarcocystis* further reveal its coccidian affinities.

## FAMILY TOXOPLASMIDAE

*Toxoplasma gondii*, the only species of the family, is a parasite of birds and mammals. Until recently, the parasites were recognized as free and encysted intracellular organisms that multiplied by a special type of schizogony known as endodyogeny. Because spores, i.e., oocysts, were unknown, *Toxoplasma* was excluded from the Sporozoa. Recent observations based on life history studies, however, have shown that *Toxoplasma* has a sexual stage and reproduces by schizogony and gametogony in the intestinal epithelium of cats with the eventual production of oocysts that are passed in the feces. This discovery and verification of this type of reproduction has led to a reconsideration of the taxonomic relationship of *Toxoplasma*. It is now regarded by some workers as a coccidian of the suborder Eimeriina in the order Coccidia. Others feel that further work is necessary to clarify a number of points before the questions of its relationship are clarified. Although the evidence points strongly to an affinity of *Toxoplasma* with the Eimeriina, it is being left in the order Toxoplasmida along with the *Sarcocystis*.

*Toxoplasma* is similar morphologically to but different biologically from species of *Isospora* of cats and dogs. It resembles *Isospora bigemina* more closely than other species of the genus. Perhaps what has been called the small race of *I. bigemina* is *Toxoplasma gondii*.

The family and genus are characterized as follows: Both schizogony and gametogony occur in the intestinal epithelium of cats; oocysts passed in the feces contain, when sporulated outside the host, two sporocysts each with four sporozoites. Trophozoites* occur in a variety of cells, mainly in the brain and muscles, in many species of mammals and birds; each trophozoite

___

* Trophozoites refer to organisms resulting from endodyogenous multiplication in intermediate hosts and merozoites to those originating from schizogony in the intestinal epithelium of cats. Both develop originally from sporozoites.

multiplies asexually by endodyogeny and produces a thick-walled cyst containing numerous naked trophozoites.

### *Toxoplasma gondii*
Nicolle and Manceaux, 1908 (Plate 36)

Minute crescent- or banana-shaped bodies (Plate 36, A; Fig. 20, C) found in the gundi (*Ctenodactylus gondi*), a small North African rodent, were named *Toxoplasma gondii*. Subsequently similar forms were discovered in many different kinds of birds and mammals. Numerous species were named, based on the different hosts in which they occurred. Similarity of structure and nonspecificity for hosts, as shown by the ease with which cross-infections occurred, led to the conclusion presently held that only one species, *Toxoplasma gondii*, exists.

*Toxoplasma* is a prevalent parasite. It is primarily intracellular, occurring in the epithelial cells of the intestine in the sexual stage and in reticuloendothelial cells of the lungs, liver, and spleen, leukocytes, voluntary and cardiac muscles, the retina, and brain in the asexual stage of the life cycle.

Because of the serious effects resulting from infection of humans and domestic animals with *Toxoplasma*, extensive investigations have been conducted on the nature of the parasite and the disease it causes. These studies led to the recent discovery of its most unusual life cycle, together with new information on the epidemiology of toxoplasmosis.

**Description.** Five separate forms of the parasite appear in the life cycle. Two, the trophozoites and cysts, are in intermediate hosts, and three, the schizonts, gametocytes, and oocysts, are in the cat final host.

Trophozoites are banana-shaped crescents 4 to 7 $\mu$ long by 2 to 4 $\mu$ in diameter (Plate 36, A). Ultrastructurally, they possess a polar ring, conoid, toxonemes (micronemes), subpellicular fibrils, nucleus, Golgi apparatus, mitochondria, and endoplasmic reticulum (Fig. 20, A, C). In acute infections, they are abundant in the peritoneal exudate as free individuals. Cysts are round to oval intracellular bodies with a thick, resilient wall and packed with banana-shaped trophozoites (Plate 36, I, h', i'). Cysts represent the chronic stage and are in the brain, retina, striated muscles, and elsewhere.

Oocysts (Plate 36, a–d) which appear in the feces are oval, smooth, with a double wall, and of a light greenish hue when alive. They measure 9 to 11 $\mu$ by 11 to 14 $\mu$, with a mean of 10 by 12 $\mu$. Sporulated oocysts contain two oval sporocysts about 6 by 8 $\mu$, each with four elongate, curved sporozoites 2 by 8 $\mu$ in size, and a residuum of cytoplasm. There is no Stieda body.

Schizonts containing bipolar merozoites and residual body develop inside the epithelial cells of the small intestine. Microgametocytes with a light-colored, loosely formed nucleus and later with microgametes arranged peripherally, and macrogametocytes with small, compact nucleus and prominently staining cytoplasmic granules are in epithelial cells of the small intestine.

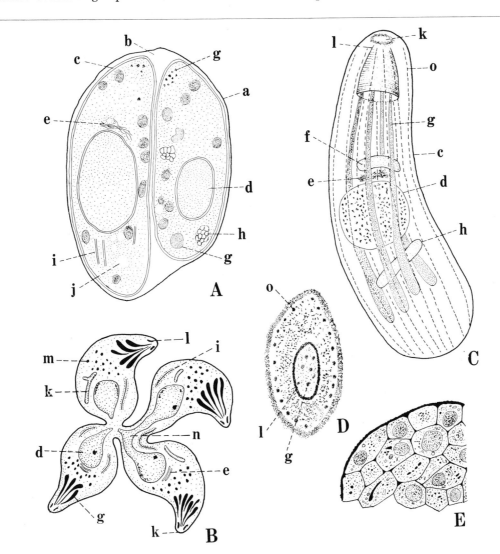

**Fig. 20.** Schematic diagrams from electron micrographs of trophozoites of *Toxoplasma gondii*. A, Longitudinal section through two toxoplasms inside host cell. B, Dividing organism inside host cell showing eight-membered rosette with four individuals attached to residual mass of cytoplasm. C, Reconstruction of mature trophozoite based on electron micrographs. D, Cross section of trophozoite through region of conoid. E, Section of cyst from mouse brain showing thick wall and numerous tightly packed trophozoites cut in different planes. **a,** Wall of vacuole in host cell containing parasite; **b,** vacuole in host cell containing parasite; **c,** double pellicular membrane of parasite; **d,** nucleus with double membrane; **e,** Golgi complex; **f,** Golgi adjunct; **g,** toxoneme or microneme; **h,** mitochondrion; **i,** endoplasmic reticulum; **j,** vacuole; **k,** apical ring; **l,** conoid; **m,** granules; **n,** residual mass of mother cytoplasm; **o,** subpellicular fibrils. A and D adapted from Gavin, Wanko, and Jacobs, 1962, J. Protozool. 9: 222; B, from Sénaud, 1967, Protistologica 3:167; E, from Wanko, Jacobs, and Gavin, 1962, J. Protozool. 9:235; D reconstructed from various sources.

## Explanation of Plate 36

A, Trophozoite from peritoneal exudate of mouse. B, Trophozoites inside macrophage. C–G, Schematic representation of multiplication of trophozoite by process of endodyogeny. C, Parent cell in resting stage. D, Formation of daughter trophozoites as two bud-like processes originated from parent nucleus. E, Growth of daughter cells, disappearance of parent nucleus, and appearance of endosome in daughter cells. F, Continued growth and forward extension of daughter cells and formation of complete nucleus in each. G, Rupture of parent cell and emergence of daughter cells but still attached to each other along longitudinal axis of body. H, Cat as primary host in whose small intestine schizogony and gametogony take place. I, Mouse intermediate host which represents all vertebrates, including humans, domestic mammals, carnivorous mammals and birds, that become infected.

1, Nucleus of trophozoite; 2, nucleus of parent cell; 3, nucleus of daughter cell; 4, nucleolus of daughter cell; 5, nucleus of host cell; 6, toxoplasm in host macrophage; 7, conoid.

a, Oocyst with sporont filling entire inner space; b, oocyst in which sporont has divided to form two sporoblasts; c, oocyst with two sporocysts; d, oocyst with one sporocyst containing four fully developed sporozoites and separate residual body and the other sporocyst with three immature sporozoites still attached to residual body and one free; e, ruptured oocyst releasing sporocysts and sporozoites; f–k, schizogenous cycle; f, sporozoite entering epithelial cell of intestinal mucosa; g, rounded trophozoite; h, young schizont; i, mature schizont containing merozoites; j, ruptured schizont releasing merozoites; k, some merozoites enter other cells to initi-

ate another cycle of schizogony; l–t, gametogenous phase; l, merozoite that develops into microgametocyte; m, young microgametocyte; n, nucleus dividing to form microgametes; o, peripheral arrangement of microgametes; p, rupture of cell and release of microgametes; q, merozoite entering epithelial cell to form macrogametocyte; r, young macrogametocyte; s, macrogamete being fertilized by microgamete; t, zygote which develops into oocyst; u, oocyst being voided with feces; v, cysts of sporozoite origin (f), containing trophozoites in brain and tissues; w, mouse intermediate host containing cysts.

A′, B′, Phases of life cycle in intermediate host. A′, proliferative or multiplicative phase; b′, trophozoite entering host cell; c′, d′, multiplication of trophozoite; e′, destruction of host cell and liberation of trophozoites to infect more cells and initiate additional cycles resulting in proliferation of trophozoites. B′, cystic phase in life cycle when cysts are formed due to immunity in which proliferation is within cysts over a period of a week; f′, infection of host cell (similar to what takes place in A′, b′); g′, early formation of cyst (also called pseudocyst) of parasite origin by multiplying cells inside; h′, i′, continued multiplication of trophozoites and enlargement of cyst within distorted and stretched host cell; j′, host cell degenerates liberating cyst which releases trophozoites; k′, after decline of immunity, liberated trophozoites enter new cells to initiate new proliferative (A′) and cystic (B′) cycles.

Figure A modified from Frenkel, Dubey, and Miller, 1970, Science 167:893; C–G adapted from Goldman, Carver, and Sulzer, 1958, J. Parasitol. 44:161; a–d, from Dubey, Miller, and Frenkel, 1970, J. Exp. Med. 132:636; A′, B′, from Frenkel and Jacobs, 1958, Arch. Ophthalmol. 59:260.

**Life Cycle.** The life cycle involves cats as definitive hosts and numerous other species of animals (birds and mammals) as intermediate hosts. The sexual phase of the life cycle develops in the epithelial cells of the small intestine of cats with production of oocysts. The asexual phase develops in various cells of the body but primarily those of the brain and striated muscles of intermediate hosts as a proliferative stage which eventually produces cysts containing numerous infective trophozoites.

Infected cats void undeveloped oocysts with the feces. Under favorable conditions of moisture, aeration, and temperature, oocysts have developed two sporoblasts in 9 to 12 hours, sporocysts in 21 to 28 hours, and sporozoites in 2 to 4 days. The resistant oocysts remain in-

fective up to 12 months under favorable conditions.

Animals become infected when fully developed oocysts are swallowed. The course of development of the sporozoites varies, depending whether they are in cats, other mammals, or birds.

In cats, both the sexual phase, consisting of gametocytes and oocysts, in the intestine and the asexual phase of the life cycle, comprising trophozoites and cysts, in the various tissues occur following ingestion of ripe oocysts.

Under the influence of the digestive juices of the small intestine of cats, oocysts rupture and liberate the sporozoites. Some of them penetrate epithelial cells of the small intestine and undergo schizogony, producing merozoites. These

Plate 36    *Toxoplasma gondii*                                                                171

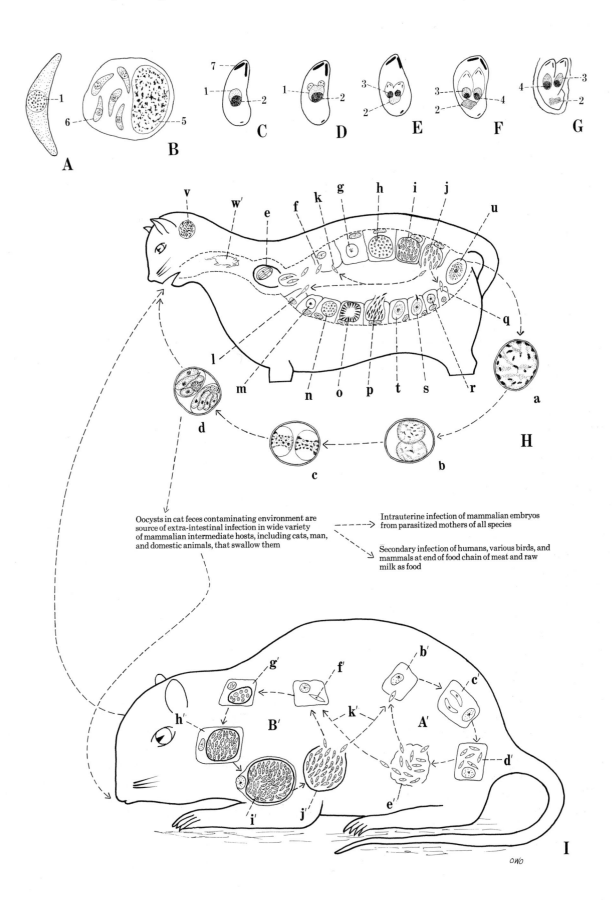

Oocysts in cat feces contaminating environment are
source of extra-intestinal infection in wide variety
of mammalian intermediate hosts, including cats, man,
and domestic animals, that swallow them

Intrauterine infection of mammalian embryos
from parasitized mothers of all species

Secondary infection of humans, various birds, and
mammals at end of food chain of meat and raw
milk as food

enter other cells to repeat several generations. Eventually merozoites with the potential to develop into gametocytes appear. They enter epithelial cells of the gut where those with male characteristics develop into microgametocytes and those with female characteristics into macrogametocytes. Following maturation of the gametocytes with formation of microgametes and macrogametes, fertilization takes place. The zygotes develop into oocysts that when mature rupture the host cells and are freed. They appear in the feces 21 to 24 days after infection.

The sexual phase of the life cycle in cats is similar to that of *Isospora bigemina*. However, *Toxoplasma gondii* differs in having evolved an asexual phase in birds and mammals which serve as intermediate hosts.

Some of the sporozoites in cats do not participate in the sexual cycle but enter into the general circulation and go to various parts of the body where they transform into trophozoites that reproduce asexually in cells as occurs in other animals.

In animals serving as intermediate hosts, only the asexual phase of the life cycle develops from ripe oocysts that are swallowed. Sporozoites freed from oocysts in the gut make their way into the bloodstream and are distributed to all parts of the body. They enter various kinds of cells, commonly those of the brain and striated muscles, both cardiac and voluntary. Early intracellular development is one of rapid multiplication by a special type of schizogony called endodyogeny (endo = internal, dyo = two, genesis = origin). Two bud-like processes appear on the anterior margin of the nucleus. As growth proceeds, the daughter cells extend anteriorly, attain greater size, and acquire a conoid, nucleus, other characteristic cytoplasmic organelles, and are enclosed by a limiting membrane. Eventually the parent cell is destroyed and the daughter cells are released, while they are still attached to each other. They soon mature and separate as individual trophozoites capable of another cycle of reproduction. Sometimes rosettes of daughter trophozoites attached to a residual body occur from successive divisions that fail to separate normally (Fig. 20, B).

This period of rapid multiplication in the tissues is known as the proliferative or acute phase (Plate 36, I, A'). During it, trophozoites appear free in the peritoneal exudate and in cells of various kinds. As the number of trophozoites increases during several cycles of proliferation, the host begins to react by producing specific antibodies that curtail the proliferative process. With the appearance of immunity, the infection passes into the chronic stage characterized by the formation of intracellular cysts packed with numerous trophozoites (Plate 36, I, i', j'; Fig. 20, E). As immunity declines, the cysts deteriorate, releasing trophozoites that enter new cells causing relapses and are the source of additional proliferation cycles (Plates 36, I). Thus a succession of proliferative and cystic phases occur in infected intermediate hosts.

Trophozoites in cysts or free in the flesh of animals are the principal source of infection for carnivores and oocysts from fecal contamination of the environment by cats for herbivores. In addition, congenital infection may take place when pregnancy and the proliferative phase of toxoplasmosis with free trophozoites in the circulation occur simultaneously.

When cats eat infected mice, or other infected flesh, the trophozoites initiate the sexual cycle in the intestinal epithelium. Oocysts appear in the feces in 9 to 11 days after cats have eaten mice with acute infection and in 3 to 5 days with chronic infections.

**Epidemiology.**    Cats are the ultimate and most important source of toxoplasmosis in populations of warm-blooded animals, including humans. Deposition of feces by cats in sandboxes, and soil around homes, barn yards, and fields results in a great store of oocysts available to people and animals, especially domestic ones and mice. Experimentally infected dogs pass oocysts.

Oocysts swallowed by animals other than cats produce infective free and cystic trophozoites in the body tissues, particularly the brain and striated muscles. Carnivorous birds and mammals readily acquire infections from their prey. The cat-mouse relationship is especially important in this respect.

All mammals may transmit infection congenitally when the acute phase of infection and pregnancy occur simultaneously. Prenatal infection is especially prevalent among sheep in England and New Zealand where toxoplasmosis results in a high rate of abortion.

In addition to oocysts from sandboxes and soil in yards where children play, humans may become infected by eating inadequately cooked meat and drinking raw milk, either from milk-producing animals or the mother, as in the case of nursing children.

**Pathology and Immunity.** Toxoplasmosis is a zoonotic infection transmitted from animals to man. Disease manifestations during the proliferating stage include fever, pneumonia, hepatitis, and involvement of lymph nodes. Prenatal infections result in stillbirths, abortion, hydrocephalus, mental retardation, and blindness.

In chronic infections, cysts occur in the eyes, heart, and central nervous system, producing severe effects. The greatest damage to these tissues occurs when large masses of cysts rupture, resulting in destruction of sizeable areas of cells. This is especially true and most frequent in the retina where the function of sight is concentrated in a small area.

Histopathological changes in the brain of infected cats include inflammation of the meninges, accumulation of lymphocytes around the capillaries, and areas of necrosis.

Natural resistance is higher in mature than in very young animals. Acquired immunity induced by cellular and humoral factors generally follows acute infection. After a marked reduction of trophozoites caused by the immune reaction, they become encapsulated in cysts that persist for long periods under the influence of infection-immunity or premunition.

## SELECTED REFERENCES

Dubey, J. P., and J. K. A. Beverley. (1967). Distribution of *Toxoplasma gondii* in the tissues of cats. II. Histopathological study. Trop. Geogr. Med. 19:206–211.

Dubey, J. P., N. L. Miller, and J. K. Frenkel. (1970). The *Toxoplasma gondii* oocyst from cat feces. J. Exp. Med. 132:636–662.

Frenkel, J. K. (1971). Toxoplasmosis. In: P. A. Marcial-Rojas, ed. Pathology of Protozoal and Helminthic Diseases. The Williams & Wilkins Co., Baltimore, pp. 254–290.

Frenkel, J. K., J. P. Dubey, and N. L. Miller. (1970). *Toxoplasma gondii* in cats: Fecal stages identified as coccidian oocysts. Science 167:893–896.

Gavin, M. A., T. Wanko, and L. Jacobs. (1962). Electron microscope studies of reproducing and interkinetic *Toxoplasma*. J. Protozool. 9:222–234.

Hirth, R. S., and S. W. Nielson. (1969). Pathology of feline toxoplasmosis. J. Small Anim. Pract. 10:213–221.

Hutchison, W. M., J. F. Dunachie, J. C. Siim, and K. Work. (1970). Coccidian-like nature of *Toxoplasma gondii*. Brit. Med. J. 1:142–144.

Jacobs, L. (1956). Propagation, morphology, and biology of *Toxoplasma*. Ann. N.Y. Acad. Sci. 64:154–179.

Jacobs, L. (1957). The interrelation of toxoplasmosis in swine, cattle, dogs, and man. Public Health Rep. 72:872–882.

Jacobs, L. (1967). *Toxoplasma* and toxoplasmosis. In: B. Dawes, ed. Advances in Parasitology. Academic Press, New York, vol. 5, pp. 1–45.

Janitschke, K., and D. Kühn. (1972). Toxoplasma-Oozysten im Kot naturlich infizierter Katzen. Berlin. München. Tierärztl. Wochenschr. 85:46–47.

Jírovec, O. (1971). Quelque remarques sur la position taxonomique de *Toxoplasma gondii*. J. Parasitol. 57(Sect. II):84.

Katsube, Y., T. Hagiwara, H. Miyakawa, T. Muto, and K. Imaizumi. (1969). Studies on toxoplasmosis. II. Distribution of *Toxoplasma* in organs of cat and dog cases of latent infection occurring naturally. Jap. J. Med. Sci. Biol. 22:317–326.

Kühn, D., W. H. Oppermann, H. Rödel, and H. Centurier. (1972). Experimentelle Infektion von Hunden mit Toxoplasma-Oozysyten. Berlin. München. Tierärztl. Wockenschr. 85:209–314.

Rommel, M., and J. Breuning. (1961). Untersuchungen uber das Vorkommen von *Toxoplasma gondii* in der Milch einiger Tierartzen und Moglichkeit der laktogenen Infektion. Berlin. München. Tierärztl. Wockenschr. 80:365–369.

Scheffield, H. G. (1970). Schizogony in *Toxoplasma gondii*: an electron microscope study. Proc. Helminthol. Soc. Wash. 37:237–242.

Scheffield, H. G., and M. L. Melton. (1968). The fine structure and reproduction of *Toxoplasma gondii*. J. Parasitol. 54:209–226.

Scheffield, H. G., and M. L. Melton. (1970). *Toxoplasma gondii*: the oocyst, sporozoite, and infection of culture cells. Science 167:892–893.

Wanko, T., L. Jacobs, and M. A. Gavin. (1962). Electron microscope study of *Toxoplasma* cysts in mouse brain. J. Protozool. 9:235–242.

### FAMILY SARCOCYSTIDAE

The family contains the single genus *Sarcocystis*. It is characterized by compartmentalized cysts located in the striated muscles of reptiles, birds, and mammals and sporulated *Isospora*-like oocysts that develop in the intestinal epithelium of cats, dogs, and people, and are voided in the feces. The cysts, known as sarcocysts, contain banana-shaped bodies similar to those occurring in *Eimeria*, *Plasmodium*, and

*Toxoplasma*. These bodies are referred to in the literature as trophozoites, spores, sporozoites, schizozoites, and Rainey's bodies. The term trophozoite will be used.

Sarcocysts commonly are elongated bodies which lie parallel to the muscle fibers. They attain a length of 10 mm or more, depending on the species. When calcified, the sarcocysts are visible to the unaided eye as opaque, white bodies. Living ones can be seen only with the aid of a microscope.

*Sarcocystis* is worldwide in distribution. Some of the common species, based on cysts in intermediate hosts, include *S. tenella* from sheep, *S. fusiformis* from cattle, *S. bertrami* from horses, *S. mierschiana* from swine, *S. leporum* from cottontails, *S. muris* from mice, *S. rileyi* from mallards, and *S. platydactyli* from geckos. A form known as *S. lindemanni* occurs in humans but only rarely. It may not be a separate species.

### *Sarcocystis tenella* Railliet, 1886 (Plate 37)

This species is prevalent in sheep and goats everywhere. It occurs most commonly in the upper and lower ends of the esophagus but is frequently found in the diaphragm, tongue, and heart.

**Description.**    The wall of mature sarcocysts comprises two layers of a granulated syncytium. The thinner outer portion 2.5 to 3 $\mu$ thick is of a spongious nature with a fine honeycomb structure. It is covered by numerous cauliflower-like projections 8 to 10 $\mu$ long that extend into the surrounding muscle tissue. This layer gives an intensive reaction for polysaccharides.

The inner layer is 4 to 5 $\mu$ thick, contains small nuclei, and gives a positive reaction for ribonucleic acid (RNA). Several rows of undeveloped trophozoites are embedded in the inner layer which extends inward as septa forming many compartments. These contain developing trophozoites in the outer chambers and mature banana-shaped ones in the deeper compartments. The central portion of the sarcocysts is usually devoid of septa and trophozoites. It is filled with fluid containing the toxin sarcocystin.

Mature trophozoites measure 6 to 15 $\mu$ long by 2 to 4 $\mu$ in diameter. They are capable of flexing and gliding movements. The body is divided into three regions based on internal morphological structures and cytochemical reactions.

The anterior third consists of the fibrillar section filled with numerous sarconemes, or micronemes. At the anterior end is a ring-like opening, the polar ring, surrounded by three concentric rings. Within the polar ring is the cone-shaped conoid from which originate about 22 tubular, subpellicular fibrils that radiate outward. Polysaccharides and glycogen are absent but RNA is present in small amounts. A micropyle occurs in this region. The middle third of the body is the granular portion. It contains large granules whose surface is osmophilic. Dispersed among the granules are particles some of which contain volutin and others RNA. The posterior third of the body is richest morphologically and cytochemically. It contains the large nucleus enclosed in a double membrane with pores. Extensions of the outer membrane connect with the endoplasmic reticulum tubules. Other organelles consist of vacuoles and small granules that surround the nucleus; large vacuoles, Golgi apparatus, one to three mitochondria with internal cristae, and glycogen masses are scattered in the cytoplasm. This region is most active chemically, being positive for deoxyribonucleic acid (DNA) in the nucleus and for polysaccharides and glycogen elsewhere.

**Life Cycle.**    While much work has been done on the life cycle of *Sarcocystis*, the true nature of it has only recently become known. Scott, in 1943, reviewed the literature dealing with observations, including his own, on the life cycle of various species. He concluded that although many erroneous statements have appeared, a basic pattern of the life cycle could be detected, especially for *S. tenella* which had been studied most extensively.

He postulated a direct life cycle in which trophozoites escaping from sarcocysts entering the bloodstream are carried to all parts of the body. Some work their way through the intestinal wall and are voided with the feces, others go through membranes of the lungs, mouth, and nasal passages to appear in their secretions. After a period in the feces outside the host, trophozoites became infective to sheep that swallow them. Based on seasonal appearance of *Sarcocystis* in lambs, Scott believed that under certain conditions at definite times infective trophozoites in feces are capable of infecting sheep by way of the alimentary canal.

Spindler *et al.*, in 1946, reported that pigs, dogs, cats, rats, mice, and chickens that have eaten swine flesh containing mature sacrocysts void stages of *Sarcocystis* in their feces and/or urine 15 days after the infectious meal that are infective to other swine. Recipient pigs fed a mixture of feces and urine from donor pigs 15 days after the latter had consumed parasitized flesh became infected. Control pigs did not become infected.

Frenkel, in 1971, stated that the literature on *Sarcocystis* includes a mixture of observations on it and *Toxoplasma*, both of which are compounded by errors of interpretation and imagination. He concluded that nothing regarding transmission of *Sarcocystis* is known and recommended critical studies on the life cycle done under tight controls be undertaken with a species in small animals such as mice which can be handled with greater precision than large ones such as sheep and swine.

Recent morphological studies conducted with the aid of the electron microscope on several species of *Sarcocystis* and biological observations with tissue cultures for an unnamed species from grackles have revealed interesting information on the life cycle. When viewed in the light of recent findings on the life cycle of *Toxoplasma*, these observations provide some insights into that of *Sarcocystis*.

*Sarcocystis*, like *Toxoplasma*, has an asexual proliferative phase in its life cycle. Multiplication occurs only in the striated muscles. This implies that infective stages may be ingested, pass through the intestinal mucosa, enter the circulation, and be carried to various parts of the body. In the case of *S. tenella*, the organisms show a predilection for muscles of the esophagus, diaphragm, tongue, and heart. In these areas, they leave the circulation and enter the muscle fibers where growth and multiplication by two separate methods take place.

In the precystic stage, the young trophozoite lies among the muscle fibrils inside the fiber where it undergoes growth and division by binary fission. There is no cyst in this stage (Plate 37, B, G–I). Young cystic stages are packed with rounded or polygonal cells. The wall of the cyst is of a granular material with internal extensions that isolate the individual cells in compartments. Young cysts exhibit variable shapes, indicating ability to move within the muscle cells at this stage of development. The outside wall of the cyst is smooth (Plate 37, C).

As growth continues, the young cysts present new external structures, probably a physiological response to nutritional requirements of the rapidly growing organism. The outer surface of the cytoplasmic membrane produces numerous hollow, finger-like villi that extend into the cytoplasm of the muscle cells. Internally, organisms called mother cells continue to multiply by binary fission in the enlarged compartments (Plate 37, D).

Mature sarcocysts have characteristic features. The external villi have ramified distally to form cauliflower-like structures that increase their absorptive surface area. Several layers of mother cells making up the outer portion of the cyst continue to proliferate by binary fission, producing several generations of daughter cells (Plate 37, E) which lack the apical ring, conoid, and associated organelles.

Daughter cells toward the center of the sarcocyst assume an elongate shape and reproduce differently by the process of endodyogeny similar to that which occurs in *Toxoplasma*. In the course of endodyogenous reproduction, a pair of cones appears in the cytoplasm toward the anterior end of the daughter cell. The nuclear membrane disappears and a pair of conoidal trunks appears on the anterior margin of the nucleus. The cones form into the anterior part of the developing trophozoites. The nucleus assumes a U-shape with the conoid trunks being located on the tip of each arm. They extend into the anterior part of the developing trophozoites. Two banana-shaped trophozoites are formed from each mother cell, each complete with poral ring, conoid, paired organelles, sarconemes, mitochondria, endoplasmic reticulum, Golgi apparatus, and other characteristic organelles. The mother cell ruptures, releasing them (Plate 37, J–M). Banana-shaped trophozoites that escape from the sarcocysts into the tissues are not phagocytized but migrate in the bloodstream to other cells, presumably to develop into more sarcocysts by repeating the entire process described above. Lymphocytes, polymorphonuclear cells, and giant ameboid cells invade the empty sarcocysts which appear caseous at first and later become calcified.

A sexual phase in the life cycle of *Sarcocystis* is suggested by the type of development obtained by Fayer in tissue cultures of trophozoites from grackles. When cultured in embryonic kidney cells of bovine, canine, and turkey origin and in embryonic chicken muscle cells,

## Explanation of Plate 37

A, Mature banana-shaped sarcosporidian cell, the trophozoite. B, Young trophozoite beginning first division in muscle cell of sheep esophagus. C, Dividing trophozoite showing individual new cells in compartments separated by septa that originate from granular membrane. D, Intermediate stage of cyst in muscle cell with finger-like villi that develop from granular layer and extend into cytoplasm of surrounding muscle cell to provide nutriment. E, Mature cyst with branched villi providing increased surface area for absorption of nutriment from surrounding layer of muscle tissue; undeveloped trophozoites occur in outer layer and mature trophozoites in compartments toward center. F, Segment of esophagus of sheep showing large oval sarcocysts. G–I, Binary fission of young trophozoites. G, Early stage of constriction of cell and elongation of nucleus. H, Constriction almost complete; nucleus divided. I, Division complete with two daughter cells. J–M, Endodyogenous division (a special type of multiplication) of banana-shaped trophozoite (endodyocyte). J, Formation of cones representing banana-shaped trophozoites and fibrillar section of body; nucleus without membrane. K, Further growth of banana-shaped trophozoites and horseshoe formation of nucleus preparatory to division. L, Banana-shaped trophozoites fully formed and separated from each other but still retained in intact mother cell. M, Mother cell ruptures releasing banana-shaped trophozoites which may undergo successive divisions of endodyogeny. N–R, Stages appearing in tissue culture. N, Entire tissue cell containing four banana-shaped trophozoites 1 hour after inoculation of culture. O, Portion of tissue cell cytoplasm with enlarged, oval, uninucleate trophozoite in host cell 24 hours. P, Binucleate schizont in vacuole 48 hours after inoculation. Q, Multinucleate schizont in vacuole 48 hours in cell. R, Cyst-like body (oocyst?) containing two oval forms after 48 hours in host cell.

1, Fibrillar region (anterior third of body); 2, granular region (middle third of body); 3, nuclear and mitochondrial region (posterior third of body); 4, three surface rings surrounding apical pore; 5, conoid; 6, sections of sarconemes, or micronemes, that fill fibrillar region (1); 7, paired organelles; 8, granules of granular region (2); 9, mitochondrion; 10, nucleus enclosed in double porous membrane, the outer of which extends tubules to connect with endoplasmic reticulum (12); 11, Golgi apparatus; 12, endoplasmic reticulum; 13, glycogen vacuoles; 14, micropyle; 15, muscle fibers; 16, muscle fibrils; 17, young trophozoite in process of first binary division with beginning of formation of septum; 18, newly formed compartment; 19, granular layer of cyst; 20, septum formed by extension from granular layer; 21, layer of finger-like, hollow villi extending from granular layer into muscle cytoplasm for increasing surface area to supply nutritional requirements of rapidly growing cyst; 22, cross sections of simple hollow villi; 23, connective tissue layer surrounding sarcocyst; 24, highly branched, cauliflower-like villi provide for greater demands of cyst for nutriments; 25, young trophozoites in layers next to granular cyst wall; these divide by binary fission (G–I); 26, mature banana-like trophozoites which divide by endodyogeny (J–M); 27, beginning of constriction of mother cell in process of dividing by binary fission; 28, mother trophozoite beginning endodyogenous division; 29, cones with conoid of banana-shaped trophozoites; 30, fibrillar area (1) of banana-shaped trophozoite forming on anterior margin of nucleus of mother cell; 31, growing banana-shaped trophozoite; 32, residual nucleus of mother trophozoite; 33, ruptured membrane of mother trophozoite freeing banana-shaped trophozoites; 34, portion (cytoplasm) of host cell in tissue culture; 35, oval, uninucleate trophozoite; 36, binucleate schizont-like body; 37, parasitophorous vacuole in cytoplasm of host cell; 38, multinucleate schizont-like body; 39, cyst-like body resembling isosporan oocyst containing two sporocysts.

Figures A–E, G–M adapted from Sénaud, 1967, Protistologica 3:167; N–R redrawn from photomicrographs by Fayer, 1970, Science 168:1104.

---

stages comparable to trophozoites, schizonts, and oocysts appeared.

Within 1 hour after placing the banana-shaped trophozoites in the tissue cultures, they had entered the cells and were moving about in the cytoplasm (Plate 37, N). At the end of 24 hours, they had formed oval, uninucleated bodies lying in a cytoplasmic parasitophorous vacuole (Plate 37, O). By 48 hours in the cells, the ovoid bodies were in various stages of development. Some were binucleate (Plate 37, P) and others multinucleate (Plate 37, Q), suggesting phases of schizogony. During the same period, the appearance of cyst-like bodies containing two oval forms (Plate 37, R) suggests similarity to an immature *Isospora*-like oocyst such as occurs in *Toxoplasma*. These stages appearing in tissue cultures are reminiscent of schizogenous and

**Plate 37**    *Sarcocystis tenella*                                                    177

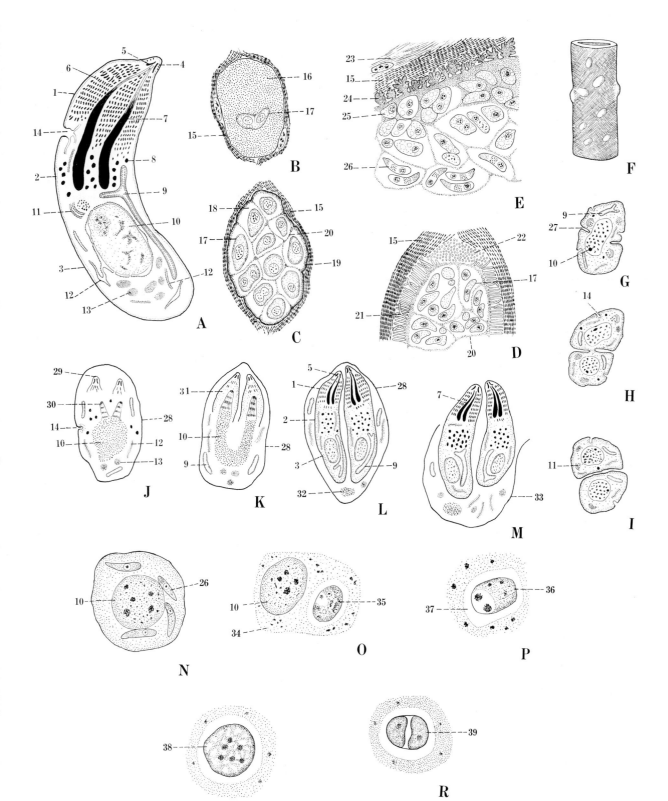

gametogenous phases with the production of oocysts. Fayer noted that the changes observed in these cultures of *Sarcocystis* are similar to those seen by him and Hammond in the transformation of sporozoites of *Eimeria bovis* under similar conditions. If this tentative interpretation is correct, the embryonic tissue culture cells might represent a host comparable to the cat in *Toxoplasma* (Plate 36). One aspect of transmission would take place through ingestion of flesh containing mature sarcocysts and the other by swallowing oocysts. With the high incidence of sarcocystosis in sheep, the dog-sheep relationship, like the cat-mouse combination in *Toxoplasma* (Plate 36), has significant epizootiological implications in regard to their respective roles in the life cycle. Such a canine host, either domestic or wild, eating sheep would be a source of infection through dissemination of oocysts available to grazing herbivorous animals.

Recent feeding experiments in Germany using three common species of *Sarcocystis* resulted in coccidia-like infections in the small intestine of cats, dogs, and humans.

When fed to cats, sarcocysts of *Sarcocystis tenella* from the esophagi of sheep produced sporocysts similar to those of *T. gondii*. Fully sporulated sporocysts with two sporozoites each and a Stieda body appear in the feces on days 12 to 13 and continue from 4 to 53 days. Sporocysts measure 10.8 to 13.9 $\mu$. Oocysts, when present, are thin-walled, regular to irregular in shape, and contain two sporocysts. Dogs did not become infected.

Both dogs and cats become infected when fed sarcocysts of *Sarcocystis fusiformis* from cattle. Sporulated sporocysts measuring 13.9 to 17 $\mu$ by 6.2 by 10.8 $\mu$ appear in the feces in 9 to 10 days and continue for 8 to 10 weeks.

Sarcocysts of *Sarcocystis fusiformis* from cattle swallowed by human volunteers produced sporulated sporocysts measuring 14.7 by 9.3 $\mu$ in 2 days, whereas *Sarcocystis miesheriana* from swine resulted in sporocysts measuring 12.6 by 9.3 $\mu$ in 10 days. The sporocysts produced by these two species of sarcocysts are probably the same and are what has been called *Isospora hominis* from man.

*Sarcocystis muris* from mice fed to cats produce Toxoplasma-like oocysts in their feces. Conversely, when the oocysts from the cats are fed to mice, sarcocysts of *S. muris* appear in the muscles.

**Toxicity and Pathology.**   The central portion of mature sarcocysts of *S. tenella* is devoid of septa and trophozoites but is filled with fluid. It contains a true animal toxin capable of rapidly killing rabbits when injected in small amounts. The chief symptom is a cholera-like diarrhea. Guinea pigs, rats, and mice are nonsusceptible to it. The toxic agent gives some protein reactions and is heat labile and dialyzable. It is a true toxin since it gives rise to antitoxins.

Pigs with heavy infections of 40 sarcocysts or more per gram of diaphragm showed symptoms of unthriftiness, weakness of loins, muscular stiffness, and temporary posterior paralysis.

Lesions include enlargement and paleness of kidneys, and inflammation of the mucosa of the stomach and intestines. Meat that is heavily infected may be condemned as unfit for human consumption.

## EXERCISE ON LIFE CYCLE

Flesh containing living sarcocysts of *S. tenella* from sheep, *S. rileyi* from mallards, or other species available should be fed to puppies and kittens free of *Isospora* and protected against natural infection. Examine their feces for isosporan oocysts. If they appear, allow them to sporulate and feed them to various animals to determine if sarcocysts appear in the striated muscles. Protect both experimental and control animals from natural infection.

## SELECTED REFERENCES

Brand, T. von. (1952). Chemical physiology of endoparasitic animals. Academic Press, New York, p. 64.

Doflein, F., and E. Reichenow. (1953). Lehrbuch der Protozoenkunde. 6th ed. Gustav Fischer, Jena, p. 1042.

Fayer, R. (1970). Sarcocystis: Development in cultured avian and mammalian cells. Science 168:1104–1105.

Fayer, R., and D. M. Hammond. (1967). Development of first-generation schizonts of *Eimeria bovis* in cultured bovine cells. J. Protozool. 14:764–772.

Frenkel, J. K. (1971). Protozoal diseases of laboratory animals. In: R. A. Marcial-Rojas, ed. Pathology of Protozoal and Helminth Diseases. Williams & Wilkins Company, Baltimore, pp. 358–361.

Heydorn, A.-O. and M. Rommel. (1972). Beiträge zum Lebenszyklus der Sarkosporidien. II. Hund und Katze als Überträger der Sarkosporidien des Rindes.

Berlin. München. Tierärztl. Wockenschr. 85:121–123.

Heydorn, A.-O. and M. Rommel. (1972). Beiträge zum Lebenszyklus der Sarcosporidien. IV. Entwicklungsstadien von *S. fusiformis* in der Dünndarmschleimhaut der Katze. Berlin. München. Tierärztl. Wockenschr. 85:333–336.

Ludvik, J. (1956). Vergleichende elektronenoptische Untersuchungen an *Toxoplasma gondii* und *Sarcocystis tenella.* Zentralbl. Bakteriol. Parasitenk. Abt. I. Orig. 166:60–65.

Ludvik, J. (1958). Elektronenoptische Befunde zur Morphologie der Sarcosporidien (*Sarcocystis tenella* Railliet, 1886). Zentralbl. Bakteriol. Parasitenk. Abt. I. Orig. 172:330–350.

Ludvik, J. (1960). The electron microscopy of *Sarcocystis meischeriana* Kuhn, 1865. J. Protozool. 7:128–135.

Ludvik, J. (1963). Electron microscopic study of some parasitic protozoa. In: J. Ludvik, J. Lom, and J. Vávra, eds., Progress in Protozoology, Proc. First Int. Congr. Protozool., Prague, 1961. Academic Press, New York, p. 387.

Mahrt, J. L. (1973). Sarcocystis in dogs and its probable transmission from cattle. J. Parasitol. 59:588–589.

Rommel, M. and A.-O. Heydorn. (1972). Beiträge zum Lebenzyklus der Sarcosporidien. III. *Isospora hominis* (Railliet und Lucet, 1891) Wenyon, 1923, eine Dauerform der Sarkosporidien des Rindes und des Schweins. Berlin. München. Tierärztl. Wockenschr. 85:143–145.

Rommel, M., A.-O. Heydorn, and F. Gruber. (1972). Beiträge zum Lebenszyklus der Sarkosporidien. I. Die Sporozyste von *S. tenella* in den Fäces der Katze. Berlin. München. Tierärztl. Wockenschr. 85:101–105.

Scott, J. W. (1943). Life history of Sarcosporidia, with particular reference to *Sarcocystis tenella.* Univ. Wyoming Agr. Exp. Sta. Bull. 259, pp. 1–63.

Sénaud, J. (1967). Contribution a l'étude des sarcosporidies et des toxoplasmes (Toxoplasmea). Protistologia 3:167–232.

Simpson, C. F. (1966). Electron microscopy of *Sarcocystis fusiformis.* J. Parasitol. 52:607–613.

Spindler, L. A., H. E. Zimmerman, Jr., and D. S. Jaquette. (1946). Transmission of *Sarcocystis* to swine. Proc. Helminth. Soc. Wash. 13:1–11.

Vusse, F. J. V. (1967). Sarcocystis infections in relation to age of Iowa cottontails (Protozoa; Sarcocystidae). Proc. Iowa Acad. Sci. 72:524–528.

Wallace, G. D. (1973). Sarcocystis in mice inoculated with Toxoplasma-like oocysts from cat feces. Science 180:1375–1377.

## CLASS CNIDOSPORIDA

This group is characterized by having resistant spores with one to four polar capsules and one or more sporoplasms. The spore covering may consist of two or three valves (order Myxosporida) or a single piece (order Microsporida). One or more polar filaments are coiled inside each capsule. The Cnidosporida are tissue parasites of invertebrates and lower vertebrates, primarily fish. The life cycles are direct.

Spore-like organs with a coiled polar filament have been considered as related forms which are placed together in a single group, the Cnidosporida. Recent studies on the ultrastructure of Myxosporida and Microsporida have led to the speculation that they may not be so closely related taxonomically as formerly thought. They are treated here as orders of the class Cnidosporida. Comparisons of the two orders, as given by Lom and Corliss, are presented in Table 2.

### ORDER MYXOSPORIDA

The spore covering consists of two valves united in a distinct sutural plane and contains one to four polar capsules, each with a coiled extensible filament. The polar capsules are grouped at one end of the spore designated as the anterior, or the spore may be elongated and have one capsule at each end, as in the family Myxidiidae. Representatives of the order appear on Plate 37.

In addition to the polar capsules, each spore contains a protoplasmic body known as the sporoplasm. While immature, it contains two haploid nuclei which fuse before or after the sporoplasm escapes.

Kudo (1920) prepared a key to the genera and species of Myxosporida in which nine families are listed. Hoffman, Putz, and Dunbar published a key, together with illustrations, of the species of *Myxosoma.*

The species of Myxosporida are cosmopolitan in distribution.

**Basic Life Cycle.**    Much diversity of opinion exists on the nature of the life cycles of these parasites. This is due to the difficulties of interpreting the stages in fixed tissues. When techniques for growing them *in vitro* are available, better understanding will be forthcoming. Only the basic aspects of the life cycle as generally accepted at present are given here.

Infection takes place when free spores are ingested. Upon reaching the intestine of a specific host, the action of the digestive juices on

**Table 2. Comparison of Myxosporida and Microsporida**

| Character | Myxosporida | Microsporida |
|---|---|---|
| Organization of trophozoites before sporogenesis | Complex; part of nuclei becomes generative cells that produce spores; remainder provides food and spore, then degenerates | Simple; nuclei and cells equal, becoming sporoplasms |
| Cytoplasmic organelles | Many mitochondria, Golgi, ergastoplasm, fibrils, microtubles, vesicles and granules; no centriole; pinocytosis | Few, only ergastoplasmic lamellae and smooth-walled vesicles; no pinocytosis |
| Complexity of spore | Sporoplasm protected by sister cells that produce protective spores | Sporoplasm produces protective shell and enclosed organelles for extrusion of filament |
| Extrusible apparatus | Filament within polar capsule | Filament + polaroplast + polar cap + posterior vacuole |
| Function of filament | Attachment of sporoplasm to gut epithelium | Injection of sporoplasm into host tissue |
| Means of extrusion of filament | Pressure developed inside spore, extrusion by urea | Osmotic pressure in polaroplast and posterior vacuole; no chemical triggering |
| Chemical composition of shell | Proteinaceous | Chitinous |

the spores cause them to open, allowing the sporoplasms to escape. The ameboid sporoplasms migrate through the intestinal epithelium where presumably they enter the circulation and are carried to various parts of the body. In organs or tissues specific to their physiological needs, the sporoplasms leave the blood vessels and begin growth as trophozoites. The nucleus of each trophozoite multiplies repeatedly. Individual nuclei are surrounded by a covering of cytoplasm and grow into uninucleate sporonts that will produce spores. Several nuclear divisions occur in each sporont; the number depends on whether one or two spores occur in each sporont. If a single spore develops, it is termed a monosporoblastic sporont, if two it is disporoblastic. The disporoblastic sporont, or mother sporoblast, is called a pansporoblast. Continued formation of sporonts results in the production of myriads of spores in the tissues.

### FAMILY MYXOSOMATIDAE

The spores of this family are characterized by two or four polar capsules and a sporoplasm that does not have a iodinophilous vacuole. Walliker reported that iodinophilous glycogen vacuoles are present in both *Myxosoma* of the Myxosomatidae and *Myxobolus* of the Myxobolidae and

therefore cannot be used as a taxonomic character. He proposed that the genus *Myxosoma* be synonomized with the genus *Myxobolus*. Since a unanimous opinion on the correct generic status of *Myxoxoma* has not materialized, Lom and Hoffman preferred the present status which is followed here.

### *Myxosoma cerebralis*
(Hofer, 1903) (Plate 38)

This species is a common parasite of the cartilaginous parts of the skeleton of salmonid fish, particularly of the head and vertebral column from vertebra 26 back in young trout. Infection leads to destruction of the parasitized parts, resulting in deformation. The mandible and opercula shorten, the mouth gapes, a humped back develops just behind the head, and the tail becomes twisted or bent and dark in color. The dark discoloration of the tail is due to damaged nerves that control the melanophores, allowing them to expand.

Infection causes the fish to swim in circles, thus leading to the common names of tail-chasing or whirling disease. Swimming is laborious and, after efforts to move, the fish sink to the bottom exhausted.

**Description.**    Mature spores are broadly

oval, sometimes slightly elongated, and rarely completely circular. In side view, they are lenticular in shape with both valves considerably vaulted. A mucus covering encloses the posterior portion of each spore which consists of two valves joined along a sutural ridge. A deep depression borders each side of the sutural ridge. At the anterior end of the sutural ridge are one or two openings through which the coiled filaments emerge. Each filament consists of five to six coils. The posterior end of the spore is covered by an irregular network of intertwined fibers (Fig. 21, B–D). Spores measure 8.7 (7.4 to 9.7) μ long by 8.2 (7 to 10) μ wide.

The internal structure of *Henneguya* is characteristic of the group. The young sporocyst (Fig. 21, A) has two polar capsule cells at the anterior end, each with a polar capsule containing a coiled polar filament and a nucleus. The large sporoplasm is located centrally. It contains various organelles, including two nuclei

with masses of ribosomes arranged on the inner side of the double membrane, mitochondria with cristae that extend outward into the nucleoplasm as well as inward, a rich network of nongranular endoplasmic reticulum, liposome vacuoles, and ribosomes. Two large valve shell cells with large nuclei are located in the posterior and lateral parts of the spore. They form the two valves.

**Life Cycle.** The life cycle has not been determined experimentally because of unsuccessful attempts to establish infection by feeding spores or infected tissue to susceptible fish. Hence, certain parts of the cycle are based on a number of observations and some assumptions (Plate 38, N–Q).

It is assumed that infection takes place when spores from the bottom of ponds are ingested by young fish at the time they begin to feed. Upon reaching the stomach or intestine, the valves of the spores open and release the sporoplasms. The haploid nuclei fuse, if they have not

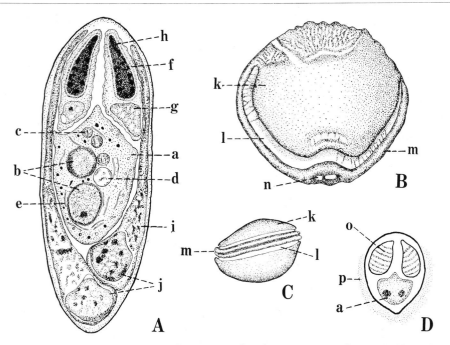

Fig. 21. Myxosporidian spores. A, Internal morphology of young spore of *Henneguya pinnae*. B, Surface view of mature spore of *Myxosoma cerebralis*. C, Lateral view of mature spore of *M. cerebralis* showing furrows, sutural ridges, and vaulted valves. D, Spore of *M. cerebralis* surrounded by mucus layer which is characteristic of this species. a, Sporoplasm; b, nuclei; c, mitochondrion; d, liposome; e, endoplasmic reticulum; f, polar capsule cell; g, nucleus of polar capsule cell; h, polar capsule showing cross sections of filament; i, shell valve cells; j, nucleus of shell valve cells; k, surface view of valve; l, furrow; m, sutural line; n, openings (sometimes single) through which polar filaments extend; o, polar capsule with coiled filament; p, mucus covering. A adapted from electron micrograph by Shubert, 1968, Z. Parasitenk. 30:57; B and C adapted from electron scanning micrographs and D from phase contrast photomicrograph by Lom and Hoffman, 1971, J. Parasitol. 57:1302.

## Explanation of Plate 38

**A, B,** Mature spores of *Myxosoma catostomi* from skin of catfish. **A,** Flat surface of spore. **B,** Edge or sutural view of spore. **C,** Immature spore of *M. catostomi*. **D, E,** *Ceratomyxa blennius,* a disporous species from the urinary bladder of blenny. **D,** Two mature spores still in cytoplasmic mass in which they developed. **E,** Free mature spore. **F, G,** *Thelohanellus notatus* from the subdermal tissue of minnows (shiners and fatheads). **F,** Flat surface of spore. **G,** Sutural view. **H, I,** *Henneguyia exilis* from skin of catfish. **H,** Flat surface of spore. **I,** Sutural view of spore. **J,** *Chloromyxum trijugum* from gall bladder of crappies (*Pomoxis*). **K,** *Myxobolus notemigoni* from skin of golden shiner, life cycle, and pathology. **L,** *Myxidium lieberkühni* from urinary bladder of pickerel (*Esox*). **M,** *Myxidium aplodinoti* from gall bladder of sheepshead (*Aplodinotus*). **N-Q,** Life cycle of *Myxosoma cerebralis*. **N,** Life cycle as presumed to occur. **O,** Intact cartilage containing a multinucleate sporont (see 1). **P,** Lesion caused by parasites in cartilage as it is being enclosed in bone. **Q,** Lesion containing ripe spores (see m).

1, Spore wall; 2, suture; 3, polar capsule; 4, polar filament; 5, sporoplasm; 6, haploid nuclei of sporoplasm; 7, vacuole of iodinophilous body of sporoplasm; 8, iodinophilous body; 9, nuclei of polar body of immature cyst; 10, nuclei of valves of spore.

1'-15', Life cycle *Myxobolus notemigoni* of golden shiner. 1', surface view of mature spore, showing two haploid nuclei of sporoplasm and vacuole left in it by iodinophilous body; 2', sutural view of spore, showing iodinophilous body; 3', zygote, called a sporont, originating from sporoplasm by union of haploid nuclei; 4'-7', developing sporonts; 4', binucleate stage; 5', tetranucleate stage; 6', 7', hexanucleate sporoblasts, containing the number of nuclei necessary to form a single spore and its parts in the monosporous species; 8', golden shiner, showing effect of *M. notemigoni;* 9', section of body wall of golden shiner, showing sporonts under scale; 10', infection causes scales to rise; 11', epidermis; 12', scale; 13', cyst of parasite; 14', corium; 15', muscle.

a, b, Sutural and surface views of mature spores freed from dead infected fish (n); c, spore being swallowed; d, expulsion of polar filaments under influence of digestive juices; e, escape of sporoplasm with haploid nuclei from spore; f, sporoplasm free in lumen of gut, nuclei come together; g, sporont forms and penetrates gut wall to enter postcaval vein; h, sporont reaches heart; i, sporont passing through capillaries of gills to dorsal aorta; j, sporont leaving dorsal aorta through arteries and capillaries to cartilaginous parts of skeleton; k, sporont in cartilage; l, multinucleate sporont in cartilage; m, spores in lesions in cartilage; n, fish dead of whirling disease or blacktail, will decompose and liberate spores; o, kingfishers and other piscivorous animals may release spores by digestion of infected fish and subsequently distribute spores in the feces; p, intact cartilage; q, cartilage cell; r, lesion caused by parasite; s, multinucleate sporont (same as l); t, mature spores in lesions; u, bone; v, cartilaginous debris; w, cellular debris.

Figure A–C adapted from Kudo, 1926, Arch. Protistenk. 56:90; D, E, from Noble, 1944, Quart. Rev. Biol. 19:213; F, G, from Kudo, 1934, Ill. Biol. Monogr. 13:3; H–J, from Kudo, 1920, Ill. Biol. Monogr. 5:245; K, from Lewis and Summerfelt, 1964, J. Parasitol. 50:388; N, from Hoffman, Dunbar, and Bradford, 1962, Fish Wildl. Serv. Spec. Sci. Rep. Fish. No. 427.

---

already done so, to form the mature sporoplasms. These penetrate the intestinal mucosa and presumably are carried in the circulatory system to various parts of the body. Those entering cartilaginous tissues, particularly of the head and posterior part of the spine, begin development.

Extremely young parasites are difficult to recognize in the cartilage. Symptoms of whirling and blacktail occur, however, as early as 35 days after newly feeding fish are placed in ponds with infected ones, but verification of infection has not been made by demonstrating parasites in histological sections.

Multinucleate trophozoites are present in cavities in the cartilage 40 days after newly feeding fish are exposed to infection in a pond with parasitized fish. At 3 months of age, ameboid trophozoites 5 by 5 $\mu$ to 30 by 8 $\mu$ in size and with at least 18 nuclei are present in cavities measuring 300 by 100 $\mu$ in size. The smaller cavities may be only cross sections of elongated ones.

At 4 months of age, the trophozoites have grown greatly, attaining a diameter of 1 mm. The nuclei have continued to divide and now appear in groups of 12 to 14 known as pansporoblasts, each of which will develop into two spores. The process of nuclear division is one of multiple fission, beginning when the sporoplasm enters the cartilage. Through it the nuclei for the fu-

**Plate 38**   *Myxosporidian Types*                                                      183

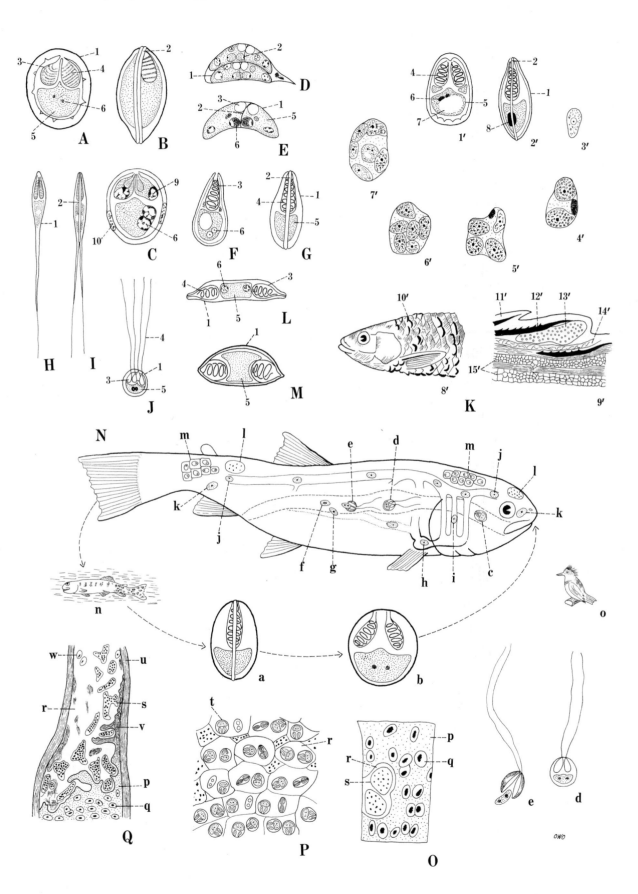

ture spores are formed. They appear in groups of at least six enclosed in a small amount of cytoplasm. Two of the nuclei form the shell, two the polar capsules, and the remaining two by a process of reduction division form two haploid nuclei with two chromosomes for the sporoplasm. The haploid nuclei may unite before the sporoplasm escapes from the spores, but it is thought generally to occur afterwards.

By 8 months after infection, the pansporoblasts have produced spores which fill the cavities. The spores generally remain in the cavities or "lesions," as they are called, where they occur in fish up to 3 years of age, possibly older. Some workers believe that the spores may escape from the "lesions," enter the circulation, and be carried to the intestinal wall whence they enter the lumen of the gut. The spores may escape from the host by several possible ways. When carried by the blood to the intestinal wall, they may pass through it into the lumen and be voided with the feces. Death and disintegration or crushing of infected fish would release them. Spores have been reported from the intestine of kingfishers, presumably from infected fish eaten and digested by them. All other piscivorous animals, including fish, could serve as a rapid means of releasing and disseminating spores from infected fish eaten by them. The fact that infected fingerlings are unable to swim well makes them easy prey and thereby a selective means in favor of spreading the parasite.

Other species of *Myxosoma* include *M. catostomi* of the common sucker (*Catostomus commersoni*) which infects muscles and subcutaneous connective tissue. The spores are broadly oval and measure 13 to 15 μ long by 10 to 11.5 μ wide. Extremely severe infections occur in very young suckers in which numerous whitish pustule-like cysts appear on the skin of the head and body. Upon examination, they are filled with spores.

*Myxosoma cartilaginis* Hoffman, Putz, and Dunbar of blue gills (*Lepomis macrochirus*), green sunfish (*L. cyanellus*), and largemouth bass (*Micropterus salmoides*) parasitizes the cartilaginous parts of the skeleton. Its life cycle appears similar to that of *M. cerebralis*. The symptoms of whirling and darkening of the body are lacking.

Mature spores are oval in frontal view, of regular shape, and quite uniform in size, measuring 11.25 (10.7 to 11.7) μ long by 9.5 (9 to 10) μ wide. There are usually seven coils of the

polar filament. A mucus envelope surrounding the spore and an iodinophilous vacuole are lacking.

## FAMILY MYXOBOLIDAE

The spores have one, two, or four polar capsules at the anterior end and a sporoplasm with an iodinophilous body. They are histozoic. *Myxobolus notemigoni* Lewis and Summerfelt of the golden shiner is representative of this family in North America.

**Description.**   Fresh spores average 11.8 μ long by 8.9 μ wide. The polar capsules average 4.1 μ long by 3.3 μ wide. The outer surface of the spore valves is smooth, but small spine-like projections appear on the inner surface of the posterior end. The spores occur in aggregations of cysts 0.9 to 3 mm in diameter under the scales scattered over the body, except on the head and fins.

**Life Cycle.**   This is a monosporous species in which each sporont produces a single spore containing six nuclei. Two of these form the valves, two the polar capsules, and two, by a process of reduction division, the sporoplasm with two haploid nuclei (Plate 38, K; Fig. 21, A). Infection of fish is probably by ingestion of spores and migration of the sporoplasm to the dermal tissue via the circulatory system. Inasmuch as the cysts are under the scales, it is likely that spores may be liberated from ruptured cysts throughout the period of infection of the shiners.

Other common genera are *Henneguya* (Fig. 21, A) with two polar capsules and an elongated process on the posterior end of each valve, and *Thelohanellus* with one polar capsule. For a listing of species, with keys and figures, the reader should consult Kudo (1920).

## FAMILY MYXIDIIDAE

The valves are elongated, giving a fusiform appearance. There are two polar capsules, one at the distal end of each valve (Plate 38, L, M). An iodinophilous body is lacking.

*Myxidium lieberkühni* Bütschli occurs in the urinary bladder of the pike (*Esox* spp.) and is widely distributed. Spores measure 18 to 20 μ long by 5 to 6 μ wide.

*Myxidium serotinum* Kudo and Sprague occurs in the gall bladder of frogs and toads in North America. Spores are 16 to 18 μ long by 9 μ wide, with two to four longitudinal and 10 to

13 transverse ridges. It is both di- and polysporous with two or several cysts developing in each pansporoblast.

*Sphaeromyxa* spp. occur in the gall bladder of marine fish.

The life cycles of the Myxidiidae presumably are basically similar to those of the Myxobolidae.

### FAMILY CERATOMYXIDAE

The spores are markedly prolonged laterally (Plate 38, D, E). There are two polar capsules in the anterior margin with the sutural plane running between them.

*Ceratomyxa blennius* Noble from the gall bladder of blennies, or butterfly fish, is a disporous species in which two cysts develop in each pansporoblast (Plate 38, D). Development of the sporont proceeds as described for the other species. In addition, cytoplasmic growth with much nuclear division (nucleogony) occurs, followed by budding off of uninucleate bodies that develop into pansporoblasts to form spores.

### EXERCISE ON LIFE CYCLE

Life cycles of the Myxosporidia have not been worked out in their entirety. Infection of hosts by feeding spores or tissue infected with them has not been accomplished.

For material to work with, seek such fish as minnows, suckers, or catfish with infections in the skin or on the gills. Since spores open in the stomach or intestine of fish, efforts should be made to get them to open in digestive juices, both natural and artificial. Try to determine whether fusion of nuclei takes place before or after escape of the sporoplasm from the spore.

Histological sections of infected tissue will reveal the various stages of development, including growth of the cytoplasm, nuclear multiplication, and formation of pansporoblasts and spores.

### SELECTED REFERENCES

Doflein, F., and E. Reichenow. (1953). Lehrbuch der Protozoenkunde. 6th ed. Gustav Fischer, Jena, p. 967.

Grassé, P. P. (1953). Traité de Zoologie. Protozoaires. Masson et Cie, Paris. Vol. 1, Fasc. 2, p. 1009.

Hoffman, G. L., and R. E. Putz. (1969). Host susceptibility and the effect of aging, freezing, heat, and chemicals on spores of *Myxosoma cerebralis.* Prog. Fish-Cult. 31:35–37.

Hoffman, G. L., C. E. Dunbar, and A. Bradford. (1962). Whirling disease of trouts caused by *Myxosoma cerebralis* in the United States. U.S. Fish Wildl. Serv. Spec. Sci. Rep. No. 427, 15 pp.

Hoffman, G. L., R. E. Putz, and C. E. Dunbar. (1965). Studies on *Myxosoma cartilaginis* n. sp. (Protozoa: Myxosporidia) of centrarchid fish and a synopsis of the *Myxosoma* of North American freshwater fishes. J. Protozool. 12:319–332.

Kudo, R. R. (1920). Studies on Myxosporidia. A synopsis of genera and species of Myxosporidia. Ill. Biol. Monogr. 5:239–503.

Kudo, R. R. (1926). On *Myxosoma catostomi* Kudo, 1923, a myxosporidian parasite of the sucker, *Catostomus commersonii.* Arch. Protistenk. 53:191–214.

Kudo, R. R. (1943). Further observations on the protozoan, *Myxidium serotinum,* inhabiting the gall bladder of North American Salientia. J. Morphol. 72:263–277.

Lom, J., and C. O. Corliss. (1967). Ultrastructural observations on the development of the microsporidian sporozoan *Plistophora hyphessobryconis* Schäperclaus. J. Protozool. 14:141–152.

Lom, J., and G. L. Hoffman. (1971). Morphology of the spores of *Myxosoma cerebralis* (Hofer, 1903) and *M. cartilaginis* (Hoffman, Putz, and Dunbar, 1965). J. Parasitol. 57:1302–1308.

Noble, E. R. (1944). Life cycles in the Myxosporidia. Quart. Rev. Biol. 19:213–235.

Schäperclaus, W. (1954). Fischkrankheiten. Akademie, Berlin, p. 379.

Schubert, G. (1968). Elektronenmikroskopische Untersuchungen zur Sporenentwicklung von *Henneguya pinnae* Schubert (Sporozoa, Myxosporidia, Myxobolidae). Z. Parasitenk. 30:57–77.

Walliker, D. (1968). The nature of the iodinophilous vacuole of myxosporidian spores and a proposal to synonymize the genus *Myxosoma* Thelohan, 1892 with the genus *Myxobolus* Bütschli, 1882. J. Protozool. 15:571–575.

Wenyon, C. M. (1926). Protozoology. William Wood and Co., New York (Reprinted by Hafner Publishing Co., New York, 1965), Vol. 1, p. 716.

### ORDER MICROSPORIDA

The Microsporida are especially numerous parasites both as individuals and species in invertebrates and fish. Extremely resistant, minute spores are probably responsible in part for their cosmopolitan distribution and frequency in occurrence. The prevalence of suitable hosts also plays an important part in their abundance.

In general, the Microsporida infect specific tissues of their hosts. Some occur in protozoan or helminth parasites within a host. Many are host-specific. For example, two species of mosquito larvae in a single pool do not share each other's microsporidian parasites.

There are two suborders based on the number of polar filaments. The Monocnidina are characterized by a single filament and the Dicnidina have two, one at each end of the spore. Some common representatives appear on Plate 38, A–E.

**Description.**    The spores are small and composed of a single chitinous piece varying greatly in shape, being spherical, pyriform, and cylindrical. Living specimens viewed under the light microscope are refractile. Each spore contains a clear, vacuole-like structure at the anterior end and in some species there is one at the posterior end. A long, slender, polar filament coils about a nucleated sporoplasm. The wall of the spore consists of three basic layers: a thin inner one surrounding the organism, a thick middle one which forms the solid part of the case, and a thin outer one composed of several exceedingly thin layers, the outermost being smooth or wrinkled on the surface. The contents of the spore consist of four main parts. They are 1) the polaroplast (same as anterior vacuole) occupies the anterior part of the cell; it consists of lamellae probably originating from the endoplasmic reticulum. It has great swelling power in the presence of water. 2) A long, slender polar filament attached to the anterior end of the spore extends backwards through the polaroplast and then forms a number of coils inside the spore. The filament is a hollow tube filled with electron dense material and composed of two layers, the outer of which is probably derived from the Golgi material. The proximal end of the filament which is attached to the spore may or may not be covered by a polar cap at the anterior end of the spore. 3) The sporoplasm is a thin, granular, uni- or binucleate cytoplasmic body around the middle third of the inside of the spore. In addition to the nucleus, the sporoplasm contains numerous ribosomes. The Golgi apparatus appears as a complex system which probably plays an important role in the formation of the polaroplast, certain membranous investments of the polar filament, and part of the posterior vacuole. 4) The posterior vacuole shows no distinct structure beyond a flocculated appearance, due probably to inadequate fixation. It is hygroscopic, as is the polaroplast, and functions in extrusion of the polar filament and forces the sporoplasm through it.

**Suborder Monocnidina**

This suborder has the greater number of species. It consists of the families Nosematidae with oval or pyriform spores, Coccosporidae with spherical spores, and Mrazekiidae with long, cylindrical, or bowed spores. In each case, the spore contains a coiled, hollow, polar filament.

### FAMILY NOSEMATIDAE

This is the largest family, having seven genera. They are based on the number of spores produced by each sporont. *Nosema* has 1 spore; *Glugea*, 2; *Gurleyia*, 4; *Thelohania*, 8; *Stempellia*, 1, 2, 4, or 8; *Duboscquia*, 16; and *Plistophora*, variable but often more than 16. Recent writers include *Perzia* as a subgenus of *Glugea* since the basic difference between them is that the latter causes hypertrophy of the infected cells and the former does not.

*Nosema* includes such well-known species as *N. apis* of honeybees and *N. bombycis* of silkworms. Both species are highly pathogenic to their hosts. Other species of *Nosema* are pathogenic, being responsible for the destruction of large numbers of other kinds of insects, including mosquito larvae. This genus has been chosen as the representative of the suborder and family.

*Nosema apis* Zander, 1909 (Plate 39)

This species is primarily a parasite of the epithelial cells of the ventriculus, or midgut, of honeybees in many parts of the world. It causes a disease known commonly as bee sickness, bee dysentery, winter losses, spring dwindling, May sickness, and scientifically as nosemiasis. The parasite is of economic importance because it curtails the activities of bees and destroys entire colonies. In addition to honeybees, it can develop in bumble bees, although it is not a common parasite of them.

**Description.**    The spores of *N. apis* are oval, measuring 4 to 6 $\mu$ long by 2 to 4 $\mu$ wide. The length of the polar filament is given as 250 to 400 $\mu$. The internal structure and that of the spore case is basically similar to that given above.

**Life Cycle.**    Bees become infected by swallowing spores in the feces deposited in the hive by parasitized members of the colony. Upon passing through the esophagus and proventriculus, the spores are deposited inside the peritrophic membrane at the anterior end of the ventriculus. Inside the ventriculus of normal live bees, some unknown substance dissolves the polar cap and permits intestinal fluids to enter the spore. The polaroplast reacts by swelling and creating pressure inside the spore, forcing the polar filament to evert through the pore. It quickly extends full length, being rigid and capable of piercing the peritrophic membrane and membranes of the epithelial cells of the ventriculus. When the polar filament has everted, swelling of the posterior vacuole begins, creating pressure that forces the sporoplasm, or germ, anteriorly into (Fig. 22) and through the polar filament (Plate 39, F–I) and into the cells where subsequent development occurs (Plate 39, f). Within 30 minutes after the spores have entered the ventriculus, the polar filaments have everted and the sporoplasms have been injected into the epithelial cells. Initially, the most active development of sporoplasms takes place in cells at the anterior and posterior parts of the ventriculus. Numerous intracellular granules of calcium phosphate in the middle portion of the ventriculus appear to be inimical to the development of the sporoplasms at the outset. As the sporoplasms develop, the granules gradually break down and the pH level in the cells becomes less basic so that the entire ventriculus becomes infected.

Inside the host cell, the sporoplasm responds to the influence of its new environment to complete the development necessary for production of spores. The first changes are elongation of the sporoplasm and migration of the nuclei to the opposite ends of the cell. Each nucleus divides once to form the first schizont which is tetranucleate. It develops into a long chain of uninucleate individuals. Upon reaching maturity, the chain breaks up into separate schizonts. The nuclei are small, centrally located, and contain densely stained clumps of chromatin located on the periphery of the nucleoplasm.

Each individual cell from the first schizogonic chain in turn produces a second schizogonic chain of uninucleated forms. The nuclei of this stage are rich in densely staining chromatin material. They may divide synchronously, producing quadrinucleate individuals or asynchronously resulting first in trinucleate and then quadri- or hexanucleate forms. The tetranucleate

individuals divide once, producing two binucleate cells called diplokaryons which are in reality sporoblasts. The nuclei of the diplokaryons remain near each other during subsequent growth leading directly to the formation of spores.

Sometimes heavily infected ventricular cells contain binucleate forms characteristic of both the first and second schizogonic stages without intermediate stages. This suggests that binucleate forms of the first schizogony might develop directly under unknown circumstances into

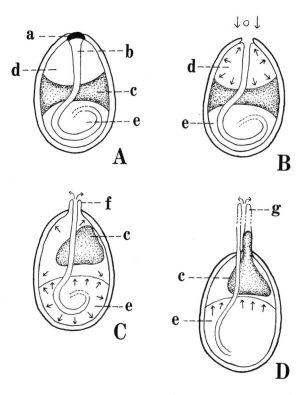

Fig. 22. Diagrammatic representation of the mechanism of extrusion of the polar filament of a microsporidan spore. A, Intact mature spore. B, Spore under influence of outer stimuli indicated by substance and arrows. C, Swelling of polaroplast and posterior vacuole creates sufficient pressure necessary to start eversion of hollow polar filament and dislodgement of sporoplasm which moves anteriorly. D, Pressure from swelling of posterior vacuole forces sporoplasm anteriorly and into lumen of polar filament through which it will be driven. a, Polar or McManus cap; b, hollow polar filament; c, sporoplasm; d, polaroplast; e, posterior vacuole; f, polar filament begins to evert by turning inside out from internal pressure; g, extruded polar filament with sporoplasm attenuating while being forced into and through it by intrasporal pressure. A–D adapted from Lom and Vávra, 1963, Acta Protozool. 1:81.

## Explanation of Plate 39

A, Longitudinal section of spore of *Plistophora hyphessobryconis* from electron micrograph. B, Reconstruction of spore of *Thelohania californica* from electron micrographs. C, Mature spore of *Nosema apis* of honeybees. D, Spore of *Nosema bombycis* of silkworms. E, Spore of *Telomyxa glugeiformis* showing two polar filaments. F–I, Sequence of movement of binucleate sporoplasm of *Nosema apis* through the hollow polar filament and expulsion from it. F, Sporoplasm beginning journey through long polar filament. G, Sporoplasm near tip of polar filament. H, Part of cytoplasm of sporoplasm emerging from tip of polar filament. I, Entire binucleate sporoplasm outside polar filament but still attached to it. J, Spore of *Nosema apis* with extremely long polar filament. K, L, Spores of *Mrazekia argoisi* before and after extrusion of polar filament. M, Spore of *Cocconema* with extruded polar filament. N, Mosquito larva (*Anopheles crucians*) infected with *Thelohania legeri*. O, Portion of ventriculus of honeybee, showing cells filled with spores of *Nosema apis*. P, Brain and cranial nerves of angler fish (*Lophuris piscatoris*) infected with *Nosema lophurii*. Q, Section of body of stickleback (*Gasterosteus*) infected with *Glugea anomala*, showing large glugea cysts in muscles. R, Smelt (*Osmerus*), showing numerous cysts on viscera caused by *Glugea hertwigi*. S, Honeybee (*Apis mellifera*), showing life cycle of *Nosema apis*.

1, Spore case; 2, polaroplast; 3, tubular polar filament; 4, polar, or McManus, cap; 5, polar granules; 6, sporoplasm with single elongate nucleus (B) or two spherical nuclei (D); 7, binucleate sporoplasm; 8, posterior vacuole; 9, sporoplasm passing through hollow polar filament (F, G), emerging (H), and outside but still attached (I); 10, masses of spores inside body of mosquito larva; 11, spores in epithelial cell of ventriculus of honeybee; 12, epithelial cell; 13, nucleus of epithelial cell; 14, nest of developing epithelial cells; 15, basement membrane; 16, outer muscular layer; 17, eye; 18, brain; 19, lobulated tumors on cranial nerves of angler fish caused by *Nosema lophurii*; 20, cyst of *Glugea anomala* in body muscles of stickleback; 21, intestine; 22, cysts of *Glugea hertwigi* on viscera of smelt.

a, Ingestion of ripe spores by adult bee; b, spore in esophagus; c, spore with polar filament extruded and sporoplasm discharged in lumen of ventriculus where it is digested; d, sporoplasm surrounded by droplet of viscous fluid; e, spore with extruded filament that has pierced peritrophic membrane (u) and entered epithelial cell of ventriculus; f, sporoplasm free in cytoplasm where it was implanted from tubular polar filament; g, binucleate sporoplasm; h, elongation of sporoplasm to form schizont preparatory to first nuclear division; i, first nuclear division with elongation of schizont; j, long chain of binucleated schizonts formed after many nuclear divisions; k, binucleate schizonts resulting from break-up of chain; l, beginning of second schizogonic chain; m, second schizogonic chain with six darkly staining nuclei; n, developed secondary schizogonic chain; o, production of binucleated diplokaryons from second schizogonic chain; p, developing spores; q, mature spores enclosed in host cell; r, epithelial cell ruptured releasing mature spores; s, spores in lumen of hindgut; t, spore voided with feces; u, peritrophic membrane lining ventriculus; v, mature spore in feces outside host.

Figure A redrawn from electronmicrograph by Lom and Corliss, 1967, J. Protozool. 14:141; B adapted from Kudo and Daniels, 1963, J. Protozool. 10:112; C adapted from Fantham and Porter, 1912, Ann. Trop. Med. Parasitol. 6:163; D adapted from Grassé, 1953, Traité de Zoologie, Vol. 1, Fasc. 2, p. 1045; F–I redrawn from photomicrographs by Kramer, 1960, J. Insect Pathol. 2:433; E, J–N, P–R redrawn from Kudo, 1924, Ill. Biol. Monogr. 9:79; O redrawn from White, 1919, U.S. Dep. Agr. Bull. 780.

---

diplokaryons and finally into spores by omitting the quadrinucleate stage.

The entire cycle is completed in 4 to 7 days. Pressure created by the developing and mature spores destroys the infected host cells. Infective spores are released into the lumen and voided with the feces.

*Nosema bombycis* Naegli is parasitic in the caterpillars of silk moths (*Bombyx mori*) and related species of lepidopterous larvae. Practically all tissues of the body, including the intestinal epithelium, are invaded. The parasite is especially dangerous to the caterpillars of silk moths because the crowded conditions in which they are reared favor constant exposure and heavy infection. Parasitized larvae show brown spots over the body, causing a peppery appearance which suggested the name pebrine disease. In severe infections, mortality is extremely high. Under favorable conditions, the life cycle, which is basically similar to that of *N. apis*, may be completed in 4 days.

*Nosema helminthorum* Moniez parasitizes the cestode *Moniezia expansa* of sheep. Spores in-

**Plate 39**   *Microsporidian Types, Nosema apis*                    189

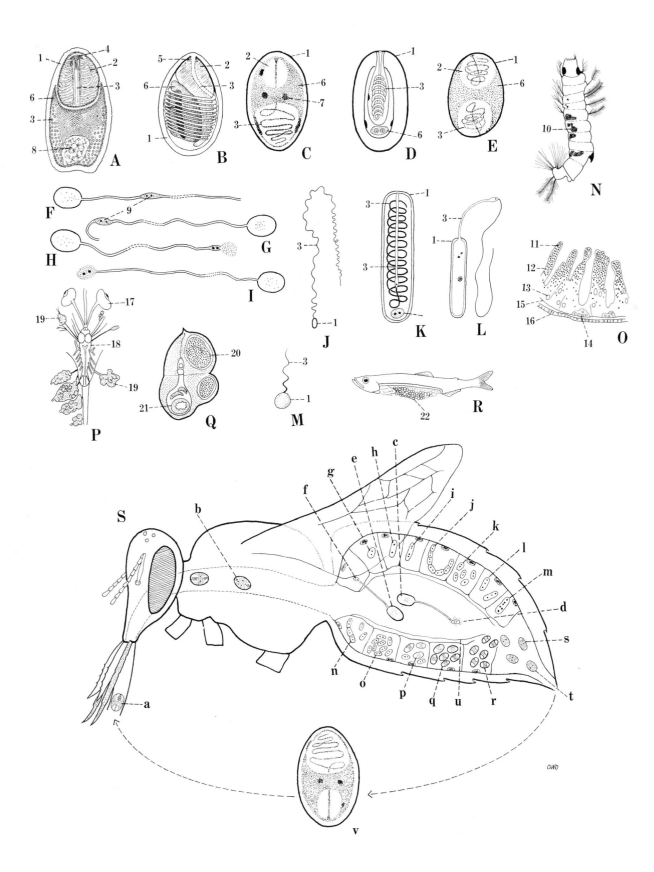

gested by sheep respond to conditions in the ovine intestine and evert the polar filament. It penetrates the thick cuticle of the tapeworms and deposits the sporoplasm inside the parenchymous tissue. Spores are disseminated upon decomposition of proglottids passed in the sheep feces. A number of species of intestinal protozoan and helminth parasites are infected by microsporidians.

*Glugea anomala* (Moniez) is parasitic in the muscles of sticklebacks (*Gasterosteus*), producing large cysts in the body wall. *Glugea hertwigi* infects the intestinal wall of smelt (*Osmerus*), forming sizable cysts that destroy the intestinal epithelium. *Glugea gasti* is pathogenic to the boll weevil (*Anthonomus grandis*). Infection probably begins in the mesenteron and spreads to all parts of the body. Development is similar to that seen in *Nosema apis*. Binucleate sporoplasms initiate the first schizogonic phase characterized by mono- and binucleated schizonts. These produce the second schizogonic phase with bi- and tetranucleate schizonts which through repeated division form numerous diplokaryons. For a comprehensive discussion of Microsporidia consult Kudo.

### Suborder Dicnidina

The Dicnidina contains a single family, the Telomyxidae. It is characterized by spores with a polar filament at each end.

*Telomyxa glugeiformis* occurs in the fat bodies of may flies (*Ephemera*). Infected nymphs are sluggish and chalky white, owing to the accumulation of spores inside. Such heavy infections are fatal. The spores are elliptical and in groups of 8, 16, or more.

**Symptoms and Pathology.**    Symptoms of nosemiasis among bees are abdominal distention, copious defecation, sluggishness, and impaired ability or complete inability to fly. Workers are unable to forage effectively, and their lifespan is shortened. The rate of brood rearing is severely depressed during April, May, and June. The effects of parasitism decrease during July, August, and September but it is already too late in the season for the bees to harvest a normal crop of honey. Greatest mortality in parasitized colonies occurs in late winter or in the spring. Egg production in queen bees declines or even ceases.

The ventricular epithelium of heavily infected bees is destroyed. In lightly infected individuals, the intestinal cells are not damaged seriously but the peritrophic membrane detaches and may disappear completely.

## EXERCISE ON LIFE CYCLE

Spores can be collected from feces deposited by bees flying over pieces of window pane placed horizontally near the entrance to hives. Mix the feces with two or three drops of water on a slide and examine under a microscope.

Extrusion of the polar filament can be induced by allowing the water suspension of feces or fragments of spore-laden tissue smeared on coverslips to desiccate for about 3 hours at room temperature. Add a layer of neutral distilled water to each dried smear which is prepared as a hanging drop so that water does not touch the slide. Examine it under bright illumination for spores with extruded polar filaments.

Expose the smears to the vapors of osmium tetraoxide for 2 to 3 minutes and stain the preparation in Giemsa's blood stain for 24 to 48 hours. The filament stains a grayish violet and inside it at various locations along its length are two bright red dots, the nuclei of the binucleate sporoplasm traveling through the hollow polar filament (Plate 39, F–I).

Another method of inducing eversion of the polar filament is to feed spores to paramecia. In the food vacuoles, the filament extrudes as a rigid rod that penetrates the cytoplasm and even projects through the pellicle.

## SELECTED REFERENCES

Bailey, L. (1955). The infection of the proventriculus of the adult honeybee by *Nosema apis*. Parasitology 45:86–94.

Bailey, L. (1959). Infectious diseases of the honeybee. Rep. Rothamstead Exp. Sta. p. 204.

Doflein, F., and E. Reichenow. (1953). Lehrbuch der Protozoenkunde. 6th ed. Gustav Fischer, Jena, p. 1004.

Fanthan, H. B., and A. Porter. (1912). The morphology and life history of *Nosema apis* and the significance of its various stages in the so-called "Isle of Wight" disease in bees (microsporidosis). Ann. Trop. Med. Parasitol. 6:163–195.

Grassé, P. P. (1953). Traité de Zoologie. Masson et Cie, Paris, Vol. 1, Fasc. 2, p. 1042.

Gray, F. H., A. Coli, and J. D. Briggs. (1969). Intracellular stages in the life cycle of the microsporidan *Nosema apis*. J. Invert. Pathol. 14:391–394.

Hassanein, M. H. (1953). The influence of infection with *Nosema apis* on the activities and longevity of the worker honeybee. Ann. Appl. Biol. 40:418–423.

Kramer, J. P. (1960). Observations on the emergence of the microsporidian sporoplasm. J. Insect. Pathol. 2:433–439.

Kudo, R. R. (1924). A biologic and taxonomic study of the Microsporidia. Ill. Biol. Monogr. 9:1–268.

Kudo, R. R., and W. E. Daniels. (1963). An electron microscope study of the spore of a microsporidian *Thelohania californica*. J. Protozool. 10:112–120.

Lom, J., and J. O. Corliss. (1967). Ultrastructural observations on the development of the microsporidian protozoan *Plistophora hyphessobryconis* Schäperclaus. J. Protozool. 14:141–152.

Lom, J., and J. Vávra. (1963). The mode of sporoplasm extrusion in microsporidan spores. Acta Protozool. 1:81–90.

Lom, J., and J. Vávra. (1963). Fine morphology of the spore in Microsporidia. Acta Protozool. 1:279–283.

McLaughlin, R. E. (1969). *Glugea gasti* sp. n., a microsporidian pathogen of the boll weevil, *Anthonomus grandis*. J. Protozool. 16:84–92.

Oshima, K. (1937). On the function of the polar filament of *Nosema bombycis*. Parasitology 29:220–224.

Roberts, M. D. (1967). Coprological examination for the detection of *Nosema apis* in the honeybee. J. Invert. Pathol. 9:143–144.

Showers, R. E., A. Jones, and F. E. Moeller. (1967). Cross-inoculation of the bumblebee *Bombus fervidus* with the microsporidian *Nosema apis* from the honeybee. J. Econ. Entomol. 60:774–777.

Sprague, V., and S. H. Vernick. (1969). Light and electron microscope observations on *Nosema nelsoni* Sprague, 1950 (Microsporidia: Nosematidae) with particular reference to the Golgi complex. J. Protozool. 16:264–271.

West, A. J., Jr. (1960). The biology of a species of *Nosema* (Sporozoa: Microsporidia) parasitic in the flour beetle *Tribolium confusum*. J. Parasitol. 46:747–754.

White, G. F. (1919). Nosema disease. U.S. Dep. Agr. Bull. 780, 59 pp.

## Subphylum Ciliophorea

The members of this subphylum have cilia, cirri, or other ciliary structures of locomotion. They contain two kinds of nuclei, the macro- and micronucleus, of different size and function. The subphylum includes the subclasses Holotricha and Spirotricha. Representatives of parasitic species occur in the Holotricha which are characterized by uniform ciliation over the body surface.

### FAMILY BALANTIDIIDAE

The trophozoite of this family is characterized by a cytostome at the anterior end of the body which opens into an elongate cylindrical cytopharynx with long cilia in the posterior half.

*Balantidium* Claparède and Lachman, 1857 is the only genus in the family. The species are cosmopolitan, being parasites of the intestine of arthropods (crustaceans and insects), fish, amphibians, and mammals, including humans and swine. Species infecting vertebrates occur in the cecum and colon.

Trophozoites are somewhat egg- or pear-shaped and otherwise as described above for the family. The macronucleus is a large sausage-shaped structure; a minute spherical micronucleus lies beside it. Two contractile vacuoles, one toward each end of the body, are present together with numerous food vacuoles. A rectal vacuole containing solid excreta discharges through the permanent cytopyge, or anus, at the posterior end of the body.

### *Balantidium coli*
(Malmsten, 1857) (Plate 40)

This species occurs frequently in pigs and commonly in humans, especially those associated with swine, throughout the temperate and warm climates. Other hosts include rats, mice, and cattle.

**Description.** Trophozoites, the ciliated forms, are 30 to 150 $\mu$ long by 25 to 120 $\mu$ in diameter. Encysted stages are enclosed in a double-walled, spherical or ovoid cyst, measuring 40 to 60 $\mu$ in diameter.

**Life Cycle.** Infection takes place when mature cysts, and presumably free trophozoites, are swallowed along with food and water contaminated with feces of parasitized animals. In the intestine, the young trophozoites are liberated by the action of digestive juices on the cysts. Freed trophozoites go rather quickly to the cecum and large intestine where they settle on the surface of the mucosal lining. They feed on particulate matter and bacteria. Certain kinds of bacteria appear to stimulate growth and others depress it in *in vitro* cultures. During this

## Explanation of Plate 40

A, Trophozoite of *Balantidium coli*. **B**, Cyst of *B. coli*. **C**, *B. praenucleatum* from cockroaches. **D**, *B. entozoon* from frogs. **E**, Trophozoite of *B. coli* in process of transverse binary fission. **F**, Section of colon of pig, showing trophozoites of *B. coli* in mucosal layer. **G**, Diagrammatic sketch, showing stages of life cycle in colon. **H**, Swine host.

1, Cytostome; 2, cytopharynx; 3, cytopyge, or anus; 4, macronucleus; 5, dividing macronucleus; 6, micronucleus; 7, contractile vacuole; 8, trophozoite of *B. coli* in mucosa of intestine (section); 9, inflammatory reaction; 10, submucosa of intestine with its nuclei and stroma; 11, cyst of *B. coli* entering lumen of intestine; 12, excystation; 13, trophozoite on surface of mucosa; 14, trophozoite dividing in lumen of intestine; 15, trophozoite in crypts; 16, trophozoite penetrating damaged mucosal epithelium; 17, multiplication of trophozoites by fission, showing destruction of tissue; 18, nest of trophozoites, with much destruction of tissue; 19, destruction of infected villus, releasing many daughter trophozoites into lumen of colon; 20, some daughter trophozoites reenter mucosal epithelium and multiply; 21, some trophozoites are voided with feces; 22, most of trophozoites encyst in colon before being voided with feces.

a, Ingestion of encysted *B. coli;* b, excystation (occurs in stomach or small intestine); c, trophozoite in lumen of intestine; d, division of trophozoite in lumen of colon; e, trophozoite entering crypts of colon; f, trophozoite penetrating epithelial mucosa; g, division and colonization of trophozoites in mucosal region of colon with daughter trophozoites; h, rupture of villus with release of daughter trophozoites; i, some daughter trophozoites reenter epithelial mucosa for further multiplication; j, many daughter trophozoites encyst upon being discharged from mucosa into lumen of colon; k, some daughter trophozoites leaving colon in an unencysted stage; l, cyst in feces; m, trophozoites in feces; n, cyst.

Figure A redrawn from Wenyon, 1926, Protozoology, Vol. 2, p. 1203; **C**, from Kudo, 1953, Protozoology, 4th Ed., p. 797; **D**, from Greel, 1956, Protozoologie, p. 265; **F**, from Brumpt, 1949, Précis de Parasitologie, p. 571.

---

time they multiply by transverse binary fission. Conjugation in the host apparently is rare, if it occurs at all, although it takes place in cultures.

In pigs, these ciliates are usually harmless commensals. However, injury from various causes that damage the mucosal lining enables them to invade it. Once inside the tissue, they begin to multiply. Colonies of trophozoites develop, producing submucosal ulcers comparable to the ones formed by *Entamoeba histolytica*. In humans, *B. coli* is regularly an active tissue invader, even entering the intact mucosal lining. Ulcers result from mechanical damage and enzymatic action of the trophozoites on the surrounding tissues. A case of human infection has been reported where a liver abscess caused by *B. coli* extended through the diaphragm to the lungs, causing death of the patient.

As the trophozoites multiply and the ulcers expand, adjacent ones coalesce, resulting in the formation of still larger lesions. Destruction and sloughing of the cells overlaying the ulcers allows the daughter trophozoites to escape into the intestinal lumen. As they are carried posteriorly with the flow of the intestinal contents, encystment begins as dehydration of the fecal material proceeds. Encystment appears to be a means of protecting them after they leave the host since multiplication in cysts does not occur. In the case of trophozoites that fail to encyst in the intestine, encystment may occur outside the host. Such unencysted trophozoites may survive up to 10 days in the feces and probably are infective if swallowed.

Other common species include *B. suis* which may be the same as *B. coli*, *B. caviae* from guinea pigs, *B. duodeni* and *B. entozoon* from frogs, and *B. praenucleatum* from American and Oriental cockroaches.

## EXERCISE ON LIFE CYCLE

Both ciliated trophozoites and cysts occur in the feces of pigs. In order to find them, suspend a small amount of fresh feces in water, strain through several layers of cheesecloth, and clean the filtrate by repeated sedimentation and decantation. Pour some of the cleared sediment into a glass dish and examine under a dissecting microscope for the large trophozoites and cysts.

Cultures of feces are preferable for examination because they yield a higher percentage of infection among pigs than direct fecal examinations. They also provide specimens in quantity for experimental studies. The simplest culture medium is Nelson's fecal filtrate consisting of one part cecal contents of a pig mixed with nine parts of Ringer's solution. The mixture is strained

**Plate 40**     *Balantidium coli*                                                                    193

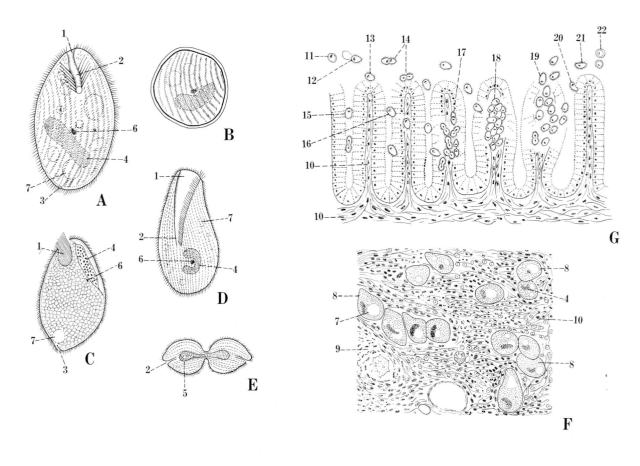

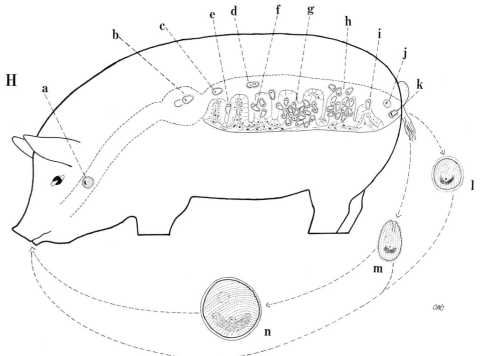

through several layers of cheesecloth and filtered through cotton in a funnel. The filtrate is the culture medium which keeps well in a refrigerator. When buffered to pH 8, the trophozoites live and multiply for 28 days.

Defined media excellent for culturing *B. coli* include Rees' modification of the Barret and Yarbrough formula, Glaser and Coria's liquid medium, Jameson's and Cox's adaptation of the HSre = S variant of the Boeck and Drbohlav medium, and Nelson's alcoholic liver or egg yolk extract. Cox found cultures of small amounts of pig feces in Boeck and Drbohlav medium, a highly sensitive means of detecting infection in a large percentage of the pigs examined.

In addition to swine, cockroaches, both American and Oriental, frogs, and salamanders are good sources for different species of *Balantidium* for experimental studies.

Stages of development occurring in the life cycle may be observed in cultures. Conjugation takes place in cultures but not in the host animals.

## SELECTED REFERENCES

Cox, F. E. G. (1963). The initiation of cultures as a means of diagnosis of *Balantidium coli*. Progress in Protozoology. Proc. First Int. Congr. Protozool. Prague, 1961, pp. 524–526.

Craig, C. F. (1948). Laboratory Diagnosis of Protozoan Diseases. Lea & Febiger, Philadelphia, p. 349.

Doflein, F., and E. Reichenow. (1953). Lehrbuch der Protozoenkunde. 6th ed. Gustav Fischer, Jena, p. 1130.

Glaser, R. W., and N. A. Coria. (1935). The partial purification of *Balantidium coli* from swine. J. Parasitol. 21:190–193.

Jameson, A. P. (1927). The behavior of *Balantidium coli* Malm. in cultures. Parasitology 19:411–419.

Krascheinnikov, S., and D. H. Wenrich. (1958). Some observations on the morphology and division of *Balantidium coli* and *Balantidium caviae* (?). J. Protozool. 5:196–202.

Kretchmer, W. (1963). Untersuchungen über den Zellbau von *Balantidium coli* und *Trichonympha agilis* zur Phylogenie der begreisselten Protozoen. Z. Tropenmed. Parasitol. 14:122–176.

Levine, N. D. (1939). Observations on *Balantidium coli* from swine in culture. J. Parasitol. 25:401–404.

Levine, N. D. (1961). Protozoan Parasites of Domestic Animals and Man. Burgess Publishing Co., Minneapolis, p. 371.

Nelson, E. C. (1940). An intestinal content cultivation medium. I. Methods of preparation and use and data obtained in the cultivation of *Balantidium coli* from the pig. Amer. J. Trop. Med. 20:731–735.

Nelson, E. C. (1947). Alcoholic extract medium for the diagnosis and cultivation of *Endamoeba histolytica*. Amer. J. Trop. Med. 27:545–552.

Rees, C. W. (1927). Balantidia from pigs and guinea-pigs: their viability, cyst production, and cultivation. Science 66:89–91.

Rostkowska, J. (1970). Biocoenotic interrelations between *Balantidium coli* (Malmsten) and the bacterial flora of the alimentary tract. Acta Parasitol. Pol. 18:377–392.

Svensson, R. M. (1955). On the resistance to heating and cooling of *Balantidium coli* in culture and some observations regarding conjugation. Exp. Parasitol. 4:502–525.

Wenger, F. (1967). Absceso hepático producido por el *Balantidium coli*. Rev. Univ. Zulia 10:27–35.

## FAMILY OPHRYOGLENIDAE

This family contains one genus that is parasitic. It is characterized by a ventrally located cytostome, sometimes inconspicuous, with an undulating membrane on the right and three membranelles on the left, and inconspicuous bordering cilia. A refractile "watch glass" body is located in the cytoplasm near the oral opening.

*Ichthyophthirius multifiliis*
Fouquet, 1876 (Plate 41)

This parasite attacks the epidermis of the skin and fins, the gill filaments, and cornea of a number of species of freshwater fish.

**Description.**    Adult trophozoites up to 1 mm long are oval and covered with numerous rows of cilia interspersed with many circular micropores about 1 $\mu$ in diameter in the pellicle which consists of an inner and outer layer. An unciliated ring-like cytostome is located at the anterior end of the body. The macronucleus is a large horseshoe-shaped body with the small micronucleus lying in the concavity of the former. Numerous contractile vacuoles are scattered throughout the cytoplasm. The contractile vacuoles together with cytoplasmic organelles form an anatomical-physiological complex of significance. Each contractile vacuole is connected through a discharge canal about 1 $\mu$ in length and diameter with an external micropore. The base of the discharge canal is closed by a double membrane consisting of the thick inner pellicular layer on the one side and the very thin vacuolar membrane on the other side (Fig. 23). These membranes rupture to empty the contractile vacuole through the discharge canal and then reform. Tubular fibrils, probably contractile in function, surround the discharge canals and cover the outer surface of the contractile vacuoles.

In a state of relaxation, or distole, the contractile vacuoles are spheres about 3.5 μ in diameter. Contracted, or systolic, vacuoles assume a multilobular figure with many fingerlike tubular extensions into the cytoplasm. Numerous minute injection canals extend from the cytoplasm to and empty through the wall of the contractile vacuoles. Granular endoplasmic reticulum and many mitochondria are intermingled with the injection canals. These highly differentiated canalicular-vesicular zones represent drainage systems from the cytoplasm into the contractile vacuoles. A permanent cytopyge is located at the posterior end of the trophozoite. Adult trophozoites on the bottom of ponds encyst, forming large gelatinous cysts with thick, clear walls.

**Life Cycle.**    Mature trophozoites are in epidermal pustules of the skin, fins, tail, and gill filaments of fish. Upon rupture of the pustules, they are liberated and swim about feebly. Upon coming to rest on aquatic plants, snail shells or other objects on the bottom of the pond, each trophozoite secretes a thick-walled, clear, gelatinous cyst about itself. Within an hour after encystment, the mother trophozoite begins to divide by simple transverse division, first with two, then four, eight and so on, until numerous daughter trophozoites, or swarmers, are produced. The number of swarmers may be as many as 1000, depending on the size of the mother trophozoite.

The swarmers are pear-shaped, translucent, ciliated protozoans 30 to 50 μ long with a spherical nucleus and single pulsating vacuole. The pointed, unciliated anterior end serves as a boring apparatus for penetrating the skin of fish.

Within 7 to 8 hours after detachment of the mother trophozoites from the fish host in water 18° to 20°C, they have encysted and completed their development and are leaving the cysts and attaching to fish. Unattached individuals die during the second day.

Upon attaching to a fish, the swarmers bore by whirling movements into and under the epidermis. Once under it, they move about, forming galleries that soon become occupied by additional swarmers so that many of them occur together, forming pustules. Division with multiplication does not occur in the epidermis.

Two to 3 days after burrowing under the epidermis, a cytostome is recognizable in the swarmers. The boring apparatus diminishes to form the center point of the ciliated area. Through ingestion of blood cells and nutritive substances in the skin, growing swarmers appear as opaque, dark-colored granules. As growth continues, the number of vacuoles in the parasites increases and the macronucleus becomes increasingly U-shaped.

**Symptoms and Pathology.**    Lesions on the skin, fins, gill filaments, and cornea of the eyes appear as grayish pustules. When present in large numbers, the pustules merge and impart a turbid appearance to the affected areas. The infection is known as skin disease, ichthyophthiriasis, or just "ich" for short. Epidermal cells slough and are replaced by a proliferation of mucus-producing cells. Trophozoites attach to the gill filaments in such numbers that exchange of oxygen and carbon dioxide through the membranes is curtailed and invasion of the filaments

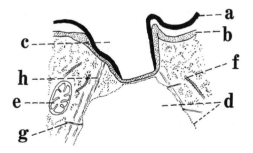

Fig. 23.    Diagrammatic figure from electron micrograph of discharge canal, excretory vacuole, and complex of associated organelles of *Ichthyophthirius multifiliis*. **a,** Outer layer of pellicle; **b,** inner granular, cytoplasmic layer of pellicle; **c,** longitudinal section of discharge canal with walls formed by outer and inner layers of pellicle (a, b) and bottom only of inner layer (b), which ruptures at intervals to release contents of contractile vacuole (d); **d,** contractile vacuole (wall dotted to differentiate it); **e,** mitochondrion; **f,** endoplasmic reticulum; **g,** injection canal from cytoplasm emptying into contractile vacuole; **h,** hollow contractile fibers that surround discharge canal and contractile vacuole. Adapted from electron micrographs from Mosevitch, 1965, Acta Protozool. 3:61.

## Explanation of Plate 41

A, Fully developed trophozoite of *Ichthyophthirius multifiliis* from pustule. B, Anterior end of fully developed trophozoite. C, Swarmer from cyst. D, E, First and second divisions of encysted trophozoite. F, Later stage of cystic multiplication. G, Cyst filled with swarmers, some of which are escaping into water. H, Section of skin of fish, showing full-grown trophozoite embedded in it. I, Section of tail of carp, showing ciliates developing in pustule. J, Infected bullhead (*Ameiurus melas*).

1, Cytostome; 2, macronucleus with nearby micronucleus; 3, longitudinal rows of cilia; 4, contractile vacuoles; 5, boring or penetration apparatus; 6, cyst; 7, dividing of macronucleus; 8, two daughter cells formed by first division; 9, four daughter cells formed by second division in cyst; 10, numerous daughter cells; 11, swarmers; 12, epidermis of fish skin; 13, pigment cell in epidermis; 14, dermis; 15, cartilaginous skeleton of tail of carp; 16, pustule containing trophozoites; 17, trophozoite under skin.

a, Pustules; b, trophozoite escaping from pustule into water; c, trophozoite free in water; d, encysted trophozoite on bottom of pond in first division, showing two daughter cells; e, cyst in second division with four daughter cells; f, cyst with many daughter cells; g, ruptured cyst liberating swarmers; h, swarmer attached to skin; i, swarmer partially embedded in skin.

Figures A, D–G, I redrawn from Kudo, 1953, Protozoology, 4th Ed., p. 709; B, C, H adapted from Schäperclaus, 1954, Fischkrankheiten, p. 334, 338.

---

damages the capillaries so the fish die of suffocation. Hemoglobin concentration declines in infected fish. Free living trophozoites are killed in salinity of 1 ppt or greater.

## EXERCISE ON LIFE CYCLE

Place infected fish in small aquaria where cysts and swarmers may accumulate on the bottom and be recovered for study.

Note the formation of the thick-walled gelatinous cyst around adult trophozoites freed from the pustules. This is followed by repeated binary fission within the cyst and the formation of swarmers with the unciliated, pointed boring end.

Transfer swarmers to a clean aquarium containing uninfected fish and observe how long it takes for pustules to appear on them. Histological sections of the skin and gill filaments containing pustules should be studied to observe the parasites in the tissue and the effect they have on it.

Young trophozoites cultured in sterile medium consisting of 0.3 gm of fresh mucus from the body surface of carp in 100 ml of filtered pond water may be observed in the various stages of development.

## SELECTED REFERENCES

Allen, K. O., and J. W. Avault, Jr. (1970). Effects of brackish water on Ichthyophthiriasis of channel catfish. Progr. Fish-Cult. 32:227–230.

Davis, H. S. (1946). Care and diseases of trout. U.S. Fish Wildl. Serv. Res. Rep. No. 12, 98 pp.

Davis, H. S. (1953). Culture and Diseases of Game Fishes. University of California Press, Berkeley, p. 209.

Doflein, F., and E. Reichenow. (1953). Lehrbuch der Protozoenkunde. 6th ed. Gustav Fischer, Jena, p. 1115.

Hłond, S. (1967). [Attempts at *in vitro* cultivation of *Ichthyophthirius multifiliis*]. Wiad. Parazytol. 13:279–282.

Kosylowski, B., and J. Antychowicz. (1964). [Anatomo- and histopathological lesions of ichthyophthiriosis (*Ichthyophthirius multifiliis*) of carp in sick fish and in the treated ones with malachite green]. Bull. Vet. Inst. Pulawy 8:136–145.

Mosevitch, T. N. (1965). Electron microscopic structure of the contractile vacuole of the ciliate *Ichthyophthirius multifiliis* (Fouquet). Acta Protozool. 3:61–71.

Reshtinkova, A. V. (1962). [Change in the blood of carp (*Cyprinus carpio*) due to *Ichthyophthirius* infection]. In: Biol. Abst. 45:80940.

Schäperclaus, W. (1954). Fischkrankheiten. Akademie, Berlin, p. 333.

**Plate 41**   *Ichthyophthirius multifiliis*   197

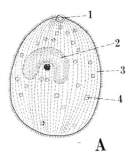
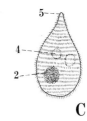
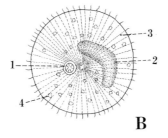
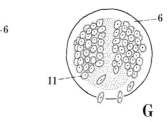

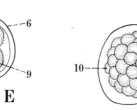
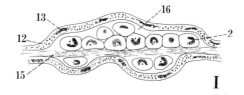
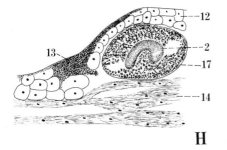
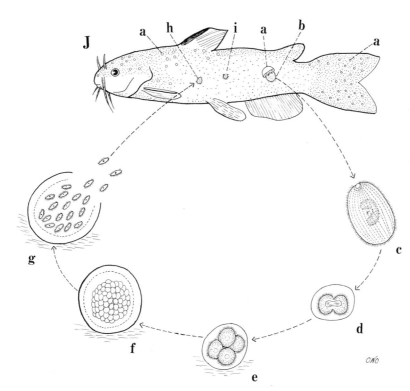

# Phylum Platyhelminthes

Parasitic representatives of the Platyhelminthes include the classes Trematoda and Cestoda. They are soft-bodied flat worms with an incomplete digestive tract in the Trematoda and none at all in the Cestoda. Since there is no body cavity, the internal organs lie embedded in the spongy parenchymatous tissue. The trematodes are monozoic, i.e., a single body, and the cestodes are monozoic or polyzoic, in which the body consists of a few to thousands of proglottids.

## CLASS TREMATODA

The Trematoda consists of three subclasses, the Monogenea, Aspidobothrea, and Digenea. Although this classification has been used for a long time, the relationship of the Monogenea to the other two groups has been questioned.

During recent years, the recognition of the larval stage known as oncomiracidium of most of the 28 families of Monogenea has provided much new information on the homogeneity of this group of worms. Studies on the development and structure of the holdfast organ and its hooks exhibit distinct differences in the oncomiracidia from the miracidia of the digenetic trematodes.

This condition has led some workers, particularly Bychowsky and Llewellyn, to regard the oncomiracidia as being more closely related to the six-hooked (hexacanth) and ten-hooked (decacanth) larvae of the Cestoda than to the hookless miracidia of the Digenea. Accordingly, Bychowsky removed the monogeneans from the class Trematoda and placed them in the class Monogenoida alongside the class Gyrocotylida of the gyrocotylidean cestodes with ten-hooked larvae. For convenience in discussing the monogeneans here, they are included in the class Trematoda.

### Subclass Monogenea

The monogenetic trematodes are the most primitive of the flukes, showing morphological relationships to free-living rhabdocoels. They are parasites of cold-blooded aquatic vertebrates, especially fish. They are primarily ectoparasitic, although a few species occur in such places as the mouth, urinary bladder, and ureters. Life cycles are direct, that is, they are completed without the need of an intermediate host.

The subclass consists of the orders Monopisthocotylea and Polyopisthocotylea. Some of the common representatives of each order are shown in Plate 42, together with the basic morphological characteristics of each.

## SELECTED REFERENCES

Bychowsky, B. E. (1957). Monogenetic Trematodes. Their Systematics and Phylogeny. (Translation of Russian Monograph.) W. J. Hargis, Jr., ed. American Institute of Biological Sciences, Washington, D. C., 627 pp.

Llewellyn, J. (1963). Larvae and larval development of monogeneans. In: B. Dawes, ed. Advances in Parasitology. Academic Press, New York, Vol. 1, pp. 287–326.

Llewellyn, J. (1968). Larvae and larval development of monogeneans. In: B. Dawes, ed. Advances in Parasitology. Academic Press, New York, Vol. 6, pp. 373–383.

Yamaguti, S. (1963). Systema Helminthum. Monogenea and Aspidocotlylea. Academic Press, New York, Vol. 4, 699 pp.

## ORDER MONOPISTHOCOTYLEA

The members of this order are characterized by a single organ of adhesion, usually in the form of a well-developed disc but sometimes sucker-like, located at the posterior extremity of the body. It is armed with one to three pairs of large hooks and 12 to 16 marginal hooklets. The

## Explanation of Plate 42

A, *Gyrodactylus cylindriformis* from *Umbra limi.* B, *Dactylogyrus extensus* from gills of *Cyprinus carpio.* C, *Ancyrocephalus aculeatus* from gills of *Stizostedion vitreum.* D, *Udonella caligorum* on *Gadus callarias* and *Myliobatis californicus.* E, *Benedenia girellae* from skin of *Girella nigricans.* F, *Acanthocotyle williamsi* from skin of "skate." G, *Heterocotyle mimima* from gills of *Squalus acanthias.* H, *Microbothrium apiculatum* from *Squalus acanthias.* I, *Rajonchocotyloides emarginata* from gills of *Raja clavata.* J, *Polystomoides cornutum* from North American turtles. K, *Diclidophora caulolatili* from gills of *Cautolatilus princips.* L, *Neohexastoma euthynni* from gills of *Euthynnus alletteratus.* M, *Mazocraes macracanthum* from gills of a mackerel. N, *Microcotyle spinicirrus* from gills of *Aplodinotus grunniens.* O, *Octomacrum lanceatum* from gills of *Catostomus commersoni.*

1, Glandular prohaptor; 2, external paired sucker-like prohaptor; 3, internal buccal suckers; 4, oral sucker; 5, mouth; 6, pharynx; 7, intestine; 8, egg; 9, first embryo; 10, second embryo; 11, ovary; 12, seminal receptacle; 13, genital pore; 14, vitellaria; 15, testis; 16, opisthaptor; 17, anchors with transverse connecting bar; 18, spines; 19, marginal hooklets; 20, sucker; 21, sickle-shaped hook in sucker; 22, opisthaptoral appendage; 23, clamps; 24, detailed view of clamp of Diclidophoridae; 25, clamp of Hexastomatidae; 26, clamp of Mazocraeidae; 27, clamp of Microcotylidae; 28, clamp of Discocotylidae; 29, haptoral hook.

Figures A, B, C redrawn from Van Cleave and Mueller, 1932, Roosevelt Wildl. Ann. 3:93; D, F–H, from Price, 1938, J. Wash. Acad. Sci. 28:185; E, from Hargis, 1941, J. Parasitol. 41:48; I, from Price, 1939, Proc. Helminthol. Soc. Wash. 7:76; J, from Price, 1939, Ibid. 6:83; O, from Price, 1943, Ibid. 10:12; K–M, from Meserve, 1938, Allan Hancock Pacific Exped. 2:43; N, from Remeley, 1942, Trans. Amer. Micr. Soc. 61:141.

---

eggs are nonoperculate. The life cycles of representatives from two families will be considered.

### FAMILY DACTYLOGYRIDAE

These are small elongate, oviparous flukes on the gills of fish. They have two or more pairs of head organs, an opisthaptor (posterior sucker) with one or two pairs of large hooks, and 14 marginal hooklets.

Species of the genus *Dactylogyrus* occur primarily on the upper half of the gills of fish. Currently about 100 species are known, distributed among four families of fish in North America. Six species occur on Catostomidae, four on Centrarchidae, one on Gasterosteidae, and the remainder, around 90, on Cyprinidae.

### *Dactylogyrus vastator*
Nybelin, 1924 (Plate 43)

These flukes occur commonly on the gills of carp, where they are especially dangerous to the health of young fish.

**Description.**    Adults measure 0.8 to 1.15 mm long by 0.15 to 0.25 mm wide. There are two prominent head lobes, each with a well-developed sticky gland and two pairs of eyespots in front of the pharynx. The opisthaptor is 0.12 mm in diameter and bears one pair of large hooks with bifurcate roots and seven pairs of small marginal hooklets. There is a single, oval testis and a cirrus that is 54 $\mu$ long. The ovary is more or less oval and located near the middle of the body. The vaginal opening is toward the right margin of the body, slightly preequatorial.

**Life Cycle.**    Adult flukes on the gill filaments lay unembryonated eggs that fall off the fish and sink to the bottom of the pond. They are characterized by being flattened on one side and bearing a bifurcated stalk-like structure on one end. Under summer temperatures, development of the embryo is rapid, and fully formed gyrodactylid larvae, or oncomiracidia, appear in 2½ days. By pressing from inside the egg, the larva forces off one end of the egg shell and escapes, leaving the embryonic membrane inside.

The larvae bear a tuft of cilia on the anterior and posterior extremities and on each side near the middle of the body. There are four eyespots and a large opisthaptor bearing a pair of large hooks and 14 small hooklets. Larvae swim about actively, attaching to the skin of carp when coming in contact with them. Having attached to the fish host, they migrate over the body toward the gills, being attracted by the mucus from them.

Sexual maturity is attained about 10 days after reaching the gills. The peak of infestation on the fish is reached about the middle of July, after which the flukes begin to disappear, being gone by fall. During the last phase of oviposition, winter eggs are laid; these remain in a resting stage during the cold weather. With the arrival

**Plate 42** *Types of Common Monogenetic Trematodes* 201

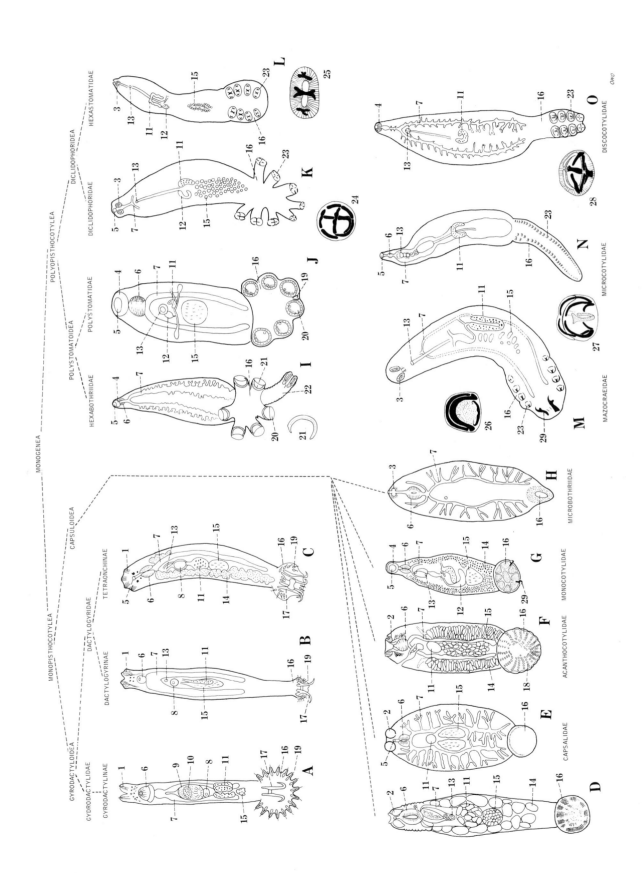

## Explanation of Plate 43

A, Adult fluke. B, Fully developed and recently hatched larva. C, Young larva having lost its cilia. D, Freshly laid and undeveloped egg. E, Partially developed embryo. F, Egg with fully developed larva. G, Larva escaping from egg. H, Carp (*Cyprinus carpio*) host with opercular cover removed to show gill filaments.

1, Four head organs; 2, eyespots; 3, pharynx; 4, cyclic intestine; 5, testis; 6, vas efferens; 7, male genital pore; 8, ovary; 9, oviduct; 10, genitointestinal canal; 11, unembryonated egg in uterus; 12, female genital opening; 13, vitelline gland; 14, vitelline duct; 15, opisthaptor; 16, anchor; 17, transverse bar; 18, marginal hooks; 19, anterior tuft of cilia on a pad of cells; 20, median tuft of cilia; 21, caudal tuft of cilia on a pyramid of cells forming so-called tail; 22, unembryonated egg in water; 23, yolk material; 24, partially developed larva; 25, embryonic membrane; 26, larva hatching.

a, Adult fluke attached to gills of carp lays eggs; b, unembryonated egg falling free from gills; c, egg on bottom of pond; d, developing larva; e, fully developed larva in egg; f, hatching egg with larvae escaping into water; g, swimming larva comes in contact with carp; h, larva has shed ciliated pads and enters mucus layer; i, larva increases in size and continues to migrate toward gills; j, larger larva reaches gills and matures in about 10 days.

Figure A adapted from various sources and Figs. B–G, from Gröben, 1940, Z. Parasitenk. 11:611.

---

of spring and the warming of the water, development of the eggs begins and hatching soon takes place. Young carp especially are infested, often fatally when large numbers of flukes are present on the gill filaments. Eggs laid during the summer hatch after a short period of incubation, resulting in an abundance of oncomiracidia with superinfection of the fish. Under these conditions, an immunity develops in the carp that reduces the infestation for about 2 months.

*Dactylogyrus extensus* infests the upper half of the gill filaments of carp simultaneously with *D. vastator* on the lower half and persists considerably longer. Eggs develop in 5 days at 20°C and the worms mature in 6 to 10 days. The life cycle is completed in 18 to 20 days at temperatures of 14° to 15°C and 17° to 19°C, showing that this species is adapted to thrive at both low and high temperatures.

*Dactylogyrus macracanthus* and *D. anchoratus* have life cycles basically similar to that of *D. vastator*.

Species of *Gyrodactylus* (Gyrodactylidae), also parasites of the gills of fish, are ovoviviparous. Their eggs develop and hatch within the uterus. Before birth, a second larva appears inside the first, a third inside the second, and even a fourth inside the third. After birth, the primary larva attaches to the gills of a fish and the sequence of growth and birth of the larvae contained within it ensues. Parasites attached to the gill filaments cause hypertrophy of the cells and stimulate excessive secretion of mucus. Respiration is hampered markedly.

## EXERCISE ON LIFE CYCLE

Eggs for experimental purposes may be obtained by two methods. Those deposited by flukes on naturally infested carp kept in small aquaria sink to the bottom and may be found by concentrating the sediment and recovering them from it. Adult flukes removed from the gill filaments and placed in dishes of water deposit their eggs.

Observations on the development of the eggs may be made by placing them on microscope slides and watching them through a microscope at frequent intervals for 2 to 3 days. Observations may be made on the reaction of larvae to a small drop of mucus from the gills of carp. This experiment gives some information on chemical attractants that help larvae find the gills.

Infection experiments may be conducted by placing larvae in a small aquarium with goldfish. Determine how long it takes from the time of exposure of the fish to larvae until eggs appear in the aquarium.

Ascertain whether eggs laid in the fall hatch in the usual time or whether they are winter eggs and hatch the following spring. Also determine whether flukes remain on the gills over winter.

## SELECTED REFERENCES

Bychowsky, B. E. (1957). Monogenetic Trematodes. Their Systematics and Phylogeny. (Translation of Russian Monograph.) W. J. Hargis, Jr., ed. Ameri-

**Plate 43**    *Dactylogyrus vastator*                    203

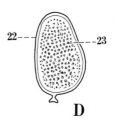
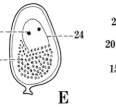
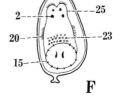
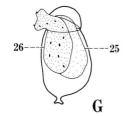
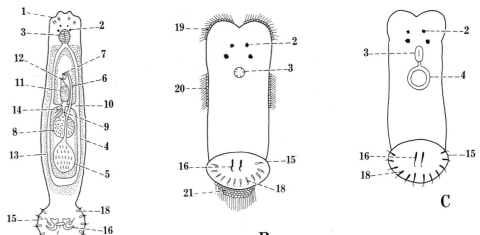
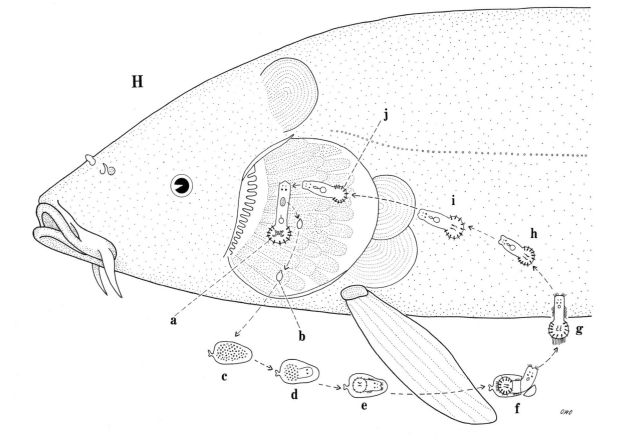

## Explanation of Plate 44

A, Adult fluke. B, Anchor (large hook). C, Unembryonated egg. D, Egg hatching. E, Young larva just hatched. F, Slightly older larva. G, Larva 16 days old. H, Salamander (*Necturus maculosus*) host. I, Salamander host.

1, Oral sucker (prohaptor); 2, opisthaptor; 3, sucker; 4, hooklet of sucker; 5, marginal hooklets; 6, anchor; 7, pharynx; 8, intestine; 9, testes; 10, vas deferens; 11, seminal vesicle; 12, common genital pore; 13, ovary; 14, oviduct; 15, genitoinestinal canal; 16, uterus; 17, egg in uterus; 18, excretory vesicle; 19, shell of unembryonated egg; 20, yolk material; 21, larva escaping from egg; 22, primordium of digestive tract.

a, Adult worm on gills of salamander; b, unembryonated egg falling from gills; c, embryonated egg on bottom of pool; d, egg that has just hatched with larva free in water; e, newly hatched larva on skin migrating toward gills; f, older larva on gills; g, 15-day-old larva on gills continues development reaching maturity about 2 months after attaching to gills.

Adapted from Alvey, 1936, Parasitology 28:229.

can Institute of Biological Sciences, Washington, D.C., 627 pp.

Dawes, B. (1956). The Trematoda. Cambridge University Press, London, p. 66.

Gröben, G. (1940). Beobachtungen über die Entwicklung verscheidener Arten von Fischschmarotzern aus der Gattung *Dactylogyrus*. Z. Parasitenk. 11:611–636.

Hyman, L. (1951). The Invertebrates. McGraw-Hill Book Co., New York, Vol. 3, pp. 232, 239.

Izumova, N. A., and N. I. Zelenstov. (1969). [Observations on the development of *Dactylogyrus extensus* Müller and Van Cleave, 1932.] Parazitologiya 3:528–531.

Kathariner, L. (1904). Ueber die Entwicklung von *Gyrodactylus elegans* v. Nrdm. Zool. Jahrb., Suppl. 7:519–550.

Kollman, A. (1966). *Dactylogyrus extensus* Müller et v. Cleave (Trematoda, Monogenoidea auf Karpfen (*Cyprinus carpio* L.) zum ersten Male in Westdeutschland nachgewiesen. Zool. Anz. 177:426–434.

Llewellyn, J. (1963). Larvae and larval development of mongeneans. In: B. Dawes, ed. Advances in Parasitology. Academic Press, New York, Vol. 1, pp. 287–326.

Llewellyn, J. (1968). Larvae and larval development of monogeneans. In: B. Dawes, ed. Advances in Parasitology. Academic Press, New York, Vol. 6, pp. 373–383.

Mizelle, J. D., and H. D. McDougal. (1970). Studies on monogenetic trematodes. XLV. The genus *Dactylogyrus* in North America. Key to the species, host-parasite and parasite-host lists, localities, emendations, and description of *D. kritskyi*. Amer. Midl. Natur. 84:444–462.

Paperna, I. (1963). Dynamics of *Dactylogyrus vastator* Nybelin (Monogenea) populations on the gills of carp fry in fish ponds. Bamidgeh Israel. 15:31–50.

Vladimirov, V. L. (1971). [Immunity of fish against *Dactylogyrus*.] Parazitologiya 5:51–58.

## ORDER POLYOPISTHOCOTYLEA

The members of this order are characterized by the opisthaptor being composed of a number of suckerlets or clamps borne on a disc-like process or on the ventral surface of the posterior end of the body. They are parasites on fish, amphibians, reptiles, and sometimes in the eyes of marine mammals.

### FAMILY POLYSTOMATIDAE

The species of this family are parasites of the gills, buccal and nasal cavities, pharynx, esophagus, urinary bladder of amphibians and reptiles, and the eyes of marine mammals.

An oral sucker forms the prohaptor, while a disc bearing one to three pairs of cup-like muscular suckers forms the opisthaptor.

### *Sphyranura oligorchis*
### (Alvey, 1933) (Plate 44)

This parasite infects the gills of necturus (*Necturus maculosus*) in the eastern part of the United States and possibly wherever the host occurs.

**Description.** Adult worms are 2.5 mm long. The opisthaptor consists of two large hooks 260 $\mu$ long and two large cup-like muscular suckers. In addition, there are 16 hooklets arranged with seven on each side of the opisthaptor and one in each sucker. There are six testes arranged in a row, one behind the other between the branches of the intestine. The ovoid ovary is located intercecally toward the left side just anterior to the middle of the body. Vitellaria extend from near the middle of the body to the posterior end. The flukes are oviparous, producing unembryonated, operculate eggs.

**Life Cycle.** Adult flukes on the gill filaments lay their unembryonated eggs which drop

**Plate 44**    *Sphyranura oligorchis*                                                    205

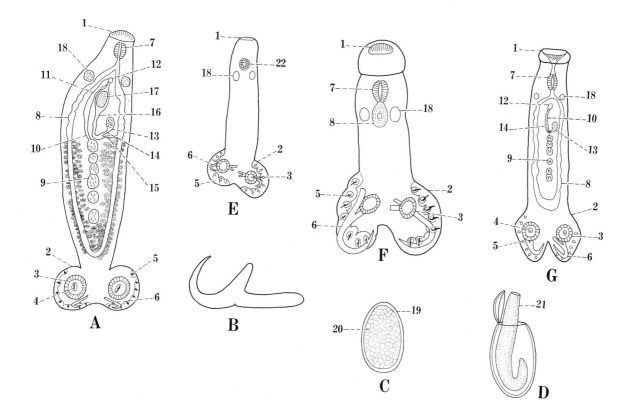

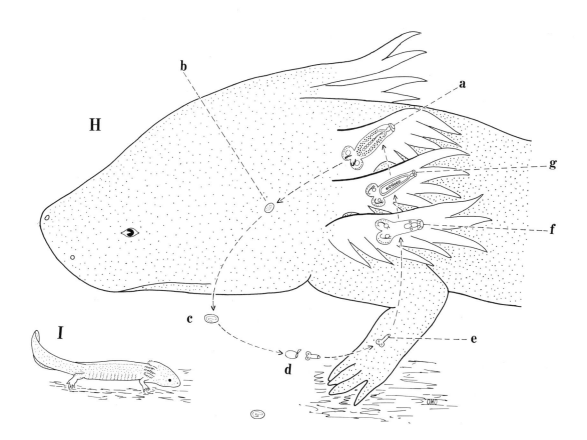

## Explanation of Plate 45

A, Adult bladder fluke. B, Adult branchial fluke. C, Embryonated egg of *P. integerrimum.** D, Dactylogyroid type larva of *P. integerrimum.** E, Adult toad (*Hyla versicolor, H. cinerea*) host of bladder generation. F, Tadpole of *Hyla* which is host of branchial generation. G, Metamorphosing *Hyla*.

1, Mouth (prohaptor); 2, opisthaptor; 3, opisthaptoral sucker; 4, anchors; 5, pharynx; 6, intestine; 7, sperm; 8, vas deferens; 9, common genital pore; 10, ovary; 11, oviduct; 12, Mehlis' gland; 13, ootype; 14, uterus; 15, genitointestinal canal; 16, vagina; 17, vaginal duct; 18, vitellovaginal duct; 19, excretory pore; 20, testis; 21, vitelline gland; 22, vitelline duct; 23, eyespot; 24, tuft of cilia; 25, esophageal gland; 26, marginal hooks; 27, excretory tubule.

a, Adult fluke of bladder generation in urinary bladder of toad; b, unembryonated eggs laid in urinary bladder are voided with urine; c, development of eggs begins in water; d, fully developed larva; e, empty egg shell; f, gyrodactyloid larva in water enters gill chamber of tadpoles by way of spiracle, thus ending the urinary bladder generation and initiating the branchial generation; g, larvae attach to gills of tadpoles and mature in about 22 days; h, unembryonated eggs laid on gills of tadpoles are washed from gill chamber through spiracle into the water; i, fully developed larva, identical with those of the urinary generation; j, empty egg shell; k, gyrodactyloid larva free in water; l, larva enters anus of metamorphosing toad, going to urinary bladder; m, young flukes entering the bladder initiate the urinary bladder generation which reaches sexual maturity in about 2 years.

Figures A–B adapted from Paul, 1938, J. Parasitol. 24:489; C, from Zeller in Claus and Sedgwick, 1884, Elementary Textbook of Zoology, Macmillan and Co., New York, Fig. 259 a; D, from Gallien, 1935, Trav. Sta. Zool. Wimereux 12:1.

---

* According to Paul (1938), these stages in *P. nearcticum* are identical with those of *P. integerrimum*.

---

off and settle to the bottom of the pond. Development is slow at room temperature, as hatching does not occur until between days 28 and 32 of incubation. Newly hatched worms swim by means of the caudal disc, or creep. Larvae are very active for a few hours after hatching, then become exhausted, sink to the bottom, and die. If, however, they come in contact with a necturus during the period of activity, they attach and migrate to the gills, no growth taking place in the meantime. Once on the gills where a rich supply of blood and protection are available, development is completed in about 2 months.

## EXERCISE ON LIFE CYCLE

Fertile eggs may be collected from the bottom of aquaria in which infected necturus are kept. When placed in Syracuse watch glasses for ready availability, individual eggs may be removed by means of a fine pipette, placed on a microscope slide, and studied daily by microscopic examination for progressive development and hatching.

Place uninfected necturus in an aquarium with fully developed eggs that are ready to hatch. Observe development of the flukes on the necturus at intervals of about 10 days from the time of attachment until sexual maturity.

## SELECTED REFERENCE

Alvey, C. H. (1936). The morphology and development of the monogenetic trematode *Sphyranura oligorchis* (Alvey, 1933) and the description of *Sphyranura polyorchis* n. sp. Parasitology 28:229–253.

*Polystoma nearcticum*
(Paul, 1935) (Plate 45)

*Polystoma nearcticum* occurs in the urinary bladder of tadpoles and adults of tree toads (*Hyla versicolor*) and on the gills of their tadpoles. There are two forms, the branchial form on the gills and the bladder form in the urinary bladder.

Paul considered the American flukes to be a subspecies of *P. integerrimum* from the urinary bladder of *Rana temporaria* in Europe and named it *P. integerrimum nearcticum*. Price regarded it sufficiently different to constitute a valid species and designated it *P. nearcticum* (Paul, 1935).

**Description.**    The bladder forms range from 2.5 to 4.5 mm long by 0.9 to 1.5 mm wide. The cordiform caudal disc bears six muscular, cup-shaped suckers and one pair of large hooks. The testis is multilobate and the ovary comma-shaped. Eggs are unembryonated and measure 300 by 150 $\mu$.

**Plate 45**     *Polystoma nearcticum*                                                                  207

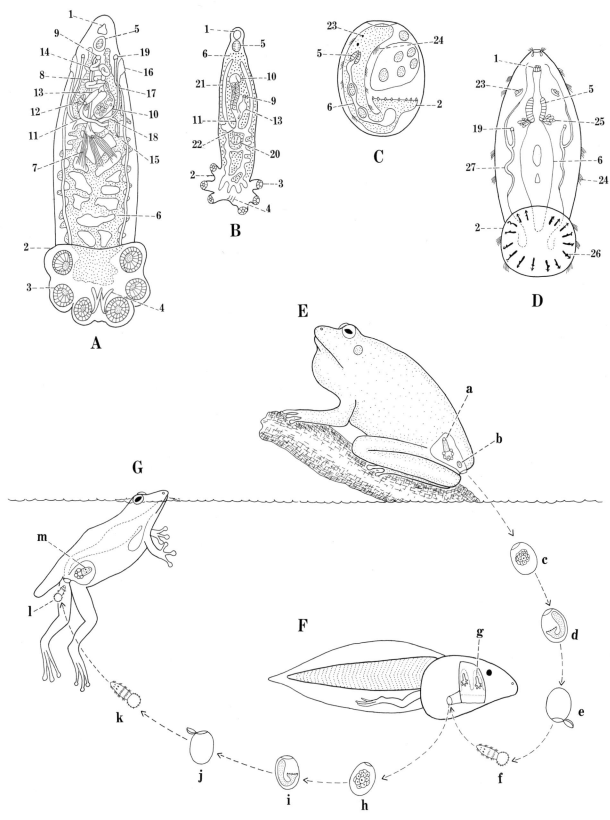

The branchial forms are 1.64 to 5 mm long and 0.29 to 0.76 mm wide. The cordiform opisthaptor bears six pedunculate suckers. The hooks are rudimentary when present. The testis is spherical and the ovary elongate. Eggs are indistinguishable from those of the bladder form.

In *Polystoma integerrimum*, some details of anatomy are known that are not reported for *P. nearcticum*. The larvae have a true body epithelium whereas the adults are covered by a cuticle. The body musculature from the outside inward consists of an outer circular, a middle diagonal, and an inner longitudinal layer. Two large hooks have a complicated musculature seen only with polarized light. In adults, the intestine is lined with a layer of epithelium. The oviduct, yolk ducts, and reproductive part of the genitointestinal ducts are lined with ciliated epithelium while the uterus is unciliated (Fig. 24).

**Life Cycle.**   Adult bladder forms of the flukes begin to lay eggs at the same time the toads become sexually active in the spring. The

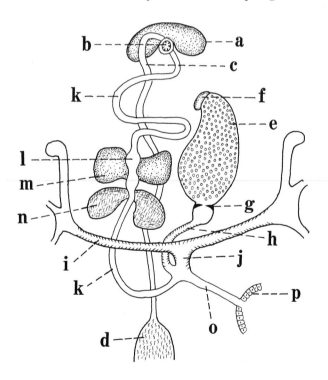

**Fig. 24.**  Schematic representation of reproductive system of *Polystoma integerrimum*. **a**, Cirrus gland; **b**, cirrus; **c**, vas deferens; **d**, testis; **e**, ovary; **f**, ovarian cap; **g**, septa; **h**, oviduct; **i**, yolk duct; **j**, yolk reservoir; **k**, uterus; **l**, ootype; **m**, mucosal part of Mehlis' gland; **n**, serosal portion of Mehlis' gland; **o**, genitointestinal canal; **p**, intestine. Adapted from Kohlmann, 1961, Z. Parasitenk. 20:495.

eggs are voided with the urine. In water, they are fully developed after 11 to 13 days at room temperature. Hatching occurs between days 12 and 13. The larvae are very active up to 20 hours, after which they slow down, sink to the bottom, and die if they do not attach to tadpoles.

Larvae that attach to the gills of tadpoles develop to sexually mature branchial flukes in 22 days. Only eggs that pass through the spiracle into the water are viable. Those that are swallowed and pass through the intestinal tract do not hatch. It is not known whether secondary infection of the gills by larvae from branchial flukes occurs.

Evidence indicates that the free-swimming larvae of the branchial flukes enter the cloaca of tadpoles by way of the anus, going to the urinary bladder as soon as it is formed. Larvae continue to enter through the anus during the process of metamorphosis of the tadpoles and young toads. At the time of atrophy of the gills during metamorphosis of the tadpoles, the branchial flukes die. Larvae entering the bladder develop into bladder forms. Inasmuch as oviposition of the flukes in the bladder coincides with that of the toads at 3 years of age, maturity of them is not attained until they, too, are 3 years old. Thus, there is an alternation of a generation of branchial flukes that requires 3 weeks to mature followed by one of bladder flukes that requires 3 years to mature. Conceivably both the branchial and bladder generations could be on and in the same tadpole. Whether this actually occurs is not known.

Experimental studies on *Polystoma integerrimum* from *Rana temporaria* of Europe show differences from those reported by early workers and that given for *P. nearcticum*. In *P. integerrimum*, the reproductive system matures in 19 months and the egg laying begins at 2 years of age. This is 1 year sooner than the spawning of the frogs. There is no evidence that gonadotropins and sexual hormones of the frogs influence reproduction in the flukes.

In sexually mature flukes, egg laying begins in the spring following a period of inactivity of both the host and parasite during the low winter temperature. Egg production in the flukes can be started artificially by subjecting host and parasite to sudden and prolonged chilling.

Yolk granules from the vitelline glands and eggs from the ovary accumulate simultaneously in a common reservoir from which they pass to the ootype. Mehlis' gland consists of two groups of cells. Mucosal cells in the anterior portion produce a lubricating substance with a pH 6.75 to 6.99. Serosal cells in the posterior part provide a thin fluid with pH 4.93 only shortly before and during egg production. It is probable that the serosal secretion upon oxidation in the ootype by a polyphenoloxydase produces the egg shell or capsule. Leftover reproductive material passes through the genitointestinal canal to the gut where it is digested.

Full grown flukes produce 900 to 1000 eggs per season. The time required for development in the natural environment ranges from 10 to 40 days or longer, depending on the temperature of the water. Unhatched larvae survive up to 60 days in water 0° to 3°C.

Adult frogs may become infected by eating young frogs harboring flukes.

## EXERCISE ON LIFE CYCLE

When a source of infected tree toads is available for conducting a life history study, special effort must be made to have a large supply of toad eggs on hand for providing tadpoles throughout the course of the investigation. This is accomplished by collecting an adequate number of eggs in advance of the experiment and storing them under sufficiently low temperature to allow their survival without development.

Infected toads isolated in jars partially filled with water serve as a source of fluke eggs. The eggs settle to the bottom of the container upon being voided with the urine. Collection of them should be at frequent intervals.

Development may be observed over a period of about 2 weeks at room temperature, at which time hatching begins. Fully incubated eggs or larvae placed with young gill-bearing tadpoles will provide all the stages of the branchial forms over a period of 3 weeks.

Using parasite-free tadpoles, ascertain whether larvae hatching from eggs laid by branchial forms will develop into similar neotenic adults on the gills or whether they will develop only in the urinary bladder. This point has never been clarified.

Try to determine whether tadpoles become infected with bladder forms by larval flukes entering the anus, another point that needs confirmation.

## SELECTED REFERENCES

Cameron, T. W. M. (1956). Parasites and Parasitism. John Wiley and Sons, New York, p. 98.

Gallien, L. (1935). Recherches expérimentales sur le dimorphisme évolutif et la biologie de *Polystoma integerrimum* Fröhl. Trav. Sta. Zool. Wimereux, Vol. 12, 181 pp.

Gorshkov, P. V. (1964). [Independence of ontogensis and annual reproductive cycle of *Polystoma integerrimum* from hormonal activity of its host (*Rana temporaria* L.).] Zool. Zhurn. 43:272–274.

Kohlmann, F. W. (1961). Untersuchungen zur Biologie, Anatomie und Histologie von *Polystoma integerrimum* Frölich. Z. Parasitenk. 20:495–524.

Paul, A. A. (1938). Life history studies of North American fresh-water polystomes. J. Parasitol. 24:489–510.

Price, E. W. (1939). North American monogenetic trematodes. IV. The family Polystomatidae. Proc. Helminthol. Soc. Wash. 6:80–92.

### Subclass Aspidobothrea

This is a small group of trematodes that parasitizes mollusks, fish, and turtles. They are characterized morphologically by the large compartmentalized adhesive organ that encompasses almost the entire ventral surface of the body. They resemble the Monogenea, on the one hand, in that none of them has asexual generations, and the Digenea, on the other hand, in that some of them have an alternation of hosts. These biological characteristics are considered justification for placing them in a position intermediate between the Monogenea and the Digenea.

The subclass consists of the families Aspidogasteridae and Stichocotylidae. The genera *Aspidogaster*, *Cotylaspis*, and *Cotylogaster* of the Aspidogasteridae are among the best known members of the subclass.

### FAMILY ASPIDOGASTERIDAE

These are small to medium-sized trematodes with the ventral surface of the body forming a large alveolated adhesive organ; they occur commonly in freshwater clams and snails.

*Aspidogaster conchicola*
von Baer, 1827 (Plate 46)

This is a common parasite of clams of the family Unionidae. It occurs in the pericardial and renal cavities, causing extensive renal metaplasia. In addition to clams, it may occur occasionally in snails and rarely in the stomach of turtles, pre-

## Explanation of Plate 46

A, Adult fluke. B, Embryonated egg. C, Egg in process of hatching with first stage larva escaping. D, Recently hatched first stage larva. E, Second stage larva. F, Third stage larva (drawing reconstructed). G, Fourth stage larva. H, Mussel (*Anadonta grandis*) host. I, Turtle which may serve as a temporary host.

1, Mouth sucker (surrounded by muscular tissue in larval stages); 2, ventral sucker divided into alveoli in adult and late fourth stage larva but not in first, second, and third stage larvae; 3, pharynx; 4, intestine; 5, testis; 6, vas efferens; 7, cirrus pouch; 8, cirrus papilla; 9, common genital pore; 10, ovary; 11, oviduct; 12, Laurer's canal; 13, descending limb of uterus; 14, terminal portion of uterus (intervening part omitted); 15, vitelline gland; 16, transverse vitelline duct; 17, common vitelline duct; 18, eggshell; 19, operculum; 20, first stage larva; 21, tail; 22, groove in ventral sucker.

Two possibilties of the life cycle are given. The first is shown in a–f and the second in g–l. a, Adult fluke in pericardial cavity of mussel; b, embryonated eggs laid in pericardium; c, egg hatching with first stage larva escaping; d, second stage larvae; e, third stage larva; f, fourth stage larva (all larvae in pericardial cavity where it is postulated that they develop to maturity, thus completing the development in a single host); g, probability that eggs may leave pericardial opening and enter kidney; h, egg passing through kidney into suprabranchial cavity; i, egg being carried from branchial cavity through excurrent siphon; j, egg hatching in the water; k, free first stage larva being drawn into branchial cavity through incurrent pore; l, first stage larva in position to enter renal pore and migrate through kidney to pericardial cavity; m, flukes liberated from infected mussels eaten by turtles may survive in the stomach of these poikilotherms which serve as a temporary second host in an otherwise simple one-host type of life cycle.

Figure A adapted from Stafford, 1896, Zool. Jahrb. Abt. Anat. 9:477; B, C, from Faust, 1927, Trans. Amer. Microsc. Soc. 41:113; D–G, from Williams, 1942, J. Parasitol. 28:467.

---

sumably persisting from a meal of infected mollusks.

**Description.**     Full grown worms are 2.5 to 2.7 mm long by 1.1 to 1.2 mm wide. The large ventral portion of the body is almost entirely covered by an alveolated sucker consisting of four rows of quadrangular sucking grooves. The narrower anterior neck-like portion, consisting of about one-fifth the total length of the worm, bears a terminal mouth. There is a single median testis and an ovary somewhat anterior and toward the right side (Fig. 25). The operculate eggs are embryonated when laid and measure 128 to 130 $\mu$ long by 48 to 50 $\mu$ wide.

**Life Cycle.**     The entire life cycle occurs in the clam host. Adult worms in the pericardial or renal cavities lay eggs containing well-developed first stage larvae. Development from the earliest larval stage to the adult is one of gradual transformation.

There are four developmental stages of larvae. The first stage is unciliated, very immobile, and 130 to 150 $\mu$ long. It consists of three main body divisions set off by constrictions. They are the anterior terminal sucker, middle section, and posterior subterminal sucker followed by a blunt rudimentary tail-like appendage. Each section contains primordial cells representing some future structure. The second stage is larger, being 150 to 275 $\mu$ long, has lost the constrictions, and

shows considerable activity, but is unable to swim. Third stage larvae are larger than the preceding one, being 880 $\mu$ long and greatly modified in shape. The oral sucker is at the end of a long neck-like structure, and the ventral sucker has assumed much larger proportions, foreshadowing that of the adult. Fourth stage larvae are about twice the size of the third, measuring 1.2 to 1.4 mm long. In general, the shape is similar to that of the adult. The large ventral sucker is well-developed but lacks the alveoli.

All of these stages except the first have been found in the pericardial cavity of clams. This condition led to the opinion (Williams) that the entire development might take place within a single clam. Dissemination of the eggs from the clams into the water, however, where they might hatch seems to be the more likely manner in which early development occurs. This would provide a means of dispersal of the parasite among the clams. A natural route exists for the escape of eggs from the pericardium. Eggs in it could easily enter the renopericardial pore, pass through the kidney and out through the kidney pore into the suprabranchial chambers of the gills. Upon being expelled by the water currents through the excurrent siphon, they would settle to the bottom of the pond where favorable conditions for hatching or embryonation exist. Infection of the pelecypods might take place by drawing first

**Plate 46**  *Aspidogaster conchicola*                                                                                          211

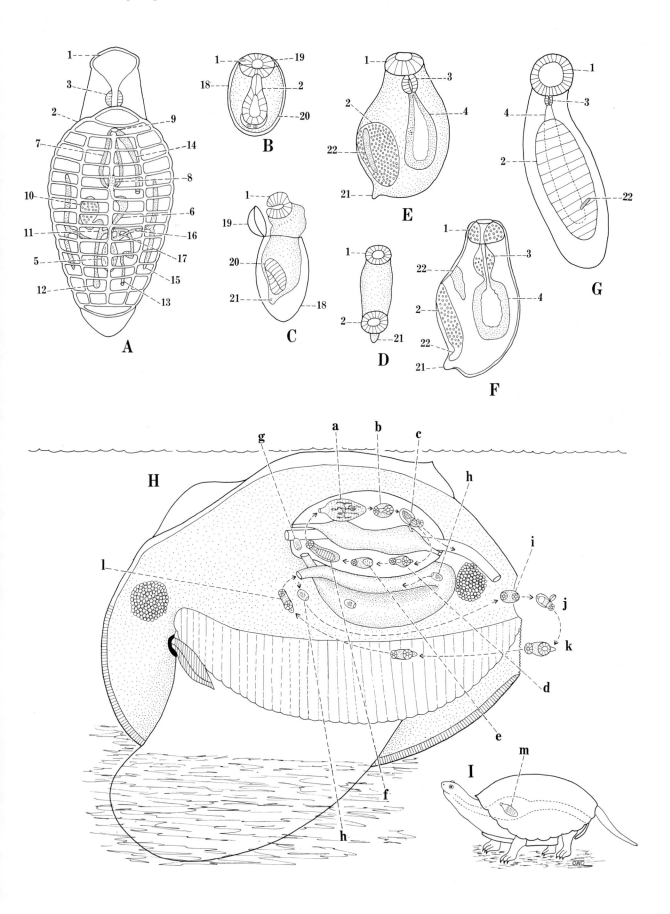

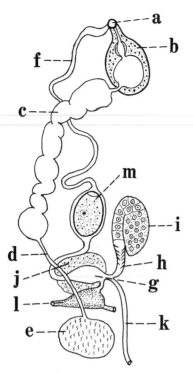

Fig. 25. Schematic representation of the reproductive system of *Cotylaspis insignis*. **a**, Common genital pore; **b**, cirrus pouch; **c**, seminal vesicle; **d**, vas efferens; **e**, testis; **f**, uterus; **g**, ootype; **h**, oviduct; **i**, ovary; **j**, Mehlis' gland; **k**, Lauer's canal; **l**, common yolk reservoir opening into oviduct; **m**, operculate egg. Adapted from Hendrix and Short, 1965, J. Parasitol. 51:561.

stage larvae through the incurrent siphon into the suprabranchial chambers where they would be in a position to enter the renal pore and migrate through the kidney into the pericardium and develop to adults. Or clams might swallow embryonated eggs that hatch and the larvae migrate to the pericardial cavity. Eggs of the flukes eaten by the snail *Viviparus malleatus* are passed in feces. Snails become infected by eating embryonated eggs, which is probably the only means of infection, as postulated above for the clams. While only the second, third, and fourth stage larvae and adults have been demonstrated in the pericardial cavity, this fact provides a basis for the postulate that the eggs are expelled into the water and hatch there. The habits, physiology, and morphology of the clams combine to provide an ideal means of infection in the manner suggested.

The principal source of nutrition is from blood obtained in the pericardial and renal cavities where it is present in abundance.

Adult *Aspidogaster conchicola* have been found in the digestive tract of turtles. Their presence probably resulted from the turtles eating infected clams. The interesting fact exists, however, that these worms which are normally parasites of mollusks are able to survive in the alimentary canal of a reptile. The question may be raised as to whether this is a step toward the introduction of a vertebrate host into a trematode cycle, as occurs in the subclass Digenea.

*Cotylaspis insignis*, also a parasite of clams, is widespread in the river systems of the United States. It locates at the junction of the foot and inner gills, within the gills and the suprabranchial cavities. Since the blood cells of clams are not confined to vessels but wander through the tissue, they would be available to these flukes for nutrition.

Alkaline phosphates in the minor excretory vessels show clearly in specimens and sections stained with Gomori's calcium cobalt. This procedure makes it easy to identify and trace the tubules in their ramifications and convolutions.

The reproductive system, recently redescribed, is typical of that occurring in the digenetic flukes (Fig. 25).

*Lophotaspis vallei* of the stomach of loggerhead turtles (*Caretta caretta*) appears to carry the evolution of trematodes a step further. Larval stages of the parasite occur in the flag conch (*Fasciolaria gigas*) which serves as food for the turtles. The larvae of *L. vallei* are ciliated and readily swim, as well as crawl, thus enabling them to reach the conchs. This cycle is significant in that it suggests another connecting link based on biological evidence in the evolution of the trematodes.

## EXERCISE ON LIFE CYCLE

More serious studies should be conducted on the biology of this and related species of Aspidobothrea.

Infected clams placed in aquaria could be used to determine if eggs are passed from the body, as postulated above. The large size of eggs and first stage larvae make them easy to find in the containers where infected clams are kept. If first stage larvae can be obtained, efforts should be made to determine whether infection takes place by their being drawn into the gill chambers through the incurrent siphon. If a sufficient number of first stage larvae can be introduced into the gill chambers through the incurrent siphon, it would be possible to determine, on the basis of numbers and the stage of development, whether they were entering the pericardium via the suprabranchial chamber and kidney.

## SELECTED REFERENCES

Brickler, D. A. (1968). Histochemical demonstration of alkaline phosphatase in the excretory system of *Cotylaspis insignis* (Trematoda:Aspidobothria). Trans. Amer. Microsc. Soc. 87:128.

Hendrix, S. S., and R. B. Short. (1965). Aspidogastrids from northeastern Gulf of Mexico river drainages. J. Parasitol. 51:561–569.

Huehner, M. K., and F. J. Ethes. (1972). Experimental transmission of *Aspidogaster conchicola* von Baer, 1827. J. Parasitol. 58:109.

Pauley, G. B., and C. D. Becker. (1968). *Aspidogaster conchicola* in mollusks of the Columbia River system with comments on the host pathological response. J. Parasitol. 54:917–920.

Stafford, J. (1896). Anatomical structure of *Aspidogaster conchicola*. Zool. Jahrb. Abt. Anat. 9:477–542.

Ward, H. B. (1918). Parasitic Flatworms. In: H. B. Ward and G. C. Whipple, eds. Freshwater Biology. John W. Wiley and Sons, New York, p. 380.

Wharton, G. W. (1939). Studies on *Lophotaspis vallei* (Stossich, 1899) (Trematoda: Aspidogastridae). J. Parasitol. 25:83–86.

Williams, C. O. (1942). Observations on the life history and taxonomic relationships of the trematode *Aspidogaster conchicola*. J. Parasitol. 28:467–475.

Yamaguti, S. (1963). Systema Helminthum. Monogenea and Aspidocotylea. Academic Press, New York, Vol. 4, 699 pp.

### Subclass Digenea

The digenetic trematodes comprise a large group of endoparasites variable in size, shape, and habitat. They are hermaphroditic except for one family in which the sexes are separate. They conform biologically in having an asexual phase of the life cycle in mollusks, usually gastropods but occasionally pelecypods, as the first intermediate host (one exception is known where a polychaete annelid replaces the molluscan intermediate host), and a sexual phase in vertebrates, the definitive host (a few exceptions occur where water beetles harbor sexually mature flukes). Many species have a second intermediate host in which early development of the sexual phase takes place.

Life cycles of digenetic trematodes are complicated. The asexual phase includes eggs that produce ciliated miracidia. Continued development of this larval stage is possible only in the bodies of mollusks where it transforms into a mother sporocyst. Germinal cells within its body develop into either rediae or daughter sporocysts but never both in the same species of trematode. There may be more than one generation of rediae or sporocysts. The final product of the asexual phase in mollusks is cercariae which appear in a variety of types (Plate 48). (For excellent references, see Dawes, and Hyman.)

With the appearance of the cercariae, the sexual phase of the life cycle is initiated but it is in an undeveloped stage. The cercariae normally escape from the molluscan host into the water. Some of them enter the definitive host directly from the water, as in the case of the blood flukes, and develop to maturity in the vascular system. The remainder encyst on objects or in the bodies of animals (second intermediate hosts), transforming into metacercariae which develop to the infective stage. Metacercariae infect the definitive host when swallowed by it. They develop to sexual maturity in various organs, according to the requirements of each species. The general plan of the different life cycles is shown in Plate 83.

The classification of the Digenea has been revised recently (LaRue) on the basis of whether the excretory bladder of the cercariae develops from the existing thin-walled excretory tubules (superorder Anepitheliocystidia) or from mesodemal cells (superorder Epitheliocystidia). This plan is followed. The basic outline of the classification is given together with the families of digenetic trematodes treated in this chapter, except where it is stated that certain groups will not be discussed.

Superorder Anepitheliocystidia
  Order Strigeatida
    Suborder Strigeata
      Superfamily Strigeoidea
        Family Strigeidae
        Family Diplostomatidae
      Superfamily Clinostomatoidea
        Family Clinostomatidae
      Superfamily Schistosomatoidea
        Family Schistosomatidae
        Family Spirorchiidae
        Family Sanguinicolidae

## Explanation of Plate 47

A, Gasterostome (*Bucephalus elegans*). B, Strigeid or holostome (*Alaria canis*). C, Monostome with muscular oral sucker (*Quinqueserialis quinqueserialis*). D, Amphistome (*Megalodiscus temperatus*). E, Monostome without muscular sucker (*Cyclocoelium* sp.). F, Echinostome (*Echinostoma revolutum*). G, Schistosome (*Schistosoma haemotobium*). H, Distome (*Plagiorchis muris*).

1, Oral sucker; 2, mouth without muscular sucker; 3, ventral sucker; 4, common genital pore; 5, adhesive or tribocytic organ; 6, ovary; 7, testes; 8, cirrus pouch.

Figures adapted from various sources.

Suborder Azygiata
  Superfamily Azygioidea
    (No members discussed)
  Superfamily Transversotrematoidea
    (No members discussed)
Suborder Cyclocoelata
  Superfamily Cyclocoeloidea
    (No members discussed)
Suborder Brachylaemata
  Superfamily Brachylaemoidea
    Family Brachylaemidae
  Superfamily Fellodistomatoidea
    (No members discussed)
  Superfamily Bucephaloidea
    Family Bucephalidae

Order Echinostomida
  Suborder Echinostomata
    Superfamily Echinostomatoidea
      Family Echinostomatidae
      Family Fasciolidae
  Suborder Paramphistomata
    Superfamily Paramphistomatoidea
      Family Paramphistomatidae
      Family Diplodiscidae
    Superfamily Notocotyloidea
      Family Notocotylidae
Order Renicolida
  Suborder Renicolata
    Superfamily Renicoloidea
      (No members discussed)

Superorder Epitheliocystidia

Order Plagiorchiida
  Suborder Plagiorchiata
    Superfamily Plagiorchioidea
      Family Dicrocoeliidae
      Family Plagiorchiidae
      Family Prosthogonimidae
    Superfamily Allocreadioidea
      Family Allocreadiidae
      Family Gorgoderidae
      Family Troglotrematidae

Order Opisthorchiida
  Suborder Opisthorchiata
    Superfamily Opisthorchioidea
      Family Heterophyidae
      Family Opisthorchiidae
  Suborder Hemiurata
    Superfamily Hemiuroidea
      Family Halipegidae

## General Characteristics of Digenea

An understanding of some of the characteristics of the digenetic flukes is helpful for purposes of general recognition of groups and comprehension of patterns of the life cycles. A few of the basic principles are set forth in the following pages.

### Adult Flukes (Plate 47)

Morphologically, adult digenetic trematodes fall into seven basic types that are easily recognized. They are 1) the gasterostomes with only a muscular oral sucker, and that on the midventral surface of the body (Plate 47, A); 2) the strigeids, or holostomes, as they are sometimes called, with a transverse equatorial constriction that divides the body into a forebody with a holdfast, or tribocytic organ, and a more or less cylindrical hindbody containing the gonads (Plate 47, B); 3) the monostomes with or without a muscular oral sucker at the anterior extremity of the body and without a ventral sucker (Plate 47, C, E); 4) the amphistomes with the oral sucker and ventral sucker at the anterior and posterior extremities of the body, respectively, with the latter much the larger (Plate 47, D); 5) the echinostomes with a spine-bearing collar around the anterior end (Plate 47, F); 6) the schistosomes with separate sexes and, in some forms, a long ventral groove, the gynecophoric groove, in the male wherein the female is held (Plate 47, G); and 7) the distomes with the ventral sucker relatively near the oral sucker and without the characteristics that distinguish the other groups (Plate 47, H).

**Plate 47** *Morphological Types of Adult Digenetic Trematodes* 215

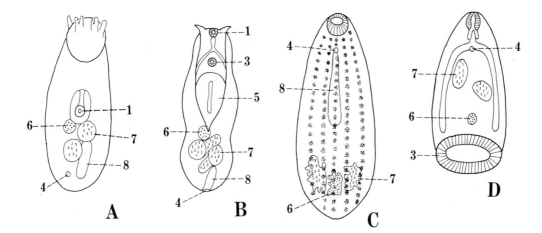

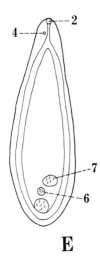

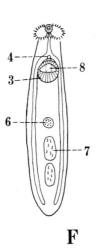

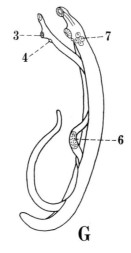

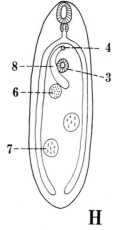

owo

## Explanation of Plate 48

A, Amphistome cercaria. B, Monostome cercaria. C, Gymnocephalus cercaria. D, Gymnocephalus cercaria of pleurolophocercous type. E, Cystophorous cercaria. F, Trichocercous cercaria. G, Echinostome cercaria. H, Microcercous cercaria. I, Xiphidiocercous cercaria. J, Ophthalmoxiphidiocercous cercaria.

K–O, Furcocercous types of cercariae. K, Gasterostome cercaria. L, Lophocercous cercaria. M, Apharyngeate furcocercous cercaria. N, Pharyngeate furcocercous cercaria. O, Apharyngeate monostome furcocercous cercaria without oral sucker. P, Cotylocercous cercaria. Q, Rhopalocercous cercaria. R, Cercariaea. S, Rattenkönig or rat-king cercariae.

Figures adapted from various sources.

---

### Cercariae (Plate 48)

Since cercariae are relatively abundant and easy to collect, they serve as a clue to the trematode fauna in a region. For recognition of the cercariae, attempts have been made to develop schemes of classification based on morphological characters. These have been concerned primarily with the number and arrangement of the suckers, presence or absence of collar spines or of stylets in the oral sucker, and morphology of the tail. Even though the shortcomings of such schemes for showing relationships among flukes are recognized, they have great practical utility for recognizing and cataloging cercariae.

LaRue recognized two basic types of cercariae, according to the development of the excretory bladder. He noted that it develops either by fusion of the excretory tubules to form a wall of thin, flat cells (Plate 49, R), or from a mass of mesodermal cells to form a thick wall of cuboidal cells (Plate 49, S). The superorders are Anepitheliocystidia and Epitheliocystidia, respectively, and represent two natural groups of digenetic trematodes.

Where life history studies have been completed, they often show relationships among adult flukes previously unsuspected on the basis of body structure alone.

For convenience in recognizing cercariae commonly encountered in surveys, the following 13 different morphological types, as recognized by Dawes, are shown on Plate 48.

1. Amphistome cercariae (Plate 48, A). These are characterized by the ventral sucker, which is the larger, being located at the posterior end of the body. They are the largest of the cercariae, develop in rediae, are born in a relatively immature stage, and complete development in the tissues of the snail hosts. Encystment is on objects in water. There are two types: Pigmenta with stellate melanophores and Diplocotylea with abundant pigmentation in the anterior region of the body.

2. Monostome cercariae (Plate 48, B). These cercariae have a muscular oral sucker at the anterior end of the body but lack a ventral sucker. They develop in rediae, are born in an immature stage, and complete development in water. Encystment is on objects in water. There are two types: Ephemera with three eyespots and Urbanensis with two.

3. Gymnocephalous cercariae (Plate 48, C, D). The cercariae of this group are without spines or stylets around the anterior end of the body or in the oral sucker, respectively. The tail is long and straight and with or without a longitudinal fin. Cercariae develop in rediae and encyst on objects in the water or in fish. Forms without a tail fin (Plate 48, C), such as the Fasciolidae, encyst on objects. Those with caudal fins and two eyespots (Plate 48, D), known also as pleurolophocercous or parapleurolophocercous cercariae, encyst in fish and are representative of the Opisthorchiidae and Heterophyidae.

4. Cystophorous cercariae (Plate 48, E). The tail has a chamber at the anterior end into which the cercariae may withdraw. It is variable in both shape and size and constitutes the morphological basis for recognizing three groups, the cystophorous cercariae, the cysticercariae, and the macrocercous cercariae.

Cystophorous cercariae have a peculiar short tail with five appendages attached, each different from the other. Development is in sporocysts in *Planorbis*. Adults are Halipegidae. Cysticercariae have a short, flat tail with a pair of large, flat, clapper-like appendages at the end. Development is in rediae in *Limnaea*. Some of these cercariae attain a length of 6 to 7 mm. They develop into Azygiidae. Macrocercous forms have a long, simple, cylindrical tail, develop in sporocysts in *Sphaerium*, and mature into Gorgoderidae.

5. Trichocercous cercariae (Plate 48, F). These are largely marine forms although a few occur in fresh water. They are characterized by having a long, slender tail that bears numerous

**Plate** 48    *Cercarial Types*                                                                  217

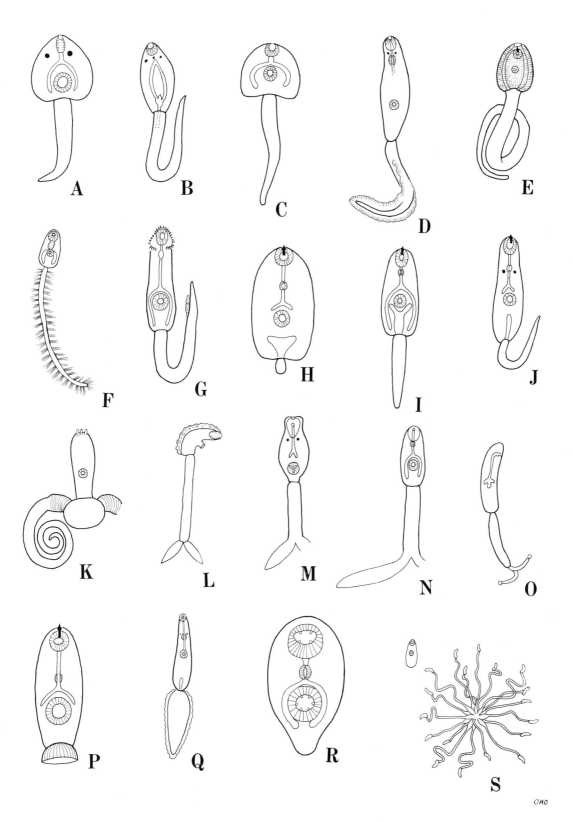

bristles, often in tufts. There are two groups, those with eyespots and those without. Some of the group develop into Allocreadiidae.

6. Echinostome cercariae (Plate 48, G). This group of cercariae is readily distinguished by the presence of a head collar bearing a ring of spines around its margin. They have a long, slender tail. The cercariae develop in rediae and encyst in snails, sometimes in the species in which they developed. They are Echinostomatidae.

7. Microcercous cercariae (Plate 48, H). These are small cercariae with a minute, knob-like tail. The oral sucker bears a stylet in the anterior margin. The cercariae develop in rediae and encyst in invertebrates. They belong to the family Troglotrematidae.

8. Xiphidiocercariae (Plate 48, I, J). This group of cercariae is characterized by a stylet in the anterior margin of the oral sucker and long tails (this feature distinguishes them from the microcercous cercariae). The cercariae develop in sporocysts and encyst in invertebrates, rarely in fish, or they may encyst in sporocysts of other trematodes. Plagiorchiidae is characteristic of this group.

Four subgroups generally are recognized. They are 1) Cercariae Microcotylae, which includes small forms with ventral sucker postequatorial, tail without a fin and equal to the body in length; 2) Cercariae Virgulae with the ventral sucker smaller than the oral, tails without a fin, the excretory vesicle V-shaped, and the presence of virgula organ which is two pyriform sacs fused in the median line, with pointed ends directed forward, and located near the posterior margin of the oral sucker; 3) Cercariae Ornate includes species with a tail fin-fold; and 4) Cercariae Armate are without a virgula organ, oral and ventral suckers are unequal in size, and the excretory bladder is Y-shaped.

9. Furcocercous cercariae (Plate 48, K–O). This is a large and varied group of cercariae with forked tails into which the body is not retractable. Some groups develop in sporocysts and others in rediae. The cercariae may penetrate the definitive host actively without prior encystment (blood flukes) or they may encyst in vertebrates.

Five groups are recognized on the bases of morphology and/or biology. They are: 1) *Bucephalus* group, which includes the gasterostome cercariae in which the oral sucker is on the midventral surface as in the adults. The tail lacks a stem but has two long furcae arising from a large bulbous structure. The cercariae develop in sporocysts (Plate 48, K). 2) Lophocercous group consists of apharyngeate, monostome cercariae. Some species have a dorsal fin extending the full length of the body. The cercariae do not encyst but penetrate the definitive host (Plate 48, L). 3) Apharyngeate or Ocellata group in which the cercariae lack a pharynx and bear two pigmented eyespots. They develop in sporocysts and penetrate directly through the epidermis of the definitive host without prior encystment. These cercariae are representative of the Schistosomatidae (Plate 48, M). 4) The pharyngeate, nonocellate cercariae that develop in sporocysts or rediae and always penetrate a vertebrate host for encystment. The strigeids (Strigeidae, Diplostomatidae) develop in sporocysts and form tetracotyle, diplostomulum, or neascus types of metacercariae. The Clinostomatidae develop in rediae and penetrate fish for encystment (Plate 48, N). 5) Suckerless apharyngeate cercariae of blood flukes of the family Sanguinicolidae that parasitize fish (Plate 48, O).

10. Cotylocercous cercariae (Plate 48, P). These are marine species with broad, short, cuplike tails that serve as adhesive organs. They belong to the Allocreadioidea.

11. Rhopalocercous cercariae (Plate 48, Q). The tail is broad, even wider than the body. They develop in rediae in fingernail clams. *Allocreadium isosporum* is a representative in freshwater fish.

12. Cercariaea (Plate 48, R). This group includes cercariae in which the tail is undeveloped. The Mutabile and *Helicis* groups develop in rediae, whereas the *Leucochloridium* group develops in peculiarly branched sporocysts which are banded with blue-green or brown colors. The root-like branches are deep in the viscera of the snails and the colored sac-like portion extends into the antennae, causing them to enlarge greatly. The cercariae in the sacs are infective to birds when eaten by them. They belong to the Brachylaemidae.

13. Rattenkönig or rat-king cercariae (Plate 48, S). These marine cercariae occur in writhing masses with the tips of the tails being attached to a body of protoplasm.

## EXERCISE ON CERCARIAE

Cercariae are obtained by isolating snails or fingernail clams in vials of pond water overnight.

Tap water may contain sufficient chlorine to kill the mollusks unless it is allowed to stand in an open container for several days.

The contents of the vials are examined by holding them toward a darkened object in a manner that light will shine through them and illuminate cercariae that might be present. They appear as minute whitish objects moving about. Their presence may be confirmed by examination under a dissecting microscope.

For studying cercariae, living specimens shed naturally by the mollusks may be observed for activity in open dishes. Internal details are seen best in specimens placed on a slide under a cover-slip in a very weak solution of neutral red or Nile blue and examined under a microscope. Regulation of cover-slip pressure is necessary to see the flame cells, as they are clearest just before the cercariae rupture from pressure.

Expose various aquatic arthropods to cercariae to find intermediate hosts that they will penetrate. Only cercariae that have emerged naturally from the mollusks should be used, as those obtained by dissection may not be mature.

## SELECTED REFERENCES

Dawes, B. (1956). The Trematoda. Cambridge University Press, New York, 644 pp.

Hyman, L. H. (1951). The Invertebrates. McGraw-Hill and Co., New York, Vol. 2, pp. 219–311.

LaRue, G. R. (1957). The classification of digenetic Trematoda; a review and a new system. Exp. Parasitol. 6:306–349.

Marten, W. E. (1968). *Cercaria gorgonocephala* Ward, 1916, a zygocercous species in northwestern United States. Trans. Amer. Microsc. Soc. 87:472–476.

Schell, S. C. (1970). The Trematodes. Wm. C. Brown, Dubuque, Iowa, 355 pp.

Yamaguti, S. (1958). Systema Helminthum. Vol. 1: Digenetic Trematodes of Vertebrates, Parts 1 and 2. Interscience Press, New York, 1075 pp., 106 pls.

## Morphology and Life History
## Stages of Digenetic Trematodes (Plate 49)

The digenetic trematodes comprise a large group of endoparasites variable in size, shape, and habitat. They conform, however, in having basic patterns of body structure and life history stages, as shown on Plate 49.

The adults are hermaphroditic, except for the blood flukes of the family Schistosomatidae, and generally similar in structure (Plate 49, A). The stages in the course of development include 1) an egg, 2) miracidium, 3) mother sporocyst, 4) daughter sporocyst or redia, 5) cercaria, and 6) metacercaria (Plate 49, B–O).

Eggs generally have a cap, or operculum, at one end (Schistosomatidae eggs are nonoperculate) and are either unembryonated or contain a fully formed miracidium when laid.

Miracidia are complex ciliated organisms that continue their development nearly always in the body of mollusks. The miracidium has penetration glands, flame cells, and associated tubules which constitute the excretory system. A mass of tissue known as germ cells provide the cellular lineage for succeeding forms appearing in the life cycle. A few miracidia are progenetic, containing later stages such as redia before they hatch, as in the case of some amphistomes and echinostomes.

Upon entering the molluscan intermediate host, the miricidia shed the ciliated covering and transform directly into a sac-like body, the mother sporocyst. Germ cells within the hollow body of the mother sporocyst develop into either daughter sporocysts without mouth or intestine (Plate 49, H, I) or rediae with pharynx and simple intestine (Plate 49, J), but not both in the same species of trematode. There may be more than one generation of sporocysts or rediae.

Germ cells in the sporocysts or rediae produce cercariae (Plate 49, K). These are sexual forms and constitute the end product of asexual reproduction of the sporocysts and rediae in the molluscan intermediate host. Upon escaping from the mollusk, the cercariae are free in the water for a short time. Except for a few families, the cercariae promptly encyst on objects in the water or in the bodies of invertebrate or vertebrate animals (second intermediate hosts) and transform into metacercariae which after a variable period of development become infective to the definitive hosts (Plate 49, L). In the case of some of the blood flukes, however, the cercariae penetrate the definitive host and develop directly to adults without going through a metacercarial stage.

In the strigeids, there are three different kinds of metacercarial forms. They are 1) tetracotyles (Plate 49, M); 2) neascus (Plate 49, N); and 3) diplostomulum (Plate 49, O). In some strigeids, there is an additional stage between the cercarial and metacercarial stages known as

## Explanation of Plate 49

A, Diagrammatic drawing of a typical digenetic trematode, showing parts. B, Egg of *Haplometrana utahensis* which is embryonated when laid and hatches only when eaten by snail host. C–F, Hatched miracidia. C, *Trichobilharzia cameroni*. D, *Bucephalus elegans*. E, *Postharmostomum helicus*. F, Progenetic miracidium of *Stichorchis subtriquetrus*, showing redia inside. G, Miracidium of *Stichorchis subtriquetrus* without ciliated epithelium, showing epidermal plates. H, Portion of branched sporocyst of *Postharmostomum helicus*. I, Simple sac-like sporocyst of *Haplometrana utahensis*. J, Redia of *Echinostoma revolutum*. K, Cercaria of *Haplometrana utahensis*. L, Metacercariae of *Paragonimus kellicotti*, which is characteristic of most digenetic trematodes. M–O, Metacercariae of strigeid flukes. M, Tetracotyle. N, Neascus. O, Diplostomulum. P, Mesocercaria of *Alaria canis*. $Q_{1-6}$, Types of excretory bladders. $Q_1$, Short club-shaped bladder of *Macroderoides typicus*. $Q_2$, Long club-shaped bladder of *Alloglossidium corti*. $Q_3$, V-shaped bladder of Lecithodendriidae. $Q_4$, Y-shaped bladder of Styphlodorinae with collecting tubules originating from sides of arms. $Q_5$, Y-shaped bladder of *Plagiorchis* with short arms and collecting tubules originating from ends of arms. $Q_6$, Y-shaped bladder of *Haplometra cylindricea* with long arms and collecting tubules originating from ends of arms. $R_{1-5}$, Diagrammatic sketches of development of excretory bladder in a furcocercous cercaria of the Anepitheliocystidia. $R_{1-2}$, Separate collecting tubules. $R_{3-4}$, Fusion of collecting tubules to form excretory bladder. $R_5$, Excretory bladder formed. $S_{1-5}$, Diagrammatic sketches of development of excretory bladder in a xiphidiocercous cercaria of the Epitheliocystidia. $S_1$, Separate collecting tubules. $S_2$, Excretory tu-

bules separate but a mass of mesodermal tissue has appeared between them. $S_3$, Excretory tubules are surrounded by mesodermal tissue. $S_4$, Mesodermal tissue forming excretory bladder. $S_5$, Excretory bladder has formed from mesodermal tissue.

1, Oral sucker; 2, ventral sucker; 3, preesophagus; 4, pharynx; 5, esophagus; 6, intestine; 7, testis; 8, vas efferens; 9, vas deferens; 10, cirrus pouch; 11, seminal vesicle; 12, prostate glands; 13, ejaculatory duct; 14, common genital pore; 15, protruding cirrus; 16, ovary; 17, oviduct; 18, Laurer's canal; 19, seminal receptacle; 20, Mehlis' gland; 21, descending limb of uterus; 22, ascending limb of uterus; 23, metraterm; 24, vitelline glands; 25, lateral vitelline duct; 26, vitelline reservoir; 27, common vitelline duct; 28, flame cell; 29, capillary tubule; 30, accessory collecting tubule; 31, anterior collecting tubule; 32, posterior collecting tubule; 33, common collecting tubule; 34, excretory bladder; 35, excretory pore; 36, apical gland; 37, penetration glands; 38, germ balls; 39, redia; 40, cephalic plate with bristles; 41, spine; 42–45, first, second, third, and fourth rows of plates, respectively; 46, cercariae; 47, birth pore; 48, lateral sucker; 49, holdfast organ; 50, mesodermal mass from which excretory bladder arises.

Figures A, B, I, K, from Olsen, 1937, J. Parasitol. 23:13; C, from Wu, 1953, Can. J. Zool. 31:351; D, from Woodhead, 1929, Trans. Amer. Micros. Soc. 48:25; E, H, from Ulmer, 1951, Trans. Amer. Micros. Soc. 70:319; F, from Skrjabin, 1949, [Trematodes of Animals and Man], Vol. 3, pl. 58; G, original; J, from Johnson, 1920, Univ. Calif. Publ. Zool. 19:335; L, from Ameel, 1934, Amer J. Hyg. 19:279; M, from Hughes, 1929, Trans. Amer. Micros. Soc. 48:12; N, Ibid, 1927, 46:248; O, from Bosma, 1956, Trans. Amer. Micros. Soc. 53:116; P, from Pearson, 1956, Can. J. Res. 34:295.

---

the mesocercaria (Plate 49, P). These are in reality cercariae, which upon entering the second intermediate host do not develop to the metacercarial stage. When the intermediary containing them is eaten by a definitive host, the mesocercariae migrate from the intestine via the hepatic portal vein, liver, and heart to the lungs where they transform into diplostomulae. From here, they proceed up the trachea and to the small intestine.

## Molluscan Intermediate
## Hosts of Digenetic Trematodes

Mollusca, especially the Gastropoda, or snails, serve as the first intermediate hosts of the digenetic trematodes. Both the land and aquatic

snails are utilized. Some clams, particularly the fingernail clams, act as intermediate hosts for a small number of species.

Illustrations are given of representatives of some of the common families of North American freshwater (Plate 50) and land snails (Plate 51) as an aid to recognizing them. A brief classification of the class Gastropoda is included to show the major groups and their relationships.

I. Sublass Streptoneura. Visceral part of body is coiled so that the two visceral nerve commissures are twisted, as suggested by the name, into the form of a figure-8. The majority of the gastropods belong to this group. A. Prosobranchiata. Gills located anterior to

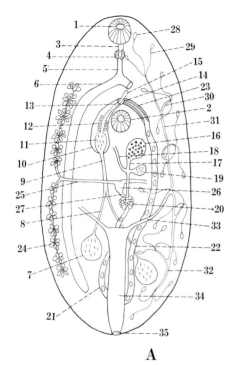

**A**

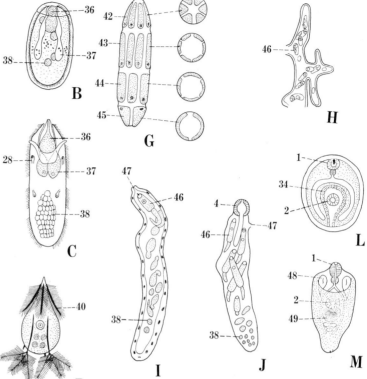

**B**

**G**

**H**

**C**

**D**

**I**

**J**

**L**

**M**

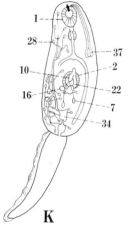

**K**

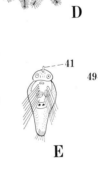

**F**

**E**

**N**

**O**

**P**

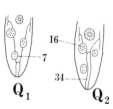

**Q₁**

**Q₂**

**Q₃**

**Q₄**

**Q₅**

**Q₆**

**R₁**

**R₂**

**R₃**

**R₄**

**R₅**

**S₁**

**S₂**

**S₃**

**S₄**

**S₅**

owo

## Explanation of Plate 50

1, Diagram of spired operculate snail with dextral whorls. 2, Diagram of spired nonoperculate snail with sinistral whorls. 3, Diagram of discoidal snail with sinistral whorls. 4–6, Physidae (pouch snails). 4, *Physa sayi.* 5, *Physa integra.* 6, *Physa gyrina.* 7–13, Lymnaeidae (pond snails). 7, *Lymnaea haldemani.* 8, *Pseudosuccinea columella.* 9, *Fossaria modicella.* 10, *Stagnicola caperata.* 11, *Bulimnea megasoma.* 12, *Lymnaea stagnalis.* 13, *Lymnaea auricularia.* 14–19, Planorbidae (orb or ram's horn snails). 14–14$_a$, *Helisoma.* 15, *Planorbula.* 16, *Segmentina.* 17–17$_a$, *Planorbis.* 18–18$_a$, *Gyraulus circumstriatus.* 19, *Gyraulus deflectus.* 20, *Menetus exacuous.* 21, 22, Ancylidae (limpets). 21, *Ferrisia.* 22, *Ancylus.* 23, 24, Viviparidae. 23, *Viviparus.* 24, *Campeloma.* 25, 26, Pleuroceridae (river snails). 25, *Pleurocerca.* 26, *Goniobasis.* 27–31, Amnicolidae. 27, *Cincinnatia.* 28, *Pomatiopsis.* 29, *Bithinia.* 30, *Amnicola.* 31, *Flumnicola.* 32, Valvatidae (round-mouthed snails). *Valvata.* 33, Ampullaridae (apple-snails). *Ampullaria.*

a, Apex; b, spire; c, operculum; d, columella; e, lip; f, suture; g, umbilicus; h, aperture; i, body whorl.

Figures redrawn from various sources.

---

the heart. Sexes are separate. An operculum is present.

1. Freshwater snails
   a. Amnicolidae*,† (Plate 50, Figs. 27–31)
      *Amnicola, Cincinnatia, Pomatiopsis, Bithinia, Flumnicola*
   b. Pleuroceridae
      *Pleurocerca, Goniobasis*
   c. Viviparidae† (Plate 50, Figs. 23, 24)
      *Viviparus, Campeloma*
   d. Valvatidae (Plate 50, Fig. 32)
      *Valvata*
2. Brackish and saltwater snails
   a. Potamidae*
      *Pirenella*
   b. Neritidae
      *Neritina*
   c. Littorinidae*
      *Littorina*
   d. Nassariidae*
      *Nassarius*
3. Land snails
   a. Helicinidae (Plate 51, Fig. A)
      *Helicina, Hendersonia*
II. Subclass Euthyneura. Visceral commissures straight, as the name states, not twisted as in the Streptoneura.
   A. Order Opisthobranchiata. Gills behind heart, as given by the name, and visceral nerve commissures not twisted; hermaphroditic. Marine, living near shore.
      1. Akeridae*
   B. Order *Haminoae.* Mantle cavity serves as a lung, and gills are absent. Shell, when present usually is a simple, regular spire. A mass of hardened mucus may close the aperture of the shells of many species of land snails. They are nonoperculate and hermaphroditic, usually ovoviparous. Mostly freshwater and land snails.
      1. Suborder Basommatophora. One pair of contractile tentacles with eyes at base. Shell conical, discoidal, or patelliform. Male and female genital tracts open separately.
         a. Freshwater snails
            1. Planorbidae*,† (Plate 50, Figs. 14–19)
               *Planorbis, Helisoma,* etc.
            2. Lymnaeidae*,† (Plate 50, Figs. 7–13)
               *Lymnaea, Fossaria, Stagnicola, Galba,* etc.
            3. Physidae*,† (Plate 50, Figs. 4–6)
               *Physa*
            4. Ancylidae*,† (Plate 50, Figs. 21, 22)
               *Ancylus, Ferrissia*
      2. Suborder Stylommatophora. Two pairs of tentacles with eyes at tips of dorsal pair, tentacles capable of being everted and inverted. Land snails with visible shells and slugs with rudimentary shells buried in tissue.
         a. Land snails
            1. Helicidae*
               *Helix*

---

\* Intermediate hosts of trematodes of humans.
† Intermediate hosts of trematodes of animals other than humans.
‡ Intermediate hosts of nematodes.
§ Intermediate hosts of cestodes.

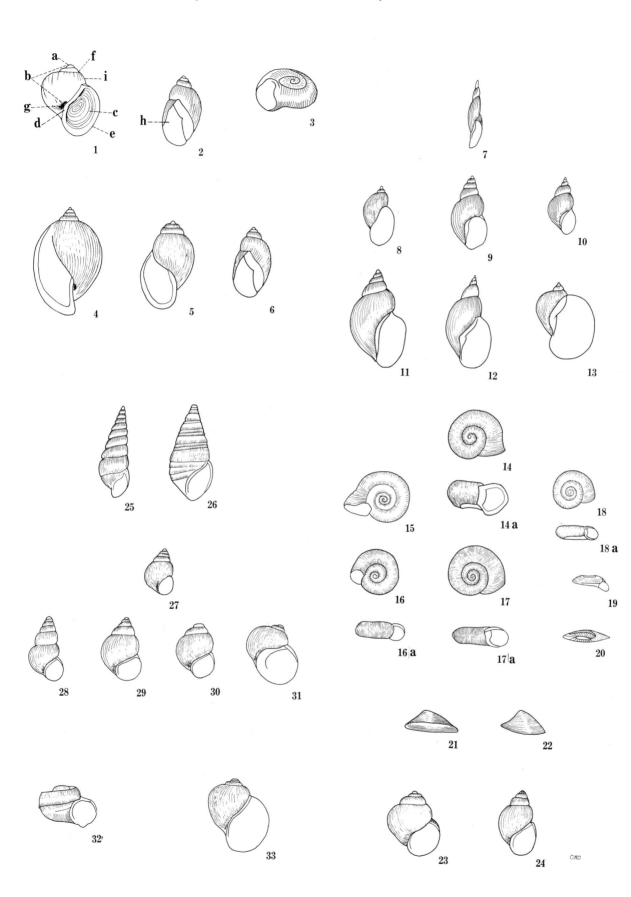

## Explanation of Plate 51

I. Order Prosobranchiata.

A, Helicinidae (*Hendersonia occulata*).

II. Order Pulmonata.

B, Polygyridae (**B₁**, *Polygyrus albolabris;* **B₂**, *Mesodon thryoides;* **B₃**, *Praticolella mobiliana*). C, Bulimulidae (*Bulimus dealbatus*). D, Zonitidae (**D, D₁**, *Ventrides ligerus.*) E, Endodontidae (**E, E₁**, *Anguispira alternata*). F, Haplotrematidae (**F, F₁**, *Haplotrema concavum*). G,

Pupillidae (**G, G₁**, *Gastrocopta procera*). **H,** Strobilopsidae (**H, H₁**, *Strobilops affinis*). **I,** Vallonidae (**I, I₁**, *Vallonia costata*). **J,** Cionellidae (**J, J₁**, *Cionella lubrica*). **K,** Succineidae (**K, K₁**, *Succinea retusa*). **L, M,** Limacidae (**L,** *Deroceras gracile;* **M,** *D. agreste*). **N,** Philomycidae (*Philomycus carolinianus*). **O,** Ellobiidae (**O, O₁**, *Carychium exiguum*).

Figures A, B, D–O redrawn from Baker, 1932, Fieldbook of Illinois Land Snails; others from various sources.

---

2. Helicellidae
   *Helicella, Cochlicella*
3. Bradybaenidae
   *Bradybaena*
4. Polygyridae*,‡ (Plate 51, Figs. B, B₁₋₃)
   *Polygyra*
5. Endodontidae†,‡ (Plate 51, Figs. E, E₁)
   *Anguispira, Helicodiscus*
6. Succineidae†,‡ (Plate 51, Figs. K, K₁)
   *Succinea*
7. Bulimulidae (Plate 51, Fig. C)
   *Bulimulus*
8. Cionellidae† (Plate 51, Figs. J, J₁)
   *Cionella*
9. Zonitidae‡ (Plate 51, Figs. D, D₁)
   *Zonitoides*

b. Slugs

1. Limacidae†,‡,§ (Plate 51, Figs. L, M)
   *Limax, Deroceras*
2. Philomycidae (Plate 51, Fig. N)
   *Philomycus*
3. Arionidae†,‡,§
   *Arion*

3. Suborder Systellommatophora. Tropical slugs. Two pairs of contractile tentacles, dorsal pair with eyes at tips. Neither internal nor external shell.

   a. Veronicellidae*,‡
      *Veronicella*

While space does not permit detailed descriptions of families of snails, a brief account of the shells is given as a guide to preliminary recognition of material under consideration. Critical identification of species is based on the anatomy

of the soft parts, such as the presence of gills or a mantle cavity that serves as a lung, and on anatomic details of the radula and internal and external parts of the reproductive organs. Details of the parts of the shell for identification are given in Plate 50, Figs. 1–3.

*Freshwater Snails*

Representatives of families of freshwater snails of North America are shown on Plate 50.

Planorbidae (orb or ram's horn snails) (Plate 50, Figs. 14–19). The shells are mainly discoidal. In *Planorbis*, they are sinistral, i.e., coil counterclockwise, and in the *Helisoma* they are dextral. The animals of all genera are sinistral because the genital organs are on the left side. Some common genera are *Planorbis, Gyraulus, Tropicorbis, Australorbis, Segmentina, Helisoma*, and *Planorbula*. The members of this family serve as intermediaries for many species of trematodes.

Lymnaediae (pond snails) (Plate 50, Figs. 7–13). These have a dextral shell with a distinctly attenuated spire. The tentacles are flat and triangular. Size of shells varies greatly. Some common genera are *Lymnaea, Pseudosuccinea, Bulimnea, Acella, Fossaria*, and *Stagnicola*.

Ancylidae (limpets) (Plate 50, Figs. 21–22). The ancylids have a small caplike or patelliform shell. The animal is sinistral or dextral. Common genera are *Ferrissia* and *Ancylus*.

Physidae (pouch snails) (Plate 50, Figs. 4–6). The shells are spired, sinistral, thin, and have large body whorls. The aperture is large. The animal is sinistral with long, slender, cylindrical tentacles. Genera include *Physa* and *Aplexa*.

Amnicolidae (Plate 50, Figs. 27–31). The shell is small, dextral, conical, with four to eight whorls, and operculate. The aperture is round. Common genera include *Amnicola, Oncomelania, Pomatiopsis*, and *Bulimus* (= *Bythinia*).

**Plate 51** *Some Families of Common Land Snails of North America* 225

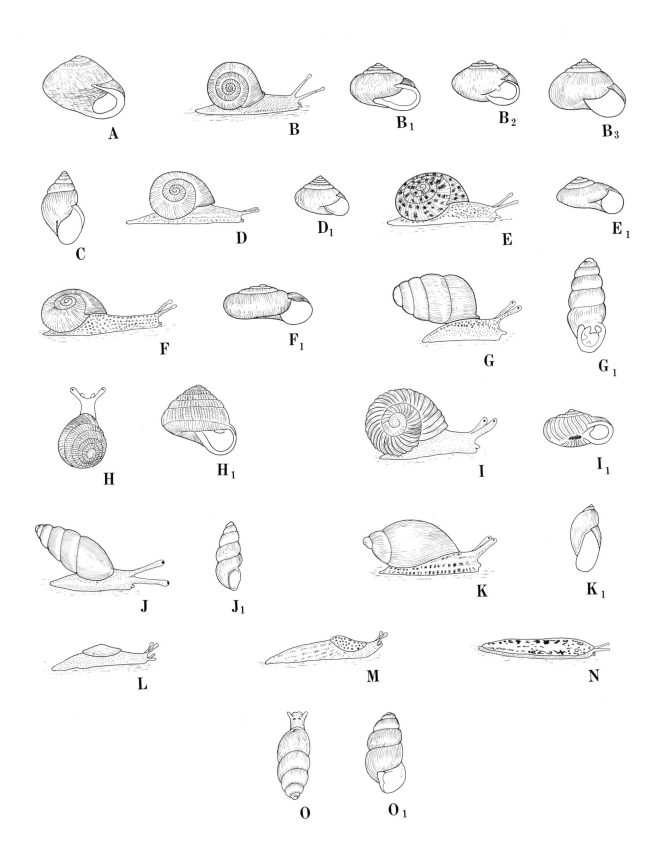

A

B

B₁

B₂

B₃

C

D

D₁

E

E₁

F

F₁

G

G₁

H

H₁

I

I₁

J

J₁

K

K₁

L

M

N

O

O₁

Pleuroceridae (river snails) (Plate 50, Figs. 25–26). Pleurocerids have a shell that is thick, with whorls that have flattened sides, and a long tapering spire. They are operculate. Genera include *Goniobasis*, *Io*, *Pleurocerca*, and *Semisulcospira*.

Ampullaridae (apple snails) (Plate 50, Fig. 33). The shell is dextral, spired, very large, and has a big aperture. Both a gill and lung are present. The proboscis is divided into two tentacle-like structures. Genera include *Ampullaria* and *Pomacea*.

### Land Snails

Figures of representative families of North American land snails appear on Plate 51. The shells are dextral.

Helicinidae (Plate 51, A). These are operculate snails largely from tropical regions. The genera *Hendersonia* and *Helicina* occur in North America.

Polygyridae (Plate 51, B, B$_{1-3}$). Shells are large, with a low spire, and in adult specimens the outer lip is reflected. *Polygyra* is the common genus with many species.

Helicidae. Shells are large and with five to seven whorls. The general shape is similar to *Praticolella*. *Oreohelix* is a common representative.

Bulimulidae (Plate 51, C). The shells are large, have an elongated spire, six rounded whorls, and deeply indented sutures. *Bulimus* is representative of the family.

Zonitidae (Plate 51, D, D$_1$). Members of this family have yellowish, horn-colored or clear, somewhat flattened shells with a shining surface. They range from minute (1.5 mm) to large (25 mm) in size. Genera include *Mesomphix*, *Ventridens*, *Retinella*, *Zonitoides*, *Paravitrea*, *Euconulus*, *Hawaiia*, and *Striatura*.

Strobilopsidae (Plate 51, H, H$_1$). The small shells are dome-shaped, sculptured with oblique ribs, and have 4.5 to 6 slowly enlarging whorls. *Strobilops* is the representative genus.

Vallonidae (Plate 51, I, I$_1$). The shells are minute (up to 2.75 mm), the spires are only slightly elevated, and provided with elevated ribs over the surface. *Vallonia* is representative of the family.

Cionellidae (Plate 51, J, J$_1$). The shell is about 6 mm long, cylindrical, smooth, shining, and horn-colored. The spire is considerably longer than the aperture. *Cionella lubrica* is the only species in North America. It is commonly listed as a member of the Cochlicopidae.

Succineidae (Plate 51, K, K$_1$). Shells are thin, amber-colored, oval, and with a very large aperture and small spire. The animal is too large for the shell. *Succinea* is the only genus.

Endodontidae (Plate 51, E, E$_1$). The shells are subconical, never smooth and shining but usually brown in color and ribbed. The lip is thin and never expanded. Size varies from minute (1.5 mm) to large (25 to 28 mm). *Anguispira*, *Goniodiscus*, *Helicodiscus*, and *Punctum* are representative of the family.

Haplotrematidae (Plate 51, F, F$_1$). The shells are large, only slightly spired, being almost discoidal, and without spotting. The lip is smooth, thin, and without protuberances. The animal is long and narrow. *Haplotrema* is the only genus.

Pupillidae (Plate 51, G, G$_1$). The shells are elongated, oval, and turreted. The outer lip is usually provided with a variable number of internally projecting teeth. The shells never exceed 5.5 mm in length. Representative genera are *Vertigo*, *Gastrocopta*, *Pupoides*, *Pupilla*, *Pupisoma*, and *Columella*.

Limacidae (Plate 51, L, M). These are the slugs. Contrary to the other land snails with coiled shells, the slugs have a rudimentary one completely buried in the tissue. The long, narrow body is humped toward the middle and bears a short rounded projection called the mantle, which covers the lung. A breathing pore is present on the right posterior part of the mantle. *Derocercas* is representative of the family. *Agriolimax* is a synonym. Species are about 2 inches long.

Philomycidae (Plate 51, N). The Philomycidae are tropical slugs with a mantle that covers three-fourths or more of the body length. The respiratory pore is on the right anterior part of the mantle. *Philomycus* and *Pallifera* are representatives of the family that occur in the United States.

Ellobiidae (Plate 51, O, O$_1$). They are largely marine forms. A few species are terrestrial. The shell is elongated. The aperture has one or more folds on the columella and frequently on the outer lip. The foot is broad and slightly indented anteriorly. It is divided ventrally by a transverse line into a short anterior and a long posterior part. *Carychium* (1.7 mm long) is representative of the family.

## EXERCISE ON REARING SNAILS

It is important to have parasite-free snails in any experimental studies on the life cycles of trematodes. While the surest way of having "clean" specimens is to rear them in the laboratory where infection with flukes is precluded, it is not always easy to do. Many aquatic snails can be reared in balanced aquaria with or without aeration. Unglazed clay pots are useful in hot climates where water in glass aquaria becomes too warm for snails to survive well.

Food consists of microflora that accumulates on the sides of the aquaria, especially on the clay containers. Dried tree leaves soaked in water to remove tannic acid are good for some snails. Fresh or boiled lettuce is good food for snails and dried powdered lettuce is excellent food for very young snails. Chopped feed used for domestic livestock provides nutritious food for snails. It is prepared by placing a few handfuls of the feed in a pail of water and allowing it to go through a fermentation process. After the water clears, the sediment provides the food that is readily eaten and does not foul the water of the aquarium. Powdered calcium or pieces of chalk may be supplied. A diet used in many laboratories where large colonies of snails are maintained consists of cerophyl (dehydrated cereal grasses), 10 gm; powdered wheat germ, 5 gm; and sodium alginate, 5 gm. It is mixed in a blender with 500 ml of water at 50°C. The viscous mixture is forced from a 2-liter suction flask by compressed air into a tray of 2 percent $CaCl_2$ solution, where it forms a continuous strand of insoluble calcium alignate food. After washing the food in water, it can be frozen and stored until needed.

Amphibious, land, and marine snails require different procedures for rearing. Amphibious snails are reared in containers tipped so as to expose some of the soil or sand at one end above the water line. Food and snail feces accumulating on the bottom of the balanced aquaria may be removed by aspiration. Land snails and slugs may be reared in terrariums in which the bottom is covered with a thick layer of moist soil overlaid with moss and dead leaves. Lettuce is good food. Marine snails can be reared in trays or aquaria of aerated sea water. Satisfactory food consists of frozen shrimp, fresh yeast, baker's yeast, and calcium carbonate. Snails may be removed from the aquaria and fed in finger bowls about twice weekly. This procedure avoids contaminating the aquaria with decaying food.

For further details on rearing snails, consult Malek and Galtsoff *et al.* In addition to mollusks, the latter reference contains procedures for rearing many different kinds of invertebrates.

## SELECTED REFERENCES

Baker, F. C. (1911). The Lymnaeidae of North and Middle America. Chicago Acad. Sci. Sp. Publ. No. 3, 539 pp., 58 pls.

Baker, F. C. (1928). The Fresh Water Mollusca of Wisconsin, Part I: Gastropoda. Wisc. Acad. Sci., Arts, and Letters, 507 pp., 28 pls.

Baker, F. C. (1939). Fieldbook of Illinois Land Snails. State of Ill., Nat. Hist. Surv. Div. Manual 2, 166 pp.

Baker, F. C. (1945). The Molluscan Family Planorbidae. University of Illinois Press, Urbana, 530 pp., 41 plates.

Chamblerlain, R. V., and D. T. Jones (1929). A Descriptive Catalog of the Mollusca of Utah. Bull. Univ. Utah, Vol. 19, 203 pp.

Eddy, S., and A. C. Hodson. (1957). Taxonomic Keys to the Common Animals of the North Central States, exclusive of the Parasitic Worms, Insects, and Birds. Burgess Publishing Co., Minneapolis, 141 pp.

Galtsoff, P. S., F. E. Lutz, P. S. Welch, and J. G. Needham. (1937). Culture Methods for Invertebrate Animals. Reprinted by Dover Publications, New York, p. 519.

Goodrich, C. (1932). The Mollusca of Michigan. Univ. Mus. Univ. Mich. Handb., Ser. 5, 120 pp.

Henderson, J. (1936). Mollusca of Colorado, Utah, Montana, Idaho, and Wyoming. Univ. Colo. Stud. 23: 81.

*Malek, E. A. (1962). Laboratory Guide and Notes for Medical Malacology. Burgess Publishing Co., Minneapolis, 154 pp.

Pennak, R. W. (1953). Fresh-water Invertebrates of the United States. Ronald Press Co., New York, p. 667.

Pilsbry, H. A. (1937–1948). Land Mollusca of North America (North of Mexico). Phil. Acad. Nat. Sci. Monogr. 1(½):1–994; 2(½):1–1113.

Pilsbry, H. A. (1940). Land Mollusca of North America. Monogr. Acad. Nat. Sci. Philadelphia; No. 5, Parts 1 and 2.

---

* This book is excellent for all aspects of malacology as it pertains to parasitology. It has basic information on morphology, classification, keys, illustrations, and procedures. The list of references is comprehensive and organized under a variety of headings in which mollusks are related to parasitic helminths.

## Explanation of Plate 52

A, Adult fluke showing internal anatomy. B, Adult fluke showing external configuration. C, Entire mother sporocyst containing germ cells, cercariae, and tetracotyles. D, Cercaria. E, Ventral view of encysted tetracotyle (metacercaria). F, Lateral view of excysted tetracotyle. G, Duck definitive host. H, Snail (*Stagnicola* and *Lymnaea*) first intermediate host. I, Snail (*Stagnicola, Lymnaea, Physa, Helisoma*) second intermediate host.

1, Forebody; 2, hindbody; 3, oral sucker; 4, pharynx; 5, esophagus; 6, intestine; 7, ventral sucker; 8, dorsal holdfast organ; 9, ventral holdfast organ; 10, adhesive gland; 11, posterior testes; 12, seminal vesicle; 13, common genital pore; 14, ovary; 15, oviduct; 16, Mehlis' gland; 17, ascending portion of uterus; 18, descending portion of uterus; 19, Laurer's canal; 20, vitelline glands; 21, daughter sporocyst; 22, developing cercaria; 23, tetracotyle; 24, penetration gland; 25, unpigmented eyespot; 26, flame cell; 27, excretory bladder; 28, collecting tubule of excretory bladder; 29, genital primordium; 30, tail; 31, cyst of tetracotyle; 32, holdfast organ; 33, orifice of lateral sucker of tetracotyle; 34, cleft of lateral sucker; 35, external meatus of holdfast organ; 36, urinary bladder; 37, genital primordium.

a, Adult fluke attached to mucosa of small intestine of duck; b, eggs laid in intestine; c, unembryonated egg passed in feces; d, egg embryonates in water and hatches; e, miracidium attacks snails of the genera *Stagnicola, Lymnaea, Physa,* and *Helisoma;* f, miracidium transforms into mother sporocysts; g, daughter sporocyst with developing cercariae; h, fully developed cercaria escaping from snail intermediate host; i, cercaria free in water, hanging in characteristic position; j, cercaria attacking snail second intermediate host, dropping its tail in the process; k, cercaria in tissues of snail; l, tetracotyle encysted in tissues of snail; m, tetracotyle encysted in redia of a different species of fluke, a case of hyperparasitism; n, tetracotyle encysted in the sporocyst of another species of trematode, another case of hyperparasitism; o, infection of definitive host occurs when snails containing tetracotyles are swallowed and released from snails by the digestive processes; p, encysted tetracotyle freed from snail tissues; q, excystation of tetracotyle in intestine of duck where it attaches to the epithelium and develops to sexual maturity after 4 days.

Figures A–B adapted from Van Haitsma, 1931, Papers Mich. Acad. Sci. Arts Lett. 13:447; C, from Hussey, Cort, and Ameel, 1958, J. Parasitol. 44:289; D, from Cort and Brooks, 1928, Trans. Amer. Microsc. Soc. 47:179; E, from Hughes, 1929, Papers Mich. Acad. Sci. Arts Lett. 10:495; F, from Faust, 1918, Ill. Biol. Monogr. 4:1.

---

Walker, B. (1918). A Synopsis of the Classification of the Fresh-Water Mollusca of North America, North of Mexico and a Catalogue of the More Recently Described Species, with Notes. Univ. Mich. Mus. Zool. Misc. Publ. No. 6.

Walker, B. (1928). The Terrestrial Shell-Bearing Mollusca of Alabama. Univ. Mich. Mus. Zool. Misc. Publ. 18, 180 pp.

Ward, H. B., and G. C. Whipple (1918). Fresh-Water Biology. John Wiley and Sons, New York, p. 957; (revised 1959, p. 1117).

### Superorder Anepitheliocystidia

In this group of trematodes, the primitive thin-walled excretory bladder of the cercariae is retained and there is no stylet in the oral sucker.

### ORDER STRIGEATIDA

The cercariae of this order are fork-tailed. The group contains four suborders, all of which show this gross characteristic. They are parasites of birds and mammals.

### Suborder Strigeata

The cercariae are fork-tailed and usually have two suckers, an oral and a ventral.

### Superfamily Strigeoidea

The cercariae usually have a pharynx and a long tail with a slender stem. The metacercariae are of three types, consisting of the tetracotyle, diplostomulum, and neascus. The body of the adults is divided into a forebody and a hindbody. The life cycle often includes one or more paratenic hosts.

### FAMILY STRIGEIDAE

These are parasites of the intestine of fish-, frog-, and snail-eating birds or mammals. The forebody of the adults is cup-shaped with the ventral sucker inside. The metacercariae are in fish, frogs, or snails.

### *Cotylurus flabelliformis*
(Faust, 1917) (Plate 52)

These are parasites of the small intestine of ducks in North America.

**Description.**   Mature flukes measure 0.56 to 0.85 mm in length; the forebody is 0.2 to 0.28 mm and the hindbody 0.36 to 0.57 mm long. The esophagus is nearly as long as the pharynx.

**Plate 52**    *Cotylurus flabelliformis*                                    229

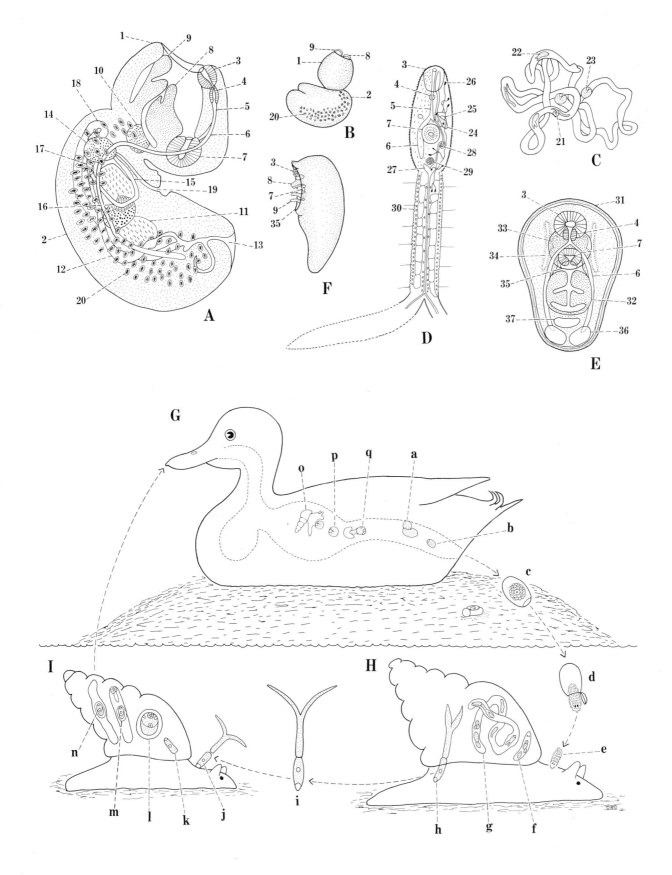

An excretory bladder is lacking. The testes are bean-shaped and a Mehlis' gland is present. Eggs are 100 to 112 $\mu$ long by 68 to 76 $\mu$ wide.

**Life Cycle.**    Adult worms in the small intestine lay unembryonated eggs which leave the host in the feces. Development takes place in the water where the eggs hatch in about 3 weeks and the ciliated miracidia swim about. Upon coming in contact with the snails *Lymnaea stagnalis* and *Stagnicola emarginata* of the family Lymnaeidae, they penetrate readily and develop normally. The ciliated epithelium is shed and the miracidium transforms into a mother sporocyst. As development progresses, the germ cells within the mother sporocyst develop into daughter sporocysts which escape and migrate to the digestive gland where they grow into slender vermiform bodies. From the germ cells in the sporocysts, fork-tailed, pharyngeate cercariae develop. When fully formed, they escape from the sporocyst and eventually from the snail into the water about 6 weeks after infection.

The cercariae are active swimmers but rest after short periods of activity. Upon coming in contact with snails, even those in which they developed, they penetrate the soft tissues and migrate to the hermaphroditic gland. Further development of them to the tetracotyle stage, a type of metacercaria characteristic of Strigeidae, is dependent upon several factors. The tetracotyles develop normally and encyst in approximately 6 weeks in the tissues of the same species of snail in which they develop as cercariae, that is, in *Lymnaea stagnalis* or *Stagnicola emarginata*. If, on the other hand, they enter physid or planorbid snails, no development takes place unless these snails are infected already with some species of trematodes. In these cases, the cercariae of *C. flabelliformis* enter the sporocysts or rediae of the trematodes already present and develop into encysted tetracotyles somewhat sooner than in *Lymnaea* or *Stagnicola*. In this case, they are hyperparasites, that is, one parasite infecting another and developing in it. Some species of *Cotylurus* encyst in leeches and fish of various kinds. Large European *Lymnaea stagnalis* are known to shed over 500,000 cercariae per day under favorable conditions of temperature and darkness.

Infection of ducks takes place when snails harboring encysted tetracotyles are eaten. Sexual maturity is attained in about 1 week.

Life histories of some other Strigeidae that have been studied show considerable variation. Several examples illustrate this. In *Apatemon gracilis*, another parasite of ducks, the cercariae develop in the snail *Helisoma antrosa*, and the tetracotyles (metacercariae) in the leech *Erpobdella punctata*. Sexual maturity is attained after 4 days in ducks and the worms are lost after 13 days.

*Apharyngostrigea pipientis* develops in pigeons under experimental conditions. The first intermediate host is *Planorbula armigera* and the second is tadpoles of *Rana pipiens*, which the cercariae presumably enter by way of the branchial or intestinal epithelium. The tetracotyles develop slowly.

*Strigea elegans* from great horned owls and snowy owls has a more complicated life cycle in which four hosts are involved. The cercariae develop in the snail *Gyraulus parvus* and penetrate the tadpoles of *Bufo americanus*, *Rana sylvatica*, and *R. clamitans*, where they develop to mesocercariae. When mesocercariae are eaten by animals other than the definitive host, such as water snakes and ducks, they develop into tetracotyles that locate in the subcutaneous tissues. These animals harboring the tetracotyles are the third intermediate host in the life cycle. A fourth host, the great horned and snowy owls, is necessary for the tetracotyles to develop to adults. An extensive summary of the life cycles of strigeid trematodes is given by Olivier.

## EXERCISE ON LIFE CYCLE

Collect specimens of *Lymnaea stagnalis*, *Stagnicola emarginata*, *Physa* spp., and *Helisoma* spp. from marshes where ducks and other animals live. Isolate specimens of them in glass jars to obtain strigeid cercariae and separate those of *Cotylurus flabelliformis*. After having determined which individuals are infected, some of them should be dissected to obtain sporocysts for study. Expose uninfected *L. stagnalis* and *S. emarginata* to cercariae of *C. flabelliformis*. Uninfected specimens may be obtained from habitats where ducks do not occur. Examine exposed specimens at intervals of 4 to 5 days for a period of 3 to 4 weeks in order to follow the development and encystment of the tetracotyles in the digestive gland.

Isolate specimens of *Physa* and *Helisoma* in order to obtain individuals infected with flukes of any kind. Expose the infected specimens to the cercariae of *C. flabelliformis*. Sacrifice some individuals at intervals of 3 to 4 days for development of the tetracotyles in the sporocysts of rediae of the flukes already in them.

Feed tetracotyles to ducklings, day-old chicks, or sparrows and determine the length of time required for eggs to appear in the feces and for the flukes to disappear spontaneously. The eggs may be used as a source of miracidia for infecting young *L. stagnalis* or *S. emarginata* in order to obtain the developmental stages of the mother and daughter sporocysts.

## SELECTED REFERENCES

Basch, P. F. (1969). *Cotylurus lutzi* sp. n. (Trematoda: Strigeidae) and its life cycle. J. Parasitol. 55:527–539.

Cort, W. W., and S. T. Brooks. (1928). Studies on holostome cercariae from Douglas Lake, Michigan. Studies on the trematode family Strigeidae (Holostomidae). No. VIII. Trans. Amer. Microsc. Soc. 47:179–221.

Cort, W. W., S. Brackett, and L. J. Olivier. (1944). Lymnaeid snails as second intermediate hosts of the strigeid trematode, *Cotylurus flabelliformis* (Faust, 1917). J. Parasitol. 30:309–321.

Cort, W. W., L. J. Olivier, and S. Brackett. (1941). The relation of physid and planorbid snails to the life cycle of the strigeid trematode, *Cotylurus flabelliformis* (Faust, 1917). J. Parasitol. 27:437–448.

Dubois, G., and R. L. Rausch. (1950). A contribution to the study of North American strigeids (Trematoda). Amer. Midl. Natur. 43:1–31.

Hughes, R. C. (1928). Studies on the trematode family Strigeidae (Holostomidae). No. XVII. *Tetracotyle flabelliformis* Faust. Papers Mich. Acad. Sci. Arts Lett. 10:495–508.

Hussey, K. L., W. W. Cort, and D. J. Ameel. (1958). The production of cercariae by a strigeid mother sporocyst. J. Parasitol. 44: 289–290.

Miller, G. C., R. Harkema, and A. Harris. (1965). Life history notes on *Strigea elegans* Chandler and Rausch, 1947 (Trematoda: Strigeidae). J. Parasitol. 51(2/2):22.

Odening, K. (1967). Die Lebenszyklen von *Strigea falconispalumbi* (Viborg), *S. strigis* (Schrank) und *S. sphaerula* (Rudolphi) (Trematoda, Strigeida) in Raum Berlin. Zool. Jahrb. Syst. 94:1–67.

Odening, K., T. Mattheis, and I. Brockhardt. (1969). Status und Lebenszyklus des Trematoden *Cotylurus platycephalus*. Angew. Parasitol. 10:76–80.

Olivier, L. (1940). Life history studies on two strigeid trematodes of the Douglas Lake region, Michigan. J. Parasitol. 26:447–477.

Olson, R. E. (1970). The life cycle of *Cotylurus erraticus* (Rudolphi, 1809) Szidat, 1928 (Trematoda: Strigeidae). J. Parasitol. 56:55–63.

Pearson, J. C. (1959). Observations on the morphology and life cycle of *Strigea elegans* Chandler and Rausch, 1947 (Trematoda: Strigeidae). J. Parasitol. 45:155–174.

Raišyte, D. (1968). [On the biology of *Apatemon gracilis* (Rud., 1819), a trematode parasitic in domestic and wild ducks.] Acta Parasitol. Lit. 7: 71–84.

Stunkard, H. W., C. H. Willey, and Y. Robinowitz. (1941). *Cercaria burti* Miller, 1923, a larval stage of *Apatemon gracilis* (Rudolphi, 1819) Szidat, 1928. Trans. Amer. Microsc. Soc. 60:485–497.

Van Haitsma, J. P. (1931). Studies on the family Strigeidae (Holostomidae) XXII. *Cotylurus flabelliformis* (Faust) and its life-history. Papers Mich. Acad. Sci. Arts Lett. 13:447–482.

Vojtec, J., V. Opravilová, and L. Vojtkova. (1967). The importance of leeches in the life cycle of the order Strigeidida. (Trematoda). Folia Parasitol. Praha. 14:107–109.

## FAMILY DIPLOSTOMATIDAE

The Diplostomatidae are parasites of the intestine of fish- and frog-eating birds and mammals. The body is generally divided into a fore and hind region. The forebody is spatulate or foliate in shape, the posterior cylindrical. An oral sucker and pharynx are present.

### *Uvulifer ambloplitis*
(Hughes, 1927) (Plate 53)

*Uvulifer ambloplitis* infects kingfishers in North America. The metacercariae, or neascus larvae, appear as small black spots in the skin of bass, rock bass, perch, and sunfish.

**Description.** Adult flukes are 1.8 to 2.3 mm long; the posterior half, or thereabouts, of the forebody forms a necklike structure. The forebody is bowl-shaped with the oral sucker inside but near the anterior margin, the ventral sucker near the center, and the holdfast organ just posterior to it. The hindbody is about 3 times the length of the forebody and about 4.5 times as long as the diameter of the small testis. Vitelline follicles extend posteriorly almost to the level of the copulatory bursa. Eggs measure 90 to 99 $\mu$ long by 56 to 66 $\mu$ wide.

**Life Cycle.** Unembryonated eggs are

## Explanation of Plate 53

A, Adult fluke. B, Miracidium. C, Mother sporocyst. D, Daughter sporocyst. E, Pharyngeate furcocercous cercaria. F, Encysted neascus (metacercaria). G, Neascus in process of excystment. H, Neascus type of metacercaria. I, Belted kingfisher (*Streptoceryle alcyon*) definitive host. J, Snail intermediate host (*Helisoma trivolvus*). K, Fish second intermediate host (*Micropterus dolomieui*).

1, Forebody; 2, hindbody; 3, oral sucker; 4, ventral sucker; 5, holdfast organ; 6, prepharynx; 7, pharynx; 8, esophagus; 9, intestine; 10, posterior testis; 11, vas efferentia; 12, seminal vesicle; 13, ejaculatory duct with sphincter; 14, genital cone; 15, ovary; 16, oviduct; 17, Mehlis' gland; 18, Laurer's canal; 19, proximal portion of uterus; 20, distal portion of uterus; 21, common genital pore; 22, vitelline glands; 23, part of vitelline duct; 24, apical papilla; 25, apical gland; 26, eyespot; 27, ganglion; 28, penetration gland; 29, germ cells; 30,

cercaria; 31, excretory bladder; 32, tail; 33, furca; 34, metacercarial cyst.

a, Adult fluke in small intestine; b, eggs laid in intestine; c, unembryonated eggs pass from intestine in feces; d, eggs develop and hatch in water; e, miracidium free in water; f, miracidium attacks and penetrates soft tissue of snail; g, mother sporocyst; h, daughter sporocyst with germ balls and cercaria; i, free-swimming cercaria in characteristic resting position; j, cercaria attacking fish second intermediate host, dropping its tail in the process; k, encysted neascus in fish intermediate host; l, infection of definitive host occurs upon swallowing of infected fish intermediary; m, digestion of fish releases cysts from its tissues; n, cyst free in digestive tract; o, neascus released from cyst by action of digestive juices, allowing it to proceed to proper site in small intestine and develop to sexual maturity.

Adapted from Hunter and Hunter, 1935, Suppl. 24th Annu. Rep. N.Y. State Conserv. Dep. 1934, (IX):267.

---

passed in the droppings of kingfishers. Under favorable temperatures, they hatch in water in about 3 weeks. The miracidia penetrate ram's horn snails (*Helisoma trivolvis* and *H. campanulata*), shed the ciliated epithelium, and transform into mother sporocysts, which are characterized by retaining the eyespots of the miracidia. Sporocysts produced by the mother sporocyst invade the digestive gland and liver. When mature, they are about 2 mm long, have constrictions of the body and a birth pore. Cercariae appear in sporocysts in approximately 6 weeks after infection of the snails. They are filled with germ balls and cercariae; the latter are in various stages of development.

When mature, the fork-tailed cercariae escape into the water. They swim for short intervals and then rest, hanging in the water with the furcae spread and the fore part of the body folded upon itself. When coming in contact with bass, perch, and sunfish, the cercariae attach to the fish and penetrate the skin, dropping the tail in the process. Upon entering the skin, they transform into neascus-type metacercariae. These secrete a roomy hyaline cyst about themselves. By the end of week 3, the host has deposited black pigment around the cysts. They are now called black spots or black grubs. Heavy infection of the fish causes a bulging of the eyes from the socket, a condition known as popeye. Fatalities occur in fry.

Kingfishers become infected by eating fish with black grubs. Development to sexual maturity in the fledgling birds requires about 27 days.

*Crassiphiala bulboglossa*, also of kingfishers, is closely related to *U. ambloplitis* and has a similar life cycle. It produces black spots in 11 species of fish from the families Cyprinidae, Cyprinodontidae, Esocidae, Etheostomidae, Percidae, and Umbridae. *Helisoma anceps* and *H. trivolvus* serve as the snail hosts.

A total of six species of strigeid flukes are reported to produce black spots in fish. Trout have black spots that are caused by Heterophyidae metacercariae of *Apophallus imperator* from gulls and loons and *A. brevis* from herons and mergansers.

## EXERCISE ON LIFE CYCLE

Locate an area where kingfishers occur along streams or near ponds. Fish and ram's horn snails from these waters in all probability will be parasitized with strigeid larvae that cause black spots. Collect ram's horn snails from waters where infected fish occur and isolate them in small wide mouth bottles to obtain the fork-tailed cercariae.

Uninfected fish, at least those without visible black spots, should be exposed to fork-tailed cercariae from ram's horn snails. Fish should be

**Plate 53**    *Uvulifer ambloplitis*                                                    233

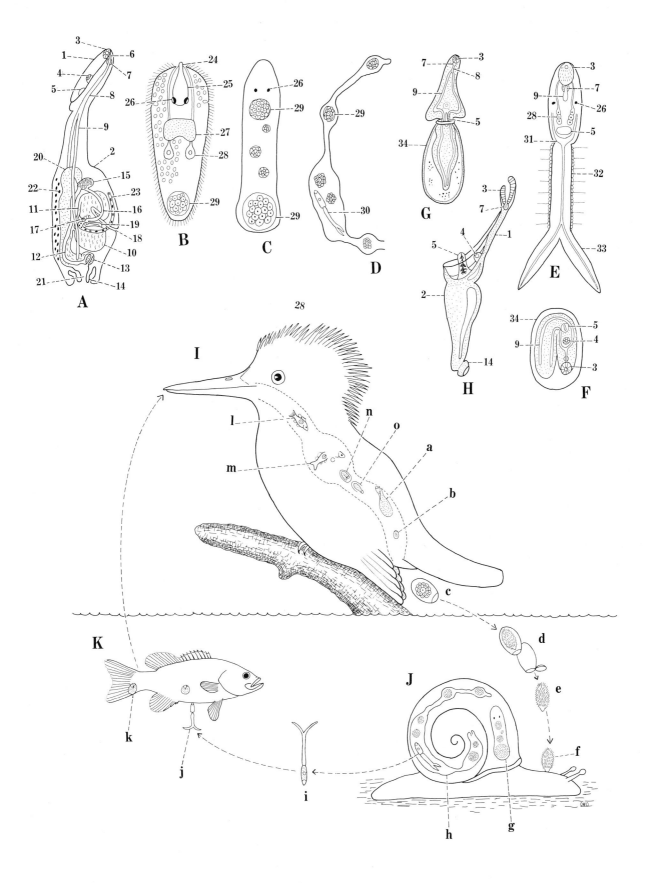

## Explanation of Plate 54

A, Ventral view of adult fluke. B, Sinistral view of adult fluke. C, Miracidium. D, Young mother sporocyst. E, Mature mother sporocyst. F, Mature daughter sporocyst. G, Pharyngeate fork-tailed cercariae. H, Diplostomulum 5 days old. I, Diplostomulum 9 days old. J, Dextral view of mature diplostomulum. K, Ventral view of mature diplostomulum. L, Ventral view of mature diplostomulum with pseudosuckers everted. M, Mallard duck definitive host. N, Snail (*Stagnicola* spp.) first intermediate host. O, Cutaway view of stickleback second intermediate host.

1, Forebody; 2, hindbody; 3, pseudosucker; 4, oral sucker; 5, ventral sucker; 6, adhesive gland; 7, pharynx; 8, esophagus; 9, intestinal crura; 10, ovary; 11, ootype; 12, testis; 13, seminal vesicle; 14, vitelline glands; 15, egg in uterus; 16, genital pore; 17, apical gland; 18, eyespot; 19, excretory canal; 20, flame cell; 21, daughter sporocysts developing in mother sporocyst; 22, germinal mass; 23, fully developed daughter sporocyst; 24, birth pore; 25, developing cercaria in daughter sporocyst; 26, penetration glands; 27, tail of cercaria; 28, excretory bladder; 29, excretory tubule of tail; 30, furca of tail; 31, everted pseudosucker; 32, genital anlagen.

a, Adult fluke in small intestine of definitive host; b, eggs laid in intestine; c, undeveloped eggs passed in feces; d, eggs develop in water where miracidia hatch; e, miracidium penetrates snail intermediate host; f, miracidium sheds ciliated covering and transforms into young mother sporocyst; g, mature mother sporocyst with germ balls and developing daughter sporocyst; h, mature daughter sporocyst with developing cercaria; i, mature fork-tailed cercaria escaping from snail into water; j, cercaria in water, hanging in characteristic position; k, cercaria penetrating gills of stickleback, dropping its tail in the process; l, cercaria in afferent branchial artery; m, cercaria escaping from afferent branchial artery; n, cercaria entering optic lobe of brain; o, metacercariae crowded in optic lobe; p, cercaria that has strayed into muscle tissues; q, metacercaria in olfactory lobe; r, metacercaria in cornea, also retina and posterior chamber; s, infection of duck host when infected sticklebacks are swallowed; t, metacercaria being freed from fish by action of digestive juices; u, metacercaria free in stomach; v, metacercaria escaping from cyst; w, young fluke attaches to wall of small intestine and grows to maturity in three to four days.

Adapted from Hoffman and Hundley, 1957, J. Parasitol. 43:613.

---

examined at intervals of 4 to 5 days to follow the development of the metacercariae and the deposition of pigment by the host. Metacercariae can be collected readily by digesting infected fish in an artificial solution of pepsin and hydrochloric acid.

A fledgling kingfisher can be used for infection experiments without harm to it. Eggs passed in the feces should be used for procuring miracidia to be used in infecting snails. By careful planning, the entire life cycle can be demonstrated.

## SELECTED REFERENCES

Hoffman, G. L. (1956). The life cycle of *Crassiphiala bulboglossa* (Trematoda: Strigeida). Development of the metacercaria and cyst, and effect on the fish hosts. J. Parasitol. 42:435–444.

Hoffman, G. L., and R. E. Putz. (1965). The black-spot (*Uvulifer ambloplitis*: Trematoda: Strigeoidea) of centrarchid fishes. Trans. Amer. Fish. Soc. 94:143–151.

Hugghins, E. J. (1959). Parasites of fishes in South Dakota. S. Dak. Agr. Exp. Sta. and S. D. Dep. Game Fish Parks Bull. 484, p. 30.

Hunter, G. W. III. (1933). The strigeid trematode, *Crassiphiala ambloplitis* (Hughes, 1927). Parasitology 25:510–517.

Hunter, G. W., and W. S. Sanborn. (1930). Contribution to the life history of *Neascus ambloplitis* Hughes 1927. J. Parasitol. 17:108.

Hunter, G. W., and W. S. Sanborn. (1935). Further studies on fish and bird parasites. In: A biological survey of the Mohawk-Hudson watershed. 24th Annu. Rep. N.Y. State Conserv. Dep. (1934) Suppl., p. 267–283.

Hunter, W. S., and G. W. Hunter III. (1935). Studies on *Clinostomum*. II. The miracidium of *C. marginatum* (Rud.). J. Parasitol. 21:186–189.

Krull, W. H. (1932). Studies on the development of *Cercaria bessiae* Cort and Brooks, 1928. J. Parasitol. 19:165.

Krull, W. H. (1934). *Cercaria bessiae* Cort and Brooks, 1928, an injurious parasite of fish. Copeia (2):69–73.

### Diplostomum baeri eucaliae
Hoffman and Hundley, 1957 (Plate 54)

This is normally a parasite of the anterior third of the small intestine of ducks in North America, and possibly many other parts of the world.

**Description.**    Adult flukes reared in baby chicks are similar to those recovered from mal-

**Plate 54** *Diplostomum baeri eucaliae* 235

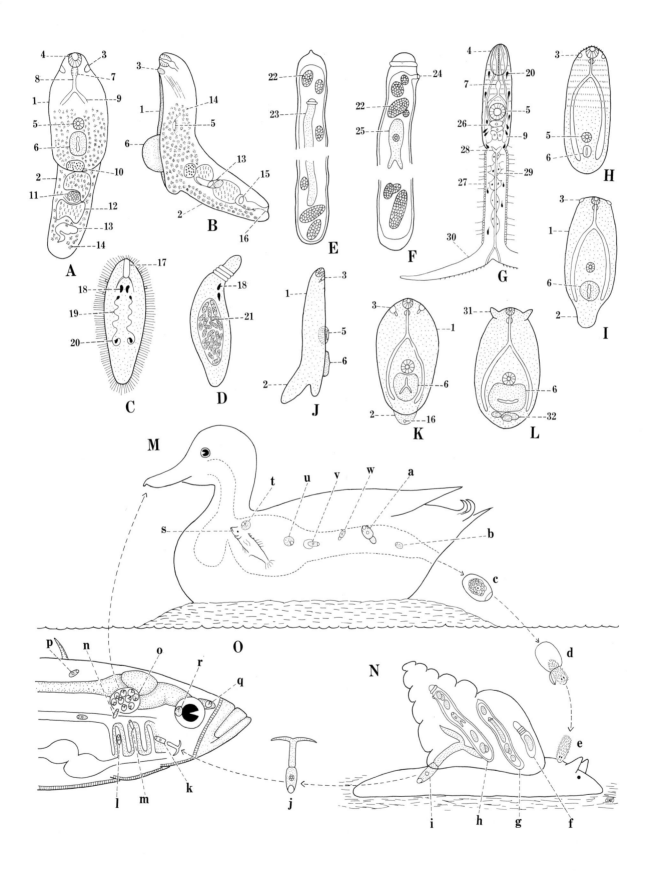

lard ducks. They are 1.35 to 1.8 mm long and about 0.48 mm at the point of greatest width. The hindbody is 0.75 to 0.96 mm long and extends dorsoposteriorly at an angle of 65° to 80°, probably the normal position. The anterior testis is 139 by 172 $\mu$ and the posterior one 163 by 209 $\mu$ in size. Eggs measure 92 to 111 $\mu$ by 54 to 64 $\mu$.

**Life Cycle.** Unembryonated eggs develop and hatch in 12 days at room temperature. The miracidia penetrate the snails *Stagnicola palustris* and *S. p. elodes* in which asexual development takes place. Upon entering the tissues of these snails, the miracidia transform near the point of penetration into mother sporocysts. By the end of 3 weeks, they are fully developed and contain daughter sporocysts which escape and migrate to the liver, reaching it by day 26. The sporocysts have matured by day 30 and are releasing cercariae. They emerge from the snails in small numbers throughout the day and night. In water having a temperature of 24° to 27°C, survival is up to 35 hours. Snails shedding cercariae at the onset of winter lose their infections during the course of hibernation.

Cercariae attach to and penetrate the skin of brook sticklebacks (*Eucalia inconstans*). Apparently, they enter blood vessels and are carried to the cephalic regions. They penetrate the optic lobes, optic nerves, cornea, retina, muscles, and gills, and transform into a diplostomulum type of metacercaria characterized by pseudosuckers, one on each side of the oral sucker. Two hundred or more cercariae may be fatal to fish by causing hemorrhage in the brain and viscera. The larvae reach the brain as early as 3 days after infecting the fish and are in the optic lobes after 5 days. Development appears to be completed by day 11 and the metacercariae are infective on day 13. Metacercariae survive in fish for 1 year and possibly more, attaining a stage of precocious development.

Ducks become infected by eating sticklebacks that harbor fully developed diplostomulae. The prepatent period for mature diplostomulae is 3 days in unfed baby chicks experimentally infected either by feeding or by injecting the metacercariae into the peritoneal cavity.

This is the only strigeid metacercariae recorded from the brain of fish in North America. Four other species are known from Europe and one from Argentina.

Trout which serve as a second intermediate host of *Diplostomum flexicaudum* normally become infected during warm weather by penetrating cercariae shed by snails. During cold weather when cercariae are not being shed, the fish acquire metacercariae by eating snails (*Lymnaea palustris* and *Physa propinqua*) that harbor precocious metacercariae.

Metacercariae of *D. flexicaudum* survive up to 4 years in the bream which is the natural host, but slightly less in the roach.

## EXERCISE ON LIFE CYCLE

Inasmuch as *Diplostomum baeri eucaliae* has been found only in brook sticklebacks, these fish should be examined as a source of experimental material. While ducks may provide eggs of the flukes or the mature parasites, they are not always as readily obtained as the sticklebacks.

Diplostomulae from the brains of fish develop readily when administered to unfed newly hatched chicks. Determine the prepatent period.

Eggs obtained from experimentally infected chicks should be incubated and the miracidia studied. Infect *Stagnicola palustris* by placing them in a small dish with miracidia. Ascertain the course of development of the sporocysts and cercariae in the mollusks.

When cercariae appear, note the time of day when emergence from the snails occurs, their behavior in the water and toward light, and postulate the relationship of it to infection of the fish under natural conditions. Naturally infected snails should be sought in waters inhabited by ducks and sticklebacks.

Expose sticklebacks to cercariae and observe the cercariae, the effect on the behavior of the fish during the period of penetration, and later when the metacercariae are developing in the brain, the route of migration to the brain, and the process of encystment. Examine the fish carefully for damage to the brain by the diplostomulae. Sections of the brain will reveal the relationship of the larval parasites to the different parts of it and some of the pathology caused by them.

## SELECTED REFERENCES

Becker, C. D., and W. D. Brunson. (1966). Transmission of *Diplostomum flexicaudum* to trout by ingestion of precocious metacercariae in molluscs. J. Parasitol. 52:829–830.

Dönges, J. (1969). *Diplostomum phoxini* (Faust,

1918) (Trematoda). Morphologie des Miracidium sowie Beobachtungen an weitern Entwicklungsstadien. Z. Parasitenk. 32:120–127.

Hoffman, G. L., and J. B. Hundley. (1958). The life-cycle of *Diplostomum baeri eucaliae* n. subsp. (Trematoda: Strigeida). J. Parasitol. 43:613–627.

Rees, F. G. (1955). The adult and diplostomulum stage (*Diplostomulum phoxini* (Faust)) of *Diplostomulum pelmatoides* Dubois and an experimental demonstration of part of the life cycle. Parasitology 45:295–312.

## *Alaria canis*
### LaRue and Fallis, 1937 (Plate 55)

*Alaria canis* in a parasite of Canidae in Canada. The adult flukes are in the first third of the small intestine where the eggs are laid.

**Description.**    Adult worms measure 2.5 to 4.2 mm long with the forebody 1.6 to 2.6 mm and the hindbody 0.68 to 1.6 mm. The forebody has foliaceous margins that join posteriorly near the level of the body constriction. There is a conical tentacular appendage of each side of the oral sucker. The holdfast organ is oval, has a longitudinal median depression, and extends from near the ventral sucker to the level of the constriction; the posterior part is more or less covered by the foliaceous margins of the forebody. The ovary is near the level of the constriction and located toward the right. The testes are lobed, with the posterior one being much the larger. Eggs are 107 to 133 $\mu$ long by 77 to 99 $\mu$ wide and unembryonated.

**Life Cycle.**    Unembryonated eggs laid in the intestine are voided in the feces of dogs or other canids. Development of them is rapid at room temperature, with hatching in 2 weeks. The fusiform miracidia are active swimmers. They penetrate the snails *Helisoma trivolvis*, *H. campanulata* and *H. duryi*, the red ram's horn snail from Florida used in aquaria. Once inside the tissues of the snail, they shed the ciliated covering, make their way to the renal veins, and transform to mother sporocysts within 2 weeks. A birth pore is not present. They persist in this stage up to 14 months, which is the usual life span of the snail host. Presumably they produce daughter sporocysts throughout their lives. Giant sporocyst-like bodies are possibly mother sporocysts that produce cercariae instead of daughter sporocysts.

Daughter sporocysts develop from germinal cells within the mother sporocysts, escaping when fully developed in 1 to 2 weeks. They migrate to the hemocoel and over the digestive gland. They may be differentiated from mother sporocysts by the subterminal birth pore. As the mother and daughter sporocysts attain an age of about 1 year, they become yellowish in color. Cercariae are fully developed in 2 to 3 weeks after the daughter sporocysts mature. They escape one at a time, oral sucker first, through the birth pore and leave the snail in intermittent bursts at intervals of several days or even weeks between 9 A.M. and 4 P.M. They are positively phototropic, accumulating on the light side of the container. They hang in a resting position in the water, slowly sinking. These rests are interrupted regularly by swimming upward.

When tadpoles of *Rana pipiens*, *R. sylvatica*, *Pseudacris nigrita*, or *Bufo americanus* swim by, creating currents in the water, the cercariae become active and attach to them, dropping their tails as they enter the integument.

Within the body of the tadpoles, the cercariae develop into mesocercariae, an unencysted form intermediate between the cercarial and diplostomula stages. They survive metamorphosis of the tadpoles. The mesocercariae are fully developed and infective in 2 weeks after entering the second intermediate host.

Activity of the cercariae and growth of metacercariae of *Bolbophorus confusus*, a strigeid, is influenced by temperature. Maximum cercarial invasion occurs when the water is between 18° and 30°C and is markedly reduced between 12° and 18°C. While metacercariae develop rapidly at 20°C, there is no growth at 14° to 15°C. When newly infected fish are kept at 14° to 15°C for a month and then returned to 20°C, development proceeds normally. Water temperature apparently plays a major role in the distribution of the metacercariae.

Infection of the definitive host may take place by two quite different means. When infected tadpoles or frogs are eaten by canids, the mesocercariae, upon being freed from the tissues, leave the gut and undergo a somatic migration. They are found first in the liver and abdominal cavity. From the latter location, they migrate forward, pass through the diaphragm, and into the lungs by the end of 2 to 3 weeks. Here they transform into diplostomulae in about 5 weeks. Presumably the diplostomulae migrate up the trachea where they are swallowed. Development is completed quickly in the small intestine, with

## Explanation of Plate 55

A, Ventral view of adult fluke. B, Miracidium. C, Adult mother sporocyst. D, Daughter sporocyst. E, Fork-tailed cercaria. F, Mesocercaria showing some internal organs. G, External view of mesocercaria. H, Diplostomulum. I, Fox definitive host. J, Planorbid snail (*Helisoma*) first intermediate host. K, Tadpole (*Rana, Bufo*) second intermediate host. L, Paratenic hosts (snakes, frogs, mice).

1, Forebody; 2, hindbody; 3, lappet or pseudosucker; 4, holdfast organ; 5, oral sucker; 6, pharynx; 7, cecum; 8, ventral sucker; 9, testes; 10, ovary; 11, eggs in uterus; 12, vitelline glands; 13, common genital pore; 14, cilia; 15, eyespot; 16, flame cell; 17, germ cells; 18, excretory opening; 19, daughter sporocyst with germ balls or developing cercariae; 20, birth pore; 21, cercaria; 22, penetration glands; 23, duct of penetration glands; 24, genital primordium; 25, excretory bladder; 26, forked tail; 27, body spines.

a, Adult fluke in small intestine; b, egg passing out of body in feces; c, unembryonated egg; d, embryonated egg; e, egg hatching; f, miracidium penetrating *Helisoma* snail; g, young mother sporocyst; h, mature mother sporocyst; i, daughter sporocyst; j, cercaria free in water in characteristic resting position; k, cercaria penetrating tadpole, casting tail as it enters; l, mesocercaria; m, mesocercaria in snake, frog, and mouse paratenic hosts; n, infection of definitive host by swallowing infected tadpole second intermediate host; o, infection of definitive host by swallowing infected paratenic host; p, mesocercariae migrate through gut wall into celom; q, mesocercariae enter hepatic portal vein but it has not been shown that they reach the lungs via the blood; r, mesocercariae pass through the diaphragm and penetrate the lungs; s, in the lungs, the mesocercariae transform to a diplostomulum stage; t, diplostomulae migrate up trachea; u, diplostomulae are swallowed, go to small intestine, and develop to maturity (a) in 5 to 6 weeks.

Figure A adapted from LaRue and Fallis, 1936, Trans. Amer. Microsc. Soc. 55:340; B–H, from Pearson, 1956, Can. J. Res. 34:295.

---

eggs appearing in the feces 34 to 37 days after ingestion of the mesocercariae.

The second method of infection involves a third host which serves as a collector of mesocercariae in its body. Water snakes, which feed on tadpoles, function especially well in this role. When infected tadpoles are eaten and digested by the snakes, as well as by frogs, mice, and possibly other animals, the mesocercariae migrate from the intestine into the tissues where they accumulate without further development. Canids eating these snakes, or other animals harboring mesocercariae, acquire large numbers of them, which undergo a somatic migration in the same manner as described for those from tadpoles and frogs. *Neodiplostomum spathoides* and *N. attenuatum* from raptors, likewise, infect by both means.

The collector, or paratenic host, serves to bridge the gap between the aquatic phase of the life cycle in tadpoles and frogs by its food habits in making the mesocercariae more readily available to the terrestrial host which normally would not eat the tadpoles because of their being in the water.

Paratenic parasitism is an adaptation by the parasite to survive due to a shortage of definitive hosts to provide a suitable habitat for development of the sexual stage rather than to a resistance on the part of the intermediaries.

*Alaria marcianae* develops in *Helisoma trivolvus* and *H. campanulata* as the first intermediate host, tadpoles of *Rana pipiens* as the second, mice, rats, and chicks as paratenic hosts, and domestic cats as the definitive host.

Pearson (1956) reviewed the literature on four other species of *Alaria* whose life cycles have been worked. They are essentially the same as that of *Alaria canis*. It is 1) the snail first intermediate host, 2) a vertebrate second intermediate host that harbors the mesocercariae, 3) a collector, or paratenic host, that makes the mesocercariae more readily available to the definitive host because of its habitat and food habits, and 4) the definitive host in which the diplostomulae and adults develop.

## EXERCISE ON LIFE CYCLE

Eggs must be obtained from adult specimens of known species of *Alaria* as the starting point, and the cycle carried out step by step.

Incubate the eggs at room temperature to obtain miracidia. The red ram's horn snails, *Helisoma duryi*, available in pet shops, make ideal hosts for *A. canis* but not for some other species (see Pearson, 1956, for a list of molluscan hosts). Expose snails to miracidia and dissect individuals at intervals of 1 to 2 days to follow the development of the sporocysts and cercariae.

**Plate 55** *Alaria canis* 239

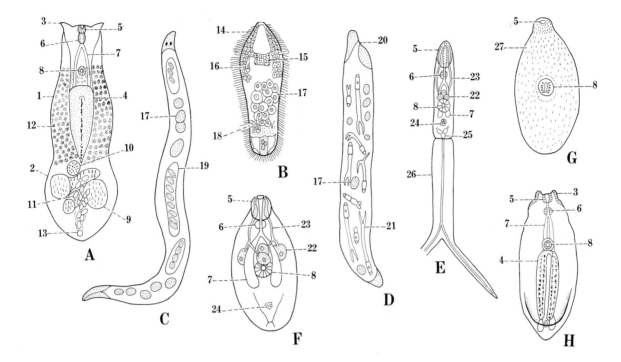

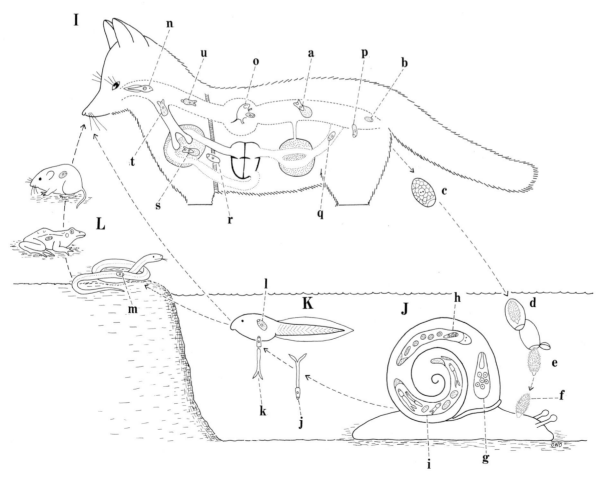

Tadpoles collected from a place where they are unlikely to be infected, as demonstrated by examination of an adequate sample, or those reared from eggs in the laboratory, should be exposed to cercariae. Examine specimens at intervals of 2 to 3 days to follow the migration of the mesocercariae.

When infected tadpoles are available, feed some to dogs and begin examining the feces after 3 weeks for eggs of flukes. If dogs are available for examination, observations should be made for mesocercariae in the body and thoracic cavities 24 hours after infection, and again 14 days later for mesocercariae and diplostomulae in the lungs.

Feed infected tadpoles to water or garter snakes, preferably specimens reared in the laboratory, to accumulate mesocercariae in the tissues. Try rats or mice to determine whether they will serve as paratenic hosts.

## SELECTED REFERENCES

Bosma, N. J. (1934). The life history of the trematode *Alaria mustelae* Bosma, 1931. Trans. Amer. Microsc. Soc. 53:116–153.

Cuckler, A. C. (1940). Studies on the migration and development of *Alaria* sp. (Trematoda: Strigeata) in the definitive host. J. Parasitol. 26 (Suppl.):36.

Johnson, A. D. (1968). Life history of *Alaria marcianae* (LaRue, 1917) Walton, 1949 (Trematoda: Diplostomatidae). J. Parasitol. 54:324–332.

LaRue, G. R., and A. M. Fallis. (1936). Morphological study of *Alaria canis* n. sp. (Trematoda: Alariidae) a trematode, *Alaria intermedia*. Trans. Amer. Microsc. Soc. 59:490–510.

Odening, K. (1965). Der Lebenszyklus von *Neodiplostomum spathoides* Dubois (Trematoda, Strigeida) in Raum Berlin nebst Beiträgen zur Entwicklungsweise verwandter Arten. Zool. Jahrb. Syst. 92:523–624.

Odening, K. (1968). Paratenischer Parasitismus als Indikator der Endwirtsspezifität bei einigen Trematoden. Helminthologia 8-9:429–435.

Odlaug, T. O. (1940). Morphology and life history of the trematode, *Alaria intermedia*. Trans. Amer. Microsc. Soc. 59:490–510.

Olson, R. E. (1966). Some experimental fish hosts of the strigeid trematode *Bolbophorus confusus* and effects of temperature on cercariae and metacercariae. J. Parasitol. 52:327–334.

Pearson, J. C. (1956). Studies on the life cycle and morphology of the larval stages of *Alaria arisaemoides* Augustine and Uribe, 1927 and *Alaria*

*canis* LaRue and Fallis, 1936 (Trematoda: Diplostomidae). Can. J. Zool. 34:295–387.

Pearson, J. C. (1959). Observations on the morphology and life cycle of *Strigea elegans* Chandler and Rausch, 1947 (Trematoda; Strigeidae). J. Parasitol. 45:155–174.

### Superfamily Clinostomatoidea

A single family, the Clinostomatidae, containing about a dozen species is recognized. They are parasites of fish- and frog-eating reptiles, birds, and mammals throughout the world. Adults and the precocious metacercariae have large bodies in which the oral sucker is surrounded by a collar-like fold. The intestinal ceca are long and with or without lateral branches. Testes are tandem with the ovary between them which is an unusual arrangement among the trematodes.

### *Clinostomum complanatum*
(Rudolphi, 1819) (Plate 56)

Adult flukes are in the mouth (not the intestine) of herons. The large metacercariae, known as yellow grubs, are embedded subcutaneously or intramuscularly in 30 or more species of fish, yellow perch, black bass, and sunfish being among the most common hosts in North America.

**Description.** Adult worms and the precocious metacercariae are about equal in size, being 3 to 8 mm long. When the oral sucker is retracted, it is surrounded by a collar-like fold. The acetabulum is the larger of the suckers and is near the junction of the anterior and middle thirds of the body.

The ceca extend to near the posterior end of the body and have numerous short outpocketings from the level of the acetabulum caudad. Testes are tandem and lie in the middle of the postacetabular part of the body with the ovary between them and the uterus anterior to them. The eggs are large and variable in size, being 104 to 140 $\mu$ long by 66 to 73 $\mu$ wide.

**Life Cycle.** Adult flukes lay their eggs in the mouth of the heron host. They may be washed into the water during the feeding activities when the birds dip their heads into the water to capture fish, or swallowed and voided with the droppings. Both embryonated and unembryonated eggs are laid. The developed ones hatch almost immediately upon reaching the water, while the undeveloped ones require about 19 days to hatch.

The active miracidia are covered with cilia,

have three pigmented eyespots arranged in a triangular form, and contain a single germ ball. Upon coming in contact with ram's horn snails (*Helisoma campanulata* and *H. antrosum*), they burrow into them. Inside the snail intermediate host, the miracidia shed the ciliated epithelium and migrate to the digestive gland or liver. They are thin-walled, sac-like creatures with three eyespots, but lack a birth pore. Inside them are several germ balls which develop into rediae.

The first generation rediae escape from the mother sporocysts and locate in the digestive gland or liver. The digestive tract is composed of a prominent pharynx, a thin esophagus, and a long wide gut. Each redia contains 3 to 15 developing daughter rediae. The second generation, or daughter, rediae are recognizable by the cuticular folds, masses of developing cercariae, and a birth pore in the anterior fourth of the body.

Cercariae, when fully developed, escape from the rediae and snail into the water. They are pharyngeate and brevifurcate, i.e., have short furcae at the end of a short tail stem, two eyes, and a longitudinal fin over the dorsal side of the body. Upon coming in contact with fish, the cercariae attach to the skin and burrow through it. In the subcutaneous tissue and muscles, they encyst, developing in about 20 weeks into large precocious metacercariae known as yellow grubs.

When infected fish are eaten by herons, the metacercariae are released from the tissues, excyst quickly, probably in the proventriculus, and migrate to the buccal cavity. Metacercariae are freed from the fish tissues by digestion in 1 to 0.5 percent pepsin in Ringer's solution adjusted to pH 2.5 with HCl and excyst in this medium within 30 minutes. In unfed, day-old chicks, metacercariae excyst in the acid pepsin of the proventriculus, are in the esophagus within 10 minutes, and the mouth in 20. They attain sexual maturity in the mouth of herons in 3 days, remain there about 2 weeks, and then are lost.

A second species, *C. attenuatum* Cort, 1913, utilizes frogs, instead of fish, as the second intermediate host. The adults are in the mouth cavity of bitterns but not herons.

## EXERCISE ON LIFE CYCLE

*In vitro* studies may be conducted on the metacercariae which are prevalent in the fish hosts. Digest cysts from the fish tissue with 1 to 0.5 percent pepsin in Ringer's solution adjusted to pH 2.3 with HCl and kept at 35° to 37°C. Observe them escape from the thin metacercarial cysts.

Unruptured cysts placed in the stomach of baby chicks will open and the metacercariae appear quickly in the mouth. Excysted metacercariae will develop in the chorioallantois of chick embryos at 35° to 37°C and produce eggs.

Metacercariae introduced surgically into the celom of mice develop to sexual maturity in 4 to 5 days. They persist up to 8 days at which time a precipitate containing leukocytes accumulates around them and degeneration follows.

If young herons in rookeries are available for examination, adult worms may be obtained by lavaging the mouths into a pan without harm to the birds. Adult worms placed in water will shed eggs. Hatching of embryonated eggs occurs quickly in water.

Planorbid snails (*Helisoma campanulata* and *H. antrosa*) become infected when exposed to newly hatched miracidia. Experimentally infected snails should be dissected at 3-day intervals to obtain mother sporocysts and the two generations of rediae. Daughter rediae are recognizable by the cercariae they contain.

Cercariae obtained from experimentally or naturally infected snails collected in lakes containing parasitized fish, will penetrate and encyst in a number of species of fish. Guppies and certain tropical fish are suitable hosts.

## SELECTED REFERENCES

Dowsett, J. A., and G. A. Lubinsky. (1966). Maturation of *Clinostomum complanatum* (Trematoda) in laboratory mice. Can. J. Zool. 44:496.

Fried, B., and D. A. Foley. (1970). Development of *Clinostomum marginatum* (Trematoda) from frogs in the chick and on the chorioallantois. J. Parasitol. 56:332–335.

Fried, B., D. A. Foley, and K. C. Kern. (1970). *In vitro* and *in vivo* excystation of *Clinostomum marginatum* (Trematoda) metacercariae. Proc. Helminthol. Soc. Wash. 37:222.

Hopkins, S. H. (1933). Note on the life history of *Clinostomum marginatum* (Trematoda). Trans. Amer. Microsc. Soc. 52:147–149.

Hunter, G. W., III, and H. C. Dalton. (1939). Studies on *Clinostomum*. V. The cyst of the yellow grub of fish (*Clinostomum marginatum*). Proc. Helminthol. Soc. Wash. 6:73–76.

Hunter, G. W., and W. S. Hunter. (1934). Studies

## Explanation of Plate 56

A, Adult fluke. B, Miracidium. C, Mother sporocyst. D, Daughter redia. E, Cercaria. F, Metacercaria. G, Great blue heron (*Ardea herodias*). H, Snail (*Helisoma campanulata, H. antrosa*) first intermediate host. I, Fish (*Perca flavescens*) second intermediate host.

1, Oral sucker; 2, ventral sucker; 3, esophagus; 4, intestine; 5, testis; 6, ovary; 7, oviduct; 8, uterus; 9, vitelline glands; 10, apical papilla; 11, apical gland; 12, penetration glands; 13, large nucleated cells; 14, germinal mass; 15, eyespots; 16, lateral papilla; 17, pharynx; 18, cercaria; 19, prepharynx; 20, tail; 21, short furca; 22, excretory tubule of tail; 23, Mehlis' gland.

a, Adult fluke in mouth of heron; b, unembryonated eggs laid in mouth are washed into water as heron strikes at fish and some are swallowed and expelled in feces; c, eggs develop in water and hatch; d, free-swimming miracidium with single germ ball inside; e, miracidium penetrating snail intermediate host; f, mother sporocyst with developing mother rediae; g, mature mother redia with developing daughter rediae; h, daughter redia containing cercariae in various stages of development; i, cercaria resting in characteristic position in water; j, cercaria penetrating fish, leaving tail behind; k, cercaria migrating in subcutaneous tissues; l, metacercaria encysted in muscles of fish; m, heron becomes infected upon swallowing fish harboring metacercariae, which are released in stomach or proventriculus by digestive juices; n, metacercaria freed from tissues; o, young fluke escaping from cyst migrates anteriorly through esophagus and pharynx into oral cavity where it develops to maturity in 3 days and remains for about 2 weeks, at which time it is lost.

Figures A, F adapted from Cort, 1913, Trans. Amer. Microsc. Soc. 32:169; B–D, G, from Hunter and Hunter, 1935, Suppl. 24th Annu. Rep. N.Y. State Conserv. Dep., p. 267; E, from Krull, 1934, Proc. Helminthol. Soc. Wash. 1:34.

---

on fish and bird parasites. In: A biological survey of the Racquette watershed, 23rd Annu. Rep. N.Y. State Conserv. Dep. (1933) (Suppl. VIII):248.

Hunter, G. W., and W. S. Hunter. (1935). Studies on Clinostomum. IV. Notes on the penetration and growth of the cercaria of *Clinostomum marginatum*. J. Parasitol. 21:411–412.

Krull, W. H. (1934). Some observations on the cercaria and redia of a species of Clinostomum, apparently *C. marginatum* (Rudolphi, 1819) (Trematoda: Clinostomidae). Proc. Helminthol. Soc. Wash. 1:34–35.

Nigrelli, R. F. (1936). Some tropical fishes as hosts for the metacercaria of *Clinostomum complanatum* (Rud. 1814) (= *C. marginatum* Rud. 1819). Zoologica 21:251–256.

Osborn, H. L. (1911). On the distribution and mode of occurrence in the United States and Canada of *Clinostomum marginatum*, a trematode parasitic in fish, frogs, and birds. Biol. Bull. 20:350–366.

Osborn, H. L. (1912). On the structure of *Clinostomum marginatum*, a trematode parasite of the frog, bass, and heron. J. Morphol. 20:189–228.

Ukoli, F. M. A. (1966). On *Clinostomum tilapiae* n. sp. and *C. phalacrocoracis* Dubois, 1931 from Ghana and a discussion of the systematics of the genus *Clinostomum* Leidy, 1856. J. Helminthol. 40:187–214.

## Superfamily Schistosomatoidea

Members of this superfamily are the blood flukes. The adults are monoecious or dioecious, living in the blood vessels of fish, reptiles, birds, and mammals. Both the adults and cercariae are apharyngeate. The cercariae enter the definitive host directly through the skin. The eggs are non-operculated and embryonated when laid, except in the Sanguinicolidae where hatching takes place in the gills of the fish host. The cercariae are apharyngeate and brevifurcate.

### FAMILY SCHISTOSOMATIDAE

This family consists of dioecious flukes parasitic in birds and mammals throughout the world. Several species are parasites of humans residing in the tropical regions.

The mature males may have a longitudinal ventral groove (gynecophoric groove) in which the more slender female is held. The ceca join to form a single terminal limb, reaching almost to the posterior extremity of the body. There are four or more testes; the ovary is oval or spirally curved and located anterior to the union of the ceca.

The cercariae of blood flukes of ducks and muskrats frequently attack people bathing or wading in water infested with them, causing a severe dermatitis which subsides after a few days. The papules resulting from the penetrating cercariae are known as swimmers itch, water itch, or cercarial dermatitis.

Two subfamilies, the Schistosomatinae and Bilharziellinae, are recognized.

**Plate 56** *Clinostomum complanatum* 243

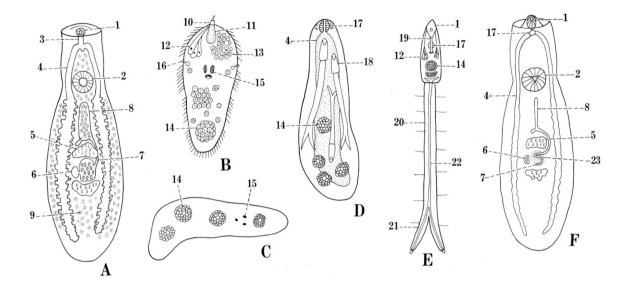

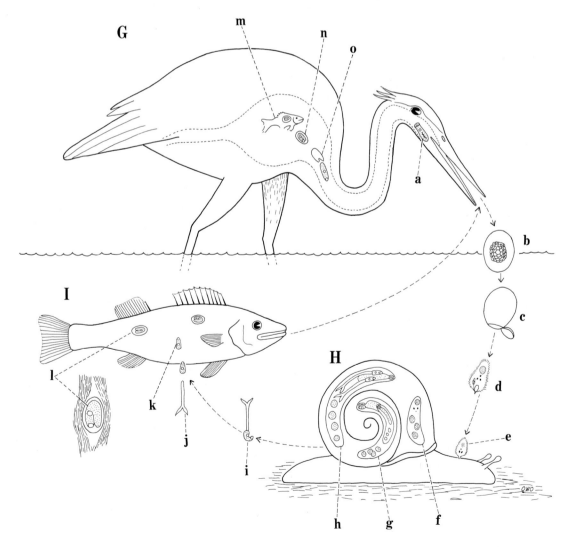

## Explanation of Plate 57

A, Adult male from hepatic venules. B, Adult female. C, Reproductive system of male. D, Reproductive system of female. E, Embryonated egg. F, Miracidium. G, Mother sporocyst. H, Daughter sporocyst. I, Outline of cercaria. J, Anterior end of cercaria, showing spination. K, Details of internal anatomy of cercaria taken from dorsal side. L, *Microtus pennsylvanicus*, definitive host of *Schistosomatium douthitti*. M, *Stagnicola*, one of the snail first intermediate hosts.

1, Oral sucker; 2, ventral sucker; 3, mouth; 4, fold of gynecophoric groove; 5, esophageal glands; 6, intestine; 7, commissures between intestinal ceca; 8, testes; 9, vas deferens; 10, cirrus pouch; 11, genital pore; 12, ovary; 13, seminal receptacle; 14, oviduct; 15, Mehlis' gland; 16, uterus filled with eggs; 17, vitelline glands; 18, common vitelline duct; 19, unhatched egg containing miracidium; 20, lateral papilla; 21, ciliated dermal plates (tiers 1, 2, 3, 4); 22, apical gland; 23, penetration glands; 24, brain; 25, flame cells; 26, germ cells; 27, eyespots; 28, ducts of penetration glands; 29, genital primordium; 30, excretory bladder; 31, common collecting tubule; 32, developing daughter sporocysts; 33, developing cercaria.

a, Adult flukes in copula in intestinal branches of hepatic portal vein; b, masses of eggs accumulating in capillaries and tissues of intestinal wall; c, eggs passing through intestinal wall into lumen of intestine; d, eggs incorporated in fecal pellets; e, embryonated egg free in water; f, egg in process of hatching; g, miracidium burrowing into foot of snail; h, miracidium shedding ciliated epithelium and beginning transformation to mother sporocyst inside snail; i, mother sporocyst in neck and esophageal region, containing germ balls and developing daughter sporocysts; j, daughter sporocyst in digestive gland, with germ balls and developing cercariae; k, cercaria freed from daughter sporocyst and escaping from snail; l, cercaria swimming in water; m, cercaria in resting position attached by ventral sucker to under side of surface film of water; n, cercaria penetrating foot of wading meadow vole; o, p, cercaria, or schistosomatula, migrating toward heart in lymphatic vessels; q, schistosomatula passing through pulmonary vein toward lungs; r, schistosomatula in capillaries of lungs; s, schistosomatula leaving capillaries for lung tissue and alveoli; t, schistosomatula in alveoli; u, adult worms in copula in lungs; v, masses of eggs in lungs; w, schistosomatula migrating from lung into pleural cavity on way to liver; x, schistosomatula migrates along esophagus and postcaval vein through pleural cavity to diaphragm, burrowing through it into peritoneal cavity; y, schistosomatula migrating through diaphragm into peritoneal cavity; z, schistosomatula in peritoneal cavity.

a', Schistosomatula penetrating liver capsule (also spleen and pancreas which are not shown); b', young fluke in liver; c', masses of eggs in liver; d', adult flukes in copula in liver.

Figures A–K redrawn from Price, 1931, Amer. J. Hyg. 13:685.

---

### Subfamily Schistosomatinae

The Schistosomatinae are those blood flukes of birds and mammals whose testes are located anterior to the union of the intestinal ceca. They include the well-known genus *Schistosoma* of humans and a number of large mammals, particularly cattle, horses, and swine in warm parts of the world. *Heterobilharzia* and *Schistosomatium* are the only blood flukes of mammals in North America, being distributed in the southern and northern regions, respectively. The vertebrate hosts are small mammals.

### *Schistosomatium douthitti*
(Cort, 1915) (Plate 57)

Adults of these flukes occur naturally in the blood vessels of the liver, mesenteries, and small intestine of voles (*Microtus pennsylvanicus*) and muskrats (*Ondatra zibethica*) in the northern part of North America. Other small mammals such as deer mice (*Peromyscus*), house mice, rabbits, and others can be infected.

**Description.**    Mature males range from 1.9 to 6.3 mm in length. The forbeody is flattened for about the first two-fifths of the body length but the remainder is broad and forms a temporary gynecophoric groove by folding the sides ventrally over the female. There are 15 to 36 testes located at the anterior end of the hindbody. The genital pore opens near the margin of the left gynecophoric fold. The ceca of both sexes have many small diverticula and in each case some unite to form a bridge between the two ceca near their posterior union.

The females are slender and shorter than the males, being 1.1 to 5.3 mm long. The ovary is in the anterior half of the body with the uterus extending anteriorly from it as a straight tube to the level of the cecal bifurcation. The nonoperculate eggs are unembryonated when

**Plate 57** *Schistosomatium douthitti* 245

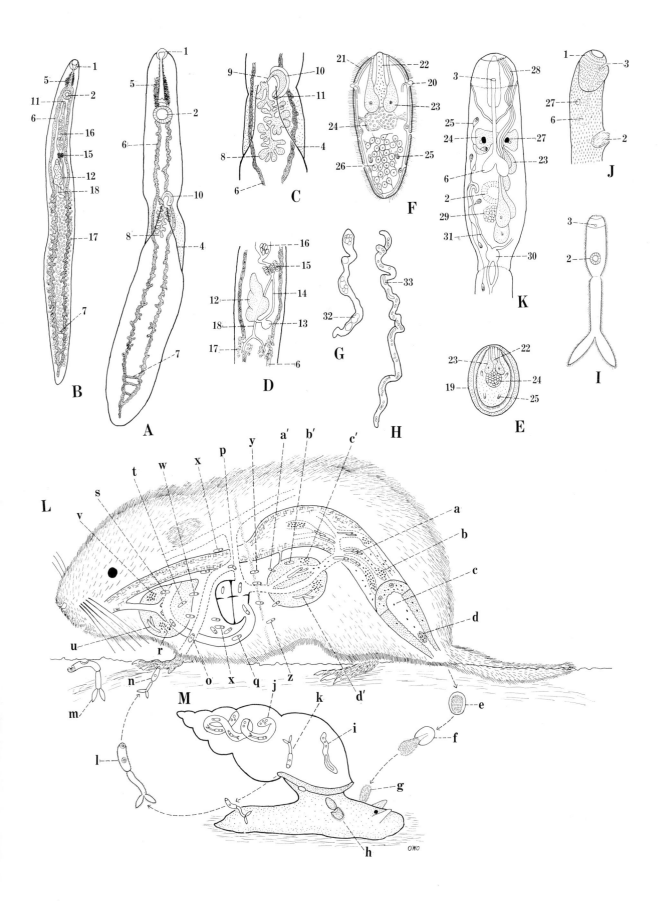

laid. The vitelline follicles extend from the ovary to the posterior extremity of the body, filling the intercecal space and extending laterally.

The cercariae are apharyngeate and brevifurcate and develop in sporocysts.

**Life Cycle.**    Mature flukes normally occur in the portal veins of the liver and intestine, including the cecum and mesenteric veins. Occasionally, they are in the lungs.

Mating presumably takes place while the females are in the gynecophoric groove of the males where they are held until ready to oviposit. At this time, they separate from the males, migrate deeper into the blood vessels, and deposit their unembryonated eggs which accumulate in masses. As the eggs penetrate the wall of the blood vessels and intestine, development proceeds, with embryonation being completed by the time they reach the lumen of the gut. They are mixed with the intestinal contents and eventually voided with the feces. Hatching occurs through a longitudinal slit soon after the eggs reach the water. The slit probably results from the combined pressure of absorbed water and the vigorous movements of the miracidium. Upon escaping, the miracidia swim swiftly and aimlessly through the water. Being short-lived, it is important that they find and enter a snail host soon lest they perish.

Natural snail intermediate hosts are species of Lymnaeidae, including *Stagnicola palustris, S. exilis, S. reflexa, S. emarginata, Lymnaea stagnalis,* and the Physidae *Physa parkeri* and *P. gyrina.*

Upon coming in contact with one of these snails, the miracidia burrow through the epidermis into the tissues. Inside, they shed the ciliated epidermal plates and transform into elongated mother sporocysts in the head and neck regions.

Germ cells within the body of the mother sporocysts give rise to germ balls which develop into daughter sporocysts, usually attached to the body wall until late in their development. By 21 days after infection of the snails, young daughter sporocyts have escaped from the mother sporocysts and reached the liver by day 36. Daughter sporocysts mature and produce cercariae in 8 to 18 days after reaching the liver, which is damaged extensively by their activities. As the sporocysts grow, the body cavity appears in which germ balls form. When these reach a given stage of development, they break apart

and each fragment develops into a cercaria. It is estimated that as many as 40,000 to 60,000 cercariae may be produced from a single miracidium. All cercariae developing from a single egg produce adults of the same sex. Females in unisexual infections produce eggs in which miracidia developed, presumably parthenogenetically, as there was no evidence of any female producing spermatozoa, or the presence of them in any of the genital ducts.

Newly formed cercariae possess six pairs of penetration glands but the anterior one disappears during migration through the snails. The cercariae emerge from the snails soon after dark. At first, they are somewhat quiescent. After a brief period of swimming, they rise and adhere to the surface film of the water by means of the ventral sucker. When swimming, the tail lashes vigorously, causing the body to rotate. Such movement is followed by a period of rest, as if the activity tired the cercaria. Movement on the bottom is in measuring-worm fashion.

Infection of voles takes place through the feet or other lightly haired or bare parts of the body exposed to infested water by wading or swimming animals. The cercariae penetrate directly through the skin or enter by way of canals of the hair shafts, dropping their tails in the process. Inflamed papillae arise at the point of entry within 5 to 10 minutes in white mice, reaching a diameter of 4 mm after an hour. They subside and disappear by day 3. Inside the definitive host, the cercariae migrate through the lymph and blood vessels to the lungs, going by way of the right side of the heart and pulmonary artery. Arrival in the lungs is as soon as 30 minutes after exposure of the host, indicating rapid transmission such as might occur in blood vessels.

The presence of the cercariae in the pulmonary blood vessels results in great damage to them and extensive congestion of the alveoli with blood and fluid. The young worms leave the blood vessels and migrate through the lung stroma for 3 or 4 days and then break into the thoracic cavity, where large numbers of them accumulate for a short time. They pass through the diaphragm, especially along the esophagus and postcaval vein, into the peritoneal cavity. A similar route is reported for *Schistosoma haematobium, S. mansoni,* and *S. japonicum,* all of humans. From here, they enter the liver by burrowing through the capsule from the outside.

By day 8 to 11 after infection, males begin to leave the liver and appear in the mesenteric veins. The females linger in the liver for a short time but soon follow the males. By day 13, nearly all worms have left the liver and are in the mesenteric veins. Sexual maturity may occur in the lungs as early as 10 days after infection but it is later in the liver and intestinal wall. By 22 to 23 days, eggs appear in the liver parenchyma, accumulating in large masses. Their presence results in much damage, producing lesions that undergo progressive development followed by regression, becoming completely fibrous by 243 days. Eggs appear in the intestinal mucosa and submucosa of mice 34 days after infection, occurring in masses in the villi. There is a general inflammatory response, with abscesses. Eggs in the intestinal wall work through it into the lumen and are voided with the feces.

Longevity of the worms varies, depending on the experimental hosts involved. In deer mice, males lived for 255 days and females for 468 to 484 days. In rats, females disappeared by 79 days, whereas a few males persisted for 136 days. This longevity is in contrast to the 28 years reported for *Schistosoma haematobium* in a human.

*Heterobilharzia americana* Price, 1929 is naturally parasitic in the raccoon, nutria, rabbit, opossum, and armadillo. There are 70 to 83 testes arranged in two irregular rows and the cuticle is covered with small tubercles. The lymnaeid snail *Fossaria cubensis* is the natural intermediate host.

Heavy infections of either species may be fatal to its host. In dogs, *Heterobilharzia americana* causes emaciation, intermittent diarrhea, eczema, loss of appetite, and terminal coma accompanied by bloody diarrhea.

## EXERCISE ON LIFE CYCLE

*Schistosomatium douthitti* is an ideal blood fluke for experimental studies because both the snail and mammalian hosts are readily available and easily maintained in the laboratory.

Snail hosts of choice are *Stagnicola palustris* and *Lymnaea stagnalis*. When collected from around muskrat houses in ponds, there is a likelihood of finding naturally infected individuals to supply cercariae for experimental studies. They may be reared successfully in balanced aquaria of 5- to 10-gallon capacity. White enamel trays and finger bowls are useful for small numbers of snails. When kept at room temperature, the snails lay eggs from which young may be hatched for experimental infections. Aquarium plants should be provided to furnish places for the snails to attach their eggs. Tap water should be freed of chlorine by aeration in large containers before using it.

Food may be supplied as fresh green leaves of lettuce but care should be taken to remove decomposing pieces. Dried maple leaves soaked a few weeks in several changes of water before adding to the aquaria are excellent food. Vitamin, a commercial product, is very good. Powdered calcium carbonate or pieces of blackboard chalk should be added to provide calcium for maintenance of the shells.

Fluke eggs in the feces of infected mice hatch quickly in water, providing a supply of miracidia for infection experiments. Young snails should be exposed for about 24 hours to a few miracidia in shallow, wide-mouthed bottles containing a small volume of water so that they can be watched under a dissecting microscope. The incubation period in snails ranges from 37 to 52 days.

Mice are good experimental definitive hosts. Exposure may be by placing mice in small jars of shallow water containing cercariae or by placing them on a clipped spot of a restrained animal. Cercariae may be transferred from the surface of the water in the container by means of a hair loop to the clipped spot. Infection is accomplished in about 30 minutes.

Migration through the host begins promptly after entering the skin. They are in the lungs in less than 1 hour after infection, the thoracic cavity in 2 to 4 days, and the liver in 8 to 10 days. By day 11, they begin to leave the liver and migrate to the mesenteric veins, where the majority of them are by the end of 2 weeks. Sexual maturity occurs in the lungs 10 to 12 days after infection and in the liver by 16 days. Nearly all worms are mature by 20 days. These times should be confirmed by examining infected mice at appropriate periods after infection.

## SELECTED REFERENCES

Batten, P. H., Jr. (1956). The histopathology of swimmers' itch. I. The skin lesions of *Schistosomatium douthitti* and *Gigantobilharzia huronensis* in the unsensitized mouse. Amer. J. Pathol. 32:363–377.

Batten, P. H., Jr. (1957). Histopathology of swimmers' itch. II. Some observations on the pulmonary lesions in the unsensitized mouse. Amer. J. Pathol. 33:729–735.

Cort, W. W. (1914). Some North American larval trematodes. Ill. Biol. Monogr. 1:451–532.

Cort, W. W. (1936). Studies on schistosome dermatitis. I. Present status of the subject. Amer. J. Hyg. 32:349–371.

Cort, W. W., D. J. Ameel, and L. Olivier. (1944). An experimental study on the development of Schistosomatium douthitti (Cort, 1914) in its intermediate host. J. Parasitol. 30:1–17.

Cort, W. W., and S. B. Talbot. (1936). Studies on schistosome dermatitis. III. Observations on the behavior of the dermatitis-producing schistosome cercariae. Amer. J. Hyg. 23:385–396.

Cort, W. W., D. J. Ameel, and A. Van der Woude. (1953). Further studies on the early development of the daughter sporocysts of Schistosomatium douthitti. Proc. Helminthol. Soc. Wash. 20:43–49.

El-Gindy, M. S. (1951). Post-cercarial development of Schistosomatium douthitti (Cort, 1914) Price 1931 in mice, with special reference to the genital system (Schistosomatidae-Trematoda). J. Morphol. 89:151–185.

Farley, J. (1962). The effect of temperature and pH on the longevity of Schistosomatium douthitti miracidia. Can. J. Zool. 40:615–620.

Kagan, I. G., and D. R. Merange. (1958). The histopathology of experimental infections in mice with Schistosomatium douthitti. Amer. J. Trop. Med. Hyg. 7:285–294.

Kagan, I. G., R. B. Short, and M. M. Nez. (1954). Maintenance of Schistosomatium douthitti (Cort, 1914) in the laboratory (Trematoda: Schistosomatidae). J. Parasitol. 40:424–439.

Malek, E. A. (1966). The epizootology of Schistosomatium douthitti and Heterobilharzia americana the two mammalian schistosomes in the United States. Proc. 1st Int. Congr. Parasitol. Rome 2:741–742.

Penner, L. D. (1939). Experimental studies on Schistosomatium douthitti (Cort) in mouse, rat, muskrat, guinea pig, and snow-shoe hare. J. Parasitol. 25(Suppl.):8–9.

Price, E. W. (1929). A synopsis of the trematode family Schistosomatidae with descriptions of new genera and species. Proc. U.S. Nat. Mus. 75(18):1–39.

Price, H. F. (1931). Life history of Schistosomatium douthitti (Cort). Amer. J. Hyg. 13:685–727.

Short, R. B. (1951). Hermaphroditic female Schistosomatium douthitti (Trematoda: Schistosomatidae). J. Parasitol. 37:547–555.

Short, R. B. (1952). Sex studies on Schistosomatium douthitti (Cort, 1914) Price, 1931 (Trematoda: Schistosomatidae). Amer. Midl. Natur. 47:1–54.

Short, R. B. (1952). Uniparental miracidia of Schistosomatium douthitti (Trematoda: Schistosomatidae). Amer. Midl. Natur. 48:55–68.

Tanabe, B. (1923). The life history of a new schistosome Schistosomatium pathlocopticum Tanabe, found in experimentally infected mice. J. Parasitol. 9:183–198.

### Subfamily Bilharziellinae

The males and females are similar in shape. The male lacks a gynecophoric groove. The cecal branches unite anterior to the equator of the body and the testes are caudad from the intestinal union. Common genera of the subfamily include *Trichobilharzia* of ducks, *Bilharziella* of ducks and herons, *Dendrobilharzia* of ducks and pelicans, and *Gigantobilharzia* of gulls. *Trichobilharzia cameroni* is discussed as a representative of this subfamily.

### *Trichobilharzia cameroni* Wu, 1953 (Plate 58)

This is a parasite of wild ducks in North America. The adults are in the peripheral blood vessels of the intestine and occasionally in the liver. The cercariae cause swimmer's itch in people who come in contact with them.

**Description.** Males measure 3.18 to 5.71 mm long. The gynecophoral fold is well developed and thickly set with spines. The acetabulum is larger than the oral sucker. There are 80 to 110 testes arranged in a single row, beginning immediately behind the gynecophoral fold and extending to near the tip of the intestinal cecum.

Females are 3.83 to 4.94 mm long. The ceca unite at about the equator of the seminal vesicle. The ovary is tubular, stout, and makes several loops; it measures 287 to 358 $\mu$ long. There is never more than a single uterine egg present. The vitellaria extend from just behind the seminal receptacle to the posterior end of the body. Eggs are spindle-shaped, 147 to 212 $\mu$ long by 57 to 73 $\mu$ wide, and embryonated when laid.

**Life Cycle.** Adult flukes are in the veins of the small and large intestine of ducks. When ready to oviposit, the female leaves the gynecophoric groove of the male where she normally rests, moves into the smaller venules, and deposits the spindle-shaped eggs. The eggs gradually work through the wall of the blood vessels and intestine into the lumen of the latter and pass in the droppings in a fully developed stage. Some eggs are carried to the liver via the hepatic portal vein. With dilution of the feces, the eggs begin to absorb water, increasing in size until they rupture after 2 to 3 hours of exposure, releasing the miracidia which swim rapidly. They have a life span of 2 to 13 hours.

Upon coming in contact with specimens of the snail *Physa gyrina* soon after hatching, the miracidia attach to them and quickly burrow into the soft parts of the body. Ability to enter the snails gradually decreases 2 to 3 hours after hatching. Upon entering the tissues, the miracidia soon shed the ciliated epithelium and transform to mother sporocysts near the point of entry into the mantle, foot, or viscera. They migrate to the liver and grow into greatly elongated tubular mother sporocysts as soon as 1 week after infection, at which time daughter sporocysts appear in them. They contain motile daughter sporocysts, germ balls in various stages of development, and single germ cells. There is no birth pore.

Daughter sporocysts are in the liver. They differ from the mother sporocysts in being smaller and in having a terminal birth pore that is surrounded by spines during the early part of their life. Sporocysts mature and produce cercariae in about 21 days, which is 28 to 32 days after infection of the snails.

Cercariae with eyes but lacking a pharynx escape from snail hosts into the water where they alternately swim and rest. Being positively phototropic, they are attracted toward the light, a factor that brings them to the surface of the water in a favorable location to make contact with the vertebrate definitive hosts.

As ducks swim about, the active cercariae attach to their feet and, with the aid of five pairs of penetration glands, burrow through the skin into blood vessels, dropping their tails as they enter. The larval flukes are carried with the blood to the right side of the heart, through the pulmonary artery to the lungs, and back to the left side of the heart via the pulmonary vein. From here, they go out through the dorsal aorta to the various parts of the body. Those entering the mesenteric arteries reach the walls of the

intestine where development to sexual maturity takes place in 12 to 14 days. Infections in ducks persist for at least 4 months, which suggests that the flukes may be carried great distances by migrating birds.

The life cycles of a number of other species of *Trichobilharzia* of ducks whose cercariae cause dermatitis in humans have been studied. They are basically the same as that of *T. cameroni*. They include *T. oregonensis* in Oregon by Macy *et al.*, *T. adamsi* in British Columbia by Edwards and Jansch, *T. physellae*, *T. ocellata*, and *T. stagnicolae* by McMullen and Beaver in Michigan, and *T. physellae* in Colorado by Hunter.

## EXERCISE ON LIFE CYCLE

*Physa gyrina* in ponds frequented by wild ducks may be infected with *T. cameroni*. Other species of *Physa* and *Stagnicola* may harbor different species of related blood flukes of ducks or other water fowl.

Cercariae obtained from naturally infected snails and determined to be those of *Trichobilharzia* may be used for experimental infections. Canaries, pigeons, and domestic ducklings are favorable hosts, especially the last species. Submerge the feet of the experimental birds in water containing a known number of freshly emerged cercariae. Determine the number that penetrated the feet by the remainder in the container at the conclusion of the exposure.

Ascertain the prepatent period of the flukes by observing the time when eggs appear in the feces. Some birds should be sacrificed at appropriate intervals to demonstrate the larval flukes in their lungs and the route of migration from the lungs to the mesenteric veins of the intestine. This point is not clear.

Snails to be used for experimental infections should be collected from areas where ducks do not occur, or reared for experimental infection at the time eggs appear in the feces of the birds that were exposed to cercariae.

When eggs appear in the feces, they should be washed free from them for hatching. Expose experimental snails of the same species from which the cercariae were obtained originally. Individuals should be dissected at intervals to determine the course of development of the mother sporocysts, daughter sporocysts, and cercariae.

## Explanation of Plate 58

A, Anterior end of adult female fluke. B, Anterior end of adult male. C, Hatched miracidium. D, Mature mother sporocyst. E, Mature daughter sporocyst. F, Cercaria showing certain details of internal morphology. G, Duck definitive host. H, Snail (*Physa gyrina*) first intermediate host. I, Cercaria attacking human host, causing dermatitis.

1, Oral sucker; 2, ventral sucker; 3, esophagus; 4, bifurcated portion of intestine; 5, single posterior limb of intestine; 6, ovary; 7, seminal receptacle; 8, oviduct; 9, Mehlis' gland; 10, uterus; 11, vulva; 12, egg; 13, vitelline gland; 14, vitelline duct; 15, testes; 16, vas deferens; 17, seminal receptacle; 18, cirrus pouch; 19, apical papilla; 20, apical gland; 21, ganglion; 22, penetration glands; 23, duct of penetration gland; 24, flame cell; 25, germinal mass; 26, daughter sporocyst; 27, birth pore; 28, developing cercaria; 29, tail; 30, eyespot.

a, Adult flukes in blood vessels of intestinal wall; b, eggs laid in blood vessels; c, eggs penetrate tissues, passing from blood vessels into lumen of gut; d, eggs mixed with feces; e, embryonated eggs pass out of body with feces; f, egg absorbs water; g, egg ruptures, releasing miracidium; h, miracidium penetrates snail; i, mature mother sporocyst with developing daughter sporocysts; j, mature daughter sporocyst with developing and mature cercariae; k, mature cercaria free in water; l, cercariae attack humans, causing cercarial dermatitis, but they do not develop to maturity; m, cercaria penetrates foot of duck, dropping tail in the process; n, cercaria reaches blood vessels and is carried toward the heart; o, cercaria approaching heart; p, cercaria passing through right side of heart; q, cercaria passes through blood vessel of lungs; r, cercaria having passed through left side of heart approaches dorsal aorta; s, cercaria in dorsal aorta; t, cercaria passing through anterior mesenteric artery to enter blood vessels of intestinal wall where maturity is attained; u, worms occur in liver on occasion.

Adapted from Wu, 1953, Can. J. Zool. 31:351.

---

## SELECTED REFERENCES

*Cort, W. W. (1936). Amer. J. Hyg. 23:349 (I); Talbot, S. B. (1936). Ibid. 23:372 (II); Cort and Talbot. (1936). Ibid. 23:385 (III); Cort. (1936). Ibid. 24:318 (IV); Brackett, S. (1940). Ibid. 31:49 (V); Brackett. (1940). Ibid. 31:64 (VI); Cort, D. B. McMullen, L. Olivier, and Brackett. (1940). Ibid. 32:33 (VII); Brackett. (1940). Ibid. 32:85 (VIII); McMullen and P. C. Beaver. (1945). Ibid. 42:128 (IX).

Edwards, D. K., and M. E. Jansch. (1955). Two new species of dermatitis producing schistosome cercariae from Cultus Lake, British Columbia. Can. J. Zool. 33:182–194.

Macy, R. W., D. J. Moore, and W. S. Price, Jr. (1955). Studies on dermatitis-producing schistosomes in the Pacific Northwest, with special reference to *Trichobilharzia oregonensis*. Trans. Amer. Microsc. Soc. 74:235–251.

Neuhaus, W. (1952). Biologie und Entwicklung von *Trichobilharzia szidati* n. sp. (Trematoda, Schistosomatidae), einen Erreger von Dermatitis beim Menschen. Z. Parasitenk. 15:203–266.

Wu, L.-Y. (1953). A study of the life history of *Trichobilharzia cameroni* sp. nov. (Family Schistosomatidae). Can. J. Zool. 31:351–371.

---

* This is a series of papers by Professor Cort and his students on the biology of some species of North American blood flukes whose cercariae produce dermatitis.

## FAMILY SPIRORCHIIDAE

The Spirorchiidae are monoecious blood flukes of turtles with the adults in many parts of the body. They are lanceolate in shape, with an oral sucker, but the acetabulum may or may not be present. The esophagus is surrounded by gland cells and the two ceca reach almost to the posterior end of the body. There are one to many testes located intercecally; they may be either pre- or postovarian.

### Spirorchis parvus
(Stunkard, 1932) (Plate 59)

Mature worms occur in the brain, spinal cord, gut wall, spleen, lungs, heart, free in the tissues, and inside arterioles but not in veins of the painted turtle in the central and eastern parts of the United States and possibly throughout its entire range.

**Description.**    Adult flukes are 1.07 to 2.04 mm long, very thin, transparent, and covered, at least along the margins, with sensory papillae. Needle-like spines cover the anterior end for the length of the oral sucker. There is no pharynx. The esophagus is surrounded by gland cells. There are four to five lobed testes lying in series between the ceca, beginning near the middle of the body and extending caudad. The ovary lies

**Plate 58**   *Trichobilharzia cameroni*                                                                 251

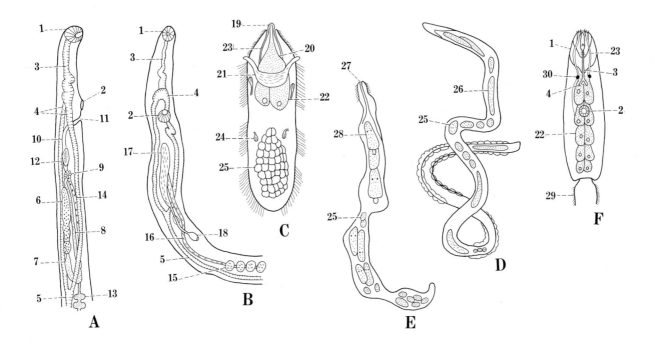

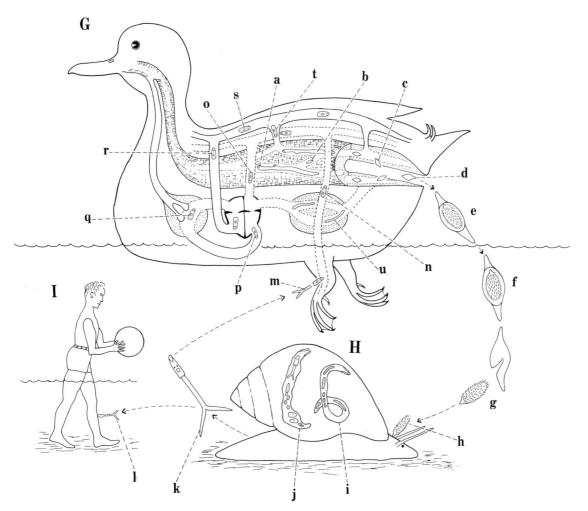

## Explanation of Plate 59

A, Adult fluke. B, Embryonated egg. C, Miracidium. D, Mature mother sporocyst. E, Daughter sporocyst with developing cercariae. F, Cercaria. G, Turtle (*Chrysemys picta*) definitive host. H, Snail (*Helisoma trivolvus, H. campanulata*) intermediate host.

1, Oral sucker; 2, esophagus, showing glandular cells; 3, intestine; 4, vitelline glands; 5, testes; 6, seminal vesicle; 7, common genital pore; 8, ovary; 9, oviduct; 10, seminal receptacle; 11, excretory bladder; 12, excretory pore; 13, apical gland; 14, penetration gland; 15, brain; 16, flame cell; 17, germ cells; 18, egg shell; 19, eyespot; 20, apical papilla; 21, lateral papilla; 22, excretory duct; 23, developing daughter sporocysts; 24, immature cercaria; 25, dorsal crest; 26, ventral sucker; 27, genital primordium; 28, tail; 29, furca.

a, Adult worm in arterioles of intestinal wall; b, eggs laid in blood vessels pass through tissues into lumen of intestine; c, embryonated eggs in intestinal contents; d, eggs passed in feces; e, eggs hatch in water; f, miracidium penetrates snail host; g, mother sporocyst with developing daughter sporocyst; h, daughter sporocyst with cercariae; i, furcocercous cercaria escaping from snail; j, cercaria free in water; k, cercaria penetrating soft tissues of turtle, dropping its tail in the process; l, tailless cercaria in blood vessels; m, cercaria in precaval vein; n, cercaria entering right auricle of heart and passing through ventricle into pulmonary artery; o, cercaria in pulmonary artery; p, cercaria having passed through lungs enters left auricle of heart; q, cercaria entering dorsal aorta; r, cercaria passing through mesenteric artery into intestinal wall; s, cercaria in arteriole where it grows to maturity in about 3 months at summer temperature and passes eggs in feces about 2 weeks later.

Adapted from Wall, 1941, Trans. Amer. Microsc. Soc. 60:221.

---

at the posterior end of the seminal vesicle. Vitelline glands surround the ceca the entire length, filling the space between them except for that occupied by the gonads. The uterus is very short and usually devoid of eggs which are 54 by 38 $\mu$ and nonoperculate. The genital pore is at the level of the posterior margin of the ovary.

**Life Cycle.** Colorless adults in the arterioles of the muscular layer of the pyloric stomach and small intestine deposit their unembryonated eggs. In cases of heavy infection, egg masses form a solid coating over large sections of the stomach and intestine of small turtles, often causing death. Two weeks are required for the eggs to work through the intestinal wall into the lumen. When voided with the feces into water, they swell considerably, at which time the miracidia become very active. Hatching occurs in 4 to 6 days between the hours of 2 A.M. and 5 A.M. through a slit in the side of the egg, and the miracidium swims away. Survival in the water is up to 15 hours at summer temperature.

Miracidia begin to attack ram's horn snails (*Helisoma trivolvus, H. campanulata*) 30 to 60 minutes after hatching and require 10 to 60 minutes to penetrate. Only young snails become infected. The miracidia transform into mother sporocysts in the mantle of the snail. Development is completed in 18 days, at which time daughter sporocysts have developed and are ready to leave. They escape from the anterior end of the mother sporocyst and migrate through the lymph spaces of the mantle and body into the digestive gland, at which time the cercariae begin to develop. Upon reaching maturity in an additional 7 to 75 days, the cercariae, which are large, apharyngeate, distomate, and bear a longitudinal fin over the dorsal side of the body, leave the snails intermittently. Being positively phototropic, they congregate at the surface of the water where they swim and rest.

Upon coming in contact with turtles, especially young ones, the cercariae penetrate the soft membranes around the anus, eyes, between the toes, in the flanks, and in the nostrils and mouth, leaving the tail on the outside. The young flukes enter the blood vessels and are carried to the right side of the heart, through the lungs and left side of the heart, and into the general circulation, going to many parts of the body. They develop to sexual maturity in 3 months during the summer. Eggs from those in the arterioles of the stomach and intestine appear in the feces within two weeks after attaining maturity.

Other species include *S. elephantis* (Cort, 1917) and *S. scripta* Stunkard, 1923. On the basis of life history studies using pure lines established by single miracidial infections from *Menetus dilatatus* and young painted turtles, *S. elegans* and *S. picta* are considered as synonyms of *S. scripta*.

*S. elephantis* has a life cycle similar to that of *S. parvus*. *Spirorchis scripta* matures in 66 days and eggs appear in the feces by day 107. The

**Plate 59**    *Spirorchis parvus*                                                          253

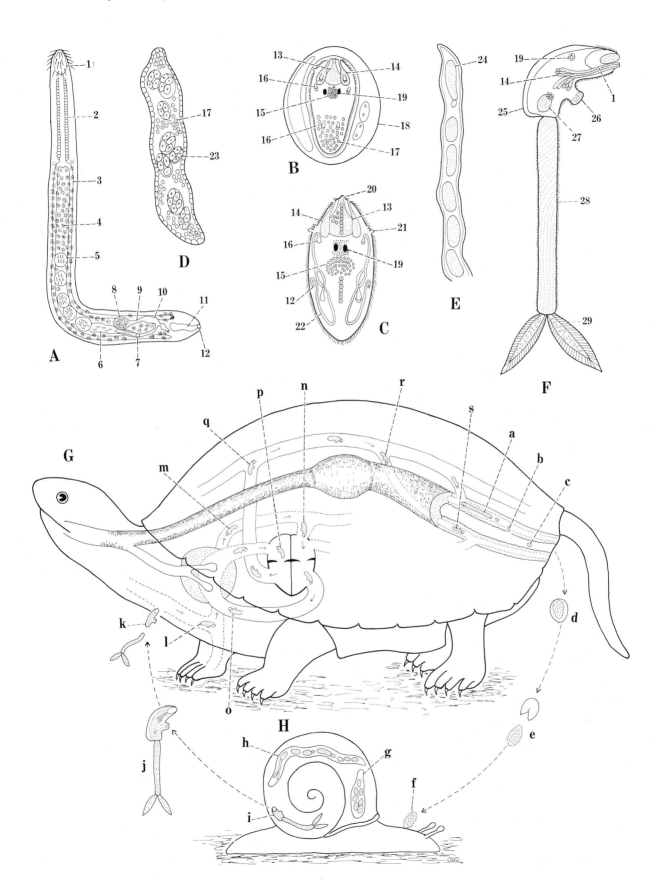

worms are in arteries, veins, heart, esophagus, pericardial cavity, abdominal cavity, and connective tissue, having a wider distribution in the definitive host than any other spirorchiid.

When infected turtles are kept in a temperature of 24° to 27°C, the worms produce more eggs than at 21°C, indicating that development of eggs and rate of oviposition are influenced by the temperature.

The parasites are detrimental to their hosts. Snails (*Menetus dilatatus*) infected with miracidia show reduction in the number of eggs laid by day 3 and have discontinued laying by the end of 1 week. Heavy infections of worms with the resulting masses of eggs in the tissues cause varying degrees of debilitation and even death of turtles. Severe neurological effects are manifested by paralysis of one side of the body as a result of damage to the brain.

## EXERCISE ON LIFE CYCLE

Eggs are obtained easiest from infected turtles by homogenizing the lungs and stomach separately in a food blender and screening the homogenate through cheesecloth to remove debris. Eggs passed through the cloth are collected in finger bowls, washed, and hatched in tap water. Expose clean, young snails to the miracidia. Infected snails must be dissected at intervals of 2 to 4 days to follow the development of the intramolluscan stages.

Large, brevifurcate, nonpharyngeate cercariae with a longitudinal fin over the dorsal surface of the body obtained from naturally infected *Helisoma trivolvus, H. campanulatum,* and *Menetus dilatatus* in ponds with infected painted turtles are those of *Spirorchis.*

Cercariae obtained from naturally infected snails and identified as those of *Spirochis* should be placed in a small aquarium with young painted turtles. Observe the penetration of the cercariae into the turtles and their reaction to the attack. By making dissections, verify that the larval flukes go through the lungs a few days after entrance into the turtles. Determine the migratory route from the lungs to the stomach wall.

## SELECTED REFERENCES

Cort, W. W., D. J. Ameel, and A. Van der Woude. (1954). Germinal development in sporocysts of the blood flukes of turtles. Proc. Helminthol. Soc. Wash. 21:85–96.

Fried, B., and W. Fee, Jr. (1964). Technique for obtaining eggs of *Spirorchis* (Trematoda) from painted turtles. J. Parasitol. 50(Suppl.):28.

Goodchild, C. G., and E. S. Dennis. (1967). Comparative egg counts and histopathology in turtles infected with *Spirorchis* (Trematoda: Spirorchiidae). J. Parasitol. 53:38–45.

Goodchild, C. G., and V. L. Martin. (1969). Speciation in *Spirorchis* (Trematoda: Spirorchiidae) infecting the painted turtle *Chrysemys picta.* J. Parasitol. 55:1169–1173.

Holliman, R. B., and J. E. Fisher. (1968). Life cycle and pathology of *Spirorchis scripta* Stunkard, 1923 (Digenea: Spirorchiidae) in *Chrysemys picta picta.* J. Parasitol. 54:310–318.

Holliman, R. B., and J. E. Fisher. (1971). Studies on *Spirorchis parvus* (Stunkard, 1923) and its pathological effects on *Chrysemys picta picta.* J. Parasitol. 57:71–77.

Hosier, D. W., and C. G. Goodchild (1967). Suppressed egg-laying by snails infected with *Spirorchis scripta* (Trematoda: Spirorchiidae). J. Parasitol. 56:302–304.

Wall, L. D. (1941). *Spirorchis parvus* (Stunkard), its life history and the development of its excretory system (Trematoda: Spirorchiidae). Trans. Amer. Microsc. Soc. 70:173–184.

Wall, L. D. (1941). Life history of *Spirorchis elephantis* (Cort, 1917), a new blood fluke from *Chrysemys picta.* Amer. Midl. Natur. 25:402–412.

## FAMILY SANGUINICOLIDAE

This is a group of monoecious digenetic flukes that lives in the blood vessels of fish. The family Aporocotylidae is now considered as a synonym of Sanguinicolidae. Species of the genus *Cardicola* have a single large testis, whereas those of *Sanguinicola* have several arranged in two regular rows. Worms are in the heart and efferent arteries of the gill arches. They cause serious damage to the gills, often resulting in death.

### *Cardicola davisi* (Wales, 1958) (Plate 60)

This species was described as *Sanguinicola davisi* from the gill arteries of trout on the Pacific coast of the United States. They occur in epizootic numbers in hatcheries, causing severe losses through death.

**Description.** Fully developed individuals are flattened and spindle-shaped flukes, having a length of 8.5 mm and a width of 0.21 mm. There is no oral sucker, acetabulum, or pharynx. The mouth is subterminal and leads into a long, slender esophagus that ends in an X-shaped cecum with four short, rounded lobes. The single testis is large, irregular in shape, and approximately equatorial in position; the sperm duct leads to a crook-like thick-walled cirrus sac. The bilobed ovary located at the posterior margin of the testis opens by means of a short oviduct into a gourd-shaped ootype. Genital pores open separately near the posterior end of the body with the female pore being anterior to the male pore. Yolk glands fill the entire body. Eggs are oval, nonoperculate, and measure 63 by 35 $\mu$.

**Life Cycle.** The adult flukes normally reside in the main gill arteries, lying parallel with the gill cartilages. One egg is produced at a time and carried by the blood into the capillaries of the gill filaments. Here development and hatching takes place. The active miracidia work through the epithelium to the surface of the gill filaments where a lobule forms. Upon rupture of the lobule, the ciliated miracidia are liberated and swim away. They develop in the snail *Oxytrema* (*Goniobasis*) *circumlineata* and possibly other species of the genus. Infected snails contain numerous sporocysts and rediae. While experimental data are lacking on the part of the life cycle in the snails, it is presumed that the miracidia transform into mother sporocysts, followed by daughter sporocysts and cercariae. Rediae have not been found in the species of this genus. The cercariae are brevifurcate. Like the adult flukes, they lack suckers and a pharynx and have a cecum closely resembling that of the mature worm. Upon coming in contact with fingerling trout, they soon penetrate the tissue of the fins, dropping their tails. The cercariae are active in the blood vessels of the fins and presumably migrate through the veins to the heart and from there to the gill arteries. No information is available on the movements in the host or on the prepatent period.

*Cardicola klamathensis* (Wales, 1958) is a closely related but much smaller species, occurring in trout of the same general area with *C. davisi*. Immature flukes occur in all parts of the circulatory system but adults are only in the efferent renal veins. The larval stages develop in the snail *Flumnicola seminalis*. The eggs are spherical and the cercariae have a longitudinal fin over the dorsal side of the body and the furcae have arrowhead-shaped tips. The flame cell pattern is 2(2 + 2).

*Sanguinicola inermis* Plehn is common in carp in Europe but has not been reported from North America. The worms vary in location in the carp according to the reproductive state of the fish. Before the fish begin to spawn, the flukes are located in the heart, but when egg-laying begins they move into the efferent arteries of the gill arches which become filled with eggs. Blockage of the arteries by worms, their eggs, and the proliferation of endothelial cells due to acute sanguinicoliasis results in the death of many fish, primarily by suffocation.

The tricornered eggs hatch in the capillaries of the gills and the miracidia eventually escape into the water, where they penetrate *Lymnaea stagnalis* and *L. auricularia*, the two snail hosts. The cercariae of this species and those of *Cardicola klamathensis* are similar in that both have a longitudinal, dorsal fin over the body.

*Sanguinicola huronis* of black bass and *S. occidentalis* of walleyed pike occur in North America, but their life cycles are not known.

## EXERCISE ON LIFE CYCLE

When infections of sanguinicolid flukes are found, a study of the life cycles should be pursued to fill the present gaps in our knowledge concerning them.

Since the morphology of the cercariae is so characteristic, they can be identified with accuracy. They may be obtained from naturally infected snails in the area where the parasitized fish occur. When the species of snails are recognized, the experiments may be started by exposing them to miracidia obtained from infected fish. Also fish may be exposed to cercariae obtained from naturally infected snails.

By careful planning, the entire life cycle can be completed. Ascertain the intramolluscan development, including the stages that occur in the cycle. Likewise, follow the development in the fish host.

## SELECTED REFERENCES

Fischthal, J. H. (1949). *Sanguinicola huronis* n. sp. (Trematoda: Sanguinicolidae) from the blood system of the largemouth and smallmouth basses. J. Parasitol. 35:566–568.

Land, J. van der. (1967). A new blood fluke (Trematoda) from *Chimaera monstrosa* L. Proc. Kon. Ned. Akad. Wetensch. Ser. C 70:110–120.

## Explanation of Plate 60

A, Adult fluke. B, Developing embryo in egg. C, Egg containing miracidium in advanced stage of development. D, Free-swimming miracidium. E, Fully developed cercaria. F, Gill filament of rainbow trout with adult fluke and developing eggs in it. G, Rainbow trout, definitive host. H, Snail [*Oxytrema* (*Goniobasis*) *circumlineata*] intermediate host.

1, Spined anterior end; 2, mouth; 3, esophagus; 4, intestine; 5, longitudinal rows of spines; 6, vitelline glands which fill entire body; 7, testis; 8, cirrus pouch; 9, male genital pore; 10, ovary; 11, ootype; 12, single egg in ootype; 13, female genital pore; 14, membrane of developing miracidium; 15, developing eyespot of miracidium; 16, cilia of miracidium; 17, refringent bar; 18, eyespot; 19, tail; 20, furcae; 21, spherical bodies at end of furcae; 22, gill filament of rainbow trout; 23, cartilaginous bar of gill filament; 24, afferent artery of gill filament; 25, egg developing in gill filament; 26, adult fluke in afferent artery of gill.

a, Adult fluke in afferent gill artery of rainbow trout; b, gill filament; c, efferent gill artery; d, dorsal aorta; e, developing eggs in tissue of gill filament; f, miracidium that has escaped from gill filament into gill chamber; g, free-swimming miracidium; h, miracidium penetrating snail intermediate host; i, mother sporocyst with germ balls; j, sporocyst with redia and germ balls; k, redia with furcocercous cercariae in various stages of development; l, free-swimming cercaria; m, cercaria penetrating pectoral fin has dropped tail; n, cercaria being carried in blood through brachial vein toward heart; o, cercaria in auricle; p, cercaria entering conus arteriosis from ventricle; q, cercaria entering afferent artery of gill where it matures and lays eggs.

Adapted from Wales, 1958, Calif. Fish Game 44:125.

---

Lucký, Z. (1964). Contribution to the biology and pathogenicity of *Sanguinicola inermis* in juvenile carp. In: R. Ergens and B. Rysăvý, eds. Parasitic Worms and Aquatic Conditions. Proc. Symp. Prague, Czechoslovak Acad. Sci. pp. 153–156.

Meade, T. G. (1967). Life history studies on *Cardicola klamathensis* (Wales, 1958) Meade and Pratt, 1965 (Trematoda: Sanguinicolidae). Proc. Helminthol. Soc. Wash. 34:210–212.

Scheuring, L. (1922). Der Lebenszyklus von *Sanguinicola inermis* Plehn. Zool. Jahrb. 44:265–310.

Van Cleave, H. J., and J. F. Mueller. (1934). Parasites of Oneida Lake fishes. Part III. A biological and ecological survey of the worm parasites. Bull. Wildl. Annu. 3:161–334.

Wales, J. H. (1958). Two new blood flukes of trout. Calif. Fish Game. 44:125–136.

## Suborder Brachylaemata

This suborder contains the families Brachylaemidae, Fellodistomatidae, and Bucephalidae whose adult stages are vastly different in appearance. The relationship of these flukes is shown by the distomate cercariae which develop in sporocysts and have a forked tail of moderate or small size, or no tail. The protonephridia (flame cells in the early stage of development of the cercariae) are stenostomate (with a narrow opening). Main excretory vessel extends anteriorly to level of oral sucker or pharynx, reverses its direction, and receives anterior and posterior collecting tubules near middle.

## Superfamily Brachylaemoidea

Members of the Brachylaemoidea are parasites of the cloaca of birds and cecum of mammals. They are usually elongate but sometimes oval in shape. The genital pore is located at the front margin of the anterior testis or caudad from it. The cercariae 1) develop in branching sporocysts in aquatic or terrestrial snails, 2) have a functional or rudimentary tail, or none at all, and 3) possess a V-shaped excretory bladder with long arms.

### FAMILY BRACHYLAEMIDAE

The flukes of this family are usually elongate and have a well-developed oral and ventral sucker and pharynx, but a very short esophagus. The ceca extend to the posterior end of the body and the genital pore is in the posterior half of the body. The main tubules of the short excretory bladder extend to near the anterior end of the body and loop back to near the middle before dividing.

### *Postharmostomum helicis*
(Leidy, 1847) (Plate 61)

*Postharmostomum helicis* is a natural parasite of the cecum of the white-footed mouse (*Peromyscus leucopus*) in the United States.

**Description.**   The oblong body is 2.8 mm long by 1.4 mm wide, with the oral sucker somewhat larger (410 to 500 $\mu$) than the acetabulum.

**Plate 60**    *Cardicola davisi*                                                                257

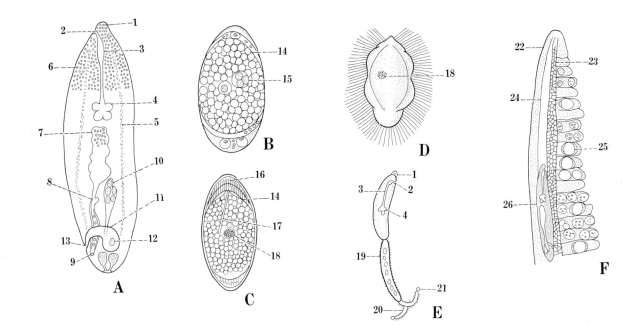

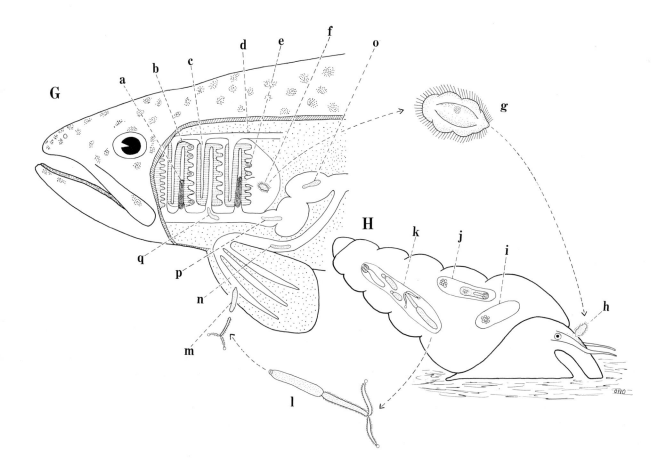

## Explanation of Plate 61

**A,** Adult fluke. **B,** Embryonated egg. **C,** Miracidium. **D,** Branched daughter sporocyst. **E,** Cercaria. **F,** Metacercaria. **G,** Cross-section of intestine showing cilia. **H,** Definitive host (*Peromyscus*). **I,** Snail first intermediate host (*Anguispira alternata*). **J,** Snail second intermediate host (*Anguispira alternata*).

1, Oral sucker; 2, ventral sucker; 3, pharynx; 4, esophagus; 5, cecum; 6, testes; 7, seminal vesicle; 8, common genital pore; 9, ovary; 10, oviduct; 11, ascending arm of uterus; 12, descending portion of uterus; 13, vitelline gland; 14, vitelline duct; 15, opening of Laurer's canal; 16, excretory bladder; 17, arm of excretory bladder; 18, anterior portion of body; 19, posterior portion of body; 20, spine; 21, glandular structure; 22, band of cilia; 23, germ cell; 24, cercaria; 25, germinal mass; 26, penetration glands; 27, tubules of penetration glands; 28, flame cell; 29, germinal primordium; 30, vitelline membrane; 31, papillae of suckers; 32, tail; 33, intestinal wall; 34, cilia.

a, Adult worm in cecum; b, embryonated eggs laid in cecum; c, eggs voided in feces; d, fully developed egg in feces on ground; e, eggs eaten by land snails; f, eggs hatch in digestive tract, where miracidium is free; g, miracidium passes through intestinal wall; h, mother sporocyst; i, young daughter sporocyst inside; j, mature daughter sporocyst containing cercariae; k, cercaria escaping from snail first intermediate host; l, mucus trail left by traveling snail; m, cercaria in mucus trail; n, cercaria entering pore of primary ureter; o, cercaria moving up primary ureter; p, kidney; q, cercaria passing through renopericardial canal into pericardial cavity; r, cercaria remain unencysted in pericardial cavity; s, heart; t, mice become infected by removing the snails from their shells and eating them; u, cercaria freed from pericardial cavity of snail by digestive juices; v, cercaria free in intestine; w, cercaria entering cecum where it attaches to the mucosa and grows to maturity in 8 days, when eggs first appear in the uterus; they appear in the feces of mice on days 19 to 20 after infection. Adult flukes live longer than 140 days.

Adapted from Ulmer, 1951, Trans. Amer. Microsc. Soc. 70:319.

---

The pharynx is about one-half the diameter of the oral sucker. Undulating ceca extend to the posterior end of the body. Reproductive organs are in the caudal end of the body, about equal in size, and arranged in a triangle with the ovary (right) and the anterior testis (left side) at the same level. The genital pore is median and at the level of the anterior margin of the ovary. Vitelline glands are lateral to the intestines and extend from the pharynx to the ovary. Eggs measure 29 by 18 $\mu$.

**Life Cycle.**    Adult worms in the cecum lay eggs that are embryonated when passed in the feces. Hatching takes place when they are eaten by the land snail *Anguispira alternata*. The miracidia migrate through the intestinal wall into the surrounding connective tissue or into the hepatic gland, where they transform into large multibranched mother sporocysts. Daughter sporocysts appear in 7 to 10 days. They localize in the hepatic gland and grow into large branched forms similar to the mother sporocysts. Some of the branches protrude from the hepatic gland into the mantle cavity. Short-tailed cercariae develop in about 12 weeks during the summer and escape through birth pores at the tips of the branches into the mantle cavity. Production of cercariae continues for longer than a year and possibly for the life of the snail. They escape from the snails via the respiratory pore and appear in the slime secreted by the host.

Snails (*Anguispira*, *Polygyra*, *Derocercas*) crawling over the slime become infected when the cercariae in it enter the pore of the primary ureter. They migrate through the ureters into the kidney and through the renal canal into the pericardial chamber. Snails containing daughter sporocysts are resistant to infection by cercariae, hence only unparasitized ones become infected as the second intermediate host. Upon reaching the pericardial chamber, the cercariae are designated as metacercariae. The short tail is lost by the end of 10 days. No encystment occurs and the infective stage is reached in about 6 months. Up to 90 percent or more of the snails are naturally infected. Chipmunks and white-footed mice become infected by eating the snails. The metacercariae reach the cecum and begin feeding on blood as early as 6 hours after being swallowed. Sexual maturity is attained in 8 days, as indicated by the appearance of eggs in the uterus, but they do not appear in the feces until days 19 to 20 after infection. A minimum of 39 weeks is required for completion of the life cycle, which

**Plate 61**   *Postharmostomum helicus*                                             259

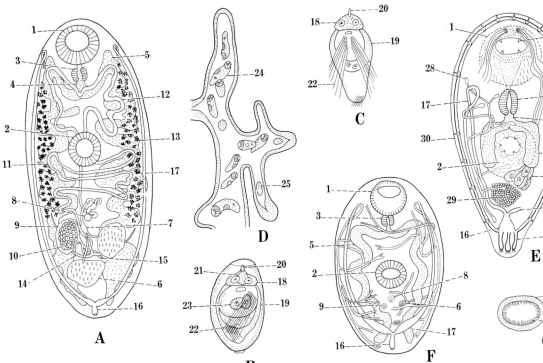

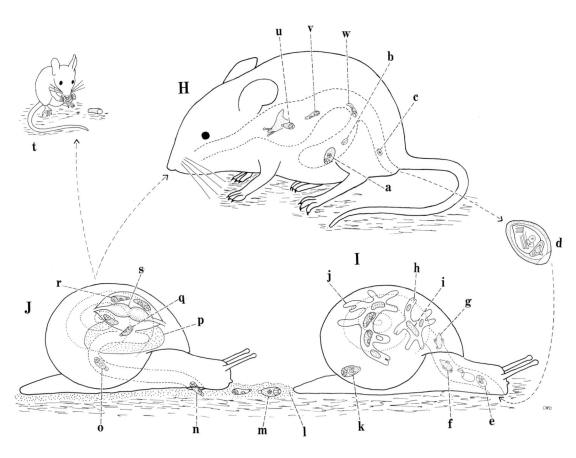

is longer than for other members of the family where known. Infections in mice persist for as long as 150 days and possibly longer.

Infection of the snail *Anguispira alternata* with sporocysts ultimately produces adverse effects on the mollusks. For the first month, growth of the snails is normal but thereafter they are retarded and usually die in about 10 months after infection. In experiments, 20°C is the optimal temperature for growth. At 30°C, movements of the snails are abnormal and death comes early. At 10°C, growth of both snails and sporocysts ceases and no cercariae are produced. When removed to a temperature of 22°C, cercariae appear in 7 weeks. Snails harboring sporocysts fail to oviposit.

Other species of Brachylaemidae, whose life cycles are known, utilize two land snails as the first and second intermediate hosts. *Brachylaemus virginianus* of opposums develops in *Polygyra thyroides*; *Postharmostomum gallinum* of chickens in *Eulota similaris* (sporocysts), *Subulina octoma*, *Praticolella griseola*, *Zachrysia auricoma*, and *Oleacina* sp. (metacercariae); *Panopistus pricei* of shrews in *Zonitoides arboreus*, *Agriolimax agrestis*, and *Ventridens ligerus* (sporocysts and metacercariae both in a single specimen); and *Leucochloridiomorpha constantiae* of ducks in *Campeloma decisum* (sporocysts and metacercariae). *Campeloma* is a gilled aquatic snail and the others are terrestrial forms.

## EXERCISE ON LIFE CYCLE

Both the cercariae and metacercariae of the several genera of the Brachylaemidae can be identified without difficulty. Land snails harboring such flukes make a good starting point for life history studies. Unencysted metacercariae found in the pericardial chamber should be fed to appropriate vertebrate hosts. Proper identification of the metacercariae will indicate the vertebrate host to which it should be fed.

When eggs become available, they should be fed to snails of the same species from which the metacercariae came in order to obtain the branching sporocysts.

If cercariae are available, uninfected snails of the same species should be exposed to them for obtaining metacercariae. Determine the development of the various stages in each snail host.

## SELECTED REFERENCES

Alicata, J. E. (1940). The life cycle of *Postharmostomum gallinum*, the cecal fluke of poultry. J. Parasitol. 26:135–143.

Allison, L. N. (1943). *Leucochloridiomorpha constantae* (Mueller) (Brachylaemidae), its life cycle and taxonomic relationships among digenetic trematodes. Trans. Amer. Microsc. Soc. 62:127–168.

Elwell, A. S., and M. J. Ulmer. (1968). Effects of *Postharmostomum helicis* sporocyst infections on *Anguispira alternata*. Trans. Amer. Microsc. Soc. 87:125.

Groschaft, J. M., T. Del Valle, and N. Lorenzo Hermandez. (1969). *Postharmostomum gallinum* Witenberg, 1923 (Trematoda: Brachylaemidae), un trematodo de las gallinas cubanas, con notas sobre ciclo evolutivo. Torreia n. s. (12):3–12.

Krull, W. H. (1935). Some observations on the life history of *Brachylaemus virginiana* (Dickerson) Krull N. 1934. Trans. Amer. Microsc. Soc. 54:118–134.

Krull, W. H. (1935). Studies on the life history of *Panopistus pricei* Sinitsin, 1931 (Trematoda). Parasitology 27:93–100.

Reynolds, B. D. (1938). Developmental stages of *Panopistus pricei* Sinitsin in *Agriolimas agrestis*. Parasitology 30:320–323.

Ulmer, M. J. (1951). *Postharmostomum helicis* (Leidy, 1847) Robinson 1949, (Trematoda), its life history and a revision of the subfamily Brachylaeminae. Part I. Trans. Amer. Microsc. Soc. 70:189–238.

Ulmer, M. J. (1951). *Postharmostomum helicis* (Leidy, 1847) Robinson 1949, (Trematoda), its life history and a revision of the subfamily Brachylaeminae. Part II. Trans. Amer. Microsc. Soc. 70:319–347.

Vilella, J. B. (1954). *Ventridens ligera*, a new first intermediate host of *Panopistus pricei* Sinitsin, 1931 (Trematoda: Brachylaemidae). J. Parasitol. 40:470–472.

### Superfamily Bucephaloidea

This superfamily is characterized by the mouth being on the midventral side of the body. There is a holdfast organ at the anterior end where the oral sucker normally is located. The adult worms are parasites of the gut of marine and freshwater fishes.

The cercariae, like the adults, are gasterostomate. The stem of the tail is short and bulbous with very long, motile furcae. They develop in branched sporocysts in clams. The protonephridia are mesostomate.

## FAMILY BUCEPHALIDAE

There is but a single family in this group. Its members occur in both marine and freshwater fishes.

### *Bucephalus elegans*
Woodhead, 1930 (Plate 62)

The adult flukes occur commonly in the cecal pouches of the rock bass (*Ambloplites rupestris*) in the Great Lakes region of North America. While a relatively large percentage of the rock bass are infected, the number of parasites per individual is small, rarely exceeding three.

**Description.**    Fully developed flukes are small spiny, cylindrical forms measuring 658 $\mu$ long by 192 $\mu$ wide. A truncate anterior end bears 7 extensile appendages. The oral sucker is on the midventral surface of the body; it opens into an esophagus directed forward which enters an oval gut that turns posteriad. The ovary and two testes are behind the gut. A long cirrus pouch extends posteriorly and opens through the genital pore near the hind end of the body. The eggs average 48 $\mu$ long by 21 $\mu$ wide and have a prominent operculum.

**Life Cycle.**    Adult worms containing up to 200 eggs in various stages of development occur in the cecal pouches of a high percentage of the rock bass. Eggs near the genital opening contain fully formed miracidia ready to hatch upon reaching the water. Hatching takes place quickly when they reach the water. The miracidium forces the operculum from the egg, emerges, and swims rapidly away. Upon striking an object, it adheres, never seeming to turn away.

Infection of the rainbow clam first intermediate host (*Eurynia iris*) has not been observed. It is likely, however, that the miracidia are swept through the incurrent siphon into the branchial chamber, where they attach to the gonad and burrow into it. Up to 6 percent of the clams from some areas are naturally infected. Branching sporocysts develop in gonads, often completely filling them. While only a single generation has been reported, it seems probable that a mother sporocyst and daughter sporocysts occur. Early development of the cercariae in the sporocysts is rapid but the time required for reaching maturity is unknown. They emerge from the gonads into the branchial chamber. Currents of water flowing from the excurrent siphon carry them away from the clams. Upon reaching the outside, they spread the long furcae and hang with the body downward, swimming slowly.

When the fish swim by, the furcae become entangled with the fins, holding the cercariae in a position that enables them to attach by the anterior holdfast and burrow into the tissues. As they work their way rapidly into the fins, the tail is discarded. Having entered the tissues, they move about, enlarging the space around them, and within an hour are enclosed in a hyaline cyst of parasitic origin.

Infection of rock bass takes place when fish harboring the cysts are eaten. Upon being released by the action of the digestive juices, the metacercariae migrate into the cecal pouches and mature in 30 days.

Woodhead reported three generations of sexual reproduction in *Bucephalus elegans*, involving sporocysts, rediae, and adults. His observations have not been confirmed nor his views accepted (Cort *et al.*).

The life cycle of *Bucephalus papillosus* Woodhead, from walleyed pike and pickerel, is essentially the same as that of *B. elegans*.

In warm water, lesions opened in the skin of fish by penetration of hordes of cercariae and the associated secondary infections result in high mortality of fish.

Marine bucephalids developing in oysters reduce the total fats, probably being hydrolyzed by lipases of parasitic origin. Neutral and fatty acids accumulate in the developing cercariae.

Larval stages of *Bucephalus cuculus* developing in oysters are hyperparasitized by the microsporidian *Nosema dollfusi*. The spores are mainly in the lumen of the sporocysts and probably are pathogenic to the developing flukes.

## EXERCISE ON LIFE CYCLE

The eggs of *Bucephalus* and miracidia contained in them are so characteristic that identification of them in the feces of fish hosts can be made with confidence. Likewise, the ox-head type of cercariae from clams (*Eurynia* and *Elliptio*) are readily recognized. With these advantages, the task of studying the life cycle is made easier.

Walleyed pike and pickerel are frequently infected with *B. papillosus*, and rock bass with *B. elegans*. Eggs from adult worms or the feces of infected fish provide material to observe hatching and for studying the action of miracidia. In

## Explanation of Plate 62

A, Adult fluke. B, Egg in early stages of cleavage. C, Miracidium. D, Branched sporocyst containing cercariae. E, Cercaria free in water. F, Morphology of mature cercaria. G, Metacercaria. H, Encysted metacercaria. I, Definitive host (*Ambloplites rupestris*). J, Clam first intermediate host (*Eurynia iris*). K, Fish second intermediate host (*Lepomis macrochirus*).

1, Fimbria; 2, anterior adhesive organ; 3, oral sucker; 4, pharynx; 5, esophagus; 6, intestine; 7, testes; 8, seminal vesicle; 9, cirrus sac; 10, cirrus extended; 11, ovary; 12, vitellaria; 13, excretory vesicle; 14, excretory pore; 15, germ cell; 16, yolk cell; 17, cephalic plate; 18, birth pore; 19, cercaria; 20, eyespot; 21, vitelline gland; 22, cirrus or perhaps anlagen of testes; 23, bulbous tail stem; 24, furca; 25, subcuticular gland; 26, genital pore; 27, cyst.

a, Small adults embedded in cecal pouches; b, eggs passed in feces; c, partially developed egg; d, egg with fully developed miracidium; e, miracidium escaping from egg; f, miracidium free in water; g, miracidium swept into mantle cavity of clam through incurrent pore; h, branched sporocyst in digestive gland (preceding stages unknown); i, developing cercaria; j, ox-head cercaria ready to escape from digestive gland; k, cercaria free in water, having been swept from mantle cavity through excurrent pore; l, cercaria attacks bluegill sunfish (experimental) when disturbed by movements of water caused by swimming fish; m, cercaria drops tail as it penetrates fins or base of them; n, encysted metacercaria under skin of fin; o, infected bluegill or other infected fish swallowed by rock bass; p, encysted cercaria released from fish, when latter is digested in the alimentary canal of rock bass; q, metacercaria escapes from cyst and migrates to the cecal pouches, where development to maturity occurs in about 30 days.

Adapted from Woodhead, 1929, Trans. Amer. Microsc. Soc. 48:25; 1930, 49:1; 1931, 50:169.

---

small clams the means of infection can be followed. Determine whether one or two generations of sporocysts occur. Follow the entire intramolluscan development of the parasite. Ascertain the time required for development of the cercariae.

Using cercariae obtained from either naturally or experimentally infected clams, study the action of them in the water and manner in which they infect the fish second intermediate host. By the use of stains, determine whether the cyst surrounding the metacercaria has the same chemical characteristic as the glandular material in the cercariae. This would provide a clue as to whether the cyst is of parasite or host origin.

By feeding encysted metacercariae to appropriate fish, determine the time required for the fluke to reach sexual maturity.

## SELECTED REFERENCES

Cheng, T. C. (1965). Histochemical observations on changes in lipid composition of the American oyster *Crassostrea virginica* (Gmelin), parasitized by the Trematode *Bucephalus* sp. J. Invert. Pathol. 7:398–407.

Corkum, K. C. (1967). Bucephalidae (Trematoda) in fishes of the northern Gulf of Mexico: *Bucephalus* Baer, 1827. Trans. Amer. Microsc. Soc. 86:44–49.

Cort, W. W., D. J. Ameel, and A. Van der Woude. (1954). Germinal development in the sporocysts and rediae of digenetic trematodes. Exp. Parasitol. 3:185–216.

Kindelin, P. de, P. Besse, G. Jolinet, and G. Tuffery. (1967). Rôle pathogène des cercaires de *Bucephalus polymorphus* (Baer, 1827) (Trematode, Bucephalidae) sur le peuplement pisicole du bassin de la Seine. C. R. Acad. Sci. Ser. D. 264:2321–2324.

Sprague, V. (1964). *Nosema dollfusi* n. sp. (Microsporidia, Nosematidae), a hyperparasite of *Bucephalus cuculus* in *Crassostrea virginica*. J. Protozool. 11:381–385.

Woodhead, A. E. (1929). Life history studies on the trematode family Bucephalidae. Trans. Amer. Microsc. Soc. 48:256–275.

Woodhead, A. E. (1930). Life history studies on the trematode family Bucephalidae. No. II. Trans. Amer. Microsc. Soc. 49:1–17.

Woodhead, A. E. (1931). The germ cell cycle in the trematode family Bucephalidae. Life history studies on the trematode family Bucephalidae. No. III. Trans. Amer. Microsc. Soc. 50:169–188.

### ORDER ECHINOSTOMIDA

This order contains approximately two dozen families of diverse morphology. They include the fasciolids, echinostomes, paramphistomes, and certain monostomes. The relationship is shown in cercariae that develop in rediae, have large bodies, strong single tails, and numerous cystogenous glands in the epidermis.

**Plate 62** *Bucephalus elegans* 263

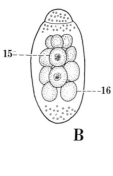

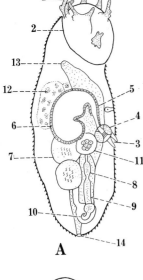

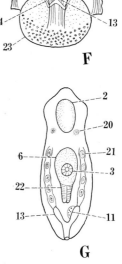

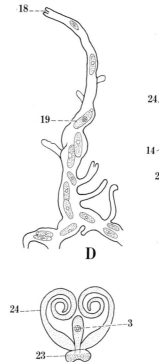

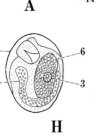

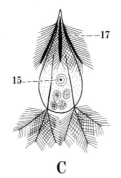

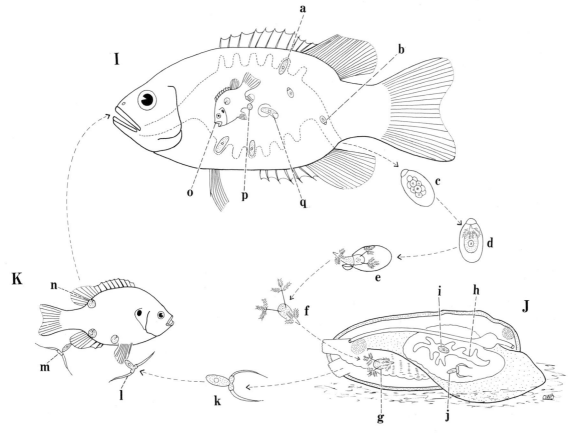

## Explanation of Plate 63

A, Adult fluke, ventral view. B, Anterior end of fluke, showing details of collar and spines. C, Embryonated egg. D, Mother redia. E, Daughter redia. F, Cercaria. G, Encysted metacercaria. H, Duck definitive host. I, Muskrat definitive host. J, Snail (*Helisoma trivolvus* and others) first intermediate host. K, Frog tadpole second intermediate host. L, Snail (*Physa*) second intermediate host.

1, Oral sucker; 2, ventral sucker; 3, pharynx; 4, esophagus; 5, intestine; 6, testis; 7, cirrus pouch; 8, common genital pore; 9, ovary; 10, oviduct; 11, Mehlis' gland; 12, uterus; 13, excretory bladder; 14, excretory pore; 15, head collar; 16, collar spines; 17, corner spines; 18, eggshell; 19, operculum; 20, miracidium; 21, apical gland; 22, flame cell; 23, excretory duct; 24, pharynx of redia; 25, gut; 26, daughter redia; 27, germinal masses; 28, birth pore; 29, cercaria; 30, prepharynx; 31, tail; 32, cyst of metacercaria.

a, Adult fluke in small intestine; b, unembryonated eggs laid in intestine; c, d, egg develops and hatches in water; e, miracidium penetrates snails; f, mother sporocyst; g, mother redia containing daughter rediae; h, daughter redia containing cercariae; i, cercaria escaping from snail; j, cercaria free in water; k, cercaria creeping into anus of tadpole; l, cercaria migrates up ureters; m, encysted metacercariae in kidney; n, cercaria penetrating snail second intermediate host; o, metacercaria encysted in snail; p, ducks and other definitive hosts become infected by eating infected tadpoles or snails; q, metacercaria released by action of digestive juices; r, young fluke escapes from cyst and develops to adult stage in 15 to 19 days.

Figures A, B, F adapted from Beaver, 1937, Ill. Biol. Monogr. 15:1; C–E, G, from Johnson, 1920, Univ. Calif. Publ. Zool. 19:335.

## Suborder Echinostomata

The cercariae are echinostomate, nonechinostomate, or show modifications of a collar and collar spines. They develop in collared rediae with stump-like lateral appendages. The life cycle usually requires two hosts, but three may occur in some species.

## Superfamily Echinostomatoidea

Members of this superfamily are trematodes with a head collar bearing a row of spines.

### FAMILY ECHINOSTOMATIDAE

The members of this family are more or less elongated distomate flukes. They have a distinct head collar armed with a single or double row of spines. They are parasites of the intestine and, occasionally, the bile duct or ureters of reptiles, birds, and mammals.

### *Echinostoma revolutum*
(Froelich, 1802) (Plate 63)

This fluke is among the most versatile as regards the variety of hosts it parasitizes. Adult flukes are able to develop in the ileum of 23 species of birds and nine of mammals, including man. Miracidia develop in at least four genera of snails (*Helisoma*, *Pseudosuccinea*, *Physa*, and *Stagnicola*) and 10 species in North America. The cercariae enter and develop to metacercariae in 16 species of pulmonate snails (many are the same ones in which they developed), two species of fingernail clams and seven species of tadpoles, together with catfish and bullheads. Because of its adaptability to mollusks and tadpoles as intermediate hosts and ducks as definitive hosts, this parasite is cosmopolitan.

**Description.**    Adult worms vary in length from 4 to 21 mm, depending on the species of definitive host in which they develop. The number and arrangement of spines in the cephalic collar are constant. There are 37 spines arranged in three general groups; they are 1) the two corner groups of five spines each, three of which extend slightly beyond the margin of the collar and two that do not; 2) the two lateral groups, consisting of six spines each arranged in a straight row, and 3) the dorsal group consisting of 15 spines arranged in two alternating rows in which seven are in the anterior and eight in the posterior row. Eggs range in size from 91 to 145 $\mu$ long by 66 to 83 $\mu$ wide, the size being influenced by the species of the host.

**Life Cycle.**    Adult worms lay an average of 2000 to 3000 unembryonated eggs per day. Energy for such large scale production comes from feeding on the mucosal lining, desquamated cells, and secretions of the gut. Glycogen is stored in the parenchyma, hepatopancreas, and muscles for use by the flukes and in the eggs to nourish the embryos.

Eggs develop slowly, requiring 18 to 30 days,

**Plate 63**   *Echinostoma revolutum*                                      265

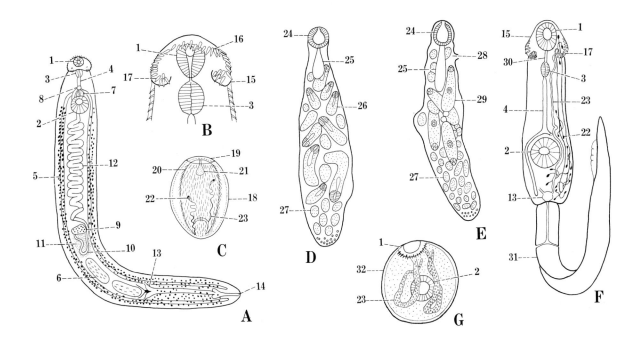

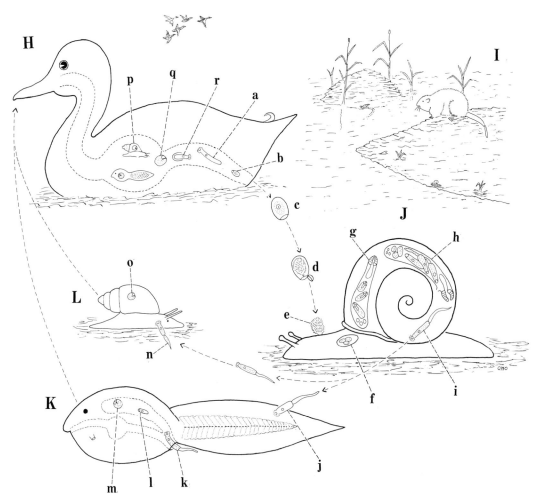

depending on the temperature of the water. Hatching takes place in light and darkness. Newly hatched miracidia swim rapidly and constantly. Upon contacting the snails *Physa gyrina, Helisoma trivolvus,* and others, they attach to the epidermis, burrow inside, and migrate through the lymphatics to the foot and mantle and transform into mother sporocysts that produce mother rediae. These migrate to the digestive gland, mature, and give birth to daughter rediae. It is the second generation of rediae, or daughters, that produce the cercariae in 9 to 10 weeks after infection of the snails by the miracidia. Large snails produce more rediae and cercariae than small ones. Redial transfers to uninfected snails will produce 42 and possibly 102 generations, indicating their prolific nature.

The cercariae have the 37 collar spines arranged in the same manner as in the adults; there is a membranous fin over the dorsal side of the tail. They are active swimmers and good crawlers. Upon coming in contact with snails and fingernail clams, they penetrate the soft parts and encyst. If they attach to tadpoles or silurid fishes, they creep into the cloaca and migrate up the ureters into the kidneys where encystment occurs.

In double infections with some species of echinostomes antagonisms toward one species by the other parasite develop that retard its development. Rediae of one species prey upon the larval stages of the other, completely eliminating all except the metacercariae.

Infection of the definitive hosts is accomplished when they eat mollusks, tadpoles of frogs and toads, or catfish and bullheads that harbor metacercariae. Upon being freed from the tissues of the second intermediate hosts, the metacercariae migrate to the ileum, attach to the mucosa, and develop to sexual maturity when 4 to 7 mm long, at which time individual flukes fertilize themselves. Growth continues as long as the flukes remain in the host. Eggs appear in the feces 18 days after infection.

The presence of flukes in the gut produces lesions as well as physiological and functional disturbances. Alkaline phosphates decrease in the duodenum, ileum, and rectum. Significant decreases in the levels of enterokinases occur in the mucosa and contents of the duodenum but increase in the rectum and droppings.

As a result of Beaver's extensive experimental studies on variation in pure strains of *E. revolu-tum,* about 15 species were declared synonyms. This is evidence for the need of studying variation in morphology as a means of determining the parameters of species of many of the parasites.

In cases where life cycles of other species of Echinostomatidae have been worked, they follow the general pattern shown by *E. revolutum.* In *Echinoparyphium recurvatum* of chickens and ducks, the cercariae develop in *Physa, Planorbis,* and *Lymnaea,* and encyst in *Lymnaea* and tadpoles; in *Himasthla quissetensis* of gulls, they develop in *Nassa* and encyst in *Mya* and several other marine clams; those of *Hypoderaeum conoideum,* a European species of ducks, encyst in *Planorbis* and *Limnaea;* and *Petasiger chandleri* of grebes utilizes *Helisoma* for development of the cercariae, and the gills or esophagus of fish and tadpoles for the metacercariae.

## EXERCISE ON LIFE CYCLE

Both cercariae and metacercariae of Echinostomatidae have an oral collar with spines which serves as an infallible character for identification. *Echinostoma revolutum* has 37 collar spines, *Echinoparyphium recurvatum,* 45; *Himasthla quissetensis,* 27; *Petasiger chandleri,* 19 to 21.

Identify any of the above genera of echinostome cercariae recovered from naturally infected snails. Following identification, expose appropriate snails, clams, tadpoles, or bullheads to the cercariae in order to obtain metacercariae. Ascertain the location of the metacercariae in the second intermediate host.

Having obtained metacercariae, feed them to some of the known definitive hosts to study the part of the life cycle in them. Eggs obtained from such infections may be hatched and snails of the same species from which the cercariae originated may be infected in order to get the intramolluscan stages of the parasites.

Since lipids localize in the excretory bladder and its tubules, staining whole mounts of flukes or cercariae with Oil Red O will delineate the morphology of the excretory system.

## SELECTED REFERENCES

Abdel-Malek, E. T. (1953). Life history of *Petasiger chandleri* (Trematoda: Echinostomatidae) from the pied-billed grebe *Podilymbus podiceps podiceps,* with some comments on other species of *Petasiger.* J. Parasitol. 39:152–158.

Basch, P. F., and K. J. Lie. (1966). Infection of single snails with two different trematodes. II. Dual infections to a schistosome and an echinostome at staggered intervals. Z. Parasitenk. 27:260–270.

Beaver, P. C. (1937). Experimental studies on *Echinostoma revolutum* (Froelich) a fluke from birds and mammals. Ill. Biol. Monogr. 15:1–96.

Dönges, J. (1971). The potential number of redial generations in echinostomatids (Trematoda). Int. J. Parasitol. 1:51–59.

Fried, B., and M. D. Kramer. (1968). Histochemical glycogen studies on *Echimostoma revolutum*. J. Parasitol. 54:942–944.

Fried, B., and L. J. Morrone. (1970). Histochemical lipid studies on *Echinostoma revolutum*. Proc. Helminthol. Soc. Wash. 37:122–123.

Fried, B., and W. Vonroth, Jr. (1968). Transplantation of *Echinostoma revolutum* (Trematoda) in normal and surgically altered chick intestine. Exp. Parasitol. 22:107–111.

Fried, B., and L. J. Weaver. (1969). Effects of temperature on the development and hatching of eggs of the trematode *Echinostoma revolutum*. Trans. Amer. Microsc. Soc. 88:253–257.

Johnson, J. C. (1920). The life cycle of *Echinostoma revolutum* (Froelich). Univ. Calif. Publ. Zool. 19: 335–388.

Lie, K. J., P. F. Basch, and D. Heyneman. (1968). Direct and indirect antagonism between *Paryphostomum segregatum* and *Echinostoma paraensis* in the snail *Biomphalaria glabrata*. Z. Parasitenk. 31:101–107.

Lie, K. J., P. F. Basch, and M. A. Hoffman. (1967). Antagonism between *Paryphostomum segregatum* and *Echinostoma barbosai* in the snail *Biomphalaria straminae*. J. Parasitol. 53:1205–1209.

Patnaik, M. M. (1968). Notes on glycogen deposits in *Lymnaea auricularia* var. *rufescens* and the parasitic larval stages of *Fasciola gigantica* and *Echinostoma revolutum*. A histochemical study. Ann. Parasitol. 43:449–456.

Patnaik, M. M., and S. K. Ray. (1966). A histopathologic study of *Lymnaea auricularia* var. *rufescens* infected with the larval stages of *Echinostoma revolutum*. Jap. J. Med. Sci. Biol. 19:253–258.

Stunkard, H. W. (1938). The morphology and life cycle of the trematode *Himasthla quissetensis* (Miller and Northrup, 1926). Biol. Bull. 75:145–164.

Tsyunene, E. (1968). [The activity of alkaline phosphatase in duck intestine during echinostomatid infections.] Acta Parasitol. Litu. 8:81–94.

Tsyunene, E. (1968). [Determination of enterokinase activity in the intestine of ducks during experimental echinostomatid infections.] Acta Parasitol. Litu. 8:95–101.

Zischke, J. A. (1967). Redial populations of *Echinostoma revolutum* developing in snails of different sizes. J. Parasitol. 53:1200–1204.

## FAMILY FASCIOLIDAE

The species of this family are parasites of mammals, commonly herbivores. They are large spiny, leaflike, distomate flukes with the suckers close to each other. The ceca are branched or simple. The testes and ovary are dendritic. Vitellaria are profuse, lateral, and confluent posteriorly.

Because of their economic importance, much research has been done on liver flukes in an effort to find ways of controlling them. The investigations have been directed toward an understanding of the biology of both the liver fluke itself and the snail intermediate host in various parts of the world in a variety of climates.

*Fasciola hepatica* Linnaeus, 1758 (Plate 64)

This is the common liver fluke of sheep and cattle in the United States and Europe, as well as other sections of the world. It is the cause of tremendous losses to growers of these animals. In addition to rendering the livers unsuitable for human consumption, it causes losses through death, reduction of meat and milk, inhibited reproduction, morbidity, and lowered resistance. Infected wild rabbits and hares serve as reservoir hosts in spreading and maintaining the parasites on grazed and ungrazed areas.

**Description.** Fully grown flukes reach a size of 30 mm long by 13 mm wide. They are shaped somewhat like a leaf with a narrow cephalic cone on the broad anterior end. The oral sucker is smaller than the ventral one which is at the level of the broad "shoulders." The highly dendritic intestinal ceca extend to the posterior extremity of the body. A branched ovary lies toward the right side a short distance posterior from the ventral sucker with the coils of the uterus between them. The testes are extensively branched, filling most of the body behind the ovary. Numerous vitelline glands extend backward along the sides of the body from the shoulders to the end of the body where they are confluent behind the testes. Eggs measure 130 to 150 $\mu$ long by 63 to 90 $\mu$ wide.

**Life Cycle.** The life cycle consists of six phases in the development of the fluke. They are 1) development to adulthood in the final

## Explanation of Plate 64

A, Adult fluke. B, Unembryonated egg. C, Miracidium. D, Young mother sporocyst. E, Mature mother sporocyst. F, Mature second generation redia. G, Cercaria. H, Sheep definitive host. I, Wild rabbits as reservoir hosts. J, Snail (*Stagnicola bulimoides, Fossaria modicella*) first intermediate host. K, Metacercaria on vegetation.

1, Oral sucker; 2, ventral sucker; 3, pharynx; 4, intestine; 5, testis; 6, vas efferens; 7, vas deferens; 8, cirrus pouch; 9, common genital pore; 10, ovary; 11, oviduct; 12, Mehlis' gland surrounding ootype; 13, uterus; 14, vitelline glands; 15, transverse vitelline duct; 16, operculum of unembryonated egg; 17, apical papilla; 18, eyespots; 19, cilia; 20, germ cells; 21, germ balls; 22, mother rediae; 23, daughter redia; 24, cercaria; 25, tail; 26, genital anlagen.

a, Adult fluke in bile duct of liver, causing great hypertrophy of the ducts; b, unembryonated eggs leave the liver by way of the common bile duct; c, eggs mixed with feces; d, eggs passed from body in feces; e, eggs develop in water and hatch, releasing miracidia; f, miracidium penetrates snail; g, miracidium transforms into a mother sporocyst; h, mature mother sporocyst containing daughter rediae; i, first generation of daughter redia with developing rediae; j, second generation redia filled with germ balls and cercaria; k, cercaria free in water; l, lateral view of metacercaria encysted below water level on grass stem; m, surface view of submerged metacercaria; n, metacercaria above water due to growth of vegetation or lowering of water, or a combination of both; o, infection of definitive host occurs when metacercariae are swallowed with forage; p, young fluke escapes from metacercarial cyst in abomasum; q, young fluke free in intestine; r, flukes burrow through intestinal wall into celom; s, flukes migrate to liver; t, flukes enter liver by penetrating capsule; u, young flukes migrate through liver parenchyma for a period and then enter bile ducts where maturity is attained in approximately 90 days after entering the host; v, metacercariae penetrating lungs.

Figures adapted from various sources.

---

host; 2) passage of eggs from the host in its feces; 3) development of the eggs; 4) free-living miracidia in the water in search of a snail host; 5) development of the parasite in the snail; and 6) emergence of the cercariae from the snails and successful encystment on plants or in the water.

Adult flukes in the biliary system of sheep lay an average of 8000 to 25,000 unembryonated eggs each per day. They pass out of the liver through the common bile duct and into the intestine to be voided with the feces. Those falling in water develop and hatch in 9 to 10 days at summer temperatures. Development is retarded at low temperatures and ceases at 10°C. In low temperatures, eggs survive for long periods and resume normal development in favorable temperatures.

Eggs hatch only when free in water as anerobic conditions in fecal masses prevent development. The miracidia are ciliated and have a proboscis-like structure at the anterior end and two semilunar eyespots. They are capable of living up to 24 hours free in the water.

Upon approaching a suitable snail intermediate host such as *Stagnicola bulimoides* or *Fossaria modicella* in North America, chemotactic receptacles direct them to substances in the mucus. Upon contacting the snails, they burrow into them, lose the ciliated coat and transform into a sac-like mother sporocyst. Each sporocyst produces numerous mother rediae that migrate to the hepatic gland where a great many daughter rediae are born. Each one gives birth to many cercariae, averaging from 9 to 649 per miracidium, beginning 5 to 7 weeks after infection. If the pools in which the snails live become dry, they burrow into the mud, surviving for long periods. When water reappears, they quickly emerge and, if infected, shed cercariae.

The cercariae are 250 to 350 $\mu$ long, the tail is double the length of the body, and the body wall is filled with cystogenous glands. In the water, the cercariae quickly attach to objects, particularly plants, or the undersurface of the water film, drop their tails, and secrete a double-walled cyst about themselves. The outer wall consists of an external layer of tanned protein that covers the dorsal and lateral sides and a thin inner one surrounding the entire worm. There is a thin, inner, transparent wall. The walls are produced from the products of the cystogenous glands and consist of various polysaccharides. Metacercariae are infective by 24 hours after encystment.

Sheep, cattle, and rabbits, all common hosts, become infected by eating forage bearing cysts. Excystment consists of two stages: activation of the young fluke and its emergence from the cyst. Activation begins in the rumen in an atmosphere of concentrated carbon dioxide, under reducing

Plate 64    *Fasciola hepatica*                                                                     269

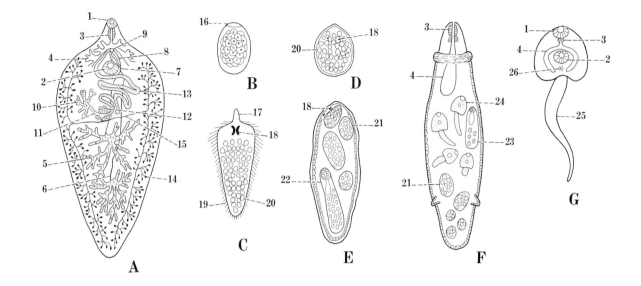

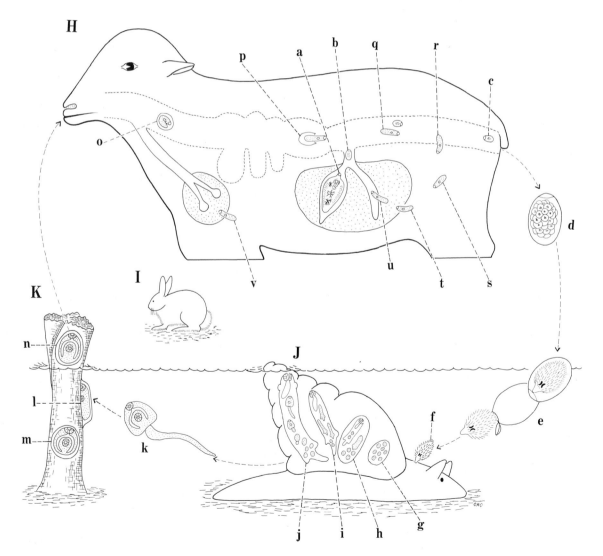

conditions, and at a temperature of 39°C. Emergence occurs in the duodenum below the opening of the bile duct where the bile triggers emergence by activating enzymes from the metacercariae that cause the emergence hole of the cyst to open. Upon escaping from the cyst, the metacercariae browse on the mucosal lining and then quickly bore through the gut wall. Most of the flukes are in the celomic cavity within 24 hours after excystation. They migrate forward to the liver, browsing along the way, and penetrate it, primarily through the ventral lobe between days 4 and 5, but the time may be much longer for some individuals. After a period of migration, feeding, and rapid growth in the parenchyma, they enter the bile ducts while still young and small enough to get into them. The flukes are self-fertilizing. Eggs appear in the feces of rabbits in about 9 weeks and of sheep and cattle in about 13 to 15 weeks. Liver flukes have great longevity, living over 11 years in sheep under conditions that precluded infection.

**Nutrition of Flukes.**    Flukes feed on tissue while on the intestinal mucosa, in the intestinal wall, while migrating in the celomic cavity and liver parenchyma, and in the bile ducts. Blood is not a part of the diet but may be ingested accidentally, especially by flukes migrating in the liver.

While flukes are burrowing in the liver, tissue and cytological changes in the wall of the bile ducts produce small papule-like epithelial outgrowths. The entire inner surface of the ducts is studded with these blebs, whose origin may be associated with the inflammatory reaction provoked by the presence of the flukes. Upon entering these bile ducts, the flukes find an ample supply of food in the hypertrophied mucosal lining which they browse off. Regenerative processes of the abnormal epithelium of the bile ducts provide a continuous supply of food for the flukes for as long as they remain in them.

**Ecology of Liver Flukes and Their Snail Intermediate Hosts.**    Inasmuch as liver flukes are entirely dependent on snails for their existence, the presence of moisture under favorable conditions is an absolute essential. Favorable temperature for development of the snails and different stages of the fluke is necessary.

The species of snails serving as intermediate hosts of the liver flukes are amphibious. They live on the mud along the margins of small pools, in marshes, and in seeps. They thrive in temporary pools where many successfully estivate in the mud for long periods when there is no water. Whenever water saturates the soil, the survivors emerge, grow, and reproduce. Being prolific, the snails quickly repopulate the areas when favorable moisture conditions return. Heavy clay soil with impervious subsoil and topography that limits drainage constitutes ideal habitat. Water is provided by heavy rain, from marshy seeps, or by floods from overflowing streams.

Final hosts, such as cattle, sheep, and rabbits, are agents for spreading fluke eggs over the pastures. Eggs and developmental stages of the flukes on the pastures face many hazards. The flukes have evolved to meet them by their prodigious powers of fecundity. With individual flukes producing up to 25,000 eggs per day for long periods (up to 11 years) and the cercarial progeny from a single miracidium being as high as 650, it is clear how heavily infested pastures may become.

Development of the larval stages in the snails is dependent on temperature of 10° to 30°C. It ceases below 10° and is impaired above 30°C.

Eggs can hatch only when freed from the fecal masses of cattle or pellets of sheep in water. However, hatching does not assure success, for miracidia have a very short existence in which to find and penetrate a specific kind of snail. Development in the snail is dependent on its survival in a temperature favorable for growth of it and the fluke. Eggs falling outside the snail habitat are lost. Miracidia failing to find and enter snails die within 24 hours.

Upon completion of development of the parasite in the snails, cercariae emerge and encyst to form metacercariae, the stage infective to the final hosts. Success of the metacercariae is dependent on being attached to herbage that is eaten by the proper animal before they die. Longevity of the metacercariae is determined by the extent to which adequate moisture is present. Upon being swallowed by an appropriate host, many of the metacercariae fail to reach the liver and attain sexual maturity. The environmental resistance and hazards take an enormous toll. The flukes have responded by evolving a high reproductive capacity. The basic ecological picture is shown in Fig. 26.

**Symptoms and Pathology.**    Symptoms differ somewhat in acute and chronic fascioliasis in sheep. Acute fascioliasis results from massive infection with simultaneous migration to and entry into the liver by young flukes. Sudden death occurs from the physical damage done by them. Damaged and dead liver tissue provides an ideal medium for the growth of the bacillus *Clostridium novyi* with production of a toxin that kills sheep suddenly by the acute condition known as "black disease." Swelling under the lower jaw, known as bottle jaw, occurs while young flukes are migrating in the liver. In chronic cases, sheep are likely to improve in condition during the summer and fall, probably from stimulation of the liver. Later, they rapidly deteriorate, losing weight and condition. The skin and mucous membranes become pale, the animals are inactive, off feed, and chew the cud less. As the composition of the blood changes,

bottle jaw and pot belly appear. Condition declines, rapid and feeble breathing develops, and there is a dejected appearance and diarrhea in the late stages just before death.

Chronic fascioliasis in cattle includes progressive cirrhosis of the bile ducts with eventual fibrosis of the liver. The bile ducts thicken and stand out prominently. They become filled with calcarious castes. The coat is ruff and infected animals lose physical condition.

Other well-known species of Fasciolidae include the liver flukes *Fasciola gigantica* and *Fascioloides magna* of ruminants, and *Fasciolopsis buski*, an intestinal fluke of humans and swine in the Orient. *Fasciola gigantica* replaces *F. hepatica* in some parts of the world, including Hawaii; *Fascioloides magna* is a natural parasite

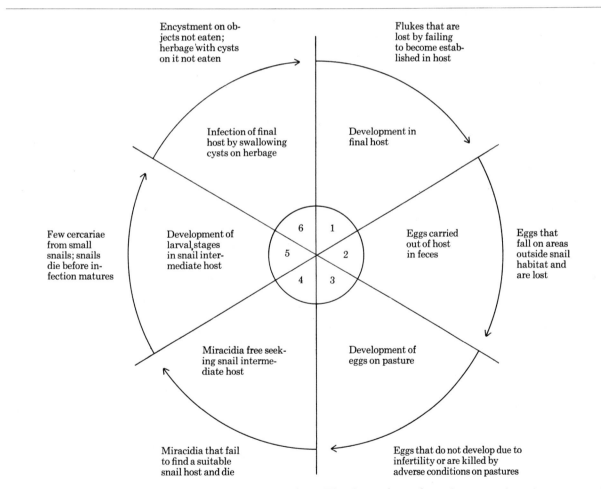

Fig. 26. Schematic representation of six phases of the ecology of the liver fluke *Fasciola hepatica* and its relatives. The stages inside the circle represent the developmental phases of the flukes on the pastures and in their intermediate and final hosts.

The hazards and environmental resistance to the various stages on the pasture and in the hosts are indicated outside the circle. Modified from Ollerenshaw, 1959, Vet. Rec. 71:957.

of Cervidae in North America which infects cattle where they occupy the same range with deer under physical conditions suitable for the snail hosts. The life cycles of these flukes are very similar to that of *F. hepatica*, except for *F. buski*, which develops in the intestine.

**Medication.** Much effort has been directed toward controlling flukes by administering fasciolicides to cattle and sheep. Carbon tetrachloride in doses of 1 ml is the earliest and most successful treatment for use in sheep but is dangerous to cattle. Hexachloroethane-bentonite-water suspension is highly efficacious and safe for use in cattle. Other drugs, usually hydrocarbons, have been developed for removing liver flukes. Some medicaments are effective against young flukes migrating in the liver. Treatment of animals to be most effective in liver fluke control must be correlated with the time when the young flukes are not present in the host and only adults occur in the bile ducts.

## EXERCISE ON LIFE CYCLE

*Fasciola hepatica* is an excellent subject for conducting the life history studies. Its life cycle was the first worked out for a digenetic trematode, having been done independently and almost simultaneously in Germany by Leuckart in 1882 and in England by Thomas in 1883.

In areas where the flukes occur, quantities of them may be procured from the condemned livers of sheep and cattle at abattoirs. Flukes placed in a dish of water lay numerous eggs that will develop and hatch in 9 to 10 days at room temperature. Snail hosts placed with the miracidia become infected. Observe the miracidia penetrate the snails and their reactions to the attacking parasites. Place specimens of *Physa*, *Helisoma*, or other nonhost species in a dish with miracidia and note whether they are attacked. Dissect infected snails at intervals of 2 to 3 days during the growth of the flukes in them to find and identify each of the stages.

When cercariae are shed by the snails, collect the encysted metacercariae from the side of the container or on a glass slide placed upright in the water, and feed them to rabbits. Examine the rabbits at intervals of 1 week after feeding the metacercariae to observe the migration and development of the flukes. Upon opening the celom of the first rabbit, flush it with water into a pan to collect the flukes that are en route from the intestine to the liver.

Put some metacercariae that have been washed in several changes of distilled water into the body cavity of a rabbit to demonstrate that the flukes are capable of migrating from it into the liver.

## SELECTED REFERENCES

Boray, J. C. (1969). Experimental fascioliasis in Australia. In: B. Dawes, ed. Advances in Parasitology. Academic Press, New York, Vol. 7, pp. 95–210.

Boray, J. C., and I. G. Pearson. (1960). The anthelmintic efficiency of tetrachlorodifluoroethane in sheep infested with *Fasciola hepatica*. Austr. Vet. J. 36:331–337. (Technique.)

Dawes, B. (1963). Hyperplasia of the bile duct in fascioliasis and its relation to the problem of nutrition in the liver fluke *Fasciola hepatica*. Parasitology 53:123–133.

Dawes, B., and D. L. Hughes. (1964). Fascioliasis: The invasive stages of *Fasciola hepatica* in mammalian hosts. In: B. Dawes, ed. Advances in Parasitology. Academic Press, New York, Vol. 2, pp. 97–168.

Kendall, S. B. (1970). Relationships between species of *Fasciola* and their molluscan hosts. In: B. Dawes, ed. Advances in Parasitology. Academic Press, New York, Vol. 8, pp. 251–258.

Krull, W. H. (1933). Infections of the white-footed mouse *Peromyscus leucopus noveboracensis* Linn. J. Parasitol. 20:99.

Krull, W. H. (1934). The intermediate hosts of *Fasciola hepatica* and *Fascioloides magna* in the United States. North Amer. Vet. 15:13–17.

Krull, W. H. (1941). The number of cercariae of *Fasciola hepatica* developing in snails infected with a single miracidium. Proc. Helminthol. Soc. Wash. 8:56–58.

Krull, W. H., and R. S. Jackson. (1943). Observations on the route of migration of the common liver fluke, *Fasciola hepatica*, in the definitive host. J. Wash. Acad. Sci. 33:79–82.

Ollerenshaw, C. B. (1959). The ecology of the liver fluke (*Fasciola hepatica*). Vet. Rec. 71:957–965.

Olsen, O. W. (1944). Bionomics of the lymnaeid snail, *Stagnicola bulimoides techella*, the intermediate host of the liver fluke in southern Texas. J. Agr. Res. 69:389–403.

Olsen, O. W. (1947). Hexachloroethane-bentonite suspension for controlling the common liver fluke, *Fasciola hepatica*, in cattle in the Gulf Coast Region of Texas. Amer. J. Vet. Res. 8:353–366.

Olsen, O. W. (1947). Longevity of metacercariae of *Fasciola hepatica* on pastures in the upper coastal region of Texas and its relationship to liver fluke control. J. Parasitol. 33:36–42.

Olsen, O. W. (1948). Wild rabbits as reservoir hosts of the common liver fluke, *Fasciola hepatica*, in southern Texas. J. Parasitol. 34:119–123.

Pantelouris, E. M. (1965). The Common Liver Fluke. Pergamon Press, New York, 259 pp.

Price, E. W. (1956). Liver flukes of cattle and sheep. In: Animal Diseases. Yrbk. U. S. Dep. Agr., pp. 148–153.

Reinhard, E. G. (1957). Landmarks in parasitology. I. The discovery of the life cycle of the liver fluke. Exp. Parasitol. 6:208–232.

Rowcliffe, S. A., and C. B. Ollerenshaw. (1960). Observations on the bionomics of the egg of *Fasciola hepatica*. Ann. Trop. Med. Parasitol. 54:172–181.

Taylor, E. L. (1964). Fascioliasis and the liver fluke. FAO Stud. No. 64, 234 pp.

Thomas, A. P. W. (1883). The life history of the liver-fluke (*Fasciola hepatica*). Quart. J. Microsc. Sci. 99:1–33.

Willmott, S., and F. R. N. Pester. (1952). Variations in faecal egg-counts in paramphistome infections as determined by a new technique. J. Helminthol. 26:147–156.

### Suborder Paramphistomata

The adult flukes are either amphistomate, i.e., the oral sucker is at the anterior extremity and the ventral sucker at the posterior extremity of the body, or monostomate, in which case only the oral sucker is present and located at the anterior end.

The cercariae have the features of the adults in the number and arrangement of the suckers. Their bodies are heavily pigmented and they possess eyespots. They develop in rediae lacking a collar and the stumpy appendages on the side of the body; encystment is on objects in the water, especially vegetation.

### Superfamily Paramphistomatoidea

These flukes, both adults and cercariae, are characterized by having a sucker at each extremity of the body.

### FAMILY PARAMPHISTOMATIDAE

The members of this family are parasites of mammals, particularly herbivores. The location of the metacercariae on vegetation makes them readily available to grazers and accounts for the frequency with which these parasites appear in cattle and sheep. The bodies are thick, the testes are usually near the middle of the body, and the ovary is post-testicular.

### *Paramphistomum cervi*
(Schrank, 1790) (Plate 65)

This is one of the rumen flukes of sheep, goats, cattle, deer, and other ruminants in almost all parts of the world.

**Description.**    The mature worms are conical in shape and pink in color during life. They are 5 to 12 mm long. The dorsal surface is somewhat convex and the ventral concave. The oral sucker is at the anterior extremity of the body and the large ventral sucker at the posterior extremity. The testes are slightly lobed, in the posterior half of the body, and anterior to the ovary. The eggs have a clear shell and are 114 to 176 $\mu$ long by 73 to 100 $\mu$ wide.

**Life Cycle.**    Life cycles of the various species of *Paramphistomum* are basically similar. The exogenous phase is similar to that of the common liver fluke in most respects. Some common snail intermediate hosts are *Bulinus liratus*, *B. mariei*, *Planorbis planorbis*, *Stagnicola bulimoides techella*, and *Pseudosuccinea columella*. The latter two species are common hosts in North America.

Eggs complete their development and hatch in water at the optimal temperature of 27°C in 12 to 17 days. Upon coming in contact with the soft parts of the snail host, the miracidia attach and quickly burrow through the epidermis and into the underlying tissues. Young snails are more susceptible than old ones.

Within 12 hours after reaching the tissues, the miracidia have lost the ciliated coat and have transformed into sporocysts 53 $\mu$ in diameter and 93 $\mu$ in length. Growth is rapid, and by day 11, they have matured and contain up to eight rediae which are liberated at this time. Following 10 days of growth (21 days after infection of the snail), the rediae are mature and measure 0.5 to 1 mm long. Each one contains 15 to 30 cercariae.

Being immature when released from the rediae, the cercariae remain in the snail's tissue for about 13 days until mature. They are released in the water around 34 to 35 days after the miracidia enter the snail. Mature cercariae have two eyespots, a long, slender tail, are brown in color, and have the arrangement of the suckers characteristic of the amphistomes. They are

## Explanation of Plate 65

A, Ventral view of adult fluke. B, Sinistral view of adult fluke. C, Free-swimming miracidium. D, Mother sporocyst containing mother rediae. E, Daughter redia (mother redia not shown). F, Cercaria. G, Bovine definitive host. H, Snail (*Stagnicola bulimoides*) first intermediate host. I, Vegetation with encysted metacercaria.

1, Oral sucker; 2, ventral sucker; 3, esophagus; 4, intestine; 5, testis; 6, vas deferens; 7, common genital pore; 8, ovary; 9, uterus; 10, vitelline glands; 11, Mehlis' gland; 12, Laurer's canal; 13, excretory bladder with excretory pore dorsal and subterminal; 14, apical papilla; 15, apical gland; 16, penetration glands; 17, germinal mass; 18, mother redia; 19, pharynx; 20, gut; 21, developing cercaria with well-formed eyespots; 22, eyespots; 23, excretory tubules; 24, tail.

a, Adult fluke in rumen; b, unembryonated eggs laid in rumen and pass out of body in feces; c, unembryonated eggs must reach water to develop; d, developed egg hatches in water; e, miracidium free in water; f, miracidium pentrating snail; g, mother sporocyst containing mother rediae and germinal masses; h, daughter redia with developing cercariae; i, cercaria escaping from snail; j, cercaria free in water; k, metacercaria encysted below water line; l, metacercaria above water line, having reached there by growing vegetation and lowering of water, or both; m, host infected by swallowing metacercariae; n, metacercaria passes through stomachs unaffected; o, metacercaria excysts in duodenum; p, young flukes enter mucosa; q, young flukes migrate forward in mucosa; r, young flukes leave mucosa and enter intestinal lumen; s, t, young flukes in lumen migrate anteriorly through abomasum, omasum, and reticulum to rumen where they grow to maturity in 2 to 4 months.

Figures A, B adapted from Fischoeder, 1903, Zool. Jahrb. Abt. Syst. 17:485; C–F, from Brumpt, 1936, Ann. Parasitol. Hum. Comp. 14:552.

---

stimulated by light to emerge from the snails, appearing during the day. Following a short period of activity in the water, they encyst on herbage and other objects.

Upon being swallowed along with forage, excystment of the metacercariae occurs in the duodenum and jejunem. The young flukes attach to the mucosal lining and burrow into it, migrating forward. In the abomasum, they return to the lumen and travel forward through the esophageal groove, passing through the omasum and reticulum to the rumen where they attach among the villi. Excysted flukes in the small intestine often wander into the bile and pancreatic ducts. In the rumen, the flukes attain maturity and eggs appear in the feces 60 to 120 days after infection of the host. Great numbers of adult worms occur in the rumen.

**Symptoms and Pathology.** During the migration of the young flukes in the small intestine, there is a severe watery and fetid diarrhea which often is accompanied by mortality of 80 to 90 percent of the infected animals. Up to 30,000 or more young flukes may be present, all attacking the mucosal lining which is destroyed. While it is generally reported that the adults are commensals, doing no harm, they cause the villi to erode and otherwise have a definite pathological effect on the rumenal tissues. Young flukes in the bile and pancreatic ducts cause marked thickening of them.

Repeated infections with small numbers of metacercariae produce a degree of immunity with marked reduction in the number of flukes, retarded growth, and limited migration.

Adult flukes can be removed effectively by drenching cattle with hexachloroethane-bentonite-water suspension.

## EXERCISE ON LIFE CYCLE

Investigation on the intramolluscan phases of the life cycle should be conducted on the same basis as given for *Fasciola hepatica*. Make careful observations to determine the stages occurring in the snail and compare them with those of *F. hepatica*. The cost of sheep and calves for experimental purposes prohibit their use in most cases.

## SELECTED REFERENCES

Bennett, H. J. (1936). The life cycle of *Cotylophoron cotylophoron*, a trematode from ruminants. Ill. Biol. Monogr. 14:1–81.

Brumpt, E. J. A. (1936). Contribution a l'étude de l'évolution des paramphistomidés, *Paramphistomum cervi* et cercaire de *Planorbis exustus*. Ann. Parasitol. 14:552–563.

Damiano, S. (1965). Contributo alla conoscenza della paramfistomosi negli animali domestici: ricerche anatomo-istopatologiche nella specie bovina. Nota 2a. Acta Med. Vet. Napoli 11:65–81.

**Plate 65**   *Paramphistomum cervi*                                                275

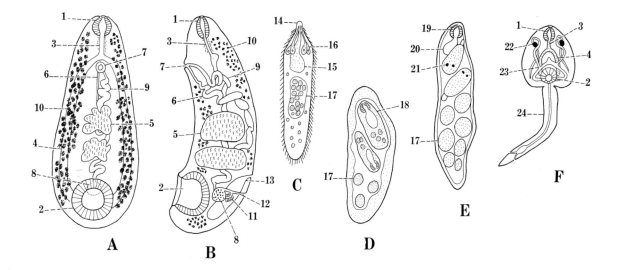

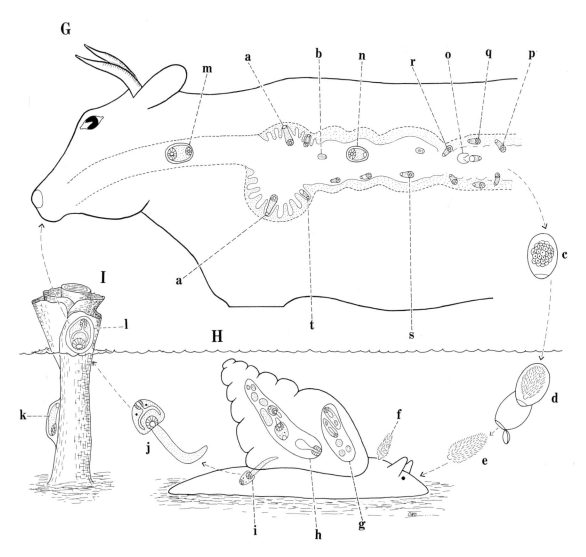

## Explanation of Plate 66

A, Ventral view of adult worm. B, Sagittal section of adult worm. C, Embryonated egg with miracidium containing a single redia. D, Miracidium. E, Miracidium without cilia showing dermal plates. F, Young mother redia just released from miracidium. G, Mature mother redia with daughter redia. H, Daughter redia with cercariae. I, Cercaria. J, Beaver (*Castor canadensis*) definitive host. K, Snail (*Fossaria parva*) intermediate host.

1, Oral sucker; 2, ventral sucker; 3, esophagus; 4, intestine; 5, testis; 6, cirrus pouch; 7, ovary; 8, seminal receptacle; 9, uterus; 10, common genital pore surrounded by muscular genital sucker; 11, vitelline glands; 12, opening of Laurer's canal; 13, miracidium in unhatched egg; 14, redia inside miracidium; 15, apical gland; 16, penetration glands; 17, excretory tubule; 18, flame cell; 19, excretory pore; 20, redia in hatched miracidium; 21, pharynx; 22, gut; 23, penetration glands; 24, first tier of six dermal plates; 25, second row of eight dermal plates; 26, third row of four dermal plates; 27, fourth row of two dermal plates; 28, pharynx; 29, germ balls; 30, birth pore; 31, young daughter redia; 32, mature cercaria escaping through birth pore of daughter redia; 33, cercaria; 34, tail.

a, Adult fluke in cecum of beaver; b, unembryonated eggs are laid in the cecum and voided in feces; c, unembryonated egg free in water; d, egg hatching; e, miracidium penetrating snail; f, young mother redia escaping from miracidium; g, mature mother redia containing daughter redia and germ balls; h, daughter redia containing cercariae and germ balls; i, cercaria escaping from snail; j, cercaria free in water; k, metacercariae encysted on vegetation or other objects in water; l, beavers become infected by swallowing infective metacercariae with food; m, metacercaria escapes from cyst under influence of digestive juices and migrates to the cecum where it develops to maturity.

Figures A, B, D, H adapted from Skrjabin, 1949, Trematodes of Animals and Man, Vol. 4, Plates 57, 58 (Russian text); C, E–G, from Bennett and Humes, 1939, J. Parasitol. 25:225.

Dinnik, J. A. (1957). Development of *Paramphistomum sukari* Dinnik, 1954 (Trematoda: Paramphistomidae) in a snail host. Parasitology 47: 209–216.

Dinnik, J. A., and N. N. Dinnik. (1954). The life cycle of *Paramphistomum microbothrium* Fischoeder, 1901 (Trematoda, Paramphistomidae). Parasitology 44:285–299.

Durie, P. H. (1953). The paramphistomes (Trematoda) of Australian ruminants. II. The life history of *Ceylonocotyle streptocoelium* (Fischoeder) Nasmark and *Paramphistomum ichikawai* Fukui. Austr. J. Zool. 1:193–222.

Durie, P. H. (1955). A technique for the collection of large numbers of paramphistome (Trematoda) metacercariae. Austr. J. Agr. Res. 6:200–202.

Durie, P. H. (1956). The paramphistomes (Trematoda) of Australian ruminants. III. The life-history of *Calicophoron calicophoron* (Fischoeder) Nasmark. Austr. J. Zool. 4:152–157.

Horak, I. G. (1967). Host-parasite relationships of *Paramphistomum microbothrium* Fischoeder, 1901, in experimentally infected ruminants with particular reference to sheep. Onderstepoort J. Vet. Res. 34: 451–540.

Horak, I. G. (1971). Paramphistomiasis of domestic animals. In: B. Dawes, ed. Advances in Parasitology. Academic Press, New York, Vol. 9, pp. 33–72.

Krull, W. H. (1934). Life history studies on *Cotylophoron cotylophoron* (Fischoeder, 1901) Stiles and Goldberger, 1910. J. Parasitol. 20:173–180.

Kurtpinar, H., and B. M. Latif. (1970). Paramphistomiasis of cattle and buffaloes in Iraq. Vet. Rec. 87:668.

Olsen, O. W. (1949). Action of hexachloroethane-bentonite suspension on the rumen fluke, *Paramphistomum*. Vet. Med. 44:108–109.

Szidat, L. (1936). Ueber die Entwicklungsgeschichte und den ersten Zwischenwirt von *Paramphistomum cervi* Zeder 1790 aus dem Magen von Wiederkäuern. Z. Parasitenk. 9:1–19.

### *Stichorchis subtriquetrus* (Rudolphi, 1814) (Plate 66)

This is a parasite of beavers throughout their range.

**Description.** The fleshy body is attenuated anteriorly and broadly rounded posteriorly. The oral sucker is large and has two pouches within the posterior wall; an acetabulum is on the ventral surface of the body near but not at the posterior extremity. A small cirrus sac and genital sucker are distinct but not strongly developed. The testes are large and lobulated, lying anterior to the small spherical ovary. Vitelline follicles are lateral and mostly posterior to the testes. Eggs measure 118 to 154 $\mu$ long by 82 to 118 $\mu$ wide.

**Life Cycle.** Unembryonated eggs laid in the cecum of beavers are expelled with the feces. Development in the water is slow but spectacu-

**Plate 66** *Stichorchis subtriquetrus* 277

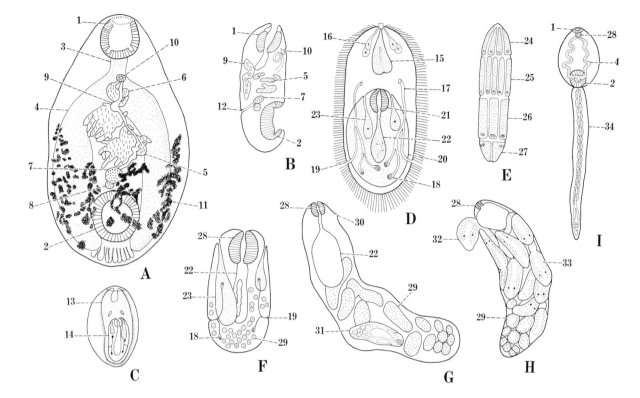

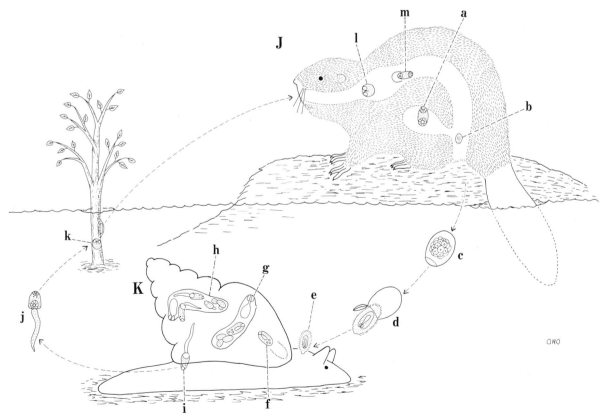

lar. By the end of 3 weeks, a typically ciliated miracidium is formed. It contains in the posterior part of its body a single fully developed mother redia. Neither germ cells nor germ balls appear in these miracidia at this time. Hatching begins at the end of week 3. In water, the miracidia may, by contractions of the posterior portion of the body, force the mother redia forward and through a rent in the wall of the anterior end. Miracidia that attack and penetrate the mantle of the amphibious snail *Fossaria parva* in North America (it develops in *Planorbis vortex*, *Bithynia tentaculata*, *Limnaea ovata*, and *Succinea putris* in Russia) liberate the redia and then disintegrate in the tissues. By the end of day 2, the mother redia has escaped from the body of the miracidium but is still in the mantle. It reaches the liver by day 17, perhaps somewhat earlier, at which time developing daughter rediae are present inside. Three weeks after infection, daughter rediae are present in the liver of the snail.

Cercariae are liberated about 35 days after infection of the snails. They swim about, beginning to encyst around 18 hours after becoming free in the water. Under optimal conditions, larval development in the snail requires 10 to 11 weeks.

Metacercariae encyst on submerged vegetation and are swallowed by the beavers along with their food. Upon liberation in the digestive tract, they migrate to the cecum. No data are available on the prepatent period or the length of time they remain in the host.

Helminths are reported to cause mortality among beavers in Russia, with *S. subtriquetrus* being responsible for much of the losses. When moving beavers from one locality to another, the Russians treat them to remove the amphistomes.

In addition to *S. subtriquetrus*, *Typhlocoelium cymbium*, a monostome fluke from the trachea of ducks, and *Parorchis acanthus*, an echinostome from the bursa Fabricii or rectum of herring gulls, have a well-developed mother redia inside the miracidium before it hatches from the egg.

## EXERCISE ON LIFE CYCLE

In localities where carcasses of beavers are available during the trapping season, the amphistomes can be obtained readily, as they are relatively common. Eggs dissected from the uterus and washed free of particles of tissue will develop and hatch in about 3 weeks. Observe the details of morphology of the miracidium and the mother redia contained inside. Also note how the redia is pushed forward in the miracidium and forced through the wall.

Snails collected from beaver ponds are a source of the amphistome cercariae and from them the metacercariae. Expose *Fossaria parva* to recently hatched miracidia. By means of sections, examine the mantle 48 to 60 hours after infection to see the miracidia and mother rediae.

Examination of snails at intervals of 1 week will demonstrate the mother and daughter rediae in the liver. Successful infection of *F. parva* is rather difficult, suggesting that perhaps other species of snails might serve as the natural intermediate host.

Note the morphological and behavioral characteristics of the cercariae together with the time required for them to develop in the daughter rediae.

Expose meadow voles to infection by feeding metacercariae to them. Guinea pigs also should be tried as an experimental host.

## SELECTED REFERENCES

Bennett, H. J., and A. G. Humes. (1939). Studies on the pre-cercarial development of *Stichorchis subtriquetrus* (Trematoda: Paramphistomidae). J. Parasitol. 25:223–231.

Romanshov, V. A. (1966). Control of helminthiasis of the beaver during acclimatization and reacclimatization. In: A. I. Yanushevich, ed. Acclimatization of Animals in the U. S. S. R. Jeruslem: Israel Program for Scientific Translations, pp. 245–246.

### FAMILY DIPLODISCIDAE

The oral sucker has a pair of posterior diverticula. The body is conical and bears a large terminal posterior sucker.

*Megalodiscus temperatus*
(Stafford, 1905) (Plate 67)

This amphistome, formerly known as *Diplodiscus temperatus*, is a common parasite in the rectum of frogs in North America.

**Description.** Fully grown flukes measure up to 6 mm long and about 2 to 2.25 mm thick, being somewhat round. The oral sucker is well developed and bears a pair of prominent contractile posterior pouches about two-thirds the length of the sucker; the posterior sucker is terminal with a diameter about equal to that of the body. The ceca begin immediately behind the pharynx and extend to the posterior extremity of the body. The testes are in the anterior half of the body and the ovary is posterior to them.

**Life Cycle.** Adult flukes in the rectum of frogs and tadpoles lay thin-shelled embryonated eggs which are deposited in the water with the feces. They hatch promptly and the miracidia penetrate young ram's horn snails, *Helisoma trivolvus, H. antosum,* and *H. campanulata,* and the limpets *Ferrissia fragilis* and *Laevaplex fuscus,* where they transform into mother sporocysts. Three generations of rediae develop in the liver. The third one produces cercariae that are released in an immature stage, and development is completed in the tissues of the snail host. Mature spermatozoa are present in the testes of the cercariae when they are shed by the snails, about 90 days after infection.

Cercariae escape daily from infected snails, the number being somewhat greater during the afternoon hours. Fewer of them appear during dull days than bright ones. These ocellated cercariae are positively phototropic, always moving toward the light areas. They encyst on the skin of tadpoles and adult frogs, especially on the dark spots of the fore- and hindlegs. They attach quickly and encyst almost immediately upon making contact. In view of their positive phototropic tendencies, the question arises as to why they select the dark spots on the skin rather than the light portions. Possibly it is due to a chemotactic response. Metacercariae are attached loosely to the skin of tadpoles and become dislodged easily, whereas they adhere tenaciously to the skin of adult frogs.

Infection of adult frogs occurs normally when they ingest the sloughed stratum corneum bearing metacercariae. Larval flukes free themselves from the cysts by vigorous movements, with excystment taking place in the rectum. Tadpoles become infected by eating encysted metacercariae or free cercariae. When cercariae are taken into the mouth of tadpoles, they encyst promptly and pass to the rectum, where excystment takes place. During metamorphosis of tadpoles, flukes that are not expelled migrate forward into the small intestine, appearing as far anterior as the stomach. As the intestine shortens after metamorphosis and the young frogs begin taking a

protein diet, the remaining flukes return to the rectum.

Maturity of the flukes may occur as early as 27 days after infection but it usually requires 2 to 3 months, depending on the temperature. When a large number of flukes crowd the rectum, development is delayed to 3 to 4 months. Flukes remain in the frogs at least from one summer to the next and possibly longer.

## EXERCISE ON LIFE CYCLE

When a source of naturally infected frogs is available for study, preparations for making the observations on the life cycle should be made in advance by obtaining a supply of frog eggs and snail eggs (*Helisoma trivolvus, H. antrosum, H. campanulata*). Laboratory-reared tadpoles, frogs, and snails are essential for making carefully controlled studies on the life history. The supply of experimental material may be kept in a refrigerator until needed.

Adult flukes freshly removed from the rectum of frogs and placed in a dish of water promptly shed fully embryonated eggs which hatch soon after reaching the water.

Snails 4 to 6 weeks old exposed to freshly hatched miracidia become infected with a high percentage of success. Examine these snails by making dissections and sections at intervals of 2 weeks or less to follow the development of the intramolluscan stages.

When cercariae are available, either from naturally or experimentally infected snails, place frogs in a dish of water with them. Observe encystment on the body. Place tadpoles in a dish with cercariae and note that they are ingested. Encystment occurs in the mouth almost at the moment cercariae enter it. By dissections of tadpoles, observe where excystation occurs. Likewise note what happens to parasites in tadpoles during and after metamorphosis.

Infection of adult frogs can be observed by watching for the sloughed stratum corneum that is eaten together with the encysted metacercariae that adhere tightly to it. Determine when eggs appear in the feces of frogs that have swallowed metacercariae.

There are other common amphistomes whose life cycles may be studied with profit when material is available. *Allassostomoides parvum* from the cloaca of snapping turtles and the urinary bladder of painted turtles and frogs develop in

## Explanation of Plate 67

**A,** Adult fluke. **B,** Miracidium. **C,** Miracidium showing epidermal plates. **D,** First generation redia. **E,** Daughter redia. **F,** Ventral view of cercaria. **G,** Dextral view of cercaria. **H,** Frog definitive host. **I,** Snail (*Helisoma trivolvus*) first intermediate host. **J,** Adult frog with encysted metacercariae on dark spots of skin. **K,** Tadpoles with metacercariae encysted on tail or eating cercariae which will encyst in mouth and be swallowed.

**1,** Oral sucker; **2,** posterior sucker; **3,** pharyngeal pouch; **4,** pharynx; **5,** intestine; **6,** testis; **7,** vas efferens; **8,** vas deferens; **9,** common genital pore; **10,** ovary; **11,** seminal receptacle; **12,** oviduct; **13,** Laurer's canal; **14,** Mehlis' gland surrounding ootype; **15,** uterus; **16,** vitelline glands; **17,** vitelline duct; **18,** penetration glands; **19,** lateral papillae; **20,** eyespots; **21,** ganglion; **22,** germ cells; **23,** anterior row of six epidermal plates; **24,** second row of eight epidermal plates; **25,** third row of four epidermal plates; **26,** fourth row of two epidermal plates; **27,** second generation redia; **28,** developing cercariae in third generation redia; **29,** germ balls; **30,** tail; **31,** excretory tubule.

**a,** Adult fluke in rectum; **b,** embryonated eggs laid in rectum; **c,** eggs passed in feces; **d,** eggs in water; **e,** hatching of egg; **f,** miracidium penetrating snail host; **g,** miracidium sheds epithelial layer and transforms into mother sporocyst; **h,** mother rediae containing daughter rediae; **i,** daughter redia with cercariae; **j,** amphistome cercaria escaping from snail; **k,** cercaria free in water; **l,** cercariae attach on dark spots of skin of frogs where they encyst; **m,** cercariae attach to tadpoles; **n,** cercariae swallowed by tadpoles encyst in digestive tract; **o,** frogs become infected by eating the sloughed stratum corneum with attached metacercariae; **p,** encysted metacercaria; **q,** young flukes escape from cyst, attach to rectal wall and grow to maturity in about three months.

Figure **D** adapted from Herber, 1938, J. Parasitol. 24:549; others from Krull and Price, 1932, Occ. Papers Mus. Zool. Univ. Mich., No. 237.

---

rediae in *Planorbis* and *Helisoma*. The cercariae encyst on vegetation, crayfish, and tadpoles. *Zygocotyle lunata* of ducks develops in rediae in *Helisoma antrosum*. The cercariae encyst on objects in the water. This species is able to infect mammals in addition to ducks.

## SELECTED REFERENCES

Beaver, P. C. (1929). Studies on the development of *Allassostoma parvum* Stunkard. J. Parasitol. 16: 13–23.

Herber, E. C. (1938). On the mother redia of *Diplodiscus temperatus* Stafford, 1905. J. Parasitol. 24:549.

Herber, E. C. (1939). Studies on the biology of the frog amphistome, *Diplodiscus temperatus* Stafford. J. Parasitol. 25:189–195.

Krull, W. H. (1933). Notes on the life history of a frog bladder fluke. J. Parasitol. 20:134.

Krull, W. H., and H. F. Price. (1932). Studies on the life history of *Diplodiscus temperatus* Stafford from the frog. Occ. Pap. Mus. Zool. Univ. Mich. No. 237, 37 pp.

Smith, R. J. (1967). Ancylid snails as intermediate hosts of *Megalodiscus temperatus* and other digenetic trematodes. J. Parasitol. 53:287–291.

Van der Woude, A. (1954). Germ cell cycle of *Megalodiscus temperatus* (Stafford, 1905) Harwood, 1932 (Paramphistomidae: Trematoda). Amer. Midl. Natur. 51:172–202.

Willey, C. H. (1941). The life cycle and bionomics of the trematode, *Zygocotyle lunata* (Paramphistomidae). Zoologica 26:65–88.

### Superfamily Notocotyloidea

This superfamily consists of monostome flukes in fish, reptiles, birds, and mammals. The oral sucker is well developed and located at the anterior end of the body. There is no pharynx and the ceca terminate near the posterior extremity of the body. Both testes and the ovary are near the caudal end of the body. Eggs are small and each end bears a long polar filament. The cercariae are monostomate, ocellated, and have very long tails.

#### FAMILY NOTOCOTYLIDAE

Notocotylidae are small flukes in the gut and ceca of birds and mammals. The cirrus pouch is well developed and the shell gland is anterior to the ovary. Vitelline follicles are lateral to the ceca, and anterior, lateral, or dorsal to the testes. The uterus, coiled transversely between the ceca, is filled with eggs. The excretory bladder is short, with two long arms that unite anteriorly.

*Quinqueserialis quinqueserialis*
Barker and Laughlin, 1911 (Plate 68)

This monostome is a natural parasite of the cecum of muskrats (*Ondatra zibethica*), meadow voles (*Microtus pennsylvanicus*), and jumping

Plate 67     *Megalodiscus temperatus*                                                    281

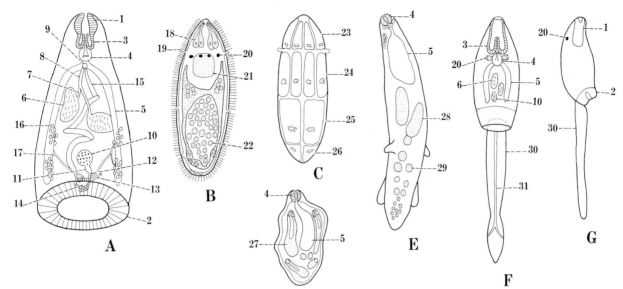

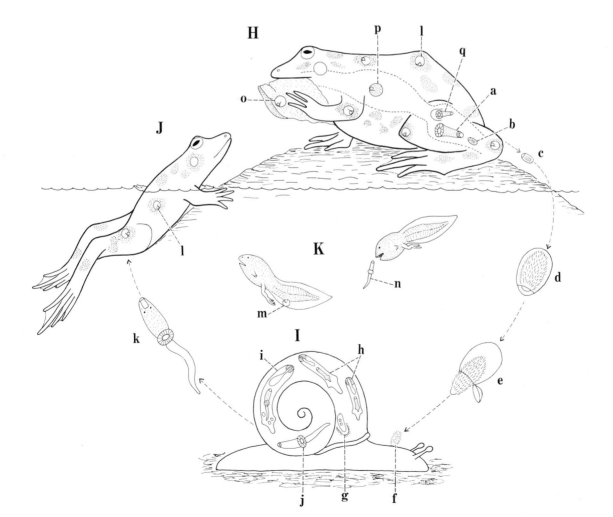

## Explanation of Plate 68

**A,** Adult fluke, showing internal anatomy. **B,** Ventral surface of adult fluke, showing five rows of glands. **C,** Embryonated filamentous egg. **D,** Young mother sporocyst in mantle of snail host. **E,** Young mother sporocyst. **F,** Mother redia. **G,** Daughter redia containing cercariae. **H,** Cercaria. **I,** Encysted metacercaria. **J,** Muskrat (*Ondatra zibethica*) definitive host. **K,** Meadow vole (*Microtus pennsylvanicus*) definitive host. **L,** Snail (*Gyraulus parvus*) intermediate host.

1, Oral sucker; 2, esophagus; 3, intestine; 4, testis; 5, vas efferens; 6, vas deferens; 7, cirrus pouch; 8, seminal vesicle; 9, common genital pore; 10, ovary; 11, Mehlis' gland; 12, seminal receptacle; 13, uterus; 14, metraterm; 15, vitelline gland; 16, vitelline duct; 17, excretory pore; 18, glands on ventral surface; 19, embryo of egg; 20, filament of egg; 21, mantle of snail host; 22, young mother sporocyst with mother rediae in mantle; 23, mother sporocyst; 24, mother redia; 25, pharynx; 26, intestine; 27, daughter redia; 28, germ balls; 29, developing cercaria; 30, eyespot; 31, excretory bladder; 32, excretory tubules; 33, tail; 34, metacercarial cyst; 35, metacercaria.

a, Adult fluke in cecum; b, embryonated filamentous eggs laid in cecum; c, eggs carried into intestine; d, eggs passed in feces; e, embryonated egg in water; f, eggs eaten by snail (*Gyraulus parvus*) intermediate host; g, eggs hatch in intestine; h, miracidium passes through wall of intestine; i, mother sporocyst containing mother rediae; j, mother redia containing daughter rediae; k, daughter redia containing cercariae; l, cercaria escaping from snail; m, cercaria free in water; n, metacercaria attached to submerged vegetation; o, definitive host becomes infected by swallowing metacercariae on food; p, metacercaria in stomach; q, excystation of metacercaria; r, young fluke enters cecum, reaching maturity in 16 days.

Adapted from Herber, 1942, J. Parasitol. 28: 179.

---

mice (*Zapus hudsonius*) in the United States and Canada. It is able to mature in 15 other species of rodents.

**Description.** Fully grown flukes are oval in shape and 2.9 mm long by 0.93 mm wide. The ventral surface bears five longitudinal rows of ventral glands with 14 to 16 in the middle row and 16 to 19 in the others. The intestinal ceca arise near the oral sucker and extend to near the posterior end of the body as irregular tubes. The two lobed testes lie at the same level outside the ceca near their ends, and the lobed ovary is located intercecally between them. The uterus appears as a series of close transverse loops between the ootype and the base of the long cirrus pouch which equals about one-third the length of the body. There are 12 to 17 groups of extracecal vitelline follicles extending from the base of the cirrus pouch to the anterior margin of the testes.

**Life Cycle.** Adult worms in the ceca lay their embryonated filamentous eggs which are mingled with the cecal contents and gradually passed from the body in the feces. They do not hatch until eaten by the snail *Gyraulus parvus*. The miracidia penetrate the gut wall of the snail host, shed the cilia, and transform into sac-like mother sporocysts containing four mother rediae in the tissue surrounding the intestine. They force their way through the body wall of the mother sporocyst and migrate to the liver, en-

tering it 4 days later, or 19 days after infection. Daughter rediae develop rapidly and are free in the liver by day 20 after infection. Within 3 days, 23 following infection, partially developed cercariae are present in the daughter rediae. They emerge through the birth pore in an undeveloped stage, requiring 3 days in the liver to reach maturity, thus making a total of 26 days from infection of the snails to the shedding of mature cercariae. Emergence from the snails is periodic, being between 9 and 11 A.M. The tail of the cercariae is about twice the length of the body and there are three eyespots in a transverse row on the ventral side of the body. When the cercariae strike an object, they attach to it and encystment is completed in 3 to 5 minutes, at which time they are infective.

Muskrats and voles feeding on vegetation in the water or on that which has emerged from it swallow the encysted flukes. Metacercariae released from the cysts go to the cecum and mature sexually in 15 days in *Microtus montanus*, 18 in *M. pennsylvanicus*, and 28 in *Ondatra zibethica*. Growth continues after sexual maturity is attained.

*Notocotylus stagnicolae* Herber, 1942 is closely related to *Q. quinqueserialis* and has a life cycle very similar to it. The snail intermediate hosts are *Stagnicola emarginata* and *Lymnaea auricularia*. They are natural parasites of willets, but chickens, ducks, and mergansers may

**Plate 68**    *Quinqueserialis quinqueserialis*                                    283

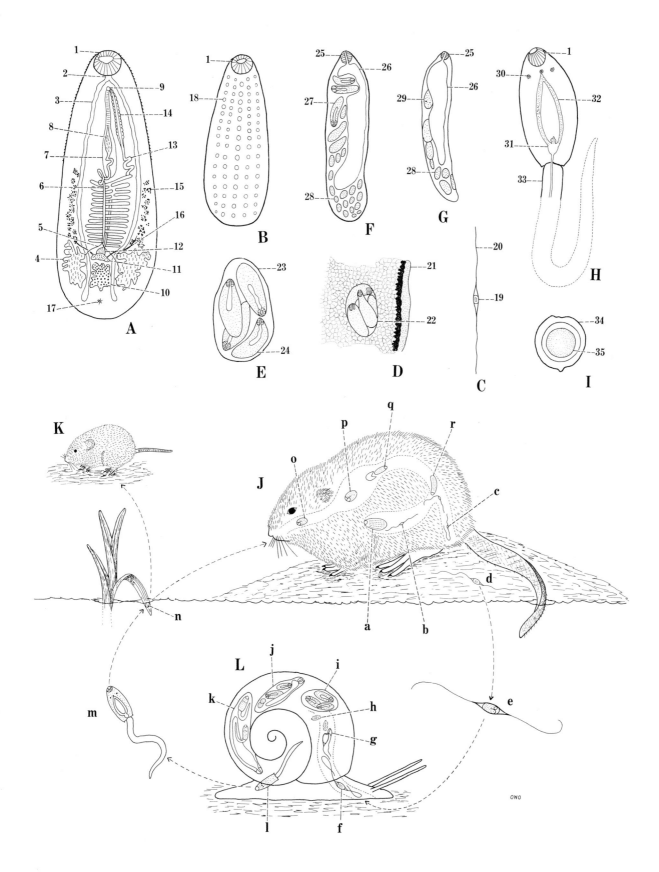

be infected experimentally. The mother sporocyst contains but one redia in contrast to four in *Q. quinqueserialis*. *Notocotylus urbanensis* is a parasite of muskrats and ducks whose cercariae develop in *Stagnicola emarginata* and *Physa parkeri*. *Notocotylus atlanticus* of eider ducks develops in the brackish water snail *Hydrobia salsa*. *Notocotylus minutus* also utilizes a marine snail intermediary. *Nudacotyle novicia* Barker, 1961 is a parasite of meadow voles and muskrats. *Cercaria marilli* occurring in rediae in the snail *Pomatiopsis lapidaria* forms metacercariae that develop into *N. novicia* when fed to voles and muskrats.

## EXERCISE ON LIFE CYCLE

Whenever muskrat carcasses are available during the trapping season, a good supply of *Q. quinqueserialis* will be assured. Filamentous eggs teased from the flukes, which will be dead if obtained from frozen carcasses, will contain fully developed and living miracidia. They are infective to *Gyraulus parvus* when eaten by them. Feed eggs to a large number of snails, preferably laboratory-reared ones, and dissect and section them at intervals of 1 day from the time of feeding to obtain the larval stages. Examine the mantle, the tissues surrounding the intestine, and the digestive gland (liver) for the various stages. Note how rapidly the cercariae encyst when freed from the snails, a function of the numerous cystogenous glands in the body wall. Feed metacercariae to baby chicks, ducklings, and voles to obtain adult flukes.

The life cycle may begin with cercariae from naturally infected snails instead of eggs which may not be available at a suitable time.

## SELECTED REFERENCES

Acholonu, A. D., and O. W. Olsen. (1967). Studies on the life history of two notocotylids (Trematoda). Proc. Helminthol. Soc. Wash. 34:43–50.

Ameel, D. J. (1944). The life history of *Nudacotyle novicia* Barker, 1916 (Trematoda: Nudacotylidae). J. Parasitol. 30:257–263.

Herber, E. C. (1940). The mother sporocysts of three species of monostomes of the genus *Notocotylus* (Trematoda). J. Parasitol. 26 (Suppl.):35.

Herber, E. C. (1943). Life history studies on two trematodes of the subfamily Notocotylinae. J. Parasitol. 28:179–196.

Herber, E. C. (1955). Life history studies on *Notocotylus urbanensis* (Trematoda: Notocotylidae). Proc. Pa. Acad. Sci. 29:267–275.

Kinsella, J. M. (1971). Growth, development, and intraspecific variation of *Quinqueserialis quinqueserialis* (Trematoda: Notocotylidae) in rodent hosts. J. Parasitol. 57:62–70.

Luttermoser, G. W. (1935). A note on the life history of the monostome, *Notocotylus urbanensis*. J. Parasitol. 21:456.

Stunkard, H. W. (1966). The morphology and life-history of *Notocotylus atlanticus* n. sp., a digenetic trematode of eider ducks, *Somateria mollissima* and the designation, *Notocotylus duboisi* nom. nov., for *Notocotylus imbricatus* (Looss, 1893) Szidat, 1935. Biol. Bull. 131:501–515.

### Superorder Epitheliocystidia

In this group of trematodes, the primitive excretory bladder of the cercariae is first surrounded by and later replaced by a layer of mesodermal cells, forming a thick epithelial wall. Caudal excretory vessels may be present or absent. The cercariae which develop in sporocysts are typically distomate in form, i.e., having two suckers near the anterior end of the body.

### ORDER PLAGIORCHIIDA

The cercariae lack the caudal excretory vessels in all stages of development but a stylet may or may not be present in the oral sucker. The adults are parasites of all classes of vertebrates.

#### Suborder Plagiorchiata

The cercariae are pharyngeate and have a horizontal stylet. They encyst in invertebrates, chiefly larval aquatic arthropods.

#### Superfamily Plagiorchioidea

The principal characteristic of this superfamily is the joining of the main collecting canals of the excretory system to the anterior end of the arms of the Y-shaped excretory bladder.

### FAMILY DICROCOELIIDAE

Dicrocoeliidae are parasites of the liver, gall bladder, pancreas, and intestine of amphibians, reptiles, birds, and mammals. The best known member of the family is the lancet fluke in the bile ducts of sheep.

Members of the family are medium-sized flukes with the oral sucker subterminal and the acetabulum in the anterior half of the body. A pharynx, short esophagus, and simple ceca of variable length are present. Testes are arranged variously but usually in the hind part of the body; the ovary is post-testicular.

### Dicrocoelium dendriticum
(Rudolphi, 1819) (Plate 69)

This species occurs in the bile ducts of sheep, goats, cattle, pigs, deer, elk, rabbits, and marmots. It is common in Europe and Asia, and limited in distribution in the northeastern part of North America and in Australasia.

**Description.** Grown flukes range from 6 to 10 mm long by 1.5 to 2.5 mm wide. The body is elongate with the widest part near the middle. The oral sucker is slightly smaller than the ventral and both are in the anterior third of the body. The large testes lie one behind the other immediately posterior to the ventral sucker. The small ovary is immediately behind the posterior testis. Loops of the uterus fill that part of the body behind the ovary. Vitellaria occupy the middle third of the lateral part of the body. Operculate eggs, which are embryonated when laid, measure 36 to 45 $\mu$ long by 22 to 30 $\mu$ wide.

**Life Cycle.** Eggs laid in the bile ducts pass through the common bile duct into the intestine and are voided with the feces. Hatching does not occur until they are eaten by land snails. Physicochemical factors in the intestine associated with low pH and reducing conditions trigger hatching. In the United States, *Cionella lubrica* is the first intermediate host. Since a number of different species of terrestrial snails other than *C. lubrica* are known to serve as intermediate hosts of this fluke in Europe, it might be expected that other species may act in this capacity in North America. As yet none has been found, although some have been exposed to experimental infection with negative results.

Upon hatching, the miracidia migrate through the gut wall to the digestive gland where they transform into mother sporocysts. These give rise to many daughter sporocysts, each with a birth pore, that produce stylet-bearing cercariae. Development of the cercariae is slow, requiring somewhere around 3 months. Upon reaching maturity, they leave the sporocysts and migrate from the digestive gland into the respiratory chamber. As the cercariae accumulate there the snails somehow are stimulated to secrete sufficient mucus to surround them. When a drop in temperature occurs, the mucus and the enclosed cercariae are expelled explosively through the respiratory pore. This is the slimeball in which the cercariae find an aquatic environment to bridge the gap between the terrestrial snail intermediary and the second invertebrate intermediate host. The outer surface of the slime mass dries, forming a tough membrane that retards loss of water from within. Only a very few slimeballs are produced by each snail during its life but they may contain up to nearly 500 cercariae each. Slimeballs and egg masses of the snails have a remarkable resemblance. Under conditions of excessive moisture, the slimeballs soon liquefy, leaving the cercariae to die.

Ants (*Formica fusca* and a variety of other species) foraging over leaves and sticks where the snails live find the slimeballs and carry them to their nests. Worker ants eat the slimeballs and cercariae, becoming infected. Up to 35 percent of the adult ants in endemic areas are naturally infected with metacercariae. When eaten by ants, the cercariae migrate through the crop wall and encyst in the gaster, thorax, and head, often attaching to the cranial nerve centers. More than 100 encysted metacercariae may occur in a single ant.

Neurophysiological effects from the assimilation by esophageal ganglia of metacercarial metabolites in the head cause infected ants to climb vegetation, where they attach by means of the mandibles and enter a stupor during the cool periods of the day and night. In this location, they are readily available to and swallowed by grazing animals. The metacercariae are released in the duodenum, probably under the influence of the pancreatic juice. Some workers believe the young flukes enter the liver by way of the bile duct, arriving within an hour after the cysts are swallowed. In sheep where the bile duct was separated surgically from the small intestine and attached to the colon, the flukes entered the blood vessels of the intestinal wall 25 to 30 cm below Vater's papilla and traveled via them to the intercellular spaces of the liver. From these locations, they went into the bile ducts and finally the gall bladder. Flukes are sexually mature in 6 to 7 weeks in lambs and eggs appear in the feces about 4 weeks later.

In view of the high incidence of natural infection of ants, large numbers of flukes may be expected in the definitive hosts. Up to 7000 have been found in the gall bladder of a single sheep and 50,000 in a liver. All the sheep in endemic areas may be infected. Damage to the liver is

## Explanation of Plate 69

A, Adult trematode. B, Sporocyst containing cercariae. C, Dorsal view of cercariae showing mucous glands. D, Ventral view of cercaria showing developing organs. E, Excysted metacercaria. F, Empty egg with operculum open. G, Definitive hosts (sheep, cattle, deer, cottontail rabbits, marmots). H, Snail (*Cionella lubrica*) first intermediate host. I, Ant (*Formica fusca*) second intermediate host.

1, Oral sucker; 2, ventral sucker; 3, pharynx; 4, esophagus; 5, intestine; 6, testis; 7, cirrus pouch; 8, common genital pore; 9, ovary; 10, uterus; 11, vitelline gland; 12, Mehlis' gland; 13, Laurer's canal; 14, excretory bladder; 15, birth pore; 16, cercaria; 17, germ ball; 18, spine; 19, mucous glands; 20, tail; 21, openings of mucous glands; 22, glands; 23, genital primordium; 24, operculum.

a, Adult fluke in hypertrophied bile duct; b, embryonated eggs laid in bile ducts pass through common bile duct into intestine; c, eggs mixed with feces voided from body; d, embryonated eggs on soil; e, eggs hatch in intestine of snail; f, miracidium passes through intestinal wall; g, mother sporocyst with developing daughter sporocyst; h, daughter sporocyst with cercariae and germ balls; i, cercaria escaping from snail; j, slimeball containing cercariae; k, ants eating slimeballs swallow cercariae; l, cercariae migrate through intestinal wall into hemocoel; m, metacercaria encysted in gaster; n, definitive host becomes infected upon swallowing infected ant; o, upon digestion of ants, metacercariae are released; p, metacercaria free in duodenum; q, through action of the pancreatic juice, young fluke escapes from cyst and enters common bile duct; r, young fluke in bile duct matures in 11 to 12 weeks after entering liver.

Figures A, B, D adapted from Neuhaus, 1938, Z. Parasitenk. 10:476; C, from Neuhaus, 1936, Z. Parasitenk. 8:431; E, from Krull and Mapes, 1953, Cornell Vet. 43:389; F, from Mapes, 1951, Cornell Vet. 41:382.

---

extensive. Rabbits and marmots, which are definitive hosts, serve to intensify the infection in local areas. Deer in their wider movements disseminate it. Shipment of infected sheep to other areas gives the flukes an opportunity to move great distances and become established because of the universality of the snail and ant intermediate hosts.

Life cycles of other dicrocoeliids, where known, are similar in general to that of *D. dendriticum*.

A species of *Lyperosomum* from the liver of grackles and meadow larks develops in the land snails *Polygyra texasiana* and *Praticollela berlandierana* which produce slimeballs. The second intermediate host is unknown (Denton). *Lyperosomum monenteron* is a common parasite in the liver of robins. Its life cycle is unknown.

*Brachylecithum americanum*, a parasite of boattail and purple grackles, meadow larks, and blue jays, develops in sporocyts in *Polygyra texasiana* and *Praticollela berlandierana* in 64 days and daughter sporocysts are present between days 64 and 70. Slimeballs containing up to 300 cercariae are expelled from the respiratory pore. Larval chrysomelid beetles serve as the second intermediate host (Denton). *Brachylecithum mosquensis* of robins utilizes the land snail *Allogona ptychophora* as the first intermediate host. Carpenter ants, *Campanotus herculeanus*, *C. pennsylvanicus*, and *C. vicinus* are the second intermediate hosts. Infected ants with metacercariae display behavioral patterns that predispose them to predation. They become obese, sluggish, and strongly photophilic, lying on bare places in bright light in contrast to unparasitized individuals that are active and photophobic. Eggs appear in the feces of robins 116 days after swallowing ants.

*Conspicuum icteridorum* from the gall bladder of purple grackles develops sporocyts in the land snail *Zonitoides arboreus*. Isopods (*Oniscus asellus* and *Armadillidium quadrifons*) serve as the second intermediate host. Metacercariae fed to grackles reach the gall bladder in 16 hours and mature in 12 weeks (Patten). *Conspicuum macrorchis* of the eastern crow develops in the snail *Bulimulus alternatus mariae*, according to Denton and Byrd. The remainder of the cycle is unknown.

*Eurytrema procyonis* occurs in the pancreas of raccoons. Its ova produce mother sporocysts in the common garden snail *Mesodon thyroides* in 70 days and daughter sporocyts in 141 days. Slimeballs expelled by the snails accumulate in grape-like clusters, adhering to the mollusks as they crawl about (Denton). *E. pancreaticum* of cattle and sheep develops in the snails *Acusta despecta* and *Bradybaena similaris*. There are two generations of sporocysts. Cercariae appear in about 5 months and are deposited on grass where the long-horned grasshoppers *Cono-*

**Plate 69**     *Dicrocoelium dendriticum*                                                                287

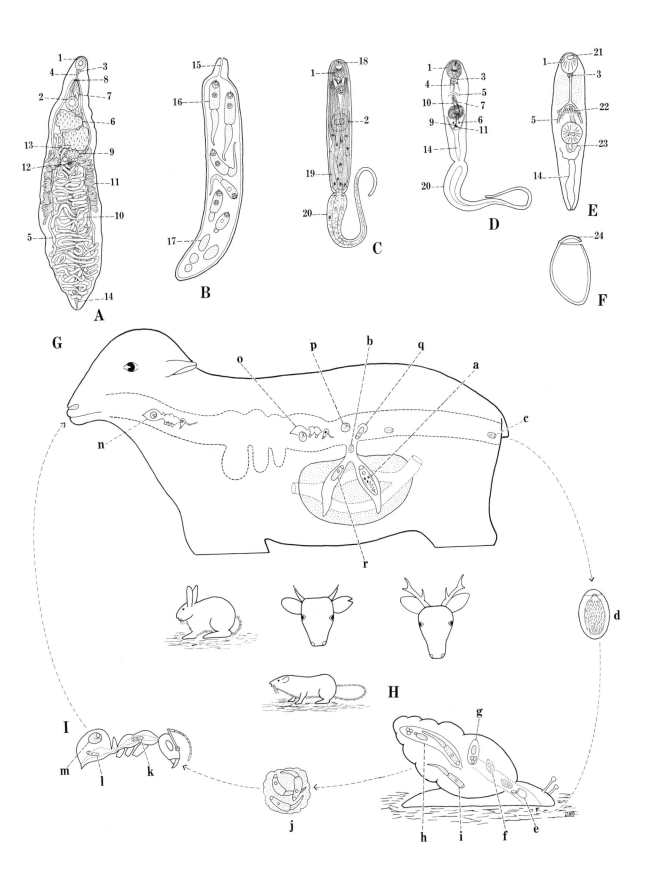

*cephalus maculatus* find and eat them, serving as the second intermediate host in Asia.

Families of mollusks which act as intermediate hosts of dicrocoeliid trematodes belong to the suborder Stylommatophora.

## EXERCISE ON LIFE CYCLE

While *Dicrocoelium dendriticum* is too limited in its distribution at this time to be useful for life history studies, the dicrocoeliid trematodes of birds and raccoons offer excellent opportunities.

A good starting point is the examination of land snails for natural infection. Stylet-bearing cercariae in them are a clue that the flukes are dicrocoeliids. The definitive hosts should be sought among the vertebrates, particularly icterids, in the same areas with the snails.

Eggs should be teased from the flukes found in the vertebrate hosts and fed to hungry snails of the species in which natural infections were found. Dissections of experimentally infected snails should be made at short intervals to recover the different generations of sporocysts.

When slimeballs are produced by the snails, they should be fed to arthropods suspected of serving as intermediate hosts because of their prevalence in the area with the naturally infected snails. Exposed arthropods should be dissected at intervals to determine the course of development in them.

Observe the habits and development of the snails. Procedures for rearing them are given in detail by Krull and in the papers on *D. dendriticum* by Krull and Mapes.

## SELECTED REFERENCES

Aliev, S. Y. (1966). [Some problems on the pathogensis of dicrocoeliasis.] Veterinariya 43:41–42.

Anokhin, I. A. (1966). [Diurnal cycle of activity and behaviour of ants (*Formica pratensis*) infected with *Dicrocoelium dendriticum* metacercariae during grazing season.] Zool. Zh. 45:687–692.

Carney, W. P. (1967). Notes on the life cycle of *Brachylecithum mosquensis* (Skrjabin and Isaitschikiff, 1927) from the bile ducts of the robin *Turdus migratorius*, in Montana. Can. J. Zool. 45: 131–134.

Carney, W. P. (1969). Behavioral and morphological changes in carpenter ants harboring dicrocoeliid metacercariae. Amer. Midl. Natur. 82:605–611.

Carney, W. P. (1970). *Brachylecithum mosquensis* infections in vertebrate, molluscan and arthropod hosts. Trans. Amer. Microsc. Soc. 89:233–250.

Denton, J. F. (1941). Studies on the life history of a dicrocoeliid trematode of the genus *Lyperosomum*. J. Parasitol. 27(Suppl.):13–14.

Denton, J. F. (1944). Studies on the life history of *Eurytrema procynis* Denton, 1942. J. Parasitol. 30: 277–286.

Denton, J. F. (1945). Studies on the life history of *Brachylecithum americanum* n. sp., a liver fluke of passerine birds. J. Parasitol. 31:131–141.

Hohorst, W. (1962). Die Rolle der Ameisen in Entwicklungsgang des Lanzettegels (*Dicrocoelium dendriticum*). Z. Parasitenk. 22:105–106.

Hohorst, W., and G. Graefe. (1966). Der Entwicklungscyklus von *Dicrocoelium dendriticum*. Int. Congr. Parasitol. (1st) Rome 2:889–890.

Jang, D. H. (1969). [Study on *Eurytrema pancreaticum*. II. Life cycle.] Korean J. Parasitol. 7:178–200.

Krull, W. H. (1937). Culture Methods for Invertebrate Animals. Comstock Publ. Co. Ithaca (Reprinted by Dover Publications, New York), p. 526.

Krull, W. H. (1955). Experiments involving potential definitive hosts of *Dicrocoelium dendriticum* (Rudolphi, 1819) Looss, 1899: Dicrocoeliidae. Cornell Vet. 46:511–525.

Krull, W. H. (1958). The migratory route of the metacercaria of *Dicrocoelium dendriticum* (Rudolphi, 1819) Looss, 1899 in the definitive host: Dicrocoeliidae. Cornell Vet. 48:17–24.

Krull, W. H., and C. R. Mapes. (1952). Studies on the biology of *Dicrocoelium dendriticum* (Rudolphi, 1819) Looss, 1899 (Trematoda: Dicrocoeliidae), including its relation to the intermediate host, *Cionella lubrica* (Müller). III. Observations on the slimeballs of *Dicrocoelium dendriticum*. Cornell Vet. 42:253–276.

Krull, W. H., and C. R. Mapes. (1952). Studies on the biology of *Dicrocoelium dendriticum* (Rudolphi, 1819) Looss, 1899 (Trematoda: Dicrocoeliidae), including its relation to the intermediate host, *Cionella lubrica* (Müller). IV. Infection experiments involving definitive hosts. Cornell Vet. 42:277–285.

Krull, W. H., and C. R. Mapes. (1952). Studies on the biology of *Dicrocoelium dendriticum* (Rudolphi, 1819) Looss, 1899 (Trematoda: Dicrocoeliidae), including its relation to the intermediate host, *Cionella lubrica* (Müller). V. Notes on infections of *Dicrocoelium dendriticum* in *Cionella lubrica*. Cornell Vet. 42:339–351.

Krull, W. H., and C. R. Mapes. (1952). Studies on the biology of *Dicrocoelium dendriticum* (Rudolphi, 1819) Looss, 1899 (Trematoda: Dicrocoeliidae), including its relation to the intermediate host, *Cionella lubrica* (Müller). VI. Observations on the life cycle and biology of *C. lubrica*. Cornell Vet. 42:464–489.

Krull, W. H., and C. R. Mapes. (1952). Studies on the biology of *Dicrocoelium dendriticum* (Rudolphi, 1819) Looss, 1899 (Trematoda: Dicrocoeliidae), including its relation to the intermediate host, *Cionella lubrica* (Müller). VII. The second intermediate host of *Dicrocoelium dendriticum*. Cornell Vet. 42:603–604.

Krull, W. H., and C. R. Mapes. (1953). Studies on the biology of *Dicrocoelium dendriticum* (Rudolphi, 1819) Looss, 1899 (Trematoda: Dicrocoeliidae), including its relation to the intermediate host, *Cionella lubrica* (Müller). VIII. The cotton-tail rabbit, *Sylvilagus floridanus mearnsi*, as a definitive host. Cornell Vet. 43:199–202.

Krull, W. H., and C. R. Mapes. (1953). Studies on the biology of *Dicrocoelium dendriticum* (Rudolphi, 1819) Looss, 1899 (Trematoda: Dicrocoeliidae), including its relation to the intermediate host, *Cionella lubrica* (Müller). IX. Notes on the cyst, metacercaria, and infection in the ant, *Formica fusca*. Cornell Vet. 43:389–410.

Mapes, C. R. (1951). Studies on the biology of *Dicrocoelium dendriticum* (Rudolphi, 1819) Looss, 1899 (Trematoda: Dicrocoeliidae), including its relation to the intermediate host, *Cionella lubrica* (Müller). I. A study of *Dicrocoelium dendriticum* and *Dicrocoelium* infection. Cornell Vet. 41:382–432.

Mapes, C. R., and W. H. Krull. (1951). Studies on the biology of *Dicrocoelium dendriticum* (Rudolphi, 1819) Looss, 1899 (Trematoda: Dicrocoeliidae), including its relation to the intermediate host, *Cionella lubrica* (Müller). II. Collection of the snail, *Cionella lubrica*, and its maintenance in the laboratory. Cornell Vet. 41:433–444.

Patten, J. A. (1952). The life cycle of *Conspicuum icteridorum* Denton and Byrd, 1951, (Trematoda: Dicrocoeliidae). J. Parasitol. 38:165–182.

Ractliffe, L. H. (1968). Hatching of *Dicrocoelium* eggs. Exp. Parasitol. 23:67–78.

Svadzhyan, P. K., and L. V. Frolkova. (1966). [Ants as first and second intermediate hosts of some trematodes and cestodes.] Zool. Zh. 45:213–219.

Tarry, D. W. (1969). *Dicrocoelium dendriticum*: the life cycle in Britain. J. Helminthol. 43:403–416.

### FAMILY PLAGIORCHIIDAE

Members of this family of flukes are parasites of all classes of vertebrates. They are variable in shape, size, and the location and arrangement of the reproductive organs. The cuticle is usually spined. Constant characters in them are 1) the genital pore is between the suckers, 2) the ovary is pretesticular, and 3) the descending and ascending limbs of the uterus pass between the testes. A Y-shaped excretory bladder is typical.

*Plagiorchis muris* Tanabe, 1922 (Plate 70)

This parasite occurs in the small intestine of rats, dogs, herring gulls, robins, sandpipers, and night hawks over a wide geographic range.

**Description.**    Adult worms average 2.67 mm long by 0.52 mm wide. The ceca extend to the posterior extremity of the body. A long C-shaped cirrus pouch extends posteriorly from the genital pore around the ventral sucker and caudad to the ovary. The testes are obliquely arranged in the third quarter of the body. The ovary is between the ventral sucker and anterior testis, lying to the right of midventral line. The uterus is long, sinuous, and extends to the posterior extremity of the body with both the descending and ascending limbs passing between the testes. It terminates as a metraterm. The excretory bladder is Y-shaped. Eggs measure 38 $\mu$ long by 19 $\mu$ wide.

**Life Cycle.**    Eggs in the morula stage are laid in the intestine and voided with the feces. Development in water requires about 24 days. Eggs of *P. muris*, along with those of *P. proximus*, hatch when swallowed by *Lymnaea emarginata*. While both flukes develop in this snail, they do not occur together. The general level of parasitism attained by each species increases with the size of the snail host. Juvenile snails do not become infected during the first summer. Mother sporocysts being attached to the intestine of the snails suggests that hatching may occur soon after the eggs are eaten. Daughter sporocysts escape from the mother sporocysts and migrate to the liver.

Cercariae, upon reaching maturity in the daughter sporocyst, bear a well-developed stylet in the anterior margin of the oral sucker and have a thick-walled, Y-shaped excretory bladder. In the process of development, they follow one of two courses. Some of them appear to be precocious and encyst within the mother sporocyst where they develop. Others escape and leave the snail to penetrate chironomid larvae (also freshwater shrimp in Japan) and other snails such as *Lymnaea stagnalis*, *L. auricularia*, and *L. pereger* and encyst. One week after encystment the metacercariae are infective to a considerable number of avian and mammalian hosts, including humans, mice, and rats.

## Explanation of Plate 70

**A**, Adult fluke from mouse (experimental infection).
**B**, Mother sporocyst attached to intestine of snail host. **C**, Daughter sporocyst. **D**, Cercaria. **E**, Encysted metacercaria. **F**, Herring gull (*Larus argentatus*) definitive host. **G**, Robin (*Turdus migratorius*) definitive host. **H**, Spotted sandpiper (*Actitis macularia*) definitive host. **I**, Nighthawk (*Chordeiles minor*) definitive host. **J**, Snail (*Lymnaea emarginata*) first intermediate host. **K**, Chironomidae larva second intermediate host. **L**, Adult Chironomidae in flight.

1, Oral sucker; 2, ventral sucker; 3, pharynx; 4, testis; 5, cirrus pouch; 6, seminal vesicle; 7, ovary; 8, Mehlis' gland; 9, descending limb of uterus; 10, ascending limb of uterus; 11, metraterm; 12, common genital pore; 13, intestine of snail; 14, mother sporocyst; 15, cercaria; 16, stylet; 17, penetration glands; 18, thick-walled, Y-shaped excretory bladder; 19, tail; 20, metacercarial cyst.

a, Adult worm in small intestine of definitive host; b, partially embryonated egg passed in feces; c, egg completes embryonation in water; d, mother sporocyst attached to intestine (a suggestion that the eggs hatch in the intestine of the snail, but this has not been observed); e, daughter sporocyst containing cercariae; f, daughter sporocyst containing encysted metacercariae; g, cercaria free in water; h, cercaria penetrating chironomid larva; i, metacercaria in chironomid larva; j, definitive hosts become infected by eating snails or chironomids harboring metacercariae; k, encysted metacercaria freed from snail host; l, metacercaria escapes from cyst, attaches to intestinal wall, and matures in 7 to 9 days.

Figures A, C–E adapted from McMullen, 1937, J. Parasitol. 23:235; B, from Cort and Olivier, 1943, J. Parasitol. 29:91.

---

Cercariae of *Plagiorchis laricola* possess four types of gland cells in the complex commonly designated as the penetration glands. Three of the cells, the ventral, dorsal, and cystogenic, function in the formation of the metacercarial cyst, and the fourth for penetration. The ventral and dorsal cells discharge their contents before the cercaria leaves the sporocyst, bathing the entire body. In the second intermediate host, the cercarial tegument forms the outer layer of the cyst wall and the contents of the cystogenic gland the inner, thin cyst wall. The dorsal and ventral glands contain acid mucosubstances and the cystogenic cells neutral mucosubstances and proteins with sulfhydral groups of tyrosine and arginine. The penetration glands produce mostly proteins with disulfide and sulfhydral groups of tyrosine, arginine, and tryptophan.

Infection occurs when encysted metacercariae in chironomids, either larval or adult forms, or in snails, are ingested by susceptible hosts. When the metacercariae are released from the snail or insect intermediary in the intestine of the definitive host, development to maturity is completed in 7 to 9 days. Infections persist about 1 month.

Eggs of *Plagiorchis noblei*, a common parasite of yellowhead and redwinged blackbirds, are embryonated in 5 days after being passed by the host. They hatch in the gut of *Lymnaea* (*Stagnalis*) *reflexa*. Infected snails produce great numbers of cercariae, up to 4000 per day. They penetrate and encyst in larval mosquitoes (*Aedes aegypti*), caddis flies, damsel flies, and midges (*Chaoborus*). Tails carried into the body cavity of midges by the entering cercariae detach and remain active for more than a week. The metacercariae become infective in 4 to 6 days and when fed to day-old chicks mature in 6 days.

The cercariae of species of *Plagiorchis* encyst in a variety of hosts. For convenience in understanding their life cycles and relationships, they are grouped into four loose categories based on the phylum in which the respective metacercariae occur. They include arthropods, mollusks, platyhelminths, and chordates. The common snail first intermediate host is included for each species of fluke. Two species of flukes utilize second intermediate hosts from widely separated phyla. 1) Cercariae from the arthropod group encyst in crustacea and larval aquatic insects. They include *Plagiorchis muris*, *P. proximus*, and *P. micracanthus* whose cercariae come from *Lymnaea emarginata*, *P. jaenschi* from *L. lessoni*, *P. maculosa* from *L. auricularia*, *P. megalorchis* from *L. pereger*, *P. parorchis* from *L. stagnalis*, and *P. arcuatus* from *Bithynia tentaculata*. 2) Cercariae of *P. noblei* from the molluscan group developing in *L. reflexa* encyst in the body tissues of *L. stagnalis*. 3) Cercariae from the platyhelminth group include precocious individuals of *P. muris* and *P. proximus*, which also penetrate insects of the arthropod group, encyst as hyperparasites in the sporocysts in which they developed without ever leaving them.

**Plate 70** *Plagiorchis muris* 291

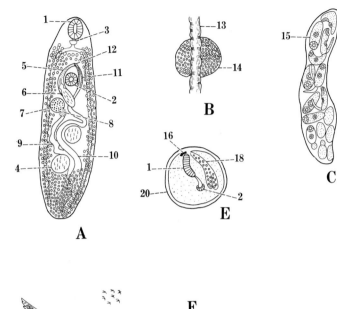

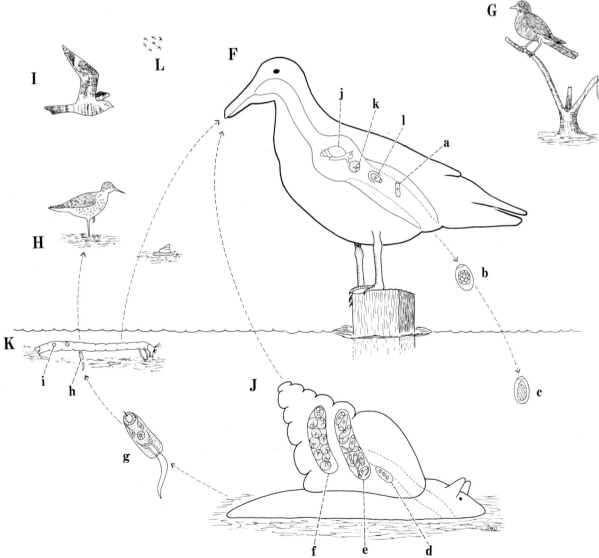

4) The chordate group includes *P. brumpti* whose cercariae come from *Planorbis planorbis* and encyst in tadpoles where the metacercariae develop to sexual maturity, producing numerous fertile eggs within 15 days.

## EXERCISE ON LIFE CYCLE

Naturally infected *Lymnaea emarginata* over wide areas of this country produce xiphidiocercariae of one or more species of *Plagiorchis*. Common species are *P. muris* of mice, *P. macracanthus* of bats, and *P. proximus* of muskrats and voles.

When infected lymnaeid snails shedding xiphidiocercariae are found, the cercariae should be placed in dishes with chironomid, caddis fly, and mosquito larvae and *Daphnia* to afford them an opportunity to encyst. Note the development of the metacercariae in the insect or crustacean. Examine the infected snails to determine whether metacercariae are present in the daughter sporocysts. Only *P. muris* and *P. proximus* are known to encyst in the sporocysts.

Since species of *Plagiorchis* are capable of developing in a rather wide range of hosts, mice, rats, and baby chicks are likely to serve as definitive hosts for experimental infection by feeding metacercariae to them. Observe the droppings of experimentally infected hosts for eggs, noting the length of the prepatent period.

Recover eggs from the feces and incubate them. Determine whether they hatch in water and penetrate snails or hatch only when eaten by snails. This point has not been determined for the American species.

When the adult worms have been recovered from the experimental hosts, they should be fixed, stained, and identified.

## SELECTED REFERENCES

Bourns, F. K. R. (1966). *Plagiorchis noblei* in nestling red-winged blackbirds. J. Parasitol. 52:974.

Buttner, A. (1951). La progénèse chez les trématodes digénètiques. Technique et recherches personnelles. III. Technique. Ann. Parasitol. 26:19–66.

Cort, W. W., and D. J. Ameel. (1944). Further studies on the development of the sporocyst stages of plagiorchiid trematodes. J. Parasitol. 30:37–56.

Cort, W. W., and L. Olivier. (1943). The development of the larval stages of *Plagiorchis muris* Tanabe, 1922, in the first intermediate host. J. Parasitol. 29:81–99.

Daniell, D. L., and M. J. Ulmer. (1964). Life cycle of *Plagiorchis noblei* Park, 1936 (Trematoda: Plagiorchiidae). J. Parasitol. 50(2/3):46.

Farley, J. (1967). The occurrence of congeneric trematode species in *Lymnaea emarginata* from Lake Ainslie, Cape Breton. Can. J. Zool. 45:1247–1254.

Johnston, T. H., and L. M. Angel. (1951). The life history of *Plagiorchis jaenschi*, a new trematode from the Australian water rat. Trans. Roy. Soc. S. Austr. 74:49–58.

Macy, R. W. (1956). The life cycle of *Plagiorchis parorchis* n. sp. (Trematoda: Plagiorchiidae). J. Parasitol. 42(Suppl.):28.

McMullen, D. B. (1937). The life histories of three trematodes, parasitic in birds and mammals, belonging to the genus *Plagiorchis*. J. Parasitol. 23:235–243.

Okabe, K., and H. Shibue. (1952). A new second intermediate host, *Neocaridina denticulata* for *Plagiorchis muris* (Tanabe); Plagiorchiidae. Jap. J. Med. Sci. Biol. 5:257–258.

Paskalskaya, M. Y. (1954). [Study on the cycle of development of *Plagiorchis arcuatus* Strom, 1924 parasite of the oviduct and bursa of Fabricius of the hen.] Dokl. Akad. Nauk. SSSR 97:561–563.

Rees, F. G. (1952). The structure of the adult and larval stages of *Plagiorchis* (*Multiglandularis*) *megalorchis* n. nom. from the turkey and an experimental demonstration of the life history. Parasitology 42:92–113.

Strenzke, K. (1952). Wirtwechsel von *Plagiorchis maculosus*. Z. Parasitenk. 15:369–391.

Williams, R. R. (1964). Life cycle of *Plagiorchis noblei* Park, 1936. J. Parasitol. 50(3/2):29.

Zajíček, D. (1971). Snails as second intermediate hosts of the trematode *Plagiorchis muris* Tanabe, 1922 (Trematoda: Plagiorchiidae. Věstn. Cesk. Spolecnosti Zool. 35:75–78.

Žďárská, Z. (1969). Gland cells of the cercaria of *Plagiorchis laricola* (Skrjabin, 1924). (Trematoda). Věstn. Cesk. Spolecnosti Zool. 33:278–286.

*Haematoloechus medioplexus*
Stafford, 1902 (Plate 71)

*Haematoloechus medioplexus* is one of the common lung flukes of adult *Rana pipiens* and *Bufo americanus* in North America. Many species of lung flukes occur in different species of frogs and toads in this and other countries.

**Description.** The flattened worms are up to 8 mm long and 1.2 mm wide. The acetabulum is near the junction of the first and second thirds of the body and about one-fourth to one-fifth the size of the oral sucker. The cuticle is densely spined. The gonads occupy a little more than the middle third of the body. The testes are large oval bodies which are much bigger than the elongate ovary. A large seminal receptacle partially overlaps the smaller ovary from the dorsal side. Limbs of the uterus pass between the testes and fill the post-testicular part of the body. A common genital pore is located at the bifurcation of the ceca. Vitellaria are arranged in groups which extend from about one-half the distance between the anterior tip of the body and ovary to behind the posterior testis. The dark brown eggs measure 22 to 29 $\mu$ long by 13 to 17 $\mu$ wide. For figures and descriptions of the different species of lung flukes of North American frogs see Cort.

Mature spermatozoa measure about 400 $\mu$ long and 1 $\mu$ in diameter. Each consists of a head 25 to 30 $\mu$, a middle piece 300 to 350 $\mu$, and a flattened, mobile, terminal portion 25 to 50 $\mu$ in length. The entire structure is surrounded by a membrane. The head and anterior part of the middle piece contains the single nucleus and mitochondrion. Inside the membrane are two sets of microtubules arranged in a row on opposite sides of the sperm and one pair of axial units. Each unit consists of nine peripheral, hollow, doublets connected by spoke-like strands of cytoplasm to a hub-like ring of 18 longitudinal

fibrils. These surround an inner, isolated core of a few small thread-like fibrils. The fibrils of the outer rim of the hub wind about the core, forming a double helix (Fig. 27).

Formation of the zygote begins when the spermatozoon comes in contact with the ovum, lying full length on the surface. The plasma membranes of both dissolve at the point of contact and the spermatozoon sinks into the ovum, leaving the peripheral microtubules lying on the underside of the plasma membrane.

**Life Cycle.**     Adult flukes in the cavity of the lungs deposit fully embryonated eggs. They are carried from the lungs up the bronchioles and through the glottis by ciliary action of the lung and bronchiole cells. Upon passing through the glottis, they are swallowed and eventually voided with the feces, sinking to the bottom of the pond. Although fully embryonated, hatching does not occur until they are swallowed by the snail *Planorbula armigera*. In lakes where the biotic and physical conditions are favorable for the snail hosts, a large percentage of them are infected with these flukes. After a short time in the stomach, the eggs pass to the intestine and hatch in 15 to 20 minutes. Eggshells and some miracidia are passed in the feces, but these miracidia soon die. Eggs hatch in the intestine of other species of snails, but the miracidia do not

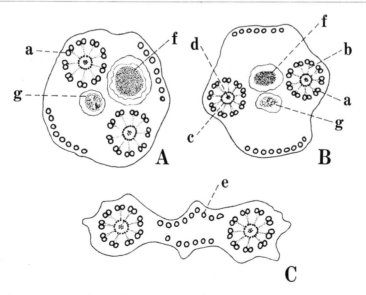

Fig. 27. Diagrammatic representation of a cross section of each of the three parts of the sperm of *Haematoloechus medioplexus* reconstructed from electron micrographs. A, Head. B, Middle piece. C, Tail. a, Nine pairs of double hollow fibrils in outer ring of axial unit; b, spoke-like plasmic strands; c, hub-like structure of axial unit; d, core of axial unit around which hub fibrils wind to form a double helix; e, microtubules; f, nucleus; g, mitochondrion. Modified from Burton, 1972, J. Parasitol. 58:68.

## Explanation of Plate 71

A, Dorsal view of adult fluke. B, Ventral view of anterior end of adult. C, Embryonated egg. D, Miracidium showing epidermal plates. E, Miracidium showing anterior and posterior plates together with internal anatomy. F, Mature daughter sporocyst. G, Cercaria. H, Lateral view of tail of cercaria, showing fin-like membrane. I, Encysted metacercaria. J, Encysted metacercaria in gill lamella of dragonfly (*Sympetrum* sp.). K, Excysted metacercaria. L, Frog definitive host (*Rana pipiens*). M, Snail intermediate host (*Planorbula armigera*). N, Naiad of dragonfly (*Sympetrum* sp.). O, Teneral of *Sympetrum* sp.

1, Oral sucker; 2, ventral sucker; 3, pharynx; 4, intestinal crura; 5, testis; 6, cirrus pouch; 7, ovary; 8, seminal receptacle; 9, descending limb of uterus; 10, ascending limb of uterus; 11, vitelline glands; 12, common collecting ducts of vitelline glands; 13, common genital pore; 14, posterior epidermal plates of miracidium; 15, anterior epidermal plates; 16, penetration gland; 17, flame cell; 18, germ cell; 19, cercaria; 20, developing cercariae; 21, germ balls; 22, stylet; 23, penetration gland; 24, excretory bladder; 25, common collecting tubule; 26, anterior collecting tubule; 27, tail; 28, fin of tail; 29, cyst of metacercaria; 30, metacercaria; 31, gill lamella; 32, trachea; 33, tracheoles of gill.

a, Adult worm in lung; b, egg deposited in lung is carried up bronchi into pharynx and then swallowed; c, embryonated egg passing through intestine to appear in feces; d, embryonated egg in water; e, unhatched eggs swallowed by snail intermediate host, *Planorbula armigera;* f, eggs hatch in intestine of snail; g, miracidium passing through wall of intestine; h, mother sporocyst; i, mature daughter sporocyst with germ balls, and cercariae in various stages of development; j, cercaria that has escaped from sporocyst; k, cercaria free in water; l, cercaria swept into gill basket of naiad of dragonfly through respiratory current attaches to gill lamella, drops the tail, penetrates, and encysts; m, metacercaria in gill lamella; n, metacercaria retained through metamorphosis of naiad to teneral; o, dragonfly naiads or adults digested in stomach of frog, releasing metacercariae; p, metacercaria released from cyst by action of digestive juices; q, metacercaria migrates up esophagus and enters bronchi, finally arriving in the lungs and developing to maturity in 37 days.

Figures A, B adapted from deFreitas and Lent, 1939, Livro de Homenagem, p. 246; C, F, from Krull, 1930, J. Parasitol. 16:207; D, E, G, H, J, K, from Krull, 1931, Trans. Amer. Microsc. Soc. 50:215; I constructed from Krull, 1930, J. Parasitol. 16:207 and 1931, Trans. Amer. Microsc. Soc. 50:215.

---

develop in them. This indicates that nonspecific digestive enzymes are capable of causing the eggs to hatch. Miracidia in *Planorbula armigera* migrate through the intestinal wall.

Mother sporocysts have not been observed. Those in the hepatic gland probably are daughter sporocysts. They mature and produce cercariae in 65 days. Up to 300 cercariae may be shed per night by a single heavily infected snail. In such heavily infected snails, the liver appears to be completely destroyed. Recently escaped cercariae swim actively at summer temperatures but they are sluggish and settle to the bottom in cold water. They survive up to 30 hours.

Naiads of the dragonfly *Sympterum obtrusum* are the second intermediate host. They respire by drawing water into and expelling it from the branchial basket by way of the anus. Cercariae swimming in the vicinity of the anus on warm days are swept into the branchial basket by the inhalent currents. Naiads show signs of great annoyance from the presence of the cercariae and attempt to expel them. Upon coming in contact with the wall and gills of the branchial

basket, the cercariae quickly pass through the chitin into the tissue of the lamellae or wall of the respiratory organ and encyst in thin transparent cysts. Metacercariae are found nowhere else. Naturally only naiads can become infected, but not before the second stage. The metacercariae are infective in 6 days after encystment and the maximum size is reached in 14 to 20 days. When the naiads metamorphose to adults, the metacercariae remain in the remnant of the branchial basket in the posterior tip of the abdomen. Adults may harbor more than 200 metacercariae from natural exposure.

Parasitism of frogs occurs through eating infected naiad or adult dragonflies. Soft tenerals resting on stems of vegetation are easily captured and eaten by the frogs. Upon being released from the dragonflies and cysts (metacercariae excyst in artificial gastric juice in 45 seconds), the young flukes migrate forward through the esophagus, pass through the glottis, and enter the lungs by way of the bronchi. They appear to move against the current created by the ciliated epithelium of the bronchi. The flukes

Plate 71    *Haematoloechus medioplexus*                                           295

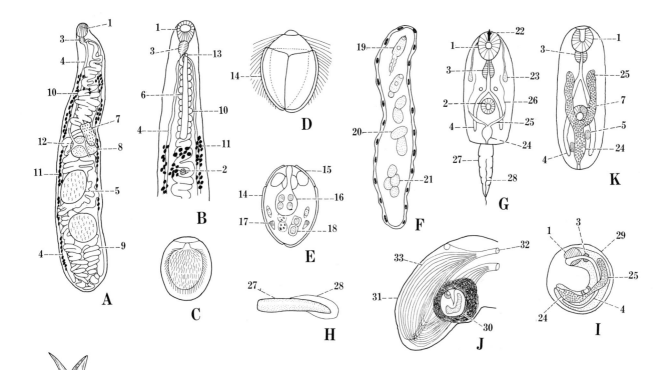

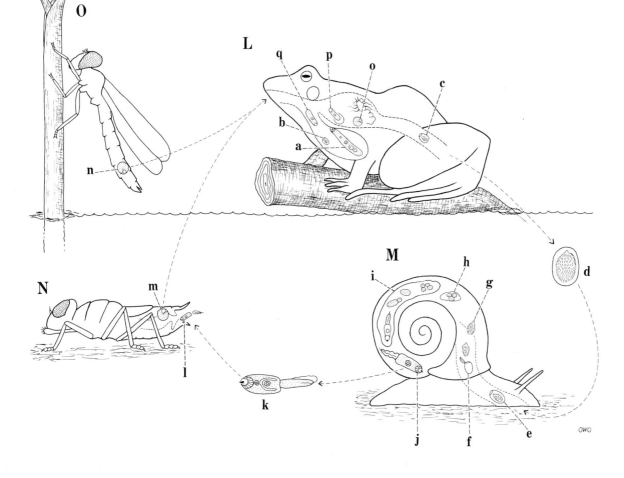

mature in 37 days during the summer. They remain in the frogs up to 15 months. After that, they are discarded and new ones take their place. Up to 75 flukes occur in individual naturally infected frogs.

The physiology of female frogs during the breeding season renders them highly refractory to infection by the flukes. Males do not react similarly and as a result have more worms in the lungs.

The life cycles of several other species of *Haematoloechus* have been studied. *Haematoloechus parviplexus* occurs in the green frog (*Rana clamitans*) and develops in the snail *Gyraulus parvus* and the dragonflies *Sympetrum rubicundium* and *S. obtrusum*. *Haematoloechus longiplexus* of bullfrogs (*Rana catesbiana*) occurs as metacercariae in damsel flies (*Lestes vigilax*). The snail host is not known. *Haematoloechus complexus* from the green frog develops in *Pseudosuccinea columella* and dragonflies (*Tetragoneuria cynosura*), and damsel flies (*Chromagrion conditum*, *Enallagma divagans*, *Lestes vigilax*, *Argia* sp.)

## EXERCISE ON LIFE CYCLE

In habitats where frogs, Odonata, and snails are present, species of *Haematoloechus* are likely to occur. Dissection of frogs from the area will show whether the flukes are present; usually a high percentage of them are infected. Flukes placed in water shed embryonated eggs which may be fed to snails. Observe the feces expelled by snails shortly after eating eggs and note the empty shells and miracidia entangled in them. For careful experimental studies, laboratory-reared snails should be used. Dissect naturally and experimentally infected snails for a study of the sequence of development of sporocysts.

Cercariae shed by *Planorbula armigera*, *Gyraulus parvus*, or *Pseudosuccinea columella* that have the characteristics of those of *Haematoloechus* should be placed in small dishes with appropriate species of dragonfly or damsel fly naiads. If the cercariae are retained in the branchial basket, with encystment, the probability is that the species is one of *Haematoloechus*. Dissect dragonfly and damsel fly naiads and examine the wall of the branchial basket for metacercariae in flattened cysts of thin hyaline material. The young flukes have a pro-nounced Y-shaped excretory bladder with long broad arms.

Feed a large number of encysted metacercariae at intervals of 2 days for 2 weeks to a frog that has been kept in the laboratory for a month in order to allow any worms that might have been acquired naturally to have matured. Sacrifice it a few hours after the last feeding and examine the intestine, stomach, esophagus, bronchi, and lungs separately to determine the route of migration and development of the young flukes.

## SELECTED REFERENCES

Burton, P. R. (1967). Fine structure of the reproductive system of a frog lung fluke. II. Penetration of the ovum by a spermatozoon. J. Parasitol. 53: 994–999.

Burton, P. R. (1972). Fine structure of the reproductive system of a frog lung fluke. III. The spermatozoon and its differentiation. J. Parasitol. 58:68–83.

Cort, W. W. (1915). North American frog lung flukes. Trans. Amer. Microsc. Soc. 34:203–240.

Hollis, P. D. (1972). Host sex influence on the seasonal incidence of *Haematoloechus medioplexus* (Trematoda: Plagiorchiidae) in *Rana pipiens*. J. Parasitol. 58:128.

Krull, W. H. (1930). The life history of two North American frog lung flukes. J. Parasitol. 16:207–212.

Krull, W. H. (1931). Life history studies on two frog lung flukes *Pneumonoeces medioplexus* and *Pneumobites parviplexus*. Trans. Amer. Microsc. Soc. 50:215–277.

Krull, W. H. (1932). Studies on the life history of *Pneumobites longiplexus* (Stafford). Zool. Anz. 99: 231–239.

Krull, W. H. (1933). Studies on the life cycle of a frog lung fluke, *Haematoloechus complexus* (Seely, 1906). Z. Parasitenk. 6:192–206.

Krull, W. H. (1934). Some additional notes on the life history of a frog lung fluke *Haematoloechus complexus*. Trans. Amer. Microsc. Soc. 53:196–199.

*Haplometrana utahensis*
Olsen, 1937 (Plate 72)

This is a parasite of the small intestine of the western frog *Rana pretiosa* in the Great Basin region of Utah and possibly over its entire range west of Rocky Mountains in the United States.

**Description.** The length of mature worms is about 5 mm, with the sides parallel and the ends rounded. The oral sucker is subterminal and the acetabulum is in the second fifth of the body. Ceca extend to near the posterior extremity of the body. Testes are in the middle fifth of the body, intercecal, subspherical, and arranged obliquely. The ovary is on the left side between the acetabulum and anterior testis. The common genital pore is near the anterior margin of the acetabulum. Vitellaria extend from the level of the ovary into the middle of the fourth fifth of the body.

Waitz compared adult specimens of Lucker, Olsen, and his own. He concluded on the basis of morphology that the entire population is represented by *H. intestinalis* Lucker, 1931. However, differences in the morphology of the cercariae and the daughter sporocysts together with their distribution in the snail host, as revealed by the life history studies of Olsen and Schell, suggest that specific differences do exist.

**Life Cycle.**    Fully embryonated eggs passed in the feces do not hatch until eaten by the snail intermediate host (*Physella utahensis, Physa ampullacea, P. gyrina*). The miracidia appear in the small intestine 75 minutes after the eggs are ingested. They migrate through the gut wall. Small sporocysts near the gut are thought to be mother sporocysts but the identity of them is not certain. Schell has found them for *H. intestinalis* in close association with the intestinal wall.

One month after infection of the snails, the youngest daughter sporocysts are nearest to the intestine and the older ones are further from it. In long established infections, the liver is heavily parasitized with mature daughter sporocysts entwined about each other in the host tissues.

Xiphidiocercariae appear during week 7 following infection of the host. They escape from the sporocysts through a terminal birth pore. There is no periodicity and up to 200 cercariae appear in a span of 24 hours. They swim aimlessly, displaying no tropisms toward or against light.

Upon coming in contact with the frog definitive host, the cercariae quickly attach and crawl rapidly over them for a short time before penetrating the epidermis, which requires about 10 minutes. Once within the epidermis, large passages are formed through which they migrate some distance before secreting a delicate cyst about themselves. The cercariae do not enter the dermis but remain in the epidermis and between it and the stratum corneum.

Infectivity of the metacercariae is attained as early as 13 hours after encystment. When the epidermis is shed, the encysted metacercariae remain attached. Inasmuch as the frogs eat the skin as it is shed, they swallow the adherent metacercariae, thus becoming infected. Hence this parasite has the unique life cycle in which the same frog may serve as both the second intermediate and definitive hosts. Upon being released from the cysts, the young flukes develop to sexual maturity in 50 days at room temperature, as indicated by small numbers of eggs in the uterus.

Schell, doing intensive studies on *H. intestinalis*, followed the development of mother and daughter sporocysts in laboratory-reared *Physa gyrina* and *P. ampullacea*. He found that the eggs hatch in the stomach and the weakly swimming miracidia penetrate the epithelium of the snail's digestive tract, coming to rest between it and the basement membrane. Here they shed the cilia and transform to mother sporocysts. Daughter sporocysts appear by day 9 and production of them ceases by the day 18. They were mature with developing cercariae by day 27 after infection of the snails. Cercariae emerged naturally on day 34. The daughter sporocysts from each mother sporocyst are held together in a mass by a surrounding membrane called the paletot, which, presumably, originates from the host tissues. The daughter sporocysts do not invade the snail tissues. The cercariae have a club-shaped excretory bladder which is different from that described for *H. utahensis*. On the basis of the cercarial anatomy of *H. intestinalis*, Schell assigned this genus to the plagiorchioid family Macroderoididae.

*Glypthelmins pennsylvaniensis* from *Hyla crucifer* has a similar life cycle in that its cercariae from *Helisoma trivolvis* encyst in the skin of frogs. The metacercariae are shed with stratum corneum which the frogs eat, becoming infected with their own flukes. *Glypthelmins hyloreus* from *Hyla regilla*, however, differs in that the cercariae develop in *Lymnaea stagnalis* and enter tadpoles through the nares. They develop in the celom into unencysted metacercariae that enter and mature in the intestine after metamorphosis of the frog.

## EXERCISE ON LIFE CYCLE

In localities where *Rana pretiosa* infected with *Haplometrana* are available, the life cycle may

## Explanation of Plate 72

A, Adult fluke. B, Embryonated egg. C, Mother sporocyst (?). D, Mature daughter sporocyst. E, Xiphidiocercaria. F, Cercaria in section of web of toes of frog. G, Metacercaria attached to shed stratum corneum. H, Adult frog definitive host (*Rana pretiosa*). I, Snail intermediate host (*Physella utahensis*). J, Frog second intermediate host (*Rana pretiosa*).

1, Oral sucker; 2, ventral sucker; 3, pharynx; 4, esophagus; 5, intestine; 6, testis; 7, cirrus pouch; 8, common genital pore; 9, ovary; 10, oviduct; 11, seminal receptacle; 12, uterus; 13, vitelline glands; 14, miracidium; 15, apical gland; 16, penetration gland; 17, germ cell; 18, birth pore; 19, cercaria; 20, spine of oral sucker; 21, penetration glands; 22, flame cell; 23, capillary tubule; 24, anterior collecting tubule; 25, posterior collecting tubule; 26, excretory bladder; 27, lateral vitelline ducts; 28, finfold on tail; 29, section of web of hind foot of frog; 30, epidermis; 31, dermis with pigment; 32, section of cercaria; 33, shed cuticle of frog; 34, metacercaria; 35, cyst of metacercaria.

a, Adult fluke in small intestine; b, embryonated egg passing through intestine; c, embryonated egg in water; d, egg swallowed by snail intermediate host; e, egg hatching in intestine of snail; f, miracidium passing through intestinal wall; g, mother sporocyst; h, daughter sporocyst with cercariae and germ balls; i, cercaria escaping from snail; j, k, cercariae swimming free in water; l, cercaria penetrating skin of frog, casting its tail in the process; m, metacercaria encysted on stratum corneum; n, metacercaria on cast stratum corneum being swallowed by the frog which sheds it; o, metacercaria being released from stratum corneum by digestive juices; p, metacercaria escaping from its cyst in the intestine to develop to sexual maturity in about 50 days.

Adapted from Olsen, 1937, J. Parasitol. 23:13.

---

be studied without difficulty. Adult worms placed in a dish of water lay embryonated eggs. Examine them for such details as can be seen through the brown shell and then hatch them in the stomach juices of snails (Schell).

Feed eggs to species of Physidae found in this area, preferably young ones taken from localities where frogs do not occur, or better still, to laboratory-reared specimens. Examine infected snails daily to find the mother sporocysts, and to follow the development of the daughter sporocysts and cercariae.

When cercariae appear, place some of them together with a piece of the web from the hind-foot of a frog in a small dish of water. Watch them penetrate the epidermis. After they have disappeared into it, place the piece of web in a fixative and prepare sections for observing the course of movements of the cercariae and their encystment in the epidermis.

Expose frogs in a small aquarium to numerous cercariae. Later, observe the shedding of the stratum corneum and the ingestion of it, together with the metacercariae. Note the development of the flukes in the frogs and the attainment of sexual maturity as demonstrated by the appearance of eggs in the feces.

## SELECTED REFERENCES

Cheng, T. C. (1961). Description, life history, and developmental pattern of *Glypthelmins pennsylvaniensis* n. sp. (Trematoda: Brachycoeliidae), a new parasite of frogs. J. Parasitol. 47:469–477.

Lucker, J. T. (1931). A new genus and new species of trematode worms of the family Plagiorchiidae. Proc. U. S. Nat. Mus. 79:1–18.

Martin, G. W. (1969). Description and life cycle of *Glypthelmins hyloreus* sp. n. (Digenea: Plagiorchiidae). J. Parasitol. 55:747–752.

Olsen, O. W. (1937). Description and life history of the trematode *Haplometrana utahensis* sp. nov. (Plagiorchiidae) from *Rana pretiosa*. J. Parasitol. 23:13–28.

Schell, S. C. (1961). Development of mother and daughter sporocysts of *Haplometrana intestinalis* Lucker, a plagiorchiid trematode of frogs. J. Parasitol. 47:493–500.

Waitz, J. A. (1959). A revision of the genus *Haplometrana* Lucker, 1931 (Trematoda: Plagiorchiidae), with notes on its distribution and specificity. J. Parasitol. 45:385–387.

## *Lechriorchis primus*
(Stafford, 1905) (Plate 73)

*Lechriorchis primus* is a parasite of the lungs of snakes of the genera *Thamnophis* and *Natrix* in North America.

**Description.**    Adult flukes are 5.5 mm long by 1.4 mm wide and have a spiny cuticle. The oral sucker is smaller than the ventral one and is located near the hind portion of the anterior half of the body. A short prepharynx, pharynx, esophagus, and intestinal ceca are pres-

**Plate 72** *Haplometrana utahensis* 299

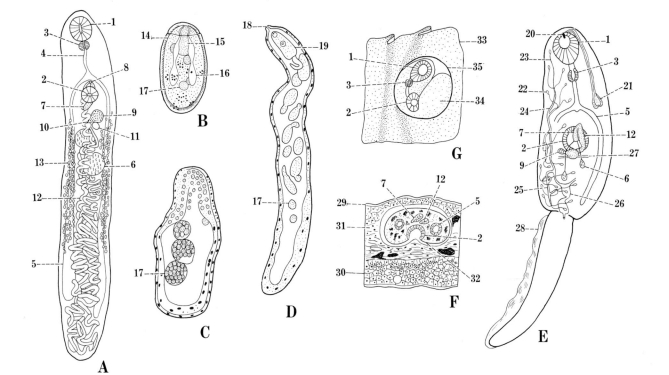

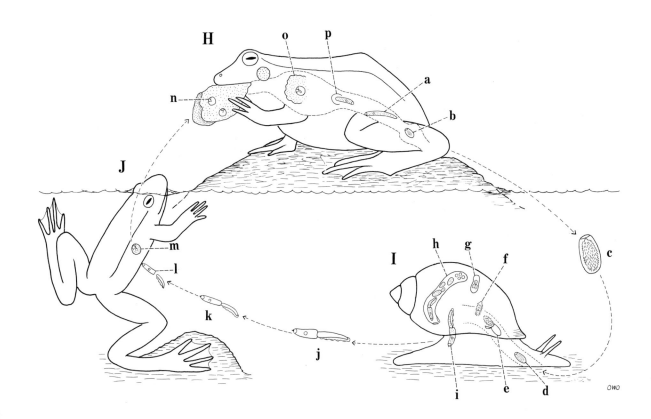

## Explanation of Plate 73

A, Dorsal view of adult fluke (experimental). B, Miracidium. C, Young mother sporocyst. D, Fully developed mother sporocyst. E, Daughter sporocyst. F, Cercaria. G, Encysted metacercaria. H, Snake definitive host (*Thamnophis sirtalis*). I, Snail first intermediate host (*Physa gyrina* or *P. parkeri*). J, Tadpole second intermediate host (*Rana clamitans*, *R. pipiens*).

1, Oral sucker; 2, ventral sucker; 3, pharynx; 4, esophagus; 5, intestine; 6, testis; 7, cirrus pouch; 8, seminal vesicle; 9, ovary; 10, oviduct surrounded by Mehlis' gland and with Laurer's canal extending free; 11, uterus; 12, metraterm; 13, vitelline glands; 14, excretory bladder; 15, apical papillae; 16, apical gland; 17, penetration gland; 18, germ cell; 19, flame cell; 20, excretory duct; 21, birth pore; 22, daughter sporocyst; 23, cercaria; 24, stylet; 25, tail; 26, anterior collecting excretory tubule (bilateral); 27, common excretory collecting tube; 28, posterior excretory collecting tubule; 29, accessory collecting tubule; 30, capillary tubule; 31, penetration glands (bilateral); 32, cyst of metacercaria; 33, metacercaria.

a, Adult fluke in lung; b, eggs laid in lungs; c, eggs carried up trachea by ciliary action of epithelial cells, enter esophagus, and are swallowed; d, eggs continue on through intestine and are voided in feces; e, eggs fully embryonated when laid but do not hatch in water; f, eggs swallowed by snail intermediate host; g, eggs hatch in intestine of snail; h, miracidium migrates through intestinal wall; i, young mother sporocyst; j, mature mother sporocyst; k, daughter sporocyst; l, cercaria escaping from snail; m, cercaria free in water; n, cercaria penetrating tadpole and casting tail at the same time; o, encysted metacercaria; p, tadpole swallowed by snake is digested, releasing encysted metacercaria; q, metacercaria escaping from cyst; r, metacercaria free from cyst; s, metacercariae migrate into small intestine where they remain several months, growing slightly; t, young flukes migrate up esophagus; u, flukes enter trachea from esophagus; v, flukes migrate down trachea to lungs where they develop to maturity at about 2 years of age.

Adapted from Talbot, 1933, Parasitology 25: 518.

---

ent. The common genital pore is at the posterior margin of the intestinal bifurcation and the cirrus pouch is 1 mm long, extending to near the middle of the ventral sucker. The oval testes are slightly oblique and are near the front end of the posterior half of the body. The small ovary is to the right of the midline and at the posterior margin of the ventral sucker. The vitellaria extend from between the anterior sucker and the acetabulum to the middle of the testes. The limbs of the uterus pass between the testes and reach to near the posterior end of the body. In fully matured flukes, it fills almost the entire body caudad from the oral sucker. Eggs measure 52 to 55 $\mu$ long by 26 to 29 $\mu$ wide.

**Life Cycle.** Adult worms in the lungs of water snakes and garter snakes lay fully embryonated eggs in brown shells. They are carried up the trachea by the ciliary action of the epithelial cells and swallowed, to be passed later in the droppings of the snake hosts. Hatching does not occur until the eggs are eaten by the snail hosts, which are *Physa parkeri* and *P. gyrina*. The miracidia escape from the eggs under the influence of the gastric juice in the stomach of the snails within an hour after being ingested. After migrating through the stomach wall, the miracidia attach to the outer side of it or the

first part of the intestine. When the miracidia are 2 weeks old, young daughter sporocysts are present, and at 3 weeks they escape through the birth pore and migrate to the liver. Five weeks after infection, cercariae have reached maturity and are escaping from the daughter sporocysts via a birth pore. They enter the lymph spaces of the liver and migrate through lymph channels to the outside for a period of about 2 hours just after dark in the evenings.

The stylet-bearing cercariae with a Y-shaped excretory bladder are distributed at random through the water, having no definite tropisms toward light or darkness. Upon coming in contact with tadpoles of frogs of several species, the cercariae penetrate the skin, migrate into the connective tissue between the muscle bundles, and encyst promptly in thin hyaline cysts. Cercariae that encyst in small perch, sunfish, and larval *Triturus* soon die. Metacercariae in frog tadpoles are infective almost immediately following encystment. After a week, the cysts become brown in color and are laminated. They persist during metamorphosis and appear in the muscles of frogs.

Snakes become parasitized by eating infected tadpoles and frogs. Excystment takes place in the stomach and the young flukes migrate to the

**Plate 73**    *Lechriorchis primus*                                                                301

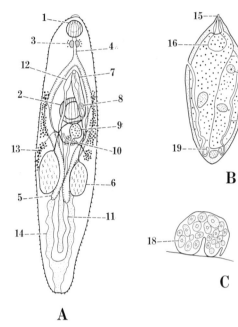

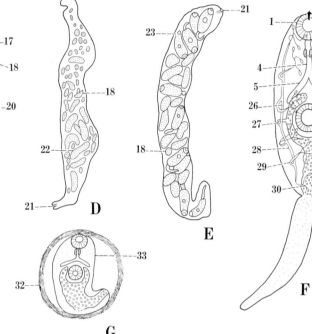

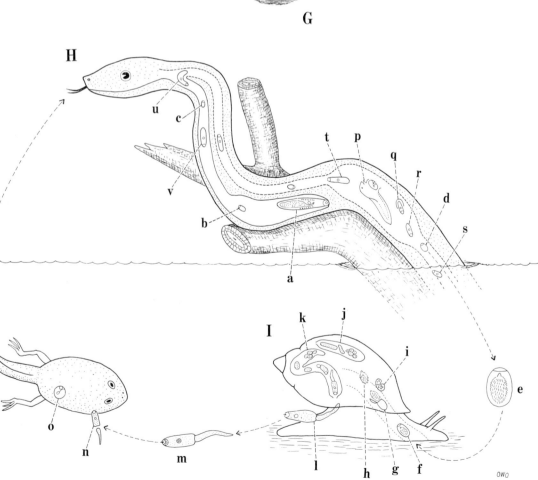

OWO

small intestine. Growth in the intestine is slow during the first 2 months. The flukes remain dormant during the first winter when the snakes are in hibernation. With the advent of spring and resumption of activity by the snakes, the young flukes begin to grow and migrate via the stomach and esophagus to the lungs. By the end of 10 months after infection, most of them have reached the lungs. Growth is rapid and eggs appear but presumably they are not fertilized until after the second winter in the snakes, at which time the flukes become mature. It is not known how long they live. Infections usually are light.

The life cycles of several related species from snakes have been worked. They are basically the same as that of *Lechriorchis primus. Lechriorchis tygarti* (Talbot), and *Zeugorchis eurinus* (Talbot), both from the lungs of garter snakes, have a cycle similar to that of *L. primus. Dasymetra villicaeca* Byrd from the intestine of *Natrix* develops in *Physa halei* and tadpoles of several species of frogs. In *Dasymetra conferta*, and possibly other species, fibroblasts form the surrounding paletot, similar to that described by Schell for *Haplometrana intestinalis*, inside which daughter sporocysts develop and give rise to cercariae. *Ochetosoma aniurum* (Leidy) of the mouth, esophagus, and lungs of *Natrix, Heterodon*, and *Lampropeltis* develops in *Physa halei* and tadpoles of frogs.

## EXERCISE ON LIFE CYCLE

Because of the long time required to complete the life cycle, it is not convenient to conduct the entirety of it. Some aspects of the cycle, however, are both interesting and informative. Eggs of these flukes from water snakes and garter snakes may be identified tentatively by being embryonated, brown in color, and with a large single yolk cell at the end opposite from the operculum. Uterine eggs measure 48 to 50 $\mu$ long by 23 to 25 $\mu$ wide.

Eggs fed to young *Physa parkeri* and *P. gyrina* hatch in the stomach and produce cercariae in 5 weeks. Hatching can be induced by placing eggs in gastric juice from any of a number of species of snails. Tadpoles of frogs should be exposed to cercariae from experimental or natural infections of snails. Note metacercariae in the muscles and how the wall of the cyst changes in color and structure after 1 week. Metacer-

cariae placed in gastric juice from snake hosts quickly excyst, an indication that excystation occurs in the stomach. After feeding infected tadpoles to snakes, note that the young flukes go to the small intestine for a long sojourn.

## SELECTED REFERENCES

Byrd, E. E. (1935). Life history studies of Reniferinae (Trematoda, Digenea) parasitic in Reptilia of the New Orleans area. Trans. Amer. Microsc. Soc. 54:196–225.

Byrd, E. E., and W. P. Maples. (1969). Intramolluscan stages of *Dasymetra conferta* Nicoll, 1911 (Trematoda: Plagiorchiidae). J. Parasitol. 55:509–526.

Cort, W. W., D. J. Ameel, and A. Van der Woude. (1952). Development of the mother and daughter sporocysts of a snake plagiorchiid, *Lechriorchis primus* (Trematoda: Reniferidae). J. Parasitol. 38: 187–202.

Price, E. W. (1936). Redescriptions of the type species of the trematode genera *Lechriorchis* Stafford and *Zeugorchis* Stafford (Plagiorchiidae). Proc. Helminthol. Soc. Wash. 3:32–34.

Talbot, S. B. (1933). Life history studies on trematodes of the subfamily Reniferinae. Parasitology 25: 518–545.

### FAMILY PROSTHOGONIMIDAE

The Prosthogonimidae are parasites of the bursa Fabricii, the oviduct, and rarely the intestine of a large number of species of birds in many parts of Europe, Asia, Africa, and North and South America.

They are relatively small, flattened, and transparent distomes somewhat narrowed anteriorly and broadly rounded posteriorly. The testes are symmetrical and postacetabular. A lobed ovary lies between the acetabulum and testes. The uterine loops pass between the testes and fill the post-testicular portion of the body. The genital pores open separately or together on the anterior margin of the body near the oral sucker. The vitellaria are sparse, grape-like clusters in the lateral fields.

*Prosthogonimus macrorchis*
Macy, 1934 (Plate 74)

This oviduct fluke occurs in chickens, ducks, pheasants, English sparrows, and crows that live near or on lakes of the Great Lakes region of North America. Being in the oviducts of laying birds, they commonly occur inside eggs. They are considered to be the cause of marked decline in egg production if not complete cessation of it.

**Description.** The flattened pyriform and transparent body is 7.56 mm long by 5.26 mm wide with the cuticle spined. The ventral sucker is near the junction of the anterior and middle thirds of the body. The ratio of size of the suckers is 1 : 1.7, with the ventral one being the larger. Large symmetrically placed testes occupy the middle third of the body. The length of the cirrus pouch is about twice the diameter of the oral sucker and lies lateral to it, opening at anterior extremity of the body. A multilobed ovary lies between the ventral sucker and testes. The ascending and descending limbs of the uterus pass between the testes and fill the post-testicular portion of the body. Vitelline glands are arranged in eight to nine lateral clusters between the anterior margin of the ventral sucker and posterior margin of the testes. Eggs average 28 $\mu$ long by 16 $\mu$ wide.

**Life Cycle.** Flukes in the oviducts or bursa Fabricii lay eggs that are expelled through the anus into the water. Hatching has not been observed, although the eggs are embryonated when laid. Upon reaching the water, they gradually sink to the bottom where they remain until swallowed by *Amnicola limnosa porata*, the snail intermediate host. Snails crawling over the bottom of the lake are in a favorable position to find the eggs. Since hatching does not take place in water, it must occur in the gut of the snail. Mother sporocysts have not been seen but simple sac-like daughter sporocysts in the digestive gland produce small cercariae bearing a stylet and a relatively short tail. These are expelled from the snail into the water. They are feeble swimmers.

As the cercariae swim by the posterior end of dragonfly naiads of the genera *Leucorrhina*, *Tetragoneura*, *Epicordulia*, *Gomphus*, and *Mesothemis*, they are drawn through the anus by the respiratory currents into the branchial basket. They attach to the wall of the rectum and the gill filaments and burrow into the body cavity. Encystment takes place primarily in the muscles of the body wall, particularly those on the ventral side toward the posterior end of the abdomen. Exposure of the naiads is during the summer in which they hatch and the following spring before they emerge from the water to molt and fly away. Development of the metacercariae is slow, requiring up to 70 days before the

characteristic thick-walled, striated cysts are formed.

Ducks appear to be the normal hosts but chickens, pheasants, English sparrows, and crows become infected. Infections follow when parasitized dragonfly naiads or adults are eaten. Naiads living in lakes are available to ducks at all times except when the water is frozen. As the young dragonflies prepare to molt to adults, they climb out of the water onto vegetation. In this position, they are easy prey to all birds. Although adult dragonflies are swift fliers, they too may be caught, especially when resting during the cool early morning hours when birds are feeding.

When the metacercariae are released from the dragonflies as the latter are digested, they migrate to the bursa Fabricii or, if the bird is laying eggs, into the oviducts. In nonlaying birds, residence of the flukes in them is correlated with the existence of the bursa Fabricii. When it atrophies, the infection is lost. Development of the flukes in chickens requires about 1 week and they are lost in 3 to 6 weeks. Development in ducks is slower, taking 3 weeks, and the flukes are retained up to 18 weeks.

*Prosthogonimus cuneata* develops in the snails *Bithynia tentaculata* and *B. leachi* in Russia. Metacercariae are in Odonata only. In *Libellula quadrimaculata*, they become infective in 70 days. Flukes mature in the bursa Fabricii of chickens in 14 days and in the oviducts in 7 days. *Prosthogonimus ovatus* infects female pigeons more readily than males. In ducks, the genital apparatus of this species develops atypically, being like that of *Schistogonimus rarus*, the only species. It is therefore considered a synonym of *P. ovatus*.

## EXERCISE ON LIFE CYCLE

Since approximately 32 species of *Prosthogonimus* are known to occur in as many species of birds from seven orders, the likelihood of finding infected snails is fairly good during surveys for cercariae. The identity of the cercariae may be suspected by their small size, stylet, short tail, thick-walled Y-shaped excretory bladder, and origin from sac-like sporocysts in *Amnicola*.

When found, such cercariae should be placed in dishes with dragonfly naiads of the species available in the pond. Observe them being drawn into the rectum. Note whether they are retained in the rectum or expelled. In the latter case,

## Explanation of Plate 74

**A,** Adult fluke. **B,** Daughter sporocyst. **C,** Cercaria. **D,** Encysted metacercaria. **E,** Excysted metacercaria. **F,** Chicken definitive host. **G,** Ring-necked pheasant and duck definitive hosts. **H,** Snail (*Amnicola limnosa*) first intermediate host. **I,** Dragonfly naiads (*Leucorrhina, Tetragoneuria, Epicordulia, Gomphus, Mesothemis*) second intermediate host. **J,** Dragonfly imago.

1, Oral sucker; 2, ventral sucker; 3, pharynx; 4, esophagus; 5, intestine; 6, testis; 7, vas efferens; 8, cirrus pouch; 9, seminal vesicle; 10, common genital pore; 11, ovary; 12, uterus; 13, vitelline gland; 14, vitelline duct; 15, developing cercaria; 16, spine of oral sucker; 17, penetration glands; 18, body of excretory bladder; 19, arms of excretory bladder; 20, tail; 21, cyst wall of metacercaria.

**a,** Adult fluke in oviduct; **b,** embryonated eggs laid in oviduct; **c,** eggs reach cloaca; **d,** eggs pass from body in feces; **e,** egg in water; **f,** eggs hatch when swallowed by snail intermediate host (this has not been observed but probably is the only place where it occurs); **g,** miracidium passes through intestinal wall; **h,** daughter sporocyst bearing cercariae and germ balls (mother sporocyst not seen); **i,** cercaria escaping from snail; **j,** free cercaria being drawn into rectal chamber of naiad; **k,** cercaria drops tail and burrows through rectal wall into hemocoel; **l,** metacercaria in hemocoel; **m,** when naiad transforms into imago, metacercariae persist in body; **n,** definitive host becomes infected by eating naiads or imagoes with cysts in their bodies; **o,** digestion of dragonfly releases encysted metacercaria; **p,** metacercaria free in digestive tract; **q,** young fluke escapes from cyst with aid of digestive juices; **r,** young flukes leave intestine and enter bursa Fabricii, where some develop to maturity while others migrate up the oviducts of laying birds and develop to maturity.

Adapted from Macy, 1934, Univ. Minn. Agr. Exp. Sta. Tech. Bull. 98.

---

either the host is not the right one or the cercariae are not those of *Prosthogonimus.*

Dissection of dragonfly naiads from lakes where infected snails originate should reveal metacercariae with the characteristic striated, thick-walled cyst.

When metacercariae are available, they should be fed to baby chicks to get the adult worms in the bursa Fabricii. Eggs obtained from experimental infections should be fed to young *Amnicola,* preferably laboratory-reared specimens, to study the development of the intramolluscan stages. Eggs placed in the open stomach or gut of the snail host may hatch under the action of the digestive juices and the miracidia be available for study. Sections should be made of snails that have eaten eggs to study hatching and movements of the miracidia. Eggs should be studied to obtain information on the morphology of the miracidia.

### SELECTED REFERENCES

Heidegger, E. (1937). *Libellula brunnea* Fonsc. und *Platycnemis pennipes* Pall., zwei neue Hilfswirte der Eileiteregel der Hühner. Arch. Wiss. Prakt. Tierheilk. 72:224–229.

Krasnolobova, T. A. (1961). [Life cycle of *Prosthogonimus cuneata* (Rudolphi, 1809), a parasite of poultry.] Helminthologia 3:183–192.

Krasnolobova, T. A. (1969). [Experimental proof of synonymy of *Prosthogonimus* and *Schistogonimus* Lühe, 1909 (Trematoda: Prosthogonimidae).] Tr. Gel'mintol. Lab. 20:79–87.

Macy, R. W. (1934). Studies on the taxonomy, morphology, and biology of *Prosthogonimus macrorchis* Macy, a common oviduct fluke of domestic fowls in North America. Univ. Minn. Agr. Exp. Sta. Tech. Bull. 98, 71 pp.

Singe, P., and B. P. Pande. (1968). Experimental *Prosthogonimus ovatus* infection in domestic pigeons. Curr. Sci. 37:500–501.

### Superfamily Allocreadioidea

The cercariae of this superfamily developing in rediae or sporocysts in snails or clams are of various types (ophthalmoxiphidiocercariae, microcercous, rhopalocercous, ophthalmotrichocercous, tailless, or muscular tails with lateral and ventral fins). Encystment is in arthropods chiefly, rarely in vertebrates.

#### FAMILY ALLOCREADIIDAE

These are parasites of fishes, principally of the digestive tract. It is a large family containing many subfamilies, genera, and species. They are small- to medium-sized flukes, having well-developed suckers with the acetabulum near middle of body. The ceca are tubular. The testes are in the posterior half of the body with the ovary between them and the acetabulum. The uterus is usually pretesticular. The excretory bladder is tubular or saccular.

**Plate 74**   *Prosthogonimus macrorchis*                                                305

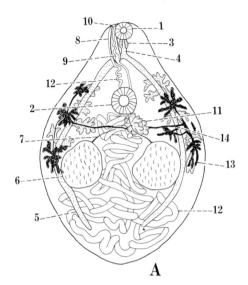

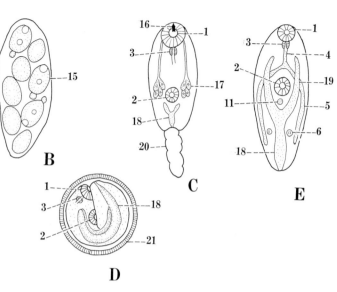

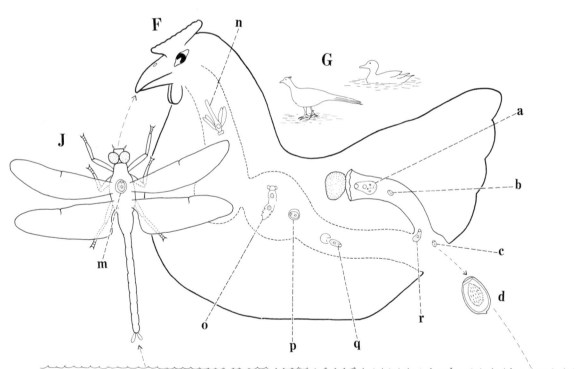

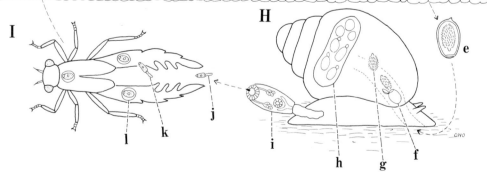

## Explanation of Plate 75

A, Adult fluke, ventral view. B, Sagittal section of adult fluke. C, Embryonated egg. D, Miracidium. E, Mother redia containing daughter rediae. F, Cercaria. G, Fish (trout and other fish) definitive host. H, Fingernail clam (*Pisidium* and *Musculium*) first intermediate host. I, Burrowing may fly (*Hexagenia*) naiad second intermediate host. J, Adult may flies retain metacercariae during metamorphosis.

1, Papillae around oral sucker; 2, oral sucker; 3, ventral sucker; 4, pharynx; 5, esophagus; 6, intestine; 7, testis; 8, cirrus pouch; 9, common genital pore; 10, ovary; 11, eggs in uterus; 12, vitelline gland; 13, embryonated egg; 14, miracidium; 15, apical gland; 16, eyespots; 17, flame cell; 18, germinal masses; 19, pharynx of mother redia; 20, daughter redia; 21, stylet of oral sucker; 22, lateral and dorsal views of stylet from oral sucker; 23, prepharynx; 24, penetration glands; 25, ducts of penetration glands; 26, posterior collecting tubule of excretory system (there is an anterior collecting tubule); 27, common collecting tubule; 28, excretory bladder.

a, Adult fluke in small intestine of fish; b, unembryonated eggs laid in intestine; c, egg developing in water; d, egg hatching; e, miracidium penetrating clam; f, mother redia with daughter rediae; g, daughter redia with cercariae (some rediae may contain both rediae and cercariae); h, cercaria free in water; i, cercaria penetrating abdomen of burrowing may fly; j, metacercariae in may fly naiad are retained throughout the life of the insect, including the adult stage; k, fish become infected by eating may flies; l, metacercaria released from may fly in stomach of fish; m, young fluke escapes from metacercarial cyst in intestine and develops to maturity.

Adapted from Hopkins, 1937, Ill. Biol. Monogr. 13:1.

---

*Crepidostomum cooperi*
Hopkins, 1931 (Plate 75)

This species occurs in the pyloric ceca of a number of species of fish, including sunfish, bass, bullheads, catfish, and trout.

**Description.**    Adult worms are elongate forms measuring up to 1.5 mm in length. The oral sucker bears six large papillae and the acetabulum is near or somewhat anterior to the middle of the body. The genital pore is between the anterior margin of the acetabulum and bifurcation of the intestine. A long slender cirrus sac containing a seminal vesicle twice its length extends from the genital pore to the anterior margin of the ovary. The testes are tandem in the posterior half of the body, with a pear-shaped ovary and the seminal receptacle between the anterior one and the acetabulum. The uterus is anterior to the testes and contains few eggs, ranging in size from 50 to 75 $\mu$ long by 30 to 55 $\mu$ wide. Vitelline glands extend from about midway between the pharynx and acetabulum to near the caudal end of the body. The excretory bladder is sac-like and extends forward to between the anterior and posterior margins of the front testis.

**Life Cycle.**    Unembryonated eggs laid by adult worms in the alimentary canal do not begin to develop until they have been deposited in the water with the feces. Development is completed at summer temperature in 7 days and hatching begins by day 10.

Miracidia penetrate fingernail clams (*Musculium transversum, Pisidium subtruncatum, P. compressum, P. adbitum, P. llijeborgi, P. nitidum*). Infections in them may run as high as 67 percent in some locations. The miracidia penetrate between the inner and outer layers of gills and the mantle where they transform. Presumably they change into mother sporocysts but this point is not clear, as only rediae have been seen. Two generations of rediae occur. Some contain only daughter rediae and others either cercariae only or both cercariae and rediae. Large rediae are in the liver. The situation concerning the rediae is not fully understood. The length of time required for the intramolluscan phase to develop is unknown.

The ophthalmoxiphidiocercariae are active swimmers, responding positively toward bright light. They attack and penetrate the naiads of may flies (*Hexagenia limbata, H. recurvata, Polymitarcys* sp.), including up to 95 percent of *H. limbata* during the summer in some areas. Upon entering the naiads, the cercariae go to the muscles, usually those of the gill-bearing filaments. At first, the cyst is thin and hyaline, being of parasite origin, but by day 4 it is surrounded by a layer of orange-brown material deposited by the host. The metacercariae are fully formed in 2 to 3 weeks. In old individuals, the internal organs appear in a well-developed stage, and numerous sperms are produced but no eggs have been seen in this species.

**Plate 75**    *Crepidostomum cooperi*                                                                307

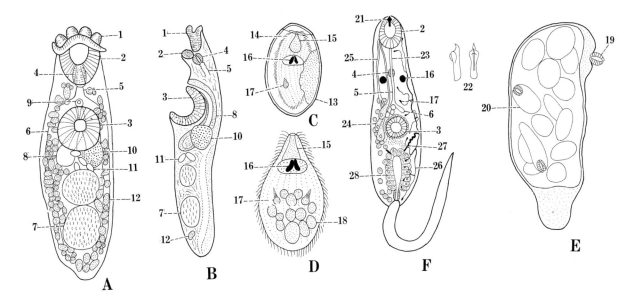

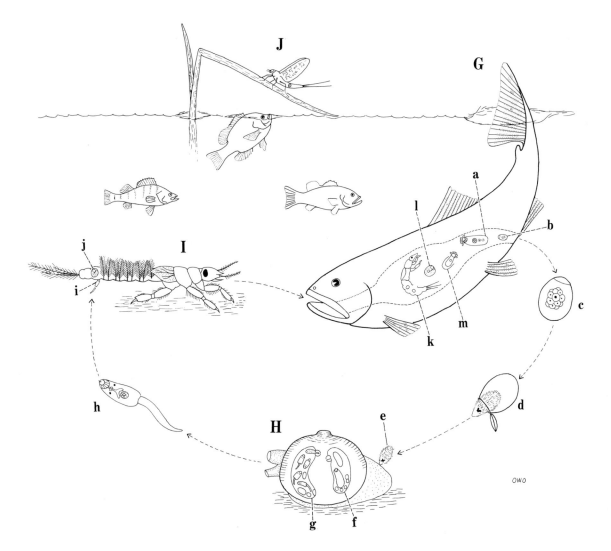

When infected may fly naiads of a number of species are eaten by susceptible fish, the metacercariae escape from the cysts and develop to maturity in 3 to 4 weeks.

Other species of *Crepidostomum* have life cycles very similar to that of *C. cooperi*. The molluscan hosts are fingernail clams and the invertebrate hosts are *Hexagenia* naiads for *C. isostomum*. The cercariae of *C. cornutum* encyst in the vicinity of the reproductive organs of crayfish; they are particularly abundant in the females, where up to 80 percent may be infected. The metacercariae are progenetic, having fully developed eggs which are deposited in the cysts. Adults occur in a variety of fishes in North America. *Crepidostomum farionis* of trout develops in *Pisidium* sp. and may fly naiads (*Ephemera* sp.) in lakes at high altitudes.

The life cycles and bionomics of *Crepidostomum metoecus* and *C. farionis* of the brown trout (*Salmo trutta*) differ in certain aspects in the same habitat. The first intermediate host of *C. metoecus* is the snail *Lymnaea pereger* and of *C. farionis* is the fingernail clam *Pisidium casertanum*. *Gammarus pulex*, the common freshwater scud, is the second intermediate host for both. *Crepidostomum metoecus* develops to sexual maturity in the pyloric ceca and upper part of the small intestine of the trout and *C. farionis* in the lower part. In a similar stream, *C. farionis* is far more common because of a greater abundance of its molluscan host and a higher percentage of infected gammarids. Cercariae are shed more abundantly during daylight and survive up to 5 days in water at 80°C. *Crepidostomum metoecus* does extensive damage to its snail (*L. pereger*) and gammarid hosts but as an adult, it seems to have no adverse effect on the fish.

Rainbow trout infected with *C. farionis* tested in a stamina tunnel performed equally well with uninfected control fish.

*Megalogonia ictaluri* from catfish has a life cycle similar to that of *C. cooperi*, except that the metacercariae are in the gills of *Hexagenia* and possibly other may fly naiads.

*Allocreadium ictaluri*, an Allocreadiidae from catfish, develops in rediae in the snail *Pleurocerca acuta*. The cercariae, which are biocellate and do not have a stylet, encyst in unionid clams. In Europe, *Allocreadium angusticolle*, a parasite of eels and Miller's Thumb, develops in the snail *Neritina fluviatilis* and the metacercariae are in gammarids (*Gammarus pulex* and *Asellus aquaticus*).

## EXERCISE ON LIFE CYCLE

With *Crepidostomum* being a very common parasite of fish in many types of habitats, it provides excellent material for studies on the life cycle of an allocreadiid trematode. When related forms are present, they can be studied, using somewhat similar approaches.

Adult flukes obtained from the pyloric ceca of fish provide a source of eggs for hatching and infection of fingernail clams. Study the development and morphology of the miracidia.

Expose fingernail clams to miracidia. Special effort should be directed toward finding the very earliest stages. Determine whether the miracidia transform into mother sporocysts, since they have not been reported.

Biocellate, stylet-bearing cercariae issuing from rediae in fingernail clams should be placed in dishes with may fly naiads, particularly those of the burrowing may fly (*Hexagenia*). Other genera, such as occur in the habitat, may be the second intermediate host. Crayfish are the hosts of some species, as are gammarids for others. Note the location of the cysts and the nature of the cyst wall. Ascertain whether the flukes are progenetic, i.e., producing eggs while still encysted.

Feed may fly naiads, or other forms containing metacercariae, to fish to complete the life cycle. Determine the time required for development to the adult stage.

## SELECTED REFERENCES

Abernathy, C. (1937). Notes on *Crepidostomum cornutum* (Osborn). Trans. Amer. Microsc. Soc. 56: 206.

Ameel, D. J. (1937). The life history of *Crepidostomum cornutum* (Osborn). J. Parasitol. 37:218–220.

Awachie, J. B. E. (1968). On the bionomics of *Crepidostomum metoecus* (Braun, 1900) and *Crepidostomum farionis* (Müller, 1784) (Trematoda: Allocreadiidae). Parasitology 58:307–324.

Choquette, L. P. E. (1954). A note on the intermediate hosts of the trematode *Crepidostomum cooperi* Hopkins, 1931, parasitic in the speckled trout (*Salvelinus fontinalis* (Mitchill)) in some lakes and rivers of the Quebec Laurintidae Park. Can. J. Zool. 32:375–377.

Crawford, W. W. (1943). Colorado trematode

studies. I. A further contribution to the life history of *Crepidostomum farionis* (Müller). J. Parasitol. 29:379–384.

Henderson, H. E. (1938). The cercaria of *Crepidostomum cornutum*. Trans. Amer. Microsc. Soc. 57: 165–172.

Hopkins, S. H. (1934). The papillose Allocreadiidae. Univ. Ill. Biol. Monogr. 13:49–124.

Klein, W. D., O. W. Olsen, and D. C. Bowden. (1969). Effects of intestinal fluke *Crepidostomum farionis* on rainbow trout, *Salmo gairdnerii*. Trans. Amer. Fish. Ass. 98:1–6.

Mathias, P. (1937). Cycle évolutif d'un trématode de la famille des Allocreadiidae Stossich (*Allocreadium angusticolle* (Hausmann)). C. R. Acad. Sci. Paris 205:626–628.

Seitner, P. G. (1951). The life cycle of *Allocreadium ictaluri* Pearse, 1924 (Trematoda: Digenea). J. Parasitol. 37:223–244.

### *Plagioporus sinitsini*
#### Mueller, 1934 (Plate 76)

This allocreadiid fluke is a common parasite of the gall bladder of cyprinid and catostomid fishes in North America.

**Description.** Adult worms reach a length of 0.67 to 1.5 mm. The acetabulum is near the center of the body and much larger than the oral sucker. The prepharynx is short, the pharynx is followed by a short esophagus, and the ceca are large, reaching to the level of the posterior testis. Testes located in the last quarter of the body are subequal in size, and obliquely arranged. The ovary is dextral and at the level of the anterior margin of the forward testis. The cirrus pouch is large, sigmoid, oblique in position, and extends to the middle of the acetabulum. The genital pore is displaced toward the left side of the body at a level between the pharynx and intestinal bifurcation. The vitellaria are lateral and extend from the pharynx to near the posterior end of the body. The eggs are few in number and large in size, measuring 67 to 75 $\mu$ long by 41 to 48 $\mu$ wide. A subterminal nodule is present on the anopercular end.

**Life Cycle.** The eggs are in the two- to four-cell stage when laid. They pass through the ductus choledochus into the intestine and are voided with the feces. Development is completed and hatching occurs 12 to 20 days after oviposition when the eggs are kept at room temperature.

The miracidia have 19 epidermal plates ar-

ranged in four tiers, an arrangement unique to this group of flukes. They swim actively for 24 hours. Upon coming in contact with the operculate snail *Goniobasis livescens* of the family Pleuroceridae, they penetrate them. Early stages of development in the snail are unknown. There are two generations of sporocysts in the liver. The first, probably the mother sporocyst, produces sporocysts. The second gives rise to unarmed cotylocercous cercariae which never leave them, apparently having lost the ability to swim. Large cercariae creep about inside the daughter sporocysts where they encyst. A thin cyst is secreted about each cercaria and the tail is slowly absorbed so that mature metacercariae are tailless.

Mature sporocysts containing metacercariae are abundant in the rectum of the snails but they may be found on rare occasions in the lymph spaces below the liver, indicating their former association with that organ. How they reach the rectum from the liver is unknown. Sporocysts with infective metacercariae emerge through the anus of the snail into the water, where they are active for about 24 hours.

Small fish such as minnows (*Nocomis biguttatus*, *Notropis cornutus*, *Lebistes reticulatus*), and fingerling suckers (*Catostomus commersoni*) readily eat the sporocysts containing the metacercariae, becoming infected as a result. At least 10 species of cyprinids in Michigan are naturally infected. The worms are free in the intestine of the fish 4 to 6 hours after the sporocysts are swallowed and in the gall bladder in 25 hours. Mature flukes with eggs appear 15 to 30 days after the fish eat the sporocysts containing metacercariae.

The life cycle of *Plagioporus lepomis* differs from that of *P. sinitsini* in which the cercariae escape from the sporocysts in the snail, *Goniobasis liviscens*, and encyst in the amphipod *Hyalella knickerbockeri*. Fish become infected by eating parasitized crustaceans.

The glandular complex of the cercariae consists of three pairs of cells: cephalic, mucoid, and caudal. Secretions of the cephalic glands produce an enzyme containing protein with sulfhydral groups that acts against the chitinous exoskeleton of the amphipod intermediate host, enabling the cercariae to enter. Mucoid glands provide a mucus covering, consisting of acid polysaccharides and a protein, for the cercariae before they leave the snails. The caudal glands produce an

## Explanation of Plate 76

A, Dorsal view of adult worm. B, Embryonated egg. C, Miracidium. D, Miracidium showing arrangement of dermal plates. E, Miracidium showing variation in arrangement of dermal plates. F, Daughter sporocyst from rectum of snail with enclosed cercariae and metacercariae. G, Cercaria. H, Metacercaria. I, Body of snail intermediate host (*Goniobasis livescens*) showing mass of sporocysts in rectum. J, Fish (*Notropis cornutus*) definitive host. K, Snail (*Goniobasis livescens*) first and second intermediate host.

1, Oral sucker; 2, ventral sucker; 3, pharynx; 4, esophagus; 5, intestine; 6, testis; 7, vas efferens; 8, vas deferens; 9, cirrus pouch; 10, seminal vesicle; 11, ovary; 12, seminal receptacle; 13, oviduct with Laurer's canal nearby to left; 14, Mehlis' gland; 15, proximal portion of uterus; 16, distal portion of uterus (intervening loops left out); 17, common genital pore far to the left of the midline; 18, vitelline glands; 19, vitelline duct; 20, excretory bladder; 21, miracidium in egg; 22, penetration glands; 23, flame cell; 24, apical papilla; 25, glandular structure; 26, excretory tubule; 27, excretory pore; 28, first tier of six dermal plates; 29, second tier of seven dermal plates; 30, third tier of four dermal plates; 31, fourth tier of two dermal plates; 32, birth pore; 33, cercaria; 34, metacercaria; 35, prepharynx; 36, genital primordia; 37, tail; 38, early stage of cyst; 39, sensory papillae of oral sucker; 40, sensory papillae of ventral sucker; 41, Laurer's canal; 42, common excretory tubule; 43, dermal papilla; 44, stomach; 45, small intestine; 46, rectum filled with sporocysts; 47, liver.

a, Adult fluke in gall bladder; b, unembryonated eggs laid in gall bladder; c, eggs pass into water with feces; d, developing egg in water; e, fully embryonated egg; f, egg hatches in water in 12 to 20 days; g, experimental infection of snail host with cercaria not successful but presumably they penetrate soft tissues; inside the snail they probably transform into mother sporocysts; h, stomach of snail; i, mother sporocyst with daughter sporocyst in liver; j, daughter sporocyst with cotylocercous cercaria in liver; k, small intestine; l, rectum filled with daughter sporocysts; m, mature daughter sporocyst containing metacercariae escaping through anus of snail; n, colored daughter sporocyst with encysted metacercariae free on bottom of pool; o, fish definitive host becomes infected when it swallows mature daughter sporocysts containing metacercariae or infected snails; p, metacercaria escapes from cyst through action of digestive juices; q, metacercaria enters common bile duct and migrates up it to gall bladder where maturity is attained in 15 to 30 days.

Adapted from Dobrovolny, 1939, Trans. Amer. Microsc. Soc. 58:121.

---

adhesive containing a mucopolysaccharide and protein.

## EXERCISE ON LIFE CYCLE

Eggs obtained from adult worms collected from the gall bladder of fish may be incubated to observe development and hatching. Note the unique number and arrangement of the epidermal plates.

Expose small *Goniobasis livescens* to infection by miracidia and determine if it can be accomplished experimentally and, if successful, try to ascertain the early developmental stages of the parasite. Observe adult snails collected from areas where infected fish occur for the appearance of the colorful sporocysts containing the cotylocercous cercariae and encysted metacercariae. Also dissect infected snails to locate the two generations of sporocysts and present evidence that both generations do occur. Note the route of exit through the snail of sporocysts containing metacercariae.

Feed sporocysts containing metacercariae

that have escaped from the snails to small cyprinids and follow the development of the parasites in them.

## SELECTED REFERENCES

Dobrovolny, C. G. (1939). Life history of *Plagioporus sinitsini* Mueller and embryology of new cotylocercous cercariae (Trematoda). Trans. Amer. Microsc. Soc. 58:121–155.

Dobrovolny, C. G. (1939). The life history of *Plagioporus lepomis*, a new trematode from fishes. J. Parasitol. 25:461–470.

Porter, C. W., and J. E. Hall. (1970). Histochemistry of a cotylocercous cercaria. I. Glandular complex in *Plagioporus lepomis*. Exp. Parasitol. 27:368–377.

### FAMILY GORGODERIDAE

The members of this family are parasites of the urogenital system of fish, amphibians, and reptiles. The body of some species is divided into a narrow anterior and a broad posterior portion. The suckers are well developed, with the ventral one much the larger and projecting prominently

**Plate 76**    *Plagioporus sinitsini*                                              311

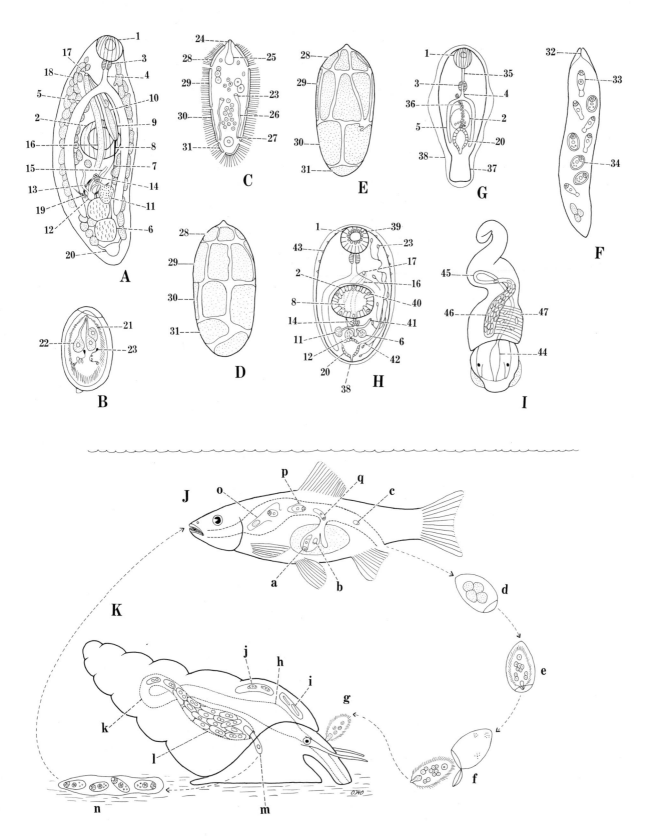

## Explanation of Plate 77

A, Adult fluke. B, Embryonated egg. C, Miracidium. D, Mother sporocyst. E, Daughter sporocyst. F, Anterior portion of cercarial tail showing chamber with cercaria enclosed. G, Excysted metacercaria. H, Frog (*Rana* spp.) definitive host. I, Clam (*Musculium partumeium*) first intermediate host. J, Snail (*Physa*) second intermediate host. K, Snail (*Helisoma*) second intermediate host. L, Tadpole (*Rana* spp.) second intermediate host (salamander larvae, *Ambystoma maculatum* also become infected). M, Crayfish (*Cambarus* sp.) second intermediate host.

1, Oral sucker; 2, ventral sucker; 3, esophagus; 4, intestine; 5, testis; 6, seminal vesicle; 7, common genital pore; 8, ovary; 9, uterus; 10, vitelline gland; 11, embryonated egg; 12, miracidium; 13, apical gland; 14, penetration glands; 15, flame cell; 16, germ cells; 17, germinal mass destined to form daughter sporocyst; 18, birth pore; 19, germinal masses; 20, cercaria escaping through birth pore of daughter sporocyst; 21, tail of cercaria; 22, chamber walls of cavity forming anterior portion of cercarial tail; 23, cercaria detached from stem of tail; 24, excretory bladder; 25, point of attachment of cercaria to its tail; 26, vas efferens; 27, excretory pore.

a, Adult fluke in urinary bladder; b, embryonated eggs deposited in urinary bladder; c, eggs expelled with urine into water hatch quickly; d, swimming miracidium drawn through incurrent siphon into gill chamber of clam penetrates tissues; e, mother sporocyst in gills; f, daughter sporocyst giving birth to fully developed cercariae; g, cercariae freed from sporocysts make their way through the tissues into the gill chamber; h, cercariae escape from clam through excurrent siphon; i, cercariae are swallowed by snails, tadpoles, and crayfish; j, cercariae pass into intestinal wall; k, encysted metacercariae in intestinal wall; l, frogs acquire the parasites by eating infected intermediate hosts; m, metacercariae released when intermediate hosts are digested; n, young fluke released from cyst by action of digestive juices migrates to urinary bladder; o, flukes travel up ureters to kidneys or up oviducts and return to urinary bladder in two weeks where they develop to maturity in 21 to 60 days.

Adapted from Goodchild, 1948, J. Parasitol. 34: 407.

---

from the ventral side of the body. The testes are two or more in number. The ovary lies between the testes and the ventral sucker. Vitellaria consist of two compact or lobed glands located behind the ventral sucker and between the ceca.

### *Gorgodera amplicava* Looss, 1899 (Plate 77)

*Gorgodera amplicava* parasitizes the urogenital system of most anurans and one urodele in North America.

**Description.**     The body is 3 to 5 mm long with the anterior portion smaller than the posterior, being about one-third as long. The ventral sucker is 2.5 to 3.0 times larger than the oral sucker. There are nine testes, with five in a row on the same side as the ovary and four on the other. The ovary lies on the left side between the testes and the posterior margin of the ventral sucker. The vitelline gland consists of two groups of follicles between the ventral sucker and ovary. The eggs are fully developed when laid.

**Life Cycle.**     Adult worms in the urinary bladder lay fully embryonated nonoperculate eggs. They are voided with the urine and hatch almost immediately or up to 1 day after reaching the water.

The miracidia are deliberate swimmers, living up to 72 hours in the water, after which death ensues if they have not entered *Musculium partumeium*, a fingernail clam, which serves as the only first intermediate host in this country. The miracidia do not seek the clams but are drawn into the mantle cavity through the incurrent siphon by the current of water flowing into it. When in contact with the gills, they attach and penetrate within 10 minutes.

Upon entering the gills, the miracidia transform to mother sporocysts. Growth is slow. Small germ balls developing in the wall are shed into the central cavity and slowly develop into a single generation of daughter sporocysts after 40 to 50 days.

Daughter sporocysts are located ventrally between the inner and outer gill lamellae of the clams and appear as large entangled masses wholly embedded in the gill tissue. Up to 16 fully developed cystocercous cercariae may be present in a single daughter sporocyst. They escape singly through a terminal birth pore. The cercariae are 12 to 17.4 mm long. They consist of an almost structureless tail with a deep chamber at one end in which the young fluke is enclosed. The stylet-bearing larval fluke itself is spindle-shaped and 0.6 to 0.78 mm long by 0.195 to 0.235 mm wide.

Cercariae are liberated from the gills of the

**Plate 77**    *Gorgodera amplicava*                                                313

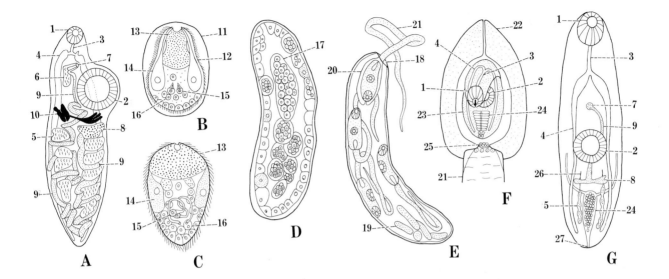

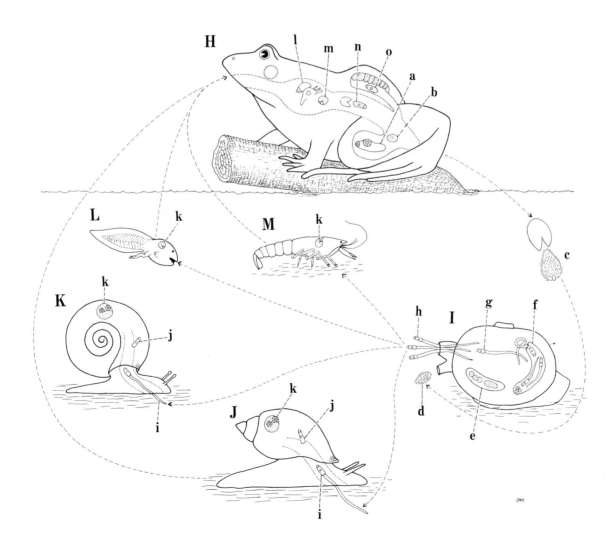

clams during the late afternoon and early evening, being expelled through the excurrent pore into the water. The tail is smooth and glistening when first expelled but soon becomes sticky and adheres to objects in the water. The wriggling cercariae attract tadpoles of frogs, larvae of salamanders (*Ambystoma maculatum*), snails (*Physa halei, P. parkeri, Lymnaea traskii, Helisoma trivolvis, H. antrosa, Pseudosuccinea columella*), and crayfish (*Cambarus* sp.), which eat them. The young distomes emerge from the anterior chamber of the tail in the mouth, stomach, and intestine of these second intermediate hosts. They creep over the surface of the intestine for 10 or 15 minutes, after which they burrow into the wall and encyst within it. The metacercariae are infective to frogs after they have been in tadpoles approximately 24 hours. Dragonfly naiads do not become infected even though they eat the cercariae.

Frogs become parasitized when they swallow infected second intermediate hosts. Upon digestion of the hosts in the stomach, the metacercarial cysts are liberated. Encystment may occur in the stomach but more often in the small intestine. The larval flukes migrate posteriorly through the intestine, arriving in the cloaca in 8 hours or less. They seek out the openings of the excretory and reproductive ducts and migrate up them, remaining about 2 weeks. After this time, they return to the urinary bladder where changes leading to the adult worms take place. Some flukes become sexually mature in 21 days and all of them are fully developed by the end of 2 months.

Ultrathin sections of the integument show no physical evidence that an interchange of substances takes place through it.

Other species of Gorgoderinae, whose life cycles have been worked, have a pattern similar to that of *Gorgodera amplicava*. The first intermediate host is a clam and the second an anamniotic vertebrate.

*Gorgoderina attenuata* from frogs (*Rana*) and newts (*Triturus viridiscens*) develop in sporocysts in *Sphaerium occidentalis*. The cercariae encyst in tadpoles when eaten by them. *Gorgoderina vitelliloba* from the bladder and kidneys of *Rana temporaria* transplanted to *Bufo bufo* and *B. viridis* develop normally and survive up to 8 months. Specimens from the kidneys migrate back to them for further development.

*Phyllodistomum solidum* from the urinary bladder of the dusky salamander (*Desmognathus fuscus*) develops in sporocysts in the fingernail clam *Pisidium abditum*. Odonatan naiads (*Ischnura verticalis, Argia* sp., *Enallagma* sp., *Libellula* sp.) serve as the second intermediate hosts. The metacercariae encyst in the thoracic hemocoel.

## EXERCISE ON LIFE CYCLE

Fully embryonated eggs may be obtained from infected frogs by grasping the frogs behind the front legs and squeezing quickly. They react by struggling and forcefully emit urine which may be caught in a container. Note that the eggs are nonoperculate. Study the miracidia inside the eggs. With a micropipette, transfer some of the eggs from the urine to water and note the rapidity with which they hatch. Determine the point of dilution of the urine with water at which eggs will hatch.

Put a specimen of *Musculium partumeium* in a small dish of water together with miracidia. Note whether they react to the mollusk. By means of a micropipette, place some miracidia in the current of water going into the inhalent siphon and observe how they are carried by it. Pipette miracidia onto the gills of an open clam and observe their reaction and penetration into the tissue. After they have completely entered the gills, fix them for sectioning to observe the location and condition of the miracidia.

Cercariae from naturally infected clams may be used for study. Find the daughter sporocysts by dissecting the mollusks. Watch the cercariae escape through the birth pore. Place fully developed cercariae in separate dishes with snails, tadpoles, and small crayfish and observe them being eaten. Dissect specimens of each species of intermediate host shortly after they have eaten the cercariae to observe the migration into the intestinal wall and encystment.

Feed several infected tadpoles or snails to each of a dozen uninfected mature frogs. Examine the alimentary canal, ureters with associated ducts, kidneys, and urinary bladder of four frogs at intervals of 2 hours, beginning 6 hours after giving the metacercariae to them. Note the course of migration before the flukes enter the urinary bladder the first time. Examine one of the remaining eight frogs every third day to determine migrations of the metacercariae in the body prior to their returning to the urinary bladder where final development to maturity takes place.

## SELECTED REFERENCES

Burton, P. R. (1966). The ultrastructure of the integument of the frog bladder fluke *Gorgoderina* sp. J. Parasitol. 52:926–934.

Cort, W. W. (1912). North American frog bladder flukes. Trans. Amer. Microsc. Soc. 31:151–166.

Goodchild, C. G. (1943). The life-history of *Phyllodistomum solidum* Rankin, 1937, with observations on the morphology, development and taxonomy of the Gorgoderinae (Trematoda). Biol. Bull. 84:59–86.

Goodchild, C. G. (1948). Additional observations on the bionomics and life history of *Gorgodera amplicava* Looss, 1899 (Trematoda: Gorgoderidae). J. Parasitol. 34:407–427.

Krull, W. H. (1935). Studies on the life history of a frog bladder fluke, *Gorgodera amplicava* Looss, 1899. Pap. Mich. Acad. Sci. Arts Lett. 20:697–710.

Mitchell, J. B. (1966). Successful transplantation of a frog bladder fluke from one host species to another. Nature (London) 211:305.

Rankin, J. S. (1939). The life cycle of the frog bladder fluke *Gorgoderina attenuata* Stafford, 1902 (Trematoda: Gorgoderidae). Amer. Midl. Natur. 21:476–488.

Schell, C. S. (1967). The life cycle of *Phyllodistomum staffordi* Pearse, 1924 (Trematoda: Gorgoderidae Looss, 1901). J. Parasitol. 53:569–576.

### FAMILY TROGLOTREMATIDAE

Troglotrematidae are small to medium-sized fleshy flukes with spinous cuticle and dense vitellaria. They are parasites of the intestine, lungs, nasal cavities, frontal sinuses, and subcutaneous tissues in birds and mammals in many parts of the world.

### *Paragonimus kellicotti*
Ward, 1908 (Plate 78)

This parasite occurs as pairs in cysts in the lungs of cats, dogs, pigs, muskrats, rats (experimental), wildcats, and mink from North America. On the basis of the frequency of natural infections, mink are believed to be the natural host.

**Description.** Adult flukes have thick oval-shaped bodies 7 to 12 mm long, 4 to 6 mm wide, and 3 to 5 mm thick. They are covered with scale-like spines. The ventral sucker is slightly anterior to the middle of the body. Unbranched ceca extend to near the posterior end of the body. The two testes are irregularly lobed, slightly oblique, and in the posterior third of the body. There is no cirrus pouch, but a prostate gland is present. The ovary is lobed, larger than the testes, and on the right side near the level of the ventral sucker. A genital pore is at the posterior margin of the ventral sucker. Extensive vitelline glands fill the lateral fields for almost the full length of the body. Eggs measure 75 to 118 $\mu$ long by 48 to 65 $\mu$ wide.

**Life Cycle.** Adult flukes in cysts in the lungs of the definitive hosts lay unembryonated eggs which are transported up the trachea into the pharynx, swallowed, and voided in the feces. Development at summer temperatures requires at least 2 weeks, at the end of which some hatch. Others are very retarded in development and do not hatch for at least 4 months.

The miracidia are very active just prior to hatching time. Their movements probably aid in forcing the operculum off the egg. In swimming about, they come in contact with *Pomatiopsis lapidaria*, *P. cincinnatiensis*, and *Oncomelania nosophora*, the snails which serve as the first intermediate hosts, and penetrate the head and neck of the young snails. Old ones are refractory.

*Pomatiopsis lapidaria* is a freshwater snail of the eastern half of the United States, occurring in marshy seepages and along the banks and flood plains of streams. They are dormant during the winter and hot period of the summer, hiding under vegetation and objects with the operculum closed. Males are about 7 mm long and females 8.5. Eggs are laid singly during the spring and hatch in late September.

Upon entering the snail hosts, the miracidia accumulate around the esophagus, stomach, and intestine where they transform into inactive sac-like mother sporocysts. They mature in 2 weeks and begin producing short, stumpy mother rediae, each with a well-developed collar-like thickening around the anterior end.

The mother rediae escape from the sporocysts about 29 days after infection of the snails. They come to rest in the lymph channels adjacent to the liver and develop to maturity in approximately 34 days, i.e., 63 days after infection. The number of daughter rediae present at one time is small, averaging about 12 per individual.

Upon escaping from the mother rediae, the daughter rediae migrate into the lymph system of the liver. They are fully developed and con-

316    ANIMAL PARASITES

## Explanation of Plate 78

A, Adult fluke. B, Miracidium. C, Mature mother sporocyst. D, Mother redia. E, Daughter redia. F, Cercaria. G, Encysted metacercaria. H, Mink (*Mustela vison*) definitive host. I, Snail (*Pomatiopsis lapidaria*) first intermediate host. J, Crayfish (*Cambarus* spp.) second intermediate host.

1, Oral sucker; 2, ventral sucker; 3, pharynx; 4, intestine; 5, testis; 6, vas efferens; 7, common genital pore; 8, ovary; 9, oviduct; 10, uterus; 11, vitelline glands; 12, vitelline duct; 13, excretory bladder; 14, apical papilla; 15, lateral papilla; 16, flame cell; 17, excretory tubule; 18, anterior and first row of six epidermal plates; 19, second row of six epidermal plates; 20, third row of three epidermal plates; 21, fourth and last row of a single epidermal plate; 22, mother redia; 23, daughter redia; 24, cercaria; 25, oral spine; 26, prepharynx; 27, penetration glands; 28, ducts of penetration glands; 29, common collecting tubule of excretory system; 30, cyst of metacercaria; 31, encysted metacercaria.

a, Two adult flukes in cyst in lungs; b, unembryonated eggs laid in cysts pass up trachea and are swallowed; c, eggs mixed with feces; d, eggs leave body in feces; e, eggs develop in water where hatching occurs; f, miracidium free in water; g, miracidium penetrates snail intermediate host; h, mature mother sporocyst with mother rediae inside; i, mother redia with daughter rediae; j, daughter redia with cercariae; k, cercaria free in water; l, cercaria penetrates soft parts of crayfish and migrates to heart; m, metacercaria encysts on heart of crayfish; n, definitive host acquires parasites by eating infected crayfish; o, metacercaria is freed from crayfish when latter is digested; p, metacercaria free in digestive tract; q, young fluke escapes from metacercarial cyst in intestine; r, young fluke passes through intestinal wall into celom; s, fluke migrates forward in celom; t, young fluke migrates through diaphragm into pleural cavity; u, flukes burrow into lungs where cysts connecting with the air ducts are formed and development proceeds to maturity.

Figure A somewhat schematic from several sources; B–G, from Ameel, 1934, Amer. J. Hyg. 19: 279.

---

tain mature cercariae as soon as 15 days later, or 78 days after infection of the snails. In some snails, however, development may be delayed another 2 weeks. There is an average of around 22 cercariae per daughter redia at any one time.

The cercariae are spiny, microcercous, stylet-bearing forms that emerge in the late afternoon and early evening, a time that coincides with the activity of the crayfish hosts. The cercariae enter crayfish (*Cambarus propinquis, C. robustus, C. virilis, C. diogenes, C. rusticus*), through the soft membranous chitin between the segments. All parts of the crayfish are vulnerable to them during ecdysis. Upon entering the body, the cercariae go to the cardiac region, arriving within 12 hours.

Encystment takes place on the outside of the heart. The metacercariae generally are aligned in a transverse band across the dorsal side of the heart between the two ostia. A thin cyst appears around each fluke by day 5 and they are infective by day 46. On the average, there are 20 to 30 cysts per crayfish but 147 have been reported. Old cysts are thick-walled, hyaline, and have a reddish color.

Infection of the definitive hosts occurs when they eat parasitized crayfish. The metacercariae are freed from their cysts in the small intestine and the young flukes migrate through the intestinal wall into the peritoneal cavity as early as 5 hours after being swallowed. They travel forward to the diaphragm and burrow through it into the pleural cavity by 96 hours at the earliest, but a few are still in the abdominal cavity 259 days later. Some reach the lungs in 14 days after entering the host. The flukes find each other and normally occur as pairs in lung cysts; occasionally there may be more per cyst. In small mammals, the cysts appear as nodules projecting above the surface of the lungs, whereas in large mammals they may be deep below the surface. The cysts are lined with bronchiolar-like epithelium surrounded by a fibrous capsule whose stroma is largely collagenous. Sexual maturity is attained in 22 to 24 weeks. Infection in cats persists at least 27 months, and perhaps longer.

*Paragonimus westermani*, the oriental lung fluke which some workers believe to be conspecific with *P. kellicotti*, infects man as well as a number of domestic and wild animals. It is a fluke of great medical importance in oriental countries because of the frequency with which it occurs and the extent of the pathology it produces in humans. Its life cycle is very similar to that of *P. kellicotti*, except that crabs in addition to crayfish serve as intermediate hosts.

**Plate 78**   *Paragonimus kellicotti*                                                                317

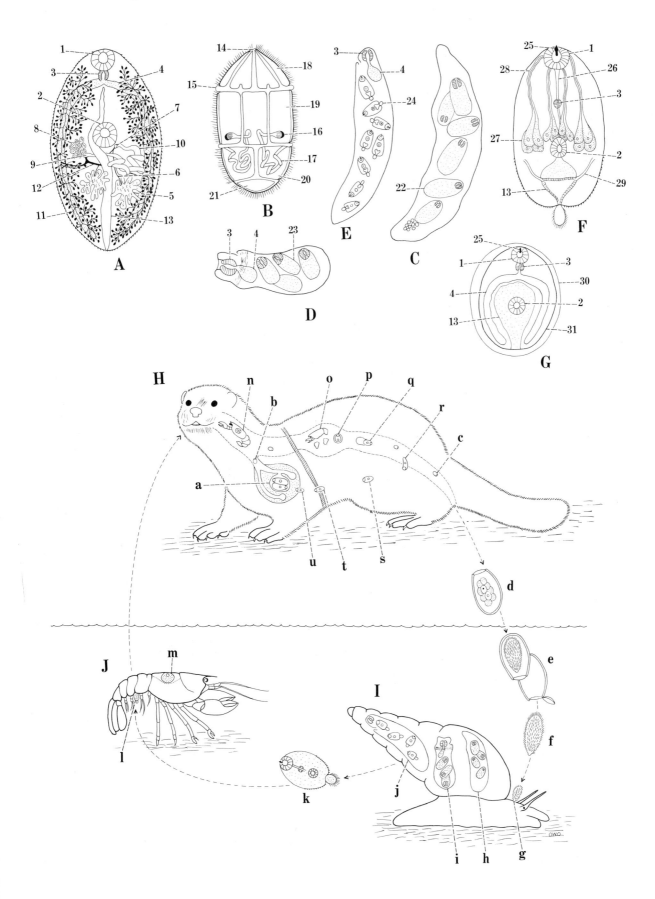

## EXERCISE ON LIFE CYCLE

The quickest and easiest way to determine whether *Paragonimus kellicotti* occurs in a region is to examine the hearts of crayfish for the metacercariae whose location and morphology are characteristic.

Upon finding a source of metacercariae, infection experiments with white rats or cats may be conducted. Using rats, feed at least half a dozen of them simultaneously with upward of 25 metacercariae each. Examine one rat after 6 hours to check for flukes in the small intestine and the peritoneal cavity. After 96 to 125 hours, examine a second one for flukes in both the peritoneal and pleural cavities. Sixteen days after infection, examine the lungs and body cavities of a third rat for flukes. If none appear in the lungs at this time, examine a fourth one. Keep the two remaining ones to determine the prepatent period by means of fecal examinations for the appearance of eggs.

Examine *Pomatiopsis lapidaria* from the same area where the infected crayfsh originated for natural infections with sporocysts, rediae, and cercariae. Note the different sizes of the two generations of rediae. Observe the spiny microcercous cercariae. Expose small crayfish to the cercariae. Examine infected individuals at 1-hour intervals for a period of 8 hours after infection to determine when the metacercariae reach the pericardial cavity. Sacrifice one of the remaining individuals daily for the next 14 days to ascertain when encystment occurs.

If eggs are available, incubate them at room temperature and note when hatching occurs. Expose young *Pomatiopsis lapidaria* to miracidia in order to follow the intramolluscan development. Find the sporocysts around the esophagus, stomach, and intestine, and the rediae in the liver. Note the time when cercariae appear. Carcasses of wild mink when available should be examined for eggs to be used in infecting snails.

## SELECTED REFERENCES

Ameel, D. J. (1934). *Paragonimus*, its life history and distribution in North America and its taxonomy (Trematoda: Troglotrematidae). Amer. J. Hyg. 19: 279–317.

Ameel, D. J., W. W. Cort, and A. Van der Woude. (1951). Development of the mother sporocyst and rediae of *Paragonimus kellicotti* Ward, 1908. J. Parasitol. 37:395–404.

Basch, P. F. (1959). Two new molluscan intermediate hosts for *Paragonimus kellicotti*. J. Parasitol. 45:273.

Chen, P. D. (1937). The germ cell cycle in the trematode *Paragonimus kellicotti* Ward. Trans. Amer. Microsc. Soc. 56:208–236.

Dundee, D. S. (1957). Aspects of the biology of *Pomatiopsis lapidaria* (Say). (Mollusca: Gastropoda: Prosobranchia). Univ. Mich. Misc. Publ. Mus. Zool. No. 100, 37 pp.

Lumsden, R. D., and F. Soganderes-Bernal. (1970). Ultrastructural manifestations of pulmonary paragonimiasis. J. Parasitol. 56:1095–1109.

Soganderes-Bernal, F. (1965). Studies on American paragonimiasis. I. Age immunity of snail host. J. Parasitol. 51:958–960.

Soganderes-Bernal, F. (1966). Studies on American paragonimiasis. IV. Observations on the pairing of adult worms in laboratory infections in cats. J. Parasitol. 52:701–703.

### *Nanophyetus salmincola*

Chapin, 1927 (Plate 79)

*Nanophyetus salmincola* is a parasite of the small intestine of dogs, foxes, coyotes, raccoons, rats, otters, skunks, weasels, mink, great blue herons, and mergansers in the Pacific northwest of the United States. It transmits *Neorickettsia helminthoeca* to dogs and other canines which is highly fatal to them.

**Description.** Adult flukes are 0.9 to 2.5 mm long by 0.3 to 0.5 mm wide. The suckers are nearly equal in size with the ventral one near the middle of the body or slightly anterior. The pharynx and esophagus are about equal in length and the ceca extend near or to the posterior margin of the testes. The testes are large oval bodies arranged symmetrically at the sides in the posterior half of the body. A large cirrus pouch is located to the right of the midline and behind or somewhat dorsal to the ventral sucker; it contains a large seminal vesicle divided by a constriction into two subequal parts. The round ovary is considerably smaller than the oral sucker and located near or adjacent to it on the left of the midline. The common genital pore is median and near the posterior margin of the ventral sucker. Vitellaria consist of irregular follicles scattered over the dorsal part of the body except that portion cephalad from the cecal bifurcation. Eggs measure 82 to 97 $\mu$ long by 38 to 55 $\mu$ wide.

**Life Cycle.** Unembryonated eggs are passed in the feces of the vertebrate host. They develop slowly, requiring a minimum of 85 days at room temperature; many take 185 to 200 days to hatch.

Newly hatched miracidia swim at random, not being attracted to snails. However, upon coming in contact with the snail intermediate host (*Oxytrema silicula*), they burrow into them. The immediate development of the miracidia upon entering the snails is unknown. At least no sporocysts have been observed. Only rediae are known. They are small in size, taper posteriorly, and contain a few small germ balls; medium-sized ones have only a few cercariae inside; and large mature individuals are filled with 74 to 76 cercariae. They are liberated intermittently, often by the thousands, in long strands of mucus where they live up to 48 hours. Small fish may even become entangled in the mucus threads.

Larval stages of the parasite do much damage to the snail host by increased pressure from sheer numbers, destruction of hepatic tubules by feeding on them, and by the added burden of disposing of the metabolites of the parasites. Uptake of lipids and glycogen is by mouth and by absorption across the body wall. Alkaline phosphatases of the fluke function in glycogen metabolism.

The cercariae show no reactions upon coming close to fish. If, however, they touch the fish at any point, they attach and quickly penetrate, entrance being completed in 30 seconds to 2 minutes. They soon enter the blood vessels and are carried to all the organs of the body, including the cornea and retina. In addition to salmonid fish, naturally infected fish include sculpins, eels, and the red-sided bream. The Pacific giant salamander is naturally infected. Species susceptible to experimental infection are goldfish, suckers, bluegills, sticklebacks, and gambusias.

Young salmonids migrating down rivers to the sea become infected en route. Metacercariae in the muscles and internal organs remain viable during the marine sojourn and are infective upon return of the fish to the rivers on their way back to the ancestral spawning ground.

Being transported to all the internal organs where encystment occurs, the cercariae cause high mortality in small fish. Different species of salmonids ranging from 30 to 37 mm in length vary greatly in their ability to withstand the effects of infection by cercariae. The following number of cercariae cause mortality in 50 percent of the fish: 58 in kokanee, 74 in Yellowstone cutthroat trout, 110 in Atlantic salmon, 200 in silver salmon, 295 in rainbow trout, and 430 in Columbia River cutthroat trout. The metacercariae are inside the fin rays, blood vessels, and internal organs within 1 hour after penetrating the skin. If the cyst ruptures, the flukes are able to reencyst. Metacercariae over 10 days of age are infective. When fish harboring them are eaten by mammalian hosts, the young flukes are released in the intestine and grow to maturity in 6 to 7 days.

Both the flukes and their eggs carry the rickettsial pathogen *Neorickettsia helminthoeca* that causes salmon-poisoning disease. Symptoms appear 6 to 10 days after swallowing metacercariae of *Nanophyetus salmincola*. There is fever and loss of appetite, followed by a pus-like discharge from the eyes, severe vomiting, and profuse diarrhea that may contain blood. Death follows in 50 to 90 percent of the infected dogs. Recovered animals are immune to reinfection. Flukes penetrating the mucosa of the small intestine produce much damage, resulting in severe enteritis that leads to bleeding.

Because of the high incidence of infection of snails, it is believed that mink and raccoons, which eat fish and defecate in the water, are the most common natural definitive hosts.

*Nanophyetus schikhobalowi*, believed by some workers to be *N. salmincola*, infects most of the dogs, cats, and people in some of the far eastern Russian villages. About 16 percent of the snail intermediary, *Semisulcospira laevigata*, and practically all the graylings and red salmon are infected.

*Sellacotyle mustelae* Wallace, a closely related minute form, occurs in the small intestine of mink. The cercariae develop in rediae in *Campeloma rufum* and encyst in fish, mainly bullheads (*Ameiurus melas*). The metacercariae are infective after 8 days and develop to maturity in mink by day 5.

### EXERCISE ON LIFE CYCLE

The characteristic cercariae are easily recognized by their short tails, stylet, and large mucous gland on the ventral side near the posterior end of the body. Observe the strands of mucus emitted by the snail with the cercariae entangled in them.

Expose small fish to a large number of cercariae and observe their reaction to the infection.

## Explanation of Plate 79

A, Adult fluke. B, Mother redia. C, Daughter redia. D, Lateral view of cercaria. E, Encysted metacercaria. F, Excysted metacercaria. G, Dog definitive host. H, Snail (*Oxytrema silicula*) first intermediate host. I, Fish (*Salvelinus salmo, Oncorhynchus* and others) second intermediate host.

1, Oral sucker; 2, ventral sucker; 3, prepharynx; 4, pharynx; 5, esophagus; 6, intestine; 7, testes; 8, vas efferens; 9, vas deferens; 10, seminal vesicle; 11, common genital pore; 12, ovary; 13, oviduct; 14, uterus; 15, egg in uterus; 16, Laurer's canal; 17, vitelline glands; 18, vitelline duct; 19, common vitelline duct; 20, developing daughter redia; 21, birth pore; 22, cercaria; 23, germinal masses; 24, oral spine; 25, penetration glands; 26, genital anlagen; 27, excretory bladder; 28, mucous gland; 29, tail; 30, cyst of metacercaria; 31, metacercaria; 32, ruptured cyst of metacercaria.

a, Adult fluke embedded in wall of small intestine; b, unembryonated eggs laid in intestine; c, eggs pass from body in feces; d, eggs develop in water and hatch; e, miracidium free in water; f, miracidium penetrates snail; g, mother sporocyst presumably occurs but has not been reported; h, mother redia; i, daughter redia; j, cercaria having escaped from daughter redia is free in tissues of snail; k, cercaria free in water; l, cercaria attaches to salmonid fish, entering muscular tissue; m, encysted metacercaria in fish; n, dog becomes infected by eating fish carrying metacercariae; o, digestion of fish releases metacercaria; p, young fluke escapes from metacercarial cyst and attaches to the wall of the intestine where maturity is attained in 5 to 6 days.

Adapted from Bennington, 1951, M. S. Thesis, Oregon State Univ., and Bennington and Pratt, 1960, J. Parasitol. 46:91.

---

When the fish die, fix them and section various organs and part of the musculature to ascertain the location of the cercariae. In fish less heavily exposed, observe the metacercariae and distribution of them in the body. Six days and again 10 days after infection, feed metacercariae to hamsters in order to determine first, the approximate time required for the metacercariae to attain infectivity, and second, the prepatent period of the flukes. Place eggs in a bottle of water and observe them weekly for a period of 6 to 7 months to follow the rate of development and time required for hatching. Expose small snails to miracidia and search for sporocysts; also observe development of the rediae and cercariae.

## SELECTED REFERENCES

Baldwin, N. L., R. E. Millemann, and S. E. Knapp. (1967). "Salmon-poisoning" disease. III. Effect of experimental *Nanophyetus salmincola* infection on the fish host. J. Parasitol. 53:556–564.

Bennington, E., and I. Pratt. (1960). The life history of the salmon-poisoning fluke *Nanophyetus salmincola* (Chapin). J. Parasitol. 46:91–100.

Chapin, E. A. (1926). A new genus and species of trematode, the probable cause of salmon-poisoning in dogs. North Amer. Vet. 7:36–37.

Filimonova, L. V. (1963). [Life-cycle of the trematode *Nanophyetus schikhobalowi*.] Tr. Gel'mintol. Lab. Akad. Nauk. SSSR 13:347–357.

Gebhardt, G. A., R. E. Millemann, S. E. Knapp, and P. A. Nyberg. (1966). "Salmon-poisoning" disease. II. Second intermediate host susceptibility studies. J. Parasitol. 52:54–59.

Millemann, R. E., G. A. Gebhardt, and S. E. Knapp. (1964). "Salmon-poisoning" disease. I. Infection in a dog from marine salmonids. J. Parasitol. 50:588–589.

Nyberg, P. A., S. E. Knapp, and R. E. Millemann. (1967). "Salmon poisoning" disease. IV. Transmission of the disease to dogs by *Nanophyetus salmincola* eggs. J. Parasitol. 53:694–699.

Philip, C. B. (1955). There's always something new under the "parasitological" sun (the unique story of helminth-borne salmon-poisoning disease. J. Parasitol. 41:125–148.

Porter, C., I. Pratt, and A. Owczarzak. (1967). Histopathological and histochemical effects of the trematode *Nanophyetus salmincola* (Chapin) on the hepatopancreas of its snail host *Oxytrema silicula* (Gould). Trans. Amer. Microsc. Soc. 86:232–239.

Schlegel, M. W., S. E. Knapp, and R. E. Millemann. (1968). "Salmon-poisoning" disease. V. Definitive hosts of the trematode vector, *Nanophyetus salmincola*. J. Parasitol. 54:770–774.

Wallace, F. G. (1935). A morphological and biological study of the trematode *Sellacotyle mustelae* n. g., n. sp. J. Parasitol. 21:143–164.

Witenberg, G. G. (1932). On the anatomy and systematic position of the causative agent of so-called salmon-poisoning. J. Parasitol. 18:258–263.

## ORDER OPISTHORCHIIDA

The cercariae of this order have caudal excretory vessels during development. An oral stylet is never present.

**Plate 79**   *Nanophyetus salmincola*                                                                                          321

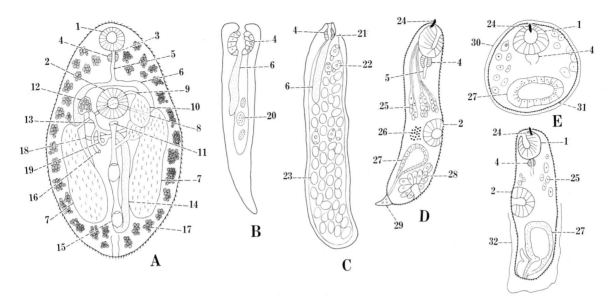

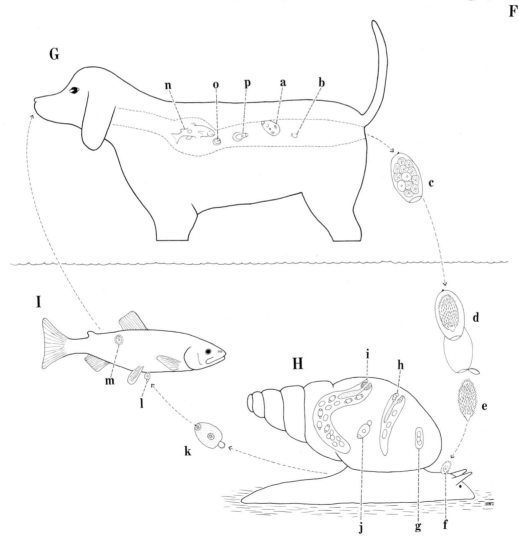

## Explanation of Plate 80

**A**, Adult fluke. **B**, Embryonated egg. **C**, Mother redia. **D**, Young daughter redia. **E**, Sinistral view of cercaria. **F**, Ventral view of cercaria without tail. **G**, Encysted metacercaria. **H**, Heron (*Ardea herodias*) definitive host. **I**, Cats, together with dogs and raccoons serve as definitive hosts. **J**, Snail (*Goniobasis livescens*) first intermediate host. **K**, Bullheads (*Ameiurus melas*) are the second intermediate host.

1, Oral sucker; 2, ventral sucker; 3, pharynx; 4, esophagus; 5, intestine; 6, testis; 7, vas efferens; 8, vas deferens; 9, seminal vesicle; 10, common genital pore; 11, ovary; 12, seminal receptacle; 13, uterus; 14, vitelline glands; 15, vitelline duct; 16, vitelline reservoir; 17, miracidium; 18, germ cells; 19, cercaria in redia; 20, eyespots; 21, genital anlagen; 22, tail; 23, prepharynx; 24, penetration glands; 25, ducts of penetration glands; 26, dorsal membranous tail fin; 27, ventral membranous tail fin; 28, excretory bladder.

**a**, Adult fluke embedded in mucosa of intestine; **b**, embryonated eggs laid; **c**, eggs pass out of intestine in feces; **d**, eggs drop in water but do not hatch; **e**, embryonated eggs eaten by snails; **f**, eggs hatch in intestine and miracidia migrate through intestinal wall; **g**, mother sporocyst presumably forms, producing mother redia; **h**, mother redia containing daughter redia; **i**, daughter redia with cercaria; **j**, cercaria escaping from snail host; **k**, cercaria free in water; **l**, cercaria penetrating bullhead host, casting its tail in the process; **m**, metacercariae encysted in muscles, reaching infectivity in about 4 weeks; **n**, infection of definitive hosts occurs when they ingest infected fish; **o**, digestion of fish host releases metacercariae; **p**, young fluke escapes from cyst and penetrates intestinal mucosa, developing to maturity in about 1 week.

Adapted from Cameron, 1936, Can. J. Res. D. 14:59.

---

### Suborder Opisthorchiata

These are medium- to small-sized flukes with weakly developed musculature. They lack a cirrus pouch and the testes are behind the ovary. A seminal receptacle is present.

The primary pores of the V-shaped or globular excretory bladder of the cercaria open on the lateral margins of the tail near its base. The tail is pleuro- or parapleurolophocercous, i.e., bearing a finlike membrane. Encystment is in the lower vertebrates, chiefly fish.

### Superfamily Opisthorchioidea

The general description is the same as for the suborder. The eggs are small, thick-shelled, and operculate.

#### FAMILY HETEROPHYIDAE

The species of this family are small to minute, ovoid or pyriform distomes having weak suckers with the ventral one often in a genital sinus. The cuticle is covered with large scales. A genital sucker (gonotyl) or sinus is present and near the acetabulum. The loops of the uterus lie in front of the testes. The vitellaria are lateral.

### *Apophallus venustus*
(Ransom, 1920) (Plate 80)

These small flukes are parasites in the ileum of dogs, cats, raccoons, and great blue herons in eastern Canada and the northeastern section of the United States.

**Description.** Mature specimens have elongated oval to pyriform bodies 0.95 to 1.4 mm long by 0.25 to 0.55 mm wide. The anterior half of the body is covered with scale-like spines and the hind fourth is smooth. The ventral sucker is slightly smaller than the oral one and is situated at the posterior margin of the genital sinus near the middle of the body. The pharynx and esophagus combined are about one-fourth the length of the body, and the ceca reach to the posterior extremity of it.

The testes are arranged obliquely in the posterior half of the body and a large S-shaped seminal vesicle opens into the genital sinus. The ovary is partially overlapped by the seminal vesicle. A large transverse seminal receptacle lies just posterior from the ovary. Vitelline glands extend laterally from the intestinal bifurcation to the posterior end of the body. The excretory bladder is Y-shaped with the stem extending between the testes. Eggs measure 26 to 32 $\mu$ long by 18 to 22 $\mu$ wide.

**Life Cycle.** Fully embryonated eggs passed in the feces do not hatch until eaten by the freshwater snail *Goniobasis livescens*. The miracidia migrate through the intestinal wall into the tissues. Although mother sporocysts have not been seen, they do occur in other members of the family and on that basis they are presumed to be present in this species.

Small rediae filled with spherical cells appear about 16 days after infection of the snails. Large

**Plate 80**    *Apophallus venustus*    323

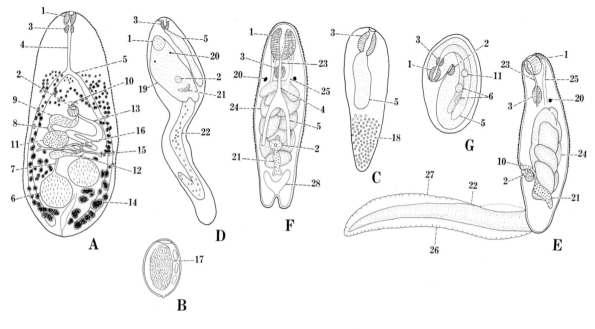

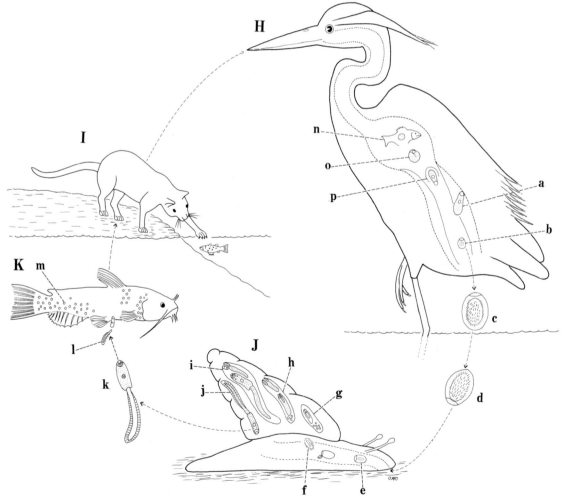

ones containing long-tailed cercariae appear later. Possibly these are mother and daughter rediae, respectively, although their identity is not certain.

The cercarial tails are about 1½ times the length of the body and bear a complete dorsal and a partial ventral membranous fin. There are two eyespots and a rudimentary ventral sucker. Sixteen large glands practically fill the body. The cercariae are fully developed when shed daily by the snails between the hours of 8 and 10 P.M. While they lash about rapidly in the water, progress is largely in an up and down direction. Upon coming in contact with one of about a dozen common species of fish (bullheads, catfish, garpike, common suckers, carp, dace, pickerel, pike, smallmouth bass, sunfish, bowfins), they quickly penetrate the fins, enter the blood vessels, and go to the muscles where encystment occurs. An adventitious cyst formed by the fish surrounds that secreted earlier by the parasites. The metacercariae grow slowly, becoming infective in about 4 weeks.

Infection of the definitive host takes place when fish harboring metacercariae in their flesh are eaten, developing to maturity in 7 days. The flukes are lost from the host in a few months.

*Apophallus imperator* encysts in trout in Quebec. It develops in experimentally infected cats and pigeons. The other hosts are unknown. Presumably its life cycle is similar to that of *A. venustus*. *A. muehlingi* from central Europe, which occurs in fish-eating birds (gulls, pelicans, cormorants, etc.) and mammals (dogs, foxes, cats, mink). Rediae and pleurolophocercous cercariae develop in the hydrobriid snail *Lithoglyphus naticoides*. Cercariae penetrate and develop to metacercariae in a number of species of small aquarium cyprinids, including several barbs, the zebra danio, and goldfish.

*Cryptocotyle lingua* is a common heterophyiid trematode of gulls, terns, cormorants, kittiwakes, murres, loons, bitterns, grebes, herons, and other fish-eating birds, as well as cats, rats, and guinea pigs under experimental conditions. Cercariae develop in rediae in the marine snail *Littorina littorea* and encyst in the skin of cunners. Birds and mammals become infected by eating parasitized fish.

Developing larval flukes inhibit movements of infected snails. In comparison with healthy periwinkles, parasitized ones move more slowly, have a more limited distribution, and are less able to make the normal seasonal migrations to and from deep water. A higher percentage of the large snails (11 mm and more long) are parasitized.

Rediae have many epithelial folds extending into the lumen of the gut. They entrap and phagocytize food globules which sink into the intestinal cells as vesicles. The body integument forms numerous folds which engulf nutrients by pinocytosis. The integument contains mitochondria, membrane-bound vesicles, and glycogen-like granules. There is acid phosphatase activity in the body wall, pharynx, and cecum. Rediae use the host's free amino acids rather than products of protein hydrolysis.

The excretory bladder is a multinucleate syncytium bearing luminal projections. Mitochondria, secretory granules, Golgi complex, and granular endoplasmic reticulum are present.

The high percentage of infected cunners is attributable to a variety of factors. Two of them are the concentration of cercariae in areas where colonies of inactive parasitized periwinkles localize and the sluggish habits of the cunners which idle in one place for long periods. This state of inactivity enables large numbers of the weak, nonacetabulate cercariae to attach over the entire body and penetrate, especially in the fins and tail.

The metacercarial cyst is spherical, about 360 μ in diameter, and composed of a thick, outer fibrous wall and a thin, smooth inner one. The occurrence of parasitized fish over a widespread area indicates the presence of many infected birds and periwinkles.

*Heterophyes heterophyes*, another member of the family, is well known because it parasitizes humans in the Near and Far East where raw and freshly salted fish are eaten. Metacercariae are common in mullets from brackish water. Eggs deposited among the villi of the small intestine of vertebrate hosts (cats, dogs) work into the wall and enter the blood vessels. Upon reaching the coronary circulation of the heart, their presence results in serious damage to that organ.

## EXERCISE ON LIFE CYCLE

When adult flukes are available from cats or herons, eggs may be obtained for infecting *Goniobasis livescens*. Infect a sufficient number of small snails to examine one or two daily for several weeks to follow the complete development of the parasite, namely, the mother sporocyst, the mother and daughter rediae, and the cercariae.

If naturally infected snails occur in the area, they may be used to conduct experimental infections of bullheads. The pleurolophocercous cercariae are characteristic and strongly suggest the identity of the general group of flukes. Fish, especially bullheads, from the same area should already be naturally infected. By grinding some of the fish and digesting them in artificial gastric juice, metacercariae may be obtained for study and feeding experiments.

By feeding infected bullheads to cats, the prepatent period can be determined and adult flukes obtained for study.

## SELECTED REFERENCES

Cameron, T. W. M. (1936). Studies on the heterophyid trematode, *Apophallus venustus* (Ransom, 1920) in Canada. Part I. Morphology and taxonomy. Can. J. Res. 14:59–69.

Cameron, T. W. M. (1937). Studies on the heterophyid trematode, *Apophallus venustus* (Ransom, 1920) in Canada. Part II. Life cycle and bionomics. Can. J. Res. 15:38–51.

Cameron, T. W. M. (1937). Studies on the heterophyid trematode, *Apophallus venustus* (Ransom, 1920) in Canada. Part III. Further hosts. Can. J. Res. 15:275.

Krupa, P. L., A. K. Bal, and G. H. Cousineau. (1967). Ultrastructure of redia of *Cryptocotyle lingua*. J. Parasitol. 53:725–734.

Krupa, P. L., G. H. Cousineau, and A. K. Bal. (1968). Ultrastructure and histochemical observations on the body wall of *Cryptocotyle lingua* rediae (Trematoda). J. Parasitol. 54:900–908.

Krupa, P. L., G. H. Cousineau, and A. K. Bal. (1969). Electron microscopy of the excretory vesicle of a trematode cercaria. J. Parasitol. 55:985–992.

Lambert, T. C., and J. Farley. (1968). The effect of parasitism by the trematode *Cryptocotyle lingua* (Creplin) on zonation and winter migration of the common periwinkle, *Littorina littorea* (L.). Can. J. Zool. 46:1139–1147.

Lyster, L. L. (1940). *Apophallus imperator* sp. nov., a heterophyid encysted in trout, with a contribution on its life history. Can. J. Res. 18:106–121.

Odening, K. (1970). Der Entwicklungszyklus von *Apophallus muehlingi* (Trematoda: Opisthorchiida: Heterophyidae) in Berlin. Z. Parasitenk. 33:194–210.

Sekhar, S. C., and W. Threlfall. (1970). Infection of the cunner, *Tautogolabrus adsperus* (Walbaum), with metacercariae of *Cryptocotyle lingua* (Creplin, 1825). J. Helminthol. 44:189–198.

Sheir, Z. M., and M. El-S. Aboul-Enein. (1970). Demographic, clinical, and therapeutic appraisal of heterophyasis. J. Trop. Med. Hyg. 76:148–152.

Stunkard, H. W. (1930). The life history of *Cryptocotyle lingua* (Creplin), with notes on the physiology of the metacercariae. J. Morphol. Physiol. 50:143–191.

Watts, S. D. M. (1970). The amino acid requirements of the rediae of *Cryptocotyle lingua* and *Himasthla leptosoma* and of the sporocyst of *Cercaria emasculans* Pelseneer, 1900. Parasitology 61:491–497.

## FAMILY OPISTHORCHIIDAE

These are parasites of the gall bladder and bile ducts of all classes of vertebrates in many parts of the world. They generally are medium-sized and weakly muscled flukes, but some are very long, with poorly developed suckers. A cirrus pouch is usually absent and the tubular seminal vesicle is free in the parenchyma. The many-coiled uterus winds between the ovary and genital pore which is just pre-acetabular. The excretory bladder is Y-shaped.

### Metorchis conjunctus
### (Cobbold, 1860) (Plate 81)

*Metorchis conjunctus* occurs in the liver of dogs, foxes, cats, raccoons, mink, muskrats, and occasionally humans in North America, particularly in the eastern half of Canada.

**Description.**    There is great variation in the dimensions of this species, depending on the host in which it developed; larger flukes occur in the larger hosts. Adults range from 1 to 6.6 mm long by 0.59 to 2.6 mm wide. The suckers are about equal in diameter, which is roughly equal to the length of the pharynx. The genital pore lies at the anterior margin of the acetabulum. Voluminous ceca extend almost to the posterior extremity of the body. The testes are tandem or slightly oblique, and lie in the third quarter of the body. There is no cirrus pouch and the seminal vesicle is an enlargement of the vas deferens. The ovary is a spherical, oval, or trilobed body located a short distance in front of the anterior testis and partially overlaid by a seminal receptacle larger than it in size. The uterus occupies nearly all the intercecal space anterior to the ovary. Vitelline glands are lateral, never crossing the middle line of the body, and

## Explanation of Plate 81

A, Adult fluke. B, Mother sporocyst. C, Mother redia. D, Daughter redia. E, Pleurolophocercous cercaria. F, Encysted metacercaria. G, Dog definitive host. H, Mink and fox definitive hosts. I, Human definitive hosts. J, Snail (*Amnicola limosa porata*) first intermediate host. K, Fish (*Catostomus commersoni*) second intermediate host.

1, Oral sucker; 2, ventral sucker; 3, pharynx; 4, intestine; 5, testes; 6, ovary; 7, seminal receptacle; 8, uterus filled with eggs; 9, vitellaria; 10, developing mother rediae; 11, gut; 12, germinal masses; 13, developing cercariae; 14, eyespots; 15, prepharynx; 16, penetration glands; 17, excretory bladder; 18, hair-like process; 19, tail; 20, finfold; 21, encysted metacercaria; 22, inner cyst; 23, outer cyst.

a, Adult fluke in bile ducts of liver; b, embryonated eggs laid in bile ducts pass out of body with feces; c, eggs in water do not hatch; d, eggs hatch in intestine of amnicolid snail when eaten; e, young mother sporocyst; f, mature mother sporocyst filled with developing rediae; g, mother redia containing embryonic daughter rediae; h, daughter redia with developing cercariae; i, cercaria escaping from snail; j, cercaria free in water; k, cercaria penetrating sucker, casting its tail in the process; l, metacercariae encysted in musculature; m, infection of dog takes place when suckers containing viable metacercariae are eaten; n, digestion of fish releases metacercariae from muscles; o, pancreatic juices dissolve outer cyst and young fluke ruptures inner cyst to escape; p, young flukes enter liver by migrating through common bile duct, maturing in about 28 days.

Adapted from Cameron, 1944, Can. J. Res. 22:6.

---

extend from the intestinal bifurcation to the ovary. Eggs measure 22 to 32 $\mu$ long by 11 to 18 $\mu$ wide.

**Life Cycle.** Embryonated eggs laid in the bile ducts or gall bladder enter the intestine and are voided with the feces. Hatching occurs in the intestine of the snail *Amnicola limosa porata*. The sparsely ciliated miracidia penetrate the gut wall, enter the liver, and transform into mother sporocysts. There is one generation of rediae and it produces typical pleurolophocercous cercariae. The anterior half of the cercarial body is covered with minute spines and the posterior half bears eight pairs of long, protoplasmic, hair-like processes. Upon reaching the water, cercariae exhibit positive heliotropism and are able to survive 60 to 75 hours.

When contact is made with the common sucker (*Catostomus commersoni*), the cercariae burrow into them, encysting in the lateral muscles from the level of the dorsal fin to the tail. The metacercariae are enclosed in a double cyst. They are able to survive freezing in the fish.

Definitive hosts infect themselves by eating suckers that harbor viable metacercariae. These are freed in the stomach as the fish are digested. The outer cyst is dissolved in the pancreatic juices and the fluke escapes from the inner one, presumably as a result of muscular activity rather than from the effect of digestion. The young flukes migrate through the ductus choledochus to the liver where development proceeds to maturity. Eggs appear in the feces in 28 days. The flukes are known to live 5 years in cats but how long they can live beyond this time is unknown. Sledge dogs commonly become infected from frozen suckers fed to them as the main item of food. Young dogs frequently die as a result of infection. Infection in dogs produces cystic dilation of the bile ducts and hyperplasia of the mucosa of the gall bladder. In cases of heavy and prolonged infection, only small islands of liver parenchyma remain. Since canine hepatitis and *Metorchis conjunctus* have been found associated in dogs, these flukes have been suggested as transport hosts.

Species of *Metorchis* (*M. intermedius, M. orientalis, M. taiwanensis, M. xanthosoma*) of ducks have a similar life cycle, with fish as the second intermediate host. *Metorchis intermedius* in eastern Europe develops in the snail *Bithynia*. *Opisthorchis tonkae* from muskrats encyst in the shiner *Notropis stramineus*. Another species of this family is *Clonorchis sinensis*, the Oriental liver fluke which is of great public health importance in many parts of Asia. The cercariae encyst in cyprinid fish and people become infected by eating them raw.

These cycles are representative of those species of the family known to have pleurolophocercous cercariae that develop in rediae and the metacercariae encyst in fish.

## EXERCISE ON LIFE CYCLE

If pleurolophocercous cercariae are obtained from naturally infected snails, they should be identified and allowed to encyst in fish similar

Plate 81    *Metorchis conjunctus*                                     327

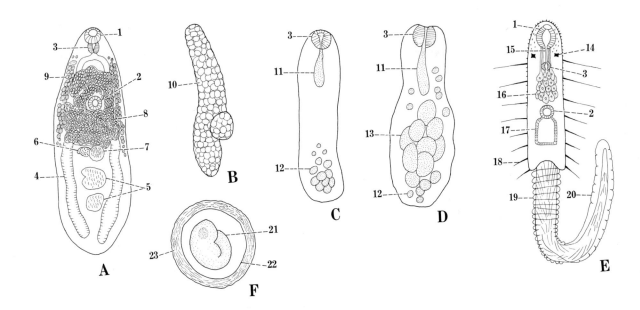

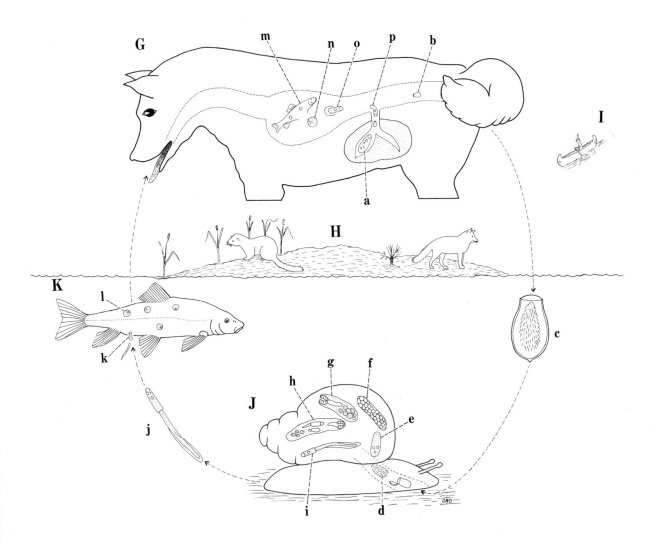

to those found in the same habitat. When infected fish are fed to a variety of hosts such as ducks, rats, guinea pigs, and cats, it is likely that adult worms will be obtained from some of them.

Eggs recovered from any of these hosts can be used in experiments to infect snails and study the development of the intramolluscan stages.

If only adult opisthorchiids are obtained from a vertebrate host, eggs dissected from them or obtained from the gall bladder or feces of the host should be fed to common snails, preferably species of genera known to serve as the intermediate host. In the event eggs are available, an entire life cycle may be worked out by careful planning. If the adult flukes are present in an area, the other stages are likely to be there.

## SELECTED REFERENCES

Cameron, T. W. M. (1944). The morphology, taxonomy, and life history of *Metorchis conjunctus* (Cobbold, 1860). Can. J. Res. 22: 6–16.

Heinemann, E. (1937). Ueber den Entwicklungskreislauf der Trematodengattung Metorchis sowie Bermerkungen Systematik dieser Gattung. Z. Parasitenk. 9:237–260.

Mills, J. H. L., and R. S. Hirth. (1968). Lesions caused by the hepatic trematode, *Metorchis conjunctus* Cobbold, 1860. A comparative study in Carnivora. J. Small Anim. Pract. 9:1–6.

Mongeau, N. (1961). Hepatic distomatosis and infectious canine hepatitis in northern Manitoba. Can. Vet. J. 2:33–38.

Wallace, F. G., and L. R. Penner. (1939). A new liver fluke of the genus *Opisthorchis*. J. Parasitol. 25:437–440.

Vishkvartseva, N. V. (1969). [Morphology of developmental stages of *Metorchis intermedius* (Opisthorchiidae) from the cormorant.] Parazitologiya 3:346–353.

### Suborder Hemiurata

The cercariae are of the cystophorous type or a modified form of it. They have a sac-like or cylindrical excretory bladder with the main collecting tubules fused anteriorly. The eggs are small, often with a filament opposite the operculum, and contain a nonciliated spinose miracidium with large spines on the anterior end. The cercariae develop in rediae and encyst in copepods and ostracods.

### Superfamily Hemiuroidea

These flukes are elongate, nonspinous, and with the suckers widely separated. The testes are in the posterior part of the body and anterior to the ovary. The vitellaria are sparse.

### FAMILY HALIPEGIDAE

The Halipegidae occur under the tongue and in the eustachian tubules of frogs. Vitellaria are divided into two compact masses located at the posterior extremity of the body.

#### *Halipegus eccentricus*
Thomas, 1939 (Plate 82)

This is a comparatively rare fluke of the eustachian tubes of adult frogs in Michigan. Other species occur in Massachusetts, California, Colorado, Oregon, Canada, and Mexico, as well as other parts of the world.

**Description.**    Fully grown worms are 6 to 6.5 mm long by 1.8 mm wide, the body being almost cylindrical. The oral sucker is subterminal and the ventral one near the middle of the body or caudad from it. The esophagus is extremely short and the ceca extend to the posterior end of the body. The testes are extracecal and obliquely arranged in the last third of the body. The ovary is median in position and posterior to the testes, being very near the caudal end of the body. The few vitelline glands, which consist of four to five follicles each, are between the ovary and the hind end of the body. Eggs measure 67 $\mu$ long by 27 $\mu$ wide; the opercular end is broad while the other terminates as a long filament.

**Life Cycle.**    Adult flukes deep in the eustachian tubes lay fully embryonated eggs which work downward through the canal into the mouth, are swallowed, and eventually voided with the feces. Hatching takes place in the intestine of snails (*Physa gyrina*, *P. sayii*, *P. parkeri*, *Helisoma trivolvus*). The nonciliated miracidia have a spiny cuticle and eight large pen-shaped spines at the anterior end. They burrow through the intestinal wall by means of the spines, and possibly with the aid of histolytic secretions.

Once inside the tissue of the snail, the miracidia slowly change into mother sporocysts about which little is known. They produce three to eight rediae characterized by punctate orange-yellow bands within a month after entering the snail hosts. Only a single generation of rediae has been reported. In about 1 month, they produce cercariae that are enclosed within a bulbous tail. The bulb-like structure is complicated in morphology. It contains the elongated and somewhat coiled cercaria with a long fluted excretory appendage attached to the posterior end. There are two long ribbon-like appendages between which is a conical delivery tube.

Cercariae lying on the bottom of the pond thrust out the excretory tubules whose movement appears to attract cyclops (*Cyclops vulgaris*, *Mesocyclops obsoletus*). The crustaceans attack and ingest the cercariae. Upon entering the intestine, the cercarial body detaches from the bulbous cyst-like structure and begins migrating posteriorly along the gut. It soon burrows through the wall into the hemocoel.

Within the body cavity of the cyclops, the cercaria transforms into a quite different type of unencysted metacercaria. It is cylindrical with the oral sucker terminal and the ventral sucker near the posterior end. A tuft of villi evert from the excretory bladder. Metacercariae become infective within 2 to 3 weeks.

Infection of the frogs takes place when the tadpoles eat infected cyclops. Upon being freed from the cyclops in the stomach, the flukes remain in the cardiac end of it until metamorphosis. During this time, the villi of the excretory bladder are inverted. As the tadpoles transform into frogs, the flukes migrate up the esophagus into the mouth, and finally into the eustachian tubes of fully developed frogs where maturity is attained.

The life cycle of *Halipegus occidualis* is similar in most respects to that of *H. eccentricus*. *Helisoma antrosa* is the snail intermediate host, and *Cyclops vernalis*, *C. serrulatus*, and the ostracod *Cypridopsis vidua* have been infected experimentally as the second intermediate hosts. Metacercariae have been found in mature dragonflies (*Libellula incesta*). Presumably dragonfly naiads become infected by eating cyclops harboring metacercariae which transfer to them, continuing on through metamorphosis. Hence frogs could become infected by eating dragonflies as naiads, tenerals, or mature individuals. Frogs have been infected experimentally by placing metacercariae in the mouth.

*Halipegus amherstensis* produces sluggish sporocysts and rediae in the digestive gland of *Physa gyrina*. The metacercariae develop in *Cyclops viridus*. Adult flukes occur in the stomach of tadpoles, and the mouth and eustachian tubes of frogs (*Rana catesbiana* and *R. clamitans*) in Massachusetts.

## EXERCISE ON LIFE CYCLE

Cystophorous cercariae obtained in surveys of snails for cercariae in ponds where frogs occur may be suspected as being those of *Halipegus*. Feed the cercariae to species of *Cyclops* known to serve as intermediate hosts for these flukes. Movement of the cercariae in the gut, their penetration into the hemocoel, and development into the metacercariae can be observed in living specimens.

Flukes in the eustachian tubes of frogs can be detected by holding them toward the light. Infected frogs serve as a source of eggs for study and infecting species of *Physa* or *Helisoma* to observe development of the sporocyst, rediae, and cercariae.

Feed metacercariae to tadpoles and note the location and stage of development in the alimentary canal before, during, and after metamorphosis. Place some free metacercariae in the mouth of an adult frog and observe what happens to them.

## SELECTED REFERENCES

Ameel, D. J., W. W. Cort, and A. Van der Woude. (1949). Germinal development in the mother sporocyst and redia of *Halipegus eccentricus* Thomas, 1939. J. Parasitol. 35:569–578.

Krull, W. H. (1935). Studies on the life history of *Halipegus occidualis* Stafford, 1905. Amer. Midl. Natur. 16:129–143.

Macy, R. W., and W. R. Demott. (1957). Ostracods as second intermediate hosts of *Halipegus occidualis* Stafford, 1905 (Trematoda: Hemiuridae). J. Parasitol. 43:680.

Rankin, J. S. (1944). A review of the trematode genus Halipegus Looss, 1899, with an account of the life history of *H. amherstensis* n. sp. Trans. Amer. Microsc. Soc. 63:149–164.

Thomas, L. J. (1934). *Cercaria sphaerula* n. sp. from *Helisoma trivolvis* infecting cyclops. J. Parasitol. 20:285–290.

Thomas, L. J. (1939). Life cycle of a fluke, *Halipegus eccentricus* n. sp., found in the ears of frogs. J. Parasitol. 25:207–221.

## Explanation of Plate 82

A, Ventral view of adult worm. B, Embryonated egg. C, Free miracidium. D, Young mother sporocyst. E, Young redia. F, Cystophorous cercaria recently escaped from snail. G, Cercaria with excretory tubule extended. H, Metacercaria from cyclops. I, Frog intermediate host. J, Snail (*Physa*) intermediate host. K, Cyclops second intermediate host. L, Tadpole definitive host.

1, Oral sucker; 2, ventral sucker; 3, pharynx; 4, intestine; 5, testis; 6, ovary; 7, vitelline glands; 8, uterus; 9, spines; 10, apical gland; 11, penetration glands; 12, germ cells; 13, spined cuticle; 14, gut; 15, germinal masses; 16, orange-yellow colored band; 17, bulbous portion of tail; 18, cercaria; 19, excretory appendage; 20, excretory bladder; 21, delivery tube; 22, streamers; 23, developed excretory bladder; 24, everted villi of excretory bladder.

a, Adult fluke in eustachian tube; b, embryonated eggs pass down eustachian tube into pharynx; c, eggs pass down intestine; d, eggs enter water in feces; e, eggs hatch only when eaten by snail host; f, miracidium migrates through wall of intestine; g, spined cuticle is cast off; h, mother sporocyst; i, banded redia with developing cercariae; j, cystophorous cercaria escaping from snail; k, cercaria with everted excretory tubule free in water; l, when eaten by cyclops, cercaria detaches from tail and passes through intestinal wall into hemocoel; m, metacercaria in hemocoel; n, tadpole becomes infected by eating cyclops; metacercariae remain in stomach without further development until tadpole metamorphoses into adult at which time they migrate into the eustachian tubes and develop to maturity.

Adapted from Thomas, 1939, J. Parasitol. 25: 207.

## CLASS CESTODA

The cestodes are flat worms parasitic in freshwater oligochaetes and all classes of vertebrates, both aquatic and terrestrial. In some forms, the body consists of a single segment and is designated as monozoic, while in others, the body, or strobilus, is a chain of segments called proglottids, and are referred to as polyzoic forms. There is no digestive tract. For the most part, the adult cestodes are parasites of the digestive tract or liver.

Lacking an alimentary canal, cestodes have a body wall modified morphologically and physiologically for absorption of nutrients from the host gut, protection against digestion, adhesion to the intestinal wall, and synthesis and transport of proteins (Fig. 28).

Morphologically, the body wall consists of two layers; the outer tegumental layer (t) is a trilaminated cystoplasmic membrane said to be syncytial by some workers and of squamous-like cells by others. Its outer margin is formed into numerous, minute, finger-like cytoplasmic projections, the microvilli (mi). Beneath the microvilli is a wide zone of vacuolated ground substance through which pore canals (pc) pass. The inner portion of the tegument is dense and contains many mitochondria (m). It is bordered internally by a well-defined basement membrane (bm) which continues through the pore canals and around the large cells much as the intercellular substance surrounds cells in other animals.

The subtegumental layer consists primarily of three muscle layers, the circular (c), longitudinal (l), and diagonal. Below this layer and located deep in the parenchymal substance of the body are large electron dense (d) and smaller electron light cells. The dense cells clearly are associated with the tegument through several connecting tubules (ct) that have protoplasmic connections between the walls and mitochondria in them. The cells themselves have a large, double-membraned nucleus (n) whose outer wall connects with the large, complex endoplasmic reticulum (er). In addition, they contain mitochondria, protein crystalloids (p), and fat or glycogen droplets (f).

The arrangement of the tegument, with its outer covering of cytoplasmic microvilli and dark cells, suggests an inverted intestine.

The morphology of the tegument and its associated cells point to its probable physiological functions of absorption and secretion. Microvilli may serve several purposes. As structures of attachment, they prevent expulsion of the worms from the gut. Their pointed ends abrade the intestinal epithelium, thereby allowing a high concentration of nutritive cellular sap to ooze into the immediate vicinity of the worm, where the myriad microvilli provide a greatly increased absorptive surface.

The living nature of the covering tegument resists digestion by the host. It is readily renewable from the large cells buried deeply in the parenchyma of the worm's body. The presence of numerous vacuoles in the outer portion of the

**Plate 82**     *Halipegus eccentricus*                                                                331

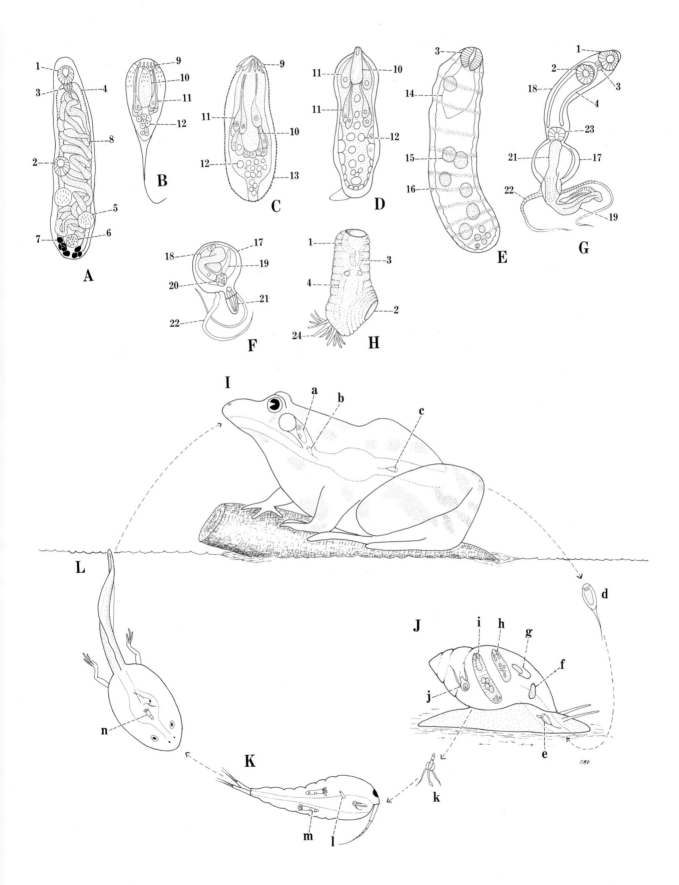

## Plate 83

tegument suggests pinocytosis of nutriments and their transportation. The occurrence of mitochondria, endoplasmic reticulum, and crystalline storage bodies further suggests that the cells synthesize adsorbed substances and either store or transport them to the parenchyma or the outer tegument.

About a dozen orders of cestodes are recognized. Of these, four occur in selachians. The remainder are in freshwater and terrestrial hosts and are the ones most frequently seen.

Common species of cestodes from four orders parasitizing freshwater and terrestrial vertebrates have been chosen to depict the basic patterns of morphology and life cycles. These include the orders Caryophyllidea from oligochaete annelids and fish, Pseudophyllidea from all classes of vertebrates, Proteocephala from fish mainly, and Cyclophyllidea from birds and mammals. The basic life cycles of the Pseudophyllidea, Proteocephala, and Cyclophyllidea are presented in Plate 100.

### Morphological Types (Plate 84)

The four common orders of cestodes found in freshwater and terrestrial hosts have morphological features characteristic of each of them.

The Caryophyllidea (Figs. A–G) are monozoic. The sexual stages are considered as progenetic larval forms. The operculate eggs are unembryonated when laid. In the genus *Archigetes* (Figs. A–C), the sexual stage is in the celom of aquatic annelids. It resembles the tailed procercoid of the Pseudophyllidea. The life cycle is direct. In *Caryophyllaeus* (Figs. D–G) and related genera, the sexual stage is in the intestine of catostomid fish. It resembles a plerocercoid of the Pseudophyllidea. Aquatic annelids serve as the first intermediate host, harboring the procercoid.

The Pseudophyllidea (Figs. H–N) and remaining orders are polyzoic. Members of this order are parasites of fish, amphibians, reptiles, birds, and mammals. Size varies greatly, ranging from small to large, even exceeding 40 feet in length.

The scolex is variable in shape but always bears a pair of longitudinal slits or bothria (Figs. H, N). The proglottids are wider than long, with the genital openings on the midventral surface. A uterine pore is present, opening separately from but near those of the male and female organs. The gravid uterus is a long tube filled with eggs and folded into loops in the center of the proglottid between the ovary and uterine pore. The operculate eggs are unembryonated when laid.

The testes and vitelline follicles are numerous, small bodies scattered throughout the proglottid except for the space occupied by the ovary and uterus. The vitelline follicles are in a layer toward the ventral side of the proglottid and the testes in a layer toward the dorsal side. The eggs are laid separately and the embryos hatching from them are ciliated hexacanths known as

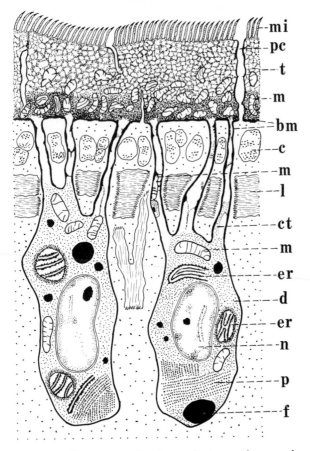

Fig. 28.   Representation from electron micrographs of the tegument of the cestode *Dipylidium caninum.* **bm,** base membrane; **c,** circular muscles; **ct,** connecting tubules; **d,** dark cells; **er,** endoplasmic reticulum; **f,** fat or glycogen; **l,** longitudinal muscles; **m,** mitochondrium; **mi,** microtrich; **n,** nucleus, **o,** opposed membranes; **p,** protein crystalloid; **pc,** pore canal; and **t,** tegument. Adapted from Threadgold, 1962, Quart. J. Microsc. Sc. Ser. 3, 103:135.

**Plate 83** *Some Representative Life Cycles of Digenetic Trematodes* 333

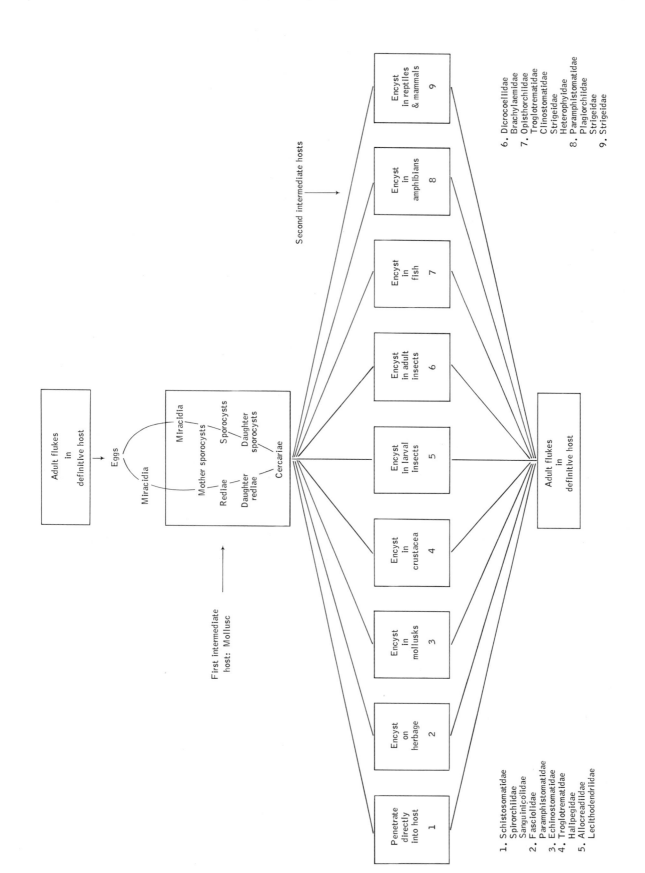

## Explanation of Plate 84

A–G, Caryophyllidea. A–C, *Archigetes sieboldi*. A, Procercoid-like adult. B, Embryonated operculate egg. C, Hexacanth larva. D–G, *Glaridacris catostomus*. D, E, Anterior and posterior ends of plerocercoid-like monozoic adult cestode. F, Unembryonated operculate egg. G, Procercoid.

H–N, Pseudophyllidea. H, Scolex of *Diplyllobothrium latum*, a polyzoic cestode, with two bothria. I, Mature proglottid. J, Gravid proglottid. K, Undeveloped operculate egg. L, Coracidium. M, Procercoid. N, Plerocercoid.

O–S, Proteocephala. O, Scolex of *Proteocephalus ambloplitis*. P, Mature proglottid. Q, Hexacanth embryo. R, Procercoid. S, Plerocercoid.

T–L', Cyclophyllidea. T–B', Taenioid cestodes. T, Scolex of hooked forms (*Taenia, Hydatigera, Multiceps, Echinococcus*). U, Scolex of nonhooked form (*Taeniarhynchus*). V, Mature proglottid. W, Gravid proglottid (*Taenia solium*). X, Taenioid egg. Y–B', Different kinds of larval forms. Y, Cysticercus (*Taenia, Taeniarhynchus*). Z, Strobilocercus (*Hydatigera*). A', Coenurus (*Multiceps*). B', Hydatid (*Echinococcus*).

C'–K', Nontaenoid cyclophyllidean cestodes. C'–F', Types of proglottids. C', *Hymenolepis* with single lateral genital pore and single set of reproductive organs. D', E', *Mesocestoides* with midventral genital pore. D', Mature proglottid. E', Gravid proglottid with paruterine organ, a specialized type of uterus. F', *Dipylidium* with double genital pores and reproductive organs (*Anoplocephala* and *Thysanosoma* also have double sets of reproductive organs and double pores). G'–I', Types of eggs. G', Normal or common type represented by *Hymenolepis*. H', Pyriform type of Anoplocephalidae. I', Packet type of *Dipylidium*; the fringed tapeworm, *Thysanosoma*, produces a similar type. J', K', Larval types. J', Tailed cysticercoid. K', Nontailed cysticercoid. L', Cross section of proglottid of a cestode, showing structure.

1, Adhesive organ; 2, bothrium; 3, sucker; 4, apical sucker; 5, rostellum; 6, oncospheral hooks; 7, tail or cercomer; 8, common genital pore; 9, cirrus pouch; 10, testes; 11, ovary; 12, uterus; 13, paruterine organ (gravid uterus); 14, vitellaria; 15, vagina; 16, dorsal excretory or osmoregulatory canal; 17, ventral excretory canal; 18, cortex; 19, medulla; 20, circular muscles; 21, internal longitudinal muscles; 22, transverse muscles; 23, external longitudinal muscles; 24, cortical cells; 25, cuticle; 26, brood capsule; 27, scolex.

Figures adapted from various sources.

---

coracidia. The other larval stages are a procercoid in entomostraca and a plerocercoid in fish.

The Proteocephala (Figs. O–S) are parasites of fish primarily; a few species occur in amphibians and reptiles. They are different from the other polyzoic cestodes considered in having the vitelline follicles arranged in longitudinal bands along each lateral margin of the proglottids. The cirrus pouch and vagina open in a common atrium located on the lateral margin of the proglottid. A uterine pore is lacking. The unarmed scolex bears four muscular in-cupped suckers. A fifth sucker, or remnant of one, appears at the apex of the scolex. The gravid uterus is a large sac-like structure filling the intervitelline space. Uterine eggs contain fully developed, unciliated hexacanth larvae. Eggs are released when proglottids voided from the intestine disintegrate in the water. Procercoids occur in entomostraca and plerocercoids in the same crustaceans or in fish, varying according to the species of tapeworm.

The Cyclophyllidea (Figs. T–L') include the largest number of families of polyzoic cestodes of any of the orders. The scolex bears four in-cupped muscular suckers and in most species an armed rostellum. The genital ducts open into a common atrium located on the side of body. In some species, a double set of reproductive organs is present, in which case an atrium is located on each side of the proglottid.

As a group, they are readily recognized from all others by the compact vitelline gland lying immediately caudad from the ovary. For convenience in consideration, they may be separated into two groups, the Taeniidae, or taenioid cestodes, and the remaining families, or nontaenioid cestodes.

The taenioid cestodes (Figs. T–B') are characterized by several features, both morphological and biological. Most of these worms are large, muscular forms with an armed rostellum. The gravid uterus consists of a median stem with lateral branches. The eggs have thick shells, which in optical section appear to be striated. The larval stages consist of four kinds of bladders filled with fluid. They are 1) a cysticercus with a single invaginated scolex, 2) a strobilocercus in which the bladder is small and there is a long everted strobilus bearing an evaginated scolex, 3) a coenurus which is a single bladder containing numerous scolices attached to the inside of the

Plate 84     *Morphological Types of Cestodes*                                              335

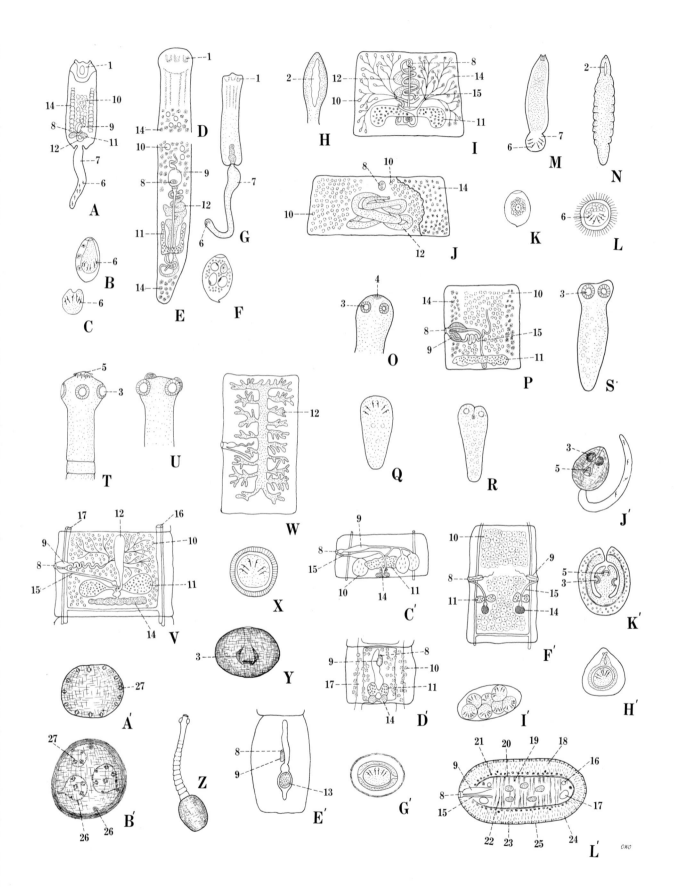

wall, and 4) a hydatid consisting of a large capsule containing many smaller ones called brood capsules, each being attached by a slender stalk to the germinal layer of the mother cyst. The brood capsules produce scolices by an internal budding process from the wall. The larval stages of taenioid cestodes always develop in mammals that have swallowed eggs.

The nontaenioid cyclophyllidean cestodes (Figs. C'–K') comprise an assembly of families varying in structural details but having a common type of larval form known as a cysticercoid. Its body consists of tissue rather than the fluid-filled, thin-walled bladder of the taenioid cestodes, and contains a retracted scolex. It develops in invertebrates and fish but not in mammals. In addition to the cysticercoid, there are other kinds of larvae. Members of the family Mesocestoididae have a larval form known as a tetrathyridium and those of the Paruterinidae have a plerocercoid. The eggs vary but never have the thick, striated shell characteristic of the Taeniidae. The majority of the eggs are simple, oval, and thin-shelled. Those of the Anoplocephalidae have an inner membrane, with two finger-like processes, that encloses the embryo and is called a pyriform apparatus. The scolex usually bears an armed rostellum but in some families (Anoplocephalidae, Mesocestoididae) it is absent. The gravid uterus may be a network of tubules, may be sac-like, may break up into small bits, forming packets containing one to several eggs each, or may form a paruterine organ in which it becomes surrounded by a thick fibrous layer (Paruterinidae, Mesocestoididae).

## SELECTED REFERENCES

Lumsden, R. D. (1966). Cytological studies on the absorptive surfaces of cestodes. I. The fine structure of the strobilar integument. Z. Parasitenk. 27:355–382.

Rybicka, K. (1966). Embryogenesis in cestodes. In: B. Dawes, ed. Advances in Parasitology. Academic Press, New York, Vol. 4, pp. 107–186.

Smyth, J. D. (1963). The biology of cestode life cycles. Commonw. Bur. Helminthol. Tech. Commun. No. 34, pp. 1–38.

Smyth, J. D., and D. D. Heath. (1970). Pathogenesis of larval cestodes in mammals. Helminthol. Abst. Ser. A. 39:1–22.

Threadgold, L. T. (1962). An electron microscope study of the tegument and associated structures of *Dipylidium Caninum*. Quart. J. Microsc. Sci., Ser. 3, 103:135–140.

Voge, M. (1967). The post-embryonic developmental stages of cestodes. In: B. Dawes, ed. Advances in Parasitology. Academic Press, New York, Vol. 5, pp. 247–297.

Wardle, R. A., and J. A. McLeod. (1952). The Zoology of Tapeworms. Univ. Minn. Press, Minneapolis. 780 pp.

Yamaguti, S. (1959). Systema Helminthum. The Cestodes of Vertebrates. Academic Press, New York, Vol. II, 860 pp.

## ORDER CARYOPHYLLIDEA

Caryophyllidea are small monozoic cestodes parasitic in the body cavity of aquatic oligochaetes and the intestine of fish throughout the world.

The genera *Archigetes* and *Glaridacris* are chosen for discussion because their life cycles are known best and they present an interesting sequence of development in relation to the two sexually undeveloped larval stages of the more advanced pseudophyllidean cestodes in which there is a procercoid stage in a crustacean first intermediate host and a plerocercoid stage in a fish second intermediary. The plerocercoid develops into a sexually mature strobilate worm in the intestine of a vertebrate definitive host.

In *Archigetes*, the procercoid is neotenic, producing eggs in aquatic oligochaetes and fish. In *Glaridacris*, procercoids in the oligochaete first intermediate host remain normal but upon entering fish, they develop into neotenic plerocercoids that produce eggs.

### FAMILY CARYOPHYLLAEIDAE

These small cestodes are characterized by a scolex with frilled margins or one to three pairs of suctorial grooves. They are parasitic in freshwater oligochaetes and fish.

### *Archigetes sieboldi*
Leuckart, 1878 (Plate 85)

The life cycle is direct, with the neotenic procercoids normally in the oligochaete *Limnodrilus hoffmeisteri* and rarely in *Tubifex tubifex*.

**Description.**     Gravid worms from Wisconsin measure 1.4 to 1.7 mm long, whereas European forms are 2.5 to 6 mm long. The scolex consists of a well-defined median bothrium and lateral loculi on the dorsal and ventral sides. A long tail bears six embryonic hooks near the posterior end. About 90 to 110 testes 30 by 60 $\mu$ in size are arranged in dorsal and ventral layers anterior to the genital pore and between the two lateral bands of vitelline follicles. An external, oval seminal vesicle 50 by 80 $\mu$ in size is separated from the small round cirrus pouch 80 to 120 $\mu$ in diameter by a length of vas deferens. The numerous elliptical vitelline follicles 20 by 50 $\mu$ in size extend as two lateral bands almost the full length of the body, being both pre- and postovarian. There is a dumbbell-shaped ovary with follicular wings 110 to 120 $\mu$ wide. Coils of the glandular uterus are pre- and postovarian. Operculate eggs 32 to 37 $\mu$ in diameter and 46 to 60 $\mu$ in length accumulate under the cuticle and begin embryonation which is completed outside in the water.

**Life Cycle.** Sexually mature worms escape from the celom of the oligochaetes through a rent in the weakened body wall. Upon death and disintegration of the annelid host, the eggs are freed into the water where embryonation is completed in about 16 days. No further development occurs before hatching. When annelids swallow the embryonated eggs, they hatch in the posterior part of the intestine. The oncospheres burrow through the gut wall into the celom and slowly migrate anteriorly to the seminal vesicles which they enter. After 120 days, they have matured to egg-laying neotenic, tailed procercoids which leave the seminal vesicles, returning to the celom. From 1 to 15 worms occur in a single annelid. A high percentage of the worms may be infected.

Gravid parasites occur only during the spring and early summer; only immature procercoids are present at other times. Eggs are produced in 120 days after infection.

*Archigetes sieboldi* is normally a parasite of oligochaetes, as indicated by the high incidence of infection. However, it is able to infect carp but only during the spring and then somewhat rarely. *Archigetes limnodrili*, likewise, appears to be parasitic in oligochaetes where it occurs in abundance but has not been found in fish. On the other hand, *A. iowensis* normally requires carp in which the procercoids complete development and produce eggs. It develops only in *Limnodrilus hoffmeisteri*.

The reasons for both the annelid and fish hosts of these neotenic procercoids are not clear. One postulate suggests that it is a biological response to the absence of suitable fish hosts.

## EXERCISE ON LIFE CYCLE

Oligochaetes may be obtained from ponds or gently flowing streams whose bottoms are covered with rich organic ooze. Infected individuals can be recognized by the bulging body wall. *Archigetes* and *Cryptobothrius*, related genera, may be recognized by being sexually mature individuals filled with undeveloped operculated eggs and thereby differentiated from the sexually immature procercoids of *Glaridacris*, *Caryophalleus*, and related genera occurring in annelids and which parasitize suckers.

Eggs that are obtained by dissecting gravid worms and permitted to embryonate in water in a glass container at room temperature are infective to oligochaetes of the genera *Limnodrilus* and *Tubifex*.

Oligochaetes to be infected should be kept in the laboratory several weeks in order to allow any natural infections to mature and be recognized. Infected worms should be discarded. Expose uninfected annelids by placing eggs in the dish of water with them, either mixed with the ooze or free in a clean container. After they have eaten the eggs, dissect them at intervals of 1 or 2 days to observe the growth and migration of the parasites. Note how long it takes for the worms to reach sexual maturity at room temperature. Observe the procercoid characteristic of *Archigetes*.

## SELECTED REFERENCES

Calentine, R. L. (1964). The life cycle of *Archigetes iowensis* (Cestoda: Caryophyllaeidae). J. Parasitol. 50:454–458.

Calentine, R. L., and B. L. DeLong. (1966). *Archigetes sieboldi* (Cestoda: Caryophyllaeidae) in North America. J. Parasitol. 52:428–431.

Kennedy, C. R. (1965). The life-history of *Archigetes limnodrili* (Yamaguti) (Cestoda: Caryophyllaeidae) and its development in the invertebrate host. Parasitology 55:427–437.

Wisniewski, L. W. (1930). Das Genus Archigetes R. Leuck. Eine Studie zur Anatomie, Histogenese, Systematik und Biologie. Mem. Acad. Polon. Sci. Lett. Cl. Sci. Math. Natur. Ser. B Sci. Natur. 2:1–160.

*Glaridacris catostomi*
Cooper, 1920 (Plate 86)

*Glaridacris catostomi* parasitizes the stomach and small intestine of suckers (*Catostomus commersoni* and related forms) in North America.

## Explanation of Plate 85

A, Ventral view of sexually mature nongravid worm. B, Ventral view of gravid worm. C, Lateral view of gravid worm. D, *Limnodrilus hoffmeisteri* with three fully developed worms. E, Embryonated egg. F, Hatched oncosphere. G, Cross section through scolex. H, *Limnodrilus* with *Archigetes* escaping through slit in body wall. I, Infection of *Limnodrilus*.

1, Bothrium or loculus; 2, common genital pore; 3, cirrus pouch; 4, testes; 5, ovary; 6, uterus; 7, vitelline glands; 8, tail with hooks of oncosphere; 9, gravid uterus; 10, *Limnodrilus* or *Tubifex*; 11, intestine of oligochaete; 12, three *Archigetes* in celom; 13, egg shell; 14, operculum; 15, oncosphere; 16, vitelline cells; 17, hooks of oncosphere.

a, Sexually mature *Archigetes* escaping from celom of oligochaete through break in body wall; b, recently escaped worm alive on ooze; c, dead and decomposing worm liberates eggs in the ooze on the bottom of the pond; d, free unembryonated egg; e, embryonated egg; f, oligochaete swallows eggs; g, h, eggs pass through intestine; i, shell of egg that has hatched in posterior part of intestine; j, oncosphere in intestinal lumen; k, oncosphere passing through intestinal wall into celom; l, developing oncosphere; m, young procercoid continues traveling anteriorly; n, well-developed procercoid approaching region of gonads; o, gravid neotenic procercoids associated with gonads.

Adapted from Wisniewski, 1930, Mem. Acad. Polo. Sci. Lett. Cl. Sci. Math. Nat. Ser. B Sci. Natur. 2:1.

**Description.**    The worms measure up to 25 mm long and are somewhat cylindrical in cross-section. The scolex is short, broad, and chisel-shaped, with three sucker-like depressions of each side. The testes are few and not completely surrounded by the vitellaria. The ovary is lobular and H-shaped. The vitelline glands are in both the anterior and posterior parts of the body. The eggs have a boss on the anopercular end and are undeveloped when laid.

**Life Cycle.**    Eggs voided with the feces of the fish host sink to the muddy bottom of the ponds and become fully embryonated in about 2 weeks. When swallowed by the freshwater oligochaetes *Limnodrilus udekemianus, L. hoffmeisteri*, or *Tubifex tempeltoni*, they hatch in the intestine. Young oligochaetes are more susceptible to infection than old ones. The liberated, six-hooked, unciliated embryos burrow through the intestinal wall into the celomic cavity where they develop into procercoids with well-developed scolex and reproductive organs and are infective to fish in 15 to 25 days.

Fish become infected by eating parasitized annelids. Procercoids released in the small intestine of the fish develop into neotenic plerocercoids that produce fertile eggs. Except for the neotenic condition, these plerocercoids are comparable to those of the diphyllobothriids.

*Glaridacris confusa, Hunterella nodulosa, Monobothrium hunteri*, and *M. ingens* have life cycles basically similar to that of *G. catostomi*. Of these species, *H. nodulosa* is the only one whose procercoids prevent sexual development of the annelid intermediate host.

Procercoids of *Biacetabulum infrequens* and *B. macrocephalus* of white suckers, *Catostomus commersoni*, develop in *Tubifex tempeltoni* but not in *Limnodrilus*. They are mature and infective to fish in 62 days. Only single procercoids occur in naturally infected oligochaetes and these die in about 120 days.

## EXERCISE ON LIFE CYCLE

Aquatic annelids of the genus *Limnodrilus* in waters where suckers occur are infected commonly with *Glaridacris* and related forms. Infected annelids generally can be recognized readily by the bulge of the body wall caused by the larval cestodes in the celom. After they have been examined inside the annelid, they may be dissected for further study.

Adult worms removed from the intestine of fish and placed in a dish of tap water shed eggs which when incubated at room temperature develop to six-hooked larvae. Aquatic annelids placed in the dishes with the embryonated eggs ingest them and become infected. Hatching in the gut and development in the celom may be followed by dissecting infected annelids at appropriate intervals.

Suckers may be infected by allowing them to eat annelids that have been infected experimentally. Fingerlings kept in balanced aquaria are the preferred size because of the difficulty of keeping larger fish alive sufficiently long to complete the experiments.

Suckers may be reared in the laboratory by stripping eggs and milt from adult fish and

Plate 85    *Archigetes sieboldi*                                                                    339

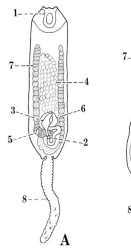

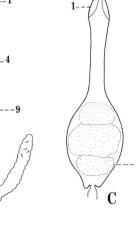

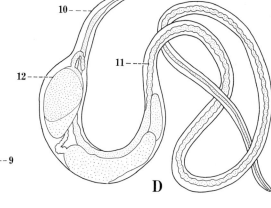

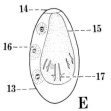

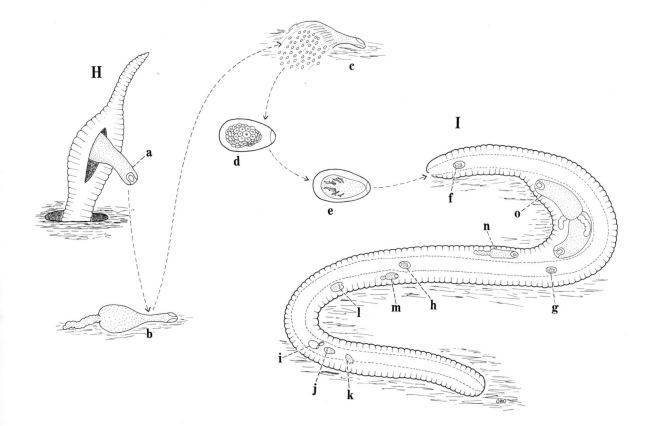

## Explanation of Plate 86

A, Anterior end of adult. B, Posterior portion of adult worm. C, Unembryonated egg. D, Fully embryonated egg. E, Large and small hooks of embryo. F, Fully developed embryo from annelid. G, Young worm from intestine of sucker. H, Sucker (*Catostomus commersoni*) definitive host. I, *Limnodrilus udekemianus*, the intermediate host.

1, Loculus or sucker; 2, opening of male reproductive system; 3, cirrus pouch; 4, seminal vesicle; 5, sperm duct; 6, testes; 7, opening of female reproductive system; 8, vagina; 9, ovary; 10, oviduct and Mehlis' gland; 11, uterus; 12, vitelline glands; 13, internal longitudinal muscles; 14, operculum; 15, germ cells of egg; 16, yolk granules; 17, embryo; 18, cercomer with larval hooks at tip.

a, Adult worm in intestine of sucker; b, eggs pass out of intestine with feces; c, unembryonated egg in water or mud; d, embryonated egg; e, eggs eaten by aquatic annelids; f, eggs hatch in intestine; g, larvae pass through wall of intestine into celom; h, fully developed and infective larvae; i, annelid digested in stomach of sucker; j, larval cestode freed from annelid; k, larvae shed cercomer and grow to neotenic plerocercoids in the intestine of suckers.

Adapted from McCrae, 1960, Dissertation, Colorado State Univ.

---

placing them together in an aquarium in which the water is constantly aerated. Dog biscuits provide suitable food for rearing the fish.

## SELECTED REFERENCES

Calentine, R. L. (1965). The biology and taxonomy of *Biacetabulum* (Cestoda: Caryophyllaeidae). J. Parasitol. 51:243–248.

Calentine, R. L. (1965). Larval development of four caryophallaeid cestodes. Proc. Iowa Acad. Sci. 72: 418–424.

Calentine, R. L., and D. D. Williams. (1967). Larval development of *Glaridacris confusa* (Cestoda: Caryophyllaeidae). J. Parasitol. 53:692–693.

McCrae, R. C. (1961). Studies on the Caryophyllaeidae (Cestoda) of the white sucker, *Catostomus commersoni* (Lacepede) in northern Colorado. Diss. Abst. 21:2835–2836.

## ORDER PSEUDOPHYLLIDEA

The pseudophyllideans are polyzoic cestodes, occurring in the intestine of all the classes of vertebrates over the world, especially the piscivorous species. The worms range in size from small to over 10 m in length, being among the largest of the cestodes. The scolex has a dorsal and ventral groove or lobes. Testes are small, numerous, and scattered throughout each proglottid. Genital pores open separately and unembryonated, operculate eggs are laid.

### FAMILY DIPHYLLOBOTHIIDAE

The scolex bears a dorsal ventral groove, followed by a neck and distinctly segmented body. The genital openings are on the midventral surface of the proglottids. Both the testes and the vitellaria are scattered throughout the proglottids, and the tubular gravid uterus loops into a rosette shape.

### *Diphyllobothrium latum*
(Linnaeus, 1758) (Plate 87)

This is a member of the family Diphyllobothriidae and is commonly known as the broad fish tapeworm of humans. It has been selected as a typical representative of this order for study of the life history. It parasitizes man and other fish-eating mammals such as dogs and bears. While worldwide in distribution, it is more prevalent in the northern hemisphere than elsewhere.

**Description.** Fresh worms are white to ivory in color and up to 10 m long. The scolex is almond-shaped, with two elongated grooves. Proglottids are wider than long. There is a separate uterine and common genital pore on the midventral surface of each segment. Flame cells consist of 50 to 80 flagella arising from a compact basal plate. The flagella are hexagonal in cross section, are closely packed, and have the classic ciliary morphology of nine pairs of peripheral and one pair of central longitudinal fibrils. Testes and vitelline glands are scattered throughout the proglottid. Coils of the gravid uterus form a centrally located rosette. The eggs are operculate and undeveloped when laid.

**Life Cycle.** Eggs laid singly in the intestine of the definitive host incubate when discharged with feces into the water. They develop at variable rates, depending on the temperature of the water as determined by the weather as well as the depth to which they descend. Upon hatching, the sluggish, ciliated, six-hooked coracidia swim aimlessly about. When eaten by crustaceans of the genera *Diaptomus* and *Cyclops*, particularly the former, the coracidium sheds

**Plate 86**    *Glaridacris catostomi*                                    341

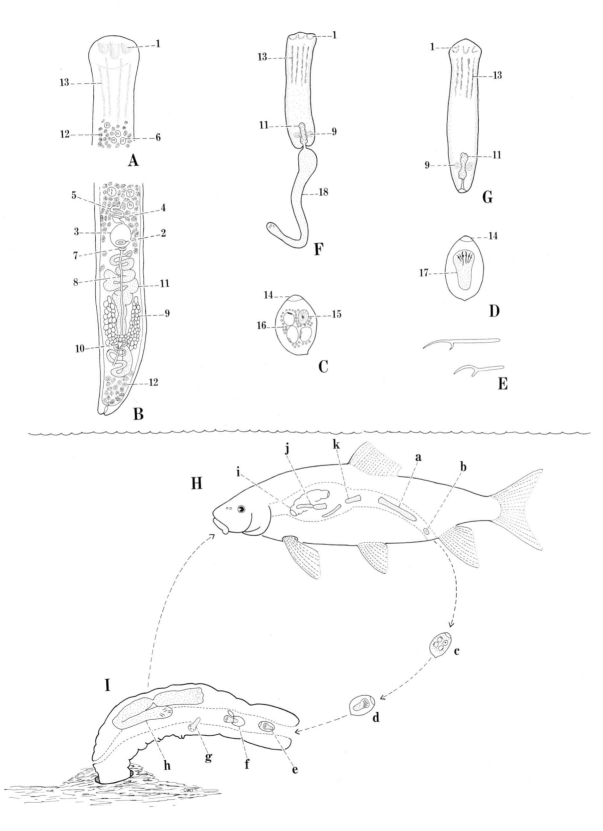

## Explanation of Plate 87

A, Scolex of adult worm. B, Cross section of scolex. C, Mature proglottid. D, Ripe proglottid. E, Procercoid. F, Plerocercoid. G, Definitive host. H, Copepod first intermediate host. I, Walleyed pike (*Stizostedion vitreum*) second intermediate host.

1, Bothrium; 2, common genital atrium; 3, male genital pore; 4, female genital pore; 5, uterine pore; 6, bilobed ovary; 7, Mehlis' gland surrounding ootype; 8, vitelline duct; 9, proximal portion of vagina; 10, oviduct; 11, vitelline glands; 12, vagina; 13, vas deferens; 14, testes; 15, uterus; 16, cercomer with oncospheral hooks.

a, Adult in intestine of definitive host; b, egg passing out of intestine with feces; c, unembryonated egg; d, embryonated egg; e, hatched egg; f, ciliated six-hooked coracidium free in water; g, coracidium eaten by crustacean first intermediate host; h, coracidium sheds ciliated covering; i, coracidium migrates through intestinal wall into hemocoel; j, procercoid; k, infected crustacean is swallowed by fish second intermediate host and the procercoid is liberated by the digestive enzymes; l, procercoid passes through intestinal wall into celom and finally into muscles of fish where development continues to infective stage; m, plerocercoid in muscles; n, infection of definitive host occurs when infected fish are eaten; o, plerocercoids are liberated and develop to adults.

Figures adapted from various sources.

---

the ciliated epithelium in the intestine and migrates through the gut wall into the hemocoel where it develops into a procercoid. When freshwater fish—including pike, perch, trout, and salmon, as a few representatives—ingest infected crustaceans, the liberated procercoids migrate from the intestine to the muscles where they develop into plerocercoids. Natural selective processes play an important role in the infection of small fish with plerocercoids. Young fish still feeding on plankton acquire infections primarily during the summer when the greatest number of tapeworm eggs are hatching in the warmed water and are being eaten by the active and abundant crustacean intermediate hosts. Growth and maintenance of the procercoids place high demands on the infected copepods for providing nutriment and disposing of metabolites of the parasites, rendering them sluggish and easily captured by the fish. If small fish infected with plerocercoids are eaten by large ones, the larval tapeworms, having paratenic tendencies, migrate to the muscles of the new host and and reencyst, a process that may be repeated. Once encysted in the muscles of fish, either directly from the crustacean host or from infected fish, they develop to the infective stage for the definitive host.

Infection of mammalian hosts occurs when fish harboring plerocercoids are eaten. Upon being released from the muscle tissue in the intestine by the digestive processes, the parasites attach to the intestinal wall and develop to sexual maturity in 5 to 6 weeks. Bile appears to be a necessary stimulant for development to sexual maturity.

Nutritive substances are absorbed through the body surface of cestodes, and the demands by the various stages of development are reflected in the structure of the integument. Procercoids and plerocercoids with limited growth and nutrition requirements are covered with short microvilli about 1.5 $\mu$ long. Upon going from a cold-blooded intermediate host to a warm-blooded definitive host, great growth and enormous egg production by the worm, together with general maintenance needs, require prodigious amounts of nutritive substance. To meet these demands, the absorptive surface is increased many times by the greater size of the worm and its covering of microvilli now 4 $\mu$ long (Fig. 28).

*Spirometra mansonoides* (Mueller, 1935) is a parasite of cats in the eastern half of the United States whose life cycle is typical of the diphyllobothriids. Under optimal conditions, development may be completed in about 50 days. Eggs hatch in 10 days and procercoids develop in the warm water copepod *Cyclops vernalis* in 14 days. When copepods harboring mature procercoids are swallowed by mice or water snakes (*Natrix*) infective plerocercoids appear in the muscles and fascia in 2 weeks. They mature in the small intestine of cats in 11 days, as manifested by eggs in their feces.

If plerocercoids are taken into the body of animals other than cats, either cold or warm blooded, where sexual development is not possible, they migrate through the gut wall and into the tissues where growth in size continues. Plerocercoids may be transferred indefinitely in this manner, or experimentally, thus prolonging

**Plate 87**   *Diphyllobothrium latum*                                               343

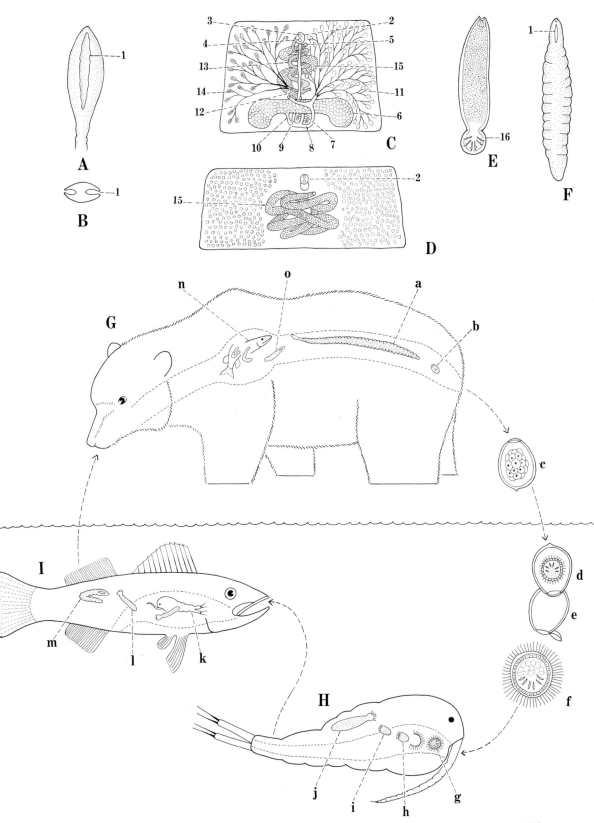

the time required for sexual development of the parasite in a suitable host capable of providing the necessary stimuli.

The extensive work of Mueller on this species provides a wealth of information on host-parasite relationships in the diphyllobothriids.

Species of diphyllobothriids present a good example of a zoonosis which is defined as an infection transmitted between animals and man. Being normally parasites of wild animals, especially those of amphibians, birds, and mammals, these worms are especially suitable for transmission in food when appropriate conditions are met.

Adult worms occur in man when plerocercoids originating from wild animals are ingested with inadequately cooked fish. Six species of *Diphyllobothrium*, *D. latum*, *D. dendriticum*, *D. lanceolatum*, *D. ursi*, and *D. dallae*, all parasites of wild animals in Alaska, occur as adults in humans. Spargana (migrating plerocercoids, especially unknown species) occur in humans when plerocercoids incapable of developing to sexual maturity in them enter the body. Plerocercoids of *Spirometra mansonoides* entering the human gut migrate to the body tissues and appear as spargana. Those of *S. erinacei*, a cosmopolitan species often designated as *S. mansoni*, probably of felids, occurs in amphibians. They migrate into the tissues of people from the muscles of frogs used as poultices on the eyes in some parts of the Orient. Spargana of *S. mansonoides* may get into humans under natural conditions when infected copepods are swallowed with drinking water or by eating insufficiently cooked meat from pigs foraging in marshes and woods where they acquire spargana.

## EXERCISE ON LIFE CYCLE

Eggs of *D. latum* are difficult to obtain in most localities but those of related pseudophyllidean cestodes of fish, amphibians, birds, or mammals may be obtained without too much difficulty. When available, they should be incubated in dishes of tap water at room temperature. Immediately upon hatching, place species of *Cyclops* and *Diaptomus* in the dish with them. Observe the crustaceans with the aid of a dissecting microscope to ascertain whether they ingest the coracidia. Dissect procercoids from the crustaceans and study them. Infected crustaceans may be fed to small fishes in an effort to obtain plerocercoids. If successful, appropriate definitive hosts may be infected by feeding the fish to them. Also dissect plerocercoids from the muscles of the fish for study. Feed them to other animals and observe if they migrate into the tissues or develop into adults.

## SELECTED REFERENCES

Becker, C. D., and W. D. Brunson. (1967). *Diphyllobothrium* (Cestoda) infections in salmonids from three Washington lakes. J. Wildl. Manage. 31:813–824.

Bogomolova, N. A., and L. I. Pavlova. (1961). [Vitelline cells in *Fasciola hepatica* and *Diphyllobothrium latum* and their role in the formation of egg-shell and nutrition of the embryo.] Helminthologia 3:47–59.

Bonsdorff, C. H. von, and A. Telkka. (1966). The flagellar structure of the flame cell in fish tapeworm (*Diphyllobothrium latum*). Z. Zellforsch. Mikrosk. Anat. 70:169–179.

Bråten, T. (1968). An electron microscope study of the tegument and associated structures of the procercoid of *Diphyllobothrium latum*. Z. Parasitenk. 30:95–103.

Bråten, T. (1968). The fine structure of the tegument of *Diphyllobothrium latum* (L.). A comparison of the plerocercoid and adult stages. Z. Parasitenk. 30:104–112.

Cameron, T. W. M. (1945). Fish-carried parasites in Canada. (1) Parasites carried by fresh-water fish. J. Comp. Med. 9:245–254; 283–286; 302–311.

Corkum, K. C. (1966). Sparganosis in some vertebrates of Louisiana and observations on human infection. J. Parasitol. 52:444–448.

Furukowa, T. (1968). Growth and survival of *Diphyllobothrium erinacei* in bileless dogs. Jap. J. Parasitol. 17:419–423.

Janicki, C., and F. Rosen. (1918). Le cycle évolutif du *Dibothriocephalus latus* L. Recherches expérimentales et observations. Bull. Soc. Neuchatel. Sci. Natur. (1916–1917). 42:19–53.

Klekowski, R. Z., and A. Guttowa. (1968). Respiration of *Eudiaptomus gracilis* infected with *Diphyllobothrim latum*. Exp. Parasitol. 22:279–287.

Meyer, M. C. (1967). Hatching of *Diphyllobothrium sebago* eggs in nature. Trans. Amer. Microsc. Soc. 86:239–243.

Meyer, M. C. (1970). Cestode zoonoses of acquatic animals. J. Wildl. Dis. 6:249–254.

Mueller, J. F. (1938). The life history of *Diphyllobothrium mansonoides* Mueller, 1935, and some considerations with regard to sparganosis in the United States. Amer. J. Trop. Med. 18:41–66.

Mueller, J. F. (1966). Host-parasite relationships as

illustrated by the cestode *Spirometra mansonoides*. In: J. E. McCauley, ed. Host-Parasite Relationships, Proc. 26th Annu. Biol. Colloq., Oreg. State Univ., pp. 15–58.

Mueller, J. F. (1966). Human sparganosis in the United States. Proc. 1st Int. Congr. Parasitol. (Rome) 2: 815–816.

Rausch, R. L., and D. K. Hilliard. (1970). Studies on the helminth fauna of Alaska. XLIX. The occurrence of *Diphyllobothrium latum* (Linnaeus, 1758) (Cestoda: Diphyllobothriidae) in Alaska, with notes on other species. Can. J. Zool. 48:1201–1219.

Rausch, R. L., E. M. Scott, and V. R. Rausch. (1967). Helminths in Eskimos in western Alaska, with particular reference to *Diphyllobothrium* infection and anaemia. Trans. Roy. Soc. Med. Hyg. 61:351–357.

## ORDER PROTEOCEPHALA

This group of cestodes is parasitic in fish almost exclusively. The great majority of cestodes of freshwater fish are species of the genus *Proteocephalus*. The adults are located in the small intestine. They are world wide in distribution.

### FAMILY PROTEOCEPHALIDAE

The scolex bears four in-cupped muscular suckers and sometimes a fifth apical one. The genital pores are lateral and the vitellaria are in two lateral bands, one on each side of the proglottid. The gravid uterus is saccular.

### *Proteocephalus ambloplitis*
Leidy, 1887 (Plate 88)

*Proteocephalus ambloplitis* is a parasite of bass (*Micropterus dolomieui, Huro salmoides*), yellow perch (*Perca flavescens*), bowfins (*Amia calva*), and other freshwater fish. Due to the pathology caused by the plerocercoids to the gonads of bass, especially the females, reproduction in infected fish is curtailed if not completely eliminated.

**Description.**    The scolex has four in-cupped muscular suckers and a vestigial apical sucker. The genital pore is lateral. The ejaculatory duct consists of a large number of coils, and the testes are enclosed laterally by the vitellaria and posteriorly by the ovary. The vagina is surrounded by a very large sphincter near the genital pore. Vitellaria are arranged in lateral bands extending from the anterior to the posterior ends of the proglottid. The gravid uterus has 15 to 20 lobes on each side and is filled with somewhat dumbbell-shaped eggs which are fully embryonated.

**Life Cycle.**    As the cestodes mature in the definitive hosts, the terminal proglottids become gravid or "ripe" and detach from the strobilus. They pass from the intestine with the feces. Upon absorbing water, they rupture and release large numbers of eggs, each containing a fully developed oncosphere. Eggs eaten by copepods (*Cyclops, Eucyclops*) hatch in the intestine and the nonciliated oncospheres migrate into the hemocoel where they develop into procercoids, bearing suckers characteristic of the adult. When infected copepods are eaten by fry of bass or pumpkinseed sunfish, which serve as the second intermediate hosts, the procercoids migrate through the intestinal wall into the celom and transform into plerocercoids, the next larval stage and the one that is infective to the definitive hosts. Infection of the fish definitive host with adult cestodes takes place when fish containing fully developed plerocercoids are eaten.

If, on the other hand, fry harboring immature plerocercoids are eaten by larger fish, the parasites are unable to develop to sexual maturity even though they are in the intestine of a favorable host. Instead, they migrate to the celom again, frequently entering the gonads, where development of the plerocercoid stage is completed. Infected gonads are destroyed by the parasites, resulting in sterile fish. These plerocercoids are capable of development to sexually mature tapeworms if eaten by bass or other appropriate hosts.

The demand for free-tissue carbohydrates differs in the various developmental stages and in parts of the body of *Proteocephalus ambloplitis*. They increase during migration of the plerocercoid from the visceral organs to the intestinal lumen. There is no change during development in the intestine. In adult specimens, the concentration of glycogen is much greater in the mature and gravid proglottids due to the high demands of sexual reproduction than in the scolex and immature segments.

When bass and rainbow trout occur simultaneously in lakes, especially shallow ones, the latter become infected with plerocercoids of *Proteocephalus ambloplitis*. While trout do acquire intestinal infections of adult worms, they are not compatible hosts. Without the presence of bass in the lake, the parasite cannot maintain itself in trout alone and soon dies out.

## Explanation of Plate 88

A, Scolex. B, Mature proglottid. C, Gravid proglottid. D, Young procercoid. E, Older procercoid. F, Plerocercoid from lumen of intestine of fish. G, Plerocercoid encysted on mesenteries of fish. H, Bass (*Micropterus*) definitive host. H', Rock bass (*Ambloplites*) definitive host. H", Bowfin (*Amia*) definitive host. I, Crustacean (*Cyclops, Eucyclops, Macrocyclops*) first intermediate host. J, Bass fingerling or fry intermediate host. K, Mature bass paratenic host.

1, Scolex of mature cestode; 2, sucker; 3, rudiment of apical sucker; 4, common genital pore; 5, cirrus pouch; 6, vas deferens; 7, testes; 8, vagina with strong terminal sphincter; 9, ovary; 10, vitellaria; 11, gravid uterus; 12, hooks of oncosphere on procercoid; 13, cyst wall of plerocercoid.

a, Adult cestode in intestine; b, gravid proglottid shed from terminus of strobilus; c, gravid proglottid free in water ruptures and releases embryonated eggs; d, embryonated eggs free in water; e, embryonated egg held by crustacean for eating; f, oncosphere escapes from egg shell in intestine; g, oncosphere migrates through gut wall into hemocoel; h, procercoid in hemocoel; i, procercoid liberated from cyclops by digestive processes in stomach of bass fry will migrate through intestinal wall into celom and encyst; j, encysted plerocercoid attaches to mesenteries or enters gonads (one of two courses follow from this point, depending on whether the fry have procercoids in the digestive tract from recently digested cyclops, or underdeveloped plerocercoids in the celom when eaten by large bass, or other hosts); k, procercoid released from intestine of fry penetrates wall of intestine of bass to continue development in celom; l, immature plerocercoid from body cavity of fry passing through intestinal wall into celom to continue development as a plerocercoid; m, plerocercoid in celom; n, plerocercoid in gonads; o, infection of definitive host with adult cestodes takes place when fish harboring fully developed plerocercoids (J, K) are eaten; p, fully developed plerocercoid free in digestive tract; q, plerocercoid attaches in crypts of small intestine and develops to maturity; r, infected gonads.

Adapted from Hunter and Hunter, 1929, 16th Annu. Rep. N. Y. State Conserv. Dep. Suppl. p. 198.

---

The life cycles of several additional species of *Proteocephala* have been worked. In the cases of *Ophiotaenia perspicua* from snakes, and *Proteocephalus pinguis* and *Corallobothrium parvum* from fish, the life history is similar to that of *P. ambloplitis*, wherein the procercoids are in copepods and the plerocercoids in fish. In *P. tumidocollis* from trout, the life cycle differs in that both the procercoids and plerocercoids develop in succession in a single copepod.

### EXERCISE ON LIFE CYCLE

Adult cestodes of *Proteocephalus ambloplitis* and related species are readily available almost wherever bass and yellow perch occur. Likewise the copepod intermediate hosts (*Cyclops vulgaris* and *Eucyclops agilis*) are ubiquitous.

Place eggs of the cestodes in glass dishes containing pond water, a sprig of water plant, and copepods. Observe the crustaceans eat the eggs. Examine the intestinal contents of the copepods for oncospheres shortly after they have eaten the eggs, and the body cavity 4 to 5 hours later for young procercoids.

Allow fry of bass, perch, or pumpkinseed sunfish to eat the infected copepods. Examine the liver, mesenteries, and gonads of fish after 5, 10, and 15 days for plerocercoids. At the end of this time, infect hungry yearling bass or yellow perch by allowing them to eat fry containing young or fully developed plerocercoids and observe the course of development of each age group. Examine the intestine, liver, mesenteries, and gonads of the yearling bass at 3, 30, 60, and 90 days, if possible to keep them that long, for various stages and location of the parasites. It may be impractical or impossible to have yearling bass for the final step in the life cycle.

Species of *Proteocephalus* other than *P. ambloplitis* may be available for study. The life cycle would be studied in the manner described above with possible variations in the species of copepods and fish.

### SELECTED REFERENCES

Becker, C. D., and W. D. Brunson. (1968). The bass tapeworm: a problem in northwest trout management. Prog. Fish-Cult. 30:76–83.

Hunter, G. W. III. (1928). Contributions to the life history of *Proteocephalus ambloplitis* (Leidy). J. Parasitol. 14:229–242.

Hunter, G. W. III. (1929). Life-history studies on *Proteocephalus pinguis* LaRue. Parasitology 21:487–496.

Hunter, G. W. III, and W. S. Hunter. (1929). Further studies on the bass tapeworm, *Proteocephalus ambloplitis* (Leidy). In: A biological sur-

Plate 88    *Proteocephalus ambloplitis*                                                347

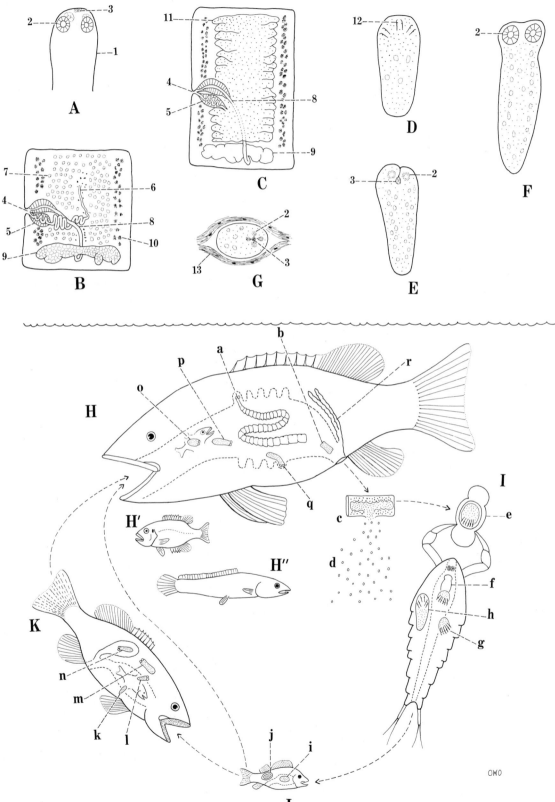

### Explanation of Plate 89

A, Scolex and neck. B, Large and small rostellar hooks. C, Mature proglottid. D, Gravid proglottid. E, Egg. F, Invaginated cysticercus. G, Evaginated cysticercus. H, Human definitive host. I, Swine intermediate host.

1, Scolex; 2, sucker; 3, armed rostellum; 4, large rostellar hook; 5, small rostellar hook; 6, common genital pore; 7, cirrus pouch; 8, sperm duct; 9, testes; 10, vagina; 11, seminal receptacle; 12, ovary with accessory median lobe; 13, oviduct; 14, vitelline gland; 15, Mehlis' gland; 16, immature uterus; 17, longitudinal excretory canal; 18, lateral branch of gravid uterus; 19, striated shell of egg; 20, oncosphere or hexacanth; 21, bladder of cysticercus.

a, Adult cestode in intestine of human; b, detached gravid proglottid; c, proglottid rupturing to release eggs; d, eggs free; e, eggs swallowed by swine; f, eggs hatch; g, oncospheres penetrate intestinal wall and enter hepatic portal vein; h, oncospheres passing through liver; i, oncospheres in right side of heart; j, oncospheres in lungs; k, oncospheres in left side of heart; l, oncospheres enter general circulation; m, oncospheres entering skeletal muscles; n, cysticercus in skeletal and cardiac muscles; o, cysticercus in ham; p, cysticerci in sausages; q, infection of human by swallowing cysticerci in infected pork; r, cysticerci in piece of pork; s, cysticercus released from meat has evaginated; t, gravid proglottid carried to stomach by reverse peristalsis where eggs are released; u, eggs in intestine hatch; v, oncospheres in intestine; w, oncospheres in hepatic portal vein; x, oncospheres in right side of heart; y, oncospheres in lungs; z, oncospheres in general circulation.

a', Oncospheres leaving blood vessel in brain; b', cysticerci in brain; c', cysticerci in skeletal muscles (also in eyes).

Figures adapted from various sources.

---

vey of the Erie-Niagara system. Annu. Rep. N. Y. State Conserv. Dep. (1928), Suppl., pp. 198–207.

Larsh, J. E., Jr. (1941). *Corallobothrium parvum* n. sp., a cestode from the common bullhead, *Ameiurus nebulosus* Le Sueur J. Parasitol. 27:221–227.

Marra, M., and G. W. Esch. (1970). Distribution of carbohydrates in adults and larvae of *Proteocephalus ambloplitis* (Leidy, 1887). J. Parasitol. 56:398–400.

Thomas, L. J. (1934). Further studies on the life cycle of the frog tapeworm *Ophiotaenia saphena* Osler. J. Parasitol. 20:291–294.

Wagner, E. D. (1954). The life history of *Proteocephalus tumidocollus* Wagner, 1953 (Cestoda), in rainbow trout. J. Parasitol. 40:489–498.

### ORDER CYCLOPHYLLIDEA

The cyclophyllidean cestodes are common parasites of birds and mammals throughout the world. Most species are inhabitants of the small intestine as adults, although a few occur in the bile ducts. They are characterized as sucker-bearing tape worms with one or two compact vitelline glands located posterior to the ovary.

Generally an intermediate host is involved in the life history of these parasites although at least one species may complete its cycle with or without the succor of an intermediary. Two basic types of life cycles are known. The species of Taeniidae utilize mammals as intermediate hosts and the larval stage is a vesicular cysticercus, strobilocercus, coenurus, or hydatid cyst. In the other families, the larvae develop in invertebrates or vertebrates and are nonvesicular cysticercoids, tetrathyridia, or plerocercoids.

For convenience of discussion, the cyclophyliidean cestodes are divided into two groups: 1) the family Taeniidae, and 2) other families.

### FAMILY TAENIIDAE

Five genera of this family are common parasites as adults of humans, dogs, and cats. They include *Taenia*, *Taeniarhynchus*, *Hydatigera*, *Multiceps*, and *Echinococcus*. The larval stages occur in humans, cattle, sheep, swine, deer, rabbits, hares, rats and others.

The first four of these genera have been transferred to the genus *Taenia* by some workers on the basis of similarity of the adults. Considering the larval characteristics and certain adult features as reliable and valid generic characters, the classification that includes the various genera formerly accepted is followed here.

**Description.**    They are small to large muscular cestodes with distinct segmentation. The scolex bears four muscular in-cupped suckers, and may have a double or single crown of rostellar hooks, or none. The genital pore is lateral and irregularly alternate. The gravid uterus consists of a medium stem and lateral branches filled with embryonated eggs having thick, striated shells, or embryophores.

Species commonly found as adult worms in man are *Taenia solium*, *Taeniarhynchus saginatum*, and *T. confusum*.

### *Taenia solium* Linnaeus, 1758 (Plate 89)

*Taenia solium* is known as the pork tapeworm of humans. It occurs only in people as the adult

Plate 89     *Taenia solium*                                                                                   349

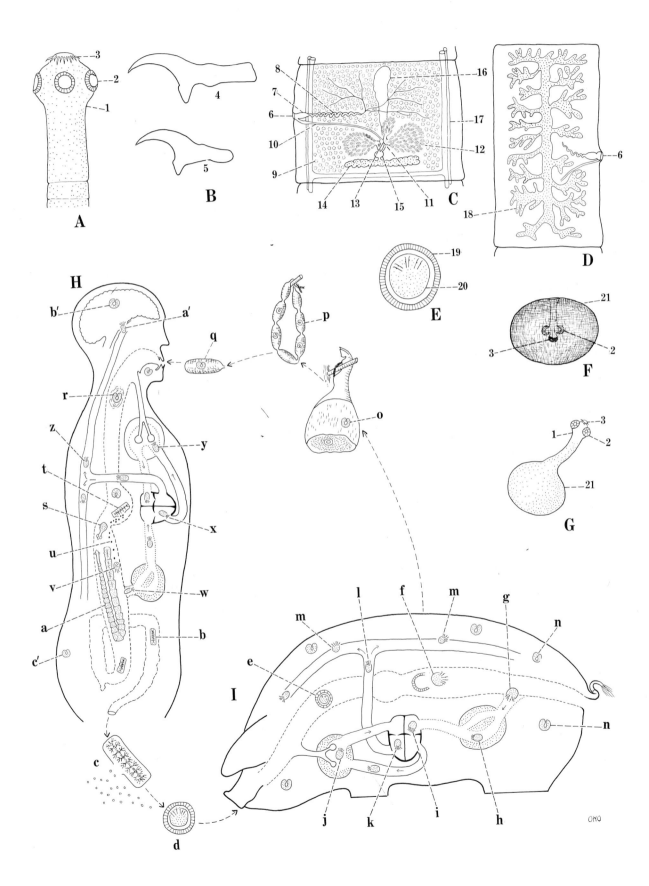

A

B

C

D

E

F

G

H

I

worm and in swine as the larval cysticercus, but cysticerci also may appear in humans infected with adult worms. Wherever pork is eaten in a cured or inadequately cooked condition and where human feces are left for pigs to eat, these worms normally are prevalent.

**Description.**    Fully developed worms are up to 7 m long and with less than 1000 proglottids. There is a double crown of 22 to 32 rostellar hooks consisting of two sizes, measuring 160 to 180 $\mu$ and 110 to 140 $\mu$ long. Testes number 150 to 200 and are scattered throughout the proglottid. The ovary has two large lateral lobes and a median accessory one. The gravid uterus has 7 to 13 lateral branches on each side.

**Life Cycle.**    Gravid proglottids detached from the strobilus in the intestine of humans are eliminated from the body in the feces. Through muscular activity, the embryonated eggs are forced from the anterior end of the proglottid through ruptured branches of the uterus. When swallowed by swine or humans, the preliminary action of the gastric juices followed by that of the intestinal juices cause a rapid breakdown of the embryophore and liberation of the oncosphere. The rate of hatching of the eggs and migration of the oncospheres in humans is uncertain. Eggs of *Hydatigera fasciolaris* injected into ligatured sections of the intestine of rats hatch in 15 minutes and the oncospheres begin penetration, mostly in the duodenum, in 30 minutes. By 2 hours, they are in the deeper parts of the mucosa and in the submucosa. Oncospheres are in the blood in 24 hours, possibly sooner.

Upon entering the hepatic portal vein, they are carried through the liver, heart, and lungs into the general circulation. They leave the blood vessels and develop into cysticerci in the skeletal and cardiac muscles. Infection of humans occurs when viable cysticerci are swallowed with pork insufficiently cooked to kill them.

Inasmuch as development of *T. solium* in the intestine is not known in detail, that of *Taenia hydatigena* is given as a pattern of what may take place in humans. The liberated scolex first migrates posteriorly in the small intestine and then returns to the duodenal region to begin growth. Strobilization occurs in 7 days, testes are present in 20 days, mature eggs appear in the uterus in 48 days, and proglottids are passed in the feces in 50 days. An average of around 30,000 eggs develops in a single proglottid.

When gravid proglottids are carried into the stomach of humans parasitized with the worms from the small intestine by reverse peristalsis, the eggs are released, hatch, and the oncospheres enter the circulation. They migrate through the human body and undergo development in the muscles as in the case of swine that swallow them. Cysticerci developing in vital organs such as the brain and eyes cause serious results. This is internal autoinfection and occurs only in *T. solium* among the taenioid cestodes. External autoinfection can take place from swallowing eggs and may be more common than internal autoinfection.

Adult worms in the intestine generally cause no serious damage. Weakened persons suffer from a number of discomforts, including loss of appetite and weight, nervous disorders, chronic indigestion, and persistent diarrhea, or alternating diarrhea and constipation.

## SELECTED REFERENCES

Banerjee, D., and K. S. Singh. (1969). Studies on *Taenia fasciolaris*. I. Studies on the early stages of infection in cysticerciasis in rats. Indian J. Anim. Sci. 39:149–154.

Esch, G. W., and J. T. Self. (1965). A critical study of the taxonomy of *Taenia pisiformis* Bloch, 1780; *Multiceps multiceps* (Leske, 1780); and *Hydatigera taeniaeformis* Batsch, 1786. J. Parasitol. 51:932–937.

Faust, E. C., P. R. Russell, and R. C. Jung. (1970). Craig and Faust's Clinical Parasitology. 8th ed. Lea & Febiger, Philadelphia, p. 529.

Featherstone, D. W. (1969). *Taenia hydatigena:* I. Growth and development of adult stage in dog. Exp. Parasitol. 25:329–338.

Gönnert, R., G. Meister, R. Struffe, and G. Webbe. (1967). Biologische Probleme bei *Taenia solium*. Z. Tropenmed. Parasitol. 18:76–81.

Gönnert, R., G. Meister, and H. Thomas. (1968). Das Freiwerden der Eier aus Taenia-Proglotten. Z. Parasitenk. 31:282–288.

Verster, A. (1967). Redescription of *Taenia solium* Linnaeus, 1758 and *Taenia saginata* Goeze, 1782. Z. Parasitenk. 29:313–328.

Verster, A. (1969). A taxonomic revision of the genus *Taenia* Linnaeus, 1758, s. str. Onderstepoort J. Vet. Res. 36:3–58.

Webbe, G. (1967). The hatching and activation of taeniid ova in relation to the development of cysticercosis in man. Z. Tropenmed. Parasitol. 18:354–369.

*Taeniarhynchus saginatum*
(Goeze, 1782) (Plate 90)

This is the beef tapeworm of humans, commonly designated as *Taenia saginatum*. It is the largest of the taenioid cestodes and is prevalent whenever beef is eaten and human sanitation is lax, leaving feces available for cattle to eat or used to fertilize pastures.

**Description.**    Adult worms are up to 25 m long (average, 4 to 5), with 1000 to 2000 proglottids. A rostellum and hooks are lacking and the ovary is without an accessory lobe. There are 300 to 400 testes per proglottid. A well-developed vaginal sphincter is present. When gravid, the uterus has 15 to 20 lateral branches on each side.

**Life Cycle.**    The life history of this species is similar to that of *Taenia solium* except that cattle instead of swine serve as the intermediate host and internal autoinfection does not occur in the definitive host.

Eggs swallowed by cattle hatch in the duodenal region. They migrate into the lymphatics and venules of the hepatic portal system and are distributed via the bloodstream throughout the body. In the voluntary muscles, especially the masseters, they develop into the oval, opalescent, infective cysticerci (*Cysticercus bovis*) 7.5 to 10 mm long by 4 to 6 mm in diameter in 60 to 75 days. Other natural intermediate hosts include bison, giraffe, and llama. External autoinfection may occur if eggs are swallowed by humans. About 3 months are required for cysticerci to grow into sexually mature worms in the human intestine.

Cattle become infected when their feed, corrals, or pastures are contaminated with the feces of humans infected with *T. saginatum*. Sludge from sewage plants used as fertilizer on pastures is a source of viable tapeworm eggs to grazing cattle. Eggs remain infective in moist situations similar to what occurs on pastures for 60 to 70 days at 20°C and up to 180 days at 10°.

In the United States, 12,000 to 16,000 infected cattle are discovered per year (1959 through 1967) in abattoirs. Of this number 72.6 percent occurred in California in 1967. In man, the concentration of *T. saginatum* is in the far western and northeastern parts of the United States. Of 1,852,764 human fecal samples examined for parasites, 429, or 23 per 100,000 had *Taenia*. Since the species are not distinguishable on the basis of eggs it is not known which was represented, although *T. saginatum* is far more prevalent in the United States than *T. solium*.

These large worms block the digestive tract, give off toxic metabolic substances that are absorbed, and cause diarrhea and loss of weight. Single detached proglottids crawling from the anus are a source of discomfort. Worms survive many years in the intestine unless removed by medication.

*Taeniorhynchus confusum* has been found occasionally in humans in parts of the United States; it occurs commonly in East Africa. The cysticerci develop in calves fed eggs. Adults and cysticerci resemble *T. saginatum* in being without rostellar hooks which has led some workers to believe it is a variant of that species. I obtained a specimen of *T. confusum* from a person in Colorado who had been engaged recently in wildlife investigations in Kenya for several years.

**SELECTED REFERENCES**

Enigk, K., M. Stoye, and E. Zimmer. (1969). Die lebensdauer von Taenieneiern in Gärfutter. Deutsch. Tierärztl. Wochenschr. 76:421–425.

Faust, E. C. (1930). A study of a rare tapeworm, *Taenia confusa*, with a report of the fourth case. Southern Med. J. 23:903–906.

Faust, E. C., P. F. Russell, and R. C. Jung. (1970). Craig and Faust's Clinical Parasitology. 8th ed. Lea & Febiger, Philadelphia, p. 535.

Schultz, M. G., J. A. Hermos, and J. H. Steele. (1970). Epidemiology of beef tapeworm infection in the United States. Publ. Health Rep. 85:169–176.

*Echinococcus* Rudolphi, 1801

The species of this genus are the smallest of the taenioid cestodes, with the body consisting of three to five proglottids. Adult worms live in the small intestine of dogs, wolves, coyotes, foxes, cats, and related animals. The larval stage, or hydatid cyst, occurs commonly in the liver and lungs of humans, cattle, sheep, swine, cervids, kangaroos, as well as other ruminants, and rodents. *Echinococcus* is especially important as a parasite because of the prevalence of the dangerous hydatid cysts in people who live in close association with dogs whose food consists of parts of carcasses of animals that serve as the intermediate host.

## Explanation of Plate 90

A, Scolex and neck. B, Mature proglottid. C, Gravid proglottid. D, Egg. E, Invaginated cysticercus. F, Evaginated cysticercus. G, Human definitive host. H, Bovine intermediate host.

1, Scolex; 2, suckers; 3, common genital pore; 4, cirrus pouch; 5, sperm duct; 6, testes; 7, vaginal sphincter; 8, vagina; 9, ovary; 10, oviduct; 11, uterus; 12, vitelline gland; 13, ventral longitudinal excretory canal; 14, dorsal longitudinal excretory canal; 15, transverse excretory canal; 16, striated egg shell; 17, oncosphere; 18, bladder of cysticercus.

a, Adult cestode in small intestine; b, gravid proglottid passing out of intestine; c, ruptured proglottid liberating eggs; d, eggs and proglottids mixed with hay; e, eggs on grass; f, infection of bovine by swallowing eggs; g, eggs hatch in small intestine; h, oncospheres enter hepatic portal vein; i, oncospheres pass through liver; j, oncospheres pass through heart; k, oncospheres pass through lungs; l, m, oncospheres enter general circulation; n, cysticercus in skeletal muscles; o, live cysticercus in inadequately cooked beef; p, infection of definitive host; q, cysticercus being freed from flesh in stomach; r, cysticercus evaginates in small intestine, attaches to intestinal mucosa, and grows to adult cestode.

Figures adapted from various sources.

---

*Echinococcus granulosus*
(Batsch, 1786) (Plate 91)

This cosmopolitan species produces an unilocular hydatid cyst in ruminants and people. Adult worms are in the small intestine of dogs, wolves, and coyotes.

**Description.** The body normally consists of the scolex and usually one each of an immature, mature, and gravid proglottid. Occasionally more may be present. Adult worms are 1.5 to 6 mm long. The rostellum is armed with a double row of 28 to 50 (average 30 to 36) hooks of characteristic shape and of two sizes. The large ones are 40 to 49 $\mu$ long and the small ones 30 to 42 $\mu$. Testes, 45 to 65 (average 56), are both anterior and posterior to the genital pore which is in the posterior half of the proglottid. Lateral branches occur on the gravid uterus.

The hydatid cyst consists of a thin inner syncytial germinative layer surrounded externally by a thick laminated one of connective tissue. Internal buds from the germinative layer form brood capsules containing scolices as well as daughter capsules with their own scolices.

**Life Cycle.** The life cycle is similar in most respects with that of the taenioids discussed above. Eggs in the feces of the canine hosts hatch when swallowed by one of the intermediate hosts. Oncospheres enter the venules of the hepatic portal vein and are transported by the blood to various parts of the body. Many are filtered out in the liver and lungs where development is completed, attaining an enormous size in long-lived animals. Others continue through the lungs and heart to other parts of the body where they may develop. The marrow cavity of the long bones of man is a common site of hydatids. Pressure of the growing cysts results in erosion of the bone with a tendency to fractures that fail to heal properly.

Brood capsules containing scolices from hydatid cysts obtained from the liver and lungs injected intraperitoneally into rats, mice, gerbils, and birds produce normal, fertile hydatid cysts. Serious problems associated with surgical removal of hydatid cysts from humans are the unnoticed escape into the body cavities of scolices that will develop subsequently into normal cysts, and the escape of hydatid fluid into the tissue, causing anaphylactic shock.

Infection of the definitive host takes place when hydatid cysts are eaten, the wall digested, and the scolices liberated. The scolices evaginate in the small intestine as a result of the combined influences of the peptic and tryptic juices. The evaginated scolex has numerous microtriches, or tegumental projections, on the rostellum and knoblike projections on the posterior region. In order for growth to proceed to strobilization and sexual maturity, the rostellum with its microtriches must be in contact with a solid nutritive substrate such as the intestinal mucosa. *In vitro*, the rostellum must have the surface of the solid submerged nutritive substrate roughened. The tegumental projections initially on the scolex and later over the entire strobila presumably serve as absorptive surfaces through which nutriments necessary for growth, strobilization, and reproduction pass (Fig. 28).

The highly resistant eggs remain alive and infective for 2 years and more on sodded pastures with adequate soil moisture.

*Echinococcus multilocularis* Leuckart, 1863 is primarily from the northern hemisphere. It occurs as adults mostly in foxes, dogs, coyotes,

**Plate 90** *Taeniarhynchus saginatum* 353

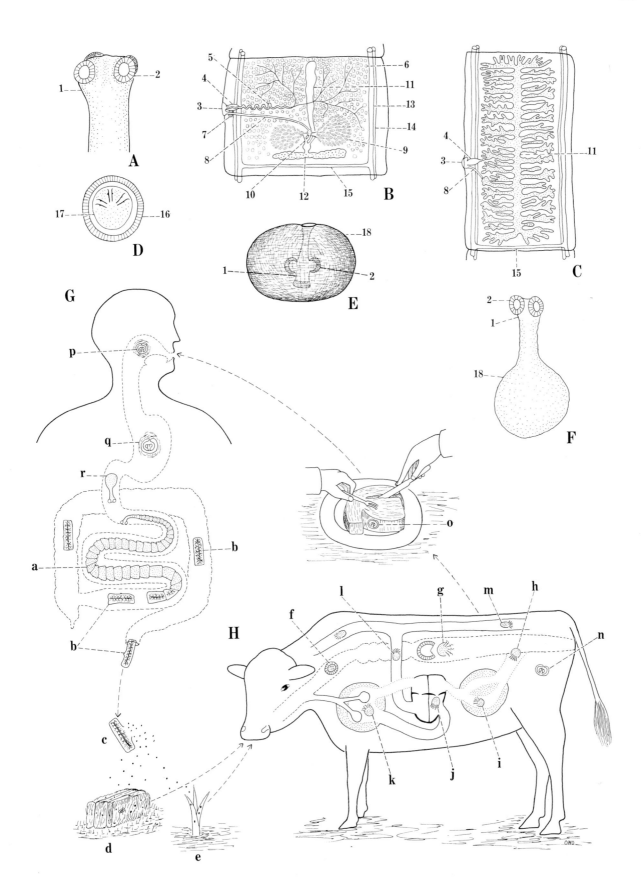

## Explanation of Plate 91

A, Adult cestode. B, Rostellar hooks. C, Segment of hydatid cyst. D, Invaginated scolex from hydatid cyst. E, Evaginated scolex. F, Dog definitive host. G, Sheep intermediate host.

1, Scolex; 2, armed rostellum; 3, suckers; 4, immature proglottid; 5, mature proglottid; 6, common genital pore; 7, cirrus pouch; 8, testes; 9, vagina; 10, uterus; 11, ovary; 12, Mehlis' gland; 13, vitelline gland; 14, gravid proglottid; 15, large rostellar hook; 16, small rostellar hook; 17, laminated wall of hydatid cyst; 18, germinal layer of mother cyst; 19, brood capsules in which scolices will develop; 20, germinal epithelium extruding to outside through rupture in cyst wall; 21, brood capsule; 22, developing scolex; 23, brood capsule; 24, daughter capsule with developing scolices; 25, fully developed scolex in brood capsule; 26, stalk of brood capsule.

a, Adult cestode in small intestine; b, eggs passed out in feces; c, characteristic taenioid type of egg; d, infection of intermediate host; e, eggs hatch in small intestine; f, oncospheres enter hepatic portal vein; g, oncospheres enter liver parenchyma from blood vessels; h, hydatid cyst; i, j, some oncospheres continue through liver and right side of heart; k, oncospheres leave pulmonary circulation to enter parenchyma of lungs; l, hydatid; m, oncospheres pass through lungs and left side of heart; n, oncospheres in general circulation; o, hydatids in other organs such as brain and marrow of hollow bones; p, cyst wall; q, brood capsule with scolices; r, exogenous cyst that originated from germinal epithelium of mother cyst; s, strip of germinal epithelium free in mother cyst; t, infection of definitive host occurs when hydatids are eaten and digested, releasing scolices; u, scolices released in intestine evert and attach to intestinal mucosa, maturing in 6 to 7 weeks.

Figure C adapted in part from Chandler, 1955, Introduction of Parasitology, 9th ed., John Wiley and Sons, New York, p. 359; D, E, from Belding, 1952, Textbook of Clinical Parasitology, 2nd ed., Appleton-Century-Crofts, New York, p. 586.

---

and cats, and as larval multilocular cysts in microtine rodents.

Morphologically it differs from *E. granulosus* in the genital pore being near the middle of the proglottid and in having few testes (17 to 26) located from the level of the genital pore and backward. The uterus is without lateral branches. The hydatid is an alveolar cyst which proliferates by external budding.

Its life cycle is basically the same as that of *E. granulosus*. Natural sylvatic infection is through foxes and microtine and, to some extent, cricetine rodents. Ground beetles feeding on infected droppings of foxes ingest eggs which when transmitted to white-footed mice (*Peromyscus maniculatus*) eating them hatch and develop into normal multilocular cysts.

In Canada and Alaska, *E. multilocularis* is restricted largely to the tundra and islands of the Bering sea. Natural transmission involves foxes, wolves, and microtine rodents. However, dogs harbor the adults and probably are the source of larval infection of the people who work and live in close contact with them. In the conterminous United States, it occurs in the north central states where in addition to foxes, coyotes are natural definitive hosts.

## SELECTED REFERENCES

Faust, E. C., P. R. Russell, and R. C. Jung. (1970). Craig and Faust's Clinical Parasitology. 8th ed. Lea & Febiger, Philadelphia, p. 541.

Gemmel, M. A. (1957). Hydatid disease in Australia. I. Observations on the incidence of *Echinococcus granulosus* (Batsch 1786) (Rudolphi 1805) in the dog in New South Wales. Austr. Vet. J. 33: 8–14.

Gemmel, M. A. (1957). Hydatid disease in Australia. II. Observations on the geographical distribution of *Echinococcus granulosus* (Batsch, 1786) (Rudolphi, 1805) in the dog in New South Wales. Austr. Vet. J. 33:217–226.

Gemmel, M. A. (1958). Hydatid disease in Australia. III. Observation on the incidence and geographical distribution of hydatidiasis in sheep in New South Wales. Austr. Vet. J. 34:269–280.

Gemmel, M. A. (1959). Hydatid disease in Australia. IV. Observations on the incidence of *Echinococcus granulosus* on stations and farms in endemic regions of New South Wales. Austr. Vet. J. 35:396–402.

Gemmel, M. A. (1959). Hydatid disease in Australia. VI. Observations on the Carnivora of New South Wales as definitive hosts of *Echinococcus granulosus* (Batsch, 1786), (Rudolphi, 1801), and their role in the spread of hydatidiasis in domestic animals. Austr. Vet. J. 35:450–455.

Gemmel, M. A. (1959). Hydatid disease in Australia. VII. An appraisal of the present position and some problems of control. Austr. Vet. J. 35:505–514.

Gemmel, M. A., and P. Brydon. (1960). Hydatid disease in Australia. V. Observations on hydatidiasis in cattle and pigs in New South Wales and the economic loss caused by the larval stage of *Echinococcus granulosus* (Batsch 1786) (Rudolphi 1801) in food animals in Australia. Austr. Vet. J. 36:73–78.

**Plate 91** *Echinococcus granulosus* 355

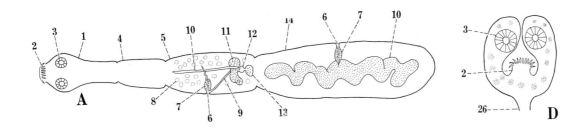

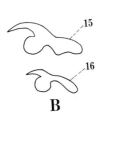

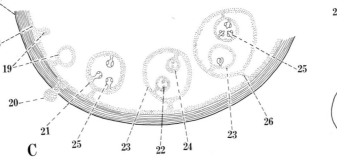

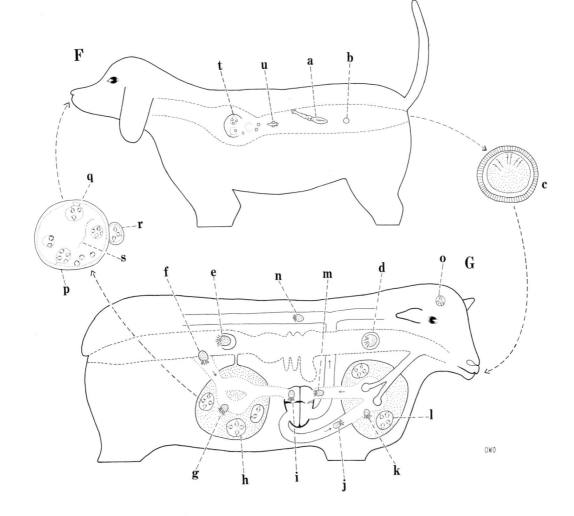

OWO

## Explanation of Plate 92

I. Definitive hosts
  A. Dogs and related Canidae.
  B. Cats and related Felidae.
II. Adult worms and eggs
  1. Dogs. *Taenia hydatigena, T. pisiformis, T. ovis, T. krabbei, Multiceps multiceps, M. seralis.*
    Cats. *Hydatigera taeniaeformis.*
  2. *Dogs. Echinococcus granulosus.*
  3. Embryonated taenioid eggs.
III. Larval types
  a. Cysticercus. *Taenia.*
  b. Coenurus. *Multiceps.*
  c. Hydatid. *Echinococcus.*
  d. Strobilocercus. *Hydatigera.*

IV. Intermediate hosts
  Rabbits and hares. **a,** *Taenia pisiformis;* **b,** Multiceps serialis.
  Sheep. **a,** *Taenia hydatigena;* **b,** *Multiceps multiceps;* **c,** *Echinococcus granulosus.*
  Swine. **a,** *Taenia hydatigena;* **c,** *Echinococcus granulosus.*
  Deer and other cervids. **a,** *Taenia hydatigena, T. Krabbei;* **c,** *Echinococcus granulosus.*
  Humans. **c,** *Echinococcus granulosus.**
  Rats. **d,** *Hydatigera taeniaeformis.*

* While larvae develop in man, normally he does not function in transmitting the parasite to dogs.

---

Heath, D. D. (1970). The development of *Echinococcus granulosus* larvae in laboratory animals. Parasitology 60:449–456.

Jha, R. K., and J. D. Smyth. (1969). *Echinococcus granulosus* ultrastructure of microtriches. Exp. Parasitol. 25:232–244.

Liu, I. K. M., C. W. Schwabe, P. M. Schantz, and M. N. Allison. (1970). The occurrence of *Echinococcus granulosus* in coyotes (*Canis latrans*) in the Central Valley of California. J. Parasitol. 56:1135–1137.

Leiby, P. D., and M. P. Niclel. (1968). Studies on sylvatic echinococcus. I. Ground beetle transmission of *Echinococcus multilocularis* Leuckart, 1863 to deer mice, *Peromyscus maniculatus* (Wagner). J. Parasitol. 54:536–537.

Leiby, P. D., W. P. Carney, and C. E. Woods. (1970). Studies on sylvatic echinococcosis. III. Host occurrence and geographic distribution of *Echinococcus multilocularis* in the north central United States. J. Parasitol. 56:1141–1150.

Morseth, J. D. (1966). The fine structure of the tegument of adult *Echinococcus granulosus, Taenia hydatigena* and *Taenia pisiformis.* J. Parasitol. 52:1074–1085.

Morseth, J. D. (1967). Fine structure of the hydatid cyst and protoscolex of *Echinococcus granulosus.* J. Parasitol. 53:312–325.

Rausch, R. (1956). Studies on the helminth fauna of Alaska. XXX. The occurrence of *Echinococcus multilocularis* Leuckart, 1863, on the mainland of Alaska. Amer. J. Trop. Med. Hyg. 5:1086–1092.

Rausch, R. (1960). Recent studies on hydatid disease in Alaska. Parasitologia 2:391–398.

Smyth, J. D. (1967). Studies on tapeworm physiology. XI. *In vitro* cultivation of *Echinococcus granulosus* from the protoscolex to the strobilate stage. Parasitology 57:111–133.

Smyth, J. D. (1969). The biology of hydatid organisms. In: B. Dawes, ed. Advances in Parasitology. Academic Press, New York, Vol. 7, pp. 327–347.

Vibe, P. P. (1968). [Survival of *Echinococcus* eggs in the environment.] Vestn. Sel'skokhoz. Nauki, Alma-Ata (7):75–76.

Williams, J. F., and C. W. Colli. (1970). Primary cystic infection with *Echinococcus granulosus* and *Taenia hydatigena* in *Merione unguiculatus.* J. Parasitol. 56:509–513.

## Life Cycles of Some Common Taenioid Cestodes of Dogs and Cats

Several species of taenioid cestodes occur commonly as adults in canines and felines. The life cycles (Plate 92) are basically the same as those species discussed for man.

The eggs of the taenioid cestodes are so similar as to be virtually indistinguishable. The species of adult cestodes, with the exception of *Echinococcus*, are difficult to identify. Larval stages, on the other hand, form four distinct groups. *Taenia* produces a cysticercus which has a single scolex invaginated in a vesicular bladder; *Hydatigera* has an evaginated larva with a single scolex followed by a series of immature proglottids, and terminated posteriorly by a small vesicular bladder; *Multiceps* is distinguished by a single

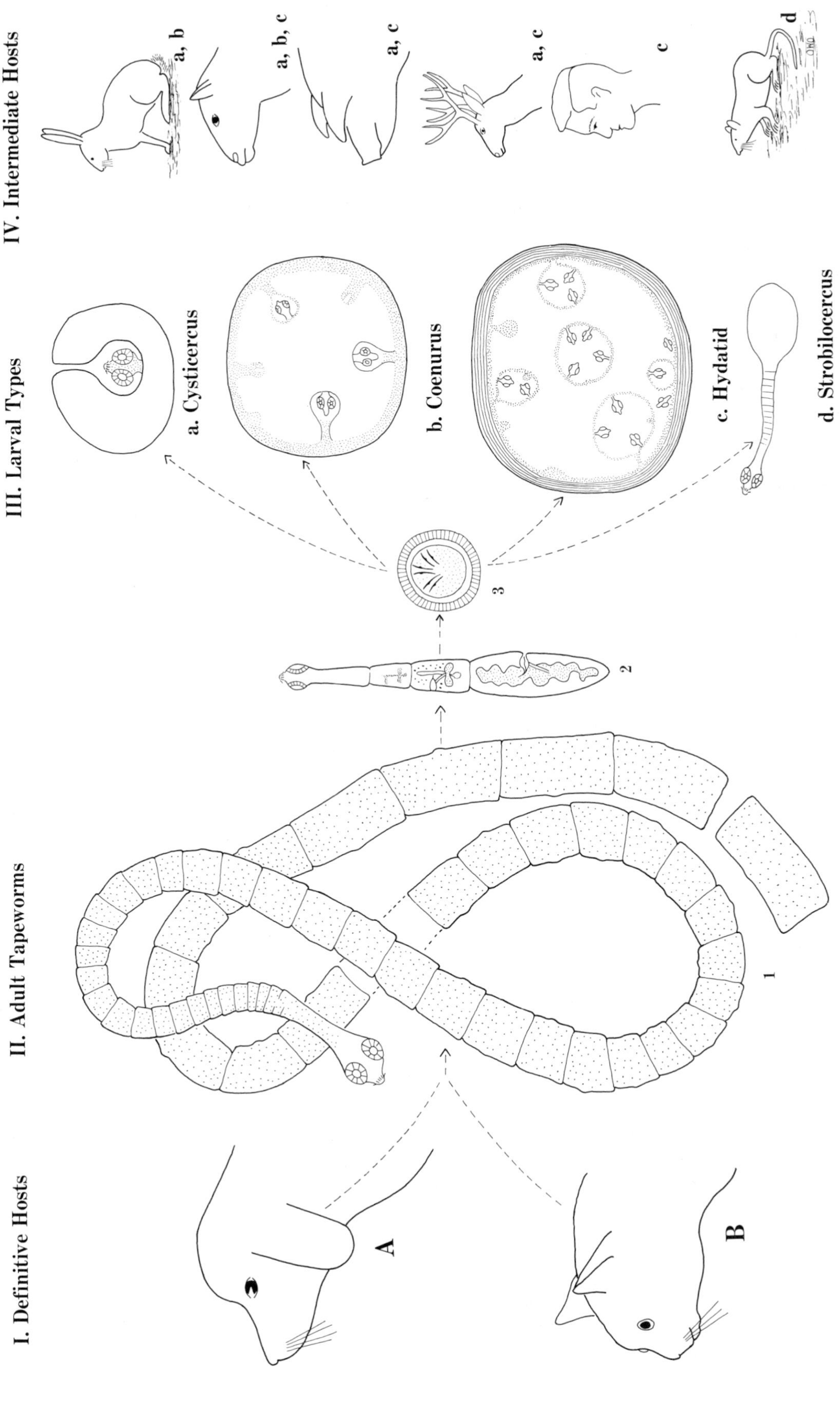

**IV. Intermediate Hosts**

a, b

a, b, c

a, c

a, c

c

d

**III. Larval Types**

a. Cysticercus

b. Coenurus

c. Hydatid

d. Strobilocercus

3

2

**II. Adult Tapeworms**

1

**I. Definitive Hosts**

A

B

vesicular bladder containing many invaginated scolices; and *Echinococcus* produces a complicated larva, the hydatid cyst which may be unilocular or multilocular, consisting of a tough external laminated wall lined with a thin layer of germinal epithelium which produces brood capsules and daughter cysts which in turn produce scolices from their own germinal epithelium.

The cysticerci of *T. pisiformis* and *T. hydatigena* occur in the visceral cavity of lagomorphs and large herbivores, respectively. Those of *T. ovis* and *T. krabbei* are in the musculature of sheep and cervids, respectively, causing a condition of the meat commonly known as measles.

Adults of *Hydatigera taeniaeformis* are in the intestine of cats. The larval stage is a strobilocercus that develops in the liver of rodents, chiefly rats and mice.

*Multiceps multiceps* and *M. serialis* are parasites of the intestine of canines. The larval stage is a coenurus in the central nervous systems of ovines and intermuscular connective tissue and abdominal cavity lagomorphs, respectively. Coenuri of both species have been found in humans.

Hydatids of *Echinococcus granulosus* occur primarily in ruminants, whereas those of *E. multilocularis* are found in rodents, chiefly microtines. Hydatids of both species infect humans.

## EXERCISE ON LIFE CYCLE

Material for the study of life cycles of several species of taenioid cestodes may be obtained readily from veterinary hospitals where dogs and cats are treated for the removal of tapeworms. These include *Taenia pisiformis* and *Multiceps serialis* from dogs, and *Hydatigera taeniaeformis* from cats.

Large numbers of eggs dissected from gravid proglottids should be fed to appropriate intermediate hosts, i.e., eggs of *Taenia pisiformis* and *Multiceps serialis* to rabbits, and those of *Hydatigera taeniaeformis* to rats or mice.

Hatching takes place in the small intestine of mammals. Give a large number of eggs to several white mice by means of a stomach tube. Examine the contents of the intestine in 15 minutes for free oncospheres and sections of the duodenum in 30 minutes for oncospheres in the mucosa and submucosa, and of the liver in one hour in the region of the venules of the hepatic portal vein. The larval stages begin development in the liver. Only *H. taeniaeformis*

remains in it; the others drop into the body cavity to complete their growth.

About a week after ingesting eggs, the larvae appear in the liver as small pearly bodies without hooks. In 3 weeks, they have attained a length of approximately 1 cm. On day 30, the larvae of *T. pisiformis* and *M. serialis* drop into the body cavity to complete their development. They are capable of infecting the definitive host several weeks later if eaten by it.

Introduce eggs and scolices of cysterci, strobilocerci, or coenuri into the peritoneal cavity of mice and observe whether they develop into new larval forms characteristic of the species used.

Dogs and cats may be infected by feeding the larval stages to them. Experimental animals should be wormed to assure freedom from the cestodes by natural infection prior to feeding the larval stages. Development of the worms to sexual maturity will be recognized when gravid proglottids appear in the feces.

## SELECTED REFERENCES

Banerjee, D., and K. S. Singh. (1969). Studies on *Taenia fasciolaris*. I. Studies on the early stages of infection in cysticerciasis in rats. Indian J. Anim. Sci. 39:149–154.

Leiby, P. D., and W. C. Dyer. (1971). Cyclophyllidian tapeworms of wild carnivora. In: J. W. Davis and R. C. Anderson. eds. Parasitic Diseases of Wild Animals. Iowa State Univ. Press, Ames, p. 174.

Singh, B. B., and B. V. Rao. (1967). Some biological studies on *Taenia taeniaeformis*. Indian J. Helminthol. 18:151–160.

Soulsby, E. J. L. (1968). Helminths, Arthropods and Protozoa of Domesticated Animals. 6th ed. Mönnig's Veterinary Helminthology & Entomology. The Williams & Wilkins Co., Baltimore, p. 114.

Tadros, G., R. Iskander, and G. Riad. (1966). Blindness in domestic rabbits in Egypt, U. A. R., due to *Coenuris serialis* (Gervais), with a redescription of this species. J. Arab Vet. Med. Ass. 26:165–180.

Williams, J. F., and C. W. Colli. (1970). Primary cystic infection with *Echinococcus granulosus* and *Taenia hydatigena* in *Meriones unguiculatus*. J. Parasitol. 56:509–513.

## FAMILIES OF CYCLOPHYLLIDEA OTHER THAN TAENIIDAE

Only about a dozen well-known families of this order, other than Taeniidae, are recognized. They are common parasites of birds and mammals. Only the life cycles of species occurring in domestic and a few species of game animals have been studied to any extent.

The infective larvae are most commonly cysticercoids or some variation of them in which the scolex is invaginated in a solid rather than a vesicular body, as in the Taeniidae. A plerocercoid-type larva occurs in the family Paruterinidae and a tetrathyridium-type in the Mesocestoididae. While invertebrates commonly serve as the intermediate hosts, vertebrates also function in this role.

### FAMILY DILEPIDIDAE

A common parasite of this family is *Dipylidium caninum*, the double-pored tapeworm of dogs and cats. It is prevalent in all parts of the world where fleas and biting lice infest these mammals. Infections in dogs can be diagnosed macroscopically by the occurrence on the feces of the active reddish-yellow gravid proglottids, which are about the size and shape of a cucumber seed.

*Dipylidium caninum*
(Linnaeus, 1758) (Plate 93)

*Dipylidium caninum* occurs in the small intestine of dogs, foxes, and cats. It is found occasionally in humans, particularly children.

**Description.**     Adult worms attain a length of 50 cm. The scolex bears an elongated rostellum armed with four to seven rows of rose-thorn shaped hooks. Mature and gravid proglottids resemble cucumber seeds in size and shape. There are two sets of reproductive organs, each opening in a common genital pore, one on each lateral margin of the proglottid. Testes are numerous and fill the space between the excretory canals. The ovaries and vitelline glands are in separate clusters with the latter posterior. Several eggs occur in individual rust-colored capsules.

**Life Cycle.**     Gravid proglottids voided with the feces, or leaving the host spontaneously, disseminate the egg capsules. These may be on the hair of the hosts or in their beds. Biting lice, *Trichodectes canis*, or larvae of the dog flea, *Ctenocephalides canis*, or the cat flea, *C. felis*, eating the eggs, become infected. Each egg consists of a distinct outer and inner layer separated by a thick cytoplasmic one containing lipids, mitochondria, and glycogen. The embryophore, the inner layer surrounding the oncosphere, is composed of two layers of rods lying

at right angles to each other. The oncosphere contains both germinative and somatic cells, six keratinaceous hooks with their associated muscles, and a penetration gland. They hatch in the digestive tract of the insects and the oncospheres migrate into the body cavity. In biting lice, they develop quickly to the infective tailed cysticercoids. Growth in fleas, however, is more prolonged and development is associated with the stages of metamorphosis of the insect host. Oncospheres develop very little in the larval flea, considerable growth takes place during the pupal stage, and final development is completed in adult fleas when the latter begin to take blood meals.

Infection of the definitive host is accomplished when infected adult fleas or biting lice are swallowed. The cysticercoids escape in the intestine and develop directly to adult cestodes in 3 to 4 weeks. Infection of children takes place in the same manner.

### EXERCISE ON LIFE CYCLE

Flea eggs may be obtained by putting an infested dog or cat in a cage under which a pan is placed to catch eggs that drop from the hair. Eggs placed in Petri dishes containing fine moist sand hatch successfully at room temperature. The larvae feed well on dog biscuits and dried blood.

Flea larvae intended for infection are isolated in suitable dishes and starved for 24 hours. Gravid proglottids are teased apart in a small amount of saline to free the egg capsules. Sufficient dried and powdered blood is added to the eggs to form a soft paste which the flea larvae readily eat. After they have ingested the eggs, the larvae must be kept on the moist sand until they become adults. Little development of the oncospheres occurs in the larvae, but considerable development takes place in the pupae, and is completed when the adult fleas begin feeding on the host. Dissection of larvae, pupae, and adults should be made to observe the development of the cysticercoids.

If dogs or cats are to be fed experimentally infected fleas, they should be treated first to remove any tapeworms that might be present in the intestine from natural infections.

Another common Dilepididae readily available and useful for life history studies is *Choanotaenia infundibulum* of chickens. Eggs from

## Explanation of Plate 93

A, Adult cestode. B, Scolex. C, Rose-thorn shaped rostellar hook. D, Mature proglottid. E, Gravid proglottid. F, Individual egg. G, Capsule with eggs. H, Tailed cysticercoid. I, Dog definitive host. J, Flea larva intermediate host. K, Flea pupa. L, Adult flea. M, Adult biting louse intermediary.

1, Scolex; 2, immature proglottids; 3, mature proglottids; 4, gravid proglottids; 5, rostellum; 6, sucker; 7, common genital pore; 8, cirrus pouch; 9, vas deferens; 10, testes; 11, vagina; 12, ovary; 13, vitelline gland; 14, longitudinal excretory canal; 15, egg capsules containing eggs; 16, egg shell; 17, oncosphere; 18, egg capsule; 19, eggs; 20, cysticercoid; 21, tail with some oncospheral hooks still attached.

a, Adult cestode in small intestine; b, gravid proglottid detached from strobilus; c, gravid proglottid outside body; d, egg; e, egg hatching in intestine of flea larva; f, oncosphere migrating through intestinal wall into hemocoel; g, oncosphere grows only slightly in larval flea; h, developing cysticercoids in pupal flea; i, cysticercoids attain full development in imago when it begins to feed on blood from vertebrate host; j, eggs hatch in gut of adult biting louse; k, oncospheres migrate from gut into body cavity; l, fully developed cysticercoid; m, cysticercoids released from insect hosts in stomach of dog; n, cysticercoids migrate to small intestine, attach to the mucosa, and develop to maturity in 3 to 4 weeks.

Figures B, D, F adapted from Faust and Russell, 1957, Craig and Faust's Clinical Parasitology, 6th ed., Lea & Febiger, Philadelphia, p. 632; H adapted from various sources.

---

gravid proglottids develop into cysticercoids when fed to houseflies, beetles, and grasshoppers. Infected arthropod intermediate hosts fed to young chicks produce adult cestodes. For details on procedure and specific intermediate hosts, read the paper by Horsfall and Jones.

## SELECTED REFERENCES

Horsfall, M. W., and M. F. Jones. (1937). The life history of *Choanotaenia infundibulum*, a cestode in chickens. J. Parasitol. 23:435–450.

Marshall, A. G. (1967). The cat flea, *Ctenocephalides felis* (Bouché, 1835) as an intermediate host for cestodes. Parasitology 57:419–430.

Pence, D. B. (1967). The fine structure and histochemistry of the infective eggs of *Dipylidium caninum*. J. Parasitol. 53:1041–1054.

Vernard, C. E. (1938). Morphology, bionomics and taxonomy of the cestode *Dipylidium caninum*. Ann. N. Y. Acad. Sci. 37:273–328.

Zimmermann, H. R. (1937). Life-history studies on cestodes of the genus Dipylidium from the dog. Z. Parasitenk. 9:717–729.

### FAMILY DAVAINEIDAE

Members of this family are parasites of the intestine of birds and mammals throughout the world. Some of them cause serious damage to poultry.

They are small- to medium-sized cestodes. The rostellum is retractable and armed with two to three rows of minute hammer-shaped hooks, and the margins of the suckers likewise may bear tiny hooks. The gravid uterus breaks up into egg capsules which contain one to several eggs each, depending on the species.

### Davainea proglottina
(Davaine, 1860) (Plate 94)

This minute cestode occurs in the duodenum of chickens, other gallinaceous birds, and pigeons in all parts of the world.

**Description.** Mature worms may be 0.5 to 3 mm long, with up to six proglottids. The rostellum is armed with 80 to 94 hooks 7 to 8 $\mu$ long, and the suckers with a few rows of minute thorn-shaped hooks along the margins. The genital pores are regularly alternating and located on the anterior corner of the proglottid. Eggs are distributed singly in capsules throughout the parenchyma of the ripe proglottid.

**Life Cycle.** Eggs released from motile proglottids expelled with the feces hatch when eaten by slugs (*Limax, Arion, Cepaea, Milax,* and *Agriolimax*) and land snails (*Polygyra* and *Zonitoides*). The newly hatched oncospheres migrate from the intestine to the tissues where development into infective cysticercoids requires 3 to 4 weeks during the summer. Fowls acquire the cestodes by eating infected mollusks. Development to sexual maturity in chickens is attained in about 2 weeks.

*Davainea tetraoensis* of ruffed grouse, *Bonasa umbellus*, has a seasonal cycle in Canada. Fully developed specimens occur in the grouse during the summer and poorly developed forms with no ripe proglottids during the winter. Environmental temperatures falling to around 0°C cause

Plate 93    *Dipylidium caninum*                                                                                                361

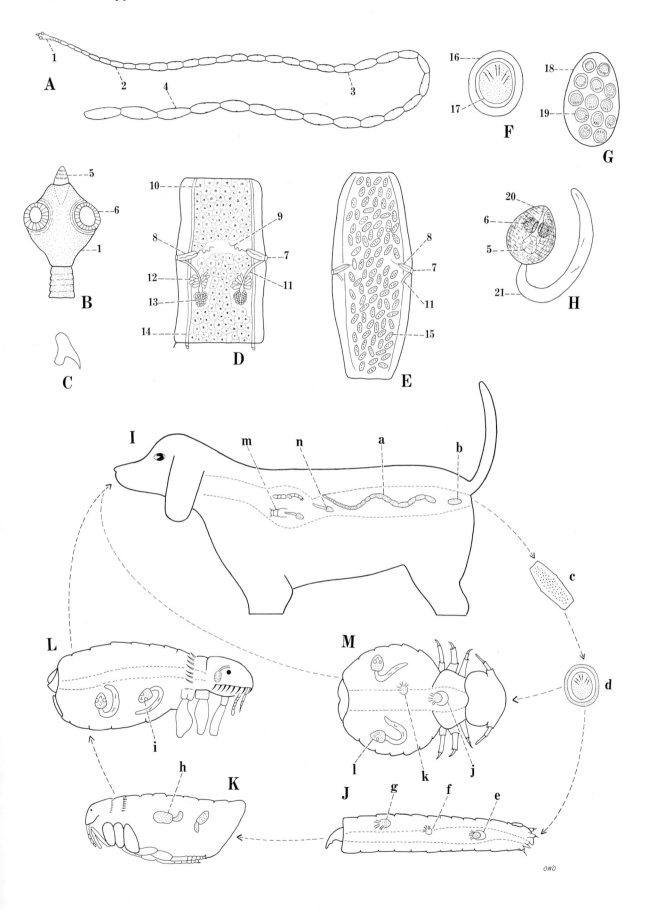

## Explanation of Plate 94

A, Entire adult cestode. B, Scolex. C, Mature proglottid. D, Ripe proglottid. E, Hammer-shaped hook from rostellum. F, Thorn-shaped hook from suckers. G, Hatched oncosphere or hexacanth. H, Invaginated cysticercoid. I, Chicken definitive host. J, Slug (*Arion, Limax, Agriolimax, Milax*) intermediate host.

1, Scolex; 2, spiny sucker; 3, rostellum; 4, rostellar hooks; 5, immature proglottid; 6, mature proglottid; 7, gravid proglottid; 8, common genital pore; 9, cirrus pouch; 10, cirrus; 11, vas deferens; 12, testes; 13, vagina; 14, seminal receptacle; 15, ovary; 16, vitelline gland; 17, embryonated eggs in uterus; 18, hammer-shaped hook from rostellum; 19, thorn-shaped hook from suckers; 20, hooks of oncosphere; 21, cells of developing embryo; 22, wall of cysticercoid.

a, Adult cestode in small intestine; b, gravid proglottid detaches from strobilus and passes to outside with feces; c, gravid proglottid; d, gravid proglottid in feces; e, slugs feeding on feces eat eggs; f, embryonated egg in intestine of slug; g, oncosphere escapes from egg; h, oncosphere migrates through intestinal wall and grows into cysticercoid; i, cysticercoid embedded in body tissues; j, definitive host becomes infected by eating slugs that harbor cysticercoids; k, cysticercoids released from slug by digestive processes; l, cysticercoid everts scolex, attaches to wall of intestine, and grows to maturity (a) in 14 days.

Figures E–G, from Wetzel, 1932, Arch. Wiss. Prakt. Tierheilk. 65:595; others from various sources.

---

the transition from the summer to the winter form of worms. The change back to the summer type is linked with the appearance of reproductive hormones of the grouse in the spring. In the land snail, *Zonitoides arboreus*, development from oncosphere to cysticercoid requires 12 days in small snails up to 2 mm in diameter and 19 days in individuals 4.5 mm or more in size.

## SELECTED REFERENCES

Abbou, A. H. (1958). The life-cycle of *Davainea proglottina* Davaine and relation between the proglottids discharged daily and the number of tapeworms in the domestic fowl. Can. J. Comp. Med. 22:338–343.

Dick, T. A., and M. D. B. Burt. (1971). The life cycle and seasonal variation of *Davainea tetraoensis* Fuhrmann, 1919, a cestode parasite of ruffed grouse, *Bonasa umbellus* (L.). Can. J. Zool. 49:109–119.

Wetzel, R. (1932). Zur Kenntnis des weniggliedrigen Hühnerbandwurmes *Davainea proglottina*. Arch. Wiss. Prakt. Tierheilk. 65:595–625.

Wetzel, R. (1938). Insekten als Zwischenwirte von Bandwürmen der Hühnervogel (Sammelreferat). Z. Hyg. Zool. 30:84–92.

### *Raillietina* Fuhrmann, 1920

The species of this genus are parasites of the small intestine of birds and mammals; at least one species is reported from reptiles. Three of them are of common occurrence in poultry and of cosmopolitan distribution.

In cases where the life cycles are known, ants and many species of beetles of the families Tenebrionidae, Scarabaeidae, and Carabidae serve as intermediate hosts. Houseflies also are intermediaries for some species, becoming infected as adults rather than as maggots.

### *Raillietina* (*Skrjabinia*) *cesticillus* (Molin, 1858) (Plate 95, A–F)

This cestode is very common throughout the world, owing, probably, to the occurrence simultaneously of chickens and many species of intermediate hosts in the same area. They include houseflies (Muscidae), ground beetles (Carabidae), dung beetles (Scarabaeidae), and meal worms (Tenebrionidae).

**Description.** They are usually about 4 cm long but may be up to 13 cm. The scolex is large and bears a broad rostellum armed with 400 to 500 small hooks. The suckers are small and unarmed. There is no distinct neck. Genital apertures are irregularly alternating. Egg capsules scattered throughout the parenchyma contain a single egg.

**Life Cycle.** Ripe, or gravid, proglottids passed in the feces liberate their embryonated eggs. When these are eaten by houseflies (*Musca domestica*), or numerous species of beetles, they hatch in the intestine and the oncospheres migrate into the body cavity. They soon develop to infective cysticercoids. Birds become parasitized by eating the infected insects.

Plate 94     *Davainea proglottina*                                                                  363

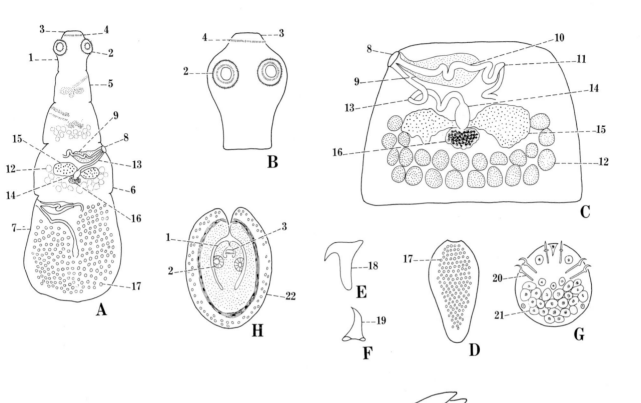

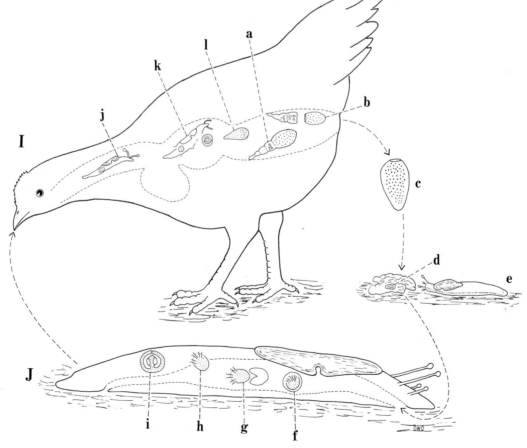

## Explanation of Plate 95

A–F, *Raillietina (S.) cesticillus*. A, Scolex. B, Hook from rostellum. C, Mature proglottid. D, Gravid proglottid. E, Egg with oncosphere. F, Invaginated cysticercoid. G–K, *R. (R.) echinobothrida*. G, Scolex. H, Hooks from rostellum. I, Mature proglottid. J, Gravid proglottid. K, Invaginated cysticercoid (cysticercoids of *R. (R.) echinobothrida* and *R. (R.) tetragona* are similar). L–O, *R. (R.) tetragona*. L, Scolex. M, Hooks from rostellum. N, Mature proglottid. O, Ripe proglottid. P, Fowl definitive host. Q, Ground beetle intermediate host. R, Ant intermediate host. S, Dung beetles (Scarabaeidae), along with ground beetles (Carabidae), and houseflies are intermediate hosts.

1, Scolex; 2, sucker; 3, rostellum with hooks; 4, genital pore; 5, cirrus pouch; 6, vas deferens; 7, testes; 8, vagina; 9, ovary; 10, vitelline gland; 11, uterine pouch or egg capsule; 12, egg shell; 13, oncosphere; 14, body of cysticercoid.

a, Adult cestode in small intestine; b, detached gravid proglottids; c, free ripe proglottid; d, egg; e, motile proglottid in feces discharges egg capsules; f, eggs hatch in intestine of intermediate hosts; g, oncosphere migrates from lumen of intestine to body cavity of intermediate hosts; h, cysticercoid; i, infection of definitive host by eating infected intermediate hosts; j, cysticercoids released from intermediate hosts by digestive processes; k, cysticercoid evaginates and attaches to mucosa of small intestine, and develops to maturity.

Figure E adapted from Reid, Ackert, and Case, 1938, Trans. Amer. Microsc. Soc. 57:65; F, from Ackert and Reid, 1936, Ibid., 55:97; K, from Horsfall, 1938, J. Parasitol. 24:409; all others from Ransom, 1904, 21st Annu. Rep. U. S. Dep. Agr., p. 268.

---

### *Raillietina (Raillietina) echinobothrida* (Megnin, 1880) (Plate 95, G–K)

Like the preceding species, this one is a cosmopolitan parasite infecting the small intestine of chickens.

**Description.**    The body length is up to 25 cm. A small rostellum is armed with about 200 small hooks arranged in two rows and the suckers have 8 to 10 rows of minute hooks. The genital pores usually are unilateral. The uterus fragments into capsules, each containing several eggs.

**Life Cycle.**    Egg capsules released from free proglottids hatch when eaten by ants (*Tetramorium and Pheidole*). Cysticercoids in the body cavity are infective to chickens.

### *Raillietina (Raillietina) tetragona* (Molin, 1858) (Plate 95, L–O)

This is another cosmopolitan parasite of the small intestine of chickens, pigeons, and guinea fowls.

**Description.**    The length of the strobilus is up to 25 cm. The tiny rostellum is armed with about 100 minute hooks and the suckers with 8 to 10 rows of them. The genital pores are unilateral. Egg capsules contain 6 to 12 ova each.

**Life Cycle.**    The cysticercoids develop in houseflies and ants (*Tetramorium* and *Pheidole*). *Raillietina (R.) loeweni* of black-tailed jackrabbits utilizes harvest ants (*Pheidole sitarches campestris* and *P. bicarinata*) as intermediate hosts in Kansas.

### EXERCISE ON LIFE CYCLE

A plentiful supply of representatives of the Davaineidae for experimental studies can be obtained from establishments where farm-raised poultry is processed for market. Species of *Raillietina* are common parasites of columbiformes.

Eggs dissected from the proglottids should be placed in a glass dish with the respective intermediate host. These must be collected from localities where fowls have not been in order that natural infections are not present to complicate the experiments, or they should be reared under controlled conditions in the laboratory.

The eggs hatch shortly after reaching the intestine of the intermediate hosts. Hence the gut of the intermediary should be examined within 1 to 2 hours after the eggs have been eaten, at which time the oncospheres will appear free in the intestinal lumen or be penetrating the epithelium. Mature cysticercoids appear in the tissues of snails and slugs or the body cavity of arthropods in 2 weeks during the warm summer. They may be found by dissecting the intermediate hosts that have eaten eggs.

Baby chicks fed infected mollusks or insects, depending on the species of cestode, pass gravid proglottids in their feces in about 2 weeks.

### SELECTED REFERENCES

Ackert, J. E., and W. M. Reid. (1936). The cysticercoid of the fowl tapeworm *Raillietina cesticillus*. Trans. Amer. Microcs. Soc. 55:97–100.

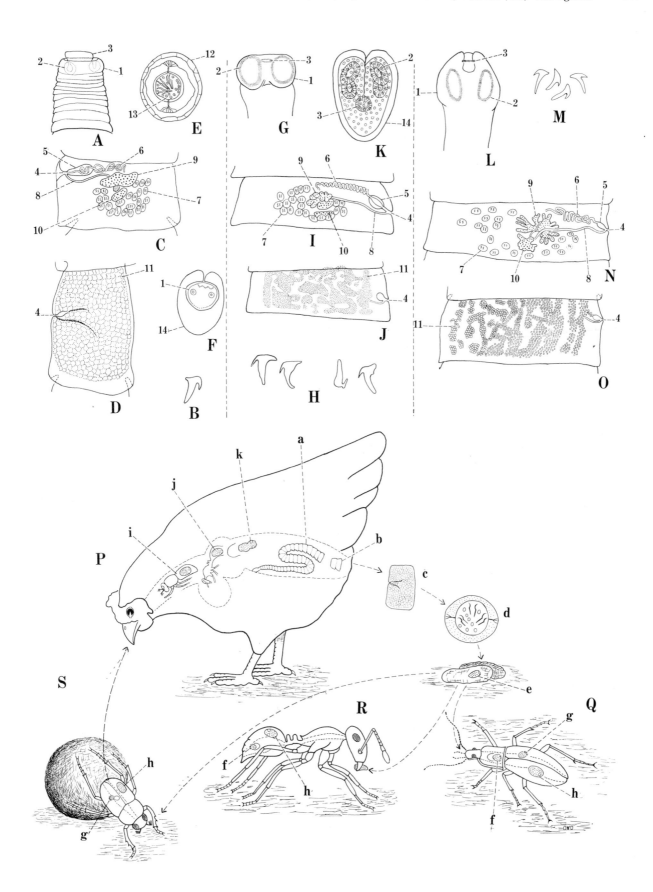

## Explanation of Plate 96

A, Scolex. B, Minute hooks from sucker. C, Immature proglottid. D, Mature proglottid. E, Gravid proglottids. F, Egg. G, Hook of oncosphere. H, Tailed cysticercoid. I, Chicken definitive host. J, Dung beetle (scarabeid) intermediate host. K, Histerid beetle intermediate host. L, Carabid beetle intermediary. M, Stable fly (larva and adult) reported as intermediate hosts but it is not certain that they are.

1, Scolex; 2, sucker; 3, unarmed rostellum; 4, genital pore; 5, long cirrus pouch leading from spherical seminal vesicle; 6, vas deferens; 7, testes; 8, vagina with enlarged seminal receptacle; 9, ovary; 10, vitelline gland; 11, longitudinal excretory canal; 12, transverse excretory canal; 13, gravid uterus containing eggs; 14, egg membranes; 15, oncosphere with hooks; 16, body of cysticercoid; 17, tail; 18, embryonic hooks.

a, Adult cestode in small intestine; b, eggs passing out of intestine; c, egg; d, egg hatching in intestine of beetle intermediary; e, hexacanth migrating through intestinal wall; f, cysticercoid; g, digestion of intermediate host releases cysticercoid; h, cysticercoid excysts, evaginates, and attaches to intestinal mucosa where it develops into an adult cestode in about 3 weeks.

Figures A–G adapted from Ransom, 1902, Trans. Amer. Micros. Soc. 23:151; H, from Jones, 1929, J. Agr. Res. 38:629.

---

Bartel, M. H. (1965). The life cycle of *Raillietina* (*R.*) *loeweni* Bartel and Hansen, 1964 (Cestoda) from the black-tailed jackrabbit, *Lepus californicus melanotus* Means. J. Parasitol. 51:800–806.

Enigk, K., and E. Sticinsky. (1959). Zwischenwirte der Hühnerbandwürmer *Raillietina cesticillus, Choanotaenia infundibulum* und *Hymenolepis carioca*. Z. Parasitenk. 19:278–308.

Horsfall, M. W. (1938). Observation on the life history of *Raillietina echinobothrida* and *R. tetragona*. J. Parasitol. 24:409–421.

Jones, M. F. (1931). On the life histories of species of *Raillietina*. J. Parasitol. 17:234.

Reid, W. M., J. E. Ackert, and A. A. Case. (1938). Studies on the life history and biology of the fowl tapeworm *Raillietina cesticillus*. Trans. Amer. Micosc. Soc. 57:65–76.

### FAMILY HYMENOLEPIDIIDAE

Species of this large family are parasites of birds and mammals throughout the world. Some occur in poultry and others in mice and men.

In general, the strobila is long and slender. Usually there is a rostellum armed with a single row of relatively few, rather large hooks. In most species, there is but a single set of reproductive organs. There are one to four large testes per proglottid. The genital pore is unilateral. A few species have a double set of reproductive organs. The uterus is generally sac-like and filled with embryonated eggs enclosed in three membranes.

The three species considered here represent variations on the basic life cycle. They include one from terrestrial birds with a land insect as an intermediate host, one from waterfowl with an aquatic arthropod intermediary, and one from mice in which the life cycle may be direct either by swallowing eggs or by internal autoinfection, or indirect with the aid of an arthropod intermediate host, as in the case of the first two.

### *Hymenolepis carioca*
(Magalhães, 1898) (Plate 96)

This is one of the commonest cestodes of chickens in the United States.

**Description.**    The thread-like body is 3 to 8 cm long with about 500 proglottids and bears a scolex with an unarmed rostellum, but the suckers are armed with minute hooks. The genital pores are near the middle of the proglottids and located on the lateral margin. The length of the cirrus pouch equals about half the width of the proglottid and is preceded by a large seminal vesicle. There are three testes arranged in a triangle. Eggs are enclosed in three membranes, characteristic of the group.

**Life Cycle.**    When eggs passed in the feces of chickens are eaten by dung beetles (*Anisotarsus terminatus, A. agilis, Aphodius granarius, A. fimetarsus, A. fossor, Choeridium histeroides, Geotrupes silvaticus, Ontophagus hectae, Sphaeridium bipustulatum, S. scarabaeoides*) of the family Scarabaedidae, the hister beetle (*Hister 14-striatus*) of the family Histeridae, and the spider beetle (*Niptus hololeucus*) of the family Ptinidae, they quickly hatch in the intestine and the oncospheres migrate through the intestinal wall into the hemocoel. The tailed cyscticercoids develop to the infective stage in the body cavity in 12 to 18 days. The tail may be contracted and short or extended to several times the length of the cyst proper.

Stable flies and ground beetles (Carabidae)

**Plate 96**    *Hymenolepis carioca*    367

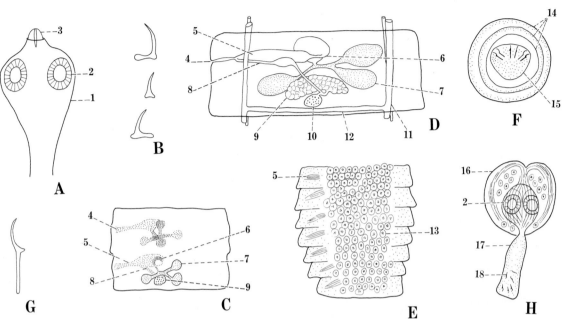

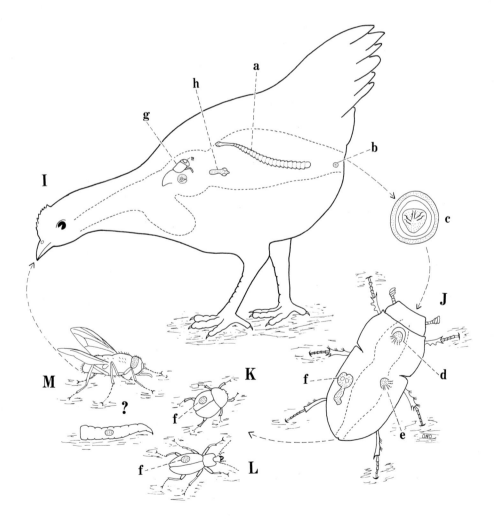

## Explanation of Plate 97

A, Scolex and anterior end of strobilus. B, Hook from rostellum. C, Mature proglottid. D, Gravid proglottid. E, Embryonated egg. F, Invaginated cysticercoid. G, Evaginated cysticercoid. H, Duck and goose definitive hosts. I, Copepod intermediate host.

1, Scolex; 2, armed rostellum; 3, sucker; 4, neck; 5, anterior proglottids; 6, handle; 7, guard; 8, blade; 9, protruding cirrus; 10, cirrus pouch; 11, seminal vesicle; 12, testes; 13, ovary; 14, oviduct; 15, vitelline gland; 16, Mehlis' gland; 17, seminal receptacle; 18, vagina; 19, developing uterus; 20, dorsal longitudinal excretory canal; 21, ventral excretory canal; 22, eggs; 23, common genital pore; 24, egg shell; 25, inner membrane; 26, oncosphere; 27, hooks; 28, body of cysticercoid; 29, tail; 30, oncospheral hooks.

a, Adult cestode in small intestine; b, eggs passing out of gut; c, egg in water; d, egg hatching in gut of copepod; e, oncosphere migrating through intestinal wall; f, cysticercoid in hemocoel; g, digestion of infected copepod releases invaginated cysticercoid; h, evaginated cysticercoid attaches to intestinal wall and grows into a mature cestode.

Adapted from Ruszkowski, 1932, Bull. Int. Acad. Pol. Sci. Cla. Sci. Math. Natur. (II), (1-4): 1–8.

---

have been reported as intermediate hosts but their status is questionable.

Following ingestion of the infected beetles and digestion of them, the cysticercoids are liberated. Further details on this aspect of the life cycle of *H. carioca* are unknown. In the case of *H. exigua* of chickens, however, invaginated cystericercoids are free from their amphipod intermediate hosts (*Orchestia platensis* in Hawaii) in the gizzard in 6 hours and evaginated ones are present in the duodenum by 20 hours. Sexually mature worms are present in the small intestine by 21 days, and possibly sooner.

Young chicks experimentally infected with 1000 cysticercoids of *H. carioca* and kept on an adequate diet developed normally when compared with uninfected controls.

## EXERCISE ON LIFE CYCLE

Adult cestodes can be obtained from almost any place where chickens that have been kept on the ground are slaughtered and dressed. Obtain gravid proglottids and feed them to dung beetles obtained from localities where chickens have not been kept to assure that they are not infected naturally. Dissect some of the infected beetles on alternate days for about 3 weeks to observe the migration and progressive development of the cysticercoids.

At the end of 3 weeks, feed some of the infected beetles to young chicks that have been kept in insect-proof pens to avoid natural infection. Examine the feces for eggs or gravid segments which should appear about 3 weeks after infected beetles have been eaten. Control chicks should be kept.

The report of Guberlet that stable flies serve as intermediate hosts should be reexamined.

## SELECTED REFERENCES

Alicata, J. E., and E. Chang. (1939). The life history of *Hymenolepis exigua*, a cestode of poultry in Hawaii. J. Parasitol. 25:121–127.

Cram, E. B., and M. F. Jones. (1929). Observations on the life histories of *Raillietina cesticillus* and *Hymenolepis carioca* tapeworms of poultry and game birds. N. Amer. Vet. 10:49–51.

Enigk, K., and E. Sticinsky. (1959). Die Zwischenwirte der Hühnerbandwürmer *Raillietina cesticillus*, *Choanotaenia infundibulum* und *Hymenolepis carioca*. Z. Parasitenk. 19:278–308.

Guberlet, J. E. (1929). On the life history of the chicken cestode, *Hymenolepis carioca* (Magalhaes). J. Parasitol. 6:35–38.

Guthrie, J. E., and P. D. Harwood. (1944). Limited tests of mixtures of tin oleate with ammonium compounds for the removal of experimental tapeworm infections of chickens. Proc. Helminthol. Soc. Wash. 11:45–48.

Jones, M. F. (1929). *Aphodius granarius* (Coleoptera), an intermediate host for *Hymenolepis carioca* (Cestoda). J. Agr. Res. 38:629–632.

Jones, M. F. (1929). *Hister* (*Carcinops*) *14-striatus* an intermediate host for *Hymenolepis carioca*. J. Parasitol. 15:224.

Jones, M. F. (1932). Additional notes on intermediate hosts of poultry tapeworms. J. Parasitol. 18:307.

Luttermoser, G. W. (1940). The effect on the growth-rate of young chickens of infections of the tapeworm *Hymenolepis carioca*. Proc. Helminthol. Soc. Wash. 7:74–76.

### *Drepanidotaenia lanceolata*
(Block, 1782) (Plate 97)

These are large Hymenolepidiidae from the small intestine of ducks and geese throughout the world.

**Description.**    The body is lance-shaped,

**Plate 97**  *Drepanidotaenia lanceolata*  369

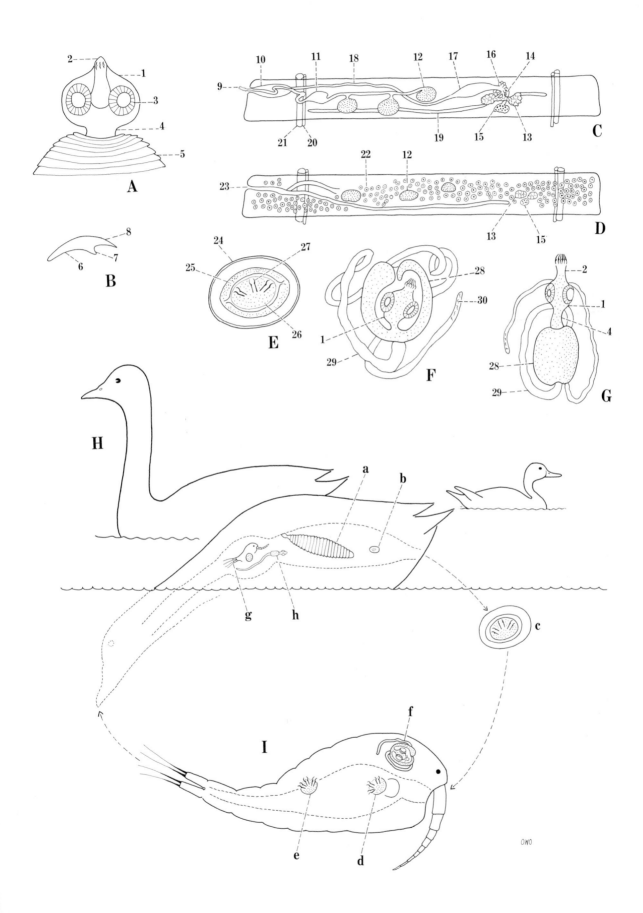

3 to 13 cm long, with very narrow anterior proglottids and posterior ones 5 to 18 mm wide. The scolex is small, the rostellum is armed with eight hooks, and the neck is very short. There are three testes on the poral side arranged in linear fashion across the proglottid. The ovary is aporal from the testes.

**Life Cycle.**    Eggs are laid singly in the small intestine and are passed in the feces. Only those falling in water are available to cyclops. When eaten by *Cyclops strennus* and *Eucyclops serrulatus*, which are especially susceptible to infection, hatching proceeds. The outer covering of the eggshell is removed mechanically in the gut by bursting and splitting. The next two coverings are digested by the cyclops, leaving the oncosphere enclosed in a thin, hyaline membrane. By means of the two pairs of broad, short, lateral hooks, the oncosphere stretches the membrane taut, tears it open with the two long, slender, median hooks, and escapes into the intestine. By means of its hooks and penetration glands, the oncosphere quickly migrates through the gut wall into the hemocoel to develop into a tailed cysticercoid.

In the process of development, a central cavity forms and the body elongates. An extended scolex forms on the anterior end on which four suckers and a rostellum with eight hooks develop. A tail, or the cercomer, up to 1 mm long with its six embryonic hooks, develops on the opposite, or posterior, end. Finally, the scolex invaginates and becomes enclosed in the body of the cysticercoid to form the mature larva, or cercocyst as the tailed forms are sometimes called.

Ducks and geese become parasitized by eating infected cyclops. Upon digestion of the crustaceans, the cysticercoids escape, evert the scolex, attach to the gut wall, and develop to maturity. Carbohydrates are hydrolyzed in the outer microtrichial membrane of the body and absorbed through it. The worms migrate anteriorly in the host's intestine so that the greatest part of the absorptive areas of their bodies are in contact with that part of the gut most favorable for their nutrition and development.

In the case of other species of hymenolepiids occurring in birds and mammals living in an aquatic environment, aquatic invertebrates are most likely to serve as the intermediate hosts, with microcrustaceans being predominant in this role.

**EXERCISE ON LIFE CYCLE**

Adult cestodes may be obtained from the intestine of wild ducks and geese during the hunting season, as well as from domestic birds when available. Feed infective eggs from the posterior proglottids of the tapeworms to copepods; dissect individuals on the day they eat them and examine the intestinal contents for oncospheres, and again at intervals of 2 days thereafter to observe the migration and development of the cysticercoids in the body cavity.

If available, ducklings may be infected by feeding copepods known to contain cysticercoids. Watch for the first occurrence of eggs or proglottids in the feces in order to determine the prepatent period.

**SELECTED REFERENCES**

Crompton, D. W. F., and P. J. Whitefield. (1968). A hypothesis to account for the anterior migrations of adult *Hymenolepis diminuta* (Cestoda) and *Moniliformis dubius* (Acanthocephala) in the intestine of rats. Parasitology 58:227–229.

Kisielieska, K. (1956). [Investigations on the development of larves of *Drepanidotaenia lanceolata* (Bloch) in the intermediate host.] Acta Parasitol. Pol. 3:397–428.

Ransom, B. H. (1904). An account of the tapeworms of the genus *Hymenolepis* parasitic in man, including several new cases of the dwarf tapeworm (*H. nana*) in the United States. Hyg. Lab., U. S. Pub. Health Marine Hosp. Serv. Bull. 18, pp. 1–138.

Ruszkowski, J. S. (1932). Le cycle évolutif du cestode *Drepanidotaenia lanceolata* (Bloch). Bull. Int. Acad. Pol. Sci. Lett. Cl. Sci. Math. Natur. Ser. B (II) (1-4), pp. 1–8.

Taylor, E. W., and J. N. Thomas. (1968). Membrane (contact) digestion in three species of tapeworm *Hymenolepis diminuta, Hymenolepis microstoma* and *Moneizia expansa*. Parasitology 58:535–546.

*Hymenolepis fraterna* Stiles, 1906 (Plate 98)

This small cestode is a parasite of rats and mice in many parts of the world. It is so similar morphologically to *H. nana* of humans that some authors consider them to be physiological strains rather than distinct species. It is chosen for discussion because the life cycle deviates markedly in certain aspects from that normally encountered among other hymenolepiid cestodes.

**Description.**    These small cestodes measure 7 to 80 mm long. The retractable rostellum is armed with 16 to 18 hooks (24 to 26 according to some authors). The genital pores are unilateral and on the left side of the proglottid. The three testes are arranged in a transverse line and separated by the ovary so that one testis is poral and two aporal.

**Life Cycle.** The life cycle of this species is perhaps the most complicated of all the cyclophyllidean cestodes. It is either direct, being completed without an intermediate host or indirect in which case transmission is through an intermediate host, including widely separated phylogenetic groups.

There are two aspects of the direct type of life cycle. One is by ingestion of eggs passed in the feces of infected animals. The other is by internal autoinfection, where eggs hatch in the gut without leaving it.

Eggs swallowed by mice hatch in the intestine. Some of the liberated oncospheres enter the villi of the intestine where development proceeds to completely formed cysticercoids. These break out of the villi, attach to the intestinal mucosa, and develop into strobilate, sexual worms. Other oncospheres enter the lacteals and blood vessels to be transported to the various organs. In the lymph nodes, liver, lungs, and pancreas, they lodge and develop into cysticercoids. It is presumed that the cysticercoids when mature migrate from these locations back to the small intestine to develop into adult worms.

Eggs laid by worms in the gut may hatch *in situ*. The oncospheres migrate to the villi and visceral organs where they develop in the same manner as described above for the cases of oral infection with eggs. Mice infected with a single cysticercoid each soon harbor an adult tapeworm. If these animals are caged individually so as to prevent infection by swallowing eggs from their own tapeworm, cysticercoids appear in the tissues followed by additional adult worms in the intestine. This is evidence of autoinfection.

Eggs injected intraperitoneally in mice hatch and the oncospheres migrate to the viscera and develop into cysticercoids. Obviously intestinal enzymes are not required for hatching.

In the indirect cycle, beetles function as the common natural intermediate host but mice harboring cysticercoids are capable of serving in this capacity. When eggs are ingested by beetles such as the confused flour beetle *Tribolium confusum* and the meal worm *Tenebrio molitor*, the outer shell is broken by the mandibles of the feeding beetles. In the midgut, the middle layers of the shell are digested, leaving the oncosphere enclosed in the thin, inner membrane. By means of its lateral hooks, the oncosphere pulls the membrane taut and tears it open with the median hooks, and escapes through the rent into the lumen of the midgut, all within a few minutes. By tearing with its hooks, muscular movement, and dissolving tissues by secretions from the penetration glands, the oncospheres migrate through the gut wall into the hemocoel. Having completed the migration, the penetration glands are depleted of their contents and the hook muscles of their glycogen. Development of the cysticercoids in the body cavity of *Tribolium confusum* proceeds through five major stages. They are 1) appearance of the oncospheres in the body cavity during hours 12 to 48; 2) formation of the central cavity and enlargement of an ovoidal body in hours 48 to 72; 3) rapid elongation of the anterior part of the body, forming the future scolex with rudiments of suckers and rostellum and the combined body and tail during hours 72 to 96; 4) withdrawal of the scolex into the central cavity of the body, forming a double wall of infolded tissue during hours 96 to 120; and 5) completion of the scolex and membrane development with complete maturation to infectivity of the cysticercoid which is during hours 100 to 140.

Infection of the definitive host occurs when cysticercoid-bearing beetles are eaten, digested, and the parasites liberated. Following evagination of the scolex, the worm attaches to the mucosal epithelium and grows to sexual maturity in a period of about 10 days. Senescence begins after 14 days of productivity, and then they begin dying.

Lymph nodes from mice harboring adult cestodes fed to clean mice produce intestinal infections with strobilate egg-producing worms. Thus mice may serve as both a definitive and an intermediate host.

Natural infections with adults of *H. fraterna* occur in colonies of mice, rats, chinchillas, and gerbils. Immunity develops that impedes development of the worms and precludes reinfection. Severe and prolonged exertion such as fighting among male mice and extreme nutritional deficiencies break down acquired immunity and render mice susceptible to reinfection.

Eggs whose shells have been broken by shaking with glass beads develop into normal

## Explanation of Plate 98

A, Scolex. B, Rostellar hook. C, Mature proglottid. D, Gravid proglottid. E, Embryonated egg. F, Cysticercoid. G, Cysticercoid in villus of intestine. H, Mouse definitive host. I, Larval intermediate beetle host (Tenebrio molitor, Tribolium confusum.) J, Adult beetles.

1, Scolex; 2, sucker; 3, armed rostellum; 4, handle of hook; 5, guard; 6, blade; 7, common genital pore; 8, cirrus pouch; 9, testes; 10, vagina; 11, seminal receptacle; 12, ovary; 13, oviduct; 14, vitelline gland; 15, longitudinal excretory canal; 16, gravid uterus; 17, shell or outer membrane of egg; 18, thick inner membrane with terminal filaments; 19, oncosphere; 20, tail of cysticercoid; 21, oncospheral hooks; 22, intestinal villus; 23, cysticercoid embedded in villus.

a, Adult worm in small intestine; b, gravid proglottid detached from strobilus; c, egg in feces passing out of intestine; d, egg free; e, egg swallowed and returned to small intestine; f, egg hatches in small intestine; g, oncosphere with penetration glands burrows into intestinal wall; h, cysticercoid develops in villus; i, cysticercoid breaks out of intestinal wall and evaginates; j–l, a, cysticercoid attaches to intestinal wall and grows to sexually mature cestode; m, some infective eggs do not leave the body, but hatch in the intestine, initiating internal autoinfection in which development proceeds as in e–l; n, eggs passed in feces are infective to arthropod intermediate hosts; o, eggs swallowed by larva of beetle or flea hatch in intestine; p, oncospheres enter hemocoel and develop into cysticercoids; q, eggs swallowed by adult beetles hatch in intestine; r, oncospheres migrate from intestine into hemocoel; s, tailed cysticercoids develop; t, infected beetles swallowed by definitive hosts; u, cysticercoids released from beetle; v, cysticercoids evaginate and shed tail; j–l, a, cysticercoids develop to adult worms.

Figures A–E adapted from Stiles, 1903, Hyg. Lab., U. S. Pub. Health Serv. Marine Hosp. Bull. 25; F, from Voge and Heyneman, 1957, Univ. Calif. Publ. Zool. 59:549; G, from Dr. W. S. Bailey, Auburn University, Alabama.

---

cysticercoids when injected subcutaneously or intramuscularly into mice, rats, rabbits, guinea pigs, golden hamsters, Peromyscus maniculatus, Microtus californicus, canaries, and lizards. No development occurs in chickens, turtles, amphibians, and fish.

Hymenolepis nana of humans in some parts of the world occurs in great numbers in some individuals. It is believed that such heavy infections result from internal autoinfection.

## EXERCISE ON LIFE CYCLE

These parasites are sufficiently common in rats and mice in most regions to provide material for life history studies. The direct cycle and that involving an intermediate host are preferred for general studies.

Gravid proglottids fed to flour beetles (Tribolium confusum) result in infections. Cysticercoids are developed in 2 to 3 weeks at room temperature. Examine the intestinal contents of beetles 4 to 5 hours after ingestion of eggs for the presence of newly hatched oncospheres and again at intervals of 1 or 2 days thereafter to observe the development of the cysticercoids in the hemocoel.

Infect mice with cysticercoids originating from worms taken from mice, and likewise rats with those from rats. Also try cross infection of rats with cysticercoids originating from mice and vice versa to determine whether differences of susceptibility occur with the use of different physiological strains of cestodes. Watch for eggs in the feces to ascertain the prepatent period.

Feed gravid proglottids directly to mice to establish infections. Examine the intestinal contents of a mouse for oncospheres 4 to 5 hours after it has eaten eggs. Examine the intestinal mucosa at intervals of two days to observe 1) the development of the cysticercoids; 2) the time they enter the lumen of the intestine, and attach to the mucosa; and 3) the course and rate of development to adults. Sections should be employed for the first two steps.

## SELECTED REFERENCES

Astafiev, B. A. (1966). [New data on the migration of eggs and larvae of Hymenolepis nana in white mice.] Med. Parazitol. 35:149–154.

Collin, W. K. (1969). The cellular organization of hatched oncospheres of Hymenolepis citelli (Cestoda: Cyclophyllidea). J. Parasitol. 55:149–166.

DiConza, J. J. (1968). Hatching requirements of dwarf tapeworm eggs (Hymenolepis nana) in relation to extraintestinal development of larval stages in mice. Z. Parasitenk. 31:376–381.

Garkavi, B. L. (1958). [Mice-intermediate host of Hymenolepis fraterna (Stiles, 1906).] Papers on

Plate 98    *Hymenolepis fraterna*                                                                    373

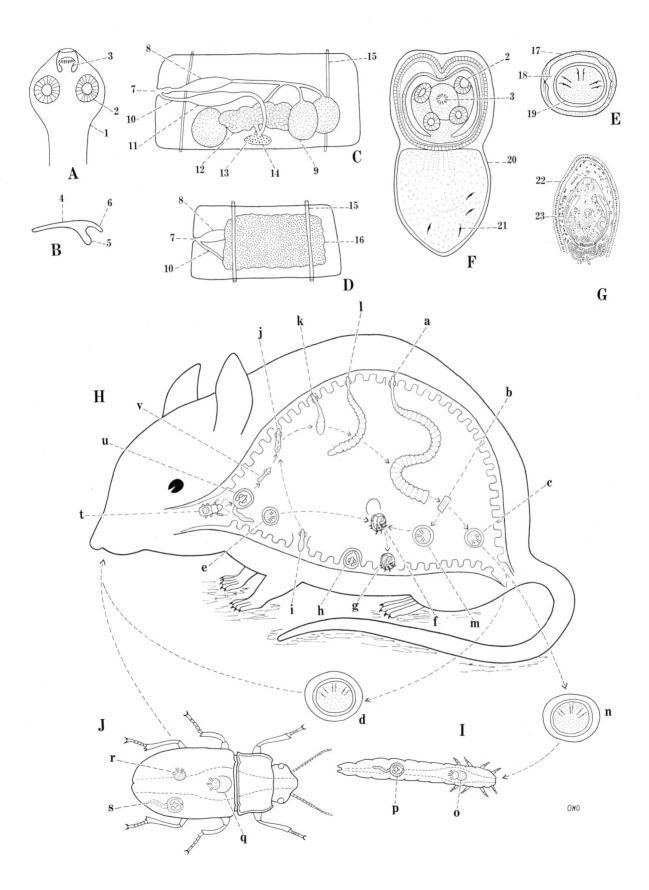

Helminthology presented to Academician K. I. Skrjabin on his 80th Birthday. Moscow. Izdatelstov Akad. Nauk SSSR, pp. 93–95.

Garkavi, B. L., and I. Y. Glebova. (1957). [Development of *Hymenolepis fraterna* (Stiles, 1906) and *Hymenolepis nana* (Siebold, 1852) in the organs of white mice.] Zool. Z. 36:986–991.

Heyneman, D. (1959). Experimental autoinfection of *Hymenolepis nana* in isolated mice restrained from coprophagy. J. Parasitol. 45(4/2):25–26.

Hunninen, A. V. (1935). Studies on the life history and host-parasite relationships of *Hymenolepis fraterna* (*H. nana* var. *fraterna* Stiles) in white mice. Amer. J. Hyg. 22:414–443.

Hunninen, A. V. (1936). An experimental study of internal autoinfection with *Hymenolepis fraterna* in white mice. J. Parasitol. 22:84–87.

Lethbridge, R. C. (1971). The hatching of *Hymenolepis diminuta* eggs and penetration of the hexacanth in *Tenebrio molitor* beetles. Parasitology 62:445–456.

Lussier, G., and F. M. Lowe. (1970). Natural *Hymenolepis nana* infection in Mongolian gerbils (*Meriones unguiculatus*). Can. Vet. J. 11:105–107.

Namitokov, A. A. (1965). [Functions of embryonic hooks of the embryo of *Hymenolepis nana*.] Med. Parazitol. 34:354–355.

Olsen, O. W. (1950). Natural infection of chinchillas with the mouse tapeworm, *Hymenolepis nana* var. *fraterna*. Vet. Med. 440–442.

Schiller, E. L. (1959). Experimental studies on morphological variation in the cestode *Hymenolepis*. I. Morphology and development of the cysticercoid of *Hymenolepis nana* in *Tribolium confusum*. Exp. Parasitol. 8:91–118.

Schiller, E. L. (1959). Experimental studies on morphological variation in the cestode genus *Hymenolepis*. II. Growth, development and reproduction of strobilate phase of *H. nana* in different mammalian species. Exp. Parasitol. 8:215–235.

Voge, M., and D. Heyneman. (1957). Development of *Hymenolepis nana* and *Hymenolepis diminuta* (Cestoda: Hymenolepididae) in the intermediate host *Tribolium confusum*. Univ. Calif. Publ. Zool. 59:549–579.

Weiman, J. C. (1968). Larval development of *Hymenolepis nana* (Cestoda) in different classes of vertebrates. J. Parasitol. 55:1141–1142.

Weiman, J. C., and A. H. Rothman. (1967). Effects of stress upon acquired immunity to the dwarf tapeworm, *Hymenolepis nana*. Exp. Parasitol. 21:61–67.

## FAMILY ANOPLOCEPHALIDAE

The members of this family are medium-sized to relatively large cestodes with the scolex devoid of a rostellum and hooks. The proglottids have one or two sets of reproductive organs with the genital pores located on the lateral margins. They are parasites of reptiles, birds, and mammals.

### *Moniezia expansa*
(Rudolphi, 1810) (Plate 99)

This is one of the largest members of the family and occurs commonly in the small intestine of sheep and cattle in most parts of the world.

**Description.**    The length is up to 600 cm and the width 1.6 cm. Proglottids are wider than long and have a double set of reproductive organs in each one. The testes are numerous and scattered throughout the proglottid. A single transverse row of hollow interproglottid glands lies at the posterior border of each proglottid. The eggs have a pyriform apparatus.

**Life Cycle.**    Fully embryonated eggs of the tapeworm passed in the feces or released from chains of ripe proglottids voided by sheep and cattle abound on pastures where these animals graze. Eggs in feces survive less than a day and exposed ones die sooner. None is able to survive over winter on pastures. Survival of the embryos is dependent on the eggs being eaten by one of a number of species of soil mites. At least 18 genera and more than 25 species of mites from eight families are known as intermediate hosts of anoplocephaline cestodes.

The mites live in the first inch of the turf and humus where they hide during the day. In the twilight hours, they come out, crawling over the soil and grass in search of food at the same time cattle and sheep are grazing. Concurrent searching for food by both mite intermediate host and ruminant definitive host provides ideal ecological conditions for perpetuation of the tapeworms at maximum levels. The most common species of the family Oribatidae involved are *Galumna virginiensis*, *G. emarginata*, and *Scheloribates laevigatus*. It is estimated that 6 to 9 million mites of each of these species occur per acre of permanent pasture. On such pastures, 3 to 4 percent of the mites are naturally infected with 4 to 13 cysticercoids each, making an average of 400,000 per acre. This figures 4 to 7 cysticercoids per 100 mites. A grazing sheep may swallow 1200 mites with each pound of grass eaten.

When eggs are eaten by foraging mites, they hatch in the intestine and the oncospheres migrate to the hemocoel where they develop to cysticercoids. Development and growth of cysticercoids in mites (*Liacarus itascensis*) is known best for *Monoecocestus americanus* and *M. variabilis*, related tapeworms from American porcupines. Upon arriving in the hemocoel, the oncospheres undergo growth, leading to a fully formed cysticercoid. The rate is dependent upon the ambient temperature. Development follows a series of definite stages. They are 1) appearance of oncospheres in the hemocoel and slight growth by elongation, becoming 18 by 26 $\mu$ in size; 2) formation of a morula 40 by 50 $\mu$ in which the surrounding membrane begins to appear; 3) development of the blastula 150 $\mu$ in diameter with a thicker membrane and capable of feeble movement; 4) the anterior wall of the blastula thickens to produce the future scolex and the thinner posterior wall of the cercomer; the length is about 300 $\mu$; 5) the cercomer completes development and constricts at the base; muscle fibers appear beneath the thickened cuticle; 6) the primordium of the scolex appears preparatory to invagination, flame cells appear, and the cyst wall is well developed; 7) invagination of the scolex by a complex process is completed; and 8) maturation of the cysticercoid which requires 50 to 80 days at room temperature and 80 to 100 days under natural conditions (Fig. 29).

In mites infected during November, the oncospheres do not develop to mature cysticercoids until the following July, about 8 months later. Those infected in July produce cysticercoids in 66 to 69 days. The environmental temperature is a major factor controlling the rate of development. *Moniezia expansa* and *M. benedeni* have a double-peaked infective period, one in the spring and the other in the autumn.

Other species of anoplocephaline cestodes where studied utilize mites as intermediate hosts in their development. Some of them are *Bertiella studeri* of primates; *Cittotaenia ctenoides* and *C. denticulata* of rabbits; *Anoplocephala perfoliata*, *A. magna*, and *Paranoplocephala mamilliana* of equines; *Moniezia benedeni* of ruminants; *Thysaniezia giardi* of sheep, and *Monoecocestus americanus* and *M. variabilis* of porcupines.

## EXERCISE ON LIFE CYCLE

Adult cestodes for experimental studies are available in abundance from abattoirs where sheep and lambs are slaughtered. Soil mites of the genera *Galumna*, *Scheloribates*, and *Scutovertex* may be obtained from sod placed in a modified Berlese funnel with a 60-watt lamp hanging over it. The heat from the light drives the mites through the soil whence they drop into the funnel and fall into the container below.

Mites placed in a glass container together with eggs of *Moniezia* eat them and become infected. Dissection of the mites and examination of the intestinal contents during the early hours of the experiment will show empty eggshells and free oncospheres. Examination of the hemocoel of mites at intervals of 1 or 2 days will reveal stages of cysticercoids during the course of development.

The life cycles of other anoplocephalids may be examined in a similar manner. Some common ones include *Cittotaenia* from wild rabbits and hares (Stunkard) and *Monoecocoestus* from porcupines (Freeman). Field mice (*Microtus*) har-

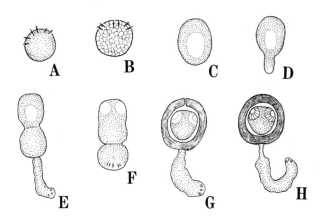

**Fig. 29.** Stages in the development of the cysticercoid of *Monoecocestus americanus* and *M. variabilis* of porcupines in the mite *Liacarus itascensis*. A, Stage 1 of oncosphere in hemocoel of mite; B, Stage 2, the morula; C, Stage 3, a blastula showing thick-walled anterior end of future scolex and thin-walled posterior end of future cercomer; D, Stage 4, the elongation phase indicating future parts; E, Stage 5, the segmented form showing future scolex, body, and well-developed cercomer; F, Stage 6, pre-invagination stage with suckers forming; G, Stage 7, invagination phase, with invagination canal still extant; and H, Stage 8, which is fully developed, darkened, infective cysticercoid. Adapted from photomicrographs by Freeman, 1952, J. Parasitol. 38:111.

## Explanation of Plate 99

A, Scolex. B, Mature proglottid. C, Egg. D, Suckered cysticercoid. E, Segmenting cysticercoid. F, Invaginated cysticercoid. G, Mature cysticercoid. H, Sheep definitive host. I, Oribatid mite intermediate host.

1, Scolex; 2, sucker; 3, immature proglottid; 4, common genital pore; 5, cirrus pouch; 6, vas deferens; 7, testes; 8, vagina; 9, seminal receptacle; 10, ovary; 11, vitelline gland; 12, longitudinal excretory canal; 13, interproglottid glands; 14, eggshell (outer membrane); 15, pyriform apparatus; 16, oncosphere; 17, scolex of cysticercoid; 18, membrane surrounding scolex; 19, segmenting tissue to form surrounding membrane; 20, tail.

a, Adult cestode in small intestine of sheep; b, proglottid passing to outside with feces; c, free proglottids; d, egg released from free proglottids; e, egg eaten by oribatid mite hatches in intestine and the oncosphere migrates to the body cavity; f, cysticercoid encysted in body cavity; g, infected mite swallowed by sheep; h, cysticercoid escapes from digested mite; i, cysticercoid evaginates, loses tail, and attaches to intestinal mucosa where it develops to a mature cestode in about 5 to 6 weeks; adults persist in a sheep for about 3 months after attaining sexual maturity.

Figures D–G adapted from Potemkin, 1948, in Spasskii, 1951, Osnovy tsestodologii (Fundamentals of Cestodology), Vol. 1, Fig. 81.

---

bor a related genus (*Paranoplocephala*) whose life cycle has not been studied.

## SELECTED REFERENCES

Edney, J. M., and G. W. Kelly, Jr. (1953). Some studies on *Galumna virginiensis* and *Moniezia expansa* (Acarina, Oribatoidea; Cestoda: Anoplocephalidae). J. Tenn. Acad. Sci. 28:287–296.

Freeman, R. S. (1949). The biology and life history of *Monoecocestus* Beddard, 1914 (Cestoda: Anoplocephalidae) from the porcupine. J. Parasitol. 38: 111–129.

Kates, K. C., and C. E. Runkel. (1948). Observations on oribatid mite vectors of *Moniezia expansa* on pastures, with a report of several new vectors from the United States. Proc. Helminthol. Soc. Wash. 15:10, 19–33.

Krull, W. H. (1939). On the life history of *Moniezia expansa* and *Cittotaenia* sp. (Cestoda: Anoplocephalidae). Proc. Helminthol. Soc. Wash. 6:10–11.

Kuznetsov, M. I. (1959). [Survival of *Moniezia* eggs on pastures in the Lower Volga steppes.] Byull. Nauch. Tekh. Inform. Vses. Inst. Gelmintol. im K. I. Skrjabin, (5):48–51.

Kuznetsov, M. I. (1970). [Development times of *Moniezia* cysticercoids in *Scheloribates laevigatus* under natural conditions.] In: E. M. Bulanova-Zakhvatkina *et al.*, eds. [Oribatids and Their Role in the Process of Soil Formation.] Vilnius, Akad. Nauk Litovskoi SSR, pp. 223–227.

Prokopic, J. (1967). Bionomische Studien über Bandwürmer der Gattung *Moniezia*. Angew. Parasitol. 8:200–209.

Stunkard, H. W. (1934). Studies on the life history of anoplocephaline cestodes. Z. Parasitenk. 6:481–507.

Stunkard, H. W. (1938). The development of *Moniezia expansa* in the intermediate host. Parasitology 30:491–501.

Plate 99    *Moniezia expansa*                                                                          377

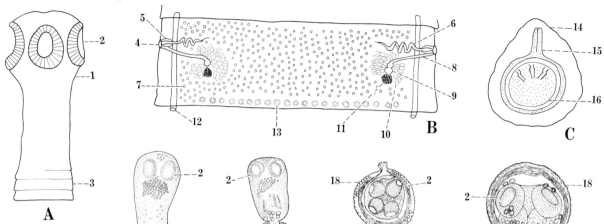

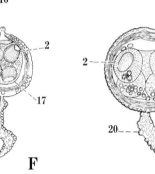

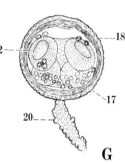

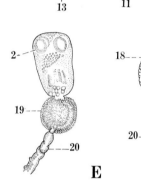

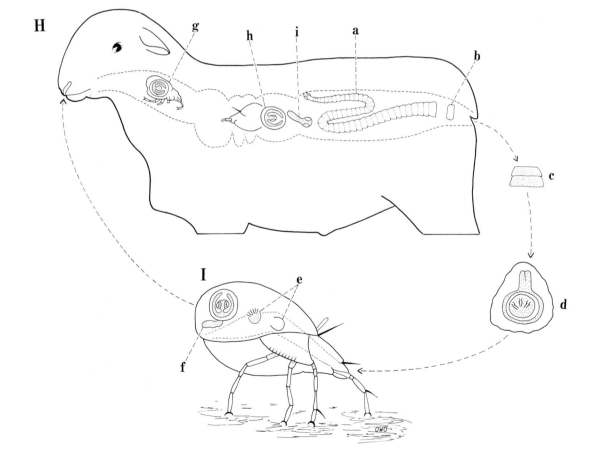

# IV

## Phylum Acanthocephala

This phylum consists of a small group of unique, cylindrical, unsegmented, parasitic worms characterized by a thorny retractable proboscis at the anterior end of the body as the most obvious and distinguishing feature. As adults, they are parasites of vertebrates. Some characteristic morphological features are presented on Plates 101 and 102.

The body is of two parts, the presoma, consisting of the proboscis and neck, and the hindbody, or trunk (Plate 101, B). The anterior part of the presoma consists of a globular or cylindrical proboscis that is almost universally covered with hooks (Plate 102, B, C). The size of the hooks varies according to their location on the proboscis, those near the base being the smallest. The arrangement of the hooks over the prosboscis is in three basic patterns. They are 1) alternating radial rows (Plate 102, B, C), 2) radial rows (Plate 102, I), and 3) horizontal rows (Plate 101, T). The last arrangement may be interpreted as spiral. In some cases a few inordinately large hooks appear, disrupting the characteristic pattern of distribution in the area around them (Plate 102, B). In some cases, it is difficult to interpret the pattern of arrangement

of the hooks. The posterior part of the presoma, which is usually unarmed, is the neck and connects with the trunk (Plate 102, C–H). The entire proboscis is retractable by special muscles into a thick-walled sac, the proboscis receptacle (Plate 101, A, Q). It is everted hydrostatically by fluids under pressure by body contraction.

The trunk consists of the large part of the body posterior to the presoma (Plate 101, A, B). It is essentially a hollow structure containing the reproductive, excretory, and nervous systems, as well as the pseudocelomic fluid (Plates 101, I; 102, H).

Acanthocephalans, like cestodes, have no alimentary tract and therefore must absorb their nutriments through the body wall. In addition to absorption through the wall, certain metabolic activity takes place in it.

Morphologically, the body wall proper consists of five layers (Fig. 30, E, C, S, F, R). Beneath these are two layers of muscles (cm, lm), each being supported by fibrous connective tissue (ct).

The epicuticle (E) is the outermost layer, consisting of a thin, irregular deposit of acid polysaccharide. The cuticle (C), formed of stabilized lipoprotein, is triple-layered and perforated by numerous pores (p), which are the openings of the canals (c) of the third, or striped, layer (S). This layer is of a homogeneous nature; the cuticular lined canals pass through it and extend into the fourth, or felt, layer (F), which consists of many hollow, fibrous strands (f). In addition, there are mitochondria (m), vesicles (v) which may be extensions of the canals (c) or sections of smooth endoplasmic reticulum, and thin-walled lacunar canals (lc). The fifth and deepest layer of the wall proper is the radial layer (R). It contains fewer fibrous strands but more and larger lacunar canals than the felt layer. Mitochondria are plentiful. Nuclei of the body wall are in this layer (Plate 102, G). The limiting plasma membrane (pm) at the inner side of the radial layer is abundantly folded. The ends of some of the folds contain what appear to be droplets of lipid.

The layer inward from the radial one consists of basement membranes (b), fibrous connective tissue layers (ct) bounding the circular (cm) and longitudinal (lm) muscle bands. Numerous roughened endoplasmic reticula (e) occur.

The physiology of the body wall is indicated by its structure. The porous nature of the cuticle suggests an absorptive function of nutrients from the lumen of the host's gut or from its microvilli when in contact with them. The canals in

## Plate 100

the striped layer serve as conduits for nutrients to the felt layer and into the lacunar canals for general transport. The matrix of the striped and felt layers probably functions as skeletal material. The radial layer, with its many mitochondria and deeply folded plasma membrane containing lipid droplets, is believed to be the most active part of the body wall in which absorbed compounds are metabolized. The folded plasma membrane may function in water and ion transportation.

The body wall of Acanthocephala is very different from that of the Trematoda, Cestoda, and Nematoda, thereby suggesting no phylogenetic relationship.

A lacunar system of dorsal and ventral, or lateral, canals extends the length of the body and are connected by a network or transverse system of small canals (Plates 101, A, B, S; 102, A, D, J, K, H).

An elongated lemniscus attached to the anterior wall of the body at each side of the proboscis receptacle hangs free in the pseudocoel, or body cavity (Plate 101, A, B, Q). Extending obliquely and posteriorly from the posterior end of the proboscis receptacle to the body wall where they attach are two ventral and one dorsal retractor muscles (Plate 101, A). The gonads are suspended in the pseudocelomic fluid by a ligament arising from the base of the proboscis

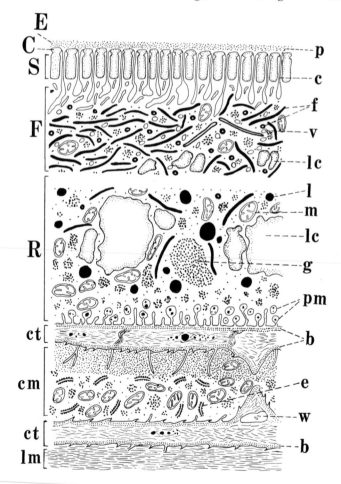

Fig. 30.    Fine structure of the body wall of *Polymorphus minutus*. E–S, body layers. E, Epicuticle. C, Cuticle. S, Striped layer. F, Felt layer. R, Radial layer. b, basement membrane; c, canals; cm, circular muscle layer; ct, connective tissue; e, endoplasmic reticulum; f, hollow fibrous strands; g, glycogen; l, lipid; lc, lacunar canal; lm, longitudinal muscle layer; m, mitochondrion; p, pores of canals in striped layer; pm, folded plasma membrane; v, vesicles, possibly smooth endoplasmic recticula or extensions of canals; w, wandering cell. Adapted from Crompton and Lee, 1965, Parasitology 55:357.

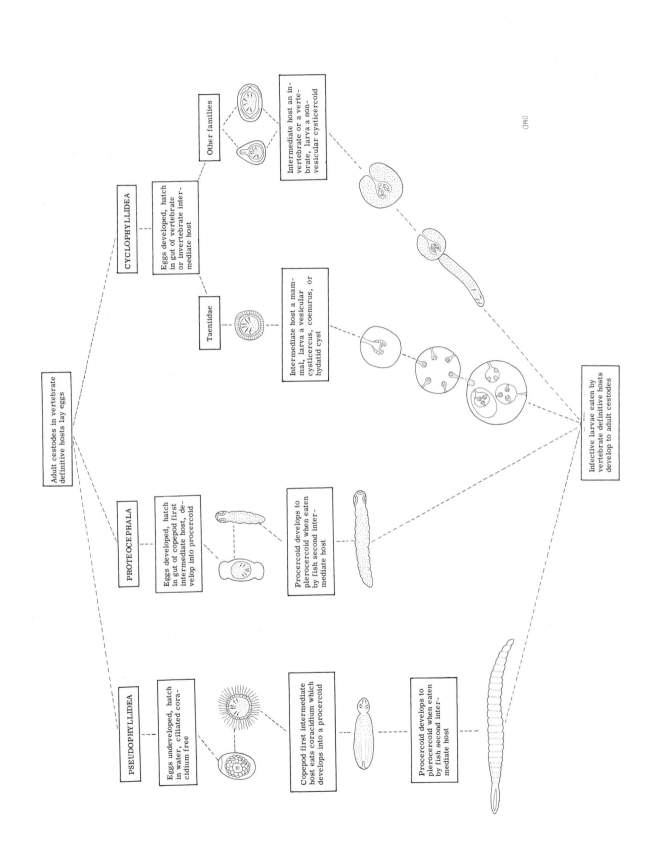

## Explanation of Plate 101

A–K, General morphological characteristics of Acanthocephala. A, Partial dissection of adult female *Neoechinorhynchus rutili*. B, Adult male of *Acanthocephalus ranae*. C–H, Six genera of Acanthocephala, showing relative development of proboscis and neck. C, *Neoechinorhynchus*. D, *Acanthocephalus*. E, *Polymorphus*. F, *Eocollis*. G, *Pomphorhynchus*. H, *Filicollis*. I, Cross section of body of *Filisoma fidium* in region of double-walled receptacle of proboscis. J, Portion of section of body wall of *Centrorhynchus pinguis*. K, Portion of surface view of body wall of *Quadrigyrus torquatus*, showing lacunar canals and giant nucleus. L–N, Eggs typical of acanthocephalans whose hosts are aquatic animals. L, *Corynosoma strumosum* from seals. M, *Echinorhynchus gadi* from cod. N, *Polymorphus* from ducks. O, P, Eggs typical of acanthocephalans whose hosts are terrestrial. O, *Moniliformis* from shrew. P, *Pachysentis* from fox. Q, R, *Corynosoma villosum* (Echinorhynchidea) from Stellar's sea lion, showing the major internal organs. Q, Male. R, Female. S–W, Basic morphological features of the order Gigantorhynchidea. S, Lacunar system of *Moniliformis*, showing dorsal longitudinal canal and encircling branches that connect dorsal and ventral longitudinal canals. T, Proboscis of *Oncicola*, showing concentric rows of hooks. U, *Hamanniella*, showing protonephridia of male. V, *Oligacanthorhynchus* female reproductive system from left side, showing dorsal and ventral ligament sacs. W, Cement glands, showing individual uninucleate glands.

1, Proboscis armed with hooks; 2, neck; 3, trunk or hindbody; 4, proboscis sheath; 5, dorsal retractor muscle of proboscis sheath; 6, ventral retractor muscles of proboscis sheath; 7, retractor muscles of anterior end of worm; 8, lemniscus with nucleus; 9, dorsal longitudinal lacunar canal with branches; 10, ventral longitudinal lacunar canal; 11, lateral longitudinal lacunar canal with branches; 12, floating ovary; 13, eggs; 14, adjacent walls of dorsal and ventral ligaments; 15, dorsal ligament sac; 16, ventral ligament sac; 17, uterine bell; 18, pockets of uterine bell; 19, opening from uterine bell into dorsal ligament sac; 20, duct from uterine bell to uterus; 21, uterus; 22, exit from uterine bell to body cavity; 23, vulva; 24, vagina; 25, testis; 26, vas efferens; 27, ejaculatory duct; 28, penis; 29, copulatory bursa; 30, cement glands; 31, cement reservoir; 32, Saefftigen's pouch; 33, genital sheath; 34, suspensory ligament; 35, giant nucleus of body wall; 36, cuticle; 37, subcuticle; 38, circular canal of lacunar system; 39, body musculature; 40, body cavity; 41, acanthor; 42, protonephridium; 43, excretory bladder.

Figures A, B, S, V redrawn from Meyer, 1933, Bronn's Klassen und Ordnung des Tierreichs 4:2; C–H, from Van Cleave, 1952, Exp. Parasitol. 1:308; L–P, from Baer, 1961, Traité de Zoologie 4:733; I, from Van Cleave, 1947, J. Parasitol. 33:487; K, from Van Cleave, 1920, Proc. U. S. Nat. Mus. 58:455; T, from Van Cleave, 1941, Quart. Rev. Biol. 16:157; W, from Van Cleave, 1949, J. Morphol. 48:431; M, from Rauther, 1930, Handbuch der Zoologie (Kükenthal and Krumbach) 2:449; U, from Kilian, 1932, Z. Wschr. Zool. 141:246.

---

receptacle (Plates 101, R, V; 102, F). In the early stages of development, it consists of a dorsal and ventral sac.

The adult male reproductive system consists of two testes (Plates 101, B, Q; 102, K), followed posteriorly by cement glands, pouches, and genital ducts. The cement glands are of three types. One consists of six to eight uninucleate glands, another of six multinucleate ones, and the last of a large single multinucleate syncytial one with an enclosed cement reservoir. The ducts of the cement glands join posteriorly with the ejaculatory duct (Plates 101, Q, W; 102, D, E, K, L). The copulatory apparatus consists of a muscular pouch containing the penis and an eversible copulatory bursa at the posterior extremity of the body (Plates 101, Q; 102, D, K). Eversion is probably by means of hydrostatic pressure and inversion by muscular action. The cement glands provide a substance

that forms a plug in the vagina following mating to prevent loss of the spermatozoa.

The female reproductive system consists of an ovary, uterine bell, uterus, vagina, and vulva. The ovary develops on the wall of the dorsal ligament sac. As development proceeds, it breaks into ovarial balls, or floating ovaries, that occur in the dorsal ligament sac in the pseudocoel in cases where the ligaments are ephemeral (Plates 101, A, R, V; 102, F).

The uterine bell is, as the name implies, a bell-shaped structure with the large opening directed anteriorly and its narrow one opening posteriorly by means of a duct into the uterus. Near the posterior end of the bell, lateral pores open into the dorsal ligament sac or into the pseudocoel when the sac is absent (Plates 101, R, V; 102, F).

Immature eggs are liberated continuously in the dorsal ligament sac or body cavity where

**Plate 101**   *Morphological Characteristics of Acanthocephala*                                    383

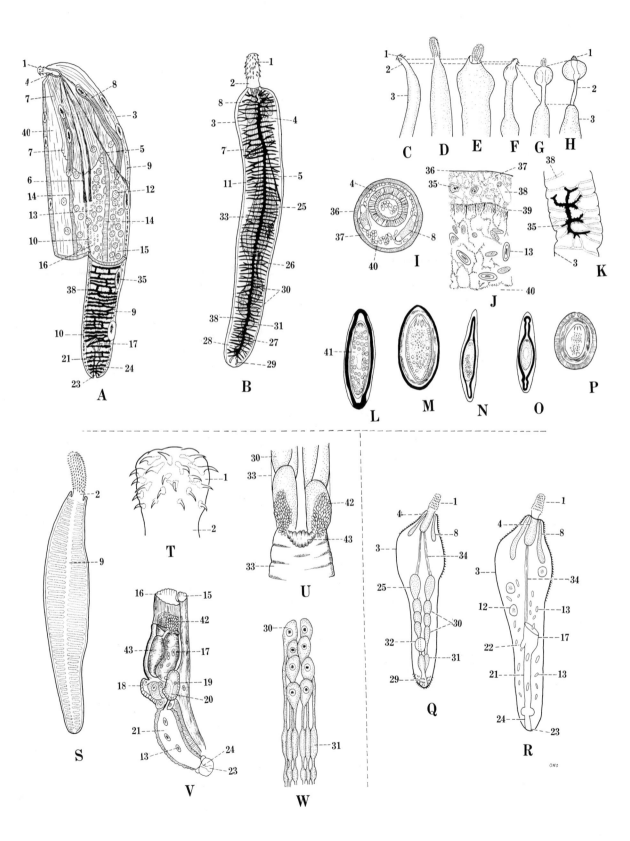

## Explanation of Plate 102

**A–H,** Basic morphological features of the order Echinorhynchidea. **A,** Lacunar system of *Centrorhynchus,* showing net-like branching. **B,** Proboscis of *Corynosoma turbidum,* showing large ventral hooks that disturb symmetrical, alternating radial arrangement of hooks. **C,** Proboscis of *Acanthocephalus dirus,* showing typical alternating radial or quincuncial arrangement of hooks. **D,** Male reproductive system of *Illiosentis furcatus.* **E,** Cement glands, showing arrangement, number, and multinucleated characteristic of the order Echinorhynchidea. **F,** Reproductive system of female *Acanthocephalus lucii,* showing suspensory ligament but no ligament sacs. **G,** Giant nuclei of body wall of *Macracanthorhynchus.* **H,** Cross section through anterior region of body of echinorhynchidian, showing remnant of ephemeral ligament and lateral lacunar channels. **I–M,** Basic morphological features of the order Neoechinorhynchidea. **I,** Proboscis of *Neoechinorhynchus,* showing characteristic number of hooks and radial arrangement of them. **J,** Cross section of anterior part of body showing dorsal and ventral ligament sacs and dorsal and ventral lacunar channels. **K,** Male reproductive system of *Gracilisentis gracilisentis,* showing syncytial type of cement gland. **L,** Cement gland characteristic of the eoacanthocephalans. **M,** *Neoechinorhynchus emydis* in process of attaching to the intestine of turtle host. **N,** Larval stages of *Macracanthorhynchus hirudinaceus* typically found in the life cycle of Acanthocephala.

1, Proboscis armed with hooks; 2, neck; 3, trunk or hindbody; 4, lemniscus; 5, dorsal longitudinal lacunar canal with branches; 6, ventral longitudinal lacunar canal; 7, lateral longitudinal lacunar canal with branches; 8, egg; 9, dorsal ligament sac; 10, ventral ligament sac; 11, remnant of ligament; 12, uterine bell; 13, uterus; 14, uterine glands; 15, sphincter of uterine bell; 16, testis; 17, vas efferens; 18, ejaculatory duct; 19, penis; 20, copulatory bursa; 21, cement glands; 22, cement reservoir; 23, Saefftigen's pouch; 24, giant nucleus of body wall; 25, body cavity; 26, intestinal wall of host.

A redrawn from Yamaguti, 1939, Jap. J. Zool. 8, Pl. 43, fig. 13; F, H, J, from Meyer, 1933, Bronn's Klassen und Ordnung des Tierreichs 4:2; D, K, from Baer, 1961, Traité de Zoologie 4:733; B, C, I, from Van Cleave, 1941, Quart. Rev. Biol. 16:157; E, L, from Van Cleave, 1949, J. Morphol. 84:431; G, from Van Cleave, 1951, Trans. Amer. Microsc. Soc. 70: 37; M, from Van Cleave, 1952, Exp. Parasitol. 1: 317; N, from Van Cleave, 1947, J. Parasitol. 33:118.

---

embryonation is completed. Hence a mixture of eggs in all stages of development is present at all times. When eggs in various degrees of development enter the uterine bell, it is believed by some to serve as a sorting apparatus by rejecting the immature ones through the lateral pores and allowing the mature ones to pass into the uterus. Since both mature and immature eggs appear in the uterus, the concept of a sorting apparatus appears untenable to others. The idea of its being a mechanism by which the worm may pass eggs, regardless of the stage of development, into the uterus for oviposition or return them to the body at will is more acceptable to them. The uterus is a simple tube, surrounded by a sphincter, that empties through a terminal or subterminal funnel-shaped vagina (Plates 101, A, R, V; 102, F). Eggs are embryonated when laid and variable in shape, being oval, fusiform, or elliptical. There are three shells of different thickness, the two outer ones being thick and the inner one membranous. The anterior end of the fusiform larva bears hooks and the body is covered with minute spines. Inside is a centrally located mass of small nuclei (Plate 101, L–P).

The excretory system consists of a pair of protonephridia located bilaterally. They may be dendritic or of a simple capsular type. Each is provided with a tuft of flickering cilia at the end of the individual dendrites or capsules. A duct from each protonephridium extends medially where several unite to form a common tubule that empties into the ejaculatory duct or uterus (Plate 101, U).

The basic pattern of the life cycle is similar for all acanthocephalans (Plate 102, N). Eggs hatch only when ingested by insect or crustacean intermediate hosts. In case of terrestrial definitive hosts, the intermediaries likewise are terrestrial, and when the definitive hosts are aquatic, the intermediate hosts also are aquatic. Infection occurs when invertebrate hosts containing fully developed acanthellas are eaten. In some cases, transport hosts are common. These are vertebrates in which the acanthellas are unable to develop to maturity. Under these circumstances, they migrate from the gut into the body cavity and encyst. A common example is the encysted forms of *Corynosoma* of seals and sea lions that occur in marine fish.

The classification of Yamaguti consisting of four orders is followed. His monograph is cur-

**Plate 102**  *Morphological Characteristics of Acanthocephala*  385

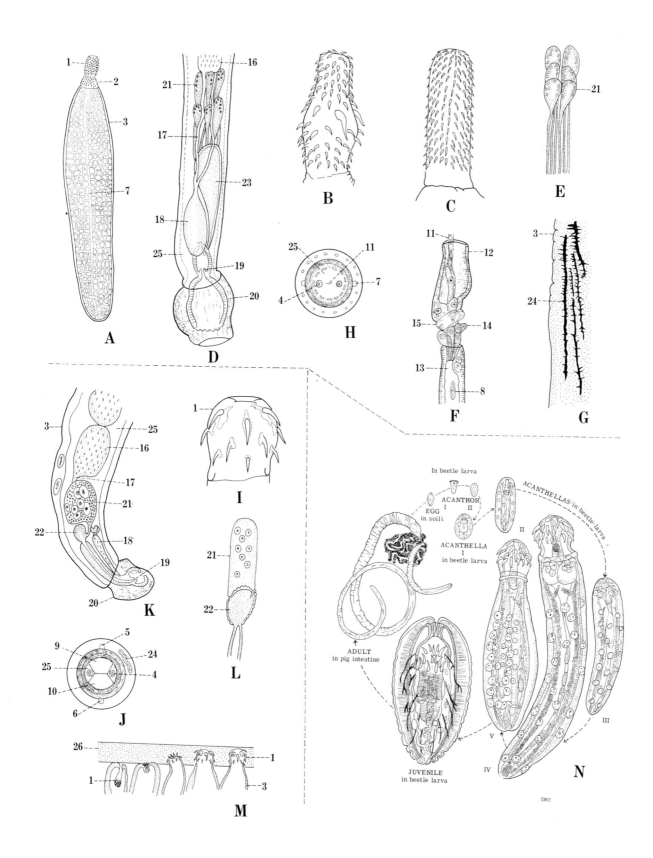

rent, complete, readily available, and contains a brief synopsis of the biology of each species where known. A few of the common families of each order, some of their genera, and the types of hosts parasitized are listed here.

I. Order Apororhynchidea: Small, subconical bodies; proboscis bulbous, covered with numerous small spines, and incapable of being withdrawn; no prosboscis receptacle. Main longitudinal lacunar vessels dorsal and ventral. Protonephridial organs absent. Lemnisci paired, long, and multinuclear. Eight compact cement glands. Eggs enclosed in three shells. Parasites of birds.

Family Apororhynchidae
*Apororhynchus;* birds.

II. Order Neoechinorhynchidea: Small bodies with prosboscis covered with relatively few hooks and retractible into single-layered receptacle. Hypodermal nuclei few and large. No protonephridial organ. Cement gland syncytial, usually single. Eggs elliptical, usually without polar elongations of middle layer of shell. Parasites of fish, amphibians, reptiles.

Family Neoechinorhynchidae
*Neoechinorhynchus;* marine and freshwater fish, batrachians, chelionians
*Octospinifer;* fish
*Paulisentis;* freshwater fish
*Eocollis;* freshwater fish
*Gracilisentis;* fish

Family Quadrigyridae
*Quadrigyrus;* fish

III. Order Echinorhynchidea: Small spinose or smooth bodies; proboscis cylindrical, armed with many hooks, retractible into double-layered receptacle. Hypodermal nuclei small and numerous. Lacunar canals lateral. Lemnisci paired, short, and club-shaped. Four to eight cement glands. Eggs elliptical to fusiform. Parasites of fish, amphibians, reptiles, birds.

Family Echinorhynchidae
*Echinorhynchus;* freshwater and marine fish
*Acanthocephalus;* fish, amphibians, reptiles
*Leptorhynchoides;* freshwater and marine fish
*Filisoma;* fish

Family Fessisentidae
*Fessisentis;* freshwater fish

Family Filicollidae
*Filicollis;* aquatic birds
*Falsifilicollis;* aquatic birds

Family Polymorphidae
*Polymorphus;* aquatic birds, occasionally mammals
*Arhythmorhynchus;* fish, amphibians, reptiles
*Corynosoma;* marine birds, mammals
*Bolbosoma;* marine mammals

Family Plagiorhynchidae
*Plagiorhynchus;* birds

Family Pomphorhynchidae
*Pomphorhynchus;* fish

IV. Order Gigantorhynchidea: Medium to large in size, proboscis armed with few to many hooks, retractible into single or double layered receptacle. Protonephridial gland present or absent. Cement glands divided into three or more compact lobes. Elliptical eggs lack polar elongations. Embryos with hooks at each end and small spines over body. Birds and mammals.

Family Gigantorhynchidae
*Gigantorhynchus;* mammals
*Mediorhynchus;* birds

Family Centrorhynchidae
*Centrorhynchus;* birds (raptores), mammals

Family Moniliformidae
*Moniliformis;* mammals

Family Oligacanthorhynchidae
*Oligacanthorhynchus;* birds
*Pachysentis;* carnivora
*Macracanthorhynchus;* mammals
*Hamanniella;* mammals (marsupials)
*Prosthenorchis;* mammals (monkeys)
*Oncicola,* mammals (carnivora)

Family Prosthorhynchidae
*Prosthorhynchus;* birds, mammals

## SELECTED REFERENCES

Baer, J. C. (1961). Embranchement des Acanthocephales. Traité Zool. 4:733–782.

Crompton, D. W. T. (1970). An Ecological Approach to Acanthocephalan Physiology. Cambridge Monogr. Exper. Biol. 17, Cambridge Univ. Press, 136 pp.

Crompton, D. W. T., and D. L. Lee. (1965). The fine structure of the body wall of *Polymorphus minutus* (Goeze, 1782) (Acanthocephala). Parasitology 55:357–364.

Hyman, L. H. (1951). The Invertebrates. Acanthocephala, Aschelminthes, and Entoprocta. McGraw-Hill Book Co., New York, Vol. 3, pp. 1–52.

Meyer, A. (1933). Acanthocephala. Bronn's Klassen und Ordnung Tierreichs 4(2/2):1.

Nicholas, W. L. (1967). The biology of the Acanthocephala. In: B. Dawes, ed. Advances in Parasitology. Academic Press, New York, Vol. 5, pp. 205–206.

Pratt, I. (1969). The biology of Acanthocephala. In: M. Florkin and B. T. Scheer, eds. Chemical Zoology. Academic Press, New York, Vol. 3, pp. 245–252.

Rauther, Max. (1930). Sechste Klasse des Cladus Nemathelminthes. Acanthocephala = Kratzwurmer. Handb. Zool. (Kükenthal u. Krumbach). Vol. 2, 10. Lief., Teil 4, Bogen 33–37, pp. 449.

Van Cleave, H. J. (1936). The recognition of a new order in the Acanthocephala. J. Parasitol. 22:202–206.

Van Cleave, H. J. (1941). Hook patterns of the acanthocephalan proboscis. Quart. Rev. Biol. 16: 157–153.

Yamaguti, S. (1963). Systema Helminthum. Acanthocephala. John Wiley and Sons, New York, Vol. 5, 423 pp.

## ORDER NEOECHINORHYNCHIDEA

Hosts of the members of this order are aquatic, usually fish but occasionally turtles.

### FAMILY NEOECHINORHYNCHIDAE

These are small to medium-sized acanthocephala with spines only on the proboscis. The wall of the proboscis receptacle is a single layer of muscle. The nuclei of the lemnisci and body wall are fixed in number, definite in arrangement, and very large, the latter producing pronounced oval elevations on the body surface. The testes are elliptical and usually contiguous. The cement gland is a single synctial mass. The Neoechinorhynchidae are common parasites of fish and occasionally of turtles throughout the world.

### Neoechinorhynchus cylindratum
(Van Cleave, 1913) (Plate 103)

This species occurs in a number of North American fish, including largemouth bass, walleyes, pickerel, bowfins, chubsuckers, carp, common eels, and possibly others.

**Description.**     Adult females are 7 to 11.2 mm long by 0.35 to 0.7 mm wide, with greatest diameter between first and second nuclei. The proboscis is globular in shape, measuring 0.1 to 0.14 mm long by 0.16 to 0.19 mm broad. It bears three circles of six hooks each. The uninucleate

lemniscus is slightly shorter than the binucleate one, which is 0.95 to 1.4 mm long. Uterine bell, uterus, and vagina total 0.49 to 0.73 mm long. The eggs are ellipsoidal and measure 50 to 60 $\mu$ long by 17 to 28 $\mu$ wide.

Adult males measure 4.7 to 6.3 mm long by 0.36 to 0.63 mm wide. The proboscis is 0.1 to 0.14 mm long by 0.15 to 0.17 mm wide. The uninucleate lemniscus is slightly shorter than the binucleate one, which is 0.84 to 1.2 mm long. The testes are arranged in tandem, followed by an elongate synctial cement gland with eight large nuclei. A cement reservoir is incorporated in the posterior part of the cement gland. There is a bell-shaped copulatory bursa.

**Life Cycle.**     Adult females in the intestine of largemouth bass lay fully embryonated eggs that are voided with the feces. Ostracods [*Cypria* (*Physacypria*) *globula*], which are common in ponds, serve as the first intermediate host. Hatching of the eggs occurs when they are eaten by the ostracods. Upon release from the eggs, the acanthors promptly burrow through the intestinal wall, arriving in the body cavity within 24 hours. When too many embryos are eaten, the ostracods die after 10 days. Growth is slow, requiring up to 30 days or more before the larvae have reached the definitive shape of the adult males and females. There is little change in size or structure during the first 6 days. On day 10 after infection, the larvae have enlarged greatly and show a few large cuticular nuclei. By the end of week 3, the size and shape are similar but a primordial body wall is apparent and the interior is filled with dividing cells. On day 28, after entering the ostracod, the body has elongated greatly, giving it a vermiform appearance. The large nuclei which will develop into the prosboscis and its various parts, such as hooks and receptacle, have appeared by this time. The precise time after week 4 when acanthellas become infective is unknown.

Bluegills serve as the second intermediate hosts. They become infected by eating ostracods harboring fully developed acanthellas. In the alimentary canal of the fish, ostracods are digested and the acanthellas freed. Those that have developed to the infective stage burrow through the intestinal wall and penetrate the liver. Inside, the worms coil up and become enclosed in a cyst composed of modified liver cells. A layer of granular material lines the inside of the cyst wall. Encysted worms are similar to

## Explanation of Plate 103

A, Adult male *Neoechinorhynchus cylindratum*. B, Proboscis, showing arrangement of hooks. C, Adult female. D, Egg. E, Embryo 24 hours after ingestion by ostracod. F, Embryo in ostracod 10 days. G, Embryo in ostracod 20 days. H, Embryo in ostracod 28 days. I, Immature male from ostracod. J, Encysted immature female from liver of bluegill. K, Immature male dissected from cyst from liver of bluegill. L, Immature female from cyst. M, Large-mouth bass (*Micropterus salmoides*) definitive host. N, Ostracod [*Cypria* (*Physacypria*) *globula*] first intermediate host. O, Bluegill (*Lepomis macrochirus*) second intermediate host.

1, Hooked proboscis; 2, hook of proboscis; 3, proboscis receptacle; 4, lemniscus; 5, retractor muscle of proboscis receptacle; 6, giant nucleus of body wall; 7, testes; 8, syncytial cement gland; 9, cement reservoir; 10, seminal vesicle; 11, copulatory bursa; 12, ovarial ball or floating ovary; 13, eggs; 14, ligament; 15, uterine bell; 16, uterus; 17, vagina; 18, outer shell of egg; 19, inner shell of egg; 20, shell-folds at each end of acanthor; 21, acanthor; 22, nuclear mass of acanthor; 23, anlagen of subcuticula.

a, Adult female with smaller male beside her, both attached to intestinal wall; b, eggs; c, embryonated egg voided with feces into water; d, egg ingested by ostracod; e, acanthor hatching from egg; f, acanthor burrowing through wall of gut; g, acanthor in body cavity; h, growing acanthor; i, young acanthella; j, ostracod in gut of bluegill; k, young acanthella escaping from ostracod as latter is digested; l, acanthella burrowing through wall of small intestine and into liver of bluegill; m, fully developed and encysted infective acanthella in liver of bluegill; n, infected bluegill eaten by bass; o, acanthella released by digestion in intestine; p, young acanthocephalans, larger female and smaller male in intestine, having been released from liver and cysts.

Figures A–L adapted from Ward, 1940, Trans. Amer. Microsc. Soc. 59:327.

---

adults except in size and degree of development of the reproductive organs. They are infective to bass when encystment is completed.

Bass become infected by eating parasitized bluegills. Little growth occurs during the first 30 days and sexual maturity is not attained until about 5 months after ingestion of encysted acanthellas. Longevity of the adult worms in the fish is unknown.

Studies on the life cycles of *Neoechinorhynchus emydis* from the map turtle and *N. rutili* from many species of fish have been done. Eggs hatch when ingested by the ostracod (*Cypria maculata*). The acanthors enter the body cavity and develop into unencysted juveniles in 21 days. When infected ostracods are eaten by the snail *Campeloma rufum*, the juvenile worms encyst in the tissues, especially in the foot. When infected snails are eaten by turtles, development of the worms takes place in the intestine. *Neoechinorhynchus rutili* has a life cycle different from the other two species in that there is no second intermediate host. In Washington State, the eggs hatch in the intestine of the ostracod *Cypria turneri*. Acanthors are free in the hemocoel of the ostracod after 6 to 12 days and complete their development to the juvenile stage in 48 to 57 days. These juveniles are infective directly to various species of freshwater fishes.

*Octospinifer macilentis* from the white sucker, *Catostomus commersoni*, develops in the ostracod *Cyclocypris serena*. Eggs hatch in the intestine of the intermediate host in about 4 hours. In the fish, males mature in 8 to 10 weeks and females in 16. A paratenic host is not required as in the case of *Neoechinorhynchus cylindratum* and *N. emydis*.

*Paulisentis fractus* from the creek chub, *Semotilus atramaculatus*, develops in the copepod *Tropocyclops prasinus*. It is the only neoechinorhynchid known to utilize a copepod instead of an ostracod intermediate host. Acanthors are in the hemocoel within 2 to 4 hours and fully developed anthellas are infective by 13 days.

## EXERCISE ON LIFE CYCLE

Infected fish or map turtles provide a source of eggs for experimental infection of ostracods. Eggs may be obtained from the feces of infected hosts but this can be a laborious task with a low yield. By rupturing the posterior region of the gravid female worms, a mass of eggs escapes into the dish. They are washed by a series of centrifugations and stored in tap water in a refrigerator for future use.

Infect ostracods by placing them in a small dish of water containing a few eggs which they will eat. Care should be taken to avoid too heavy infections lest the ostracods die as a result. Observe the development of the acanthellas in the

Plate 103    *Neoechinorhynchus cylindratum*                                    389

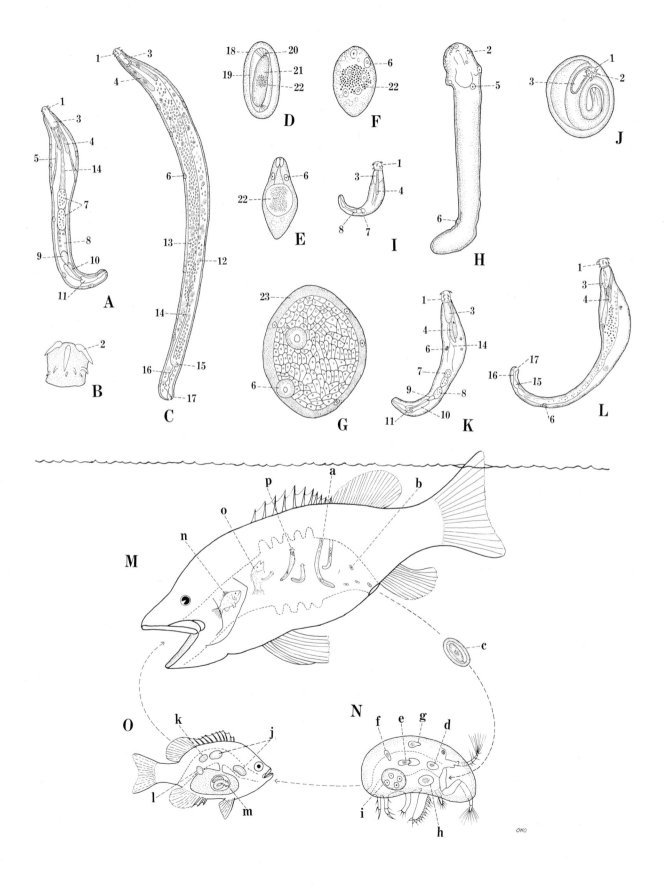

body cavity of the ostracods by dissecting them at intervals in order to obtain all of the stages. Determine the time of infectivity of the juvenile worms in the ostracods by appearance and by feeding experiments. These should be designed to determine whether infection of the definitive host may be obtained by feeding infected ostracods directly to them. If *N. rutili* is available for experimentation, determine whether the life cycle may also include a second intermediate host such as a snail or some other animal.

## SELECTED REFERENCES

Cable, R. M., and W. T. Dill. (1967). The morphology and life history of *Paulisentis fractus* Van Cleave and Bangham, 1949 (Acanthocephala: Neoechinorhynchidae). J. Parasitol. 53:810–817.

Harms, C. E. (1965). The life cycle and larval development of *Octospinifer macilentis* (Acanthocephala: Neoechinorhynchidae). J. Parasitol. 51: 286–293.

Hopp, W. B. (1954). Studies on the morphology and life cycle of *Neoechinorhynchus emydis* (Leidy), an acanthocephalan parasite of the map turtle, *Graptemys geographica* (Le Sueur). J. Parasitol. 40:284–299.

Merritt, S. V., and I. Pratt. (1964). The life history of *Neoechinorhynchus* and its development in the intermediate host (Acanthocephala: Neoechinorhynchidae). J. Parasitol. 50:394–400.

Uglem, G. L., and O. R. Larsen. (1969). The life history and larval development of *Neoechinorhynchus saginatus* Van Cleave and Bangham, 1949 (Acanthocephala: Neoechinorhynchidae). J. Parasitol. 55:1212–1217.

Ward, H. L. (1940). Studies on the life history of *Neoechinorhynchus cylindriatus* (Van Cleave, 1913) (Acanthocephala). Trans. Amer. Microsc. Soc. 59: 327–347.

### ORDER ECHINORHYNCHIDEA

The members of this order are primarily parasites of aquatic vertebrates, including fish, birds, and mammals. A few species occur in birds of prey and terrestrial mammals. Species of the family Rhadinorhynchidae are parasites of marine fishes, with exception of the genus *Leptorhynchoides* which includes forms in both marine and freshwater fish.

### FAMILY ECHINORHYNCHIDAE

Members of this family are slender with a smooth trunk and a cylindrical proboscis usually armed with numerous hooks or it may be spherical and with a small number of hooks. The lemnisci are club-shaped and rather short. The testes are oval to elliptical but never cylindrical. There are four to eight pyriform cement glands.

### *Leptohynchoides thecatus* (Linton, 1891) (Plate 104)

This is a common parasite of at least six families of fish in the United States, especially the largemouth bass (*Micropterus salmoides*) smallmouth bass (*Micropterus dolomieui*), and rock bass (*Ambloplites rupestris*).

**Description.** The proboscis, when fully extended, forms an angle with the long axis of the body. It is about 1 mm long and armed with 12 longitudinal rows of 12 to 13 hooks each. Individual hooks are surrounded throughout most of their length by an ensheathing cuticular collar. Males are 7 to 12 mm long; there are six cement glands closely compacted at the posterior border of the hind testis. Females measure 11 to 26 mm long. The embryonated eggs in the body cavity are 80 to 110 $\mu$ long by 24 to 30 $\mu$ wide.

**Life Cycle.** Eggs passed in the feces of fish are immediately infective to amphipods (*Hyalella azteca*) when ingested by them. Hatching occurs in the intestine within 45 minutes after being swallowed. The acanthors are very active and crawl over the surface of the intestinal epithelium for a short time after emergence from the eggs. They soon attach, however, and bore through the epithelium without the aid of the blade-like hooks which are lacking, coming to rest between the cells and the serosa of the intestine. Motility ceases and the acanthors enter the acanthella stage.

Growth of the acanthella begins as a bulging of the body which develops into a large spherical structure that is attached by the remains of the acanthor. The serosa ruptures and the newly formed acanthella hangs free in the celom, being anchored by means of the dwindling acanthor. Between 9 and 14 days after infection, there is extensive growth in size of the acanthella and organization of the central nuclear mass. At the end of this period, the acanthella breaks its stalk and becomes free in the hemocoel. The basic structure of the adult worm is attained by day 25, and it becomes a fully developed infective juvenile by day 32 in the amphipod.

Infection of fish is accomplished when amphipods harboring juvenile acanthellas are eaten. Digestion of the crustaceans results in release of the parasites in the stomach of the fish. They enter the pyloric ceca and attach by means of the proboscis. The rate of development is controlled somewhat by the temperature of the water. Under favorable conditions, eggs appear in the feces 8 weeks after the fish eat infected amphipods. The males are sexually mature at the end of 4 weeks.

## EXERCISE ON LIFE CYCLE

For experiments, amphipods, fish (rock bass or sunfish are preferred because they are natural hosts and are adaptable to laboratory conditions), and worm eggs are needed. Amphipods collected from habitats lacking fish hosts are suitable for experimental infections. They may be kept readily in battery jars to which are added a few willow rootlets and duckweed, in addition to bits of yeast from time to time. A stream of air bubbled through the jar will improve the environmental conditions in it. Rock bass and sunfish live well in aquaria when kept on a diet of earthworms.

Gravid females of *L. thecatus* provide eggs. The worms may be stored in water in a refrigerator for long periods without adverse effects on the viability of the eggs.

Amphipods may be infected by placing 100 of them in a jar with a large number of worm eggs. An hour of exposure is sufficient to allow four or five larvae per individual. After exposure, wash the crustaceans to free them of any eggs that might be adhering and place them in clean jars. Dissect them at frequent intervals over a period of 40 days to observe the progress of development. Examine the intestinal lumen for free newly hatched acanthors and the wall for those that have entered it. Examine the hemocoel for acanthellas.

Since fish are commonly infected naturally, uninfected ones are difficult to obtain from natural habitats. In order to assure having fish suitable for experimental feeding, it is necessary to hold them in aquaria for about 2 months. In this time, any parasites present would have developed to maturity and could be detected by eggs in the feces. Fish are infected readily by introducing a known number of infected acanthellas into the stomach by means of a glass pipette or by allowing them to eat infected gammarids. Fish should be wrapped in a wet cloth while handling in order to avoid injury. Fecal examinations should begin during week 6 after infection in order to determine the length of the prepatent period.

## SELECTED REFERENCE

DeGuisti, D. L. (1949). The life cycle of *Leptorhynchoides thecatus* (Linton), an acanthocephalan of fish. J. Parasitol. 35:437–460.

### ORDER GIGANTORHYNCHIDEA

Two common and well-known members of this order are the widely distributed large thornheaded worm *Macracanthorhynchus hirudinaceus* of the family Oligacanthorhynchidae found in the small intestine of swine and *Moniliformis dubius* of the family Moniliformidae of rats.

#### FAMILY MONILIFORMIDAE

The body is long and slender with regular constrictions, giving it a somewhat beadlike appearance. The proboscis is cylindrical and armed with small hooks, becoming fewer basally. They are parasites primarily of rodents.

*Moniliformis dubius*
Meyer, 1933 (Plate 105)

This is a common parasite of rats throughout the world, particularly in regions where cockroaches, which serve as intermediate hosts, are present.

**Description.**    The body is divided superficially, except at the extremities, into numerous bead-like pseudosegments. The proboscis is cylindrical and truncate distally; it bears 12 longitudinal rows of hooks, each row with 10 hooks. Each hook has a strongly recurved blade and a simple root directed posteriorly.

Males are up to 80 mm long; the testes are elongated. There are six cement glands crowded together immediately anterior to Saefftigen's pouch. The bursal cap has eight short digitiform rays. Females attain a length of 200 mm. The eggs have an outer shell and an inner embryonic membrane; they are 112 to 120 $\mu$ long by 54 to 64 $\mu$ wide. A fully developed embryo covered with numerous minute spines is present in the egg when it is laid.

**Life Cycle.**    The females lay fully developed eggs which are passed in the feces. When ingested by cockroaches (*Periplaneta*

## Explanation of Plate 104

A, Adult male. B, Embryonated egg containing acanthor. C, Acanthor just hatched. D, Acanthor lying between basement membrane and serosa of intestine of gammarid intermediate host. E, Formation of acanthella from acanthor in five days in intestinal wall of gammarid. F, Acanthella at 14 days of development, forming a distinct stalk between it and body of acanthor. G, Male acanthella at 17 days of development in amphipod. H, Male acanthella at 20 days of development. I, Male acanthella at 22 days of development. J, Female juvenile at 32 days of development in amphipod. K, Rock bass (*Ambloplites rupestris*) definitive host. L, Amphipod (*Hyalella azteca*) intermediate host.

1, Proboscis; 2, neck; 3, body or trunk; 4, proboscis sheath; 5, lemniscus; 6, retractor muscle of proboscis; 7, testes; 8, cement glands; 9, Saefftigen's pouch; 10, muscular cap of bursa; 11, bursa; 12, genital pore; 13, eggshell; 14, acanthor; 15, larval spines; 16, nuclear mass; 17, giant nuclei; 18, epithelial cells of intestine of amphipod host; 19, serosa of intestine; 20, basement membrane of intestine; 21, acanthella; 22, anlagen of lemniscal ring; 23, anlagen of proboscis sheath; 24, anlagen of body musculature; 25, anlagen of genital suspensory ligament; 26, anlagen of testes; 27, anlagen of cement glands; 28, anlagen of penis and bursa; 29, anlagen of bursa and genital pore; 30, cells from which proboscis is formed; 31, giant cells which form division between presoma and trunk; 32, brain; 33, genital sheath; 34, cortical layer; 35, ovary; 36, uterine bell; 37, uterus; 38, stalk.

a, Adult acanthocephalan attached deep in gastric cecum; b, embryonated eggs laid in intestine; c, eggs passed with feces into water; d, amphipod intermediate host eats eggs; e, eggs hatch in intestine of amphipod; f, acanthor enters intestinal wall; g, acanthella develops from acanthor at 5 days; h, acanthella at 14 days forms a distinct stalk; i, male acanthella at 17 days having become detached by rupture of stalk and liberated in hemocoel; j, male acanthella at 22 days free in hemocoel; k, female juvenile after 32 days of development in amphipod; l, fish become parasitized upon swallowing infected amphipods; m, juvenile acanthocephala escape when amphipod is digested; n, juvenile free in alimentary canal; o, juvenile enters gastric cecum and males develop to sexual maturity in four weeks and females begin laying eggs in 8 weeks.

Figure A adapted from Van Cleave, 1919, Bull. Ill. State Lab. Natur. Hist. 13:225; other figures from De Guisti, 1949, J. Parasitol. 35:437.

---

*americana* and probably other species since the parasite is worldwide in distribution), electrolytes in the intestine rather than osmotic pressure stimulates the acanthor to secrete chitinase that decomposes the chitinous eggshell. Hatching occurs in the midgut within 24 to 48 hours but requires only about 5 minutes *in vitro*. The newly hatched acanthors bearing six blade-like hooks work their way through the intestinal wall and appear as minute iridescent specks attached to the outer wall in 10 to 12 days. During the process of development, the acanthor passes through two stages and the acanthella six before becoming a cystacanth. In about 22 days, the acanthors are nearly spherical; after that, they begin to elongate and internal development takes place. Sometime between days 38 and 44 after infection, the structure, shape, and size have progressed to a stage intermediate between that of the acanthor in the egg and the infective stage. The shape begins to resemble that of an acanthocephalan in which the major organs can be recognized. Between days 44 and 51, the larval worms continue to elongate and the organs assume the definitive form. By day 55 after infection of the cockroaches, the acanthellas complete their development. They are enclosed in cysts with the proboscis withdrawn into the proboscis receptacle and they are designated as cystacanths.

Infection of the rats takes place when infected roaches are eaten. Upon digestion of the insect hosts, the cysts are liberated, and begin growing. As their length increases, the worms migrate forward in the intestine to bring the absorptive surface of the body into locations where the most abundant and nutritious food exists for optimal development. Maturity is attained in 5 to 6 weeks.

The tegument through which nutritive substances must pass is composed of six layers. They are 1) cuticle; 2) plasma and subplasma; 3) striated layer of branching septa separated by vesicles; 4) felt layer; 5) thick radial fibrial layer containing nuclei, mitochondria, lipid droplets, glycogen, and perhaps ribosomes; 6) basal layer consisting of basal membrane (Fig. 30). Vacuole formation on the surface of the tegument may be the mechanism of pinocytosis through which carbon particles up to 800 Å pass through the body wall into the pseudocoel. The surface structure of the lemnisci and inner sur-

**Plate 104** *Leptorhynchoides thecatus* 393

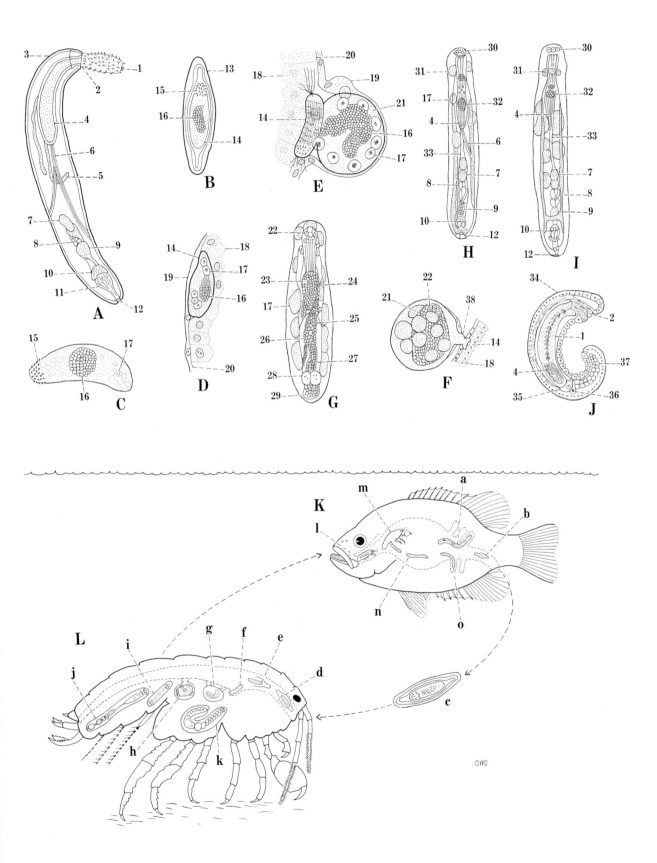

## Explanation of Plate 105

**A,** Outline of adult worm showing moniliform shape. **B,** Embryonated egg. **C,** Acanthor in process of escaping from egg shell and membranes. **D,** Acanthor dissected from gut wall of cockroach. **E,** Acanthella from hemocoel 46 days after infection. **F,** Encysted acanthella. **G,** Cystacanth free from cyst and with proboscis evaginated. **H,** Proboscis of fully developed acanthella. **I,** Rat (*Rattus*) definitive host. **J,** Cockroach (*Periplaneta americana*) intermediate host.

1, Proboscis; 2, lemniscus; 3, testes; 4, cement glands; 5, cement reservoir; 6, inverted bursa; 7, outer eggshell; 8, inner eggshell; 9, rostellar hooks; 10, central nuclear mass of acanthor; 11, body spines; 12, retractor muscle; 13, giant nuclei; 14, nuclei of nerve ring; 15, proboscis sheath; 16, brain; 17, inverter muscles; 18, hypodermis; 19, urinary bladder; 20, cyst; 21, suspensory ligament; 22, nuclei of apical ring; 23, Saefftigen's pouch.

a, Adult worm in small intestine; b, egg passing out of body with feces; c, embryonated egg in feces; d, egg swallowed by cockroach; e, egg hatches in gut; f, acanthor migrates through gut wall; g, acanthella in early stages of development; h, fully developed and infective acanthella (cystacanth); i, infected cockroach being digested in stomach of rat; j, cystacanth being released from roach; k, acanthella escaping from cyst; l, acanthella attaches to gut wall and grows into an adult worm.

Adapted from Moore, 1946, J. Parasitol. 32: 257.

---

face of the body wall are similar in structure, with many infoldings, suggesting that these organs provide greater area for absorption and transport of materials into the body cavity.

## EXERCISE ON LIFE CYCLE

In habitats occupied concurrently by cockroaches and rats, both are likely to show a high incidence of infection. Adult worms obtained from rats provide eggs for experimental feeding to roaches. Developing acanthellas can be obtained by dissecting naturally infected roaches.

For experiments on infection of cockroaches, it is necessary to either rear them or obtain them from locations where natural infections do not occur. Dissection of an adequate sample of a population will show whether they are free of natural infection and therefore suitable for experimental purposes.

Cockroaches to be used for experimental purposes may be kept readily in glass jars containing moist sand over which are placed wet paper towels. The latter maintain the moisture and provide a place for the roaches to hide. They thrive on pabulum and are fond of apples.

Acanthocephalan eggs from gravid females placed on bits of pabulum are eaten readily by roaches from which food has been kept for 24 hours. Infect a sufficient number of them so that dissections can be made at intervals of 10 days or less for a period of 2 months. This procedure will reveal all of the stages of development within the intermediate host.

## SELECTED REFERENCES

Burlingame, P. P., and A. C. Chandler. (1941). Host-parasite relationships of *Moniliformis dubius* (Acanthocephala) in albino rats, and the environmental nature of resistance to single and superimposed infections with this parasite. Amer. J. Hyg. 33(D):1–21.

Crompton, D. W. T., and P. J. Whitfield. (1968). A hypothesis to account for the anterior migrations of adult *Hymenolepis diminuta* (Cestoda) and *Moniliformis dubius* (Acanthocephala) in the intestine of rats. Parasitology 58:227–229.

Edmond, J. S. (1966). Hatching of eggs of *Moniliformis dubius*. Exp. Parasitol. 19:216–226.

Edmond, J. S., and B. R. Dixon. (1966). Uptake of small particles by *Moniliformis dubius* (Acanthocephala). Nature, London 209:99.

King, D., and E. S. Robinson. (1967). Aspects of the development of *Moniliformis dubius*. J. Parasitol. 53:142–149.

Moore, D. V. (1942). An improved technique for the study of the acanthor stage in certain acanthocephalan life histories. J. Parasitol. 28:495.

Moore, D. V. (1946). Studies on the life history and development of *Moniliformis dubius* Meyer, 1933. J. Parasitol. 32:257–271.

Moore, D. V. (1962). Morphology, life history, and development of the acanthocephalan *Mediorhynchus grandis* Van Cleave, 1916. J. Parasitol. 48:76–86.

Nicholas, W. L., and E. H. Mercer. (1965). The ultrastructure of the tegument of *Moniliformis dubius* (Acanthocephala). Quart. J. Microsc. Sci. 106:137–146.

Wright, R. D. (1970). Surface ultrastructure of the

Plate 105    *Moniliformis dubius*                                                    395

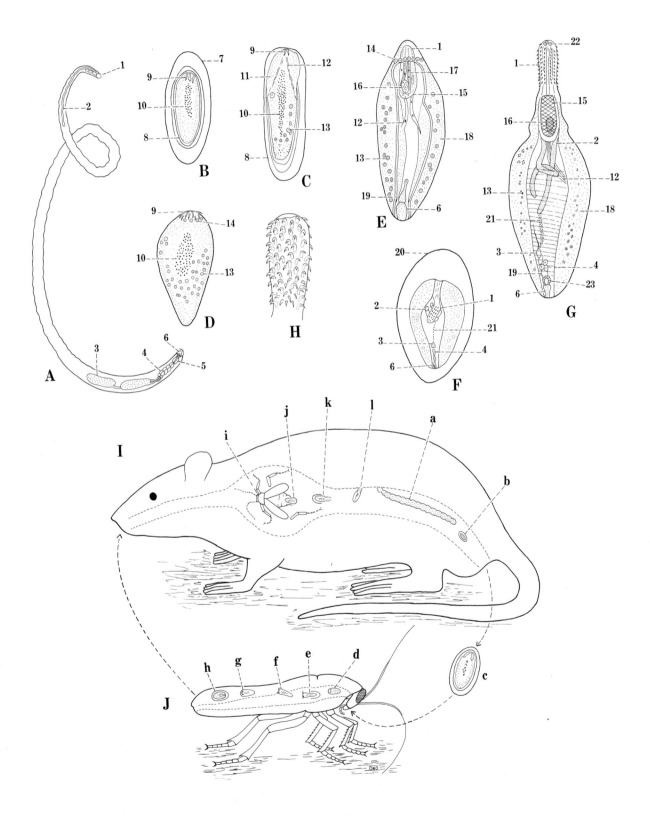

## Explanation of Plate 106

A, Outline of adult worm. B, Presoma and anterior end of trunk. C, Egg with part of shell removed to show layers and larva. D, Acanthor. E, Acanthella with proboscis extended. F, Cystacanth with proboscis retracted. G, Reproductive system of female. H, Cross section of body of female anterior to uterine bell. I, Proboscis. J, Swine definitive host. K, Beetle grub (*Cotinus* and *Phyllophaga*) intermediate host.

1, Proboscis; 2, spines; 3, anterior end of trunk; 4, sculpturing characteristic of eggs of this species; 5, egg shell and three inner membranes; 6, raphe of shell; 7, acanthor; 8, larval hooks; 9, body wall nuclei; 10, embryonic nuclear mass; 11, proboscis sheath; 12, brain; 13, lemniscus; 14, nuclei of lemniscus; 15, testes; 16, urinogenital primordium of male; 17, subcuticula; 18, ovary; 19, dorsal ligament; 20, ventral ligament; 21, protonephridium; 22, excretory bladder; 23, uterine bell; 24, opening of uterine bell; 25, uterine duct; 26, uterus; 27, vagina; 28, median canal; 29, cuticle; 30, hypodermis; 31, circular muscles; 32, longitudinal muscles; 33, lateral nerve; 34, pseudocoel; 35, eggs in various stages of development; 36, sense organ.

a, Adult worm in small intestine; b, eggs passing from intestine in feces; c, embryonated egg in soil; d, egg swallowed by grub; e, egg hatches in midgut; f, acanthor migrates through gut wall; g, acanthor attaches to wall of intestine; h, early acanthella free in hemocoel; i, j, later stages of acanthellas; k, infective acanthella (cystacanth); l, grub being swallowed by pig; m, digestion of grub and liberation of cystacanth; n, young worm ready to attach to intestinal wall and grow to maturity.

Figures C–F adapted from Kates, 1943, Amer. J. Vet. Res. 4:173; G, H adapted from Rauther, 1930, Handb. Zool. (Kükenthal u. Krumbach) 2 (10. Lief., Teil 4, Bogen 33–37, pp. 460); I adapted from Baer, 1961, Traité Zool. 4:733.

---

acanthocephalan lemnisci. Proc. Helminthol. Soc. Wash. 37:52–56.

Wright, R. D. (1971). The egg envelopes of *Moniliformis dubius*. J. Parasitol. 57:122–131.

### FAMILY OLIGACANTHORHYNCHIDAE

The members of this family are long slender worms, having slightly ringed or wrinkled bodies. The relatively small ovoid or globular proboscis bears a few circles of hooks, with those at the apex being larger than those toward the base.

### *Macracanthorhynchus hirudinaceus*
### (Pallas, 1781) (Plate 106)

These large wrinkled worms attach to the wall of the small intestine of swine by forcing the spiny proboscis into the lining of it. They do not remain attached to one spot but move about.

**Description.**    The body is more or less flattened and with numerous transverse wrinkles. The proboscis is small, globular in shape, and bears about six transverse rows of hooks. Adult males measure up to 100 mm long. Females are much larger, reaching 350 mm or more in length. The eggs have a four-layered shell; they measure 80 to 100 $\mu$ long by 46 to 65 $\mu$ wide. A fully developed and hooked embryo known as an acanthor is present when the eggs are laid.

**Life Cycle.**    Embryonated eggs laid by the females leave the body of the host in its feces. The spined acanthor is already infective to the grubs of a number of species of May and June beetles of the genera *Cotinus* and *Phyllophaga*. When swallowed by grubs, the eggs hatch within an hour and the acanthors quickly migrate through the gut wall into the hemocoel. They usually remain attached to the outer surface of the midgut for 5 to 20 days, becoming somewhat oval in shape.

As growth continues, the larvae detach from the intestinal wall, becoming free in the body cavity. They go through a period of gradual development. By 35 days, the testes are well developed but the primordia of the "egg-balls" and eggs are present as isolated cells within the ligaments. Cells which are the primordia of various other structures appear. As growth continues, the various organs develop until a fully formed acanthella, which is essentially a juvenile parasite, appears. It has an armed prosboscis, a trunk containing all the organs, and body wall nuclei. The latter are greatly attenuated and branched at this stage. These juvenile stages appear with the prosboscis extended or with it retracted into a proboscis sheath. When in the latter condition, they are referred to as cystacanths. The acanthella is fully developed and infective to swine by 65 to 90 days after entering grubs that are kept under optimal conditions for growth.

Swine eating infected grubs acquire the parasites which develop to sexual maturity in 2 to 3 months. Females lay up to 260,000 eggs per

**Plate 106**  *Macracanthorhynchus hirudinaceus*                                    397

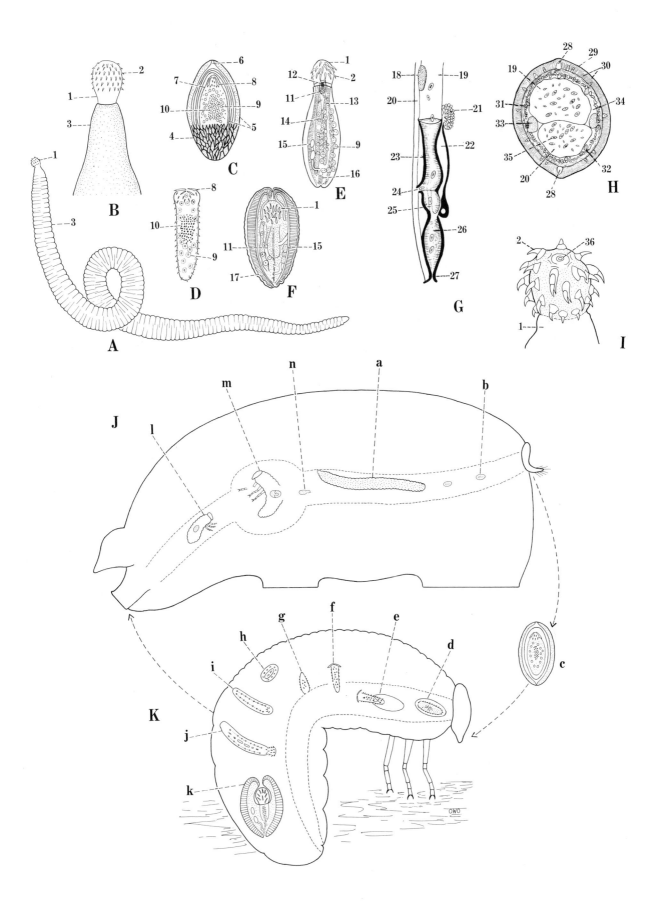

day during the peak of the reproductive period, which lasts for about 10 months.

*Prosthorhynchus formosus* from robins and flickers use terrestrial isopods, *Armadillidium vulgare*, *Porcellio laevis*, and *P. scaber*, as intermediate hosts. Under experimental conditions, the acanthellas develop to adults in chickens and turkeys.

## EXERCISE ON LIFE CYCLE

The hatching of the eggs and the development of the acanthors to the acanthella stage can be followed during the season when grubs of May and June beetles are available.

Eggs of the thorn-headed worms may be obtained from adult females collected at packing houses where swine are slaughtered. When the eggs are mixed with soil in which the grubs are kept, the larval insects ingest them. Acanthors may be found in the lumen of the intestine within an hour, and a short time later in the hemocoel. The various stages of development in the hemocoel should be sought at 15, 30, 40, 50, and 60 days after ingestion of the eggs by grubs that were kept at 70°F during the period of the experiment.

## SELECTED REFERENCES

Kates, K. C. (1942). Viability of eggs of the swine thorn-headed worm (*Macracanthorhynchus hirudinaceus*). J. Agr. Res. 64:93–100.

Kates, K. C. (1943). Development of the swine thorn-headed worm, *Macracanthorhynchus hirudinaceus,* in its intermediate host. Amer. J. Vet. Res. 4:173–181.

Schmidt, G. D., and O. W. Olsen. (1964). Life cycle and development of *Prosthorhynchus formosus* (Van Cleave, 1918) Travassos, 1926, an acanthocephalan parasite of birds. J. Parasitol. 50:721–730.

Van Cleave, H. J. (1947). A critical review of terminology for immature stages in acanthocephalan life histories. J. Parasitol. 33:118–125.

## Resume of Acanthocephalan Life Cycles

Acanthocephala appear to utilize only arthropods for the true intermediate hosts in which the infective juveniles develop. Parasites of terrestrial definitive vertebrate hosts utilize land invertebrates, primarily insects, as intermediate hosts. Those of aquatic definitive hosts similarly employ aquatic invertebrates as intermediaries. These are commonly crustaceans, both Entomostraca and Malacostraca, but insects may also serve in this capacity.

Some species parasitize animals that do not normally eat the small invertebrate intermediate hosts. In these cases, a secondary host that serves as a transport to carry the parasite to the definitive host has been included in the life cycle. When the definitive host is a land species, the transport hosts generally are a number of small animals that eat insects. Those in turn are preyed upon by the definitive host and the infection is transferred to them. In the case of species in aquatic definitive hosts, the intermediaries are commonly crustaceans but may also be aquatic insects. Snails and fish eating them act as transport hosts.

Parasites are masters in the art of adaptability, an asset they exploit to great advantage.

Phylum Nemathelminthes

## General Considerations

The phylum Nemathelminthes includes a great assemblage of species. Some of them are free-living and others are parasites of either plants or animals.

In 1933, Chitwood proposed that the phylum Nematoda be divided into two subclasses which he called the Phasmidia and Aphasmidia, based on the presence or absence of the phasmidial organs. This classification was used by Chitwood and Chitwood in 1937. Later, in 1950, they raised the subclasses to the rank of classes, using the same names. Dougherty, in 1958, noted that Phasmidia had been used much earlier as a name for walking sticks (Orthoptera) and therefore was not available for the nematodes. In order to correct this situation, he emended and brought forth the name Secernentea von Linstow, 1905 to replace Phasmidia. Following suit in the same year, i.e., 1958, Chitwood replaced the name Aphasmidia with the emendation Adenophorea von Linstow, 1905. Thus the two classes became Secernentea and Adenophorea. Chitwood and Allen in 1959 and Thorne in 1961 used Secernentea and Adenophorea as classes of the phylum Nematoda. While free-living and parasitic species occur in both classes, the great majority of those infecting plants and animals belong in the class Secernentea.

More recently, M. B. Chitwood presented a revised classification based on the earlier one of Chitwood and Chitwood. She created some new superfamilies, elevated a number of genera to family rank, and changed some families from one suborder to another. Other authors, such as Yamaguti and Hyman, follow a different basis for classification. The original plan of classification of the Chitwoods is used here because of its greater simplicity and fundamental applicability to understanding the basic relationships. Students should consult M. B. Chitwood's more complete consideration of the classification of the Nematoda.

The life cycles of many of the nematodes parasitic in animals have been studied. They range from simple to complex, the latter involving two or three hosts. The degree of complexity does not always follow the sequence of phylogenetic progression and relationship of the worms.

The species selected to illustrate representative cycles are presented in order, according to the classification scheme of Chitwood and Chitwood, as given below. Only the specific representatives discussed in the text are included in the classification. The basic types of life cycles of nematodes are given on Plates 141 and 142.

## Classification of Nemathelminthes Included in this Chapter

Class Secernentea
 Order Rhabditida
  Suborder Rhabditina
   Superfamily Rhabditoidea
    Family Strongyloididae
     *Strongyloides papillosus*
  Suborder Strongylina
   Superfamily Strongyloidea
    Family Ancylostomatidae
     *Ancylostoma caninum*
     *Uncinaria lucasi*
    Family Strongylidae
     *Oesophagostomum columbianum*
     *Stephanurus dentatus*
     *Strongylus edentatus*
     *Strongylus equinus*
     *Strongylus vulgaris*
     *Syngamus trachea*
   Superfamily Trichostrongyloidea
    Family Trichostrongylidae
     *Haemonchus contortus*
     *Ostertagia circumcincta*
     *Trichostrongylus colubriformis*

Superfamily Metastrongyloidea
Family Metastrongylidae
*Dictyocaulus filaria*
*Metastrongylus elongatus*
*Muellerius capillaria*
*Protostrongylus rufescens*
Suborder Ascaridina
Superfamily Oxyuroidea
Family Oxyuridae
*Enterobius vermicularis*
Superfamily Ascaridoidea
Family Heterakidae
*Heterakis gallinae*
*Ascaridia galli*
Family Ascaridae
*Ascaris lumbricoides*
*Toxocara canis*
*Contracaecum aduncum*

Order Spirurida
Suborder Camallinina
Superfamily Dracunculoidea
Family Dracunculidae
*Dracunculus medinensis*
Suborder Spirurina
Superfamily Spiruroidea
Family Thelaziidae
*Oxyspirura mansoni*
Family Spiruridae
*Habronema megastoma*
Family Acuariidae
*Tetrameres crami*
Family Ascaropidae
*Physocephalus sexalatus*
*Ascarops strongylina*
Family Physalopteridae
*Physaloptera phrynosoma*
*Physoloptera rara*
Superfamily Filarioidea
Family Diplotriaenidae
*Diplotriaena translucidus*
Family Onchocercidae
*Litomosoides carinii*
*Dirofilaria immitis*

Class Adenophorea
Order Enoplida
Suborder Enoplina
Superfamily Trichuroidea
Family Trichuridae
*Trichuris ovis*
*Capillaria annulata*
*Capillaria hepatica*
*Capillaria plica*

Family Trichinellidae
*Trichinella spiralis*
Suborder Dioctophymina
Superfamily Dioctophymoidea
Family Dioctophymatidae
*Dioctophyma renale*

## SELECTED REFERENCES

Chitwood, B. G., and M. W. Allen. (1959). Nematoda. In: T. W. Edmondson, ed. H. B. Ward and G. C. Whipple, Fresh-water Biology. 2nd ed. John Wiley and Sons, New York, pp. 368–401.

Chitwood, B. G., and M. B. Chitwood. (1937). An Introduction to Nematology. Monumental Publ. Co., Baltimore, 372 pp. [New, consolidated and revised reprint in preparation by University Park Press, Baltimore]

Chitwood, B. G., and M. B. Chitwood. (1950). An Introduction to Nematology, Sec. I. Anatomy, rev. Monumental Publ. Co., Baltimore, 213 pp.

Chitwood, M. B. (1969). The systematics and biology of some parasitic nematodes. In: M. Florkin and B. T. Scheer, eds. Chemical Zoology. Academic Press, New York, Vol. 3, pp. 223–244.

Dougherty, E. C. (1958). Notes on the naming of higher taxa, with special reference to the phylum (or class) Nematoda. Bull. Zool. Nomencl. 15:896–906.

Hyman, L. H. (1951). The Invertebrates: Acanthocephala, Aschelminthes, and Entoprocta. The pseudocoelomate Bilateria. McGraw-Hill Book Co., New York, Vol. 3, pp. 53, 197–520.

Levine, N. D. (1968). Nematode Parasites of Domestic Animals and Man. Burgess Publ. Co., Minneapolis, 600 pp.

Thorne, G. (1961). Principles of Nematology. McGraw-Hill Book Co., Inc. New York, 480 pp. (see p. 88).

Yorke, W., and P. A. Maplestone. (1926). The Nematode Parasites of Vertebrates. P. Blakiston's Son & Co., Philadelphia [Reprinted by Hafner Publishing Co., New York, 1962], 536 pp.

Yamaguti, S. (1961). Systema Helminthum. The Nematodes of Vertebrates. Academic Press, New York, Vol. III, Parts 1 and 2, 1261 pp.

## Morphological Characteristics of Nematodes

Nematodes are cylindrical, nonsegmented animals with a complete digestive tract and body cavity. The body is covered with a thin cuticle secreted by a noncellular hypodermis. The cuticle may be smooth, with fine transverse striations or with adornments. These include wartlike elevations, longitudinal ridges along the sides, epaulets and cordons at the anterior end, lateral expansions anteriorly and posteriorly, and spines.

The hypodermis consists of a thin layer except for thickened cords located at the dorsal, ventral, and lateral margins that extend the full length of the body, dividing it into quadrants. Bundles of longitudinal muscles in varying numbers lie between the cords. Species with about 12 bundles per quadrant as in the Ascaridoidea are designated as polymyarian, those with two to three large bundles per quadrant are meromyarian and represented by the Oxyuroidea, and forms with uniform musculature are holomyarian and include the Trichuroidea (Plate 107, F–H).

Lacking a peritoneal lining, the body cavity is designated as a pseudocoel. In males, the intestine and the reproductive tract both open into the cloaca. The female reproductive tract opens separately. The alimentary canal consists of a muscular esophagus, an intestine, and a rectum. The esophagus and rectum are lined with an inflexion of the external cuticle. The mouth may be surrounded by six, three, two, or no lips (Plate 108, $A_2$, $E_3$, $I_3$, $J_3$). In some species, the mouth is surrounded by a crown of leaflets (Plate 108, $B_1$).

### Reproductive Systems

The reproductive organs of the male consist of a muscular ejaculatory duct, followed in succession by a seminal vesicle, sperm duct, and a filamentous testis (Plate 107, B, R). In many forms, there are one or two spicules that may be extended from the anus. The gubernaculum, a small sclerotized structure of variable shape, lies dorsal to the spicules and serves as a guide to them. In some species, there is a cuticularized thickening of the ventral wall of the cloaca known as the telamon.

The males of the Strongylina have a membranous copulatory bursa supported by a system of fleshy rays (Plate 108, $C_4$) and those of Dioctophymoidea have a fleshy, bell-shaped, rayless bursa (Plate 108, $M_2$). In groups in which the males do not have a bursa, there are paired sessile or pedunculate papillae arranged in lateroventral rows on the posterior end of the body (Plate 108, $E_2$, $F_2$, $G_2$, $H_2$).

The female reproductive system consists of a basic pattern but with variations. The ventrally located vulva opens into a vagina which gives rise to one or two uteri, occasionally four. These may arise directly from the vagina and extend posteriorly, or there may be an anterior and a posterior uterus followed in each case by an oviduct, and a filamentous ovary. In some species, a muscular ovejector connects each uterus to the vagina (Plate 108, $C_3$).

Eggs vary in shape, surface markings, and structure of the shell (Plate 108, C–F, I, L, M). Some are undeveloped when laid, whereas others are embryonated.

The excretory system is basically H-shaped with variations resulting from the loss of parts. Modifications occur in the reduction of one or both anterior limbs, giving h- or inverted U-shaped structures. Reduction may include both the anterior and posterior limbs on one side. Other forms include the original H-shape but with one or two oval glands attached to the posterior side of the bar. In some species, only a single gland, known as a renette, appears. There are no flame cells. The minute excretory pore opens ventrally near the middle region of the esophagus.

The nerve system consists of an esophageal ring and an anal ring from which nerve fibers extend anteriorly and posteriorly. Six fibers extend from the nerve ring to the lips, where their branches enervate the labial papillae. In addition, one pair leads to each of the amphids, or cephalic pits, one of which lies on each side of the head. Large commissures extend through the cords, connecting the anterior and posterior parts of the body. The rectal commissure sends nerves to rectal muscles, caudal papillae, and phasmids. The phasmids are minute pits located laterally between the anus and tip of the tail in the Secernentea.

### Life Cycles

The life cycle of nematodes may be direct without the need of an intermediate host, or it may be indirect in which case an intermediary is necessary for the development of the larvae to the infective stage. In some nematodes such as *Physaloptera* and *Gnathostoma*, third-stage larvae when eaten by unsuitable hosts migrate from the intestine to the tissues where they continue to exist. If these are fortunate enough to enter a susceptible host at a later time, growth to maturity will occur.

Development in the nematodes follows a simple pattern consisting of four stages, each

## Explanation of Plate 107

A–E, Systems. A, Digestive system. B, Typical male reproductive system. C, Typical female reproductive system as shown by Trichostrongyloidea. D, Excretory system of *Rhabditis*. E, Excretory system of *Ascaris*. F–H, Types of body musculature. F, Polymyarian type of musculature of the Ascaridoidea. G, Meromyarian type of musculature of the Oxyuroidea. H, Holomyarian type of musculature of the Trichuroidea. I–S, Developmental stages of nematodes as represented by the hookworm *Ancylostoma duodenale*, except L which is *Strongyloides papillosus*. I, First stage rhabditiform larva. J, Second stage rhabditiform larva in process of shedding molted cuticle of first stage larva. K, Third stage larva enclosed in the shed cuticle of the second stage larva; it has a strongyliform esophagus characteristic of Strongylina. L, Ensheathed third stage larva of *Strongyloides papillosus* showing filariform esophagus characteristic of most superfamilies. M–Q, Development of primitive and definitive buccal capsule in third stage larva. M, Bladder-like structures of primitive buccal capsule forming around oral cavity of the third stage larva. N, Primitive oral capsule nearly complete together with teeth in base; larval oral cavity persists in center. O, Fully developed primitive capsule with teeth and beginning of bladders as forerunners of definitive buccal capsule. P, Further development of dorsal and ventral bladders. Q, Appearance of definitive buccal capsule; primitive capsule still present and attached to esophagus prior to being lost. R, S, Final or fourth stage. R, Male after final molt but still enclosed in shed cuticle of third stage larvae with adherent primitive buccal capsule. S, Female in process of undergoing final molt with primitive capsule still attached.

1, Esophagus; 2, rhabditiform larval esophagus; 3, filariform larval esophagus; 4, intestine; 5, spicules; 6, ejaculatory duct; 7, seminal vesicle; 8, vas deferens or sperm duct; 9, testis; 10, vulva; 11, vagina; 12, ovejector; 13, uterus; 14, oviduct; 15, ovary; 16, excretory pore; 17, ducts of H-shaped excretory system with glands of Rhabditoidea; 18, glands of excretory system; 19, ducts of H-shaped excretory system without glands of Ascaridoidea; 20, renette cell; 21, dorsal cord; 22, lateral cord; 23, ventral cord; 24, longitudinal muscle bundles of body wall; 25, larval buccal capsule; 26, shed cuticle from first stage larva; 27, shed cuticle from second stage larva; 28, shed cuticle from third stage larva; 29, early stage of primitive buccal capsule; 30, later stage of primitive buccal capsule with angular teeth and remnant of larval oral cavity (25); 31, fully formed primitive buccal capsule with two pairs of well-developed teeth; 32, appearance of dorsal (on left) and ventral (on right) bladders that are precursors of definitive buccal capsule; 33, advanced stage of development of dorsal (on right) and ventral (on left) bladders; 34, early stage of definitive capsule; 35, definitive capsule; 36, primitive capsule is shed with cuticle of fourth stage larva; 37, genital primordium; 38, bursa with rays.

Figures redrawn and modified from various sources.

---

separated by a molt of the cuticle and a period of growth. This procedure may be expressed by the following formula:

$$\text{Egg} \rightarrow L_1 + M \rightarrow L_2 + \boxed{M \rightarrow L_3} +$$
$$M \rightarrow L_4 + M \rightarrow \text{Adult}$$

First and second stage larvae are rhabditoid in that the esophagus is rhabditiform. The third stage larvae have a slender strongyliform or a filariform esophagus, depending on the superfamily to which they belong, and are the infective stage. Generally, they are inside the free cuticle of the second stage, which serves as an enclosing sheath (Plate 107, K, L).

Upon entering the definitive host, third stage larvae begin development. In the Strongylina, the larval buccal capsule is replaced by a primitive one suggestive of the definitive stage in the adult. As development proceeds, a dorsal and ventral vesicle called bladders develop at the base of the primitive buccal capsule. These are precursors of the definitive buccal capsule (Plate 107, Q 31, P 33). As the definitive capsule develops, the primitive one is shed along with the cuticle during the final molt which takes place at the end of the fourth stage. During the parasitic phase of the third stage larva, the genital primordium develops into the definitive male or female reproductive systems (Plate 107, R, S).

Upon maturity, gravid females deposit unembryonated eggs (Strongyloidea, Trichostrongyloidea, Ascaridoidea, Oxyuroidea, Mermithoidea, Trichuroidea), fully embryonated eggs (Spiruroidea, Metastrongyloidea, some Filaroidea), or produce first stage larvae (Camallanoidea, Dracunculoidea, most Filarioidea). Upon embryonation, the eggs may hatch, producing first stage larvae, or development may continue to the third stage before hatching

**Plate 107**  *Basic Morphology and Stages of Development of Nematodes*  403

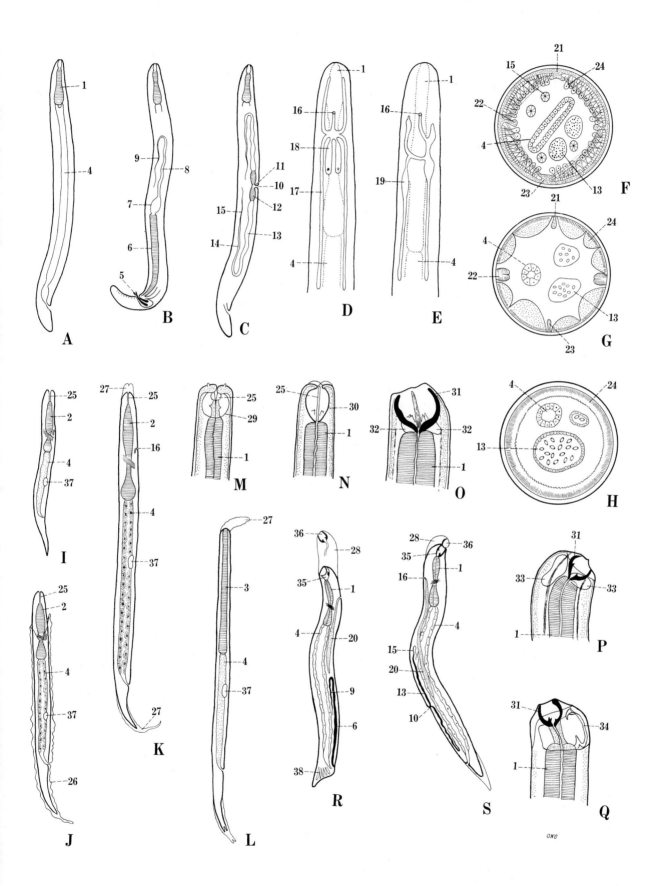

## Explanation of Plate 108

I. Order Rhabditida
  A    Suborder Rhabditina
    A. Superfamily Rhabditoidea
       *Rhabditis strongyloides:* 1, anterior end of body; 2, *en face* view; 3, right side of caudal end of male; 4, left side of caudal end of female.
  B–D Suborder Strongylina
    B. Superfamily Strongyloidea
       *Strongylus vulgaris:* 1, anterior end, showing large buccal capsule; 2, caudal end with bursa and supporting rays characteristic of suborder.
    C. Superfamily Trichostrongyloidea
       *Cooperia curticei:* 1, anterior end with rudimentary buccal capsule characteristic of superfamily; 2, caudal end of female; 3, vulva and ovejectors; 4, caudal end of male, showing rayed bursa; 5, egg in early stage of cleavage, characteristic of Strongyloidea and Trichostrongyloidea.
    D. Superfamily Metastrongyloidea
       *Metastrongylus apri:* 1, anterior end with rudimentary buccal capsule; 2, bursa with rays reduced in number and size normal for Strongylina; 3, thick-walled embryonated egg (eggs hatch in the uterus in some species.)
  E–F Suborder Ascaridina
    E. Superfamily Ascaridoidea
       *Ascaris lumbricoides:* 1, anterior end of body, showing straight esophagus and three lips; 2, caudal end of male; 3, *en face* view, showing three lips characteristic of suborder; 4, egg showing ridges over surface.
    F. Superfamily Oxyuroidea
       *Syphacea obvelata:* 1, anterior end of body, showing bulb at posterior end of esophagus, characteristic of the superfamily; 2, caudal end of male; 3, caudal end of female; 4, *en face* view, showing three lips; 5, asymmetrical egg with flattened side, characteristic of superfamily.

II. Order Spirurida
  G–H Suborder Camallanina
    G. Superfamily Camallanoidea
       *Camallanus americanus:* 1, anterior end, showing buccal plate and two-part esophagus (anterior muscular and posterior glandular portions) characterized by members of the order; 2, dextral view of tail of male, showing papillae; 3, *en face* view, showing plates.

    H. Superfamily Dracunculoidea
       *Dracunculus medinensis:* 1, anterior end of body showing swollen anterior and posterior parts of glandular portion of esophagus; 2, ventral view of posterior end of male; 3, *en face* view.
  I–J Suborder Spirurina
    I. Superfamily Spiruroidea
       *Physaloptera rara:* 1, anterior end, showing collarette and two-parted esophagus; 2, ventral view of posterior end of male, showing pedunculate papillae; 3, *en face* view showing two lips, each with a tooth-like projection; 4, thick-shelled embryonated egg of Spiruroidea.
    J. Superfamily Filarioidea
       *Dirofilaria immitis:* 1, anterior end of adult female with vagina in region of esophagus as is characteristic of this superfamily; 2, ventral view of caudal end of male, showing spicules of different length, a characteristic of the suborder Spirurina; 3, *en face* view of head.

III. Order Enoplida
    K. Suborder Dorylaimina
       Superfamily Mermithoidea
       *Paragordius robustus:* 1, anterior end; 2, ventral view of caudal end of male; 3, dorsal view of caudal end of female; 4, egg.
    L. Superfamily Trichuroidea
       *Capillaria hepatica:* 1, anterior end of adult female, showing anterior muscular part of esophagus and long posterior stichosome portion characteristic of this superfamily; vulva is near end of esophagus; 2, caudal end of male, showing single spicule in extruded spicular sheath; 3, egg characteristic of Trichuroidea.
    M. Suborder Dioctophymina
       *Dioctophyma renale:* 1, anterior end of female; 2, caudal end of male, showing bell-shaped fleshy bursa with single spicule characteristic of this suborder; 3, *en face* view; 4, egg in optical section.

a, Lips; b, dorsal lip; c, ventrolateral lip; d, mouth; e, labial papillae; f, buccal capsule; g, excretory pore; h, caudal papillae; i, phasmid; j, amphid; k, spicule enclosed in spicular sheath; l, bursa; m, bursal rays; n, buccal plates; o, muscular portion of esophagus; p, glandular portion of esophagus; q, bulb of esophagus; r, stichosome; s, vagina; t, uterus; u, ovejector; v, eggs.

Figures redrawn from various sources.

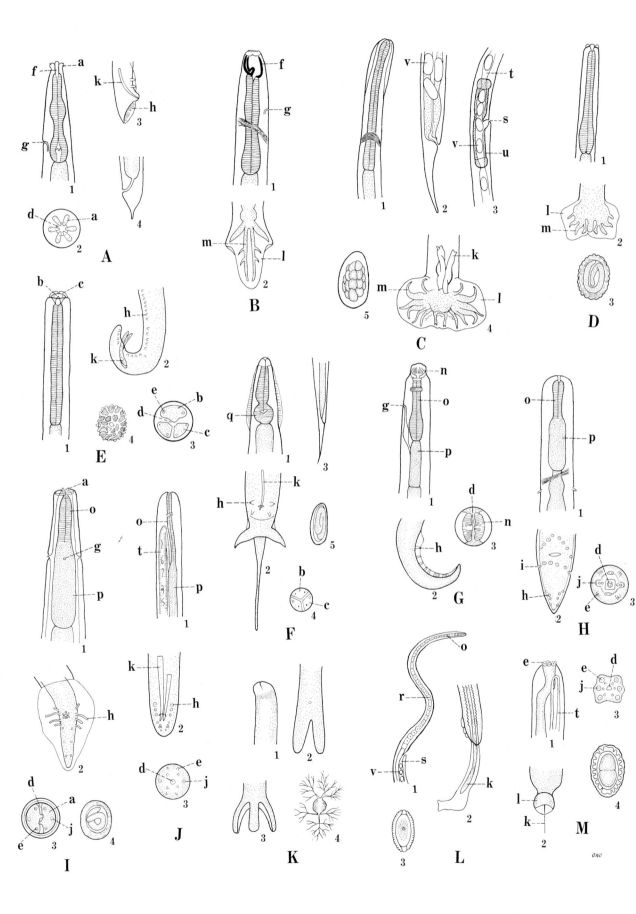

(*Nematodirus, Uncinaria*). In others, first stage larvae hatch when eaten by the intermediate hosts (Spiruroidea) or as second stage larvae upon reaching the intestine of the definitive host (Ascaridoidea).

Hatching of eggs occurs under a variety of conditions in the different groups of nematodes. In most of the Strongyloidea and Trichostrongyloidea, it takes place when the larvae are in the first stage, whereas in the genera *Uncinaria* and *Nematodirus* of the two above superfamilies, respectively, it does not occur before the larvae have developed to the third stage. In the Ascaridina, the larvae develop to the second stage in the eggs but hatch only when swallowed by the definitive host.

First stage larvae are present in the eggs of the Spiruroidea, some Metastrongyloidea, and the Filarioidea of the lungs and air sacs at the time of oviposition. There is no further development until they hatch in the intestine of an intermediate host. In the Camallanoidea, Dracunculoidea and tissue-dwelling Filarioidea, hatching occurs in the uterus of the female and first stage larvae are born.

### CLASS SECERNENTEA

Phasmids are present laterally between the anus and tip of tail but they are difficult to see in adult parasitic worms. The terminal portion of the excretory duct is lined with cuticle. Lateral canals are present. The cephalic sensory organs are papilloid. The amphids are two small pore-like openings located laterally on the lips.

### ORDER RHABDITIDA

Esophagus of adult worms is variable in shape, ranging from clavate (club-shaped) to cylindrical but showing a corpus (elongate anterior enlargement), isthmus, and bulbular posterior portion in the early larval stages.

#### Suborder Rhabditina

There is no stylet in the mouth and the latter is not surrounded by a corona radiata, i.e., one or two rows of leaflet-like structures. There are 0, 2, 3, or 6 lips.

#### Superfamily Rhabditoidea

The stoma (mouth) is usually distinct and the esophagus consists basically of a corpus, isthmus, and valved or nonvalved posterior swelling.

### FAMILY STRONGYLOIDIDAE

The members of this family are small and for the most part free-living nematodes. Some of them have adapted to a parasitic life during part of their existence. *Rhabditis strongyloides* is normally free-living but invades sores in the skin of dogs, under which circumstances it is parasitic. Members of the genus *Strongyloides*, on the other hand, have advanced their relationship with the vertebrate host so that in addition to the free-living heterogonic generation of rhabditiform males and females there is a parasitic homogonic one of only parthenogenetic females with greatly elongated filariform esophagus in the intestine of vertebrates.

*Strongyloides papillosus*
(Wedl, 1856) (Plate 109)

This species is parasitic in the small intestine of domestic and wild ruminants, and rabbits, particularly in warm moist climates.

**Description.**    Parasitic females are slender worms 4.78 to 5.85 mm long by 50 to 60 $\mu$ in diameter at the vulva. The cuticle is marked with fine striations. There are four lips, four cephalic papillae, and two large amphids. The tail is 32 $\mu$ long, tapering finger-like. There are two ovaries, one anterior and one posterior, each containing eggs. The vulva lies between the middle and posterior thirds of the body. The eggs are embryonated when laid and measure 40 to 60 $\mu$ by 32 to 40 $\mu$.

Third stage larvae are active and measure 575 to 640 $\mu$ long by 16 $\mu$ wide; the tip of the tail is trifid. The esophagus is filariform and about 40 percent of the total length of the body.

Free-living female adults measure 770 to 1110 $\mu$ long by 50 to 90 $\mu$ wide. There are two lips, each with two papillae. The esophagus is rhabditiform and 140 $\mu$ long. Ovaries are paired, one anterior and one posterior; the vulva is median. Males are 700 to 825 $\mu$ long by 50 $\mu$ wide with a single testis. The gubernaculum is 20 $\mu$ long by 2.5 $\mu$ wide and the arcuate spicules are 33 $\mu$ long.

Parthenogenetic females in the alimentary canal of the vertebrate host produce eggs with different numbers of chromosomes. They are 1) the 3N type that develops directly into homogonic filariform female larvae; 2) the 1N type that produces heterogonic free-living rhabditiform males; and 3) 2N type that produces heterogonic free-living rhabditiform females. The progeny of the 1N males and 2N females are 3N larvae that develop into parasitic parthenogenetic females upon entering the vertebrate host.

**Life Cycle.**    This parasite has two kinds of life cycles, the heterogonic and the homogonic.

In the homogonic cycle, the parthenogenetic females are parasitic, being partially embedded in the mucosa of the small intestine. The 3N eggs produce first stage rhabditiform larvae that during growth pass through a second rhabditiform stage into filariform infective third stage larvae. These are infective to sheep, entering either by mouth or through the skin. They develop into parthenogenetic females. Parasitic males have not been found.

In the heterogonic cycle, the 1N eggs develop into rhabditiform free-living males and the 2N eggs into similar free-living females. They develop to adults by going through four molts characteristic of all nematodes. Their offspring are 3N larvae which develop into infective filariform larvae, similar to those of the homogonic cycle. The preparasitic larvae molt twice in the soil before reaching the infective filariform stage. Infection of sheep takes place by two routes. Filariform larvae may enter by mouth along with contaminated food or by burrowing through the skin from moist soil. When swallowed, they molt twice in the intestine and grow to maturity. If entrance is by way of the skin, they are carried by the blood through the heart and into the lungs where one molt occurs. Leaving the pulmonary vessels, the larvae migrate up the trachea into the pharynx, and are swallowed. The fourth and final molt takes place in the small intestine. Maturity is attained in about 1 week.

*Strongyloides ratti* of rats and mice, *S. ransomi* of swine, and *S. westeri* of horses have a life cycle similar to that of *S. papillosus* except that they are also transmitted in the colostrum of the mother to her offspring. It is not known whether this mode of transmission takes place in *S. stercoralis* of humans and dogs, *S. cati* of cats, and *S. papillosus*. *S. avium* of chickens and turkeys has a life cycle consisting of parasitic and free-living adults.

*Strongyloides fülleborni* of primates is said to have a succession of free-living generations. Growth and development are dependent on semiaerobic conditions in the fecal masses. At 26°C, eggs of the free-living stage hatch in 5 to 7 hours,

the first molt occurs in 10 to 12 hours, the second in 16 to 18 hours, and the third in 22 to 24 hours. After 34 hours, adult free-living forms are present and eggs are being laid. In 8 days, second generation adults of both sexes are present. It is not known whether larvae are transmitted in the milk.

### EXERCISE ON LIFE CYCLE

The most suitable species for experimental studies is *Strongyloides ratti*, found commonly in wild rats. Freshly voided feces from infected rats contain eggs. When droppings are kept moist at room temperature for 48 to 72 hours, preferably the latter, infective filariform larvae are present. They are easily isolated by means of a small Baermann apparatus.

Infection of rats may be accomplished by 1) feeding filariform larvae, 2) placing the larvae on a shaved area of the skin in a drop of water which is permitted to evaporate, or 3) injecting them subcutaneously. Infection with a single filariform larva is done by placing it on the shaved skin of a rat or injecting it under the skin. If the infection is successful, a single parthenogenetic female will develop in the small intestine and produce eggs in about a week.

Following large doses of infective larvae given by mouth or through the skin, the rats should be sacrificed on days 2, 3, and 5 to determine the route followed by the larvae in reaching the intestine. Mince the lungs of each rat and baermannize them to determine when the larvae are migrating through them. Also examine the contents of the stomach and intestine separately to ascertain where the larvae are at a given period after infection. Determine whether both third stage filariform and rhabditiform larvae appear in the feces after incubation for 48 to 72 hours.

### SELECTED REFERENCES

Abadie, S. H. (1963). The life cycle of *Strongyloides ratti*. J. Parasitol. 49:241–248.

Basir, M. A. (1950). The morphology and development of the sheep nematode, *Strongyloides papillosus* (Wedl, 1856). Can. J. Res. 28:173–196.

Beg, M. K. (1968). Studies on the life cycle of *Strongyloides fülleborni* von Linstow, 1905. Ann. Trop. Med. Parasitol. 64:502–505.

Faust, E. C., P. R. Russell, and R. C. Jung. (1970).

## Explanation of Plate 109

A, Adult parasitic female (males unknown). B, *En face* view of adult parasitic female. C, Embryonated egg. D, First stage rhabditiform larva. E, First molt producing second stage rhabditiform larva. F, Second molt with loose cuticle enclosing third stage filariform infective larva. G, Tip of tail of third stage larva. H, Adult free-living female. I, *En face* view of free-living adult. J, Adult free-living male. K, Spicule. L, First stage rhabditiform larva of free-living parents. M, Second stage rhabditiform larva with molting cuticle. N, Third stage rhabditiform larva with loose cuticle. O, Sheep definitive host.

1, Filariform esophagus; 2, nerve ring; 3, intestine; 4, anus; 5, tail; 6, anterior ovary; 7, oviduct; 8, uterus; 9, posterior ovary; 10, oviduct; 11, egg in uterus; 12, vulva; 13, cephalic papilla; 14, amphid; 15, mouth; 16, larva in egg; 17, genital primordium; 18, loose cuticle of first stage larva; 19, cuticle of second stage larva enclosing third stage larva; 20, trifurcate tip of tail of third stage parasitic filariform larva; 21, rhabditiform esophagus; 22, testis; 23, seminal vesicle; 24, spicules.

a, Adult parthenogenetic female partially embedded in intestinal mucosa; b, eggs deposited in mucosa; c, eggs escape from mucosa into intestinal lumen; d, eggs passed in feces; e, eggs already embryonated when voided with feces; f, eggs hatch, releasing rhabditiform larva; g, free first stage rhabditiform larva of parasitic line in soil; h, second stage rhabditiform larva shedding cuticle of first stage larva; i, third stage filariform larva shedding cuticle of second stage larva; j, third stage infective larva; k, infection occurs when fully developed third stage larvae are swallowed; l, larvae pass down digestive tract; m, third stage larvae molt, producing fourth stage; n, fourth stage larvae penetrate mucosa of intestine and mature; o, infection of host may occur also through the skin, with larvae entering circulatory system; p, larva carried toward heart in veins; q, larvae pass through right side of heart toward lungs; r, in lungs, larvae pass from blood vessels into alveoli; s, larvae molt and migrate up trachea; t, larvae enter pharynx, are swallowed, and pass through stomachs; u, larvae molt fourth time; v, larvae burrow into mucosa; w, adult female.

a', Molting first stage larvae of rhabditiform free-living line; b', second stage larva molts (two other molts follow); c', adult free-living male; d', adult free-living female; e', embryonated egg from free-living female; f', egg produces rhabditiform first-stage larva; g', second stage rhabditiform larva; h', filariform third stage parasitic larva infective to sheep by mouth or by skin.

Figures adapted from Basir, 1950, Can. J. Res. (D) 28:173.

---

Craig and Faust's Clinical Parasitology. 8th ed. Lea & Febiger, Philadelphia, p. 284.

Graham, G. L. (1936). Studies on Strongyloides. I. *S. ratti* in parasitic series, each generation in the rat established with a single homogonic larva. Amer. J. Hyg. 24:71–87.

Graham, G. L. (1938). Studies on Strongyloides. II. Homogonic and heterogonic progeny of a single homogonically derived *S. ratti* parasite. Amer. J. Hyg. 27:221–234.

Graham, G. L. (1938). Studies on Strongyloides. III. The fecundity of single *S. ratti* of homogonic origin. J. Parasitol. 24:233–243.

Graham, G. L. (1939). Studies on Strongyloides. IV. Seasonal variation in the production of heterogonic progeny by single established *S. ratti* from a homogonically derived line. Amer. J. Hyg. (D) 30: 15–27.

Graham, G. L. (1939). Studies on Strongyloides. V. Constitutional differences between a homogonic and a heterogonic line of *S. ratti*. J. Parasitol. 25:365–375.

Graham, G. L. (1940). Studies on Strongyloides. VI. Comparison of two homogonic lines of singly established *S. ratti*. J. Parasitol. 26:207–218.

Graham, G. L. (1940). Studies on Strongyloides. VII. Length of productive life in a homogonic line of singly established *S. ratti*. Stud. Rockefeller Inst. Med. Res. 116:515–529.

Katz, F. F. (1969). *Strongyloides ratti* (Nematoda) in newborn offspring of inoculated rats. Proc. Pa. Acad. Sci. 43:221–225.

Lucker, J. T. (1934). Development of the swine nematode *Strongyloides ransomi* and the behavior of its larvae. U. S. Dep. Agr. Tech. Bull. 437; pp. 1–30.

Lyons, E. T., J. H. Drudge, and S. Tolliver. (1969). Parasites from mare's milk. The Blood-Horse 95: 2270–2271.

Monocol, D. J., and E. G. Batte. (1966). Transcolostral infection of new-born pigs with *Strongyloides ransomi*. Vet. Med. Small Anim. Clin. June, pp. 583–586.

Spindler, L. A. (1958). The occurrence of the intestinal threadworms, *Strongyloides ratti*, in the tissues of rats, following experimental percutaneous infection. Proc. Helminthol. Soc. Wash. 25:106–111.

Supperer, R., and H. Pfeiffer. (1967). Zum Problem der 'pränatel' Strongyloides-invasion beim Schwein. Wien. Tierärztl. Monatsschr. 54:101–103.

**Plate 109** *Strongyloides papillosus* 409

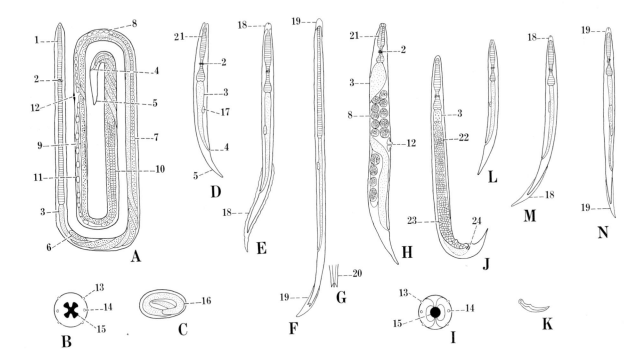

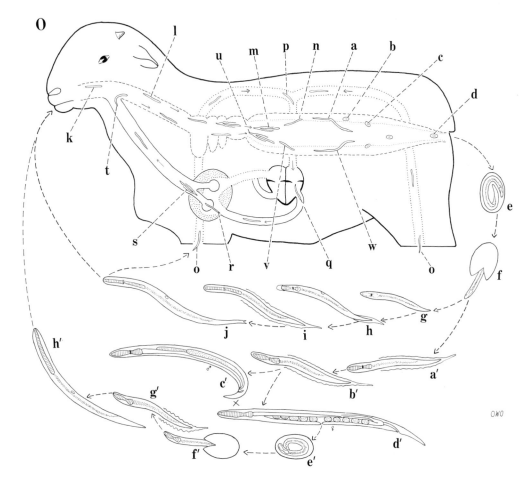

## Explanation of Plate 110

**A**, Anterior end of adult *Ancylostoma caninum*. **B**, Bursa of *A. caninum*. **C**, Anterior end of *A. braziliense*. **D**, Bursa of *A. braziliense*. **E**, Anterior end of rhabditiform second stage larva. **F**, Posterior end of second stage rhabditiform larva. **G**, Anterior end of filariform infective third stage larva. **H**, Posterior end of third stage larva. **I**, Dog definitive host. **J**, Humans susceptible to infection by larvae of *A. braziliense* hatching from eggs passed by dogs and cats.

**1**, Mouth; **2**, dorsal cutting plates; **3**, ventral cutting plates; **4**, wall of buccal capsule; **5**, esophagus; **6**, intestine; **7**, anus; **8**, copulatory bursa; **9**, dorsal lobe of bursa; **10, 11**, ventral rays (10, ventroventral rays; 11, lateroventral rays); **12–14**, lateral rays (12, externo- or anterolateral rays, 13, mediolateral rays; 14, posterolateral rays); **15, 16**, dorsal rays (15, externodorsal rays; 16, dorsal ray); **17**, rhabditiform esophagus of second stage larva; **18**, nerve ring; **19**, excretory tubule; **20**, filariform (or strongyliform) esophagus of third stage larva; **21**, retained cuticular sheath of second molt enclosing third stage larva.

**a**, Adult worms in small intestine; **b**, unembryonated eggs laid in intestine; **c**, eggs leave body in feces; **d**, development of eggs takes place outside of body; **e**, fully developed first stage larva in egg; **f**, hatching of first stage larva; **g**, first stage rhabditiform larva feeds and grows; **h**, molting of first stage larva produces the second stage rhabditiform larva; **i**, molting of second stage larva produces third stage

stage; **j**, infection of dogs occurs when infective filariform larva enclosed in loose cuticle of second third stage larvae are swallowed; **k**, larvae molt twice en route through stomach and intestine to final location where they attach and mature (a) in about 5 weeks; **l**, infective larvae penetrating skin enter blood vessels or **m**, lymph vessels and are carried to the heart in the circulation; **n**, larvae from blood vessels and lymphatics enter postcaval vein; **o**, larvae in right side of heart; **p**, larvae leave heart via pulmonary artery; **q**, larvae migrate from capillaries of lungs into alveoli of lungs; **r**, larvae migrate up trachea; **s**, larvae enter pharynx and are swallowed; **t**, two molts occur before they grow to maturity in small intestine (a); **u**, some larvae pass through the lungs in the blood vessels; **v**, they enter left side of heart and are carried into the general circulation; **w**, larvae enter dorsal aorta; **x**, larvae in dorsal aorta; **y**, some larvae enter uterine artery; **z**, in cases of pregnancy, larvae pass through the fetal membrane and are carried by the blood into the liver of the fetus where they remain until after birth, at which time migration through the lungs is completed and infection occurs in the intestine of very young pups.

**a′**, In the case of *A. braziliense*, third stage larvae penetrate the skin of humans coming in contact with them and cause a dermatitis known as creeping eruption; the larvae are eventually destroyed in the skin.

Figures A–C adapted from various sources; E, F, schematized stages of hookworm larvae; **D** adapted from Biocca, 1951, J. Helminthol. 25:1.

---

### Suborder Strongylina

The stoma of this group ranges from well-developed to rudimentary but it is never collapsed. There may or may not be a corona radiata surrounding the mouth. The males have a true or strongyloid bursa supported by four groups of well-developed muscular rays. With the exception of the dorsal ray, which is single and median, they are bilateral. These groups are 1) the single dorsal ray; 2) one pair of externodorsal rays; 3) the paired postero-, medio-, and anterolateral rays; and 4) the paired latero- and ventroventral rays.

### Superfamily Strongyloidea

The mouth may or may not be surrounded by a crown of leaflet-like structures. The stoma is well developed and it has thick walls. Adults occur in the gut, kidney, or respiratory tract of reptiles, birds, and mammals. Two families are considered.

### FAMILY ANCYLOSTOMATIDAE

This family includes the hookworms which occur in the gut of humans, dogs, cats, domestic ruminants, swine, and pinnipeds. Hookworms are common in the warm, moist climates but may reach abundant proportions in cold areas, as in the fur seals on the Pribilof Islands in the Bering Sea. In this group there is no corona radiata surrounding the mouth. The oral opening has a well-developed pair of dorsal and ventral cutting plates.

*Ancylostoma caninum*
Ercolani, 1859 (Plate 110)

*Ancylostoma caninum* is a parasite of dogs, foxes, and cats; it inhabits the small intestine, attaching by means of the buccal capsule to the mucosal lining. It occurs in warm and temperate climates, especially where there is adequate moisture.

**Description.**    Males are 10 to 12 mm and

**Plate 110**   *Ancylostoma caninum and A. braziliensis*   411

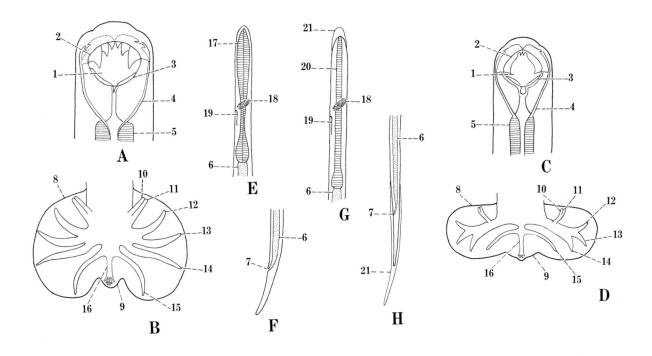

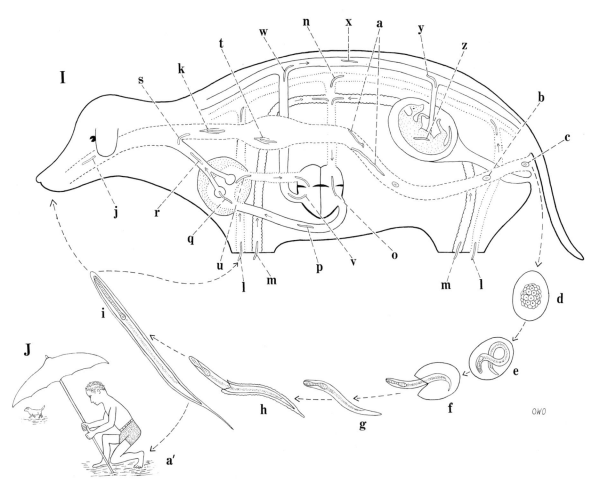

females 14 to 16 mm long. They usually have blood in the intestine. The anterior end is bent dorsad and the mouth has a pair of dorsal cuticularized plates, each with three sharp teeth, the outer being the largest; there is a pair of triangular dorsal and ventrolateral teeth inside the buccal capsule. The male bursa is well developed and the rays are arranged in a manner characteristic of the species. Spicules are 0.9 mm long. The vulva is near the junction of the middle and posterior thirds of the body. Eggs are 56 to 65 $\mu$ long by 38 to 43 $\mu$ wide and have developed to about the eight-cell stage when laid.

**Life Cycle.**    The partially embryonated eggs develop and hatch into first stage rhabditiform larvae in the soil. They feed on organic matter for a short time and then undergo the first molt, completely shedding the cuticle. After a short period of feeding again, the cuticle of the second stage rhabditiform larva loosens and forms an enclosing sheath for the third stage infective filariform larva. It differs from the two preceding rhabditiform stages in having an elongated strongyliform esophagus, i.e., one with a weak, flask-like swelling at the posterior end. The infective third stage is reached in about a week during warm weather.

Infection of dogs occurs when the infective third stage larvae are swallowed or burrow through the skin. Swallowing appears to be the common means of infection due to the eating habits of dogs. Toughness of the skin makes burrowing through it difficult, thus limiting it as a route of infection. Upon being swallowed, the larvae molt in the stomach and enter the crypts. After a short time, they migrate to the small intestine, molt the fourth and final time, and develop to maturity in about 5 weeks.

If entrance is via the skin, the larvae enter through the hair follicles, generally leaving them just above the sebaceous glands, and migrate into the dermis and on into the hypodermis with its many blood and lymph capillaries. Upon entering them, the larvae are carried to the lungs, where they leave the capillaries and go up the trachea, are swallowed, and pass on to the intestine.

Some larvae pass through the capillaries of the lungs and are carried back to the heart. These worms circulate throughout the body, finally lodging in various organs where many die and are absorbed or become calcified. Those in

the mammary glands enter the milk cisterns whence they are passed in the colostrum and milk to nursing pups for a period of 3 weeks, at which time no larvae are left. In the case of pregnant bitches, however, the larvae may enter the fetuses, thus infecting them prenatally. They remain dormant in the liver until the puppies are born, at which time they undertake the pulmonary portion of the migration, reaching the intestine and developing to maturity while the pups are still very young.

Upon reaching the intestine, the worms begin feeding by drawing tufts of mucosa into the buccal capsule. The tissue is digested by enzymatic action and swallowed to be absorbed through the microvilli of the epithelium (Fig. 31). The worms move to new sites and attach for continued feeding. Through action of a polypeptide anticoagulant produced by the cephalic glands and injected into the wound, there is prolonged bleeding into the lumen from the bites. Blood appears first in the feces by day 8 and reaches a maximum by days 12 to 16. A second peak of hemorrhage occurs during days 23 to 25. Peak losses of erythrocytes precede the onset of

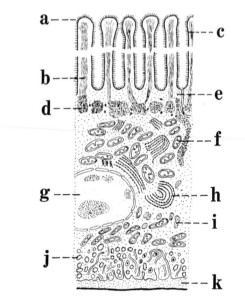

Fig. 31.   Schematic representation from electron micrographs of an epithelial cell of the intestine of the nematode *Ancylostoma caninum*. **a,** Fibrous surface coat; **b,** central core of microvillus; **c,** cell membrane; **d,** terminal web; **e,** ribosomes; **f,** mitochondrion; **g,** nucleus; **h,** endoplasmic reticulum; **i,** lipid-like droplet; **j,** vesicular reticulum; **k,** basal lamina. Adapted from Lee, 1969, Exp. Parasitol. 24: 336.

egg production, at which time the nutritional demands of the females are greatest.

The life cycles of the other species of hookworms occurring in humans, cattle, sheep, goats, equines, and swine are basically the same as that described for *A. caninum*. The point on prenatal or transcolostral infection is not known for them. Penetration through the skin is the more common route.

*Ancylostoma braziliense* (Plate 110), a parasite of cats and dogs, is of particular interest because the third stage larvae readily penetrate the skin of humans. Being in a foreign host, they are unable to pass through the skin but migrate laterally in it, causing a severe dermatitis known as creeping eruption. Beaches, yards, and playgrounds frequented by infected dogs and cats become heavily infested with larvae which attack people working or playing in contact with the soil.

## EXERCISE ON LIFE CYCLE

Eggs obtained from the feces of dogs, preferably young ones, may be incubated at room temperature in a mixture of moist sand and feces, and the different stages of larvae obtained by baermannizing portions of the material at short intervals. As a precaution, the sand should be heated prior to use in order to destroy free-living nematodes that might be in it, creating complications in identification and experimentation. Likewise, the feces should not have been in contact with the soil as that might be a means of introducing free-living nematodes. Third stage larvae in the protective sheath-like cuticle of the second stage larvae are easily procured from the culture by means of a Baermann apparatus.

The route of migration in the vertebrate host after entering by way of the mouth or the skin may be followed in mice. After feeding a large number of third stage larvae to each of five or six mice, examine the intestine, liver, and lungs of one at intervals of 24 hours to determine where the larvae are at each period. Place third stage larvae in a drop of water on the shaved skin of an equal number of mice, allowing it to dry, and examine them on the same plan presented above. Lungs of these mice but not the liver should be examined by means of a Baermann apparatus and by sections to determine route of larvae. Place a piece of mouse skin over a beaker of warm saline water (37°C) with the flesh side in

contact with the water, and pipette a known number of third stage larvae on the haired side. Examine the sediment in the beaker after several hours to determine whether larvae have passed through it. Wash the haired side in another beaker of water to recover those that did not penetrate. Determine what percentage of them burrowed through the skin. Section a portion of this skin to ascertain whether larvae were migrating through it.

## SELECTED REFERENCES

Eiff, J. A. (1966). Nature of an anticoagulant from the cephalic glands of *Ancylostoma caninum*. J. Parasitol. 52:833–843.

Enigk, K., and M. Stoye. (1968). Untersuchungen über den Infektions weg von *Ancylostoma caninum*. Med. Klin. 63:1012–1017.

Faust, E. C., P. F. Russell, and R. C. Jung. (1970). Craig and Faust's Clinical Parasitology. 8th ed. Lea & Febiger, Philadelphia, p. 297.

Foster, A. O. (1932). Prenatal infection with the dog hookworm, *Ancylostoma caninum*. J. Parasitol. 19:112–118.

Foster, A. O., and S. X. Cross. (1934). The direct development of hookworms after oral infection. Amer. J. Trop. Med. 14:565–573.

Georgi, J. R. (1968). Estimation of erythrocyte loss in hookworm infection by whole-body $^{59}$Fe counting. J. Parasitol. 54:417–425.

Kalkofen, U. P. (1970). Attachment and feeding behavior of *Ancylostoma caninum*. Z. Parasitenk. 33:339–354.

Lee, C. C. (1969). *Ancylostoma caninum*: fine structure of intestinal epithelium. Exp. Parasitol. 24:336–347.

Matsusaki, G. (1951). Studies on the life history of the hookworm. Part VI. On the development of *Ancylostoma caninum* in the normal host. Yokohama Med. Bull. 1:111–120.

Miller, T. A. (1966). Blood loss during hookworm infection, determined by erythrocyte labelling with radioactive $^{51}$chromium. I. Infection of dogs with normal and with X-radiated *Ancylostoma caninum*. J. Parasitol. 52:844–855.

Rep, B. H. (1965). The pathogenicity of *Ancylostoma braziliense*. II. Cultivation of hookworm larvae. Trop. Geogr. Med. 17:329–337.

Vetter, J. C. M. (1970). Skin penetration by hookworm larvae. J. Parasitol. 56(4/2):479.

## *Uncinaria lucasi* Stiles, 1901 (Plate 111)

*Uncinaria lucasi* is a hookworm parasitic in the lower part of the small intestine of only young

## Explanation of Plate 111

A, Dorsal view of adult showing mouth and cutting plates. B, Dextral view showing buccal capsule with pair of spine-like teeth on ventral wall. C, Dorsal view of copulatory bursa of adult male. D, Sinistral view of tail of adult female. E, Ventral view of anterior end of 81-hour-old fourth stage larva. F, Sinistral view of tail of 81-hour-old male fourth stage larva. G, Egg in early stage of development. H, First stage larva. I, Second stage larva. J, Third stage larva. K, Adult cow seal nursing pup. L, Newborn pup taking its first meal of milk. M, Small group of seals.

1, Mouth; 2, cutting plates; 3, buccal capsule; 4, spines in buccal capsule; 5, esophagus; 6, ventroventral ray; 7, lateroventral ray; 8, anterolateral ray; 9, mediolateral ray; 10, posterolateral ray; 11, externodorsal ray; 12, dorsal ray; 13, membrane of bursa; 14, anus; 15, tail; 16, buccal capsule of fourth stage larva; 17, vacuole from which buccal capsule of fifth stage will develop; 18, cuticle surrounding fourth stage larva; 19, developing bursa of fifth stage; 20, developing egg; 21, first stage larva; 22, second stage larva; 23, loose cuticle of first stage larva serving as a sheath (remains of buccal capsule inside sheath); 24, third stage larva; 25, so-called "spears" which are buccal capsule in optical section; 26, loosened cuticle of second stage larva serving as a sheath.

a, Adult worms in pups (the prepatent period is about 14 days; for convenience of illustrating the life cycle, adult worms are shown in newborn pups); b, developing egg; c, egg with first stage larva; d, egg with second stage larva; e, egg with third stage larva; f, third stage larva hatching; g, free-living third stage larva which shows but one sheath; h, free-living third stage larvae penetrate flippers of seals of all ages; i, larvae enter veins; j, larvae go to vena cava; k, larvae pass through heart; l, larvae enter capillaries of lungs and continue through them; m, larvae in pulmonary artery; n, larvae go through left side of heart; o, larvae enter dorsal aorta whence they are distributed to all parts of the body; p, larvae in blood vessels leading to ventral wall of body; q, third stage parasitic larvae accumulate in blubber of belly of all ages and both sexes of seals; r, larvae accumulate in mammary glands of pregnant cows; s, everted nipple; t, third stage parasitic larvae swallowed by newborn pups during their first meal; u, fourth stage larvae molt; v, fifth stage, which will develop to maturity in about 14 to 15 days; w, placenta of newborn pup; x, eggs from intestinal phase of adult passed in feces of only young pups; y, eggs hatch in sand of rookeries and third stage infective larvae appear; z, larvae penetrate flippers of all ages and both sexes of seals and go to tissues where they remain as third stage parasitic larvae unless ingested with the milk.

Figures E–J adapted from E. T. Lyons, 1961, Doctoral Dissertation, Colorado State Univ.

---

pups of the northern fur seal (*Callorhinus ursinus*) and Steller's sea lions (*Eumetopias jubata*). Pups of fur seals beyond 5 months of age rarely remain infected and older seals are not infected. These parasites are known to occur only on the breeding grounds, or rookeries as they are called, of the fur seals on the Pribilof and Kommandorski Islands in the Bering Sea. Probably they occur wherever sea lions breed but this point has not been ascertained. It is the only strongyle parasite of fur seals.

**Description.**     Adult males are 7.4 to 8.7 mm long and the females 12.4 to 16 mm. The lateral rays of the copulatory bursa are almost equal in length. Spicules measure 500 to 560 $\mu$ long and have transversely striated flanges along the dorsal side for almost the entire length; the flanges are fused near the distal tips of the spicules but the tips of the latter are free. The tail of the female has a small terminal spike-like tip. Eggs have transparent, three-layered shells, and measure 120 to 140 $\mu$ long by 80 to 88 $\mu$ wide. They are in the early stages of cleavage when passed in the feces.

**Life Cycle.**     The life cycle consists of three phases. They are 1) the intestinal; 2) the tissue; and 3) the free-living phases. In the intestinal phase, adult worms occur in the lower extremity of the small intestine of only young fur seal pups from 2 weeks to 4 to 5 months of age; they do not occur in any other age group of fur seals. In the tissue phase, third stage larvae occur in the tissues, especially the blubber of the belly of all age groups of seals. They are in the mammary glands of pregnant cows during the sojourn at sea and in parous cows for a short period after parturition. The free-living phase consists of third stage parasitic larvae in the soil.

Newborn pups are infected during the first few days of their lives by means of the milk which contains parasitic third stage larvae. Upon entering the alimentary canal, two molts occur

**Plate 111**   *Uncinaria lucasi*                                                                 415

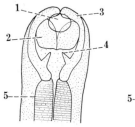

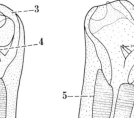

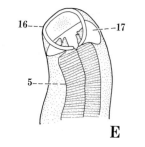

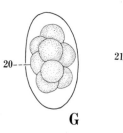

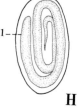

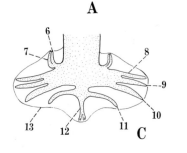

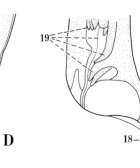

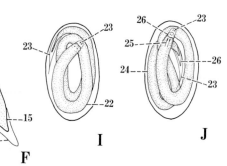

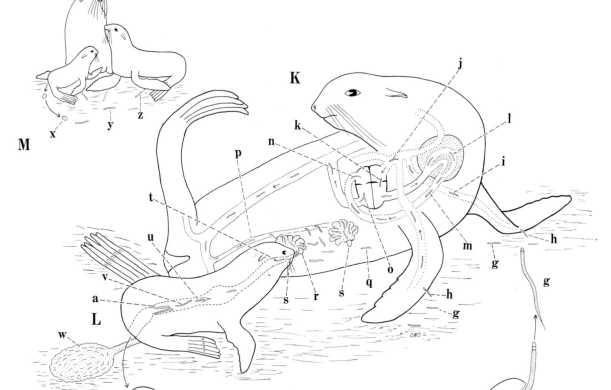

and the worms develop to maturity in the lower portion of the small intestine without migrating through the body.

Sexual maturity is attained in about 2 weeks by all of the worms, since they were acquired at one time through the colostrum, and large numbers of eggs in the early stages of cleavage are passed in bloody feces. The eggs continue development in the sand during the summer (June to September), reaching the third stage strongyliform larvae. These do not hatch, however, until late in the summer, usually some time around the beginning of September, although the time may vary somewhat. The free-living third stage larvae are hardy and may or may not survive the winter months. They may be abundant in the sand in the spring when the breeding seals return to the rookeries following their winter sojourn at sea.

Adult seals returning to the rookeries each season are exposed to the active penetration of the third stage free-living larvae in the sand. Upon entering the tissues, the larvae presumably enter the circulation and are carried to the heart, pass through the lungs, and back to the heart, whence they are distributed to all parts of the body. They accumulate in the blubber of the belly of all seals and in the mammary glands of the females, where they develop to parasitic third stage larvae.

Newborn pups are exposed to the free-living third stage larvae in the sand and to the parasitic third stage larvae in the milk of their mothers. Larvae entering the skin from the sand go to the belly blubber, remaining in the third stage, whereas those that are swallowed with the milk develop to adults in the small intestine.

Having homing instincts, together with being social and gregarious, the breeding seals assemble in great numbers in compact herds on the rookeries, which are used year after year from early June until late August. At the latter time, the social structure of the harems begins to disintegrate and the seals move to the water's edge. The social and homing instincts of the seals favor the perpetuation of the hookworms, assuring a very high incidence of infection of each generation of pups with the intestinal phase and subsequent production of eggs.

While eggs are passed onto the rookeries throughout the summer, hatching does not occur until late August or early September, when the harems begin to break up and the seals leave them for the sea and the narrow margin of land along the shore. Presumably the intense contamination of the rookeries with feces and urine inhibits hatching of the eggs until such a time that the soil begins to freshen in the absence of the seals and under the influence of the weather. The larvae infect seals that return to the harem areas during the remainder of their stay on land and when they return the following year.

Breeding animals returning to the harems year after year presumably build up intense infections in the tissue, since the larvae survive beyond a year in them.

In summary, the life cycle of *U. lucasi* is as follows: parasitic third stage larvae of the tissue phase in the mammary glands are swallowed by the pups with their first meal of milk. These larvae develop in about 2 weeks to adults which remain in the intestine of the pups until late in the fall, when they are lost completely. Eggs passed onto the rookeries in the feces during the summer begin hatching at the end of August or thereabout and infect seals that come in contact with the larvae. These go to the tissues and remain there. Larvae surviving the winter in the soil enter the flippers of the adult seals as soon as they return to the harems from the sea in the spring, and the pups as soon as they are born. Larval infections in the tissues persist for long periods.

The life cycle is unique in that the larvae of the tissue phase are transmitted to the young pups through the milk and develop to the intestinal phase as adults only in the young, never occurring as such in adult animals.

*Uncinaria hamiltoni*, a hookworm of the southern sea lion *Otaria byroni*, the California sea lion *Zalophus californicus*, and the sea elephant *Mirounga leonina* may have a life cycle similar to that of *U. lucasi*. Gibbs showed that the life cycle of *U. stenocephala* of foxes is similar to that of *Ancylostoma caninum*.

## EXERCISE ON LIFE CYCLE

While no life history studies on the hookworms of seals are possible, investigations on the biology of other hookworms should be directed toward learning whether there is a tissue phase with transmission of larvae in the milk. Prenatal infection of pups has been reported for *Ancylostoma caninum*. The possibility of infection through the milk was not ruled out.

*Ancylostoma caninum, A. braziliense,* and *Uncinaria stenocephala* may be studied in dogs and cats for transmission in the milk. *Uncinaria stenocephala* of canines apparently has a life cycle similar to that of *Ancylostoma caninum.*

## SELECTED REFERENCES

Baylis, H. A. (1947). A redescription of *Uncinaria lucasi* Stiles, a hookworm of seals. Parasitology 25: 308–316.

Olsen, O. W., and E. T. Lyons. (1965). Life cycle of *Uncinaria lucasi* Stiles, 1901 (Nematoda: Ancylostomatidae) of fur seals, *Callorhinus ursinus* Linn., on the Pribilof Islands, Alaska. J. Parasitol. 51: 689–700.

### FAMILY STRONGYLIDAE

The members of this family tend to be stout, varying in size from small to large—up to 35 mm long. The mouth, surrounded by a double or single leaf crown (corona radiata), is without cutting plates but may have toothlike spines inside the capsule which has a groove along the dorsal side. They are parasites of reptiles, birds, and mammals, especially equines. Four common species, one of sheep and three of equines, are considered.

### *Oesophagostomum columbianum*
(Curtice, 1890) (Plate 112)

This is the well-known nodular worm of sheep and goats throughout the world, particularly in warm, moist climates. The adults are in the large intestine. Larvae enter the colonic and cecal epithelium for part of their development. Some of them, however, fail to escape and are surrounded by a caseous material that forms a nodule, which gives rise to the popular names of nodular worm or pimply gut.

**Description.**    Stout worms measure 12 to 16 mm (males) and 14 to 18 mm (females) long. A groove-like ring sets the head off from the body. A ventral cervical groove is present near the level of the cervical papillae which are anterior to the middle of the esophagus. The external leaf crown has 20 to 24 elements in it and the internal one 40 to 44. Spicules are 750 to 950 $\mu$ long and the vulva is near the posterior end of the body. Eggs are 74 to 88 $\mu$ long by 45 to 54 $\mu$ wide.

**Life Cycle.**    Adults in the colon lay eggs already in the early stages of cleavage. They are capable of surviving 60 days or more on pastures where the relative humidity is 48 to 50 percent and the average temperature is 11° to 12°C. They hatch into first stage rhabditiform larvae in 15 to 20 hours after being voided in the feces. After feeding for about 24 hours, they molt to form similar stage larvae. These two stages are very sensitive to drying and die rather quickly. They likewise feed and grow for 3 days, when the cuticle loosens and forms a free investing sheath for the third stage infective larvae. Feeding ceases, but the larvae may live several months under favorable conditions. Freezing is lethal to them.

Infection of sheep is by swallowing third stage ensheathed larvae with the forage. In the abomasum, the larvae escape from the loose cuticle and migrate to the intestine, where they enter the epithelial mucosa and lie coiled next to the muscularis mucosae. After about 4 days of growth, they undergo the third molt, entering the fourth stage, following which further growth takes place. Presence of the third stage larvae in the mucosa of the small intestine and the discarded cuticular sheaths left by them results in small gritty lesions. Within another 3 to 4 days, the larvae migrate from the nodules into the lumen of the intestine, where they molt the last time and develop to adults in about 7 weeks after entering the host. When larvae leave the nodules, the latter subside and disappear. Bacterial invasion of the migratory tracts and sites of the larvae in the mucosa of the cecum and large intestine sets up an inflammatory reaction, resulting in the formation of caseous nodules that trap and eventually destroy the larvae.

In sheep, *O. columbianum* causes loss of appetite for food and water and produces diarrhea with decline of weight, beginning within 1 week after infection and becoming severe by day 5. Due to hemorrhage into the lumen of the gut, there is marked anemia. In cattle infected with *O. radiatum*, bleeding begins at day 16 and is severe by days 19 to 20 at the time of the final molt of the larvae and persists for 9 weeks. There is leakage of protein from the mucosa into the intestinal lumen, causing hypoproteinanemia.

Other species of *Oesophagostomum* common in this country are *O. venulosum* of sheep, goats, and deer, *O. radiatum* of cattle, and *O. dentatum* of pigs. The life cycles are similar except that *O. venulosum* does not cause nodules.

Biological agents function in the transfer of *O. dentatum.* Third stage larvae clinging to

## Explanation of Plate 112

A, Anterior end of adult worm. B, *En face* view. C, Ventral view of bursa. D, Lateral view of posterior end of adult female. E, Anterior end of infective third stage larva. F, Posterior end of third stage larva. G, Sheep host.

1, Crown of leaflets (corona radiata); 2, cephalic papillae; 3, cervical papillae; 4, esophagus; 5, intestine; 6, nerve ring; 7, mouth; 8, amphids; 9, ventroventral ray; 10, lateroventral ray; 11, externolateral ray (also called anterolateral); 12, mediolateral ray; 13, posterolateral ray; 14, externodorsal ray; 15, dorsal ray with terminal digitations; 16, spicules; 17, anus; 18, posterior loop of uterus; 19, ovejector; 20, vulva; 21, tail; 22, loose cuticle of second stage larva enclosing third stage larva; 23, excretory pore.

a, Adult worms in colon; b, egg being voided with feces; c, developing egg; d, embryonated egg; e, first stage larva hatching; f, molting first stage larva, giving rise to second stage larva; g, infective third stage larva enclosed in free cuticle of second stage larva; h, infective larvae on forage; i, infective larvae being swallowed; j, third stage larvae shed loose cuticle in abomasum (this is not a molt); k, larvae entering intestinal mucosa; l, nodule containing third stage larva; m, third molt in nodule; n, fourth stage larvae escape from nodules, leaving cuticle; o, fourth and final molt in lumen of intestine, after which larvae develop to adult worms.

Figures A, C, D adapted from Ransom, 1911, U. S. Dep. Agr. Bur. Anim. Ind. Bull. (127) p. 42; B, from Goodey, 1924, J. Helminthol. 2:97; E, F, from Dikmans and Andrews, 1933, Trans. Amer. Microsc. Soc. 52:1.

---

*Psychoda*, a fly that breeds in pig feces, are capable of being transported in a viable state from farm to farm. Larval oesophagostomes encapsulated in the intestinal mucosa of rats develop to sexually mature worms when the rodents are eaten by pigs.

### EXERCISE ON LIFE CYCLE

Adult females obtained from the colon of infected sheep slaughtered at abattoirs provide a source of eggs for hatching. Eggs may be available in feces obtained from infected flocks.

Eggs should be incubated at room temperature in a moist animal charcoal-feces mixture or one of sand and feces. The sand should be heated prior to using it in order to destroy free-living nematodes, because their presence complicates the problem of separating the different species. Larvae are separated from the incubation medium by a Baermann apparatus. Proper timing is necessary to obtain the first and second stages.

If the eggs are obtained from feces rather than females, there is a great probability that a mixture of species will occur, as sheep generally are host to several species of intestinal worms whose eggs are very similar. In this case, it is feasible to recognize only the third stage larvae of the oesophagostomes. In so doing, one gets excellent experience in critical observations, for the third stage larvae of the various species of nematodes of sheep are recognizable only on rather fine differences.

Known studies on the life cycles of oesophagostomes are limited to the vertebrate hosts of the respective species. No studies appear to have been made on the developmental stages as they might occur when introduced into small mammals such as rabbits, guinea pigs, mice, or rats. Such investigations might yield some interesting and valuable heretofore unknown aspects of the biology of these nematodes.

### SELECTED REFERENCES

Anataraman, M. (1942). The life-history of *Oesophagostomum radiatum*, the bovine nodular worm. Indian J. Vet. Sci. Anim. Husb. 12:87–132.

Andrews, J. S., and J. F. Madonado. (1941). The life history of *Oesophagostomum radiatum*, the common nodular worm of cattle. Puerto Rico Agr. Exp. Sta. Res. Bull. 2, 14 pp.

Brenner, K. C. (1969). Pathogenic factors in experimental bovine oesophagostomiasis. III. Demonstration of protein-loosing enteropathy with [57]Cr-albumin. Exp. Parasitol. 24:364–374.

Brenner, K. C. (1970). Pathogenic factors in experimental bovine oesophagostomiasis. V. Intestinal bleeding as a cause of anemia. Exp. Parasitol. 27:236–245.

Chhabra, R. C., and K. S. Singh. (1965). Effect of temperature on free-living stages of two species of *Oesophagostomum* (Nematoda). Indian J. Helminthol. 17:125–137.

Dhar, D. N., and K. S. Singh. (1968). Mechanism of nodule formation in the nematode *Oesophagostomum columbianum* (Curtice, 1890), Stossich, 1899. Indian J. Vet. Sci. 38:708–725.

Dikmans, G., and J. S. Andrews. (1933). A com-

**Plate 112**   *Oesophagostomum columbianum*                                                              419

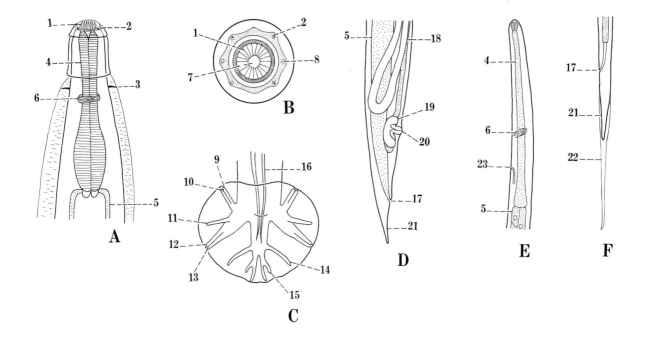

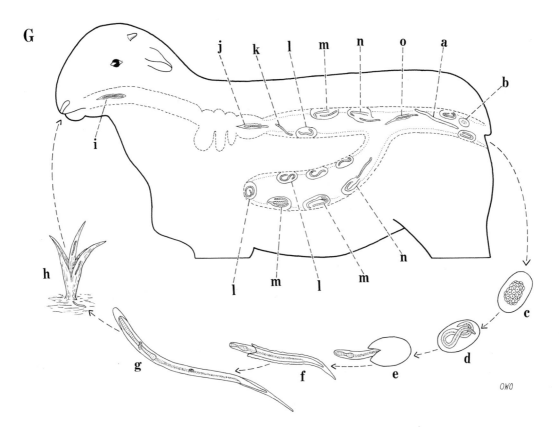

## Explanation of Plate 113

A, Anterior end of adult worm. B, Copulatory bursa of male. C, Partially embryonated egg. D, Anterior end of first stage rhabditiform larva. E, Anterior end of second stage rhabditiform larva. F, Anterior end of third stage strongyliform larva. G, Caudal end of third stage larva enclosed in investing cuticle of second stage larva. H, Swine definitive host. I, Earthworm (a collector of third stage larvae).

1, Buccal capsule; 2, leaf crown elements; 3, cuticular thickenings of rim of buccal capsule; 4, teeth in base of buccal capsule; 5, esophagus; 6, intestine; 7, anus; 8, tail; 9, sheath of third stage larva; 10, nerve ring; 11, excretory pore; 12, lateral lobe of bursa; 13, ventroventral ray; 14, externoventral ray; 15, anterolateral ray; 16, mediolateral ray; 17, posterolateral ray; 18, externodorsal ray; 19, dorsal ray with two bifid branches.

a, Adult worm in cyst in pelvis of kidney; b, adult worm encysted in wall of ureter; c, eggs in urinary bladder, having gone through the ureter; d, partially developed egg voided with urine; e, embryonated egg; f, first stage rhabditiform larvae emerging from egg; g, second stage rhabditiform larva shedding cuticle of first stage larva; h, third stage infective strongyliform larva enclosed in loose cuticle of second stage larva; i, earthworms swallow eggs which hatch, pass through the intestinal wall into the celom, where they develop to infective third stage larvae; j, infection of swine by mouth with third stage larvae; k, larvae enter mucosa of stomach; l, third molt takes place in mucosa; m, fourth stage larva leaves mucosa and goes into intestine; n, larva enters hepatic portal vein; o, larva passes through liver; p, larva in right side of heart; q, larva in lungs; r, larva in left side of heart; s, larva in aorta; t, larva in hepatic artery; u, growth and final molt occurs in liver; v, larva migrates from liver into celom; w, larva enters kidneys or ureters from celom (?) and matures; x, larva penetrates skin; y, third stage larva molts in blood vessel; z, larva goes to right side of heart and follows route to kidneys as described above (p–w).

a', Some larvae do not find their way to kidneys and ureters but become lost in the hams, loins, and perirenal fat.

Figures A, B, from various sources; C–G adapted from Alicata, 1935, U. S. Dep. Agr. Tech. Bull. (489), 73 pp.

---

parative morphological study of the infective larvae of the common nematodes parasitic in the alimentary tract of sheep. Trans. Amer. Microsc. Soc. 52:1–25.

Dobson, C. (1966). Distribution of *Oesophagostomum columbianum* larvae along the alimentary canal of sheep. Austr. J. Agr. Res. 17:765–777.

Dobson, C. (1967). The effects of different doses of *Oesophagostomum columbianum* larvae on the body weight, intake and digestibility of feed and water intake of sheep. Austr. Vet. J. 48:291–296.

Dobson, C. (1967). Pathological changes associated with *Oesophagostomum columbianum* infestations in sheep: haemotological observations on control worm-free and experimentally infected sheep. Austr. J. Agr. Res. 18:523–538.

Fourie, P. J. J. (1936). A contribution to the study of the pathology of oesophagostomiasis in sheep. Onderstepoort J. Vet. Sci. Anim. Ind. 7:277–347.

Goldberg, A. (1951). Life history of *Oesophagostomum venulosum*, a nematode parasite of sheep and goats. Proc. Helminthol. Soc. Wash. 18:36–47.

Jacobs, D. E., A. M. Dunn, and J. Walker. (1971). Mechanisms for the dispersal of parasitic nematode larvae. 2. Rats as potential paratenic hosts for *Oesophagostomum* (Strongyloidea). J. Helminthol. 45:139–145.

Singh, K. S., and R. C. Chhabra. (1965). Effect of humidity on free-living stages of two species of *Oesophagostomum* (Nematoda). Indian J. Parasitol. 17:118–124.

Spindler, L. A. (1933). Development of the nodular worm, *Oesophagostomum longicaudum* in the pig. J. Agr. Res. 46:531–542.

Tod, M. E., D. E. Jacobs, and M. A. Dunn. (1971). Mechanisms for the dispersal of parasitic nematodes. I. Psychodid flies as transport hosts. J. Helminthol. 45:133–137.

*Stephanurus dentatus*
Diesing, 1839 (Plate 113)

This is the kidney worm of pigs which is prevalent in warm climates throughout most of the world. The adults occur in cysts in the pelvis of the kidneys and the wall of the ureters from which the eggs enter the urinary bladder and are finally voided with the urine. Larvae migrating in the body are found in many places such as the perirenal fat, pancreas, loins, hams, and spinal cord.

**Description.**     The adult worms are stout, with the males 20 to 30 mm long and the females 30 to 45 mm by 2 mm wide. The buccal capsule has thick walls and is cup-shaped, with six teeth at the base. The margin bears a crown of a few small leaflet-like structures and six thickened areas. The copulatory bursa is small and with

**Plate 113**  *Stephanurus dentatus*  421

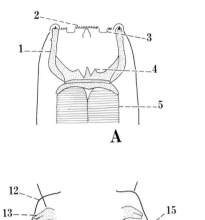

**A**

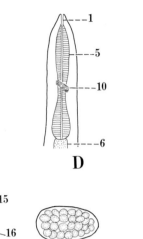

**D**

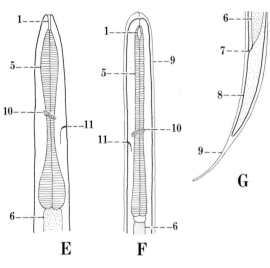

**E**  **F**  **G**

**B**  **C**

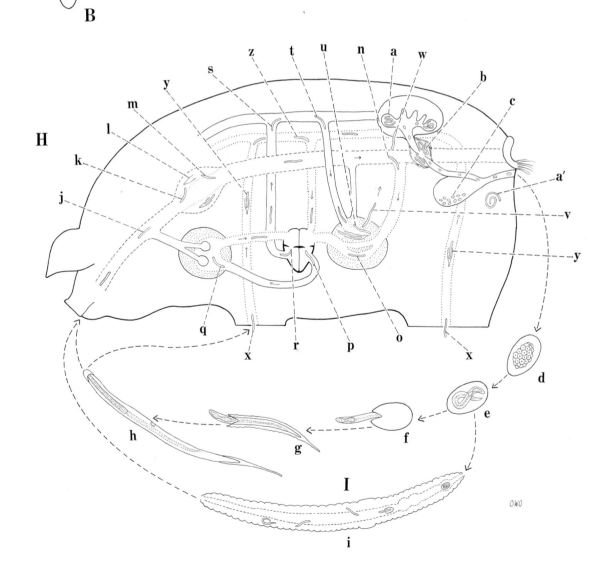

**H**

**I**

OWO

short rays, of which the ventrals, laterals, and digitations of the dorsal ray are fused distally. The spicules are equal or unequal in length and 0.66 to 1 mm long. The vulva is near the anus. Eggs are 100 by 60 $\mu$ in size.

**Life Cycle.**    Adult kidney worms oviposit in cysts in the wall of the pelvis of the kidneys or the ureters. These cysts have openings through which the eggs pass into the pelvis of the kidney or lumen of the ureters and are washed into the urinary bladder. Vast numbers of partially developed eggs are voided with the urine at each micturation. In moist, warm but shaded soil, the eggs develop and hatch usually within 48 hours. The first stage rhabditiform larvae feed and soon molt to form foraging second stage rhabditiform ones. In 3 to 5 days, the second molt takes place, with retention of the cuticle, to form the infective third stage strongyliform larvae. When embryonated eggs are swallowed by earthworms, *Eisenia foetida*, they hatch in the gut, enter the celom, and develop to infective larvae. Eggs and all stages of larvae free in the soil are quickly killed by freezing and desiccation. Third stage larvae may live several months in warm, moist soil protected by shade.

Infection of swine takes place by mouth or through the skin. Regardless of the route of entry, part of the larval development must be in the liver, where essential nutrients are available. When infective larvae free in the soil or in earthworms are swallowed, they enter the gastric epithelium and undergo the third molt. Young fourth stage larvae penetrate the intestinal wall and enter the vessels of the hepatic portal vein or mesenteric lymphatics and to a lesser degree the peritoneal cavity. In the liver, they escape from the blood vessels and wander in the parenchyma, where the final molt occurs about 70 days after infection. Upon reaching the surface, they continue wandering and developing underneath the capsule for 3 months or more. Having completed the hepatic phase of growth, they escape into the abdominal cavity, migrating at random in it. Eventually some of them reach the perirenal tissues, burrow into the pelvis of the kidneys and the wall of the ureters, encyst, and develop to sexual maturity. This requires about 6 months from the time of infection. Because of their tendency to migrate and burrow, erratic larvae occur encysted in the pancreas, loins, hams, perirenal fat, and spinal cord. Larvae in these tissues fail to develop to maturity.

Larvae burrowing into the host through the skin enter blood vessels and bypass the liver en route to the lungs where they leave the terminal vessels of the pulmonary vein. Some of them enter the air sacs, go up the trachea, are swallowed, burrow through the gut wall, and enter the celom. Of those remaining in the lung tissue, some are encapsulated and others escape to wander aimlessly in the thoracic cavity. Some of these wandering larvae burrow into major arteries and are carried to the liver via the mesenteric artery. They leave the arterioles and enter the hepatic parenchyma. These larvae and those that reached the celom via the trachea and intestine and have entered the liver by burrowing into it continue development similar to that of those entering the host per os.

In pregnant sows, prenatal infection occurs when larvae in the general circulation enter the umbilical artery and lodge in the fetal liver.

Extensive lesions resulting from destruction of liver cells by migrating larvae are replaced by widespread and massive increase of fibrous tissue.

### EXERCISE ON LIFE CYCLE

Adult females may be obtained from the kidneys and ureters of pigs at abattoirs as a source of eggs for experiments on the life history.

The three stages of free-living larvae may be obtained by incubating eggs in a mixture of animal charcoal and pig feces or in moist sand and pig feces. Care must be taken to destroy eggs and nematodes of unwanted species by heating the sand and feces prior to introducing the eggs of the kidney worms. The different stages of larvae can be procured by baermannizing portions of the medium at appropriate intervals.

Embryonated eggs fed to earthworms hatch and the larvae develop to the infective stage in the annelids. The various stages may be recovered from them.

Guinea pigs are suitable experimental animals. By exposing a series of them to infective larvae by mouth and by the skin, it is possible to follow the intramammalian routes of migration and the major stages of development as they occur in these animals. The liver, lungs, body cavity, and kidneys should be examined at appropriate intervals after exposure to third stage larvae in order to find the different stages.

## SELECTED REFERENCES

Alicata, J. E. (1935). Early developmental stages of nematodes occurring in swine. U. S. Dep. Agr. Tech. Bull. 489, 96 pp.

Batte, E. G., D. J. Moncol, and C. W. Barber. (1966). Prenatal infection with the swine kidney worm (*Stephanurus dentatus*) and associated lesion. J. Amer. Vet. Med. Ass. 149:758–765.

Schwartz, B., and E. W. Price. (1931). Infections of pigs through the skin with the larvae of the swine kidney worm, *Stephanurus dentatus*. J. Amer. Vet. Med. Ass. 79, n. s., 32:359–375.

Spindler, L. A. (1934). Field and laboratory studies on the behavior of the larvae of the swine kidney worm, *Stephanurus dentatus*. U. S. Dep. Agr. Tech. Bull. 405, 18 pp.

Spindler, L. A. (1934). Skin penetration experiments with the infective larvae of *Stephanurus dentatus*. N. Amer. Vet. 15:32–36.

Spindler, L. A. (1942). Internal parasites of swine. In: Keeping Livestock Healthy. U. S. Dep. Agr. Yearbook, 1942, Washington. pp. 745–786.

Spindler, L. A., and J. S. Andrews. (1955). The swine kidney worm, *Stephanurus dentatus*. Proc. 58th Annu. Meet. U. S. Livestock Sanit. Ass., pp. 296–302.

Tromba, F. G. (1955). The rôle of the earthworm, *Eisenia foetida*, in the transmission of *Stephanurus dentatus*. J. Parasitol. 41:157–161.

Tromba, F. G. (1958). Observations on swine experimentally infected with kidney worm, *Stephanurus dentatus*. J. Parasitol. 44(Suppl.):29.

Waddell, A. H. (1969). The parasitic life cycle of the swine kidney worm *Stephanurus dentatus* Diesing. Austr. J. Zool. 17:607–618.

### Genus *Strongylus* Müller, 1780

Several species of *Strongylus* are common parasites of equines. *Strongylus edentatus*, *S. equinus*, and *S. vulgaris* of the cecum and colon are the best known. They are referred to variously as large strongyles, red worms, and palisade worms. These parasites, particularly *S. vulgaris*, are credited with causing untoward effects in horses.

The mouth is directed anteriorly and surrounded by an external leaf crown. They have a large semiglobular buccal capsule with or without internal toothlike projections and with a thickened longitudinal ridge known as the dorsal gutter. The dorsal ray of the copulatory bursa is double for most of its length and has lateral projections. The life cycle of these three species is basically the same for the stages outside the host. They differ inside but there is not general agreement on the details for some of them.

### *Strongylus edentatus*
(Looss, 1900) (Plate 114, G)

This is the medium-sized, toothless strongyle. It occurs mainly in the ventral colon and to a lesser extent in the cecum. Yamaguti placed this species in the genus *Alfortia*.

**Description.** The buccal capsule is cone-shaped and devoid of any kind of thickenings resembling teeth. The males measure 23 to 44 mm long by 1.6 to 2.3 mm wide, and the females 33 to 44 mm long by 1.6 to 2.3 mm wide. The vulva is 9 to 10 mm from the caudal end.

**Life Cycle.** The eggs are in the early stages of cleavage when voided in the feces. They become fully embryonated and hatch in a day or so under favorable conditions of moisture and temperature. The first stage rhabditiform larvae feed for a short while and then molt, shedding the cuticle and transforming into larger, second stage larvae. When they molt to form the third stage, the cuticle of the second stage merely loosens and forms a protective sheath for the free-moving infective strongyliform larvae on about day 5.

The development of the eggs and the larvae to and including the third stage is similar for all of the horse strongyles.

Infection is passive and occurs when third stage larvae ascend blades of grass and are swallowed with the forage by grazing horses, or are otherwise picked up from the ground or stalls. This type of infection is common for all of the species of strongyles of equines.

In the large intestine, the larvae burrow through the intestinal wall to the outer layer of connective tissue, where nodules are formed in which they grow for about 3 months. After this time, they return to the intestinal wall and form more nodules in which development is continued. Eventually they leave the nodules and enter the lumen of the large intestine and cecum, where maturity is attained in about 11 months. The only migration in this species is in the wall of the large intestine and cecum.

### *Strongylus equinus*
Müller, 1780 (Plate 114, H)

This is another of the large strongyles of horses, often called the large-toothed strongyle, occurring primarily in the cecum.

## Explanation of Plate 114

A, Right side of head of *Strongylus edentatus*. B, Dorsal view of copulatory bursa of *S. edentatus*. C, Right side of head of *S. equinus*. D, Dorsal view of bursa of *S. equinus*. E, Right side of head of *S. vulgaris*. F, Dorsal view of bursa of *S. vulgaris*. G, Life cycle of *S. edentatus*. H, Life cycle of *S. equinus*. I, Life cycle of *S. vulgaris*.

1, External leaf crown; 2, internal leaf crown; 3, buccal capsule; 4, dorsal gutter; 5, dorsal tooth; 6, ventral teeth; 7, esophagus; 8, copulatory bursa of male; 9, dorsal lobe of bursa; 10, externodorsal ray; 11, trifurcated dorsal ray.

a, Adult worms of *S. edentatus*; b, unembryonated eggs passed in feces; c, egg in early stages of cleavage; d, fully developed larva; e, first stage rhabditiform larva escaping from egg; f, first stage larva molting to form second stage rhabditiform larva; g, third stage infective filariform or strongyliform larva enclosed in loose, intact cuticle of second stage larva; h, infective larvae on grass; i, larvae swallowed with forage; j, larvae of *S. edentatus* enter intestinal and cecal mucosa where they form nodules and molt; k, larvae escape from nodules, enter cecum, and develop to maturity in about 10 to 11 months.

a′, Adult worms of *S. equinus* in colon and cecum; b′, unembryonated eggs passed in feces develop in soil similar to *S. edentatus* (c–h); i′, in-fections occur when larvae are swallowed with forage; j′, larvae molt en route through intestine; k′, fourth stage larvae penetrate wall of colon; l′, nodules formed by larvae on outer surface of intestinal wall; m′, larvae break out of nodules into celom; n′, larvae burrow into liver; o′, larvae burrow into pancreas; p′, larvae emerge from liver or pancreas; q′, another molt occurs and larvae enter cecum, presumably by penetration of the wall, or some may go through pancreatic duct.

a″, Adult of *S. vulgaris* in colon and cecum; b″, unembryonated eggs passed in feces develop in soil similar to the two species discussed above (c–h); i″, infection occurs when infective larvae are swallowed; j″, larvae penetrate intestinal wall and enter hepatic portal vein; k″, larvae pass through liver; l″, larvae pass through right side of heart; m″, larvae carried through pulmonary artery; n″, larvae leave pulmonary capillaries and enter alveoli of lungs; o″, some larvae migrate up trachea; p″, upon reaching pharynx, they are swallowed and go to intestine; q″, some larvae continue through the lungs in the blood and return via the pulmonary vein to the left side of heart; r″, they pass through the aortic arch; s″, they enter the dorsal aorta; t″, larvae entering mesenteric arteries may lodge in them, causing serious damage that results in aneurysms while those that pass through re-enter the intestine and develop to maturity in the colon.

---

**Description.** The males measure 26 to 35 mm long by 1.1 to 1.3 mm wide and the females 38 to 47 by 1.8 to 2.2 mm. The buccal capsule is round and has four small toothlike projections at the base; the two ventral ones are somewhat larger and less pointed than the two dorsal ones.

**Life Cycle.** The infective larvae burrow into the walls of the colon and cecum, shedding the loose cuticle. They continue to the serosa where nodules are formed in which further development takes place. After about 11 days in the nodules, the larvae molt the third time and break through into the celom, burrow into the liver, and grow there for 6 to 7 weeks. Upon leaving the liver, they go into the celom and enter the pancreas where the fourth and final molt takes place about 4 months after infection. They are now sexually differentiated but not mature. It is not clear how they get from the pancreas into the colon and cecum, but it is thought that the pancreatic duct might be the route. The migration in this species is much more extensive than in the case of *S. edentatus*.

*Strongylus vulgaris*
(Looss, 1900) (Plate 114, I)

This is the small-toothed strongyle found more commonly in the cecum than in the colon as adults. Fourth stage larvae frequently occur in aneurysms in mesenteric arteries of the colon. It is believed that colic is the result of diminished circulation to the intestine through damaged arteries leading to the colon. Yamaguti placed this species in the genus *Delafondi*.

**Description.** The males are 14 to 16 mm long by 0.75 to 0.95 mm wide and the females 20 to 24 mm by 1 to 1.4 mm. The buccal capsule is goblet-shaped with two large ear-like projections inside at the base.

**Life Cycle.** This species has the most complicated life cycle of the three. There are several opinions as to the nature of the intra-mammalian development and migrations. Lapage has reviewed the concepts of the different workers. He concluded that probably the larvae follow several routes of migration and therefore each concept of it contains an element of truth. A

**Plate 114**    *Strongylus edentatus, S. equinus, and S. vulgaris*                                    425

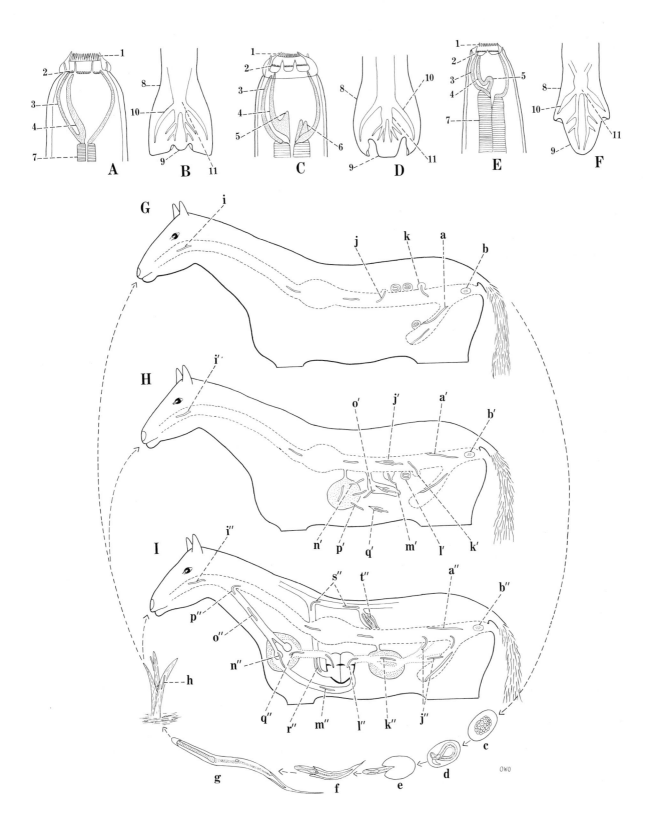

plausible explanation is the one described and illustrated. Third stage larvae entering the colon or cecum burrow into the venules of the hepatic portal vein and are carried to the lungs by way of the liver and heart. Some of them break out of the blood vessels of the lungs and migrate up the trachea to the pharynx and are swallowed. Development of these is completed in the cecum and colon. Some larvae, on the other hand, fail to escape from the blood vessels in the lungs and are carried back to the heart and into the general circulation. Such larvae reaching the anterior mesenteric artery attach to the intima and grow. Their presence in the artery and the damage done to it cause thrombi and aneurysms. This concept provides a simple, direct explanation of how the larvae may reach the anterior mesenteric artery.

Failure to find migrating larvae in the liver or lungs of experimentally infected animals led some investigators to discredit the concept of tracheal migration and the passage of larvae through the lungs back to the heart and into the general circulation. They believe that the larvae migrate from the lumen of the intestine into the wall, whence they burrow into the artery. There is no evidence of prenatal infection among these strongyles.

Of the large strongyles, *S. vulgaris* is the most pathogenic. Foals especially are susceptible and die from the effects of the parasites. The highly fatal syndrome is characterized by 1) marked elevation of body temperature, 2) loss of appetite, 3) rapid loss of body weight, 4) mental depression and loss of physical activity, 5) colic, 6) constipation or diarrhea, and 7) death in 2 to 3 weeks. These symptoms are similar to those normally associated with bacterial infections. The pathogenicity is directly related to the damage done to the mesenteric artery by the migrating larvae.

The genus *Strongylus* includes the so-called large strongyles. There are, in addition to them, eight genera containing approximately 50 species of lesser size and known collectively as the small strongyles of equines. Little is known of their life cycles except that the free-living phase is similar to that of the large strongyles.

## EXERCISE ON LIFE CYCLE

The strongyles of horses provide an abundant supply of eggs for studying the development and hatching of them together with the growth of the first three stages of larvae.

Eggs obtained from the feces of horses most probably represent a mixture of species. For ease of collecting the eggs, the feces should be comminuted in water and the coarse material removed by passing the whole mass through a series of graded screens. Further cleaning of the sample can be achieved by sedimenting and decanting until the water is clear, allowing sufficient time for the eggs to settle to the bottom of the container between each decantation.

Eggs may be recovered by placing the cleaned sediment in a cylinder filled with a saturated solution of table salt. After an hour or so, they will have risen to the surface and may be poured into a dish of water and washed free of salt. Incubation may be in a shallow dish with a small amount of water or on a moist filter paper. By examining the material at frequent intervals, the first two stages of rhabditiform larvae may be found.

A method not mentioned heretofore for collecting third stage larvae is by means of two shallow glass dishes of different diameter. In the smaller one, eggs are placed on a moist filter paper or in a mixture of animal charcoal and feces. It is placed in the second dish which is partially filled with water and the whole is covered to prevent evaporation and to maintain a film of moisture on the side of the dish. The third stage larvae creep up the sides of the small dish and into the water of the second one where they are trapped.

## SELECTED REFERENCES

Drudge, J. H., E. T. Lyons, and J. Szanto. (1966). Pathogenesis of migrating stages of helminths, with special reference to *Strongylus vulgaris*. In: E. J. L. Soulsby, ed. Biology of Parasites: Emphasis on Veterinary Parasites. Academic Press, New York, pp. 197–214.

Enigk, K. (1950). Zur Entwicklung von *Strongylus vulgaris* (nematodes) im Wirtstier. Z. Tropenmed. Parasitol. 2:287–306.

Enigk, K. (1951). Weiter Untersuchungen zur Biologie von *Strongylus vulgaris* (nematodes) im Wirtstier. Z. Tropenmed. Parasitol. 2:523–535.

Foster, A. O. (1942). Internal parasites of horses and mules. In: Keeping Livestock Healthy. U. S. Dep. Agr. Yearbook, 1942, pp. 459–475.

Lapage, G. (1956). Veterinary Parasitology. Oliver and Boyd, London, pp. 91–104.

Schwartz, B., M. Immes, and A. O. Foster. (1948). Parasites and parasitic diseases of horses. U. S. Dep. Agr. Circ. 148, rev. 54 pp.

Theiler, G. (1923). The strongyles and other nematodes parasitic in the intestinal tract of South African equines. Thèse, Faculté Sci., Univ. Neuchâtel, 175 pp.

Wetzel, R. (1940). Zur Entwicklung des grossen Palisadenwurmes (Strongylus equinus) im Pferd. Arch. Wissensch. Prakt. Tierheilk. 76:81–118.

Wetzel, R., and K. Enigk. (1938). Wandern die Larven der Palisadenwürmer (Strongylus spp.) der Pferd durch die Lungen? Arch. Wissensch. Prakt. Tierheilk. 73:83–93.

Yamaguti, S. (1961). Systema Helminthum. Interscience Publishers, New York, Vol. 3, Part 1, pp. 351, 352, 354.

## Syngamus trachea
(Montagu, 1811) (Plate 115)

*Syngamus trachea* is the gapeworm of poultry, occurring in the trachea of chickens, turkeys, pheasants, and others including some wild birds. It is widespread in many parts of the world. Young chickens and pheasants show symptoms of "gapes" by labored breathing due to the worms occluding the trachea. Turkeys are more resistant than chickens to the worms.

**Description.** Living adults are red due to the blood in the intestine. Males are 2 to 6 mm long and the females 5 to 20 mm. The male is permanently attached in copula with the female, thus giving a Y-shaped appearance to the pair. The mouth is wide and devoid of leaf crowns; the buccal capsule is thick, cup-shaped, and shallow, with 6 to 10 small teeth in the bottom. The bursa and rays of the male are short; the dorsal ray is split to the base, with each branch thick and tridigitate distally. The vulva is in the anterior third of the body. Eggs are 78 to 110 $\mu$ long by 43 to 46 $\mu$ wide, with a thickened mucoid plug at each end.

**Life Cycle.** Paired adult worms are attached to the inner surface of the trachea. Eggs in the early stages of cleavage deposited by the female under the bursa of her continuously attached male consort escape into the trachea. Gaping and coughing by the birds due to irritation caused by the worms bring the eggs up the trachea to be swallowed and voided in the feces. In moist, shaded soil between 24° and 30°C, the eggs develop. Two molts occur and a third

stage larva enclosed in both cuticular sheaths is present in 7 days. There is no development below 15°, and temperatures above 30° soon kill the embryos. While embryonated eggs survive the winter, adult worms are relatively short-lived and are eliminated from the host during the course of the winter.

In cultures and presumably in the soil, the end plugs of the eggs containing fully developed larvae begin to dissolve by day 10 of incubation and have disappeared within the next 2 days. The outermost cuticular sheath surrounding the larva is generally lost as it emerges tail first from the egg. The inner sheath is retained as a protective covering.

Hatched larvae eaten by dung-infesting earthworms (*Eisenia foetida*), slugs, and maggots of diptera such as houseflies (*Musca domestica*) and lesser houseflies (*Fannia canicularis*) burrow through the intestinal wall and encyst in the tissues or remain free in the body cavity. There is no larval development in these invertebrate reservoir hosts. Some of them accumulate large numbers of larvae which live for long periods in the favorable biological habitat provided by them.

Birds become infected by swallowing embryonated eggs or larvae while pecking food from contaminated soil or feeders, or by eating any of the parasitized reservoir hosts. Residence of larvae in earthworms appears to enhance their infectivity and success in establishing themselves in the birds.

Hatched larvae or those liberated from reservoir hosts in the gut of the birds burrow into the wall, enter the venules of the hepatic portal vein, and are carried forward through the liver and into the lungs, arriving within about 6 hours. In the lungs, they leave the vessels of the pulmonary vein for the air sacs, where the third molt occurs by the third day and a fourth stage larva with developing buccal capsule appears. A week after infection, the final molt has occurred and immature adults have already begun mating, which takes place in the lungs, and some of the pairs are in the trachea. The worms are sexually mature and laying eggs in about 3 weeks after infection.

Wild birds, such as starlings, infected with *S. trachea* flying from place to place in search of food are a source of infection to chickens and turkeys over a wide area. Strains of larval *S. trachea* from wild birds have increased infec-

## Explanation of Plate 115

A, Adult male and female worms in copula. B, Anterior end of young male. C, Copulatory bursa of young male. D, Anterior end of adult male. E, Bursa of adult male. F, Embryonated egg containing third stage larva. G, Anterior end of rhabditiform second stage larva. H, Third stage filariform or strongyliform infective larva. I, Chicken definitive host. J, Starling definitive host as a representative of wild birds that serve as hosts. K–M, Transfer hosts. K, Earthworms. L, Dung-infesting flies. M, Slugs and aquatic snails.

1, Adult male attached to female; 2, gravid adult female; 3, uterine coils; 4, buccal capsule; 5, teeth of buccal capsule; 6, esophagus; 7, anterior end of intestine; 8, copulatory bursa; 9, 10, ventral rays (9, ventroventral; 10, lateroventral); 11–13, lateral rays (11, externo- or anterolateral; 12, mediolateral; 13, posterolateral); 14, 15, dorsal rays (14, externo-dorsal; 15, dorsal ray with trifurcated branches); 16, thick shell of egg; 17, mucoid plug at each end of shell; 18, third stage infective larva; 19, cuticle of first molt; 20, cuticle of second molt, both being retained as sheaths; 21, rhabditiform type of esophagus in first and second stage larvae; 22, filariform or strongyliform esophagus of infective third stage larva; 23, esophageal nuclei; 24, excretory duct.

a, Adult worms attached to the tracheal wall; b, unembryonated eggs laid in trachea; c, eggs carried up trachea and into pharynx are swallowed; d, eggs pass from body in feces; e, undeveloped egg on ground; f, developing egg; g, egg containing second stage larva with sheath; h, fully developed third stage filariform larva encased in two cuticles resulting from previous molts; i, egg hatching with release of third stage larva; j, infection of chickens takes place when hatched or unhatched third stage larvae are swallowed; k, larvae and embryonated eggs pass down digestive tract where the latter hatch; l, larvae molt to fourth stage in small intestine; m, larvae penetrate intestinal wall and enter hepatic portal vein; n, larvae carried by blood through liver to heart; o, larvae pass through right side of heart into pulmonary artery; p, larvae migrate from blood vessels into alveoli and up trachea; q, final (fourth) molt occurs in the trachea; r, young worms attach to wall of trachea and grow to maturity; s, embryonated eggs are eaten by earthworms, houseflies, or slugs, which serve as transport hosts; t, eggs hatch in intestine; u, larvae migrate through intestinal wall; v, larvae encyst; w, digestion of transport hosts harboring encysted larvae transfers the infection to chickens.

Figure A adapted from Wehr, 1941, U. S. Dep. Agr. Leaflet 207; B, C, F–H, from Wehr, 1937, Trans. Amer. Microsc. Soc. 56:72; D, from York and Maplestone, 1962, Nematode Parasites of Vertebrates, Hafner Publishing Co., New York, p. 156; E, from Chapin, 1925, J. Agr. Res. 30:557.

---

tivity for chickens and turkeys after having been in earthworms.

In addition to *S. trachea* of poultry, other species of *Syngamus* occur in wild birds. Cattle are infected with *S. laryngeus* and sheep and goats with *S. nasicola* in some parts of the world. Species of a closely related tracheal worm of the genus *Cyathostoma* infect swans, geese, gulls, owls, and cassawaries. Their life cycles are unknown.

Damage to the lungs results from migrations of the worms, causing small hemorrhages, swelling, and even pneumonia. Worms suck blood from the tracheal mucosa and contain up to 0.013 to 0.014 ml per pair. Young birds are most seriously affected, showing the characteristic gapes in an effort to get sufficient air past the worms that occlude the small trachea. Death results from asphyxiation.

## EXERCISE ON LIFE CYCLE

In areas where gapeworms occur, a plentiful supply may be obtained from poultry-dressing plants when farm-raised chickens and turkeys are processed.

The significant thing in a study of the parasite is to see the three stages of larvae develop within the egg before hatching. In this respect, they differ from most other Strongylidae. A similar type of development occurs in *Nematodirus*, a common Trichostrongylidae parasite of sheep, in some Amidostomidae from the gizzards of ducks, and in *Uncinaria*, an Ancylostomatidae, from fur seals.

When embryonated eggs or hatched larvae are fed to houseflies, earthworms, or slugs, the larvae may be traced through the intestine and body tissues or cavities to the point of encystment.

Young chicks may be infected by feeding them embryonated eggs, hatched third stage larvae, or infected reservoir hosts such as houseflies, slugs, or earthworms. Infect a sufficient number of chicks to be able to sacrifice one each day for the first week to follow the course of migration and rate of development. Hold a few of the infected chicks until eggs of the gape-

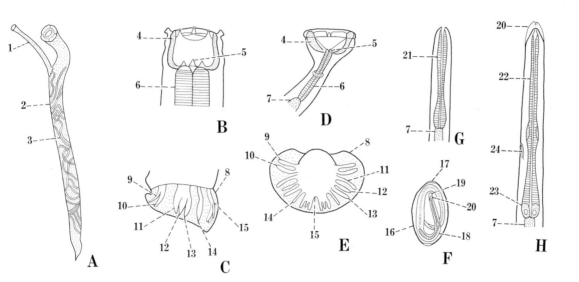

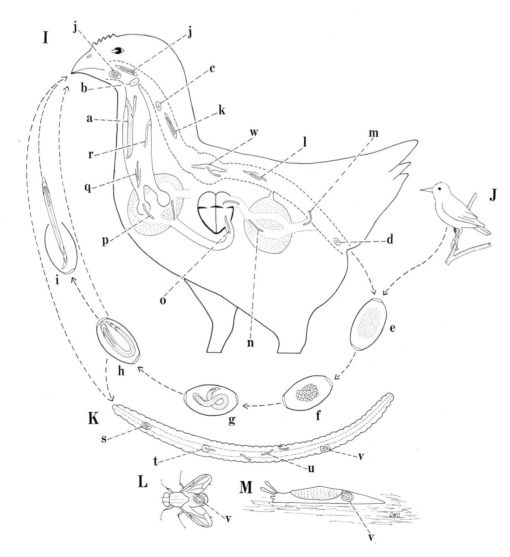

worms appear in the feces in order to determine the length of time required for them to reach sexual maturity.

## SELECTED REFERENCES

Baruš, V. (1967). Two new reservoir hosts of invasive larvae of the nematode *Syngamus trachea* (Montagu, 1811) Chapin, 1925. Helminthologia 7: 323–327.

Clapham, P. A. (1934). Experimental studies on the transmission of the gapeworm (*Syngamus trachea*) by earthworms. Proc. Roy. Soc. London, S. B. 115: 18–29.

Clapham, P. A. (1939). On flies as intermediate hosts of *Syngamus trachea*. J. Helminthol. 17:51–64.

Clapham, P. A. (1939). On the larval migration of *Syngamus trachea* and its causal relationship to pneumonia in young birds. J. Helminthol. 17:159–162.

Clapham, P. A. (1939). On a sex difference in the infection rate of birds with *Syngamus trachea*. J. Helminthol. 17:192–194.

Enigk, K., and A. Deyhazra. (1970). Zur Biologie and Pathogenität von *Syngamus trachea* (Strongyloidea, Nematoda). Deut. Tierärztl. Wockenschr. 77:521–525.

Enigk, K., H. Schanzel, and A. Deyhazra. (1970). Der Blutverlust bei Trüthuhnern durch *Syngamus trachea* (Strongyloidea, Nematoda). Z. Parasitenk. 33:225–234.

Hall, W. J., and E. E. Wehr. (1953). Diseases and parasites of poultry. U. S. Dep. Agr., Farmers' Bull. 1625, rev., 91 pp.

Morgan, D. O., and P. A. Clapham. (1934). Some observations on the gapeworm in poultry and game birds. J. Helminthol. 12:63–70.

Wehr, E. E. (1937). Observations on the development of the poultry gapeworm *Syngamus trachea*. Trans. Amer. Microsc. Soc. 56:72–78.

Wehr, E. E. (1939). The gapeworm as a menace to poultry production. Proc. 7th World's Poultry Cong., pp. 267–270.

### Superfamily Trichostrongyloidea

The members of this superfamily are relatively small, thread-like nematodes of the intestine of amphibians, reptiles, birds, and mammals. They are among the commonest and most abundant parasites of sheep, goats, cattle, and wild ruminants. Likewise, they are the most pathogenic, notwithstanding their small size.

**Description.** They are small to minute worms with the buccal capsule very small or absent and with no leaf crowns; male bursa is strongyloid, well developed, and supported by a full complement of rays similar to Strongyloidea.

### FAMILY TRICHOSTRONGYLIDAE

Only a single family is recognized but at least 14 subfamilies have been described. Of the many genera, *Haemonchus*, *Ostertagia*, and *Trichostrongylus* are well known and common representatives in sheep and cattle throughout the temperate and tropical regions of the world. Because of their great economic importance as parasites of domestic livestock, species of these genera have been studied extensively. As a result, much is known regarding the host-parasite relationships of them and the ecology of the eggs and free-living larval stages.

The external phase of the life cycle of the species of these three genera is similar to that of strongyles of horses. Infection of the host is passive, i.e., by swallowing the infective third stage larvae with forage.

### *Haemonchus contortus*
### (Rudolphi, 1830) (Plate 116, A–D)

These are the so-called twisted or barberpole stomach worms of the abomasum of sheep chiefly but commonly of goats and cattle, as well as many wild ruminants. They are more prevalent in warm, moist regions than in cold, dry ones. They are among the most pathogenic nematode parasites of sheep, being voracious bloodsuckers.

**Description.** The males are 10 to 20 mm and the females 18 to 30 mm long. The white uteri and ovaries winding around the red blood-filled intestine give a twisted or barberpole appearance. The small buccal capsule bears a curved dorsal tooth. There are two prominent lateral spike-like cervical papillae near the junction of the first and second quarters of the esophagus. The male bursa has long lateral lobes and slender rays with a flap-like dorsal lobe located asymmetrically near the base of the left lateral lobe. The spicules are 450 to 500 $\mu$ long, each with a terminal barb; the gubernaculum is navicular. Usually, the vulva is covered by an anterior thumb-like flap which may be reduced to a mere knob in some individuals. The oval eggs are 70 to 85 $\mu$ long by 41 to 44 $\mu$ wide and in the early stages of cleavage when laid. They are somewhat yellowish.

**Life Cycle.** Adult females in the abomasum lay an average of 5000 or more eggs per day. The longevity of the worms in the host is uncertain. In sheep reasonably but not exclusively protected against reinfection, Ransom found worms present 19 months after the original infection.

Eggs in the early morula stage are distributed in feces which vary greatly in moisture content. The feces are not dropped at random over the pasture but have a tendency to be concentrated by the grazing flocks. These areas are avoided for 5 to 10 days at which time the sheep regraze them. Such habits of the sheep favor continuous infection.

Pastures with ample vegetation and a mat of organic material provide a microclimate with temperature, moisture, and oxygen favorable for development of the eggs. At 26°C, development to the tadpole stage is reached in 6 hours, with hatching well underway by 14 hours, and largely completed by 48 hours. The rate of development varies under similar conditions for stomach worms from various parts of the world, presumably due to differences in races that have evolved. Lower temperature retards development of the eggs and freezing kills undeveloped ones in 24 hours.

Developed larvae initiate hatching by secreting chitinase and protease enzymes that deteriorate the eggshell and its inner membrane, allowing the larvae to hatch. First stage rhabditiform larvae in the fecal pellets feed on bacteria. At the end of their growth, they shed the cuticular covering and become second stage rhabditiform larvae. These, too, spend a period of feeding and growth in the feces. Food is stored in the cells of the intestine. The duration of the first and second larval stages is about 2 days each under optimal conditions but longer in cooler circumstances. Both occur in the feces. At completion of the second stage during days 4 to 6, the cuticular covering of the larvae loosens and is retained as an intact protective sheath enclosing a third stage larva. Being prevented from taking food by the sheath, nutriment is derived from that stored in the body, particularly in intestinal cells.

Underdeveloped eggs together with first and second stage larvae are highly susceptible to drying and cold, dying quickly when exposed to these conditions. Prehatch eggs and infective larvae are far more resistant. Third stage larvae survive 4°C for several months but most of them die when exposed to freezing temperatures in the microenvironment during the winter.

The majority of third stage larvae ascend the blades of grass during twilight hours when dew is present or on dim rainy days. Others remain near the soil in clumps of grass or in the mat but do not migrate laterally to any extent.

Infection of sheep is by swallowing third stage larvae with the forage. They must escape from the enclosing sheath before proceeding to further development. Exsheathment is triggered in the rumenal fluid with its high pH by stimulating neurosecretory cells located between the base of the esophagus and excretory pore to secrete a specific exsheathing fluid, leucine aminopeptidase, which flows from the excretory pore. The exsheathing fluid erodes a ring-like space around the 8-layered sheath at the level of the pore. In the abomasum, the anterior end of the sheath separates at the ring and the larva escapes.

Within 12 hours after being swallowed, the third stage larvae appear on the surface of the abomasal mucosa ready to penetrate. Migration into it begins and by the end of the first day nearly all of them are in the mucosa, mostly at the level of the gastric pits. After a short period of feeding and growth, the third molt takes place in the tissue. The fourth stage larvae return to the surface of the mucosa where nearly all of them are located by 40 hours after entering the host. The final molt occurs and the worms grow, becoming sexually mature in 14 to 21 days after entering the host (Plate 116, a–o).

If the host animals survive the original infection, they may undergo the phenomenon of self-cure in which the adult worms are spontaneously expelled. This was thought to result from acquired immunity. However, it appears that self-cure is brought about following massive reinfections and alteration by the exsheathing fluid of the required acid environment of a low pH to one of a high pH. Such drastic environmental changes in the pH of the abomasum prevent the larvae from becoming established.

Application of a knowledge of the biology and ecology of these parasites to principles of sheep husbandry is important in keeping losses at a minimal level. Drenching sheep at prescribed periods based on the biology of the worms with phenothiazine kills the stages in the lumen of the abomasum. Administering the drug free-choice in feed or salt maintains a low level of

## Explanation of Plate 116

A–D, *Haemonchus contortus*. A, Anterior end of adult. B, Anterior extremity of adult showing small buccal capsule and dorsal tooth. C, Copulatory bursa. D, Caudal end of female showing vulvar flap. E–H, *Ostertagia circumcincta*. E, Anterior end of adult. F, Copulatory bursa of male. G, Tail of adult female. H, Caudal end of female showing vulvar flap. I–M, *Trichostrongylus colubriformis*. I, Anterior end of adult. J, Copulatory bursa. K, Spicule. L, Gubernaculum. M, Caudal end of female. N, Sheep definitive host.

1, Circumoral papillae; 2, cervical papillae; 3, buccal capsule; 4, buccal spine; 5, esophagus; 6, intestine; 7, anus; 8, tail; 9, lateral lobe of copulatory bursa; 10, dorsal lobe of bursa; 11, ventroventral ray; 12, lateroventral ray; 13, antero- or externo-lateral ray; 14, mediolateral ray; 15, posterolateral ray; 16, externodorsal ray; 17, dorsal ray; 18, spicule; 19, gubernaculum; 20, vulvar flap; 21, vulva; 22, ovejector; 23, uterus; 24, egg in uterus; 25, ovary.

*Haemonchus contortus.* a, adult worms attached to abomasal mucosa; b, eggs passed in feces; c, egg in early stage of cleavage; d, embryonated egg; e, hatching of egg; f, first stage rhabditiform larva; g, molting and appearance of second stage rhabditiform larva; h, ensheathed third stage infective, filariform larva; i, infection of sheep takes place when third stage larvae are swallowed; j, third stage larva shedding loose cuticular sheath; k, third stage larva penetrating abomasal mucosa; l, larva in mucosa; m, larva molts; n, fourth stage larva in mucosa; o, fourth stage larva emerges from mucosa, molts to young adult stage, and develops to maturity in abomasum.

*Ostertagia circumcincta.* a', adults attached to abomasal mucosa where females lay eggs; b–j, life cycle outside of host is similar to that of *Haemonchus contortus;* k', larvae penetrate abomasal mucosa; l', larvae molt third time to form fourth stage larvae; m', larvae grow; n', larvae escape from mucosa into abomasum but remain in contact with epithelium; o', larvae undergo fourth and final molt; p', q', larvae grow to sexual maturity.

*Trichostrongylus colubriformis.* a'', adult worms attached to mucosa of small intestine; b–j, similar to *Haemonchus contortus;* k'', larvae pass through four stomachs and enter intestine, burrowing into the mucosa; l'', larvae molt third time to form fourth stage in mucosa; m'', larvae leave mucosa; n'', larvae in lumen and in contact with epithelium; o'', fourth and final molt producing last stage; p'', young nematodes grow into adults.

Figures adapted from Ransom, 1911, U. S. Dep. Agr. Bur. Anim. Ind. Bull. 127, 132 pp.

---

infection in the animals. Residual phenothiazine in the feces interferes with development and hatching of the eggs as well as killing many first and second stage larvae.

## SELECTED REFERENCES

Andrews, J. S. (1942). Stomach worm (*Haemonchus contortus*) infection in lambs and its relation to gastric hemorrhage and general pathology. J. Agr. Res. 65:1–18.

Christie, M. G. (1970). The fate of very large doses of *Haemonchus contortus* and their effect on conditions in the ovine abomasum. J. Comp. Pathol. 80:89–100.

Crofton, H. D. (1963). Nematode parasite population in sheep and on pasture. Commonw. Bur. Helminthol. Tech. Communic. No. 35, 104 pp.

Dikmans, G., and J. S. Andrews. (1933). A comparative morphological study of the infective larvae of the common nematodes in the alimentary tract of sheep. Trans. Amer. Microsc. Soc. 52:1–25.

Dinaburg, A. A. (1944). Development and survival under outdoor conditions of eggs and larvae of the common ruminant stomach worm, *Haemonchus contortus*. J. Agr. Res. 69:421–433.

Kates, K. S. (1950). Survival on pasture of free-living stages of some of the common gastro-intestinal parasites of sheep. Proc. Helminthol. Soc. Wash. 17:39–58.

Ransom, B. H. (1910). The prevention of losses among sheep from stomach worms (*Haemonchus contortus*). U. S. Dep. Agr. Annu. Rep. Bur. Anim. Ind. (1908), pp. 269–278.

Rogers, W. P. (1965). The role of leucine aminopeptidase in the molting of nematode parasites. Comp. Biochem. Physiol. 14:311–321.

Rogers, W. P. (1966). Exsheathment and hatching mechanisms in helminths. In: E. J. L. Soulsby, ed. Biology of Parasites: Emphasis on Veterinary Parasites. Academic Press, New York, pp. 33–40.

Rogers, W. P., and R. I. Sommerville. (1960). The physiology of the second ecdysis of parasitic nematodes. Parasitology 50:329–348.

Rogers, W. P., and R. I. Sommerville. (1963). The infective stage of nematode parasites and its significance in parasitism. In: B. Dawes, ed. Advances in Parasitology. Academic Press, New York, Vol. 1, pp. 109–177.

Rogers, W. P., and R. I. Sommerville. (1968). The infectious process, and its relation to the development of early parasitic stages of nematodes. In: B.

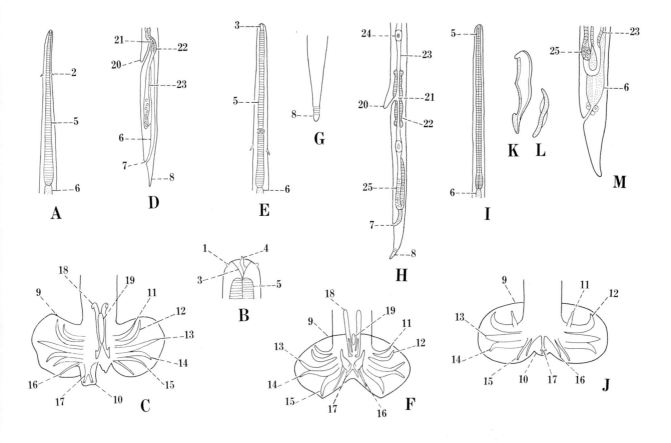

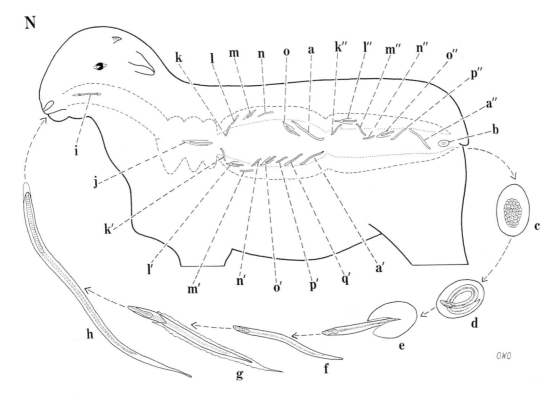

Dawes, ed. Advances in Parasitology. Academic Press, New York, Vol. 6, pp. 327–348.

Shorb, D. A. (1944). Survival on grass plots of eggs and larvae of the stomach worm, *Haemonchus contortus*. J. Agr. Res. 68:317–324.

Shorb, D. A. (1944). Factors influencing embryonation and survival of eggs of the stomach worm *Haemonchus contortus*. J. Agr. Res. 69:279–287.

Silverman, P. H., and J. A. Campbell. (1959). Studies on parasitic worms of sheep in Scotland. I. Embryonic and larval development of *Haemonchus contortus* at constant conditions. Parasitology 49: 23–38.

Stoll, N. R. (1943). The wandering of Haemonchus in the sheep host. J. Parasitol. 29:407–416.

Veglia, F. (1915). The anatomy and life-history of the *Haemonchus contortus* (Rud.). Dep. Agr. Union S. Africa, 3rd and 4th Rep. Dir. Vet. Res., pp. 347–500.

Wang, G. T. (1971). *Haemonchus contortus*: Food of preinfective larvae. Exp. Parasitol. 201–207.

## *Ostertagia circumcincta*
(Stadelmann, 1894) (Plate 116, E–H)

The species of *Ostertagia* are parasites of the abomasum of sheep, goats, cattle, and other ruminants. Because of their color, they are known collectively as the brown stomach worm. They are smaller than the twisted stomach worm and different in structure. While *Ostertagia* is widespread, it tends to reach into the colder climates beyond the optimal range of *Haemonchus. Ostertagia circumcincta* is one of the most common members of the genus occurring in sheep.

**Description.**    The males measure 7.5 to 8.5 mm and the females 9.8 to 12.2 mm long. The buccal capsule is rudimentary. Cervical papillae are located toward the posterior end of the esophagus. Spicules are slender; the distal end is bifurcated, with the branches parallel and subequal; the longer outer one has a terminal knob and the shorter one is acute. The vulva is usually in the posterior fifth of the body and generally covered by a flap. Several thickened cuticular rings occur near the tip of the tail of the female. Eggs are 80 to 100 $\mu$ by 40 to 50 $\mu$ in size.

**Life Cycle.**    The life cycle of both the preparasitic and parasitic phases is basically similar to that of *Haemonchus contortus* which thrives best in warm and mild climates. Being highly resistant to cold and desiccation as eggs and larvae, *Ostertagia circumcincta*, as well as other species of the genus, is the predominant stomach worm in cooler areas, where it replaces *Haemonchus contortus* to a great extent.

The hatching and exsheathing processes are similar to those described for *H. contortus*. About 4 days after infection, exsheathed larvae are in the gastric glands, where the third molt takes place. On day 8 after infection, most of the fourth stage larvae have left the mucosa and are attached to the surface of it. A few inhibited larvae remain as long as 60 days before emerging to resume development. The worms are mature in 15 days and eggs appear in the feces 17 days after infection. Most of them are lost from the host in 60 days. The cycle is presented on Plate 116, a′, b–j, k′–q′.

This and other species of *Ostertagia* produce clinical symptoms of anemia and diarrhea.

Other common species are *Ostertagia ostertagi* primarily of cattle but also in sheep, and *O. trifurcata* of sheep and goats. Both of them are parasites of the abomasum.

## SELECTED REFERENCES

Armour, J. F., W. F. Jarrett, and F. W. Jennings. (1966). Experimental *Ostertagia circumcincta* infections in sheep: development and pathogenesis of a single infection. Amer. J. Vet. Res. 27:1267–1278.

Furman, D. P. (1944). Effects of environment upon the free-living stages of *Ostertagia circumcincta* (Stadelmann) Trichostrongylidae. 1. Laboratory experiments. Amer. J. Vet. Res. 5:79–86.

Jayawickrama, S. D., and H. C. Gibbs. (1967). Studies on ostertagiasis in sheep. Can. Vet. J. 8: 29–38.

Rose, J. H. (1969). The development of the parasitic stages of *Ostertagia ostertagi*. J. Helminthol. 43:173–184.

Sommerville, R. I. (1944). On the histotropic phase of the nematode parasite, *Ostertagia circumcincta*. Austr. J. Agr. Res. 5:130–140.

Threlkeld, W. L. (1934). The life history of *Ostertagia circumcincta*. Va. Agr. Exp. Sta. Tech. Bull. 52, 24 pp.

Threlkeld, W. L., and T. O. Downing. (1936). A report on experimental infections of *Ovis aries* with the infective larvae of *Ostertagia circumcincta*. J. Parasitol. 22:187–201.

## *Trichostrongylus colubriformis*
(Giles, 1892) (Plate 116, I–M)

*Trichostrongylus colubriformis* is a widespread parasite of the small intestine of sheep, goats, and cattle. Its life cycle is well understood and representative of the other species of the genus occurring in mammals.

**Description.** Males are 4.5 to 5 mm and the females 5 to 7 mm long. The buccal capsule is absent. There are no cervical papillae, and a vulvar flap is lacking. Spicules are stout, consisting of a single piece with a distal enlargement; the length is 0.135 to 0.156 mm; the gubernaculum is arcuate. Eggs are 75 to 85 $\mu$ long by 34 to 35 $\mu$ wide.

**Life Cycle.** When deposited and passed in the feces, the eggs are in the morula stage. They hatch within 19 hours at 18° to 21°C and undergo the first molt 25 to 28 hours later. Optimal temperature for development is between 15° and 30°C with the greatest yield at 27°. Above that, mortality is high. The second stage persists about 40 hours, at which time the second molt occurs around 60 hours after infection.

Embryonated eggs and ensheathed third stage larvae are highly resistant to desiccation but first and second stage larvae are very susceptible. Fecal pellets on the pasture provide sufficient moisture for development and protection of the first and second stage larvae. Third stage larvae survive on pastures more successfully during fall, winter, and spring than during the summer. The microclimate under snow is conducive to survival. They live up to 12 months without loss of infectivity when stored at 4°C.

Hatching is under the influence of chitinase and proteinase enzymes, as in *Haemonchus contortus*. Stimulation of receptors of third stage larvae to produce exsheathing fluid comes from abomasal fluid instead of rumenal fluid as required by *Haemonchus contortus* and *Ostertagia circumcincta*. Exsheathing is in the anterior part of the small intestine.

Third stage larvae penetrate the mucosa of the first 6 to 8 feet of the small intestine and undergo the third molt about 4 days after infection. In 6 days, or 10 days after infection, the fourth stage larvae return to the lumen of the intestine, molt, and migrate posteriorly. Eggs in small numbers appear in the feces 17 days after infection, becoming abundant by day 25. This cycle is shown in Plate 116 by a″, b–j, k″–p″.

Heavy infections with third stage larvae alter the environment of the abomasum by causing the pH to rise to 6, which upsets the abomasal function. There is an increase in sodium in the abomasal fluid and a decrease in potassium content of dry matter. These changes are accompanied by diarrhea, rapid loss of body weight, marked terminal hemoconcentration and pepsin concentration in the abomasal fluid between weeks 3 and 4.

Other common species of the genus are *T. falcatus*, *T. vitrinus*, *T. capricola*, *T. rugatus* of ruminants; *T. tenuis* of chickens and turkeys; and *T. calcaratus*, *T. ransomi*, and *T. affinis* of rabbits. Their life cycles in general are similar to that described for *T. colubriformis*.

Differences in the three life cycles described above are slight. *Haemonchus* and *Ostertagia* are similar in having a migration in the abomasal mucosa. *Trichostrongylus* differs in having the tissue migration in the intestinal mucosa.

## EXERCISE ON LIFE CYCLE

*Nippostrongylus brasiliensis*, a common trichostrongyle nematode of rats and mice, is an excellent species for experimental studies on the developmental stages of larvae, mode of infection, and migration inside the host. The species infecting sheep are excellent for study of the larvae in their various stages. Infection of small mammals such as rats and mice with them should be undertaken. Species occurring in rabbits provide good material for study. Follow instructions outlined for *Strongylus*.

## SELECTED REFERENCES

Anderson, F. L., and N. D. Levine. (1968). Effect of desiccation on survival of free-living stages of *Trichostrongylus colubriformis*. J. Parasitol. 54:117–128.

Anderson, F. L., N. D. Levine, and P. A. Boatman. (1970). Survival of third-stage *Trichostrongylus colubriformis* larvae on pasture. J. Parasitol. 56:209–232.

Andrews, J. S. (1939). Experimental trichostrongylosis in sheep and goats. J. Agr. Res. 58:761–770.

Connan, R. M. (1966). Experiments with *Trichostrongylus colubriformis* (Giles, 1892) in the guinea pig. I. The effect of host response on the distribution of the parasites in the gut. Parasitology 56:521–530.

Gibson, T. E., and G. Everett. (1967). The ecology of the free-living stages of *Trichostrongylus colubriformis*. Parasitology 57:533–547.

Herlich, H. (1966). Effects of cold storage on the infectivity of third-stage larvae of the intestinal worm, *Trichostrongylus colubriformis* (Giles, 1892)

## Explanation of Plate 117

A–D, *Dictyocaulus filaria*. A, Anterior end of adult. B, Sinistral view of copulatory bursa of male. C, Anterior end of first stage larva. D, Caudal end of first stage larva. E–G, *Protostrongylus rufescens*. E, Ventral view of copulatory bursa. F, Anterior end of first stage larva. G, Caudal end of first stage larva. H–J, *Muellerius capillaris*. H, Sinistral view of caudal end of male. I, Anterior end of first stage larva. J, Caudal end of first stage larva. K, Sheep definitive host. L, Snail intermediate host of *Protostrongylus rufescens*. M, Snail intermediate host of *Muellerius capillaris*.

1, Cephalic knob; 2, esophagus; 3, intestine; 4, anus; 5, tail; 6, nerve ring; 7, copulatory bursa; 8, ventroventral ray; 9, lateroventral ray; 10, antero- or externolateral ray; 11, mediolateral ray; 12, posterolateral ray; 13, externodorsal ray; 14, dorsal ray; 15, spicules; 16, gubernaculum; 17, cuticularized bar.

a, Adult worm of all three species (*Dictyocaulus* and *Protostrongylus* in bronchioles and bronchi, and *Muellerius* in nodules in lung parenchyma); b, eggs; c, eggs hatching; d, larvae pass to outside in feces. *Dictyocaulus*. e, first stage larvae with characteristic cephalic knob; f, second stage larvae with retained cuticle; g, third stage larvae with two retained cuticles; h, third stage larvae when swallowed infect sheep; i, larvae molt to form fourth stage; j, larvae enter intestine; k, larvae penetrate intestinal wall, enter lacteals; l, larvae undergo final molt in lymph nodes; m, larvae enter lymph vessels and go to heart; n, larvae pass through right side of heart and go to lungs; o, larvae migrate from blood vessels of lungs into alveoli, bronchioles, and bronchi; p, some larvae go through blood vessels of lungs, pulmonary vein, and left side of heart; q, larvae enter dorsal aorta, are carried through general circulation; r, in case of pregnant ewes, some larvae enter uterine artery and fetal circulation, producing prenatal infection of lambs.

*Muellerius*. a′, First stage larvae free in feces and soil; b′, larvae penetrate land snails or slugs; c′, second stage larvae; d′, third stage larvae; e′, infection of definitive host by swallowing infected mollusks; migration in host same as in *Dictyocaulus*.

*Protostrongylus*. a″, First stage larvae; b″, larvae enter molluscan host; c″, second stage larvae; d″, third stage larvae; e′, infection of host and further development similar to that in *Muellerius*.

Figures adapted from various sources.

Ransom, 1911. Proc. Helminthol. Soc. Wash. 33: 101–103.

Mirzayans, A. (1969). The effect of temperature on the development of the eggs and larvae of *Trichostrongylus axei*. Brit. Vet. J. 125:37–38.

Mönnig, H. O. (1926). The life-histories of *Trichostrongylus instabilis* and *T. rugatus* of sheep in South Africa. Dep. Agr. Union S. Africa, 11th-12th Rep. Dir. Vet. Educ. Res., pp. 231–251.

Ross, J. G., D. A. Purcell, and J. R. Todd. (1969). Experimental infection of lambs with *Trichostrongylus axei*; investigations using abomasal cannulae. Res. Vet. Sci. 10:133–141.

Seghetti, L. (1948). The effect of environment on the survival of the free-living stages of *Trichostrongylus colubriformis* and other nematode parasites of range sheep in southeastern Montana. Amer. J. Vet. Res. 9:52–60.

Wang, G. T. (1967). Effect of temperature and cultural methods on development of the free-living stages of *Trichostrongylus colubriformis*. Amer. J. Vet. Res. 28:1085–1090.

### Superfamily Metastrongyloidea

The Metastrongyloidea are lungworms of land and marine mammals. The four genera considered represent the basic types of life cycles known for over a dozen species. The life cycle of species of *Dictyocaulus* is direct, whereas those of all of the others are indirect. Species of *Metastrongylus* and *Choerostrongylus* of swine utilize earthworms as intermediate hosts, and the remaining genera, insofar as known, develop in slugs and land snails. Some species utilize collector or storage (paratenic) hosts such as water snakes for *Crenosoma mephitidis* of skunks, and mice for *Aelurostrongylus abstrusus* of cats. The paratenic hosts acquire the parasites by swallowing infected snails or slugs. The larvae migrate to the tissues where they accumulate without further development. The definitive hosts become infected upon eating the paratenic hosts. None of the life cycles of the lungworms of marine mammals is known.

**Description.**    Strongyloid nematodes with small, rudimentary, or no buccal capsule. The copulatory bursa is usually asymmetrical, except in *Dictyocaulus*.

### FAMILY METASTRONGYLIDAE

Only one family is recognized.

### *Dictyocaulus filaria*
(Rudolphi, 1809) (Plate 117, A–D)

This is the common thread lungworm of sheep and goats in all parts of the world. The adults occur in the bronchi.

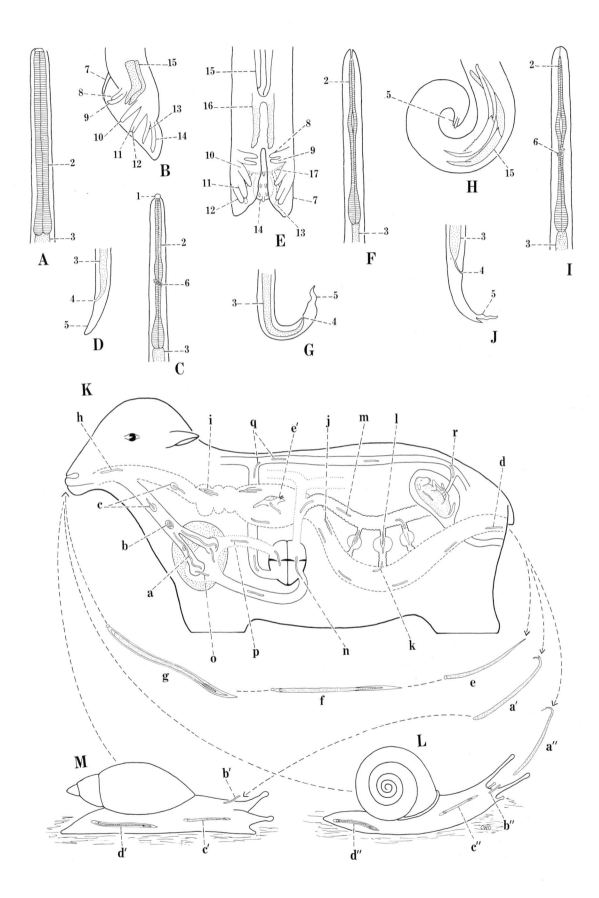

**Description.** These are white thread-like worms. Males are 30 to 80 mm long. The bursa is short, symmetrical, with the medio- and posterolateral rays fused except at the tip, and the dorsal ray is split from the base to the tip. The spicules are reticulated, dark brown, boot-shaped structures 500 $\mu$ long. Females are 50 to 100 mm long, have a pointed tail; the vulva is near the middle of the body. The uterus has an anterior and a posterior branch. First stage larvae passed in the feces are characterized by a small knob on the anterior end of the body.

**Life Cycle.** This and other species of the genus are the only lungworms of domestic animals with a direct life cycle.

Adult females in the bronchioles deposit embryonated eggs. These, together with some larvae that hatch promptly, are carried forward through the trachea by ciliary action of the epithelium or coughed up and swallowed. Some may be discharged through the mouth or nostrils with sputum or nasal secretion. Those reaching the stomach and intestine hatch and the first stage larvae are passed in the feces. Outside the host, they undergo the first molt to form the ensheathed first stage larvae in 1 to 2 days. In another 3 to 4 days, a second molt takes place, producing a third stage larva. They are enclosed in both sheaths for a time when the outer one is lost. These larvae are unable to feed during their sojourn outside the host but live on food stored in the cells. Third stage infective larvae are on the soil or blades of grass in a lethargic stage, which serves as a bridge between the free-living and parasitic phases. Both feeding and development are inhibited during this period.

Grazing sheep swallowing the infective larvae with forage open the way for further development. Stimulation of the larvae in the stomach or intestine triggers production of exsheathing fluid followed by shedding of the sheath. The larvae are now open to subsequent stimuli necessary for feeding, migration, growth, and development.

The newly freed larvae penetrate the intestinal mucosa, enter the lacteals, and go to the mesenteric lymph nodes, where the third molt occurs about 4 days after infection. The fourth stage larvae are carried in the lymph or blood vessels via the heart, always bypassing the liver, to the lungs. Being blocked against further progress by the small diameter of the capillaries forming the vascular network around the alveoli, they burrow through the tissues into them about day 8 after infection. The fourth and final molt occurs in the bronchioles about 18 days after infection. Although extensive migration occurs between the intestine and lungs, larvae appear not to wander into unborn young to establish prenatal infections. Eggs appear in the uteri 32 to 57 days after infection and larval output in the feces is at its peak 39 to 57 days following infection.

Larval lungworms require a warm and moderately wet climate for development and survival. They are very susceptible to cold but may carry over mild winters in small numbers. Desiccation kills them promptly. Survival is best on lush pastures with ample moisture. Heavily stocked irrigated pastures provide ideal environmental conditions for infection during summer months in areas otherwise highly incompatible for development and survival. Carry over from season to season is in infected animals.

Different ecological conditions encountered by the species of adult worms are responsible for differences in the amount of development of the muscles of the body wall. In *Dictyocaulus filaria* living free in the bronchioles, the muscles are well developed and function to maintain the worms in position. In *Muellerius capillaris* embedded in the parenchyma of the lungs where no effort is required to keep their location, the muscles are weakly developed.

Heavy infections of adult worms in the bronchioles, especially in young animals, cause verminous pneumonia. By 2 to 3 weeks after infection, coughing begins, accompanied by mucus discharge from the nose and mouth. Respiration becomes labored and rapid. The animals go off feed and lose weight. They may die.

*Dictyocaulus viviparus* of cattle and *D. arnfieldi* of horses have similar life cycles. *Dictyocaulus* shows morphological and biological affinities to the Trichostrongylidae and is considered by some authors as belonging to that family rather than to the Metastrongylidae.

## SELECTED REFERENCES

Deorani, V. P. S. (1965). Studies on the life-history and bionomics of *Dictyocaulus filaria* (Rudolphi, 1809) (Metastrongylidae: Nematoda). Indian J. Helminthol. 17:89–103.

Kauzal, G. P. (1933). Observations on the bionomics of *Dictyocaulus filaria* with a note on the clinical manifestations in artificial infections of sheep. Austr. Vet. J. 9:20–26.

Shaw, J. N. (1934). Lungworms (*Dictyocaulus filaria* Rudolphi) in sheep and goats. Oreg. Agr. Exp. Sta. Bull. 327, 12 pp.

Soliman, K. N. (1953). Migration route of *Dictyocaulus viviparus* and *D. filaria* infective larvae to the lungs. J. Comp. Pathol. Ther. 63:75–84.

Soliman, K. N. (1953). Studies on the bionomics of preparasitic stages of *Dictyocaulus viviparus* with reference to the same in the allied species in sheep *D. filaria*. Brit. Vet. J. 109:361–381.

Zmoray, I., and A. Gutteková. (1968). Structure of subcutaneous muscles and their ecological analysis by pneumohelminths. Biológia Bratislava 23:582–589.

### *Protostrongylus rufescens*
(Leuckart, 1865) (Plate 117, E–G)

This is the brown lungworm of sheep, goats, and deer of North America, Europe, Africa, Australia, and possibly other areas of the world. The adults are in the bronchioles.

**Description.**    The adults are thread-like brown worms. The males are 16 to 28 mm long, and have a short bursa that is strengthened dorsally by a cuticularized bar; the dorsal ray of the bursa is thick, the tubular spicules are 260 μ long, and a gubernaculum and telamon are present. Females are 25 to 35 mm long, with the vulva near the anus. First stage larvae have a characteristic undulating, pointed tail.

**Life Cycle.**    Eggs laid in the bronchioles of the lungs complete their development and hatch. The first stage larvae are carried up the trachea in the mucus by ciliary action or coughed up, swallowed, and passed with the feces. The young worms are in the mucus surrounding the fecal pellets from which they are washed onto the soil by rains.

Larval development is arrested at this stage and cannot proceed until stimuli are provided by one of the land snail intermediate hosts of the genera *Abida*, *Arianta*, *Helicella*, *Theba*, *Zebrina*, and others. First stage larvae enter the snails by way of the mucus glands of the foot and molt twice to form the infective third stage as soon as 2 weeks or as long as 50 days later. Fully developed larvae appear as dark spots in the foot of the snails, especially when the latter are attached to the wall of a Petri dish.

Further development is dependent upon snails being swallowed and digested by the sheep or other host, thus freeing the larvae. The stimulus provided by conditions and substances in the stomach of the warm-blooded host initiates physical and physiological activity leading to development and maturity. Larvae in the small intestine burrow through the mucosa, enter the lacteals, and are carried to the lymph nodes. Here the third molt occurs. Newly formed fourth stage larvae reach the lungs via the heart and blood vessels from which they enter the alveoli. In the alveoli or bronchioles, the fourth and final molt takes place and the worms develop to sexual maturity.

The effect of these parasites on the host is similar to that of *Dictyocaulus filaria*.

### SELECTED REFERENCES

Mapes, C. R., and D. W. Baker. (1950). Studies on the protostrongyline lungworms of sheep. J. Amer. Vet. Med. Ass. 116:433–435.

Gerichter, C. B. (1948). Observations on the life history of lung nematodes using snails as intermediate hosts. Amer. J. Vet. Res. 9:109–112.

Hobmaier, A., and M. Hobmaier. (1930). Life history of *Protostrongylus* (*Synthetocaulus*) *rufescens*. Proc. Soc. Exp. Biol. Med. 28:156–158.

### *Muellerius capillaris*
(Müller, 1889) (Plate 117, H–J)

These worms, known as the hair lungworms, live in the parenchyma of the lungs of sheep and goats in the United States, Europe, Australia, South Africa, and possibly other parts of the world.

**Description.**    These are delicate, white, hair-like worms. The males are 12 to 14 mm long with the pointed, nonbursate posterior end spirally coiled. The spicules are 150 μ long, curved, and with sharp points. The females are 19 to 23 mm long. The vulva has a small swelling at the posterior margin and is located near the anus. First stage larvae, with a distinct spine-like process on the dorsal side of the tail, are passed in the feces.

**Life Cycle.**    The parasites live in grayish nodules in the parenchyma of the lungs. Eggs laid in them hatch there and the larvae migrate up the trachea to the pharynx, are swallowed, and passed in the feces. The larvae are fairly resistant to drying and withstand freezing well. Like species of *Protostrongylus*, continued development is dependent on entering a terrestrial mollusk. Numerous species of snails from 20 or more genera (*Anguispira, Helix, Praticolella, Polygyra, Retinella, Succinea, Zonitoides* from North America) and the slugs *Agriolimax, Arion, Derocercus,* and *Limax* serve as intermediate hosts. Russian workers claim freshwater snails, including *Gyraulus albus, G. laevis,* and *Radix periger,* serve as natural intermediaries.

First stage larvae penetrate the mollusks, in whose bodies they survive long periods—perhaps the life span of the host. Two molts occur in the gastropods, the first in about 4 weeks and the second 3 to 4 days later under optimal conditions.

Sheep and other definitive hosts become infected by swallowing mollusks containing infective third stage larvae. Upon being released from the tissues of the gastropods in the intestine, the larvae migrate to the lungs by the same route and undergo development similar to that of *Dictyocaulus* and *Protostrongylus.* In the lungs, the larvae enter the parenchyma, where final development takes place, and the adults reside in nodules. Sexual maturity is attained in about 6 weeks. Longevity of the worms is 4 to 7 years. In cases of heavy infections, the lungs enlarge and there are purulent bronchiopneumonial abscesses.

## EXERCISE ON LIFE CYCLE

Life history studies on the lungworms of sheep may be conducted in the laboratory. Examine feces of sheep for larvae by placing freshly passed pellets in a Baermann funnel for not longer than 3 hours. It is important that the feces be fresh and uncontaminated with soil. In feces allowed to stand longer than 3 hours, eggs of intestinal nematodes hatch and those coming in contact with soil may show free-living soil nematodes. Larvae of the three species of lungworms can be identified by the morphological characters described above under each species.

Interesting experiments may be performed to demonstrate the life cycle of *Dictyocaulus filaria.* Having obtained first stage larvae from the feces, keep them in water long enough to develop to the infective third stage characterized by the two loose cuticles.

Third stage larvae fed to mice undertake the characteristic migration through the lymphatics, arriving in the lungs in about 5 days. By killing infected mice at intervals of 24 hours over a

period of 5 to 6 days, the course of migration through the lymphatics may be followed. Ascertain whether some larvae migrate by way of the hepatic portal vessel and liver.

When larvae of *Protostrongylus* or *Muellerius* are procurable in the feces of sheep, experiments using land snails and white mice may be conducted to follow the details of their life cycles. Snails known to serve as intermediate hosts collected from pastures should be examined for larvae either by crushing or digesting them. It is important to keep in mind that protostrongyline lungworms of rabbits and hares may be present and that they also infect mollusks. Snails collected from areas where sheep and lagomorphs do not occur should be uninfected. However, the preferred procedure is to rear them in laboratory terrariums.

Infect parasite-free snails with first stage larvae and examine them over a period of 4 to 6 weeks in order to follow the development of the larvae. When infective third stage larvae appear, mice may be infected by feeding snails to them. Follow the course of migration as described for *Dictyocaulus.*

## SELECTED REFERENCES

Egorov, Y. C. (1960). [Biology of *Muellerius capillaris.*] Tr. Nauch.-Issled. Vet. Inst. Minsk. 1:160–170.

Hobmaier, A., and M. Hobmaier. (1930). *Limax* and *Succinea,* zwei neue Zwischenwirte von *Muellerius (Synthetocaulus) capillaris* des Schafes und der Ziege. München. Tierärztl. Wochenschr. 81: 285–287.

Kassai, T. (1963). Uber die Lebensdauer der Protostrongyliden der Schafe. Helminthologia 4:199–205.

Ryšavý, B. (1969). Ciclo evolutivo del helminto pulmonar *Müllerius capillaris* Müller, 1889, en Cuba. Torreia, n. s. (6):3–12.

Rose, J. H. (1957). Observations on the larval stages of *Muellerius capillaris* within the intermediate hosts *Agriolimax agrestis* and *A. reticulatus.* J. Helminthol. 31:1–16.

Rose, J. H. (1957). Observations on the bionomics of the sheep lungworm, *Muellerius capillaris.* J. Helminthol. 31:17–28.

Švarc, R. (1968). Der Einfluss hoher Dosen von Invasionlarven *Müllerius capillaris* auf die Entwicklung des pathologischen Bildes der Lungengewebe. Helminthologia 8–9:541–549.

Williams, D. W. (1942). Studies on the biology of the larva of the nematode lungworm *Muellerius capillaris* in molluscs. J. Anim. Ecol. 11:1–8.

*Metastrongylus apri*
(Gmelin, 1790) (Plate 118)

The adults are slender, white worms in the bronchi and bronchioles of swine throughout the world.

**Description.**    Males are up to 25 mm long, with a small bursa in which the mediolateral and posterolateral rays are fused; the spicules are slender and 4 to 4.2 mm long, with a terminal harpoon-like barb. Females are up to 85 mm long, with the posterior end flexed ventrally upon itself; the vulva and anus open near the posterior end of the body. Eggs are 45 to 57 $\mu$ long by 38 to 41 $\mu$ wide, have thick shells, and contain first stage larvae when deposited.

**Life Cycle.**    Embryonated eggs laid in the bronchioles and bronchi are coughed up, swallowed, and voided with the feces. The thick-shelled eggs are resistant, surviving on pastures and yards for 6 to 13 months and freezing for more than 5 months. They are unable to hatch until swallowed by earthworms such as *Lumbricus terrestris, L. rubellus, Allolobophora calignosa, Dendrobaena rubida, Eisenia foetida, E. lonnbergi,* and others. Enzymes from the alimentary canal of the annelids stimulate the larvae to secrete hatching fluids that break down the egg membranes, liberating the first stage larvae. They penetrate the wall of the crop, going primarily to the lamellar sinuses of the calciferous glands which are preferred sites for localization and development. Two molts occur and the ensheathed third stage larvae appear in 10 to 30 days. They finally enter the circulatory system, accumulating in the hearts and wall of the esophagus. No further development takes place in the earthworms. Longevity of the third stage larvae is limited only by that of the annelid hosts. Hence, areas occupied by infected pigs are likely to be sources of infection for long periods.

Swine become infected by eating earthworms they root from the soil. Larvae liberated from the digested annelids exsheath and penetrate the intestinal mucosa. They enter the lacteals and go to the lymph nodes where the third molt occurs. The newly formed fourth stage larvae go via the lymph vessels, heart, and pulmonary artery to the lungs. Being stopped by the small diameter of the capillaries, they break into the air sacs throughout the lungs. The fourth molt follows and the worms go to the bronchioles and bronchi where maturity is reached in about 28 to 31 days. Egg output is at its maximum for 1

to 6 weeks immediately following patency. It ceases when pigs reach 8 to 9 months of age. The worms probably live 8 to 12 months.

The life cycles of *Metastrongylus pudentotectus* and *M. salmi,* other lungworms of pigs, are similar to that of *M. apri.*

Larval worms migrating through the lung tissues cause petechial hemorrhages and adult worms occlude bronchioles, causing verminous pneumonia. In addition to the pathological syndrome, the virus of swine influenza gains entrance to the eggs and unhatched larvae in the pigs and is introduced by these larvae to the lungs of uninfected animals in which the disease appears.

## EXERCISE ON LIFE CYCLE

Adult lungworms may be obtained from the bronchioles of swine slaughtered in abattoirs in many sections of the country. Dissect the uterus from gravid females and note the thick-shelled embryonated eggs containing first stage larvae. Mix the eggs with moist soil in a glass container and place earthworms of the genera *Helodrilus, Lumbricus, Dendrobaena,* or *Eisenia* in it. Keep the cultures at room temperature. Both soil and earthworms should be obtained from places where swine have not been kept to avoid naturally infected annelids. Examine the pharynx, esophagus, and hearts over a period of a month for larvae and note the stages of development. Also obtain earthworms from hog lots and examine them for natural infections of lungworms.

Feed infected earthworms to six white mice, white rats, or guinea pigs. Kill five of them at intervals of 24 hours after infection and examine the intestinal contents, lymph nodes, and lungs for larvae. Note the stages of development of the larvae in the different locations and correlate them with the time elapsed since infection. Leave one animal for 4 weeks or longer and examine its lungs for worms.

## SELECTED REFERENCES

Alicata, J. E. (1935). Early developmental stages of nematodes occurring in swine. U. S. Dep. Agr. Tech. Bull. 489, 96 pp.

Jericho, J. W. F., P. K. C. Austwick, R. T. Hodges, and J. B. Dixon. (1971). Intrapulmonary lymphoid tissue of pigs exposed to aerosols of carbon particles, of *Salmonella oramienberg,* of *Mycoplasma granu-*

## Explanation of Plate 118

**A–H,** *Metastrongylus apri.* **A,** Anterior end of adult. **B,** Lateral view of posterior end of mature male. **C,** Dorsal view of caudal end of male. **D,** Tip of spicule. **E,** Lateral view of posterior end of adult female. **F,** Embryonated egg. **G,** Caudal end of newly hatched first stage larva. **H,** Caudal end of third stage larva. **I–J,** *Metastrongylus pudentotectus.* **I,** Dorsal view of caudal end of adult male. **J,** Tip of spicule. **K,** *Metastrongylus salmi.* Dorsal view of bursa. **L,** Pig host. **M,** Earthworm intermediate host.

1, Mouth; 2, copulatory bursa; 3, dorsal ray; 4, externodorsal ray; 5, anterolateral ray; 6, mediolateral ray; 7, posterolateral ray; 8, externoventral ray; 9, ventroventral ray; 10, spicules; 11, tip of spicule with barb; 12, tail; 13, intestine; 14, vagina uterina with embryonated eggs; 15, vulva.

a, Adult worms in air tubes of lungs; b, embryonated eggs pass up trachea and are swallowed; c, embryonated eggs pass out of body in feces; d, embryonated eggs free in soil or feces; e, eggs swallowed by earthworms; f, eggs hatch in pharynx and esophagus; g, larvae migrate into wall of alimentary canal; h, second molt in wall of intestine; i, larvae enter hearts; j, third stage larvae in wall of esophagus; k, swine become infected by swallowing earthworms; l, larvae released from annelid intermediary in stomach; m, larvae enter intestine; n, larvae penetrate intestinal wall and enter lacteals; o, larvae enter lymph nodes and molt; p, larvae enter lymph vessels; q, larvae pass through the right side of heart; r, larvae carried in blood to capillaries of lungs; s, larvae leave capillaries, enter alveoli and molt, maturing as early as 24 days after being swallowed.

Figures A–E, I–K adapted from Gedoelst, 1922, Bull. Soc. Pathol. Exot. 16:622; F–H, from Alicata, 1935, U. S. Dep. Agr. Tech. Bull. 489, p. 37.

---

*larum,* and to an oral inoculum of *Metastrongylus apri.* J. Comp. Pathol. 81:13–21.

Kates, K. C. (1941). Observations on the viability of eggs of lungworms of swine. J. Parasitol. 27:265–272.

Polzenhagen, M., R. Buchwalder, and F. Hiepe. (1969). Untersuchungen zum Lungenwurmbefall des Schweines. 2. Mitteilung. Verlauf und Auswirkungen von *Metastrongylus*-Inokulationen bein Hausschwein unter besonderer Berücksichtigung der Lebenserwartung von *Metastrongylus apri.* Monatsh. Veterinaermed. 24:847–850.

Probert, A. J. (1969). Morphological and histochemical studies on the larval stages of *Metastrongylus* spp. (lungworms of swine) in the earthworm intermediate host, *Eisenia foetida* Savigny, 1826. Parasitology 59:269–277.

Schwartz, B., and J. E. Alicata. (1934). Life history of lungworms parasitic in swine. U. S. Dep. Agr. Tech. Bull. 456, 41 pp.

Shope, R. E. (1941). The swine lungworm as a reservoir and intermediate host for swine influenza virus. I. The presence of swine influenza virus in healthy and susceptible pigs. II. The transmission of swine influenza virus by the swine lungworm. J. Exp. Med. 74:41–68.

Shope, R. E. (1943). The swine lungworm as a reservoir and intermediate host for swine influenza virus. III. Factors influencing transmission of the virus and the provocation of influenza. J. Exp. Med. 77:111–126.

Shope, R. E. (1943). The swine lungworm as a reservoir and intermediate host for swine influenza virus. IV. The demonstration of masked swine influenza virus in lungworm larvae and swine under natural conditions. J. Exp. Med. 77:127–138.

Ustinov, I. D. (1963). [Length of life of metastrongyle eggs in the external environment, and resistance to various temperatures.] Tr. Vses. Inst. Gel'mintol. 10:56–63.

### Suborder Ascaridina

The members of this suborder are parasites of the alimentary canal of all classes of vertebrates and some terrestrial arthropods. The esophagus may have a terminal bulb or be straight; the mouth is surrounded by three, two, or no lips. The males usually have two spicules but there may be only one, the tail curves ventrally, and caudal alae may be present. Two superfamilies, Oxyuroidea and Ascaridoidea are recognized.

### Superfamily Oxyuroidea

In the oxyurids, the esophagus is terminated posteriorly by a distinct valved bulb. The stoma is cylindroid but not surrounded by esophageal tissue. There are four double or eight single lip papillae in the external circle and six minute ones in the internal circle. They are parasites of terrestrial arthropods, and all classes of vertebrates.

### FAMILY OXYURIDAE

This family contains common parasites of amphibians, reptiles, and mammals, usually in the large intestine. The tail of the female is long and slender. *Enterobius vermicularis,* a common

**Plate 118**  *Metastrongylus apri*  443

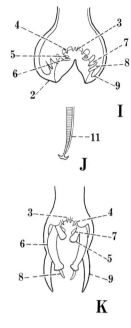

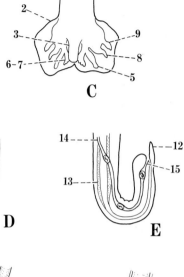

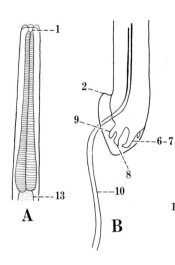
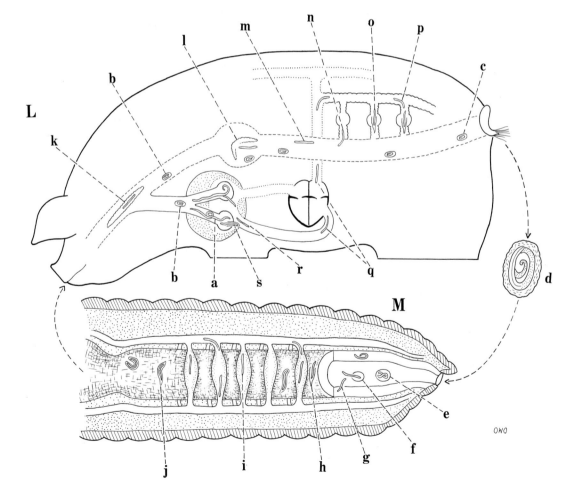

OWO

## Explanation of Plate 119

A, Adult female. B, Adult male. C, Ventral view of caudal end of male. D, Embryonated egg. E, Human host. F, Hand contaminated with eggs from anus. G, Hands contaminated from soiled underclothing. H, Bedding and sleeping garments contaminated with eggs. I, Eggs air-borne as result of sweeping, dusting, and movement of bedding.

1, Cephalic alae; 2, esophageal bulb; 3, intestine; 4, anus; 5, tail; 6, vulva; 7, vagina; 8, anterior ovary; 9, anterior uterus; 10, posterior ovary; 11, posterior uterus; 12, spicule; 13, caudal ala; 14, caudal papillae; 15, ejaculatory duct; 16, seminal vesicle; 17, sperm duct; 18, 19, testis; 20, eggshell; 21, first stage larvae.

a, Adult female in ascending colon; b, adult male in ascending colon; c, adult females in transverse and descending colon; d, gravid female making her way through rectum to anus; e, gravid female outside in perianal region to oviposit; f, eggs in perianal region; g, embryonated egg with first stage larva; h, eggs on fingers and under nails being transferred to mouth; i, embryonated egg from fingers; j, infection by swallowing eggs; k, eggs pass through stomach; l, eggs hatch in duodenum; m–o, first, second, and third molts, respectively, after which larvae continue growth to maturity; p, underwear soiled with eggs that get on hands when clothing is handled; q, eggs accumulate in bed and are stirred into air with movement of occupant or when bedding is shaken; r, eggs become airborne by sweeping and dusting; s, airborne embryonated eggs inhaled and infection results; t, eggs in perianal region hatch and larvae enter anus, producing infection by retrofection; u, larvae migrate up descending colon and develop to maturity.

---

species infecting humans, will be used to illustrate the typical life cycle of this group.

### Enterobius vermicularis
(Linnaeus, 1785) (Plate 119)

*Enterobius vermicularis* is the ubiquitous pinworm or seatworm of man. Human coprolites containing identifiable eggs and radiocarbon dated at 7837 B.C. attest the long association of man and this worm. Because of the intimate manner in which it is transmitted, infections are more prevalent in cool climates, where clothing and bed linens function in transfer of the eggs to the mouth, than in tropical regions, where bathing is frequent and raiment is less.

Upwards of 60 percent of the white children in some northern cities of the United States are infected. Whites are more commonly infected than blacks. Infection is especially prevalent in groups such as families, orphanages, and mental hospitals where a high level of personal hygiene is difficult to maintain.

**Description.**    They are small white worms. The males measure 2 to 5 mm and the females 8 to 13 mm long. The males have a single spicule 70 $\mu$ long and the caudal end of the body is curved ventrally. The tail of the female is long and pointed, equal to about one-third the total length of the body. The vulva is near the union of the anterior and middle thirds of the body and the long vagina extends caudad. When gravid, the female assumes a slight spindle shape due to the great number of eggs, which are 50 to 60 $\mu$ long by 20 to 30 $\mu$ wide, flattened somewhat on one side, and contain an embryo in the tadpole stage when laid.

**Life Cycle.**    Adult worms are in the large intestine where mating occurs, after which the males tend to disappear. As the females approach gravidity, which is 15 to 43 days after infection, they move toward the posterior end of the colon and into the rectum. When ready to oviposit, they detach from the wall, crawl out of the anus, and creep over the perianal region, leaving a trail of sticky, partially developed eggs. Each female lays from 4000 to 17,000 (average 11,000) eggs, following which she usually dies. Movement of the worms over the sensitive perianal area produces intense itching that demands attention for relief. Scratching or rubbing this area is symptomatic of enterobiasis.

The eggshell consists of two layers of chitin and an inner lipoid membrane. Being in an advanced stage of development when laid, first stage larvae are fully formed and infective in about 6 hours from the body heat of the host. In homes or institutions with infected individuals, eggs are being separated constantly from the body, underclothing, and bed linen to become airborne and settling everywhere. They are highly susceptible to drying and die within 16 hours or less. But among infected groups, infective eggs are present continuously.

Infection is by two basic routes. The more common is by swallowing eggs and the other by retrofection in which the eggs in the perianal

**Plate 119**   *Enterobius vermicularis*                                      445

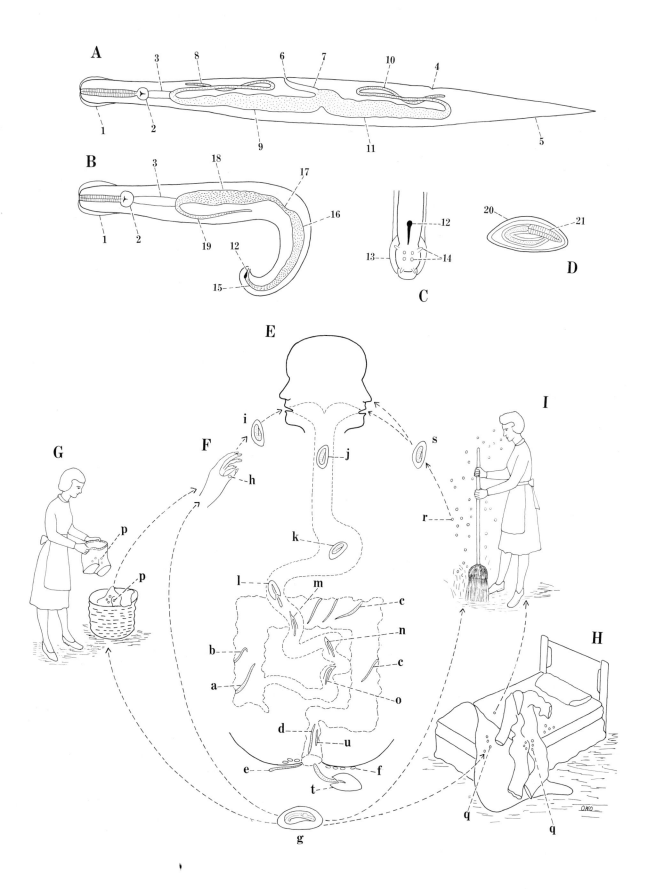

region hatch and the first stage, rhabditiform larvae crawl through the anus into the colon.

Eggs reach the host to be swallowed by several means. They are transferred to the mouth by the fingers when they become contaminated from scratching to alleviate the intense pruritis. Handling of soiled underwear and bed linen is another source of contamination of the hands to complete infection by transferring eggs to the mouth.

Airborne eggs may be inhaled through the nose or mouth and swallowed to establish infections. Air currents created from handling soiled clothing and bed linen that bear eggs set them into circulation. Movements in beds by sleeping persons raising and lowering the covers create currents that move back and forth, carrying eggs from infected persons. Sweeping and dusting, likewise, stir up eggs to be inhaled and swallowed.

Retrofection results in reinfection of the individual who harbors worms. The alkaline condition of the perianal region necessary for hatching, the relative humidity and temperature favor development, hatching, and survival of the larvae. Under these optimal conditions, the larvae creep through the anus and into the colon where development is completed.

When fully embryonated eggs are swallowed, hatching occurs in the duodenum. The larvae commence the long journey to the colon, interrupted by stops in crypts and folds of the lower part of the small intestine, molting twice en route. They attach to the mucosa of the anterior part of the colon where the fourth and final molt probably takes place. Sexual maturity is attained in 36 to 53 days after oral infection and 46 to 76 after retrofection. Adult worms survive only a few weeks but even so the epidemiology assures almost continuous reinfection.

The severe itching produced by the migrating females in the anal area causes scratching, loss of sleep, restlessness, and nervous disorders. In female patients, the worms may migrate into the reproductive tract, causing vaginitis. Larvae migrating and burrowing in the intestine cause inflammation. They often appear in excised appendices but whether they are responsible for appendicitis is uncertain.

A few other common and well-known species of oxyurids include *Passalurus ambiguus* and *Dermatoxys veliger* of rabbits, *Syphacia obvelata* and *Aspiculuris tetraptera* of rats and mice,

*Skrjabinema ovis* of sheep, and *Oxyuris equi* of equines. Their life cycles are basically the same as that of *Enterobius vermicularis*. It appears doubtful on the basis of experimental evidence available that retrofection occurs in these species.

## EXERCISE ON LIFE CYCLE

An excellent species for exercises on life cycles is *Syphacia obvelata* because of its prevalence and the ease of obtaining and handling the definitive hosts.

Like *E. vermicularis*, eggs are laid in the perianal region where they may be collected by dabbing the area with bits of Scotch tape or by washing. By collecting at appropriate time intervals, the stages of development in them may be observed. The Scotch tape method is especially good for this observation. If embryonated eggs are fed to a series of mice, the various stages of development and their location may be ascertained by killing the hosts and examining the alimentary canal at intervals of 8 to 10 hours, or even at shorter periods for the first 48 hours, and thereafter at 12 to 24 hours. Determine whether males and females have similar longevity in the gut. Try experiments to ascertain whether retrofection occurs.

## SELECTED REFERENCES

Chan, K. F. (1952). Life cycle studies of *Syphacea obvelata*. Amer. J. Hyg. 56:14–21.

Cram, E. B. (1943). Studies on oxyuriasis. XXVIII. Summaries and conclusions. Amer. J. Dis. Child. 65:46–59.

Frey, G. F., and J. G. Moore. (1969). *Enterobius vermicularis*: 10,000-year-old human infection. Science 166:1620.

Lentze, F. A. (1935). Zur Biologie des *Oxyuris vermicularis*. Zentralbl. Bakteriol. Parasitenk. I. Abt. Orig. 135:156–159.

Reardon, L. (1938). Studies on oxyuriasis. XVI. The number of eggs produced by the pinworm, *Enterobius vermicularis*, and its bearing on infection. Publ. Health Rep. 53:978–984.

Schüffner, W. (1949). Retrograde Oxyuren-Infektion, "Retrofection" IV. Mitteilung. Zentralbl. Bakteriol. I. Orig. 154:220–234.

Schüffner, W., and N. H. Swellengrebel. (1949). Retrofection in oxyuriasis. A newly discovered mode of infection with *Enterobius vermicularis*. J. Parasitol. 35:138–146.

## Superfamily Ascaridoidea

The Ascaridoidea are parasites of the various classes of vertebrates. They range from relatively small to large nematodes of the alimentary canal and are characterized by an esophagus that is straight or one that terminates in a bulb. The stoma is usually collapsed and surrounded by esophageal tissue. There are four large double papillae in the external circle on the lips.

### FAMILY HETERAKIDAE

These are common parasites of reptiles, birds, and mammals. The esophagus may or may not have a posterior bulb. The males possess a distinct precloacal sucker. The eggs are smooth, unembryonated when laid, and have thick shells.

### Heterakis gallinae
(Gmelin, 1790) (Plate 120)

This is the cecal worm of chickens, turkeys, pheasants, quails, and related upland game birds. It is extremely common and widely distributed. It is associated with the spread among birds of the flagellate *Histomonas meleagridis*, the causative organism of enterohepatitis of turkeys (Plate 1).

**Description.**    The males are 7 to 13 mm and females 10 to 15 mm long; each sex has a pointed tail. Alae, or expansions of the cuticle, extend along the lateral sides of the body. The esophagus terminates posteriorly as a distinct valved bulb. The tail of the male has wide alae, a prominent precloacal sucker, and 12 pairs of papillae. The right spicule is 2 mm and the left 0.65 to 0.7 mm long. The vulva opens directly behind the middle of the body and the eggs are 65 to 80 $\mu$ by 35 to 46 $\mu$ in size, have smooth shells, and are unembryonated when laid.

**Life Cycle.**    The life cycle is direct. Eggs are in the single-cell stage when passed in the feces. They are remarkably resistant and capable of surviving adverse conditions for long periods, up to 230 weeks in the soil and freezing for 167 days. Under average conditions such as room temperature, the larvae reach the second stage in 12 to 14 days. This is the infective stage and no further development takes place in the egg.

Infection of birds is by swallowing the eggs. While some eggs hatch in the crop, most of them do so in the gizzard and duodenum within 2 hours. They begin arriving in the ceca as soon as 6.5 hours after being swallowed without further development.

In the ceca, the larvae are closely associated with the mucosa for the first 12 days, with the peak on the third day. A few enter the tissue at first but emerge between days 2 and 5. There is no true tissue phase. Development of lumen-dwelling larvae in chicks less than 10 days old is inhibited but is optimal in birds 21 days old. This appears to be the most susceptible age. The presence of bacterial flora in the intestine is essential for development of the larvae. The role of the bacteria is unknown. It appears to be something other than related to pH and oxidation-reduction.

Earthworms (*Lumbricus terrestris, Allolobophora calignosa, Eisenia foetida*) and sow bugs (*Porcellio scaber*) serve as transport hosts. When they swallow infective eggs, the second stage larvae hatch and migrate through the gut wall into the body tissues. No development takes place but the larvae are infective to birds that eat the parasitized annelids and crustaceans.

The third molt occurs in the ceca 4 to 6 days after infection and the fourth in 9 to 10 days. Worms are fully grown in 14 days but the females do not begin ovipositing until 24 to 36 days after infection.

Worms develop more readily and produce many times more eggs in young than in old chickens and in old than in young turkeys. Chicken strains of worms thrive better in chickens than in turkeys and conversely.

The worms themselves are not particularly harmful to their hosts; however, there is an increase in eosinophils and heterophils. Their importance is the ability to transmit the protozoan *Histomonas meleagridis* which produces blackhead, or enterohepatitis, in turkey poults. The flagellate is incorporated in the embryo within the egg and transmitted by the larva to new hosts.

Other species of *Heterakis* have similar life cycles but are not known to transmit protozoan pathogens such as *Histomonas meleagridis*. *Heterakis isolonche* of chickens differs in that its second stage larvae penetrate the cecal mucosa and develop to the adult stage before returning to the lumen.

## EXERCISE ON LIFE CYCLE

These parasites are readily available from plants where farm-raised chickens are processed.

Observe the development of the eggs to the

## Explanation of Plate 120

A–C, *Ascaridia galli.* A, *En face* view of adult. B, Anterior end of adult. C, Ventral view of posterior end of adult male. D–F, *Heterakis gallinae.* D, Dorsal view of anterior end of adult. E, *En face* view of adult. F, Ventral view of posterior end of adult male. G, Unembryonated egg. H, Chicken definitive host of *A. galli* and *H. gallinae.*

1, Dorsal lip; 2, lateroventral lip; 3, ventrolateral cephalic papillae; 4, dorsolateral cephalic papillae; 5, cervical papillae; 6, cervical alae; 7, precloacal sucker; 8, papilla in precloacal sucker; 9, preanal papillae; 10, postanal papillae; 11, caudal alae; 12, spicules (markedly unequal in length in *Heterakis*); 13, buccal capsule; 14, esophagus; 15, valved bulb of esophagus; 16, intestine; 17, anus; 18, thick shell of egg.

a, Adult *Ascaridia galli* in small intestine; b, eggs in feces passing from intestine; c, unembryonated egg outside host; d, developing egg; e, egg containing first stage larva; f, egg containing second stage ensheathed infective larva; g, infection of chicken by swallowing infective egg; h–j, egg hatching in crop, gizzard, and intestine; k–n, development in lumen of intestine; k, newly hatched second stage larva; l, second molt producing third stage larva; m, third molt producing fourth stage larva attached to bottom of mucosal crypt; n, fourth molt producing final stage that matures to adult; o–r, stages with development in mucosa; o, third stage larva entering mucosa; p, third molt in mucosa forming fourth stage larva; q, fourth stage larva leaves mucosa for lumen; r, after returning to lumen, it molts fourth time, attaches to mucosa, and matures.

a′, Adult *Heterakis gallinae* in ceca; b′, eggs laid in ceca; b–j, development of eggs of *Heterakis* is similar to that of *Ascaridia*; c′, newly hatched second stage larva in small intestine; d′, second stage larva entering ceca; e′, second molt (which is the first to occur in the ceca); f′, third stage larva; g′, third molt; h′, fourth stage larva; i′, fourth and final molt after which the parasite develops to adulthood (a′).

Figures A, B, D, E adapted from various sources; C, F, from Boughton, 1937, Minn. Agr. Exp. Sta. Tech. Bull. 121, 50 pp. (spicules added in F).

---

second stage larvae enclosed in the sheath of the first molt. By feeding embryonated eggs to a series of 10 chicks and examining one postmortem each day, the movement of the larvae toward and into the ceca, and the molts that occur in the latter, may be observed. Sections of the ceca or digestion of them after the larvae have entered will provide information on whether there is migration into the epithelium.

## SELECTED REFERENCES

Baker, A. D. (1933). Some observations on the development of the caecal worm, *Heterakis gallinae* (Gmelin, 1790) Freeborn, 1923, in the domestic fowl. Sci. Agr. 13:356–363.

Clapham, P. A. (1933). On the life-history of *Heterakis gallinae.* J. Helminthol. 11:67–86.

Lund, E. E., and A. M. Chute. (1970). Relative importance of young and immature turkeys and chickens in contaminating soil with *Histomonas*-bearing *Heterakis* eggs. Avian Dis. 14:342–348.

Lund, E. E., E. E. Wehr, and D. J. Ellis. (1966). Earthworm transmission of *Heterakis* and *Histomonas* to turkeys and chickens. J. Parasitol. 52:899–902.

Lund, E. E., A. M. Chute, and S. L. Meyers. (1970). Performance in chickens and turkeys of chicken-adapted *Heterakis gallinarum.* J. Helminthol. 44:97–106.

Riley, W. A., and L. G. James. (1922). Studies on the chicken nematode, *Heterakis papillosa* Bloch. J. Amer. Vet. Med. Ass. 59, n. s. 12:208–217.

Ruff, M. D., L. R. McDougald, and M. F. Hansen. (1970). Isolation of *Histomonas meleagridis* from embryonated eggs of *Heterakis gallinarum.* J. Protozool. 17:10–11.

Spindler, L. A. (1967). Experimental transmission of *Histomonas meleagridis* and *Heterakis gallinae* by the sow-bug, *Porcellio scaber,* and its implication for further research. Proc. Helminthol. Soc. Wash. 34:26–29.

Springer, W. T., J. Johnson, and W. M. Reid. (1970). Histomoniasis in gnotobiotic chickens and turkeys: biological aspects of the role of bacteria in the etiology. Exp. Parasitol. 28:383–392.

Uribe, C. (1922). Observations on the development of *Heterakis papillosa* Bloch in the chicken. J. Parasitol. 8:167–176.

Vatne, R. D., and M. F. Hansen. (1965). Larval development of cecal worm (*Heterakis gallinarum*) in chickens. Poult. Sci. 44:1079–1085.

## *Ascaridia galli* (Schrank, 1788) (Plate 120)

These parasites occur in the small intestine of chickens, turkeys, quails, pheasants, ducks, geese, and a number of species of wild birds in most parts of the world. Unlike *Heterakis gallinae,* which has no migration in the tissue, *A. galli* has a limited one by a small percentage of

**Plate 120**     *Heterakis gallinae and Ascaridia galli*                    449

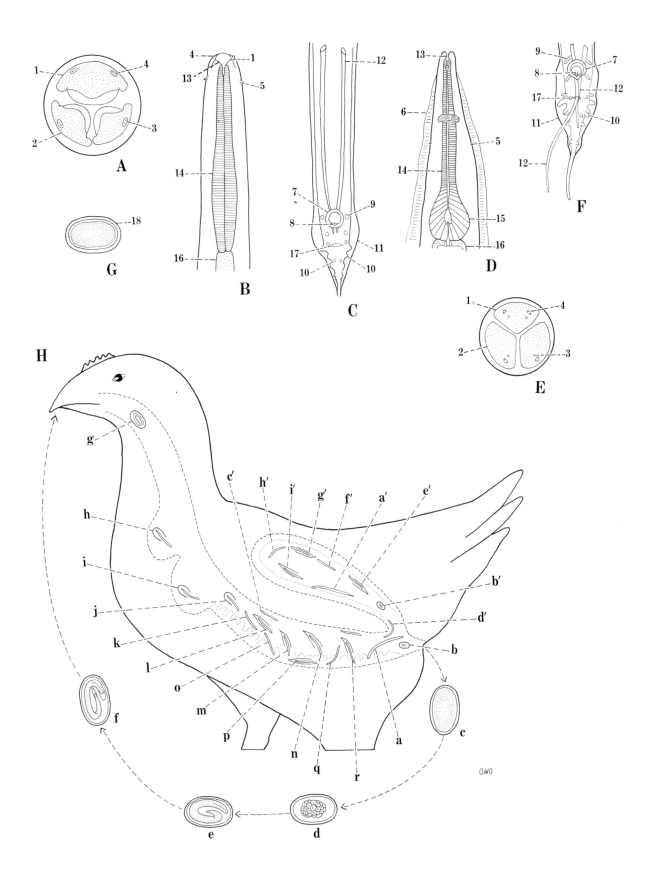

OWO

the larvae which enter the intestinal epithelium for a short period.

**Description.**    These are the largest nematodes in poultry. The males are 50 to 76 mm long and females 72 to 116 mm. The esophagus is straight, being devoid of a terminal bulb. The tail of the male bears a precloacal sucker and 10 pairs of papillae, three pairs precloacal on the ventral side, and seven pairs postcloacal and lateral.

**Life Cycle.**    Eggs are undeveloped when voided with the feces of infected chickens. When exposed to ordinary temperature, they develop and second stage larvae appear in them in 8 to 14 days. Under optimal temperature of $32°$ to $34°C$, which often prevails in poultry houses, development is complete in 5 days. There is no development above $35°C$. Eggs are killed by desiccation and heat; survival is about 1 month at $0°C$ and less than a day at $-4°$ to $-5°C$. Their infectivity decreases with age, but it may persist for 161 weeks in eggs in the soil.

Infection of birds normally takes place when infective eggs are swallowed with food or water. Eggs swallowed by earthworms and grasshoppers hatch within 15 to 60 minutes and the second stage larvae migrate to the body tissues. They are a source of infection to birds that eat them.

Hatching in bird hosts is in the proventriculus and duodenum. Larvae remain in the lumen or intervillar spaces of the duodenum 6 to 9 hours before a minority of them attach to or invade the mucosal lining, beginning at day 1 and continuing through day 26. Most of them have left the mucosa by day 8 for the lumen where the second molt takes place. The third stage larvae enter the mucosa and molt during days 14 to 19 after infection. During this time, larvae in the mucosa and lumen develop at the same rate, but after that growth of those remaining in the mucosa is inhibited. The newly formed fourth stage larvae return to the lumen and continue their growth, becoming mature between days 18 and 22 after infection. Egg laying begins from 30 to 50 days after infection. Worms mature earlier—in 5 to 6 weeks—in chicks less than 3 months old and later—in 8 weeks—in older birds. While the tissue phase is not a requisite for continued growth of the larvae, it takes place. Adult worms feed solely on the contents of the intestine, not on the tissues.

While *Ascaridia galli* from chickens or turkeys are cross infective, the chicken strain is more infective to chickens than to turkeys and conversely. *Ascaridia dissimilis* of turkeys, however, is specific for that host.

Third stage larvae migrating in the duodenal mucosa produce lesions with erosion, hemorrhage, enteritis, anemia, and diarrhea. As the infection proceeds, birds become unthrifty, emaciated, and weak. Egg production declines or even ceases. In heavy infections, the large worms obstruct the intestinal lumen. Young birds are more susceptible to damage than older ones. Dietary deficiencies, particularly vitamins, minerals, and proteins, predispose birds to heavy infections.

The life cycles of *A. dissimilis* of turkeys, *A. numidiae* of guinea fowl, and *A. columbae* of pigeons are similar to that of *A. galli*. There is a greater tendency for third stage larvae of *A. columbae* to migrate in the intestinal mucosa as some enter the bloodstream, reaching the liver and lungs as errant individuals. No development occurs and they do not migrate up the trachea.

### EXERCISE ON LIFE CYCLE

Since the life cycle of this parasite is similar to that of *Heterakis gallinae*, most of the same procedures may be followed. However, one difference exists. Since most of the larvae do not enter the intestinal mucosa, special efforts may be directed toward determining more about this phase of the cycle.

Feed each of about a dozen or more chicks 50 fully embryonated eggs. Beginning on about day 6 after infection, sacrifice one bird daily. Wash all larvae from the intestine, count them, and ascertain the stage of development. After having removed all of the worms from the lumen and those attached to the mucosa, scrape it from the intestinal wall and digest in a standard preparation of pepsin. Collect all of the worms, ascertain the stage of development, and the percentage of the total that entered the mucosa. Examine the liver, lungs, and trachea to determine whether some of them may reach the intestine by migrating through these organs.

### SELECTED REFERENCES

Ackert, J. E. (1931). The morphology and life history of the fowl nematode, *Ascaridia lineata* (Schneider). Parasitology 23:360–379.

Ackert, J. E., S. A. Edgar, and L. P. Frick. (1939). Goblet cells and age resistance of animals to parasitism. Trans. Amer. Microsc. Soc. 51:81–89.

Chute, A. M. (1970). Earthworm transmission of *Ascaridia dissimilis* Vigueras, 1931 to turkeys. J. Parasitol. 56(4/2): 56–57.

Farr, M. M. (1961). Further observations on the survival of the protozoan parasite, *Histomonas meleagridis,* and eggs of poultry parasites in feces of infected birds. Cornell Vet. 51:3–13.

Gurchenko, R. N. (1970). [Development of *Ascarida galli* in earthworms.] Veterinariya, Moscow 47:72–73.

Jacob, P. D., C. G. Varghese, P. T. Georgekutty, and C. T. Peter. (1970). A preliminary study on the role of grasshoppers (*Oedaleus abruptus* and *Spathosternum parasiniferum*) in the transmission of *Ascaridia galli* (Schrank, 1788) in poultry. Kerala J. Vet. Sci. 1:65–70.

Khouri, S. R., and B. P. Pande. (1970). Notes on histopathology in *Ascaridia galli*—an experimental study. Indian Vet. J. 47:472–474.

Khouri, S. R., and B. P. Pande. (1970). Observations on post-embryonic development in relation to tissue-phase of *Ascaridia galli* in laboratory-raised chicks. Indian J. Anim. Sci. 40:61–72.

Klimeš, B., and I. Stejskal. (1968). [Ascarid specificity in chickens, turkeys, and ducks.] Helminthologia 8–9:235–239.

Kurashvili, B. E., and N. Akhobadze. (1967). On study of feeding peculiarities of helminth *Ascaridia galli*. Zool. Anz. 179:457–463.

Todd, A. C., and D. H. Crowdus. (1952). On the life cycle of *Ascaridia galli*. Trans. Amer. Microsc. Soc. 71:282–287.

Tugwell, R. L., and J. E. Ackert. (1952). On the tissue phase of the life cycle of the fowl nematode *Ascaridia galli* (Schrank). J. Parasitol. 38:277–288.

Wehr, E. E., and J. C. Huang. (1964). The life cycle and morphology of *Ascaridia columbae* (Gmelin, 1790) Travassos, 1913 (Nematoda: Ascarididae) in the domestic pigeon (*Columbia livia domestica*). J. Parasitol. 50:131–137.

Wehr, E. E., and W. T. Shalkop. (1963). Ascaridia columbae infection in pigeons: a histopathologic study of liver lesions. Avian Dis. 7:206–211.

## FAMILY ASCARIDAE

The Ascaridae are large nematodes of the alimentary tract of all classes of vertebrates throughout the world. They have three well-developed lips. The esophagus is more or less straight and without a distinct terminal enlargement. The tail of the male bears numerous ventral papillae, bends ventrad, and is without alae; there is a pair of small spicules. The vulva is cephalad from the middle of the body. The eggs are oval and have a shell that is thick and covered with distinct ridges or pits.

### Ascaris lumbricoides
(Linnaeus, 1758) (Plate 121)

This large nematode is parasitic in the small intestine of humans and swine. Some workers consider it distinct from *A. suum* which is designated as the species in swine. Others believe that *A. lumbricoides* consists of physiological strains, one adapted to humans and the other to swine, and that there is no morphological, physiological, or biological evidence for considering two species. Cross infections between humans and pigs are possible.

**Description.** The males are 15 to 25 cm long by 3 mm wide and the females up to 41 cm long and 5 mm in diameter. There are three lips, a large dorsal one and two smaller lateroventral ones. The tail of the male bends strongly ventrad and bears a large number of preanal papillae and fewer postanal ones. The female is straight, with a rather blunt tail. The vulva is located near the posterior end of the first third of the body and joins two parallel uteri that extend posteriorly. Eggs are 50 to 75 $\mu$ long by 40 to 50 $\mu$ wide with thick shells whose surface is covered by a reticulation of ridges.

**Life Cycle.** The life cycle is direct but believed by some workers to have evolved from one with three hosts, as, indeed, some members of the family have.

The worms live over a year. Females begin producing eggs at about 2 months of age and continue for about a year when they are expelled. During this time fantastic numbers of eggs are laid, averaging around 2.7 million per day during peak production, according to some estimates. Oviposition is controlled by the circumesophageal ring and anal ganglion. Eggs are unembryonated when passed in the feces. Development takes place best between 17° and 30°C. In 10 days, the first stage larvae have appeared and molted once to form the second larvae in 13 to 18 days. However, another 2 or 3 weeks of development are required before they are capable of infecting the hosts. Eggs, both unembryonated and embryonated, are extremely resistant to adverse conditions of drying and freezing. They survive 4 years in soil under cul-

## Explanation of Plate 121

A, *En face* view of adult. B, Dorsal view of anterior end of adult. C, Lateral view of posterior end of adult male. D, Undeveloped fertilized egg. E, Embryonated egg with first stage larva. F, Egg with second stage or infective larvae. G, Second stage larva freed from egg. H, Anterior and posterior ends of second stage larva. I, Pig definitive host.

1, Dorsal lip; 2, lateroventral lip; 3, labial papillae; 4, mouth; 5, preanal papillae; 6, postanal papillae; 7, spicules; 8, esophagus; 9, intestine; 10, anus; 11, free cuticle on second stage larva; 12, shell in optical section, showing roughened surface; 13, undeveloped stage of egg; 14, first stage larva; 15, second stage larva.

a, Adult male and female worms in small intestine; b, eggs passing out of intestine with feces; c, unembryonated egg; d, egg in morula stage; e, egg with first stage larva; f, infective egg with second stage larva enclosed in cuticle of first stage larva; g, infection of host by swallowing infective eggs; h, i, eggs hatch in stomach and small intestine; j, larvae free in intestine; k, larvae enter venules of hepatic portal system; l, larvae migrate through liver; m, larvae pass through right side of heart; n, second molt occurs in veins of lungs to form third stage larva; o, third stage larvae; p, third molt to form fourth stage larva; q, fourth stage larvae enter alveoli and migrate up trachea; r, larvae enter pharynx and are swallowed; s, final molt occurs in small intestine and worms develop to adults (a); t, larvae pass through lungs, left side of heart, and into general circulation; u, larvae carried to various parts of the body, including kidneys and other organs, as well as to fetuses in pregnant sows.

Figures A, B, C adapted from various sources; D–H, from Alicata, 1935, U. S. Dep. Agr. Tech. Bull. 489, pp. 46.

---

tivation. They are capable of developing and surviving long periods in 10 percent formalin.

Hatching occurs throughout the length of the small intestine only after the eggs have passed through the stomach. Body temperature, high carbon dioxide concentration, pH near 7, and nonspecific reducing conditions stimulate the larvae to secrete hatching fluids. Chitinase secreted by the larvae passes through the vitelline membrane and reacts on one end of the chitinous shell, causing it to break down. The larva pushes its anterior end through the opening, causing the vitelline membrane to balloon out. After rupturing or dissolving the membrane, the second stage larva hatches. It secretes exsheathing fluid to break open the enclosing sheath, and emerges from it. The hatching process is completed in about 3 to 4 hours. While hatching occurs along the entire length of the small intestine, the cecum appears to be the site of maximum penetration, at least in mice.

In the intestinal wall, the mass of larvae enters the venules of the portal vein and is swept via the hepatic portal vein to the liver. Some enter lacteals, go to the mesenteric lymph nodes, and on to the heart. They are in the liver within 9 hours after infection. The second molt is in the liver 4 to 5 days after infection. Upon arriving in the liver, the gut of the larva is a solid column, consisting of seven cells with stored food in the form of glycogen and lipids. By day 3 in the liver, a lumen has developed, the second molt completed, and presumably food is taken.

On day 8 after infection, some third stage larvae are in the alveoli of the lungs and almost all are there by day 9. The third molt occurs after 5 to 6 days in the lungs. Fourth stage larvae abandon the alveoli and move into the bronchioles and bronchi, arriving en masse about 12 days postinfection. They migrate up the trachea, are swallowed, and return to the small intestine 14 to 21 days after infection. Larvae are 0.33 mm long on day 2, 0.43 on day 4, 1.54 on day 8, and 2.75 on day 15.

The fourth and final molt is about 30 days after infection and sexual maturity is attained in 49 to 62 days following ingestion of eggs. Adult worms are being expelled by 6 months and are gone from the host by 12 to 15 months.

Massive infections open the pathways through the liver and allow small, young larvae to proceed to the lungs before completing development that normally takes place in the liver.

Some of the larvae do not remain in the lungs but continue through them to the left side of the heart and into the arterial circulation. They are distributed to the tissues and organs, including the spleen and kidneys. In the case of pregnant animals, they have a tendency to accumulate in the liver of the fetuses, resulting in prenatal infection. When the young animals are born, the migration through the lungs and trachea is completed. This accounts for the early appearance of adult worms in very young animals.

It is seen from this life cycle that *Ascaris*

**Plate 121**    *Ascaris lumbricoides*                                                    453

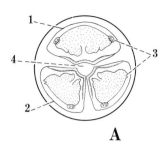

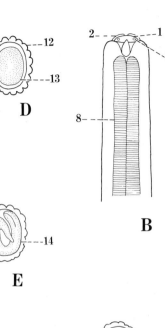

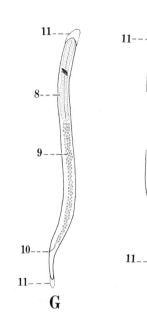

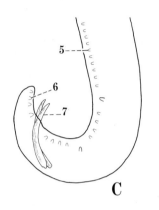

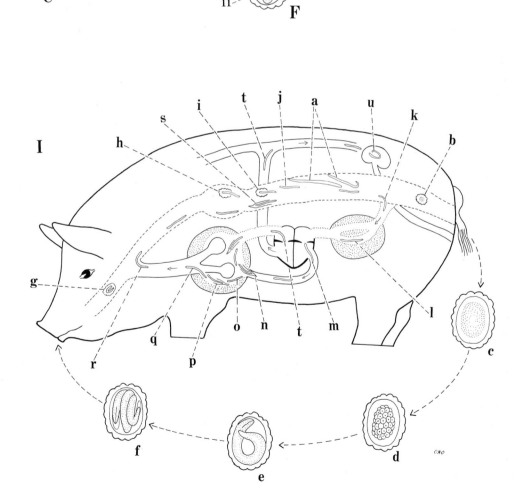

*lumbricoides* has an extended hepatopulmonary migration in the host. Only the first step of such a migration appeared in *Ascaridia galli* where but a small minority of the larvae migrated into the intestinal mucosa for a short time after which they returned to the lumen of the intestine for final development.

Ascaris is the most economically important of the helminths infecting swine. Damage to the intestine, lymph nodes, capillaries, and lungs results from the migrations of developing larvae. The extent of injury depends on the magnitude of infection. Larvae migrating in the liver, especially around the intralobular veins, cause extensive damage. Those burrowing from arterioles into alveoli of the lungs cause petechial hemorrhages, resulting in edema and other damage. In severe infections with extensive lung damage, pigs experience difficulty in breathing, verminous pneumonia characterized by labored breathing, and death in 1 to 2 weeks. Infection with *Ascaris* has a synergistic effect in connection with hog cholera by precipitating an otherwise latent infection. *Toxoplasma gondii* is not transmitted in the eggs.

Eggs swallowed by cattle hatch and the larvae migrate to the lungs, causing verminous pneumonia.

## EXERCISE ON LIFE CYCLE

Eggs are readily obtained from female worms available at abattoirs where swine are slaughtered. They are easily incubated and may be kept for long periods of time in that stage, even in a solution of 10 percent formalin. Verify that the larvae in fully developed eggs are in the second stage by observing the sheath of the first stage.

The migration route through the host can be determined by feeding large numbers of embryonated eggs to a group of about a dozen mice. One of these should be sacrificed each day for the first 6 days and one every other day thereafter. Wash out the contents of the intestine and examine them for larvae. Make sections of the intestine, liver, and lungs to ascertain when larvae are migrating through them. The trachea should be washed thoroughly and the sediment examined to determine whether larvae continue their migrations beyond the lungs.

## SELECTED REFERENCES

Alicata, J. E. (1935). Early developmental stages of nematodes occurring in swine. U. S. Dep. Agr. Tech. Bull. 489, p. 44.

Douvres, R. W., F. G. Tromba, and G. M. Malakatis. (1969). Morphogenesis and migration of *Ascaris suum* larvae developing to fourth stage in swine. J. Parasitol. 55:689–712.

Fairbain, D. (1961). The *in vitro* hatching of *Ascaris lumbricoides* eggs. Can. J. Zool. 39:153–162.

Galvin, R. J. (1968). Development of human and pig ascaris in pig and rabbit. J. Parasitol. 54:1085–1091.

Jenkins, D. C. (1968). Observations on the early migration of the larvae of *Ascaris suum* Goeze, 1782 in white mice. Parasitology 58:431–440.

Jenkins, D. C., and D. A. Erasmus. (1971). The ultrastructure of the intestine of *Ascaris suum* larvae. Z. Parasitenk. 35:173–185.

Kelly, G. W., L. S. Olsen, and A. B. Hoerlein. (1957). Rate of migration and growth of larval *Ascaris suum* in baby pigs. Proc. Helminthol. Soc. Wash. 24:133–136.

Kelly, G. W., and L. J. Smith. (1956). The daily egg production of *Ascaris suum* and the inability of low levels of aureomycin to affect egg production and embryonation. J. Parasitol. 42:587.

Kremer, M., G. Cranz, H. Fernandez, and F. Miltgen. (1969). Echec d'un essai de transmission du toxoplasme par les oeufs d'*Ascaris suum*. Ann. Parasitol. 44:527–529.

Kumar, V., and K. S. Singh. (1968). *In vivo* behaviour of *Ascaris lumbricoides* eggs and the emerging juveniles. Indian J. Med. Res. 56:1304–1308.

Lýsek, H. (1968). On the host specificity of ascarids of human and pig origin. Helminthologia 8–9:309–312.

Morrow, D. A. (1968). Pneumonia in cattle due to migrating *Ascaris lumbricoides* larvae. J. Amer. Vet. Med. Ass. 153:184–189.

Olsen, L. S., and G. W. Kelley. (1960). Some characteristics of the early phase of migration of larvae of *Ascaris suum*. Proc. Helminthol. Soc. Wash. 27:115–118.

Olsen, L. S., G. W. Kelley, and H. G. Sen. (1958). Longevity and egg production of *Ascaris suum*. Trans. Amer. Microsc. Soc. 77:380–383.

Ransom, B. H., and E. B. Cram. (1921). The course of migration of *Ascaris* larvae. Amer. J. Hyg. 1:129–159.

Ransom, B. H., and W. D. Foster. (1920). Observations on the life history of *Ascaris lumbricoides*. U. S. Dep. Agr. Bull. 817, 47 pp.

Roberts, F. H. S. (1934). The large roundworm of pigs, *Ascaris lumbricoides* L., 1758, its life history in Queensland, economic importance and control. Anim. Health Sta. Yeerongpilly Bull., 81 pp.

Rogers, W. P. (1965). The role of leucine aminopeptidase in the moulting of nematode parasites. Comp. Biochem. Physiol. 14:311–321.

Schwartz, B. (1960). Evolution of knowledge concerning the roundworm *Ascaris lumbricoides*. Smithsonian Inst. Annu. Rep. (1958–1959), pp. 465–481.

Spindler, L. A. (1940). Effect of tropical sunlight on eggs of *Ascaris suis* (Nematoda), the large intestinal roundworm of swine. J. Parasitol. 26:323–330.

Sprent, J. F. A. (1952). On the migratory behavior of the larvae of various Ascaris species in white mice. I. Distribution of larvae in tissues. J. Infect. Dis. 90:165–176.

Sprent, J. F. A. (1954). The life cycles of nematodes in the family Ascarididae. J. Parasitol. 40:608–617.

Voronyuk, N. B., and B. A. Shishov. (1968). [Study of the role of the nervous system in regulating egg-laying in pig *Ascaris* and *Ascaridia galli*.] Tr. Gel'mintol. Lab. 19:68–72.

Woodruff, A. W. (1968). Helminths as vehicles and synergists of microbial infections. Trans. Roy. Soc. Trop. Med. Hyg. 62:446–452.

## *Toxocara canis* (Werner, 1782) (Plate 122)

*Toxocara canis* is a parasite of the small intestine of dogs and foxes throughout the world. It is the largest of the ascarids occurring in canines.

While these worms are parasites of canines, the larvae may occur in the visceral organs and tissues of humans with serious effects, causing a condition known as visceral larval migrans.

**Description.**    The males measure up to 10 cm long and the females up to 18 cm. Broad cervical alae are present but caudal alae are absent. The tail of the male is abruptly reduced in diameter and bears five papillae on each side, with about 20 pairs of preanal papillae. The vulva is in the anterior fourth of the body; the eggs measure 90 by 75 $\mu$ and are covered with thick, finely pitted shells.

**Life Cycle.**    Unembryonated eggs passed in the feces are extremely resistant and durable. They withstand cold under the snow when the air temperatures fall to $-31.1°C$. Exposure to sunlight at a temperature of $55.5°C$ is lethal. Development to the second stage infective larva requires 9 to 11 days at $24°C$ and 3 to 5 days at $30°$ in the presence of adequate oxygen and a relative humidity of 75 to 100 percent.

Although the life cycle of this parasite is direct, there are extremely complex relationships between it and a series of hosts. It is another example of the importance of a thorough knowledge of the life cycle and biology of parasites in ascertaining their epizootology.

Infection begins when grown dogs, rodents, and other animals swallow embryonated eggs. Hatching occurs in the small intestine within 2 to 4 hours. The second stage larvae, 0.4 to 0.47 mm long, burrow into the intestinal mucosa. Most of them enter the venules of the hepatic portal vein and are carried to the liver, arriving sometimes as soon as 2 hours but more often in 2 days. A few enter the lacteals, go through the lymph nodes, and on to the heart via the thoracic duct. Those in the liver continue through it, arriving in the lungs via the vena cava, right side of the heart, and pulmonary artery in 1 to 4 days after infection. From this point on, the course of migration and development taken by the larvae varies, depending on whether they are in young or old dogs or in rodents and other species.

In mature dogs, the larvae reach the lungs in the second stage. The first ones arrive within a day after infection, with the peak during days 3 to 5, after which the number declines. They continue their migratory course via the pulmonary vein, left side of the heart, into the systemic circulation, and are distributed in the body muscles, liver, lungs, kidneys, and other organs. In the somatic tissues, larvae remain in the second stage without further growth. In cases where infection is by swallowing eggs, as described above, the tracheal migration is almost totally by-passed and, therefore, intestinal infections are rare, if, indeed, they occur at all.

In pups infected by swallowing eggs, the migration and development of the larvae from the lungs is different from that described above for older dogs. The larvae leave the blood vessels, enter the alveoli, and migrate up the bronchioles, bronchi, and trachea. Upon entering the pharynx, they are swallowed. The second molt takes place in the lungs, air tubes, or esophagus 1 to 5 or more days postinfection, producing third stage larvae up to 1.36 mm long. They remain in the stomach until about day 10 after infection. During this time, they molt the third time and the fourth stage larvae attain a length of 1.0 to 1.5

## Explanation of Plate 122

**A,** *En face* view of adult. **B,** Dorsal view of anterior end of adult. **C,** Lateral view of anterior end of adult. **D,** Lateral view of posterior end of adult male. **E,** Ventral view of tail of adult male. **F,** Lateral view of tail of adult female. **G,** Embryonated egg. **H,** Dog definitive host. **I,** Pup definitive host. **J,** Mouse host. **K,** Human in which larval visceral migrans occurs, as in mice.

1, Dorsal lip; 2, lateroventral lip; 3, labial papillae; 4, mouth; 5, cervical alae; 6, preanal papillae; 7, anus; 8, spicules; 9, postanal papillae; 10, esophagus; 11, ventriculus; 12, intestine; 13, anus; 14, excretory pore; 15, excretory gland; 16, nerve ring.

**a,** Adult worms relatively rare in intestine of adult dogs but common in young ones (i'); **b,** eggs voided with feces; **c,** eggs undeveloped when passed in feces; **d,** egg in morula stage; **e,** first stage larva in egg; **f,** second stage infective larva in egg; **g,** infective egg being swallowed by adult dog; **h,** eggs hatch in stomach; **i,** larvae (stippled) go into small intestine; **j,** larvae penetrate intestinal wall and enter hepatic portal vein; **k,** larvae leave capillaries in liver; **l,** some larvae accumulate in liver parenchyma; **m,** some larvae continue in blood stream through right side of heart; **n,** larvae enter lungs; **o,** larvae enter lung parenchyma; **p,** larvae in lung parenchyma; **q,** some larvae pass through lungs to left side of heart; **r,** larvae enter general circulation; **s,** larvae enter umbilical artery and eventually go into fetus; **t,** larvae in liver, heart, and lungs of fetus; **u,** larvae in kidneys; **v,** larvae in skeletal muscles.

**a,** Mouse with larvae in tissues being swallowed; **b,** larvae (unstippled) freed from mouse by digestive processes in stomach; **c,** larvae enter hepatic portal vein; **d, e,** larvae migrate to lungs via heart; **f,** larvae enter bronchioles from blood vessels; **g,** larvae migrate up trachea, are swallowed, and develop to adults, but infrequently in grown dogs.

**a',** Embryonated eggs swallowed by puppy; **b',** eggs hatch in stomach; **c'–e',** larvae (stippled) enter hepatic portal vein and migrate through liver, heart, and into lungs; **f',** larvae enter alveoli and migrate up trachea; **g',** larvae in pharynx; **h',** larvae in esophagus are swallowed; **i',** larvae develop readily to adults and lay eggs (when infected mice are eaten, the larvae (unstippled) migrate and develop in the same manner as those from eggs); **j',** infective larva passed in feces.

**a'',** Embryonated eggs swallowed by mice and other rodents; **b'',** eggs hatch in intestine; **c'',** larvae enter hepatic portal vein; **d'',** some larvae enter liver parenchyma; **e'',** larvae remain in liver; **f'', g'',** some larvae migrate through liver and heart; **h'',** some larvae enter lung parenchyma; **i'',** larvae remain in lung parenchyma; **j'', k'',** other larvae pass though heart and into arterial circulation; **l'', m'',** larvae enter central nervous system and kidneys where they remain inactive and without further development until released in stomach of dogs.

Figures A, D, E, F adapted from Mozgovi, 1953, Osnovi Nematodologii, Vol. 2, pp. 449–452; B, from Yorke and Maplestone, 1962, Nematode Parasites of Vertebrates, Hafner Publishing Co., New York, Fig. 176 A; C, from Sprent, 1958, Parasitology 48: 184.

---

mm. They spend about 13 days in the duodenum and then move on. The fourth and final molt producing the adult stage is between days 19 and 27. The worms are sexually mature and producing eggs between weeks 4 and 5. Larvae in various stages of development are passed in the feces.

In addition to becoming infected postnatally by swallowing embryonated eggs, as described above, pups almost universally are infected prenatally under natural conditions. Presumably under the influence of pregnancy hormones present in the bitches, the somatic larvae are somehow reactivated and enter the fetuses around days 42 to 43 of gestation. They accumulate in the liver, where the second molt takes place late in the gestation period, producing larvae 1 to 1.3 mm long. They are in the lungs, trachea, esophagus, and stomach within 1 week after birth when the third molt occurs. Beginning with week 2 in the intestine, the fourth stage larvae are 1.0 to 6.3 mm long when the fourth and final molt occurs. Growth is rapid and sexually mature worms are present when the pups are 2 to 3 weeks old.

Up to this point, mention has been made only of second stage somatic larvae in older dogs. These, it may be recalled, were acquired from swallowing eggs. Such dogs rarely have adult worms in the intestine. However, following parturition, reactivated somatic larvae make their way into the lumen of the intestine of bitches which were without intestinal worms prior to and during gestation. They mature in 25 to 26 days after whelping. The worms persist 8 to 108 (average 60) days when they are expelled spontaneously. Not all the second stage larvae leave the tissues during a single pregnancy, as subsequent litters are infected prenatally from bitches protected against reinfection by swallowing eggs.

Plate 122    *Toxocara canis*                                                              457

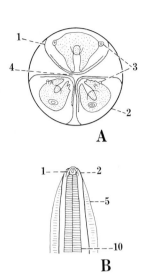

A

B

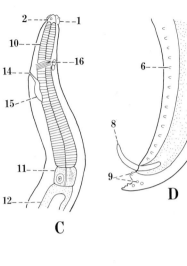

C

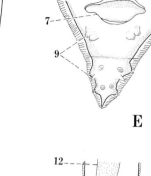

D

E

G

F

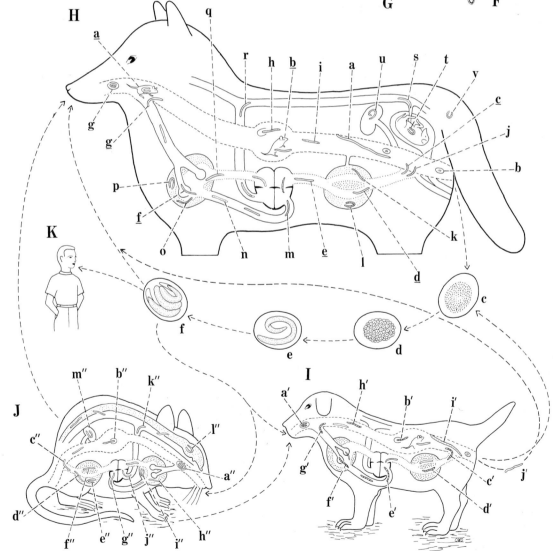

H

K

J

I

A number of animals other than dogs become infected upon swallowing eggs of *Toxocara canis*. In mice, the larvae are shunted through the lungs, going to the somatic tissues, where they remain in the second stage, just as in the case of older dogs. These mice serve as reservoir, or paratenic, hosts which, when eaten by dogs of all ages, are a source of intestinal infection with adult worms. Upon digestion of the mice, the liberated larvae undertake the tracheal migration to the esophagus and intestine, molting and developing en route. There is no somatic migration by these larvae.

Inasmuch as developing larvae are passed in the feces of prenatally infected pups, they are an additional source of infection to the mothers that lick the feces of their infected offspring. Such larvae, like those in mice, migrate to the intestine via the trachea instead of to the somatic tissues.

Infective eggs may be transmitted to dogs by houseflies and larvae in earthworms.

Infective eggs of *Toxocara canis* are a source of infection to humans who swallow them. The larvae go to the tissues and organs where they are not encapsulated but move about, producing a condition known as visceral larva migrans. They are serious when in the brain, eyes, or other vital locations.

Other closely related and common species are *Toxocara mystax* of cats and *Toxascaris leonina* of both cats and dogs. Larvae of *Toxocara mystax* hatching in the intestine of cats undergo the tracheal migration. Prenatal infection is unknown. A number of animals such as mice, lambs, chickens, cockroaches, and earthworms serve as transport hosts in which the second stage larvae are encysted in the somatic tissues. Sexual maturity is attained in the intestine in about 3 weeks. *Toxascaris leonina* has no migratory phase other than a period of about 10 days in the intestinal wall from which the larvae return to the lumen and mature. Eggs appear in the feces in about 74 days. Mice can serve as transport hosts.

## EXERCISE ON LIFE CYCLE

Obtain adult female *Toxocara canis* from a veterinary hospital where dogs are treated for worms. Dissect eggs from the proximal portions of the uteri near the vagina and incubate them in tap water or normal saline at room temperature. Note the stages of developing larvae and the time when each occurs. Recognize the second stage larva by the presence of the loose cuticle around it.

Feed eggs containing second stage infective larvae to mice. Kill the mice at intervals and examine the intestinal contents, the liver, heart, lungs, kidneys, and central nervous system to determine the course of movement of the larvae in the body and the time required for them to reach the various locations. Ascertain whether the larvae develop beyond the second stage in the mice by comparing them with larvae pressed from fully embryonated eggs.

## SELECTED REFERENCES

Douglas, J. R., and N. F. Baker. (1959). The chronology of experimental intrauterine infections with *Toxocara canis* (Werner, 1782) in the dog. J. Parasitol. 45(4/2):43–44.

Edwards, F. B. (1970). *Toxocara canis* transmitted by common house fly to the canine host. Vet. Rec. 86:581–582.

Griesemer, R. A., and J. P. Gibson. (1963). The establishment of an ascarid-free beagle dog colony. J. Amer. Vet. Med. Ass. 143:965–967.

Koutz, F. R., H. F. Groves, and W. M. Scothorn. (1966). The prenatal migration of *Toxocara canis* larvae and their relationship to infection in pregnant bitches and pups. Amer. J. Vet. Res. 27:789–795.

Okoshi, S., and M. Usui. (1968). Experimental studies on *Toxascaris leonina*. VI. Experimental infection of mice, chickens, and earthworms with *Toxascaris leonina*, *Toxocara canis*, and *Toxocara cati*. Jap. J. Vet. Sci. 30:151–166.

Olsen, L. J., N. N. Izzat, M. B. Petteway, and J. A. Reinhart. (1970). Ocular toxocariasis in mice. Amer. J. Trop. Med. Hyg. 19:238–243.

Owen, W. B. (1930). Factors that influence the development and survival of the ova of an ascarid roundworm *Toxocara canis* (Werner, 1782) Stiles, 1905, under field conditions. Univ. Minn. Agr. Exp. Sta. Tech. Bull. 71, 25 pp.

Schacher, J. T. (1957). A contribution to the life history and larval morphology of *Toxocara canis*. J. Parasitol. 43:599–612.

Scothorn, M. W., F. R. Koutz, and H. F. Groves. (1965). Prenatal *Toxocara canis* infection in pups. J. Amer. Vet. Med. Ass. 146:45–48.

Sprent, J. F. A. (1952). On the migratory behavior of the larvae of various ascaris species in white mice. I. Distribution of larvae in tissues. J. Infect. Dis. 90:165–176.

Sprent, J. F. A. (1958). Observations on the development of *Toxocara canis* (Werner, 1782) in the dog. Parasitology 48:184–209.

Sprent, J. F. A. (1961). Post-parturient infection of bitch with *Toxocara canis*. J. Parasitol. 47:284.

Sprent, J. F. A., and P. B. English. (1958). The large roundworms of dogs and cats—a public health problem. Austr. Vet. J. 34:161–171.

Warren, E. G. (1969). Infections of *Toxocara canis* in dogs fed infected mouse tissues. Parasitology 59:837–841.

Webster, G. A. (1958). On prenatal infection and the migration of *Toxocara canis* (Werner, 1782) in dogs. Can. J. Zool. 36:435–440.

### *Contracaecum* (Railliet and Henry, 1912)

*Contracaecum* contains many species infecting piscivorous fish, birds, and mammals, both freshwater and marine. The genus is ascaridoid, characterized by having lips without dentigerous ridges and two cecum-like projections at the junction of the esophagus and intestine. The esophageal cecum is the smaller; it is solid and pointed posteriorly. The intestinal cecum is almost twice as long as the other one and extends anteriorly. (*Porrocaecum* Railliet and Henry, 1912 has similar ceca but the lips have dentigerous ridges.)

Three species of this genus, one each in fish, birds, and mammals, are of interest. Their life cycles include at least one intermediate host and possibly two, but on this latter point definite information is not available.

### *Contracaecum aduncum* (Rudolphi, 1802) (Plate 123)

This species is parasitic in the intestine of the European eelpout (*Zoarces viviparus*), perch (*Perca fluviatilis*) and others, including food fishes. Numerous plankton-feeding fish serve as intermediate hosts. *Contracaecum aduncum* was chosen to illustrate the life cycle because more details concerning its stages are known than for any other species.

**Description.** Males measure 18 to 20.5 mm long by 0.434 to 0.511 mm wide. Cervical alae are 4.21 mm long. The ventriculus is 0.154 to 0.170 mm long with a ventricular cecum 0.540 to 0.528 mm long; the intestinal cecum is 0.648 to 0.930 mm long. There are 23 pairs of preanal and seven to eight pairs of postanal papillae. The spicules are 2.01 to 2.94 mm long. The females are 24 to 36 mm long with the vulva near the junction of the anterior and middle thirds of the body; the vagina divides into two uteri. Eggs are 62 to 70 $\mu$ long by 46 to 47 $\mu$ wide.

**Life Cycle.** *Contracaecum aduncum* is especially abundant in the eelpouts during the summer, virtually filling the intestine, and occurring in approximately half of the population. During the winter months, infections are relatively few and light.

The eggs are in the first cleavage when passed in the feces. At room temperature, they contain a fully developed second stage larva at the end of day 3 or 4.

Hatching apparently does not occur until the embryonated eggs are ingested by appropriate invertebrates such as the copepods *Acartia bifilosa* and *Eurytemora affinis*, the isopod *Iaera albifrons* and species of the polychaetes *Lepidonotus*, *Nereis*, *Harmothoe*, and *Gattiana*. Larvae occur in the liver and gonads of the clams *Nassa reticulata* and *Cyclonassa neritea*.

In the intestine of the first intermediate host, the larvae shed the cuticle of the first molt and migrate to the hemocoel. From this point, there is no experimental evidence that these invertebrates serve as true intermediaries. The abundance of fourth stage larvae in the tissues of plankton-feeding fish such as young herrings leads to the conclusion that invertebrates serve as the first intermediate hosts. Those found naturally infected are most likely to serve in that capacity.

While not demonstrated experimentally, it is highly probable that eelpouts become infected by eating infected herrings, since the advanced larvae in the fish and the young parasites in the intestine of the definitive host are similar, suggesting a continuous progression of growth.

*Contracaecum spiculigerum* (Rudolphi, 1809) is common in cormorants (*Phalocrocorax auritus auritus*) in the United States, and occurs in ducks, gulls, grebes, pelicans, and other fish-eating birds. The life cycle of specimens from cormorants was studied by Thomas in the United States. He found that two molts occur within the egg by the end of day 5 at room temperature, and hatching takes place the same day. The third stage larvae were very active, swimming about in the water and coiling and uncoiling within the enclosing sheaths. When eaten by guppies (*Lebistes reticulatus*), the first two enclosing cuticles were shed in the intestine and some of the larvae migrated to the body

## Explanation of Plate 123

A, *En face* view of adult. B, Sinistral view of caudal end of adult male. C, Dorsal lip. D, Caudal end of female. E, Ventral view of caudal end of male. F, Esophagointestinal junction, showing ceca. G, Egg in early stage of cleavage. H, Tadpole stage of larva. I, Fully developed first stage larva. J, Second stage larva. K, Third stage larva from intestine of copepod. L, Eelpout (*Zoarces*). M, Marine copepod (*Eurytemora*) intermediate host. N, Herring (*Clupea*).

1, Dorsal lip; 2, lateroventral lip; 3, papillae; 4, pulp of lip; 5, mouth; 6, caudal papillae; 7, spicular sheath; 8, spicule; 9, spines; 10, anal opening; 11, esophagus; 12, ventriculus; 13, anterior intestinal cecum; 14, posterior esophageal or ventricular cecum; 15, intestine; 16, rough eggshell; 17, cells of developing embryo; 18, embryonic membrane; 19, tadpole stage of larva; 20, first stage larva; 21, nerve ring; 22, cuticle of first stage larva surrounding second stage larva; 23, second stage larva; 24, third stage larva; 25, spine; 26, genital primordia.

a, Adult worm in eelpout; b, egg laid by worm; c, egg in first stage of cleavage passed in feces; d, egg in early stages of cleavage; e, larva developed to tadpole stage; f, fully developed second stage larvae; g, ensheathed second stage larva hatching from egg eaten by copepod; h, second stage larva escaping from sheath and passing through intestinal wall; i, second stage larva in hemocoel; j, second stage larva presumably is released from copepod by digestive processes in alimentary canal of plankton-feeding herring; k, larva molts (?) to third stage and burrows through wall of intestine; l, third stage larva (?) encysts in mesenteries; m, larva released from tissues of herring by digestive processes of the predacious eelpout; n, larva molts (?) fourth time, attaches to intestinal wall and matures.

Figures A–G adapted from Skrjabin *et al.*, 1951, Opredeliteli Parasiticheski Nematod, Vol. 3, Plates 205, 206; H–K, from Markowski, 1937, Bull. Int. Acad. Polo. Sci. Lett. Cl. Sci. Math. Natur. Ser. B Natur. Sci. 2:227.

---

cavity and encysted on the mesenteries. Attempts to infect largemouth bass (*Micropterus salmoides*), chicks, and ducklings were unsuccessful, presumably because they are unfavorable hosts. Natural infections with larvae occur in bluegills (*Lepomis macrochirus*), warmouth bass (*Chaenobryttus gulosus*), and largemouth bass. It is concluded without experimental evidence that small fish become infected by eating larvae, piscivorous fish by eating small infected fish, and cormorants by eating either size that contains sufficiently developed and encysted larvae.

Russian workers reported the first intermediate hosts of *Contracaecum spiculigerum* as *Cyclops strenuus*, *Macrocyclops abildus*, *M. fuscus*, *Mesocyclops leuckarti*, and *Diaptomus gracilis*. Second intermediate hosts include damsel flies (*Agrion* and *Coenagrion*) and several species of Cyprinidae.

*Contracaecum multipapillatum* from the water turkey *Anhinga leucogaster* was studied by Huizinga. At 21°C, they have developed to second stage larvae and hatched spontaneously in 5 to 7 days. The copepod *Cyclops vernalis* becomes infected by feeding on the free second stage larvae. In the gut, the anterior end of the sheath swells and a cap separates from the remainder, permitting the larva to escape. By means of a boring tooth, the larvae burrow

through the intestinal wall into the hemocoel. Five or more larvae kill the copepod but one with a single larva survives several months. There is no molt in the copepod host.

When guppies eat copepods infected with second stage larvae, they burrow through the gut wall and are in the body cavity within a few hours, encapsulating in the mesenteries, liver, intestinal wall, and pericardial sac. The larvae molt to the third stage and increase in size from 0.473 mm to 22.7 mm in 85 days. The abdomen of infected fish becomes greatly distended and growth is inhibited. Larvae occur in naturally infected bluegills, sunfish, crappies, bullheads, and mosquito fish. Largemouth bass eating infected guppies have encysted larvae on the mesenteries 7 days later. The large fish serve as transport hosts. Birds ingesting parasitized fish presumably become infected but details are unknown.

*Contracaecum osculum* (Rudolphi, 1802) is common in pinnipeds. Larvae identical with the adults occur encysted in marine and certain freshwater fish.

Larvae in the body cavity, stomach, intestine, and liver of a Miller's-thumb-like fish of Lake Baikal in Siberia are so similar to the adult worms in the freshwater Baikal seal that they are assumed to be identical. An amphipod is believed to be the first intermediate host.

**Plate 123**  *Contracaecum aduncum*  461

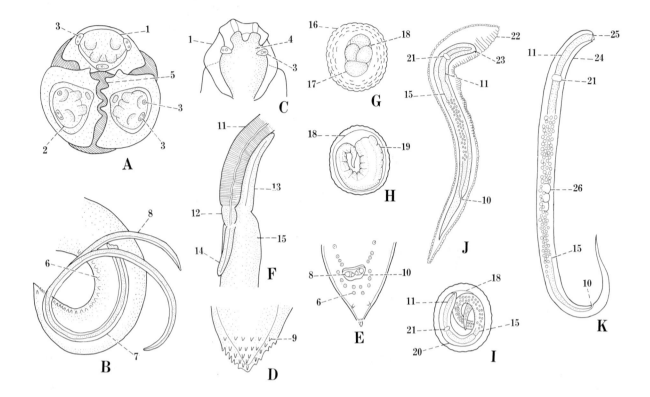

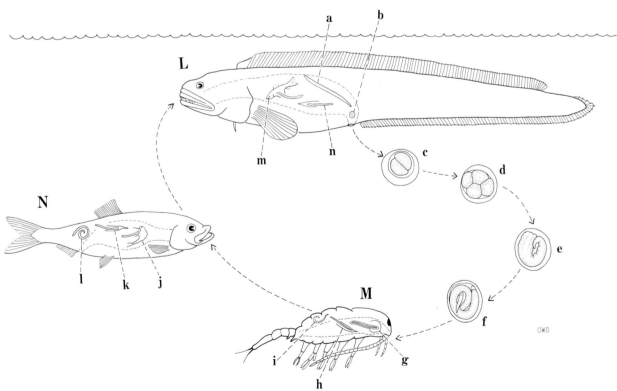

## EXERCISE ON LIFE CYCLE

Cormorants should be used to check the final step in the life cycle. This can be done only by hatching and rearing the birds in the laboratory where they will not be exposed to infection through natural food.

## SELECTED REFERENCES

Dolgikh, A. V. (1966). [Black sea molluscs in the life-cycle of *Contracaecum aduncum.*] Zool. Zh. 45: 454–455.

Huizinga, H. W. (1967). The life cycle of *Contracaecum multipapillatum* (von Drasche, 1882) Lucker, 1941 (Nematoda: Heterocheilidae). J. Parasitol. 53:368–375.

Khalil, L. F. (1969). Larval nematodes in the herring (*Clupea harengus*) from British coastal waters. J. Marine Biol. Ass. U. K. 49:641–659.

Markowski, S. (1937). Ueber die Entwicklungsgeschichte und Biologie des Nematoden *Contracaecum aduncum* (Rudolphi, 1802). Int. Acad. Pol. Sci. Lett. Cl. Sci. Math. Natur. Ser. B Sci. Natur. Bull. (II), pp. 227–247.

Mozgovoi, A. A., V. I. Shakhmatova, and M. K. Semenova. (1968). [Life cycle of *Contracaecum spiculigerum* (Ascaridata: Anisakidae), a parasite of domestic and economically important birds.] Tr. Gel'mintol. Lab. 19:129–136.

Popova, T. I., A. A. Mozgovoi, and M. A. Dmitrenko. (1964). [Biology of Ascaridata of animals from the White Sea.] Tr. Gel'mintol. Lab. 14: 163–169.

Sudarikov, V. E., and K. M. Ryzhikov. (1951). [On the biology of *Contracaecum osculatum baicalensis* —nematode from *Phoca sibirica*.] Tr. Gel'mintol. Lab. 5:59–66.

Thomas, L. J. (1937). On the life cycle of *Contracaecum spiculigerum* (Rud.). J. Parasitol. 23:429–431.

Thomas, L. J. (1944). Researches in life histories of parasites of wildlife. Trans. Ill. State Acad. Sci. 37:7–24.

Valter, E. D. (1968). [On the participation of isopods in the life-cycle of *Contracaecum aduncum* (Ascaridata, Anisakoidea).] Parazitologiya 2:521–527.

## ORDER SPIRURIDA

The members of this order are parasites of the alimentary canal, celom, subcutaneous tissues, and blood and lymph vessels of vertebrates. They are characterized by a cylindrical esophagus that basically is composed of an anterior muscular and a posterior glandular part. The mouth may be surrounded by six weak lips, a cuticular circumoral elevation, or paired lateral pseudolabia. The spicules of the males usually are unequal. The life cycle is indirect, requiring an intermediate host, usually an arthropod. Representatives from the suborders Camallanina and Spirurina are discussed.

### Suborder Camallanina

In this suborder, the esophageal glands are uninucleate and the larvae are without cephalic hooks but have large pocket-like phasmids. The intermediate hosts are copepods.

### Superfamily Dracunculoidea

The mouth is pore-like and surrounded by an inner circle of four to six papillae and an outer one of four double papillae. The amphids are lateral and posterior from the lateral papillae. The vulva is near the middle of the body; it atrophies before sexual maturity. The females are ovoviparous, with the larvae developing in copepods.

### FAMILY DRACUNCULIDAE

The esophagus consists of a short cylindrical anterior muscular portion followed by a larger and longer part that has a marked median constriction at the level of the nerve ring. The only representative of this group discussed is the guinea worm *Dracunculus medinensis*, whose life cycle is known better than any of the others.

*Dracunculus medinensis*
(Linnaeus, 1758) (Plate 124)

This is a parasite of the subcutaneous tissues of man in many parts of the tropical sections of the world, particularly in Africa and Asia. It is primarily a parasite of rural populations in dry areas where culinary water is obtained from step wells and ponds. The crustacean intermediate hosts living in these habitats become infected and are swallowed with the drinking water.

**Description.**     Adult worms are in the subcutaneous tissue. Males are up to 40 mm long and females from 60 to 80 mm, with a diameter of 1.7 to 2 mm. The mouth is triangular and surrounded by a quadrangular cuticular plate. There is an internal circle of four cephalic papillae arranged as a dorsal and ventral doublet. The external circle has four double papillae. Amphids are well developed and conspicuous. The slender, anterior muscular portion of the esophagus is followed by the larger and longer glandular part with a median constriction at the level of the nerve ring and cervical papillae. The intestine is flattened and nonfunctional in adults. The vulva, nonfunctional in adults, is approximately half way down the body. The posterior end of the males bears four pairs of preanal and six pairs of postanal papillae. A pair of large, lateral phasmids is located posterior to the anus. Spicules are subequal in length, averaging 117 mm.

**Life Cycle.**    Gravid females lie in the subcutaneous tissue with the anterior end pressing outward against the under surface of the skin. The pressure and possibly secretions produce first a papule that develops into a blister and finally into an ulcer-like sore. When the sore is wetted with water, the female is stimulated and thrusts her anterior end outward through it. Internal pressure, possibly due to water entering the pseudocoel of the worm through the membranes, causes the body wall to rupture, allowing the gravid uterus which fills the pseudocoelic cavity to prolapse. Rents appear in it and a mass of first stage larvae, up to a half million, escape into the water. The female withdraws into the subcutaneous tissue to repeat the process upon subsequent wettings of the area until her entire brood of millions of larvae has been released, following which she dies. Larvae remain active in the water up to 7 days but their ability to infect declines after day 3 and is nil by day 6.

Larvae wriggling about in the water are captured and ingested by a number of species of cyclops, *Cyclops leuckarti* and *C. hyalinus* being common intermediate hosts, together with species of *Eucyclops, Mesocyclops,* and *Macrocyclops*. Activity of larvae in the intestine together with the aid of a dorsal tooth enables them to burrow through the wall and into the hemocoel in 1 to 4 hours. Naturally infected cyclops generally contain a single larva, causing them to become relatively inactive and sink to the bottom. In temperatures above 19°C, development proceeds and two molts occur in the crustaceans. Below that temperature development ceases. Larvae are able to withstand −78°C for over 6 months and retain their infectivity for cyclops. Third stage larvae become infective to the vertebrate host in 12 to 14 days at 25°C. They are 240 to 600 $\mu$ long by 12 to 23 $\mu$ in diameter. The

tail is bi- or trifid, or with four mucrons. The intestine contains transparent globules and the genital primordium consists of six to eight cells. Lytic substances are present in larvae 21 days old or older.

Infection of the vertebrate host occurs when copepods containing fully developed third stage larvae are swallowed with water. Upon digestion of the crustaceans, the larvae are liberated in the duodenum. Within 13 hours, they are in the duodenal wall and on the abdominal mesenteries for up to 12 days. They have migrated into the thoracic and abdominal muscles by day 15. During this time, there is no appreciable growth in size of the larvae, as they average about 590 $\mu$ in length.

While it has not been demonstrated, it is likely, on the basis of what occurs in other nematodes, a final molt takes place between 15 and 22 days after the worms are swallowed by the vertebrate host.

At this time, both the males and females migrate across the subcutaneous connective tissue to the axial and inguinal regions, where they appear in 3½ months in dogs. The females have a mucus plug in the vagina, indicating that insemination has already occurred. In rhesus monkeys, the females begin moving toward the extremities of the limbs between months 8 and 10. They are full of developing eggs at month 8 and fully formed first stage larvae by month 10. Males die between months 3 and 7.

On the basis of epidemical evidence, mature females emerge and larviposit 10 to 14 months after infection of humans. A similar period is required in experimentally infected dogs. The length of time required in rhesus monkeys is 10 to 52 weeks.

Adult females in humans extending the anterior end of the body are often clamped in a split stick and slowly withdrawn from the tissues by winding the worms on the stick. This procedure of removal carries serious hazards to the persons in case the worm breaks and withdraws, resulting in secondary bacterial infection. They can be removed safely from people under aseptic conditions in hospitals.

The dracunculids occurring in North American raccoons are thought by Chandler to represent a different but very similar species which he named *D. insigne*. They occur in the subcutaneous tissue of the hind limbs. Another species, *D. ophidensis* Brackett, 1938, in garter snakes

## Explanation of Plate 124

A, *En face* view of adult female. B, Anterior end of adult. C, Ventral view of posterior end of adult male. D, Anterior end of first stage larva. E, Posterior end of first stage larva. F, Posterior end of second stage larva. G, Posterior end of third stage larva. H, Mammalian definitive host. I, *Cyclops* intermediate host.

1, Mouth; 2, inner circular circumoral elevation; 3, quadrangular elevation; 4, inner circle of six papillae; 5, outer circle of four double papillae; 6, amphid; 7, cervical papillae; 8, caudal papillae of male; 9, phasmid; 10, buccal capsule; 11, muscular portion of esophagus; 12–14, glandular portion of esophagus; 12, precorpus of glandular portion of esophagus; 13, isthmus; 14, postcorpus of esophagus; 15, nerve ring; 16, denticle; 17, esophagus of first stage larva; 18, intestine; 19, rectum; 20, phasmid cell; 21, tail of first stage larva; 22, cuticle of first stage larva; 23, tail of second stage larva; 24, freed cuticle of second stage larva; 25, tail of third stage larva; 26, anus.

a, Adult worms found first in connective tissue of esophagus; b, adult worms appear later in subcutaneous tissue; c, adult female extends anterior portion of body to skin, causing it to vesiculate; d, ulcer-like sore forms and ruptures so that the female may extend a portion of her body through the skin to the outside; e, when skin is wetted, the female is stimulated to extend the anterior end through the hole in it, at which time a rent occurs in her body wall, the uterus protrudes, ruptures, and releases a mass of first stage larvae; f, wriggling larvae attract cyclops which eat them; g, larvae migrate through intestinal wall into hemocoel; h, larvae molt first time; i, second stage larva; j, larvae molt second time; k, third stage filariform larvae infective to vertebrate host; l, vertebrate host infected by swallowing parasitized crustaceans; m, larvae freed when cyclops is digested; n, o, larvae penetrate intestinal mucosa and enter lacteals and lymph vessels; p, larvae enter postcaval vein; q–s, larvae are carried through right side of heart, lungs, and left side of heart; t, larvae in general arterial circulation; u, larvae leave blood vessels to enter subcutaneous connective tissue; v, partially grown worm in subcutaneous connective tissue (number of molts in body of definitive host unknown but there probably are two).

Figures A–C adapted from Moorthy, 1937, J. Parasitol. 23:220; D–G, from Moorthy, 1938, Amer. J. Hyg. 27:437.

---

has a life cycle apparently similar to that of *D. medinensis* but it has been studied incompletely. A tadpole has been interpolated into the life cycle of this species.

The Philometridae in the tissue of catostomid fishes have life cycles apparently quite similar to those of the Dracunculidae.

*Philometroides nodulosa* (Thomas, 1929) occurs in the head tissues of the western white sucker (*Catostomus commersoni*). Adult female worms give birth to first stage larvae. When these larvae are eaten by the crustaceans *Cyclops vernalis* and *C. thomasi*, they migrate through the gut wall into the hemocoel. The first molt occurs between days 11 and 14 and the second molt producing third stage larvae between days 18 and 20.

Suckers become infected by eating cyclops containing third stage larvae. Larvae freed from cyclops in the intestine of the fish migrate through the gut wall into the body cavity. They appear around the outside of the gut by day 9. Fourth stage larvae are in the ventral portion of the celom by day 38. The fourth molt leading to the adult stage has not been observed. Mature males occur in the head tissues of the fish by day 58 and females by day 68. Mating presumably takes place while the worms are migrating forward through the celom. Males die soon after entering the tissues. Females are in the subcutis of the head by day 92 and attain gravidity by days 126 to 142. They soon break through the skin in true dracunculid fashion, extending to the anterior part of body outside; the body wall splits, and the uterus prolapses, releasing larvae into the water. After bearing a single batch of larvae in this manner, the exhausted females die and are adsorbed by the fish tissues.

## EXERCISE ON LIFE CYCLE

*Dracunculus ophidensis* of garter snakes and *Philometroides nodulosa* of catostomid fishes provide excellent material for studying the life cycles. In areas where infected raccoons occur, *D. insigne* should be studied.

Infected animals provide a source of larvae that may be used for infecting cyclops. Ascertain the development of the larvae in the crustacean hosts. Determine whether tadpoles are essential for infection of garter snakes or whether they provide only an ecological expediency for transferring the larvae to the final host. The phases of the life cycle of *D. ophidensis* and *P. nodulosa* in the definitive hosts are inadequately known.

**Plate 124** *Dracunculus medinensis* 465

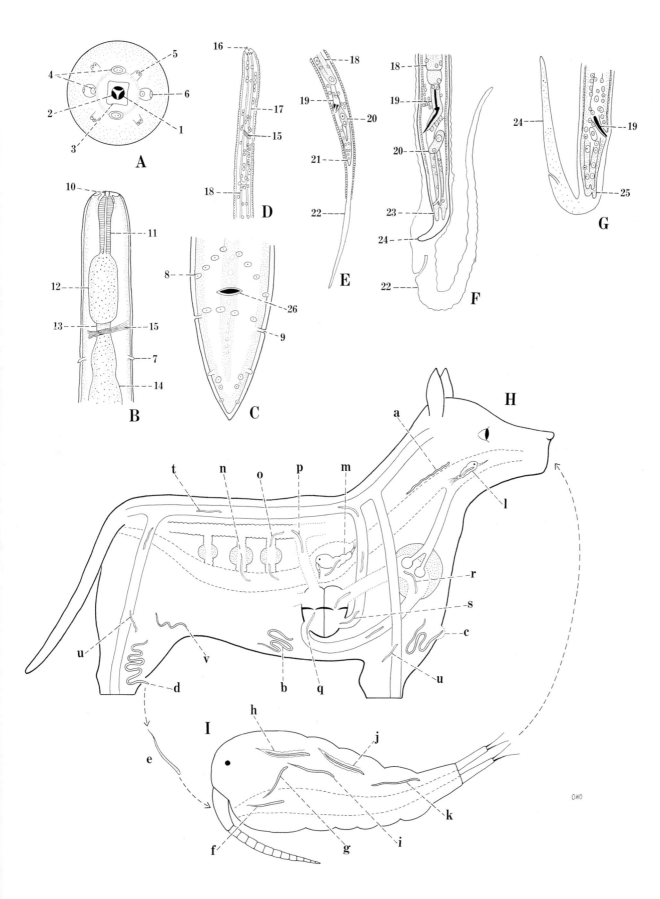

## Explanation of Plate 125

A, Anterior end of mature nematode. B, *En face* view of adult worm. C, Lateral view of posterior end of adult male worm. D, Ventral view of adult male worm. E, Lateral view of posterior end of adult female. F, First stage larva. G, Anterior and posterior ends of second stage larva, showing trifurcate tip of tail. H, Third stage larva. I, Fourth stage larva. J, Chicken definitive host. K, Cockroach (*Pycnoscelus surinamensis*) intermediate host.

1, Outer circle of cephalic papillae; 2, inner circle of cephalic papillae; 3, amphid; 4, mouth surrounded by six-lobed cuticularized ring; 5, buccal capsule; 6, esophagus; 7, esophagointestinal valve; 8, intestine; 9, nerve ring; 10, excretory pore; 11, seminal vesicle; 12, ejaculatory duct; 13, preanal papillae; 14, postanal papillae; 15, short spicule; 16, long spicule; 17, tail; 18, anus; 19, ovary; 20, uterus containing eggs; 21, vulva; 22, digitation at tip of tail of second stage larva.

a, Adult worm under eyelids of chickens and other birds; b, embryonated eggs laid in eyes enter alimentary canal by way of tear duct; c, eggs pass down esophagus into stomach; d, eggs pass through intestine; e, eggs escape from intestine in feces; f, embryonated eggs do not hatch in soil or feces; g, eggs eaten by woods cockroach; h, eggs hatch in intestine and first stage larvae pass through intestinal wall into hemocoel; i, first stage larvae molt, producing second stage larvae; j, second stage larvae molt into third stage larvae; k, third stage larvae encyst in cockroach, producing the infective stage; l, infection of birds occurs when cockroaches containing encysted larvae are swallowed; m, digestion of cockroaches releases larvae from cysts or body cavity; n, larvae escape from body of cockroach; o, larvae migrate up esophagus; p, larvae enter tear ducts and migrate into eyes where the two final molts take place, after which maturity is attained.

Figures A–E adapted from Ransom, 1904, U. S. Dep. Agr., Bur. Anim. Ind. Bull. 60; F–I, from Schwabe, 1951, Pacific Sci. 5:18.

## SELECTED REFERENCES

Brackett, S. (1938). Description and life history of the nematode *Dracunculus ophidensis* n. sp. with a redescription of the genus. J. Parasitol. 24:353–361.

Dailey, M. D. (1966). Biology and morphology of *Philometroides nodulosa* (Thomas, 1929) n. comb. (Philometridae: Nematoda) in the western white sucker (*Catostomus commersoni*). Colo. State Univ. Dissertation, 77 pp.

Moorthy, V. N. (1938). Observations on the development of *Dracunculus medinensis* larvae in cyclops. Amer. J. Hyg. 27:437–460.

Muller, R. (1971). Dracunculus and dracunculiasis. In: B. Dawes, ed. Advances in Parasitology. Academic Press, New York, Vol. 9, pp. 73–151.

Onabamiro, S. D. (1956). The early stages of development of *Dracunculus medinensis* (Linnaeus) in the mammalian host. Ann. Trop. Med. Parasitol. 50:157–166.

Thomas, L. J. (1944). Researches in the life histories of parasites of wildlife. Trans. Ill. Acad. Sci. 37:7–24.

### Suborder Spirurina

This group of the order Spirurida is characterized by multinucleate esophageal glands, larvae commonly with cephalic hooks, and phasmids that are pore-like. The intermediate hosts are arthropods but rarely copepods. The Spirurina consists of the superfamilies Spiruroidea and Filarioidea.

### Superfamily Spiruroidea

In this superfamily, the stoma is well developed and there may or may not be two large, lateral, lip-like structures. The vulva is near the middle or posterior part of the body and the eggs are fully embryonated when laid.

#### FAMILY THELAZIIDAE

These nematodes occur under the eyelids of birds and mammals in many parts of the world. They lack lips and have a small buccal capsule without thickenings in the wall.

*Oxyspirura mansoni*
(Cobbold, 1879) (Plate 125)

This is the eyeworm of chickens, turkeys, and a number of wild birds, including English sparrows, in many of the warm parts of the world, including the southeastern United States. The adults are under the nictitating membrane.

**Description.** The males are 8.2 to 16 mm long with ventrally curved tail bearing four pairs of preanal and two pairs of postanal papillae. The spicules are markedly unequal in length, one being 0.2 to 0.25 mm long and the other 3.0 to 3.5 mm. The females measure 12 to 20 mm long with the vulva 1 to 1.4 mm from the posterior end of the body. The eggs measure 50 to 65 $\mu$ long by 40 to 45 $\mu$ wide and contain a first stage larva when laid.

**Life Cycle.** Adult female worms under

**Plate 125**     *Oxyspirura mansoni*                                                                 467

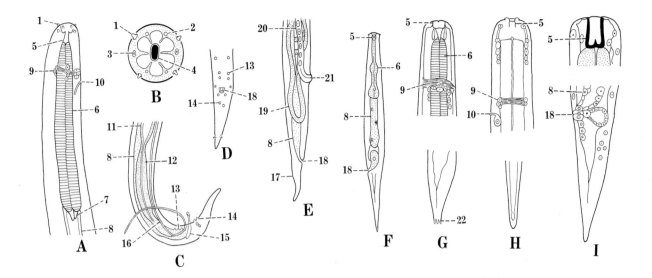

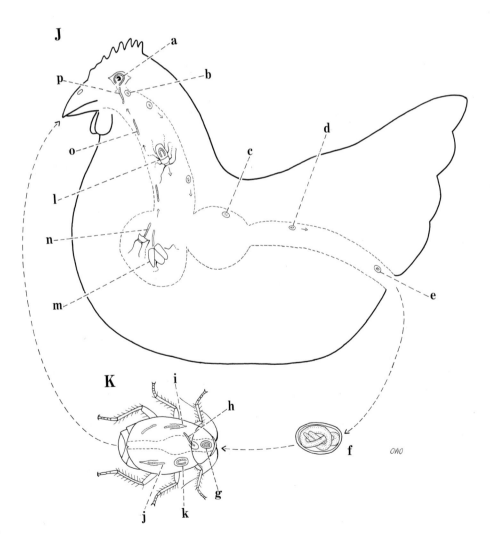

the nictitating membrane lay their eggs. These are washed through the nasolacrimal ducts into the pharynx, swallowed, and voided with the feces. Upon leaving the body, the eggs remain unhatched and without further development of the larvae until eaten by the woods or burrowing cockroach *Pycnoscellus surinamensis*, an insect of the warmer parts of the world.

Upon reaching the gut of the cockroach, the eggs begin hatching in about 48 hours, and continue for several days. First stage larvae appear in the body cavity by day 8. They are surrounded by a cyst-like structure of host origin. About 3 weeks later, the second and final molt in the roach occurs. The third stage larvae may be free or encysted in the body cavity. They are infective to birds. When roaches harboring infective larvae are swallowed by the birds, the parasites are liberated in the crop. They quickly make their way up the esophagus, into the pharynx, and through the nasolacrimal ducts into the eyes. The third molt takes place in about 2 days and the fourth and final one about 3 weeks later. Sexual maturity is attained in about 30 days after infection.

A number of species of *Thelazia*, a related genus, occurs in the eyes of sheep, goats, cattle, and other animals. *Thelazia californiensis* in sheep, deer, horses, cats, dogs, coyotes, bears, and humans in California showed development in the lesser housefly *Fannia canicularis*, a relative of *Musca*. In Russia, the flies *Musca larvipara* and *M. autumnalis* serve as intermediate hosts of *Thelazia rhodesii* in cattle. *Stomoxys calcitrans*, other bloodsucking flies, and the common housefly (*Musca domestica*) are of no importance in transmitting this parasite.

### EXERCISE ON LIFE CYCLE

In regions where eyeworms of chickens occur, interesting experiments can be performed. A colony of uninfected roaches for experimental purposes may be obtained from adult females kept in small dishes, fed brown bread, and provided with water.

Embryonated eggs obtained by macerating gravid females from the eyes of chickens are suitable for infecting laboratory-reared cockroaches. After feeding a large series of roaches embryonated eggs, examine some daily for the first 10 days and at 2 to 3 day intervals for the next 40 days. Hatching in the gut, migration of the larvae into the body cavity, and the development of the second and third stage larvae will be observed.

Upon obtaining third stage larvae, introduce some of them, freed from the roaches, into the esophagus of young unexposed chicks by means of a pipette. Check the eyes within 5 minutes by rinsing with a saline solution and repeat at similar intervals to ascertain how quickly the larvae reach them from the esophagus. Feed a dozen chicks infected roaches and examine the eyes at 2-day intervals to trace the development to adults.

### SELECTED REFERENCES

Burnett, H. S., W. E. Parmelee, R. D. Lee, and E. D. Wagner. (1957). Observations on the life cycle of *Thelazia californiensis* Price, 1930. J. Parasitol. 43:433.

Klesov, M. D. (1949). [A study of the biology of the nematode, *Thelazia rhodesi* Desmarest, 1827.] Dokl. Akad. Nauk SSSR n. s. 66:309–311.

Ransom, B. H. (1904). Manson's eye worm of chickens (*Oxyspirura mansoni*), with a general review of nematodes parasitic in the eyes of birds. U. S. Dep. Agr., Bur. Anim. Ind. Bull. 60, 72 pp.

Sanders, D. A., and M. W. Emmel. (1938). Manson's eyeworm of poultry. Fla. Agr. Exp. Sta., Press Bull. 511, 2 pp.

Schwabe, C. W. (1949). Observations on the life history of *Pycnoscelus surinamensis* (Linn.), the intermediate host of the chicken eyeworm in Hawaii. Proc. Hawaiian Entomol. Soc. 13:433–436.

Schwabe, C. W. (1950). Studies on *Oxyspirura mansoni*, the tropical eyeworm of poultry. III. Preliminary observations on eyeworm pathogenicity. Amer. J. Vet. Res. 11:286–290.

Schwabe, C. W. (1951). Studies on *Oxyspirura mansoni*, the tropical eyeworm of poultry. II. Life history. Pacific Sci. 5:18–35.

#### FAMILY SPIRURIDAE

The members of this family are parasites of the esophagus, stomach, and intestine of vertebrates. The eggs are embryonated when laid, or they may even hatch in the uterus of the female. There are two well-developed bilobed lateral lips; small interlabia, one dorsal and one ventral, may be present.

*Habronema megastoma*
(Rudolphi, 1819) (Plate 126)

This is one of the three species of related stomach worms in equines. It is the cause of stomach ulcers and granulomatous summer sores originating from wounds in the skin that are infected with larvae from the probosci of houseflies feeding on them.

The life cycle of *Habronema* in flies is very similar to that of Filarioidea in mosquitoes. It is thought that this similarity may be a connecting link between the superfamilies Spiruroidea and Filarioidea.

**Description.**    The head is separated from the body by a constriction. The males are 7 to 10 mm long. There are three pairs of preanal and one pair of postanal papillae on the ventrally curved tail. The left spicule is cylindrical and 460 μ long and the right one is flattened and 280 μ long. The females measure 10 to 13 mm long with the vulva about median. The eggs are embryonated when passed in the feces and are 40 to 50 μ long by 9 to 11 μ wide normally but they may stretch far beyond this length. Larvae of *Habronema megastoma* have a smooth cuticle, whereas those of *H. mucosae* have longitudinal cuticular ridges.

**Life Cycle.**    The elongated eggs contain a first stage larva folded upon itself when laid in the stomach. These are voided with the feces, where survival may be up to 2 weeks without further development.

When eaten by the larvae of houseflies (*Musca domestica*), the eggs hatch soon and first stage larvae 105 to 110 μ long are present in the intestine. They bear a small, inverted, V-shaped movable stylet similar to the larvae of filarioid worms. No growth takes place in the intestine of flies. The larvae enter the Malpighian tubules on day 3 where they molt and differentiate into short and thick (90 μ by 8 to 10 μ) sausage-shaped second stage larvae similar to those of the filarioids. Outlines of the esophagus, intestine, and rectum appear by day 4 when most of the larvae have entered the Malpighian tubules. The sausage stage persists into the pupal stage of the flies. As growth of the larval worms continues, the epithelial cells of the Malpighian tubules are destroyed, leaving the larvae encased in the membranes. They have grown to 500 μ long and 30 to 35 μ wide, which appears to be the only change.

On day 6, they molt the second time and transform from the sausage to the vermiform stage, which is 600 to 900 μ long and enclosed in the loose cuticle of the sausage stage. By day 7, they are 1.3 to 2 mm long. The third stage larvae

begin to migrate from the Malpighian tubules into the body cavity on day 8. They are in the thorax, head, and proboscis by days 9 to 11, being infective to the horses at this time.

Infection of horses takes place when third stage larvae escaping from the labellum of flies feeding on the moist mucus membranes of the lips are swallowed. Chilled or dead infected flies swallowed with forage also result in infections. Upon reaching the stomach, the larvae presumably enter the glands of the mucosa and molt twice, but there is no experimental evidence to verify this. As growth continues, ulcers are formed in the wall of the stomach, where adult worms occur.

When infected flies feed on the moist surfaces of wounds in the skin of horses, the larvae escape from the mouth parts and enter the wounds. Irritation caused by the larvae prevents healing, and a large pulpy mass of tissue appears that may persist throughout the summer. Since the etiology of these granulomas is larvae from infected flies, they occur during the warm season when these insects are active, hence the name summer sores. The larvae do not develop to maturity in the skin but die toward the end of the summer. *Habronema megastoma* causes cutaneous and gastric granulomata, *H. microstoma* nodular dermatitis and gastric ulceration, and *H. muscae* granular dermatitis and diffused gastritis.

The intermediate host of *Habronema muscae* is houseflies but that of *H. microstoma* is stable flies. The life cycles of these two species are basically the same as that of *H. megastoma*. In *H. microstoma*, the eggs hatch in the uterus of the parent worm, whereas those of *H. muscae* hatch in the intestine of the flies. The larvae of *H. muscae* and *H. microstoma* develop in the fat bodies of the flies instead of in the Malpighian tubules.

Some species of *Habronema* may utilize transport hosts such as appears to be the case with *H. mansioni* of falcons (*Gypaetus barbatus*) in northern China. About 60 percent of the toads (*Bufo b. asiaticus*) of the area harbor encapsulated larvae in the stomach wall. When the infected toads were fed to falcons, the larvae developed into *H. mansioni* such as occur naturally in the raptors. It might be assumed that the toads acquired the infections from insects and that the amphibians are paratenic hosts which constitute an ecological necessity in the biology

## Explanation of Plate 126

**A,** Lateral view of anterior end of *Habronema megastoma.* **B,** Lateral view of anterior end of *Habronema microstoma.* **C,** Lateral view of anterior end of *Habronema muscae.* **D,** Embryonated egg. **E,** Egg hatching. **F, G,** First stage larva. **H,** Second stage sausage-shaped larva. **I,** Third stage larva. **J,** Anterior end of third stage larva. **K,** Caudal end of third stage larva. **L,** Horse definitive host. **M,** Larval stage of housefly (*Musca domestica*). **N,** Pupal stage of housefly. **O,** Adult housefly.

**1,** Head; **2,** neck; **3,** buccal capsule; **4,** muscular portion of esophagus; **5,** intestine; **6,** rectum; **7,** anus; **8,** tail; **9,** spiny, knob-shaped tail of third stage larva; **10,** nerve ring; **11,** cervical papilla; **12,** eggshell; **13,** first stage larva; **14,** V-shaped structure at anterior end of first stage larva; **15,** cast cuticle of second stage larva; **16,** portion of Malpighian tubule, surrounding third stage larva.

**a,** Adult worms in gastric ulcers of stomach where eggs are laid; **b, c,** embryonated eggs free in stomach and being passed in feces; **d,** embryonated egg free in feces; **e,** eggs swallowed by larvae of fly hatch in intestine; **f,** larvae free in intestine; **g,** larvae enter Malpighian tubules; **h,** first molt occurs in Malpighian tubules of larval fly; **i,** second stage sausage-shaped larvae; **j,** sausage stage larvae in pupa; **k,** elongation of sausage stage; **l,** second molt in pupa or adult fly forming third stage filariform larvae; **m,** ensheathed third stage larvae; **n,** larvae leave Malpighian tubules entering thorax; **o,** larvae in thorax and proboscis; **p,** larvae escape from labellum in response to warmth of skin, when flies feed on moist mucous membranes of mouth; **q, r,** larvae deposited on lips enter mouth and are swallowed; **s,** larvae molt, probably twice, and enter glandular crypts of stomach; **t, u,** infected flies are swallowed with food; **v,** digestion of flies releases larvae; **w,** larvae from flies enter stomach wall; **x,** some larvae encyst in lungs of horse; **y,** flies feeding on open wounds deposit larvae which cause granulomatous growths known as summer sores.

Figures A–C adapted from Theiler, 1923, Thèse Faculté Sci. Univ. Neuchatel, Gov. Print. Stat. Off. Pretoria, pls. 51–53; E–K, from Roubaud and Descazeaux, 1921, Bull. Soc. Pathol. Exot. 14:471.

---

of the parasite, which has associated itself with the food chain of the falcons.

## EXERCISE ON LIFE CYCLE

If an abattoir where horses are slaughtered is accessible, adult female worms may be obtained from the stomach ulcers for a supply of eggs for experimental infections. Embryonated eggs dissected from the uterus and mixed with sterilized horse feces serve as a medium for infecting larval houseflies. These should be reared in horse feces previously sterilized by heat. Dissection of larval flies each day for 2 weeks after feeding them eggs will show the developmental stages and the time required for each to develop.

## SELECTED REFERENCES

Hsü, H.-F., and C. Y. Chow. (1938). On the intermediate host and larva of *Habronema mansioni* Seurat, 1914 (Nematoda). Chinese Med. J. Suppl. 2, pp. 433–440.

Jesus, Z. de (1963). Observations on habronemiasis in horses. Philipp. J. Vet. Med. 2:133–152.

Ransom, H. B. (1913). The life history of *Habronema muscae* (Carter), a parasite of the horse transmitted by the housefly. U. S. Dep. Agr. Bull. 163, pp. 1–36.

Roubaud, E., and J. Descazeaux. (1921). Contribution à l'histoire de la mouche domestique comme agent vecteur des habronémoses d'equidés. Cycle évolutif et parasitisme de l'*Habronema megastoma* (Rudolphi, 1819) chez la mouche. Bull. Soc. Pathol. Exot. 14:471–506.

Roubaud, E., and J. Descazeaux. (1922). Evolution de l'*Habronema muscae* Carter chez la mouche domestique et de l'*H. microstomum* Schneider chez le stomoxe (Note préliminaire). Bull. Soc. Pathol. Exot. 15:572–574.

Roubaud, E., and J. Descazeaux. (1922). Deuxième contribution a l'etude des mouches, dans leurs rapports avec l'evolution des habronemes d'equidés. Bull. Soc. Pathol. Exot. 15:978–1001.

Waddell, A. H. (1969). A survey of *Habronema* spp. and identification of third stage larvae of *Habronema megastoma* and *Habronema muscae* in section. Austr. Vet. J. 45:20–21.

### FAMILY TETRAMERIDAE

These nematodes occur in the crypts of Lieberkühn of the proventriculus of chickens, turkeys, pigeons, ducks, and many wild birds. The males are small and vermiform, whereas gravid females become greatly distended with eggs, assuming a markedly fusiform shape. In some species, the females are somewhat coiled. The anterior end of the worms is without raised looping ridges known as cordons.

Plate 126    *Habronema megastoma, H. microstoma,* and *H. muscosae*                                                471

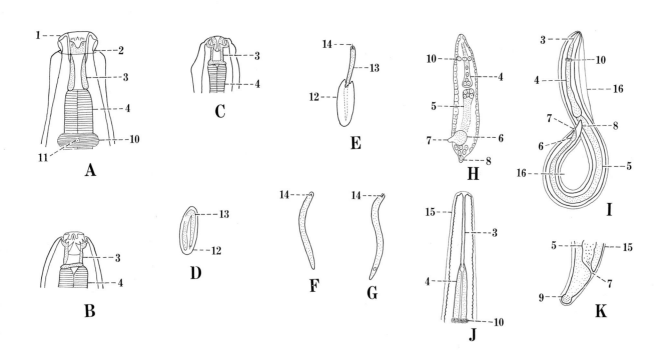

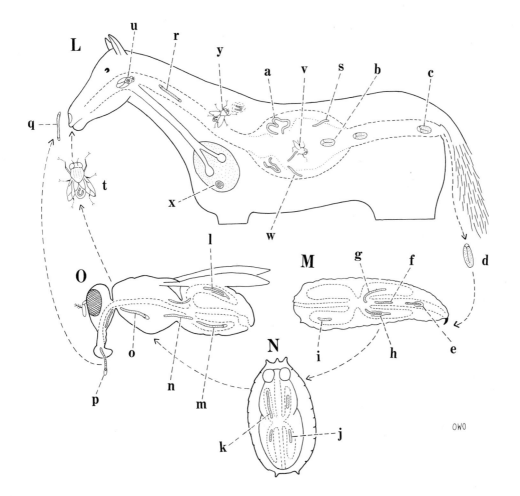

OWO

## Explanation of Plate 127

A, Anterior end of adult male. B, Posterior end of adult male. C, Gravid female. D, Embryonated egg containing first stage larva. E, Caudal end of second stage larva taken from intestine of gammarid. F, Caudal end of third stage larva from body cavity of gammarid 60 days after infection. G, Caudal end of fourth stage male from proventriculus of duck. H, Fourth stage female larva from proventriculus of duck. I, Definitive host (duck). J, Intermediate gammarid host (*Gammarus* and *Hyalella*).

1, Cephalic papilla; 2, anterior spine-like sensory organs; 3, cervical papilla; 4, buccal capsule; 5, muscular portion of esophagus; 6, glandular portion of esophagus; 7, nerve ring; 8, posterior spine-like sensory papillae; 9, genital papillae; 10, intestine; 11, anus; 12, ejaculatory duct; 13, left spicule; 14, right spicule; 15, cuticular caudal cone; 16, uterus of gravid female; 17, vulva; 18, thick shell of egg; 19, first stage larva in egg; 20, genital primordium; 21, sensory organs; 22, caudal sensory papillae; 23, primordium of male reproductive organs; 24, primordium of female reproductive organs.

a, Gravid female in crypts of Leiberkühn of proventriculus; b, embryonated eggs containing first stage larva when laid; c, eggs mixed with feces; d, embryonated eggs in water; e, gammarid intermediate host eating embryonated eggs; f, eggs hatching in intestine and liberation of first stage larvae; g, larvae pass through intestinal wall into hemocoel; h, first stage larvae molt; i, second stage larvae encyst; j, second molt takes place in cysts and third stage larvae grow; k, gammarids partially digested in crop; l, cysts released from body of crustaceans; m, larvae escape from cysts; n, loose cuticle of second molt discarded (not a molt); o, third molt; p, fourth stage larvae enter proventricular crypts; q, fourth molt not seen but presumed to take place after which development proceeds to the adult stage.

Adapted from Swales, 1936, Can. J. Res. 14:151.

---

### Tetrameres crami
### (Swales, 1933) (Plate 127)

*Tetrameres crami* is a parasite of the proventriculus of domestic and wild ducks of North America.

**Description.** The males are 2.9 to 4.1 mm long by 70 to 92 μ wide. There is a double row of spine-like sensory organs extending the full length of the body, together with one pair of cervical spines. The lateral pseudolabia are trilobed with interlabia. The right spicule is 136 to 185 μ and the left 272 to 350 μ long. The gravid females are globular to spindle-shaped, 1.5 to 3.25 mm long by 1.2 to 2.2 mm wide, and are blood red. The vulva is near the posterior end of the body. Eggs are 42 to 57 μ long by 26 to 34 μ wide.

**Life Cycle.** The eggs contain fully developed first stage larvae when laid. They hatch when eaten by freshwater gammarids (*Hyallela knickerbockeri* and *Gammarus fasciatus*). The larvae migrate into the celom and molt for the first time. Second stage larvae appear in 7 to 11 days after the eggs are eaten. They migrate to various parts of the body and generally encyst in thin-walled vesicles, commonly in the coxae, on the inner surface of the dorsal shield, and occasionally in the muscles of the walking legs.

The second molt usually occurs in the cyst with subsequent development of the third stage larvae continuing inside it. These encysted larvae acquire great length, up to 1.4 mm in 20 to 65 days. Infection of ducks occurs when gammarids containing third stage larvae are eaten. They escape from the crustaceans in the crop and make their way to the proventriculus. The third molt takes place and the fourth stage larvae are found in the crypts of Lieberkühn. While a fourth molt is indicated, it has not been observed. Male worms develop to maturity in about one month and females sometime between one and two months.

*Tetrameres americanus* of poultry and quail in North America utilizes grasshoppers (*Melanoplus femur-rubrum* and *M. differentialis*) and cockroaches (*Blatella germanica*) as intermediate hosts.

Larval development of *Microtetrameres centuri* (Barus, 1966) and *M. sternellae* Ellis, 1967 from the proventricular glands of meadowlarks takes place in species of grasshoppers of the genus *Melanoplus*. Eggs hatch in the foregut in a few hours and the first stage larvae migrate into the hemocoel and molt in 8 to 16 days. They enter the muscles and molt the second time to third stage larvae 25 days after infection. Birds become infected upon eating parasitized grasshoppers. Female worms are gravid in canaries in 72 to 184 days and in pigeons in 70 to 102 days. Adult worms contain erythrocytes in the gut.

It may be generalized that species of *Tetra-*

**Plate 127** *Tetrameres crami* 473

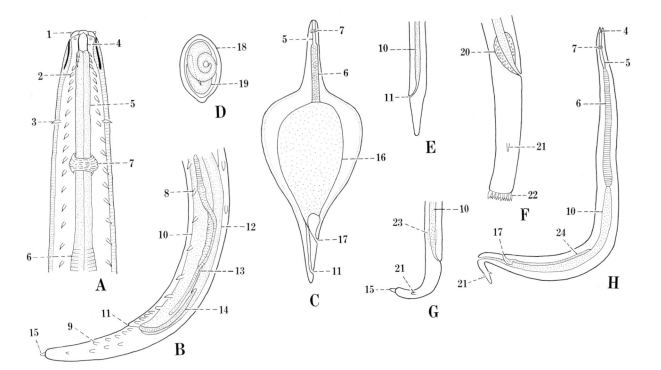

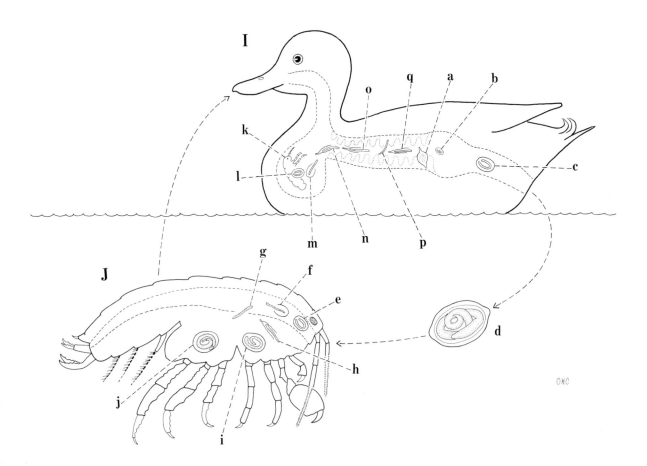

*meres* infecting terrestrial birds utilize land arthropods, namely Orthoptera, as intermediate hosts, whereas those infecting waterfowl are in aquatic crustaceans, gammarids primarily.

## EXERCISE ON LIFE CYCLE

Place eggs dissected from gravid females in small containers of water together with *Gammarus fasciatus* or *Hyalella knickerbockeri*, common gammarids in streams, lakes, and ponds of North America. Dissection of some of the crustaceans at daily intervals for 3 to 4 weeks will show the stages of development of the second and third stage larvae.

Adult worms can be obtained by feeding infected gammarids to ducklings, and at the same time the prepatent period can be determined. Eggs of *Tetrameres* from chickens should be fed to grasshoppers, preferably young ones but not necessarily only immature individuals. Dissect a series of grasshoppers at intervals after eating infective eggs in order to follow the stages of development of the larvae.

Feed grasshoppers containing infective third stage larvae to a series of chicks. Sacrifice the birds at intervals to follow the development of the worms. Check the feces for the appearance of eggs to ascertain how long it takes for the nematodes to reach sexual maturity.

## SELECTED REFERENCES

Cram, E. B. (1931). Developmental stages of some nematodes of the Spiruroidea parasitic in poultry and game birds. U. S. Dep. Agr. Tech. Bull. 227, 25 pp.

Ellis, C. J. (1967). Biology of *Microtetrameres sternellae* n. sp. (Nematoda: Tetrameridae). Dissert. Abst. 28:1725-B.

Ellis, C. J. (1969). Life history of *Microtetrameres centuri* Barus, 1966 (Nematoda: Tetrameridae). I. Juveniles. J. Nematol. 1:84–93.

Ellis, C. J. (1969). Life history of *Microtetrameres centuri* Barus, 1966 (Nematoda: Tetrameridae). II. Adults. J. Parasitol. 55:713–719.

Ellis, C. J. (1970). Pathogenicity of *Microtetrameres centuri* Barus, 1966 (Nematoda: Tetrameridae) in meadowlark. J. Nematol. 2:33–35.

Swales, W. E. (1936). *Tetrameres crami* Swales, 1933, a nematode parasite of ducks in Canada. Morphological and biological studies. Can. J. Res. 14:151–164.

## FAMILY ASCAROPIDAE

Members of this family parasitize the anterior portion of the alimentary canal. They occur in the stomach of swine, rats, and mice, in the rumen, intertwined in the mucosa and submucosa of the esophagus of ruminants, and in the crop of chickens. They are thick-bodied or filiform worms with small lips. The wall of the pharynx is provided with distinct spiral or annular thickenings (*Ascarops*, *Physocephalus*), or the anterior end of the body is covered with distinct cuticular plaques (*Gongylonema*). Irregular caudal alae are present on the males.

### *Ascarops strongylina*
(Rudolphi, 1819) (Plate 128)

*Ascarops strongylina*, together with *Physocephalus sexalatus* (Molin, 1860), a closely related species, are known as the thick stomach worms of swine. They are common in many parts of the United States.

**Description.** The pharynx appears as a spiralled rod and is 83 to 98 $\mu$ long. Males measure 10 to 15 mm long with a cervical ala on the left side; the right caudal ala is about twice the length of the left. There are four pairs of preanal and one of postanal papillae, all arranged asymmetrically. The left spicule is 2.24 to 2.95 mm long and the right 0.46 to 0.62 mm. Females are 16 to 22 mm long with the vulva slightly anterior to the middle of the body. Eggs measure 34 to 39 $\mu$ long by 20 $\mu$ wide, have thick shells with a striated plug at each end, and are embryonated when passed.

**Life Cycle.** Embryonated eggs, containing first stage larvae, passed in the feces remain infective after exposure to freezing temperatures of $-1$ to $-4°C$ for 2 weeks. They survive in water at room temperature for 1 month. The larvae are around 530 $\mu$ long and there is no further growth in the egg.

When eaten by one of 16 to 20 species of dung beetles, the eggs hatch in the midgut. Eggs fed to *Aphodius granarius* hatch and larvae are in the abdominal portion of the body cavity in 24 hours. By 13 to 15 days after infection, most first stage larvae, 158 to 160 $\mu$ long, are enclosed in spherical, thin-walled cysts located in the wall of the Malpighian tubules. In addition to the larvae, the cysts contain rounded bodies, probably fat cells. The first molt takes place between 17 and 19 days after infection. Second stage larvae begin the second molt 28 days after infection and third stage larvae 420 to 700 $\mu$ long are present 1 day later. Completely formed cysts, up to 960 $\mu$ in diameter, containing third stage larvae are usually free in the abdominal cavity, frequently with a network of superficial tracheoles. Subsequent development of the larvae is dependent upon their reaching the stomach of pigs. Third stage larvae are characterized by a smooth knob at the tip of the tail.

When dung beetles are swallowed by pigs and digested, the liberated larvae enter the mucosal layer of the fundus of the stomach. Within 5 to 9 days, they return to the lumen and migrate to the pyloric region of the stomach. The third molt occurs in both parts of the stomach 4 to 5 days after infection. The fourth and final ecdysis takes place on about day 20. Sexual maturity is attained between days 25 and 30, with the first eggs appearing in the feces 46 to 50 days after the pigs have eaten infected dung beetles. The life span of worms in pigs is about 10½ months.

Numerous animals from all classes of the vertebrates are capable of serving as reservoir hosts. Important ones include rodents, shrews, armadillos, birds, and lizards that habitually feed on insects, including dung beetles when available. Upon digestion of them, the freed third stage larvae migrate through the gut wall and reencyst in various organs in which there is no development. They may survive up to 1 year in mice and possibly as long in other species. Larvae acquired by pigs that eat reservoir hosts develop in the same manner as those gotten from beetles.

The life cycle and ecology of *Physocephalus sexalatus*, another stomach worm of swine, is similar to that of *Ascarops strongylina*. It develops in dung beetles and utilizes a number of vertebrate reservoir hosts. Third stage larvae have a digitiform knob at the tip of the tail.

*Gongylonema pulchrum*, the esophageal stitchworm of swine, along with other species of the genus, produces embryonated eggs. They hatch in many species of dung beetles as well as in the German cockroach *Blatella germanica*. Development in the cockroach is generally similar to that described above for *Ascarops strongylina*.

Larvae excyst in the stomach of pigs and promptly begin migrating forward through the gastroesophageal opening into the esophagus. Upon entering the esophagus, they burrow into the mucosa, some within 30 minutes after excystment, and begin the migration forward within it. By day 3, the van has reached the tongue. After 10 days, many are distributed throughout the length of the esophagus, as well as being in the tongue and covering of the hard palate.

They are in the fourth stage by day 12. The fourth and final molt takes place 27 to 31 days after infection, at which time the males are 11 to 12 mm long and the females 18 to 20 mm. Sexual maturity is attained between 50 and 55 days in cattle.

## EXERCISE ON LIFE CYCLE

Adult female worms procured from abattoirs where swine are slaughtered provide a source of eggs for experimental infection of dung beetles. Eggs may be obtained by chopping up gravid worms in a few drops of water and transferring the material to pieces of blotting paper which are placed in small dishes containing five or six dung beetles. Dissect some of the beetles at 12 and 24 hours after they have eaten the eggs to ascertain the location of the first stage larvae. Continue dissections at 2- and 3-day intervals to follow the development of them.

If working with *P. sexalatus*, feed beetles containing third stage larvae to young chicks to obtain larvae encysted on the wall of the alimentary canal. In the event *P. sexalatus* is not available, third stage larvae of *A. strongylina* should be fed to chicks to determine if they encyst on the viscera.

## SELECTED REFERENCES

Alicata, J. E. (1934). New intermediate hosts for heteroxenous nematodes. Proc. Helminthol. Soc. Wash. 1:13.

Alicata, J. E. (1935). Early developmental stages of nematodes occurring in swine. U. S. Dep. Agr. Tech. Bull. 489, 96 pp.

Allen, R. W., and L. A. Spindler. (1949). Note on the occurrence in farm-raised chickens of encysted third stage larvae of *Physocephalus sexalatus*, a spirurid stomach worm of swine. Proc. Helminthol. Soc. Wash. 16:1–3.

Dimitrova, E. (1966). [Distribution of *Ascarops strongylina* and *Physocephalus sexalatus* in Bulgaria and their intermediate and reservoir hosts in the Strandzha Mountain area.] Izv. Tsent. Khelminthol. Lab. Sofia 11:43–56.

## Explanation of Plate 128

A, Anterior end of adult *Ascarops strongylina*. B–D, First stage larva of *A. strongylina* (relatively similar for *P. sexalatus*). B, Sinistral view of anterior end. C, Ventral view. D, Tail. E, Tail of second stage larva. F, Tail of third stage larva. G, Encysted third stage larva. H, Anterior end of adult *Physocephalus sexalatus*. I, Tail of third stage larva of *P. sexalatus*, showing morphological characteristic. J, Swine definitive host. K, Dung beetle (*Ataenius cognatus* and *Aphodius granarius*) intermediate host.

1, Inner circle of cephalic papillae; 2, outer circle of cephalic papillae; 3, buccal capsule; 4, spiral pharynx; 5, muscular portion of esophagus; 6, cervical papillae; 7, cervical alae; 8–18, *Ascarops strongylina*; 8, cephalic hooks; 9, rows of spines; 10, tail showing terminal spine; 11, anus; 12, characteristic knoblike tip of tail of second stage larva; 13, free cuticle from first stage larva; 14, rectal glands; 15, intestine; 16, characteristic knoblike tip of tail of third stage larva; 17, third stage larva encysted in beetle host; 18, cyst wall of beetle origin; 19, spiny tip of tail of third stage larva of *P. sexalatus*.

a, Adult worm in stomach; b, embryonated eggs passed in feces; c, egg with characteristic thick shell, having striations at each end; d, eggs are eaten by dung beetles (*Ataenius* and *Aphodius*); e, eggs hatch in intestine; f, first stage larvae migrate through intestinal wall into hemocoel; g, larvae encyst and undergo first molt within cyst; h, second stage larvae with free cuticle; i, second molt in cyst with two cuticles free; j, fully developed third stage larvae in thick-walled cysts; k, pigs become infected by swallowing infected dung beetles; l, beetles digested so that cysts and larvae are liberated; m, third stage larvae penetrating gastric mucosa; n, third molt presumably takes place in mucosa; o, fourth stage larvae leave mucosa; p, fourth and final molt in stomach.

Figures A and H adapted from various sources; B–G, I, from Alicata, 1935, U. S. Dep. Agr. Tech. Bull. 489, 96 pp.

Gupta, V. P. (1970). Experimental development of the oesophageal worm of cattle in rabbit. Curr. Sci. 39:237–238.

Krahwinkel, D. J., Jr., and J. F. McCue. (1967). Wild birds as transport hosts of *Spirocera lupi* in the southeastern United States. J. Parasitol. 53:650–651.

Porter, D. A. (1939). Some new intermediate hosts of the swine stomach worm, *Ascarops strongylina* and *Physocephalus sexalatus*. Proc. Helminthol. Soc. Wash. 6:79–80.

Ransom, B. H., and M. C. Hall. (1916). The life history of *Gongylonema scutatum*. J. Parasitol. 2:80–86.

Shmitova, G. Y. (1962). [The significance of coprophagous beetles in the epizootiology of some spirurids of domestic animals.] Tr. Gel'mintol. Lab. 12:330–344.

Shmitova, G. Y. (1963). [Experimental study of reservoir parasitism in *Ascarops strongylina*.] Helminthologia 4:456–463.

Shmitova, G. Y. (1964). [Study of the ontogenetic development of *Ascarops strongylina*.] Tr. Gel'mintol. Lab. 14:288–301.

Spindler, L. A. (1942). Internal parasites of swine. In: Keeping Livestock Healthy. U. S. Dep. Agr. Yearbook 1942, pp. 745–786.

## FAMILY PHYSALOPTERIDAE

This family contains a large number of species which occur commonly in reptiles, birds, and mammals, but only rarely in amphibians. They are relatively large worms with thick bodies and attach to the mucosa of the stomach or intestine. The cuticle of the anterior end of the body is reflected forward over the pseudolabia, forming a collarette. The caudal alae are supported by long slender papillae.

### *Physaloptera phrynosoma*
Ortlepp, 1922 (Plate 129)

*Physaloptera phrynosoma* is a common parasite of the stomach of horned lizards (*Phrynosoma cornutum*) from the arid regions of the southwestern part of the United States and similar areas of Mexico.

**Description.**    Adult males are 8 to 10 mm long by 0.526 mm in diameter with a single lobed pseudolabium. The left spicule is 482 to 754 $\mu$ long and the right one 180 to 200 $\mu$. The arrangement of the ventral and lateral papillae are shown in Plate 129, D. Females are 12 to 24 mm long by 0.634 mm in diameter with two uteri (Plate 129, E). Eggs are 52 by 34 $\mu$ in size and contained in capsules.

**Life Cycle.**    Gravid females leave the lizards in the feces, die, and become dried. The eggs are contained in capsules within the uterus, where they develop and are able to survive desiccation for at least 2 years, a desirable property for living in arid regions. There are 5 to

Plate 128    *Physocephalus sexalatus and Ascarops strongylina*                                          477

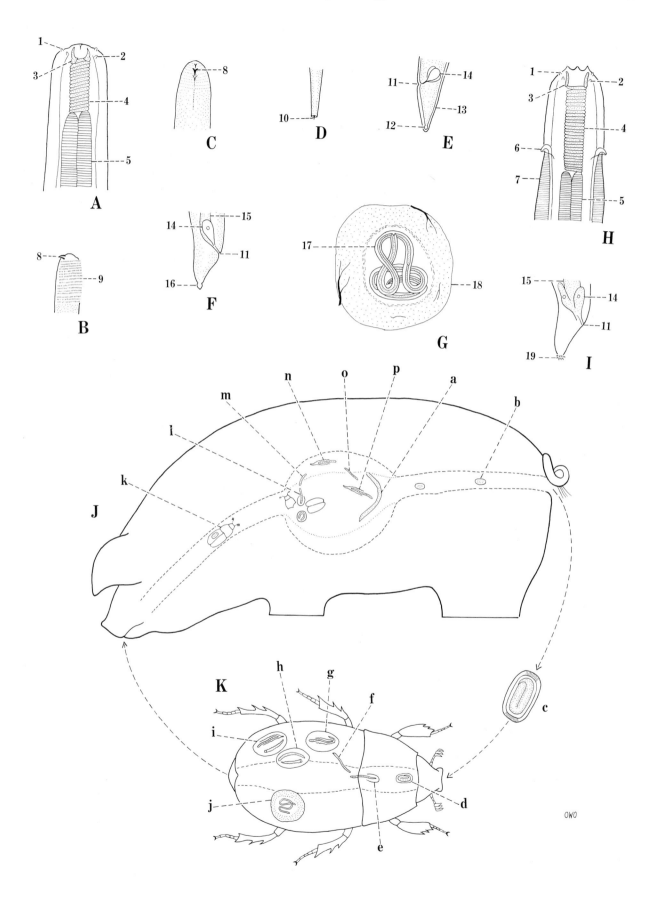

## Explanation of Plate 129

A, Lateral view of anterior end of adult worm. B, Ventral view of anterior end of adult. C, *En face* view (somewhat diagrammatic). D, Ventral view of posterior end of adult male. E, Amphidelphic uterus. F, Egg capsule containing eggs. G, Cyst from larval ant *Pogonomyrmex barbatus*. H, Cyst with larva from pupal ant. I, Cyst from gaster of callow. J, Cyst from gaster of adult ant. K, First stage larva. L, Second stage larva. M, Third stage larva. N, Horned lizard (*Phrynosoma cornutum*) definitive host. O, Worker ants (*P. barbatus*) carry gravid female worms into nest. P, Larval ant. Q, Pupal ant. R, Adult ant intermediate host.

1, Tooth; 2, amphid; 3, external circle of papillae; 4, mouth; 5, anus; 6, sessile preanal papillae; 7, sessile postanal papillae; 8, stalked lateral papillae; 9, lateral ala; 10, esophagus; 11, intestine; 12, rectum; 13, vulva; 14, vagina; 15, branch of amphidelphic uterus; 16, egg capsule; 17, embryonated eggs; 18, cyst wall; 19, fat droplets in trophocytes inside cyst; 20, larva; 21, excretory organ; 22, genital primordium; 23, female reproductive primordium; 24, nerve ring.

a, Adult worms in stomach; b, gravid female passing through intestine to outside; c, gravid female outside lizard; d, gravid females carried into nest of ants by workers; e, embryonated egg; f, eggs hatch in intestine; g, larval nematodes pass through intestinal wall into hemocoel; h, cyst containing larva and fat droplets; i, cyst in pupa; j, cyst in gaster of adult worker; k, infected worker being digested in stomach of lizard; l, cyst being liberated from ant; m, larval nematodes escape from cysts; n, larvae molt and develop to maturity (a).

Figures A–E adapted from Morgan, 1942, Lloydia 5:314; F–M, from Lee, 1957, J. Parasitol. 43:66.

---

69 eggs per capsule which is composed of five layers. Eggs become infective when second stage larvae appear in them.

Workers of agricultural ants (*Pogonomyrmex barbatus* var. *molefaciens* and possibly other species of the genus) find and carry the desiccated females of *P. phrynosoma* into the nests where they are fed to the larvae. Mature ants as such are not susceptible to infection. The eggs hatch in the intestine and the second stage larvae burrow through the wall into the hemocoel where they enter the fat body trophocytes. Cysts produced by the larval ants enclose the trophocyte and the larval nematode. As the parasite consumes the fat, it grows and molts a second time. By the time the ant has completed its metamorphosis through the pupal stage and to the callow, the parasite has consumed all of the fat together with the two molted cuticles contained in the cyst.

Adult ants harboring third stage larvae are the source of infection of the horned lizards. Upon digestion of the infected ants in the stomachs of the myrmecophagous lizards, the larvae are released and grow to adults, presumably after two more molts, making a total of four, one of which occurred in the egg, one in the ants, and two in the lizards. Sexual maturity is reached by the males in 19 days and by the females in 65 days.

## EXERCISE ON LIFE CYCLE

If unavailable naturally in a region, horned lizards may be purchased from biological supply houses in the southwest as a source of eggs of the parasite. Infective eggs are available only in gravid females that have been passed in the feces of lizards and allowed to desiccate. When lizards are kept in shoe boxes, the dried females are easily collected from the feces. The first and second stage larvae may be seen inside the eggs.

Infection of larval ants in the laboratory has not been reported and may not be possible. It should be approached by first establishing a colony of ants consisting of workers and larvae in an artificial nest. Provide dried female worms for the workers to carry into the nest to feed to the larvae. If an artificial nest cannot be established, small natural colonies away from where lizards occur might be used by placing dried female worms for the workers to find and carry into the nests. Adult ants, being resistant, possibly due to age, do not become infected. After a week or more, pupae or callows obtained from the nest may be infected. Other genera of ants should be tried to ascertain whether they are capable of acting as intermediate hosts.

Horned lizards under 50 mm long from snout to vent do not eat large agricultural ants and therefore are not infected with *P. phrynosoma*. They are useful for feeding experiments

Plate 129    *Physaloptera phrynosoma*                                                            479

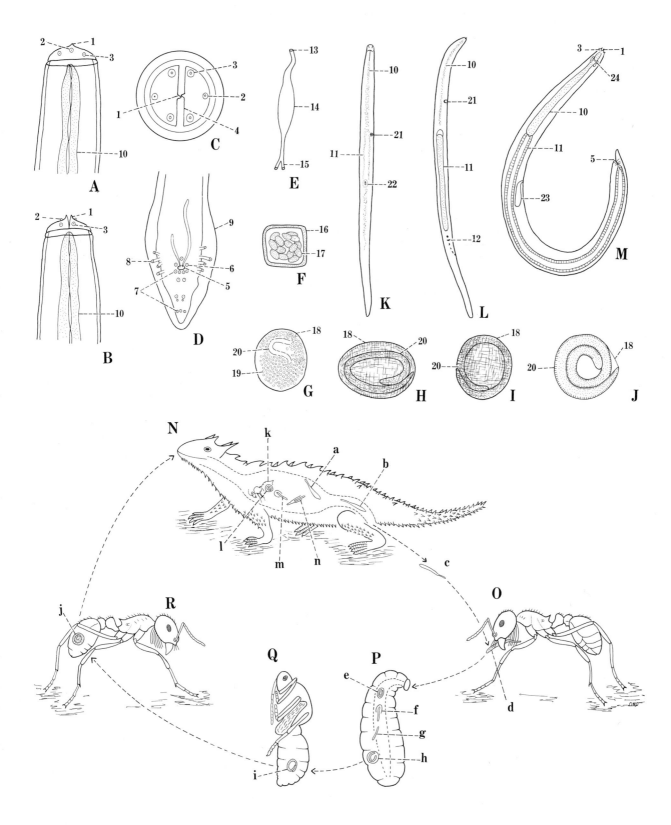

## Explanation of Plate 130

**A,** Ventral view of anterior end of adult. **B,** Ventral view of tail of adult male. **C,** *En face* view of head of adult. **D,** Lateral view of posterior end of adult female. **E,** Vagina and didelphic uterus. **F,** Embryonated egg containing first stage larva. **G,** Egg with first stage larva released by mechanical pressure. **H,** First stage larva 36 hours in intestine of cricket. **I,** First stage larva 5 days in cricket. **J,** First stage larva 11 days in cricket. **K,** First stage larva in process of undergoing first molt to form second stage larva at 10 days of age. **L,** Second stage larva molting to form third stage larva at 13 days of age. **M,** Third stage larva 21 days old removed from cyst in cricket. **N,** Dog definitive host. **O,** Cricket (*Gryllus assimilis*) intermediate host. **P,** Rodent paratenic, or collector, host, with encysted third stage larvae from insect intermediate host, eating cricket. **Q,** Snake second paratenic host with third stage larva from rodents, birds, and lizards.

1, Pseudolabia, or lips, each with three teeth; 2, mouth; 3, labial papilla; 4, amphid; 5, collar; 6, undifferentiated esophagus; 7, muscular portion of esophagus; 8, glandular portion of esophagus; 9, nerve ring; 10, caudal ala; 11, anus; 12, anal plug; 13, preanal papillae; 14, spicule; 15, postanal papillae; 16, lateral papillae; 17, ventral caudal papillae; 18, developing intestine; 19, intestine; 20, excretory pore; 21, tail; 22, first stage larva in egg; 23, first stage larva released from egg; 24, vulva; 25, vagina; 26, branches of didelphic uterus; 27, cuticle being shed in molting.

a, Adult worms in stomach and duodenum; b, egg being carried from host in feces; c, embryonated eggs with first stage larva in host's feces; d, egg swallowed by insect intermediate host; e, larva hatching from egg in intestine; f, hatched first stage larva burrowing into intestinal wall; g, first stage larva undergoing first molt; h, second stage larva; i, second stage larva molting; j, encapsulated third stage larva infective to definitive host; k, third stage larva from insect encysted in tissues of paratenic host from eating insect intermediary; l, third stage larva from rodent, bird, lizard, etc., paratenic hosts (j–l, infective to definitive host); m, invertebrate intermediate host; n, vertebrate paratenic hosts; o, third stage infective larva freed by digestion of intermediate and paratenic hosts; p, presumably a third and q, a fourth molt takes place in the stomach or duodenum.

Figures A–M after J. L. Olsen, 1971, Dissertation, Colorado State Univ.

---

to determine the phases of development in the vertebrate host.

## SELECTED REFERENCES

Lee, S. H. (1955). The mode of egg dispersal in *Physaloptera phrynosoma* Ortlepp (Nematoda: Spiruroidea), a gastric nematode of Texas horned toads, *Phrynosoma cornutum*. J. Parasitol. 41:70–74.

Lee, S. H. (1957). The life cycle of *Skrjabinoptera phrynosoma* (Ortlepp) Schulz, 1927 (Nematoda: Spiruroidea), a gastric nematode of Texas horned toads, *Phrynosoma cornutum*. J. Parasitol. 43:66–75.

### *Physaloptera rara*
### Hall and Wigdor, 1918 (Plate 130)

*Physaloptera rara* is a stomach worm of canines and felines, including domestic dogs and cats in North America.

**Description.** The general appearance is similar to that of *P. phrynosoma*. Males of *P. rara* are 25 to 29 mm long and 0.8 mm in diameter, with the left spicule 740 to 900 μ long and the right 477 to 700 μ. Females are 27 to 41 mm long by 0.9 to 1.1 mm in diameter, with the vulva located slightly anterior to the middle of the body. The uterus is didelphic, i.e., divided into two branches.

**Life Cycle.** The eggs contain first stage larvae when laid and passed in the host's feces. They are immediately infective to insect intermediate hosts such as field crickets, *Gryllus assimilis*, the German cockroach, *Blatella germanica*, and the confused flour beetle, *Tribolium confusum*, as some common ones. Of these hosts, crickets are widespread and probably the natural one. Possibly orthopterans and beetles other than crickets, cockroaches, and flour beetles serve as natural first intermediate hosts.

At room temperature, larvae hatch in the midgut of crickets and migrate into the wall of the intestine 36 hours after the eggs are swallowed. They are 223 to 270 μ long by 14 to 17 μ in diameter. By day 5, they are 249 to 307 μ long. The first stage larvae molt between days 11 and 12. The anal plug present in the first stage larva remains in position. Development and growth take place during days 13 and 14, when the body attains a length of 1.0 to 2.36 mm and a diameter of 0.45 to 1.13 mm. The second molt takes place by day 15.

**Plate 130**     *Physaloptera rara*                                                  481

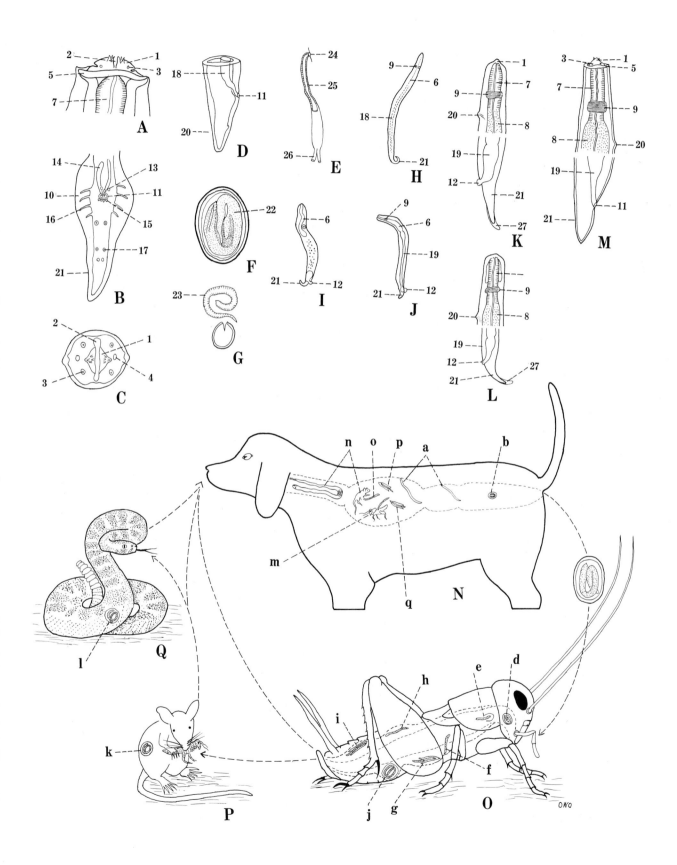

Third stage larvae live up to 152 days in crickets. Other than increase in size, up to 6 mm in length by 0.27 mm in diameter, very few morphological changes occur. Third stage larvae are infective to the definitive hosts (dogs, cats, and other related forms) and to a number of vertebrate paratenic hosts that swallow them.

In cats, third stage larvae from crickets develop to maturity, as indicated by eggs in the feces, in 79 days and from larvae in rattlesnake paratenic hosts in 75 days, according to Olsen. Widmer, on the other hand, recovered adult worms from cats fed larvae from rattlesnakes in 35 days. The short period of 35 days is difficult to explain unless the larvae may have been older and developed more rapidly.

Since the insects serving as intermediate hosts are not an important part of the food chain of the carnivorous hosts of *P. rara*, the interpolation of a small insectivorous animal normally a part of the food chain as a paratenic host is a great ecological asset to the parasite. Frogs, snakes, birds, and mice are common paratenic hosts. Upon ingesting the infected insects, the larvae are liberated and migrate through the intestinal wall into the body cavity. They reencyst in the body organs and muscles, continuing as third stage larvae without further development. In the Great Plains of the United States, rattlesnakes, *Crotalus viridis*, are commonly infected with encysted larvae. These are most likely acquired from other paratenic hosts such as lizards, birds, and mice infected from insects since the snakes themselves are not particularly insectivorous.

Development in the definitive host is the same regardless of whether the infective larvae are acquired by swallowing infected insect intermediaries or by eating parasitized paratenic hosts. While unknown at present, it is presumed on the basis of nematode life cycles that two molts occur in the stomach or duodenum, or both, where the adults reside.

*Physaloptera praeputialis* of dogs and cats produces 4500 eggs per 12-hour period. They remain viable in physiological saline at 4°C for 60 days. The cricket, *Gryllus assimilis*, is probably the natural host. Sexual maturity is attained in cats 131 to 156 days after swallowing third stage larvae.

*Physaloptera hispida* of cotton rats, *Sigmodon hispidis*, develops in German cockroaches, ground beetles, *Harpalus* sp., and earwigs, *Forficula auricularia*.

*Physaloptera turgida* of opossums develops to third stage larvae in German cockroaches, ground beetles, and earwigs. The larvae will not develop to adults in dogs, cats, guinea pigs, and chickens. However, larvae become encapsulated in the stomach wall of rats, indicating their role as a paratenic host.

## SELECTED REFERENCES

Alicata, J. E. (1937). Larval development of the spirurid nematode, *Physaloptera turgida*, in the cockroach, *Blatella germanica*. Papers on Helminth. 30th Jubl. K. I. Skrjabin and 15th Anniv. All-Union Inst. Helminthol. Moscow, p. 11.

Olsen, J. L. (1971). Life history of *Physaloptera rara* Hall and Wigdor, 1918, in definitive, intermediate, and paratenic hosts. Dissertation, Dep. Pathol., Colorado State Univ., 88 pp.

Petri, L. H. (1950). Life cycle of *Physaloptera rara* Hall and Wigdor, 1918 (Nematoda: Spiruroidea) with the cockroach, *Blatella germanica*, serving as the intermediate host. Trans. Kan. Acad. Sci. 53: 331–337.

Schell, S. C. (1952). Studies on the life cycle of *Physaloptera hispida* Schell (Nematoda: Spiruroidea) a parasite of the cotton rat (*Sigmodon hispidus littoralis* Chapman). J. Parasitol. 38:462–472.

Schell, S. C. (1952). Tissue reactions of *Blatella germanica* L. to the developing larva of *Physaloptera hispida* Schell, 1950 (Nematoda: Spriuroidea). Trans. Amer. Microsc. Soc. 71:293–302.

Widmer, E. A. (1970). Development of third-stage *Physaloptera* larvae from *Crotalus viridis* Rafinesque, 1818 in cats with notes on pathology of the larvae in the reptile (Nematoda: Spiruroidea). J. Wildl. Dis. 6:89–93.

Zago, H. F. (1958–62). Contribuicão para conhecimento do ciclo evolutivo da *Physaloptera praeputialis* von Linstow, 1889 (Nematoda: Spiruroidea). Arq. Zool. Estado Sao Paulo 11:59–98.

### Superfamily Filarioidea

These are thread-like nematodes which occur in the blood or lymphatic systems, connective tissues, nerve tissues, skin, lungs, air sacs, eye sockets, and body and nasal cavities of all classes of vertebrates except fish.

In addition to the filiform bodies and the habitats, the group is characterized further by a mouth without lips and the buccal capsule poorly developed or absent. The tail of the male is usually but not always spirally coiled, has ventral papillae, and may be with or without alae. The spicules are distinctly unequal in length. The tail of the female is straight and the vulva is near the anterior end of the body. They are ovoviviparous in most cases with the young (microfilaria) in the body fluids or tissues. Some species are oviparous and the embryonated eggs leave the host through natural openings. Arthropods, particularly insects and acarines, serve as intermediate hosts. They become infected by ingesting the microfilariae or eggs. In some species, the microfilariae are present in the peripheral blood only during certain definite periods of the day or night. They are said to show periodicity.

## FAMILY DIPLOTRIAENIDAE

The Diplotriaenidae are parasites of the air sacs and subcutaneous tissues of reptiles and birds. The family is characterized morphologically by a small mouth with a vertical opening but without lips. There are lateral cephalic cuticularized thickenings (epaulettes), or two projecting denticulate embossments, or two internal tridents. They are oviparous with thick, smooth-shelled eggs containing well-developed first stage larvae when laid.

### Diplotriaena agelaius
(Walton, 1927) (Plate 131)

This filarioid is a parasite of the air sacs of ovenbirds (*Seiurus aurocapillus*). Filarioids in the air sacs are oviparous, laying embryonated eggs that hatch after being passed in the feces of the vertebrate host and swallowed by mandibulate arthropod intermediate hosts. Their life cycle differs from that of ovoviviparous filarioids of the tissues whose hatched larvae are transmitted by bloodsucking arthropods.

**Description.**     Males are 26 to 50 mm long by 0.43 to 0.53 mm wide. The tail is straight and blunt with a subterminal circular anus. Spicules are dissimilar; the left is lance-shaped and 0.9 to 1.3 mm long, the right is twisted twice and 0.53 to 0.71 mm long; each terminates in a membranous hook-like tip. Females are 66 to 108 mm long by 0.71 to 0.81 mm in diameter. The rectum is atrophied and the anus may not be visible. The vulva opens slightly behind the junction of the muscular and glandular portions of the esophagus. There is a long vagina divided into two uteri that are packed with eggs 34 by 50 $\mu$ in size, containing short thick embryos.

**Life Cycle.**     Embryonated eggs containing first stage larvae laid in the air sacs and lungs are conveyed up the trachea into the pharynx, swallowed, and voided with the feces. Further development does not take place until the eggs are swallowed by arthropods. In the midgut of the nymphal grasshopper *Camnula pellucida*, the newly hatched first stage larvae are very active, burrowing through the wall into the hemocoel. The first molt occurs 24 days after the eggs are eaten by the grasshoppers. Molting second stage larvae are present on day 27, and the third stage larvae, both free and encysted, on day 31.

Whether grasshoppers constitute the only intermediate host is unknown. Ants (*Formica*), ground beetles (Carabidae), tenebrionids (Tenebrionidae), camel crickets (Rhaphidophorinae), field crickets (Gryllidae), and land snails (*Zonotoides* and *Discus*) are insusceptible to experimental infection. No information is available on the part of the cycle in the bird host.

The life cycle of *Diplotriaena bargusinica* of the willow thrush (*Hylocichla fuscens*) is similar with one additional bit of information that permits a postulate on the route taken by the larvae from the intestine to the lungs. Eggs hatch when eaten by several species of grasshoppers (*Melanoplus bilituratus*, *M. fasciatus*, and *Camnula pellucida*). Development of larvae takes place in the fat bodies with the first molt on day 9 and the second on days 14 to 16 at 30°C. Most third stage larvae are encapsulated. Subadult worms occur in willow and russet-backed thrushes 55 to 310 days after eating experimentally infected grasshoppers. Late fourth stage larvae found in the heart and aorta of a naturally infected nestling russet-backed thrush add another part of the life cycle.

On the basis of this evidence, migration of larvae in the vertebrate host is probably via the blood or lymph to the lungs, as illustrated in Plate 131. Since two molts normally occur in the vertebrate host when third stage larvae are swallowed, it is assumed a similar pattern follows in these filarioids. This type of life cycle probably represents the pattern of the egg-laying filarioids from the air sacs. Common genera occurring in the air sacs of birds are *Dicheilonema*, *Serratospiculum*, *Hamatospiculum*, *Monopetalonema*, *Pseudoprocta*, *Lissonema*, and *Tetra-*

## Explanation of Plate 131

A, Anterior end of adult female. B, Posterior end of adult male. C, *En face* view of adult female. D, Distal end of spicules from dorsal side. E, Thick-shelled egg containing embryo. F, First stage larva just hatched from egg in intestine of grasshopper. G, *En face* view of first stage larva showing dorsal tooth and rows of spines around anterior end of body. H, Second stage larva in process of molting. I, Third stage larva in characteristic horseshoe shape when fixed. J, Tail of first stage larva, showing serrated tip. K, Ovenbird (*Seiurus aurocapillus*) definitive host. L, Experimental arthropod intermediate host (*Camnula pellucida*).

1, Anterior muscular portion of esophagus; 2, posterior glandular part of esophagus; 3, vulva; 4, vagina; 5, uterus with eggs; 6, short right and long left spicules; 7, anus; 8, anal papillae; 9, outer and inner cephalic papillae; 10, trident; 11, amphid; 12, mouth; 13, thick, smooth shell of egg; 14, embryo; 15, nerve ring; 16, larval esophagus; 17, excretory cell; 18, intestine; 19, rectal plug; 20, tail; 21, rectum; 22, anus.

a, Adults in alveoli of lungs; b, adults in abdominal air sacs; c, eggs moving up trachea; d, eggs entering esophagus to pass through intestine; e, embryonated eggs with first stage larvae passed in feces; f, eggs eaten by grasshoppers; g, eggs hatch in intestine; h, larvae migrate from intestine into hemocoel; i, larvae molt first time to form second stage; j, third stage larva with loose cuticle; k, infection of definitive host occurs when infected intermediate hosts are eaten (probably grasshoppers are not natural hosts of *D. agelaius* since they are not an important food item of ovenbirds); l, larvae freed when arthropod intermediate host is digested; m–p, not demonstrated experimentally but it may be presumed that the hepatic portal vein or the lymphatics would be the likely route by which to reach the lungs and air sacs; m, larvae molt in intestine, burrow through wall into hepatic portal vein; n, larvae pass through liver; o, fourth stage larvae pass through right side of heart and enter pulmonary artery and lungs; p, larvae leave capillaries and enter alveoli or air sacs where they develop to maturity.

Adapted from Anderson, 1956, Can. J. Zool. 34:213; Anderson, 1957, Can. J. Zool. 35:15.

---

*cheilonema. Hastospiculum* of reptiles probably occurs in the air sacs.

## EXERCISE ON LIFE CYCLE

Diplotriaenidae occurs in a number of birds, including such common ones as grackles, redwinged blackbirds, and corvids.

Eggs obtained from female worms should be fed to insects that figure prominently in the food of the vertebrate host. Use an adequate number of each species of insect to provide enough specimens for dissection over a sufficiently long period to make the necessary observations on the development in them. Recall that in *D. agelaius* third stage larvae were found 31 days after the grasshoppers had ingested eggs.

When laboratory-reared insects are not available for feeding experiments, checks must be made on the wild ones to determine the percentage of the sample population that is naturally infected in order to give validity to the experimental results when such animals are fed infective eggs.

## SELECTED REFERENCES

Anderson, R. C. (1957). The life cycles of dipetalonematid nematodes. (Filarioidea: Dipetalonematidae). The problem of their evolution. J. Helminthol. 21:203–224.

Anderson, R. C. (1957). Observations on the life history of *Diplotriaenoides translucidus* Anderson and members of the genus *Diplotriaena*. Can. J. Zool. 35:15–24.

Anderson, R. C. (1959). Preliminary revision of the genus *Diplotriaena* Henry and Ozoux, 1909 (Diplotriaenidae: Diplotriaeninae). Parassitologia 1:195–307.

Anderson, R. C. (1959). Possible steps in the evolution of filarial life cycles. Proc. 6th Int. Congr. Trop. Med. Mal. 2:444–449.

Anderson, R. C. (1962). On the development, morphology, and experimental transmission of *Diplotriaena bargusinica* (Filarioidea: Diplotriaenidae). Can. J. Zool. 40:1175–1186.

### FAMILY ONCHOCERCIDAE

The Onchocercidae are parasites of the tissues and cavities of all classes of vertebrates except the fishes. They are characterized by a head that is smooth and round, an esophagus that may or may not be divided, and spicules that are unequal in length. Sometimes the tail of the male is weakly alate. The vulva is median or anterior but not near the mouth; the eggshells are very thin, and the microfilariae accumulate in the tissues or blood where they are ingested by bloodsucking arthropods.

**Plate 131** *Diplotriaena agelaius* 485

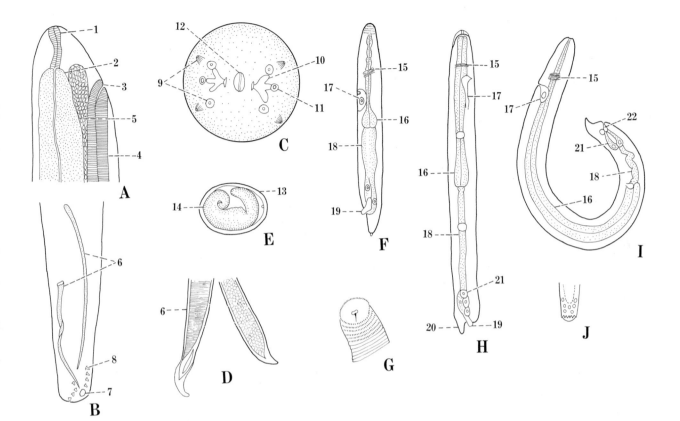

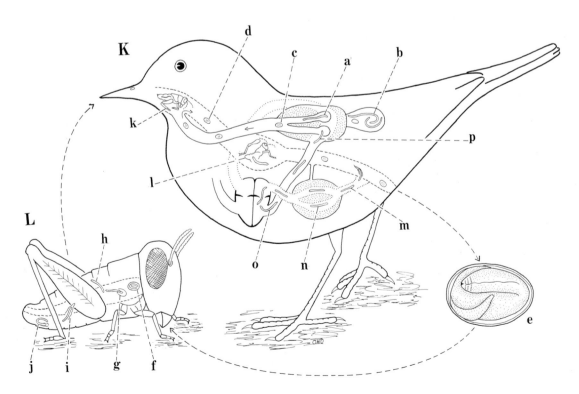

## Explanation of Plate 132

A, Anterior end of mature female. B, Posterior end of mature male. C, Right spicule of mature male. D, Posterior end of mature female. E, Vulvar region with vagina containing oviform embryos. F, Larva in embryonic sheath. G, First stage larva from intestine of mite. H, Second stage larva from hemocoel of mite. I, Third stage larva from hemocoel of mite. J, Anterior end of fourth stage larva from rat, showing development and cast cuticle. K, Posterior end of fourth stage larva, showing development and cast cuticle. L, Cotton rat (*Sigmodon hispidus*). M, Tropical rat mite (*Bdellonyssus bacoti*).

1, Mouth; 2, stoma of adult; 3, esophagus; 4, cephalic (?) glands; 5, intestine; 6, testis; 7, left spicule; 8, right spicule; 9, anus; 10, postanal papillae (three to five in number); 11, tail; 12, ovaries; 13, vulva; 14, vagina vera; 15, vagina uterina; 16, stoma of first stage larva; 17, stoma of second stage larva; 18, loose cuticle; 19, stoma of third stage larva; 20, stoma of third stage larva after it has been shed; 21, detailed view of stoma of fourth stage larva; 22, embryo; 23, embryonic sheath (egg-shell ?).

a, Adult worms in pleural cavity; b, sheathed larvae being born; c, first stage larva entering blood capillaries of lung; d, first stage larva in pulmonary vein; e, first stage larva passing through left side of heart; f, first stage larva leaving heart through aortic arch; g, first stage larva in dorsal aorta; h, first stage larva in peripheral vein of foot; i, first stage larva being ingested by mite; j, first stage larva in gut of mite; k, first stage larvae passing through intestinal wall of mite into hemocoel; l, molting first stage larvae; m, second stage larva; n, molting of second stage larva; o, third stage larva; p, third stage larvae being injected into tissues of cotton rat from where they enter veins; q, third stage larva in vein going to heart; r, s, third stage larva entering right side of heart and passing through it; t, third stage larvae in pulmonary artery on their way to lungs; u, fourth stage larvae probably enter pleural cavity from capillaries of lungs; v, peritoneal cavity.

Figures A–F adapted from Cross and Scott, 1947, Trans. Amer. Microsc. Soc. 66:1; G–J, from Scott, MacDonald, and Terman, 1951, J. Parasitol. 37:425.

---

*Litomosoides carinii*
(Travassos, 1919) (Plate 132)

This is a common parasite of the thoracic cavity of cotton rats (*Sigmodon hispidus*) of the southern United States and countries to the south.

**Description.** Males are 24 to 28 mm long and 0.13 to 0.14 mm in diameter. The tail is corkscrew-like and bears three to five postanal papillae; the right spicule measures 100 to 105 $\mu$ and the left 185 to 295 $\mu$, each with a filamentous tip. Females are 50 to 65 mm long and 0.3 to 0.32 mm wide. The vulva is about 1.25 mm from the mouth, or a distance equal to about twice the length of the esophagus. It opens into a thick muscular vagina. The uterus is filled with embryos about 94 $\mu$ long.

**Life Cycle.** Adult females in the thoracic cavity give birth to sheathed microfilariae that average 94 $\mu$ in length. They are active burrowers, going into muscles, connective tissue, and the lungs. The lungs probably constitute the most important route by which they enter the bloodstream. There is no periodicity of larvae in the blood, as they are present at all times. Microfilariae circulating in the peripheral blood are ingested by rat mites (*Bdellonyssus bacoti*) feeding on infected rats. Only those swallowed by adult mites develop. The microfilariae remain in the stomach of the mites a few hours, but have left it by the end of day 1, when they appear in the hemocoel.

During the first week, the larvae thicken in the middle region of the body and attain the sausage stage typical of late first stage filarioid larvae. By the end of the first week, they begin the initial molt and most of them have completed it by day 13. After the molt, they elongate, attaining a length of 450 $\mu$ as fully developed second stage larvae.

Beginning on day 9, the second molt commences and the majority of them have completed it by day 15. The third stage larvae grow rapidly during the first 2 days after molting, reaching an average length of 853 $\mu$. They are infective to cotton rats and white rats about 15 days after having been ingested by the mites. They migrate to the mouth parts and escape while the mites are feeding. Entry is through the wound in the skin made by the chelae of the mites.

Once inside the skin, the third stage larvae enter the lymphatic spaces and go to the regional lymph nodes within 24 hours. From these, they move to the right auricle of the heart via

**Plate 132**    *Litomosoides carinii*                                                                    487

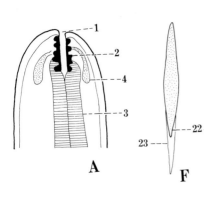

A

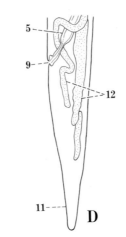

F

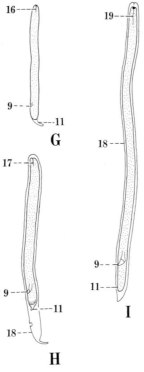

D

G

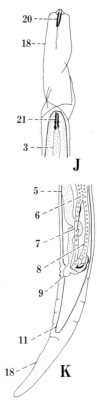

I

J

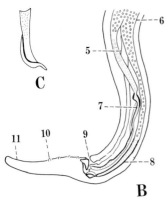

C

B

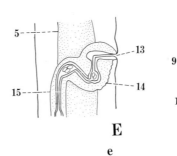

E

H

K

L

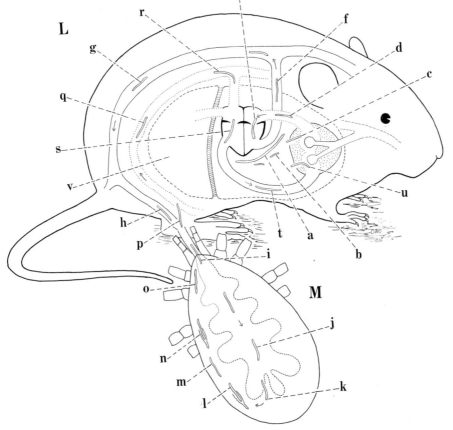

M

the lymphatics and thoracic duct, mix with the blood, and pass through the heart into the lungs, arriving by days 5 to 10 after infection. From the lungs, it is easy for them to burrow into the thoracic cavity.

Two molts occur after the third stage larvae enter the rat host. While these have been described, the place where they occur has not been given. Presumably it is in the pleural cavity. Fourth stage larvae average 1.2 mm long when the cuticle is shed by day 15. They attain an average length of 6.4 mm in the males and 8.7 in the females when fully formed. The fourth and final molt takes place on days 23 to 24 in the males and a day or two later in the females. They grow 1 to 2 mm in length during the molt.

Sexual maturity may be attained as early as 50 days after the mites have infected the rats but the prepatent period is usually 70 to 80 days. The peak of the microfilarial population in the blood occurs in 4 to 5 months and the larvae usually disappear in about 9 months. The majority of the adult worms die by the end of the first year but some may live up to about 3 years.

## EXERCISE ON LIFE CYCLE

Infected cotton rats infested with tropical rat mites captured in their natural habitat or purchased from dealers provide a source of microfilariae for study. The rats should be kept in cages containing grass and placed over sawdust. This arrangement provides places for both the rats and mites to hide and for the latter to breed.

A colony of uninfected mites should be maintained on uninfected white rats for experimental purposes. Precautions should be taken to prevent infected mites from wandering and exposing all of the experimental rats. Infected mites can be detected by examining them with the aid of a dissecting microscope and transmitted light.

Adult mites allowed to feed on infected rats provide a source of material for studying the development of the larvae through the third stage. By dissecting them in half strength Tyrode's solution at close intervals up to 15 days after feeding, development through the third stage can be followed in detail.

The migration and development of the fourth stage larvae and development of the adults can be followed in rats. Determine whether the third stage larvae migrate to the pleural cavity via the tissues or the blood. Examination of the blood of exsanguinated rats and the lungs at appropriate times after infected mites have fed on them should provide the answer to the question.

## SELECTED REFERENCES

Bertram, D. S. (1947). The period required by *Litomosoides carinii* to reach the infective stage in *Liponyssus bacoti* and the duration of the mites' infectivity. Ann. Trop. Med. Parasitol. 41:253–261.

Cross, J. B., and J. A. Scott. (1947). The developmental anatomy of the fourth stage larvae and adults of *Litomosoides carinii,* a filarial worm of the cotton rat. Trans. Amer. Microsc. Soc. 66:1–21.

Hughes, T. E. (1952). The morphology of the gut in *Bdellonyssus bacoti* (Hirst, 1913) Fonseca, 1941. Ann. Trop. Med. Parasitol. 46:54–60.

Kershaw, W. E. (1948). Observations on *Litomosoides carinii* (Travassos, 1919) Chandler, 1931. I. The development of the first-stage larva with a statistical analysis by R. L. Plackett. Ann. Trop. Med. Parasitol. 42:377–399.

Kershaw, W. E. (1949). Observations on *Litomosoides carinii* (Travassos, 1919) Chandler, 1931. II. The migration of the first stage larva. Ann. Trop. Med. Parasitol. 43:96–115.

Kershaw, W. E. (1949). Observations on *Litomosoides carinii* (Travassos, 1919) Chandler, 1931. III. The first-stage larva in the peripheral circulation. Ann. Trop. Med. Parasitol. 43:238–260.

Scott, J. A., and J. B. Cross. (1946). A laboratory infection of the rat with filarial worms. Amer. J. Trop. Med. 26:849–855.

Scott, J. A., E. M. MacDonald, and B. Terman. (1951). A description of the stages in the life cycle of the filarial worm *Litomosoides carinii.* J. Parasitol. 37:425–432.

Wenk, P. (1967). Der Invasionweg der metazyklischen Larven der *Litomosoides carinii* Chandler, 1931 (Filariidae). Z. Parasitenk. 28:240–263.

Williams, R. W. (1948). Studies on the life cycle of *Litomosoides carinii,* filariid parasite of the cotton rat, *Sigmodon hispidus litoralis.* J. Parasitol. 34:24–43.

Zein-Eldin, E. A. (1965). Experimental infection of gerbils with *Litomosoides carinii* via intravenous injection. Tex. Rep. Biol. Med. 23:530–536.

### Dirofilaria immitis
(Leidy, 1856) (Plate 133)

The adults of this filarioid are parasites mainly of the chambers of the right side of the heart and pulmonary artery of dogs, foxes, wolves, coyotes, and cats. The long, sheathless microfilariae are in the bloodstream. The worms are worldwide in distribution, being most common in mild and warm climates. In the United States, they are prevalent along the Atlantic and Gulf Coasts where the mosquito vectors abound. However, infections occur in other parts of the country.

**Description.**    Adults are long, white, thread-like worms. Males measure 12 to 16 cm long with the tail spirally coiled. It bears narrow alae and three pairs of large caudal papillae, one of which is postanal and three pairs of small ones near the tip of the tail. The left spicule is 324 to 375 $\mu$ long and the right 190 to 229 $\mu$. Females are ovoviviparous, 25 to 30 cm long, and with the vulva opening just behind the posterior end of the esophagus. The sheathless microfilariae are 218 to 329 $\mu$ long and have a long pointed tail.

**Life Cycle.**    Individual eggs developing in the uterus are enclosed in a thin vitelline membrane. As the embryo elongates, the surrounding membrane stretches to conform as an enclosing sheath. At birth, the membrane is lost and the embryo appears in the blood as a sheathless microfilaria, a characteristic of this species.

Microfilariae are deposited in the blood of the chambers of the right side of the heart and pulmonary artery. They are carried through the lungs, into the left chambers of the heart, and into the systemic circulation. There is marked nocturnal periodicity of microfilariae in the peripheral blood at which time they are commonly 5 to 10 times, occasionally even up to 50 times, more abundant in the early evening and night than during the daylight hours.

Mosquitoes, particularly *Culex pipiens*, *Anopheles quadrimaculatus*, and *A. freeborni* are effective vectors. *Aedes vexans* and others are less efficient transmitters.

When microfilariae are ingested along with the blood by the mosquito vectors, they migrate from the intestine within 24 to 36 hours into the Malpighian tubules, where further development occurs. In some species of *Culex* and *Anopheles*, the microfilariae are so severely injured in passing through the narrow buccopharyngeal armature that they die.

In the course of development, the microfilariae enter the distal cells of the Malpighian tubules, shorten and thicken to become presausage forms in 3 to 4 days and into the sausage, or first stage larvae with a complete intestine in 4 to 5 days. By days 6 to 7, they return to the lumen of the Malpighian tubules and molt to the second stage by day 9. After a period of growth, the second stage larvae break out of the shell of the Malpighian tubules into the abdominal hemocoel, where the second molt occurs. Third stage larvae about 900 $\mu$ long appear in 10 to 20 days after entering the mosquito. The cellular lining of the Malpighian tubules is completely destroyed by the developing larvae.

The third stage larvae migrate forward through the thorax along the ventral side into the head, reaching the labium in about 10 days after infection in *Culex pipiens*, 13 days in *C. quinquefasciatus*, and 14 days in *Anopheles quadrimaculatus*, *A. freeborni*, and *A. albopictus*, and 16 days in *Aedes aegypti*. There is marked retardation in development below 21°C and sustained temperatures below 15°C are lethal to the microfilariae in mosquitoes. Large numbers of microfilariae in the Malpighian tubules kill the mosquitoes. Successful natural infections are usually less than 10 microfilariae and frequently only two to five.

Infection of dogs occurs while mosquitoes with third stage larvae in the labium are feeding. The larvae escape through the labellum onto the skin. They enter it through the puncture made by the maxillae, hair follicles, or by burrowing directly through intact areas. Larvae are in the subcutaneous tissues and muscles for around 80 days, where the third molt takes place 9 to 12 days after entry. During this period in the tissues, the fourth stage larvae attain lengths up to 25 mm. They begin entering the right side of the heart shortly after the fourth molt 60 to 70 days after entering the dog. Development to maturity with males 14 to 19 cm long and females 23 to 31 cm takes 174 to 223 days, at which time microfilariae appear in the blood. The reproductive period exceeds 2 years and may extend to 5.

Microfilariae enter the amniotic fluid of developing pups. Transplacental infection occurs as a result of the boring ability of the microfilariae.

A few adult worms, up to 25, in dogs do not produce evidence of disease. Up to 60 worms in the pulmonary artery and right side of the heart cause circulatory difficulty. In cases with large numbers of worms, 100 or over, there is blockage of the pulmonary artery and right side of the

## Explanation of Plate 133

A, Anterior end of adult of female worm. B, *En face* view of head of adult worm. C, Caudal end of adult male worm. D, Ventral view of caudal end of male. E, Sinistral view of caudal end of adult female. F, Microfilaria from blood (other stages of larvae similar to those of species infecting man). G, Dog definitive host. H, Mosquito (*Aedes* spp.) intermediate host in feeding position on dog.

1, Inner circle of cephalic papillae; 2, outer circle of cephalic papillae; 3, amphid; 4, mouth without lips; 5, muscular portion of esophagus; 6, glandular portion of esophagus; 7, intestine; 8, anus; 9, nerve ring; 10, uterus containing unsheathed microfilariae; 11, preanal papillae; 12, adanal papillae; 13, postanal papillae; 14, long spicule; 15, short spicule; 16, nerve ring of microfilaria; 17, excretory cell; 18, $G_1$ cell; 19, anal space; 20, tail cells.

a, Adult worms in heart and pulmonary artery; b–d, microfilariae born in and circulating through bloodstream; e, microfilariae in peripheral blood and available to feeding mosquitoes; f, section of skin of dog; g, microfilaria being sucked up from peripheral blood vessel; h, microfilaria in stomach of mosquito; i, microfilaria entering Malpighian tubules; j, microfilaria changes to sausage stage and prepares to undergo first molt; k, second stage larva; l, sausage stage elongating and preparing for second molt; m, third stage filariform larva in Malpighian tubule; n, infective third stage larva having escaped from Malpighian tubule; o, larva migrating through thorax; p, q, larva entering and migrating down labium; r, larva escaping from labellum onto the skin; s, larva entering skin; t, larvae in subcutaneous tissue (also muscle and adipose tissues); u, larva entering peripheral vessels from tissues 85 to 129 days after infection; v, larva entering general circulation; w, larva entering heart.

Figures A, F adapted from various sources; B, from Chitwood and Chitwood, Rev. 1950, Introduction to Nematology, Sec. I, Anatomy. Monumental Printing Co., Baltimore, p. 64, Fig. A; C, D, from Ortlepp, 1924, J. Helminthol. 2:15; E, from Petrow, 1931, in Skrjabin, Shikhobalowa, and Sobolew, 1949, Opredeliteli Parasiticheskikh Nematod, Vol. 2, Plate 20.

---

heart, accompanied by interference with the heart valves. The right side of the heart becomes dilated and enlarged. Blood backs up in the liver and other parts of the body, causing general congestion and degeneration.

A second filarioid worm, *Dipetalonema reconditum*, occurs in the subcutaneous tissue of dogs in North America. The adults are about one-tenth the size of the heartworms. Microfilariae fixed with the blood in 2 percent formalin have a sharply bent tail shaped like a buttonhook, as opposed to the straight one of the heart worms. Both species occur simultaneously in dogs. The microfilariae of *Dipetalonema* develop in dog and cat fleas (*Ctenophalides canis* and *C. felis*), whereas those of *D. immitis* develop in mosquitoes. *Dipetalonema reconditum* has a midday peak and a nocturnal one about 8 to 12 hours later.

Birds are common hosts of onchocercid filarioids. The American black-billed magpie (*Pica pica hudsonia*) harbors at least four species. Midges of the genus *Culicoides* are the intermediate hosts of at least three of them. *Splendidofilaria picacardina* Hibler from the myocardium behind the aortic and pulmonary semilunar valves develops in the abdominal fat bodies of *Culicoides crepuscularis*; *Chandlerella striatospicula* Hibler from the connective tissue surrounding the splenic artery develops in the thoracic muscles of *Culicoides haematopotus*; and *Eufilaria longicaudata* Hibler from the connective tissue surrounding the esophagus develops in the thoracic muscles of *Culicoides crepuscularis* and *C. haematopotus*.

The life cycles of other onchocerid filarioids, including those in humans, where known, are similar basically to *D. immitis*. In addition to mosquitoes and fleas, other intermediate hosts for them include tabanids, simuliids, *Culicoides*, and *Stomoxys*.

An exhaustive list of filarial worms known to develop in arthropods together with both vertebrate and invertebrate hosts and references is given by Hawking and Worms.

## EXERCISE ON LIFE CYCLE

Numerous species of wild birds and mammals, also some reptiles and amphibians, are infected with filarioids, as can be ascertained first hand by looking for the microfilariae in fresh and stained blood smears.

By means of a series of blood smears taken at all hours of the day and night, determine whether there is a time when the number of microfilariae in the blood is greatest. The periodicity, if it occurs, should be established as a

**Plate 133** *Dirofilaria immitis* 491

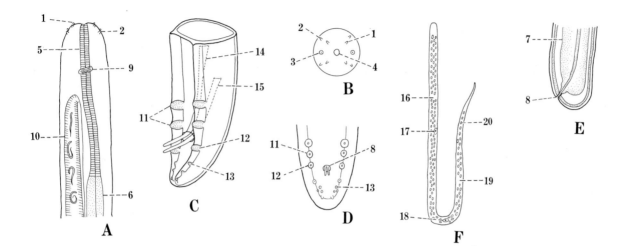

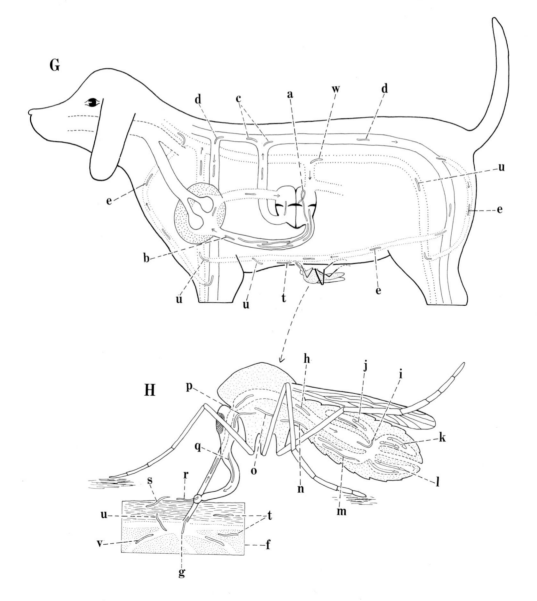

preliminary step to feeding experiments with the intermediate hosts.

Permit insects to feed when the microfilariae are at their maximal number in the blood. Dissection of them must be made at frequent intervals after feeding to determine the fate of the microfilariae ingested with the blood. Microfilariae might be found in the intestine of all dipterons that feed, but the clue to the relationship of any particular insect to the biology of the worm will be in the larvae that undergo changes leading to the sausage and filariform stages.

Having found an infected vertebrate as a source of microfilariae, attempts should be made to find the arthropod host. The choice of insect should be made on the basis of the species that feed most commonly on the vertebrate host at the time when the microfilariae are most abundant in the blood, but there is no way of determining beforehand which of these might be the intermediate host. Mosquitoes, black flies, *Culicoides*, tabanids, or stable flies and horn flies should be tested as intermediate hosts.

## SELECTED REFERENCES

Bradley, R. E., ed. (1972). Canine Heartworm Disease: The Current Knowledge. Proc. 2nd Univ. Florida Symposium on Canine Heartworm Disease. Univ. Florida Press, 148 pp.

Coluzzi, M., and R. Trabucchi. (1968). Importanza dell'armatura buccofaringea in *Anopheles* e *Culex* in realzione alle infezioni con *Dirofilaria*. Parassitologia 10:47–59.

Gubler, D. J. (1966). A comparative study on the distribution, incidence, and periodicity of the canine filarial worms *Dirofilaria immitis* Leidy and *Dipetalonema reconditum* Grassi in Hawaii. J. Med. Entomol. 3:159–167.

Hawking, F., and M. J. Worms. (1961). Transmission of filarioid nematodes. Annu. Rev. Entomol. 6: 413–432.

Hibler, C. P. (1963). Onchocercidae (Nematoda: Filarioidea) of the American magpie, *Pica pica hudsonia* (Sabine), in northern Colorado. Dissertation, Colorado State Univ., 189 pp.

Hibler, C. P. (1964). New species of Onchocercidae (Nematoda: Filarioidea) from *Pica pica hudsonia* (Sabine, 1823). J. Parasitol. 50:667–674.

Hendrikson, M. D., and R. A. Anderson. (1969). *Dirofilaria immitis* microfilariae in amniotic fluid. J. Parasitol. 55:195.

Jackson, R. F., G. F. Otto, P. M. Bauman, F. Peacock, W. L. Hinrichs, and J. H. Maltby. (1966). Distribution of heartworms in the right side of the heart and adjacent vessels of the dog. J. Amer. Vet. Med. Ass. 149:515–518.

Kartman, L. (1953). Ingestion by mosquitoes of saline and sugar suspensions of *Dirofilaria immitis* microfilariae. J. Parasitol. 39:573.

Kartman, L. (1953). On the growth of *Dirofilaria immitis* in the mosquito. Amer. J. Trop. Med. Hyg. 2:1062–1069.

Kartman, L. (1956). The vector of canine filariasis. J. Parasitol. 42(4/2):19.

Kume, S., and S. Itagaki. (1955). On the life-cycle of *Dirofilaria immitis* in the dog as the final host. Brit. Vet. J. 3:16–24.

Marquardt, W. C., and W. E. Fabian. (1966). The distribution in Illinois of filariids of dogs. J. Parasitol. 52:318–322.

Newton, W. L., and W. H. Wright. (1956). The occurrence of a dog filariid other than *Dirofilaria immitis* in the United States. J. Parasitol. 42:246–258.

Newton, W. L., and W. H. Wright. (1957). A reevaluation of the canine filariasis problem in the United States. Vet. Med. 52:75–78.

Orihel, T. C. (1961). Morphology of the larval stages of *Dirofilaria immitis* in the dog. J. Parasitol. 47:251–262.

Otto, G. F. (1969). Geographical distribution, vectors, and life cycle of *Dirofilaria immitis*. J. Amer. Vet. Med. Ass. 154:370–373.

Otto, G. F., and P. M. Bauman. (1959). Canine filariasis. Vet. Med. 54:87–96.

Stueben, E. B. (1954). Larval development of *Dirofilaria immitis* (Leidy) in fleas. J. Parasitol. 40: 580–590.

Steuben, E. B. (1954). Incidence of infection of dogs and fleas with *Dirofilaria immitis* in Florida. J. Amer. Vet. Med. Ass. 125:57–60.

Taylor, A. E. R. (1960). The development of *Dirofilaria immitis* in the mosquito *Aedes aegypti*. J. Helminthol. 34:27–38.

Wallenstein, W. L., and B. J. Tibolla. (1960). Survey of canine filariasis in a Maryland area—incidence of *Dirofilaria immitis* and *Dipetalonema*. J. Amer. Vet. Med. Ass. 137:712–716.

### FAMILY DIPETALONEMATIDAE

Members of this family are ovoviparous parasites of the body tissues, cavities, and blood of vertebrates. The microfilariae are long and slender, and devoid of spines on the anterior end.

### *Foleyella* Seurat, 1917

Species of this genus are parasites of the subcutaneous connective and muscular tissues of saurians and amphibians. They produce sheathed microfilariae which circulate in the blood. Mosquitoes are the intermediate hosts.

Adults have lateral alae extending almost the full length of the body. The spicules are unequal in length and the vulva is near the posterior end of the esophagus.

### Foleyella brachyoptera
Wehr and Causey, 1939 (Plate 134)

These filarioids are parasites of the body cavity, chiefly on the mesenteries, of the southern leopard frog (*Rana sphenocephala*) in Florida.

**Description.** The oral opening has a tooth-like projection on each side and the esophagus consists of a short narrow anterior muscular portion and long wider posterior glandular portion. Males are 15 to 18 mm long by 0.3 mm wide. Lateral alae extend from the anterior extremity to the level of the anus. There are six and possibly seven pairs of sessile postanal papillae, diminishing in size posteriorly. The long spicule is 420 to 460 $\mu$ in length, alate, and with a pointed tip; the short one is 168 to 180 $\mu$ long, alate, and with the tip bluntly rounded. Females are 60 to 70 mm long by 0.7 mm wide. The vulva is near the posterior end of esophagus. The rectum is relatively short. Microfilariae are sheathed and about 125 $\mu$ long.

The locations of anatomic structures used in the classification of microfilariae are given in percentage of body length for *F. brachyoptera* as follows: 1) nerve ring, 30; 2) excretory cell, 33; 3) beginning of inner body, 64; 4) $G_1$ cell, 74; and 5) anal pore, 87. The other G-cells ($G_2$, $_3$, $_4$) and anal pore are difficult to make out.

**Life Cycle.** Adult worms occur on the mesenteries of the alimentary canal. Upon fertilization, the eggs appear as uncleaved spherical bodies enclosed in a thick shell or membrane. As development of the larvae progresses, they are first coiled and later appear to be in three or four folds within the eggs in the vaginal portion of the uterus. As the larvae approach the size of microfilariae, the membrane stretches, becoming progressively thinner, eventually forming a thin, single-layered hyaline sheath. In fully developed microfilariae, the sheath conforms to the body shape, but extends somewhat beyond each end. Upon birth, the microfilariae penetrate tissue, finding their way into the blood vessels in which they circulate.

In the related genus *Ochoterenella*, possibly

*O. digiticaudus*, of the marine toad, *Bufo marinus*, in Colombia, there is distinct periodicity in the number of microfilariae in the blood. The number per ml of heart blood is two-thirds greater during daylight hours between 10:00 A.M. and 4:00 P.M. than the night hours from 10:00 P.M. to 1:00 A.M. During the night, the majority of the larvae are in the lungs, liver and spleen, with fewer in the kidneys, ovary, testes, gut wall, brain, striated muscles, fat, and subcutaneous tissue. Periodicity may occur in other species although it has not been reported.

Upon being swallowed by the mosquitoes, the microfilariae migrate through the gut wall into the hemocoel by the end of day 1. They are active, coiling and uncoiling. During this time, the only change is the loss of the sheath. By day 3, internal changes take place and the body becomes shortened to form the presausage stage, which is 107 by 7 $\mu$ in size. They are inactive at this time. By day 5, they have migrated into musculature of the abdominal wall and shortened still more to form the sausage stage, which is 89 $\mu$ by 9 to 14 $\mu$ in size.

During the next 3 days, internal development of the sausage stage continues, with the formation of the hypodermis and basic parts of the alimentary canal. This completes the development of the first stage larva, which measures 158 by 23 $\mu$. The first molt occurs on about day 9 in mosquitoes kept at 23°C. Second stage larvae are 456 by 26 $\mu$ in size. The esophagus consists of large cuboidal cells and is about one-half the length of the body. An anal plug projects from the anus. The second molt occurs between days 13 and 16, at which time the larvae begin the third stage. By day 18, third stage larvae have migrated from the musculature of the abdominal wall to that of the thorax. They are 88 $\mu$ long by 13 to 17 $\mu$ wide, and active. Larvae live up to 43 days after ingestion by the mosquitoes.

In *Foleyella furcata* of the chamaeleon, *Chamaeleo verrucosus*, the $R_2$ and $R_4$ cells do not divide but surround the anal plug to form the rectum. The third stage is characterized by the empty, flattened intestine and loss of the anal plug.

Infection of the frogs is presumed to take place while infected mosquitoes are feeding on them. This point, however, has not been demonstrated experimentally. Mosquitoes show reluctance to bite cold frogs. After the frogs are

## Explanation of Plate 134

A, Anterior end of adult female of *Foleyella brachyoptera*. B, Right lateral view of posterior end of adult female. C, Ventral view of adult male. D, Embryo coiled in egg membrane. E, Microfilaria partially coiled and stretching egg membrane. F, Microfilaria extended with loose-fitting membrane. G, Fully developed sheathed microfilaria in bloodstream which is stage ingested by mosquitoes. H, First stage larvae in sausage stage on day 6 after ingestion by mosquito host. I, Sausage stage on day 8 after infection of mosquito. J, K, Second stage larvae on days 11 and 13, respectively, after infection of mosquito. L, Third stage infective larva on day 18 after infection of mosquito. M, infective larva emerging from labellum of mosquito. N, Frog (*Rana sphenocephala*) definitive host. O, Mosquito (*Aedes, Culex*) intermediate host.

1, Tooth-like cuticular projections from mouth; 2, anterior muscular portion of esophagus; 3, posterior glandular position of esophagus; 4, esophagointestinal valve; 5, intestine; 6, rectum; 7, anus; 8, vulva; 9, vagina; 10, uterus; 11, right spicule; 12, caudal papillae; 13, caudal alae; 14, anal plug; 15, nerve ring; 16, egg membrane; 17, unhatched coiled microfilaria; 18, unhatched extended microfilaria; 19, sheath (elongated egg membrane) of microfilaria circulating in blood; 20, microfilaria with rows of body cells and anlage of certain organs; 21, cells of nuclear column; 22, space of nerve ring; 23, excretory pore; 24, excretory cell; 25, inner body; 26,

genital primordium; 27–30, G-cells ($G_1$, $G_2$, $G_3$, $G_4$); 31, anal cell; 32, labellum; 33, third stage larva (L) escaping from mosquito proboscis.

a, Adult worm on intestinal mesentery; b, mesentery; c, microfilariae in capillaries and subcutaneous connective tissue; d, feeding mosquito (*Aedes, Culex*); e, epidermis of frog skin; f, subcutaneous connective tissue; g, microfilaria in capillary; h, microfilaria being sucked up by feeding mosquito; i, microfilaria in foregut; j, microfilaria passing through gut wall of ventriculus into hemocoel; k, egg membrane of microfilaria; l, actively coiling microfilaria; m, microfilaria shortening in presausage stage, showing anus and excretory pore; n, musculature of abdominal body wall; o, presausage stage entering musculature of body wall; p, sausage stage (end of first stage); q, newly molted second stage larva; r, cuticle of first stage larva (sausage stage); s, full-grown second stage larva; t, recently molted third stage larva; u, cuticle of second stage larva; v, infective third stage larvae in musculature of thorax; w, infective larva entering hollow proboscis sheath; x, larva passing down proboscis sheath; y, larva through its activities ruptures thin membrane of labellum and escapes onto skin of frog during feeding activities of infected mosquitoes; z, third stage larva on skin of frog, whose fate from this point is unknown.

Figures A–C adapted from Wehr and Causey, 1939, Amer. J. Hyg. 39(D):65; D–M, from Kotcher, 1941, Amer. J. Hyg. 34:36.

---

warmed to about 38°C, however, hungry mosquitoes feed readily. During the time of feeding, larvae escape from the labellum onto the skin of the frogs, but they are unable to penetrate. Whether infection is through the skin by penetrating larvae is unknown. Infection may take place through the intestine from parasitized mosquitoes that are swallowed by frogs.

Other species of *Foleyella* occurring in American frogs include *F. ranae* Walton, 1929 with three pairs of preanal and four pairs of postanal papillae from *Rana catesbeiana*; *F. americana* Walton, 1929 with four pairs of preanal and three pairs of postanal papillae from *Rana pipiens*; and *F. dolichoptera* Wehr and Causey, 1939 with four pairs of postanal papillae from *Rana sphenocephala*.

Development of *F. dolichoptera*, *F. americana*, *F. ranae* from frogs in the United States, *F. duboisi* from frogs in Palestine, *F. philistinae* from lizards (*Agama*) in Lebanon, and *F. furcata* and *F. candezei* from chamaeleons in Madagascar

in mosquitoes is basically similar to that of *F. brachyoptera*. *Icosiella neglecta* of European frogs is transmitted by the ceratopogonid flies *Forcipomyia velox* and *Sycorax silacea*. The intermediate host of *Ochoterenella* is unknown.

In addition to serving as intermediaries for *Foleyella* in amphibians and reptiles, mosquitoes also function in a similar capacity for *Dipetalonema arbuta* in porcupines. Other arthropod intermediate hosts for dipetalonematids include *Culicoides* for *Dipetalonema streptocerca* in humans; fleas for *D. reconditum* in dogs in the United States and *D. manson-bahri* of spring hares in Kenya; hippoboscid flies for *D. dracunculoides* of dogs in Kenya; and ticks for *D. blanci* of meriones in Iran.

## EXERCISE ON LIFE CYCLE

In addition to observing the developmental stages in mosquitoes, efforts should be made to learn how infection of frogs takes place. In

**Plate 134**     *Foleyella brachyoptera*                                                           495

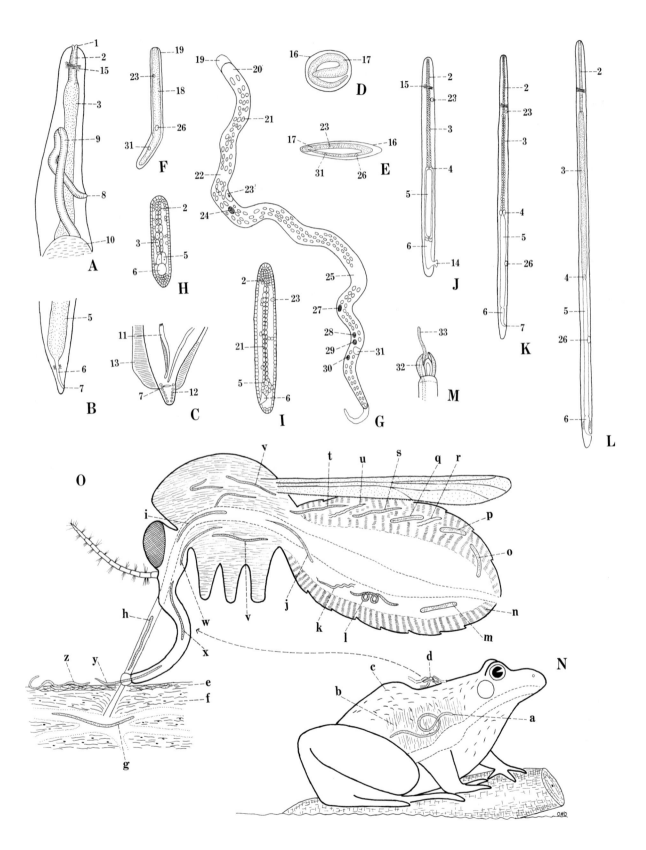

laboratory studies, mosquitoes are reluctant to bite cold frogs. After frogs are warmed to about 38°C, hungry mosquitoes will feed on them. Since the temperature of frogs under natural conditions remains far below that at which mosquitoes will feed in the laboratory, there must be circumstances under which they do feed. These conditions should be investigated.

While larvae, presumably third stage ones, emerge from the labellum of feeding mosquitoes onto the skin of frogs, they fail to penetrate. Inasmuch as the usual route of infection in the Dipetalonematidae is the skin, it would be expected to be the case for species of *Foleyella*. Experiments should be conducted to clarify the problems of mosquitoes feeding on the frogs and how infection takes place. The possibility of infection of the frogs through ingestion of parasitized mosquitoes should not be overlooked.

## SELECTED REFERENCES

Anderson, R. C. (1957). The life cycles of dipetalonematid nematodes (Filarioidea, Dipetalonematoidea): The problem of their evolution. J. Helminthol. 31:203–224.

Bain, O. (1969). Etude morphologique du développement larvaire de *Foleyella furcata* chez *Anopheles stephensi*. Ann. Parasitol. 44:165–172.

Bain, O. (1970). Etude morphologique du développement larvaire de *Foleyella candezei* chez *Anopheles stephensi* et *Aedes aegypti*. Ann. Parasitol. 45:21–30.

Baltazard, M., A. G. Chabaud, and A. Minon. (1952). Cycle évolutif d'une filaire parasite de mérion. C. R. Acad. Sci. 234:2115–2117.

Brygoo, E. (1960). Evolution du *Foleyella furcata* (von Linstow, 1899) chez *Culex fatigans* Wiedemann, 1828. Arch. Inst. Pasteur Madagascar 28:129–138.

Causey, O. R. (1939). Development of the larval stages of *Foleyella brachyoptera* in mosquitoes. Amer. J. Hyg. 30(D):69–71.

Causey, O. R. (1939). The development of frog filaria larvae, *Foleyella ranae*, in *Aëdes* and *Culex* mosquitoes. Amer. J. Hyg. 29(D):131–132.

Chardome, M., and E. Peel. (1949). La répartition des filaraies dans la region de Coquilhatville et la transmission de *Dipetalonema streptocerca* par *Culicoides grahami*. Ann. Soc. Belge Med. Trop. 29:99–119.

Desportes, C. (1942). *Forcipomyia velox* Winn. et *Sycorax silacea* Curtis, vecteurs d'*Isociella neglecta* (Diesing, 1851), filaire commune de la grenouille verte. Ann. Parasitol. 18:46–66.

Hawking, F., and M. J. Worms. (1961). Transmission of filarioid nematodes. Ann. Rev. Entomol. 6:413–432.

Highby, P. R. (1943). *Dipetalonema arbuta* n. sp. (Nematoda) from the porcupine, *Erethizon dorsatum* (L.). J. Parasitol. 29:239–252.

Katcher, E. (1941). Studies on the development of frog filariae. Amer. J. Hyg. 34:36–64.

Marinkelle, J. C. (1966). Observation on the periodicity of microfilariae. Proc. Int. Congr. Parasitol. 1st Rome 2:651–652.

Marinkelle, J. C. (1970). Observaciones sobre la periodicidad de las microfilarias de *Ochoterenella* en *Bufo marinus* de Colombia. Revista Biol. Trop. 16:145–152.

Nelson, G. S. (1961). On *Dipetalonema mansonbahri* n. sp. from the spring-hare *Pedetes surdaster larvalis*, with a note on its development in fleas. J. Helminthol. 25:143–160.

Newton, W. L., and W. H. Wright. (1957). A reevaluation of the canine filariasis problem in the United States. Vet. Med. 52:75–78.

Schacker, J. F., and G. M. Khalil. (1968). Development of *Foleyella philistinae* Schacher and Khalil, 1967 (Nematoda: Filarioidea) in *Culex pipiens molestus* with notes on the pathology in the arthropod. J. Parasitol. 54:869–878.

Walton, A. C. (1929). Studies on some nematodes of North American frogs. I. J. Parasitol. 15:16–20.

Wehr, E. E., and O. R. Causey. (1939). Two new nematodes (Filarioidea: Dipetalonematidae) from *Rana sphenocephala*. Amer. J. Hyg. 30(D):65–88.

Witenberg, G. G., and C. Gerichter. (1944). The morphology and life history of *Foleyella duboisi* with remarks on allied filariids of amphibia. J. Parasitol. 30:245–256.

Worms, M. J., R. J. Terry, and A. Terry. (1961). *Dipetalonema witei*, filarial parasite of the jird, *Meriones libycus*. I. Maintenance in the laboratory. J. Parasitol. 47:963–970.

## CLASS ADENOPHOREA

Phasmids are absent but the caudal and hypodermal glands usually are present. The terminal excretory duct generally is not lined with cuticle. The cephalic sensory organs are setose (in free-living forms) or papilloid (in parasitic forms). Amphids are circular, spiral, shepherd's crook, pocket-like shapes (free-living forms), or sometimes pore-like (parasitic forms).

## ORDER ENOPLIDA

Amphids are pocket- to pore-like, or tuboid. The esophagus may be multinucleate.

## Suborder Enoplina

The amphids usually are pocket-like. The esophagus is grossly cylindrical and the glands are uninucleate, followed by a functional intestine.

## Superfamily Trichuroidea

In this superfamily, the anterior end of the worm is longer and more slender than the posterior end. The esophagus consists of two parts. The anterior portion is the short, muscular region with a triradiate lumen. The second part, the stichosome, consists of a long chain of large glandular cells known individually as stichocytes. A pore connects each cell with the capillary esophageal tubule embedded in an infolding of the stichocytes. They contain mitochondria and abundant Golgi apparatuses, indicating secretory function. The contractile elements of the stichocytes help move the food through the capillary tubule of the esophagus. Males have one or no spicules. The females have a single ovary and the unembryonated eggs have a clear mucoid plug in each end. They are parasites of the alimentary canal of vertebrates.

### FAMILY TRICHURIDAE

The members of this family are characterized by females that are oviparous and by males that have but one spicule. They are parasites of the intestine primarily but occur in the liver and urinary bladder.

The life cycles of the species of this family present a series of patterns. There is the direct cycle in *Trichuris ovis* and *Capillaria columbae* where the eggs hatch in the intestine after being ingested and develop without migration into the tissues. In *Capillaria annulata*, the eggs hatch when ingested by oligochaete annelids and the larvae migrate to the tissues. When freed from the earthworms in the crops of chickens, the larvae penetrate its wall and develop to maturity. A somewhat similar situation occurs with *Capillaria plica*, in which case the eggs hatch when swallowed by earthworms. Larvae released in the gut of the vertebrate host migrate via the bloodstream to the kidneys, and through the ureters to the urinary bladder, where maturity is attained. *Capillaria hepatica* deposits its eggs in the liver parenchyma, from which they are freed when the host decomposes after death, or is eaten and digested by another animal and they are passed in the feces. Development to the infective stage takes place only after the eggs have been freed from the liver and exposed to the air. Upon ingestion of infective eggs by rodents, they hatch and the larvae migrate to the liver via the hepatic portal vein and develop to maturity.

### *Trichuris ovis*
(Abildgaard, 1795) (Plate 135)

This is one of the several whipworms from the cecum of sheep, cattle, and other ruminants in many parts of the world. They feed on blood.

**Description.** The anterior part of the body is much thinner than the posterior portion. Males are 50 to 80 mm long with the slender anterior part forming three-fourths of the body. The spicule is 5 to 6 mm long and the spicular sheath has an oblong spinose enlargement near the caudal end. The females are 35 to 70 mm long with the anterior end forming two-thirds or more of the body. The vulva is just posterior to the esophagointestinal junction. The unembryonated brown eggs have a transparent plug in each end and measure 70 to 80 $\mu$ long by 30 to 42 $\mu$ wide.

**Life Cycle.** The life cycles of *Trichuris* are incompletely known for any one species. For that reason, information such as is known for the various ones is included in this discussion for purposes of better understanding. Unembryonated eggs of all species are passed in the feces of the various hosts. The life cycle is direct. Development to first stage larvae is slow and hatching occurs only in the duodenum of the host.

Eggs of *Trichuris ovis* require about 3 weeks at an optimal temperature of 37.5°C to produce the first stage larvae. *T. vulpis* of dogs develops in 12 to 15 days at the same temperature. They do not survive desiccation well, dying in 15 days at room temperature and relative humidity. In wet soil, eggs develop normally at 22 to 30°C, whereas in dry soil, nearly all die in 29 days at 30°.

In the case of *Trichuris muris*, an additional period of development beyond appearance of the larva is required to reach infectivity. The thick-shelled eggs are highly resistant to unfavorable environmental conditions. Embryonated eggs of *T. suis* of pigs are able to survive and retain their infectivity for 6 years in a vacated hog lot, where grass grew for the second half of the period.

When infective eggs of *T. ovis* are swallowed

## Explanation of Plate 135

A, Gravid female. B, Caudal end of male with spicular sheath extended. C, First stage larvae pressed from egg. D, Second stage larva from intestine of rabbit. E, Third stage larva from intestine. F, Posterior end of fourth stage male larva. G, Posterior end of fourth stage female larva. H, Vulvar region of fourth stage larva. I, Sheep definitive host.

1, Slender anterior end of body; 2, broad posterior end of body; 3, mouth; 4, muscular portion of esophagus; 5, stichosome, or esophageal cells; 6, intestine; 7, anus; 8, ovary; 9, beginning of oviduct; 10, uterus; 11, vagina; 12, vulva; 13, spicular sheath; 14, spicule; 15, spear of first stage larva; 16, developing spicule; 17, developing spicular sheath; 18, developing ovary.

a, Adult female in cecum; b, unembryonated eggs laid in cecum; c, unembryonated eggs mingle with feces and pass outside of body; d, embryonated egg on ground; e, egg in two-cell stage; f, egg in four-cell stage; g, egg in six-cell stage; h, fully developed first stage larva in egg on ground; i, infection occurs when embryonated egg is swallowed; j, larva escapes from egg; k, larva enter cecum; l, first stage larva with anterior end embedded in mucosa of cecum; m, first stage larva molts; n, second stage larva attaches to mucosa, feeding and growing; o, second stage larva molts, producing third stage; p, third stage larva; q, third stage larva molts to produce fourth stage larva that is said to develop to adulthood, but an additional molt should be expected.

Figure A adapted from various sources; B–H, from Thapar and Singh, 1954, Proc. Indian Acad. Sci. Ser. B. 40:69.

---

with contaminated food and water by sheep or other ruminants, the mucoid plugs are promptly dissolved by the digestive juices of the duodenum, allowing the larvae to escape within 30 to 60 minutes after entering the host. They are 125 to 150 $\mu$ in length by 10 to 12 $\mu$ in diameter. The esophagus, the shorter part of which consists of a double row of spherical cells, occupies the major part of the body length. A buccal spear projects from the mouth. The larvae are carried directly to the cecum. Three molts, all in the cecum, are reported. The first one occurs about 3 to 4 weeks after infection. Second stage larvae are about 2.15 mm long, with the stichosome consisting of a single row of cells about two-thirds the length of the body and the rectum is very short. The larvae attach by forcing the anterior portion of the body into the mucosal layer of the cecum. They attain a length of about 5 mm, with the esophagus about 4 mm and the intestine 0.66 mm long. The second molt occurs about 3 weeks after the first. Third stage larvae are more slender than the second. In specimens 6.5 mm long, the esophagus is 5.3 mm and the intestine 1.1 mm long. The time of the third molt is unknown but it appears to be after 7 weeks of infection. A fourth molt has not been reported but should be expected. Sexual maturity is attained in 12 weeks.

In *Trichuris muris* of mice, first stage larvae are in the cecum within 1 hour after infection. By 4 hours, they have entered the glandular openings, remaining 4 to 5 days coiled around the lumen. As growth proceeds, the anterior end remains embedded while the posterior end protrudes into the lumen of the cecum. Moltings occur on days 9, 10, and 11 postinfection. Egg production during the second week of patency is 4000 to 5000 per day per female. The maximum daily output of 7500 to 7900 per day occurs during weeks 3 to 5, declining thereafter.

Eggs of *Trichuris vulpis* of dogs hatch in the small intestine within 30 minutes and the larvae are embedded in the mucosa by 24 hours. After 8 to 10 days, they return to the lumen and are carried passively to the cecum, attach, and mature in 70 to 90 days. The life span is about 16 months.

In *Trichuris suis* of pigs, the prepatent period is 41 to 45 days and the length of life 4 to 5 months.

### EXERCISE ON LIFE CYCLE

Adult females as a source of eggs may be obtained from abattoirs where sheep are slaughtered. Eggs dissected from gravid females should be washed several times to free them of tissue and then incubated in shallow water in glass dishes. Frequent changes of the water will prevent extensive growth of deleterious bacteria and fungi. Progress of embryonation should be observed by making daily examinations of the eggs.

When the first stage larvae have appeared, infection experiments may be conducted by feeding them to rabbits, or to mice and guinea pigs. Development of the larvae to the fourth

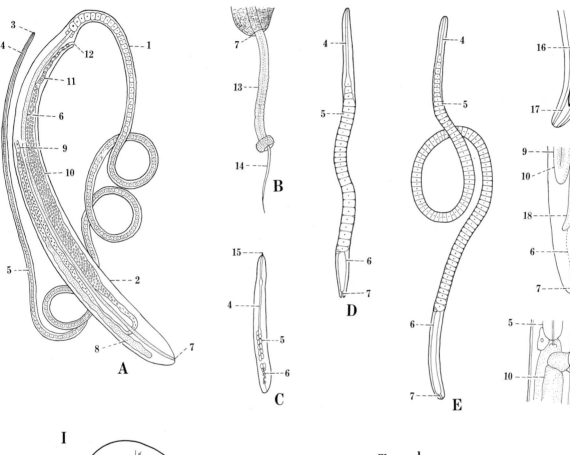

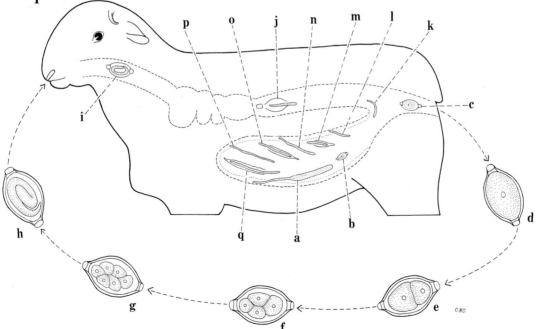

## Explanation of Plate 136

A, Anterior end of adult worm. B, Posterior end of adult male. C, Adult female. D, Adult male. E, Section of crop of chicken with adult worm embedded in it. F, Early stage larva from experimentally infected earthworm. G, Details of anterior end of larva shown in F.

1, Mouth with spear-like structure; 2, bulbous swelling of cuticle; 3, anterior muscular portion of esophagus; 4, spiny spicular sheath; 5, stichosome; 6, vulva; 7, uterus filled with eggs; 8, ovary; 9, testis; 10, portion of wall of crop; 11, adult worm embedded in mucosa; 12, egg in mucosa; 13, cephalic spine; 14, vacuolated areas; 15, intestine; 16, rectum.

a, Adult worm embedded in wall of crop; b, eggs in wall of crop; c, mucosa of crop sloughing due to damage by worms and eggs which releases trapped eggs; d, eggs passing through gizzard; e, eggs pass through intestine; f, unembryonated eggs voided in feces; g, embryonated egg; h, eggs swallowed by earthworm; i, mucoid plugs dissolved by digestive juices; j, larvae escape from eggs into intestine; k, larvae free in intestine where they may molt (but this and subsequent molts have not been seen, only assumed); l, larvae migrate through intestinal wall; m, larvae probably molt second time to form third stage; n, larvae encyst in tissues of body, becoming infective to chickens; o, earthworms disintegrate in crop of chicken; p, larvae escape from earthworms; q, larvae probably molt third time; r, fourth stage larvae (?) penetrate wall of crop; s, larvae probably molt fourth and final time, after which they grow to adulthood.

Figures A, B adapted from Ciurea, 1914, Z. Infektionsk. Hyg. Haustiere 15:49; C–E, from various sources; F, G, from Wehr, 1936, N. Amer. Vet. 17:18.

---

stage may be observed by sacrificing experimentally infected animals at 4, 7, 8, 9, and 10 weeks after having fed them infective eggs. Only three molts have been reported but four should be expected on the basis of what occurs in other nematodes.

Ascertain how long embryonated eggs will survive at room temperature when kept moist and when kept dry.

## SELECTED REFERENCES

Alicata, J. E. (1935). Early developmental stages of nematodes occurring in swine. U. S. Dep. Agr. Tech. Bull. 489, 96 pp.

Hill, C. H. (1957). The survival of swine whipworm eggs in hog lots. J. Parasitol. 43:104.

Miller, M. J. (1947). Studies on life history of Trichocephalus vulpis, the whipworm of dogs. Can. J. Res. (D) Zool. Sci. 25:1–11.

Pike, E. H. (1969). Egg output of Trichuris muris (Schrank, 1788). J. Parasitol. 55:1046–1049.

Powers, K. G., A. C. Todd, and S. H. McNutt. (1960). Experimental infections of swine with Trichuris suis. Amer. J. Vet. Res. 21:262–268.

Rubin, R. (1954). Studies on the common whipworm of the dog, Trichuris vulpis. Cornell Vet. 44:36–49.

Spindler, L. A. (1929). A study of the temperature and moisture requirements in the development of the eggs of the dog trichurid (Trichuris vulpis). J. Parasitol. 16:41–46.

Thapar, G. S., and K. S. Singh. (1954). Studies on the life history of Trichuris ovis (Abildgaard, 1795) (Fam. Trichuridae: Nematoda). Proc. Indian Acad. Sci. (B) 40:69–88.

Wakelin, D. (1969). The development of the early larval stages of Trichuris muris in the albino laboratory mouse. J. Helminthol. 43:427–436.

Wright, K. A. (1970). Ultrastructure of the esophagus of the trichuroid nematodes Capillaria hepatica and Trichuris myocastoris. J. Parasitol. (4/2):373.

### Capillaria annulata
(Molin, 1858) (Plate 136)

*Capillaria annulata* embeds in the mucosa of the esophagus and crop of chickens, turkeys and a number of other galliform birds throughout the world. In addition to domestic birds in North America, it parasitizes ruffed grouse, Hungarian partridges, ring-necked pheasants, and bobwhite quail.

**Description.** The body is of about equal diameter throughout its entire length. A bulbous bladder-like enlargement is located just behind the head. Bacillary bands occur dorsally and ventrally on the body. Males are 10 to 25 mm long by 0.05 to 0.08 mm in diameter; the tail has two lateral inconspicuous flaps. The spicular sheath is spined and usually withdrawn, and the spicule is lacking. Females are 25 to 60 mm long by 0.07 to 0.12 mm in diameter. The vulva is located near the posterior end of the stichosome. Eggs measure 60 to 66 $\mu$ long by 26 to 28 $\mu$ wide, and have a mucoid plug in each end.

**Life Cycle.** The life cycle of *Capillaria annulata* and other species of the genus which use earthworms as intermediate hosts is inade

Plate 136    *Capillaria annulata*                                                      501

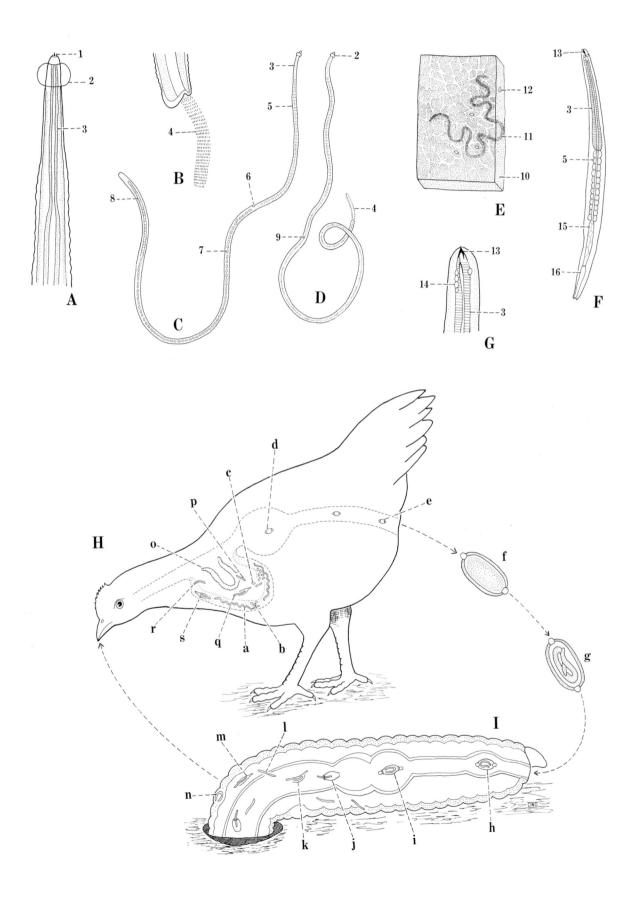

quately known. Fragmentary bits of information of the life cycles of the various species in poultry are pieced together to form a unified picture showing a basic pattern. Furthermore, it will serve as a guide to future work and clarification of the life cycles of these capillarids.

Unembryonated eggs are voided in the feces of the hosts. Development is slow. In *C. annulata*, fully developed first stage larvae 250 μ long by 10 μ in diameter appear in 24 to 32 days at 28° to 32°C. There is no further development in the egg. No development takes place at 4° to 6°C but it begins and continues normally when the eggs are transferred to room temperature. Eggs of *C. anseris* of geese are infective after 8 days of incubation at 22° to 27°C and of *C. caudinflata* of chickens after 11 to 13 days at room temperature. Eggs have great longevity. Those of *C. caudinflata* survive in a refrigerator for 344 days and unembryonated ones are more cold-hardy than embryonated ones.

Hatching occurs very quickly when fully embryonated eggs are swallowed by earthworms (*Allolobophora caliginosa, Eisenia foetida, Lumbricus terrestris*) and acted upon by the intestinal juices. Hatched larvae burrow through the gut wall into the body tissues, particularly the longitudinal body muscles. The first molt of *C. caudinflata* occurs in earthworms between days 3 and 5 after hatching. The larvae are infective to chickens in 9 days. Growth of *C. annulata* continues in the earthworms for 14 to 21 days in *Allolobophora caliginosa* and 21 to 28 days in *Lumbricus terrestris* before the larvae are infective to the bird hosts. The first molt occurs during this time. The posterior end of the first and second stage larvae of both species is bilobed.

Birds become infected by swallowing earthworms containing second stage larvae. In the case of *C. annulata*, the liberated larvae develop and burrow into the mucosa of the crop and esophagus. Nothing is known regarding molts. They attain sexual maturity in 19 to 26 days in chickens and turkeys. In *C. caudinflata*, the larvae develop in the small intestine, burrowing into the mucosa. Two molts, the second and third, occur in the definitive host. *Capillaria bursata*, which may be *C. caudinflata*, molts the second time on day 8 and the third time between days 13 and 14 in chickens and *C. caudinflata* on day 12 and again between days 16 and 17. Sexual maturity is reached by *C. caudinflata* in 22 to 24 days. Its life span is around 10 months.

*Capillaria columbae* of pigeons and *C. contorta* of chickens have a direct life cycle. The eggs of *C. columbae* develop to first stage infective larvae in 6 to 8 days at room temperature. There is no molt in the egg. They are very resistant to environmental conditions, surviving low temperatures of 4°C and exposure of 714 days in the soil. Unembryonated eggs are less hardy than developed ones.

When swallowed by birds, the eggs hatch in the small intestine, releasing larvae 135 to 169 μ long. There are three molts in the host. The first is in the second week, the second on day 14, and the third between days 14 and 21. All stages penetrate the intestinal mucosa but the larvae go deeper than the adults. Worms are sexually mature in 26 days.

## EXERCISE ON LIFE CYCLE

Specimens of *C. annulata* for experimental studies may be obtained from poultry-processing plants handling farm birds. Dissect eggs from gravid females and after washing them several times incubate them in shallow water in dishes at room temperature. Changing water daily will avoid excessive growth of bacteria which retard development of the eggs.

When first stage larvae have appeared, the eggs may be induced to hatch *in vitro* by two methods. Flush the intestine of live earthworms with water by means of a fine syringe needle placed in the mouth. Catch and filter the washings. Eggs placed in the water containing the digestive enzymes hatch in a few minutes. Embryonated eggs placed in Tyrode's, Hanks's, or Ringer's solution hatch in 16 to 18 days.

Feed embryonated eggs to earthworms for development of second stage larvae. Examination of infected annelids at daily intervals after ingestion of eggs will show the different stages of larvae, their location in the body, and time of their migrations and development.

Earthworms containing second stage larvae fed to 10 chicks which are examined at intervals of 2 days will enable the investigator to follow the development of the parasite in the vertebrate host.

## SELECTED REFERENCES

Allen, R. W. (1949). Studies on the life history of *Capillaria annulata* (Molin, 1858) Cram, 1926. J. Parasitol. 35(6/2):35.

Allen, R. W. (1950). Relative susceptibility of various species of earthworms to the larvae of *Capillaria annulata* (Molin, 1858) Cram, 1926. Proc. Helminthol. Soc. Wash. 17:58–64.

Allen, R. W., and E. E. Wehr. (1942). Earthworms as possible intermediate hosts of *Capillaria caudinflata* of the chicken and turkey. Proc. Helminthol. Soc. Wash. 9:72–73.

Cram, E. B. (1936). Species of *Capillaria* parasitic in the upper digestive tract of birds. U. S. Dep. Agr. Tech. Bull. 516, 27 pp.

Farr, M. M. (1961). Further observations on survival of the protozoan parasite, *Histomonas meleagridis,* and eggs of poultry nematodes in feces of infected birds. Cornell Vet. 51:3–13.

Morehouse, N. F. (1944). Life cycle of *Capillaria caudinflata,* a nematode parasite of the common fowl. Iowa State Coll. J. Sci. 18:217–254.

Nickel, E. A. (1953). Ein Beitrag zur Biologie und Pathogenität des Geflügelhaarwurms *Capillaria caudinflata* (Molin, 1858). Berl. München. Tierärztl. Wochenschr. 66:245–248.

Shlikas, A. V. (1965). [Biology of *Capillaria anseris* Madsen, 1945.] Tr. Gel'mintol. Lab. 15:238–240.

Shlikas, A. V. (1966). [Life-cycle of *Capillaria bursata* Freitas & Almeida, 1934.] Acta Parasitol. Litu. 6:143–147.

Shlikas, A. V. (1966). [Hatching of larvae of *Capillaria* from poultry *in vitro.*] Acta Parasitol. Litu. 6:149–153.

Shlikas, A. V. (1968). [Ontogensis of the nematodes *Capillaria bursata* Freitas & Almeida, 1934 and *C. caudinflata* (Molin, 1858) Travassos, 1915.] Acta Parasitol. Litu. 7:119–130.

Wakelin, D. (1965). Experimental studies on the biology of *Capillaria obsignata* Madsen, 1945, a nematode parasite of the domestic fowl. J. Helminthol. 39:399–412.

Wehr, E. E. (1936). Earthworms as transmitters of Capillaria annulata, the "crop-worm" of chickens. N. Amer. Vet. 17:18–20.

Wehr, E. E. (1939). Studies on the development of the pigeon capillarid, *Capillaria columbae.* U. S. Dep. Agr. Tech. Bull. 679, 19 pp.

Wehr, E. E., and R. W. Allen. (1954). Additional studies on the life cycle of *Capillaria caudinflata,* a nematode parasite of chickens and turkeys. Proc. Helminthol. Soc. Wash. 12:12–14.

Zucchero, P. J. (1942). Notes on the life cycle of *Capillaria annulata.* Proc. W. Va. Acad. Sci. 15:96–106.

## *Capillaria hepatica*
(Bancroft, 1893) (Plate 137)

*Capillaria hepatica* is a common parasite of rodents as well as other animals. It occurs in more than 20 familiar species, such as marmots, prairie dogs, ground squirrels, squirrels, gophers, beavers, muskrats, mice of several kinds, rabbits, lemmings, brown rats, dogs, peccaries, monkeys, and others. It occurs in humans, with at least 11 authentic cases.

The adult worms are embedded in the liver near the surface. Eggs are laid in the liver parenchyma in spaces around the worms and appear as yellow patches. Development of the eggs to the infective stage takes place only after they are freed from the liver and exposed to optimal temperature and moisture in the presence of sufficient oxygen.

**Description.**    Males are 17 to 32 mm long by 0.04 to 0.08 mm in diameter. The muscular esophagus is 0.32 mm long and the glandular portion 6.2 to 7.5 mm. The posterior end is blunt and bears a pair of subventral lobes. There is a well-developed though lightly cuticularized spicule 425 to 500 $\mu$ long and a membranous protrusible spicular sheath that expands distally into a funnel-shaped dilatation. Adult females attain a length of 98 mm and a diameter of 0.19 mm. The vulva is located slightly posterior to the end of the esophagus and has a protrusible membranous structure. The surface layer of the eggshell is traversed by rod-like structures, each with a knob-like enlargement on the end. Measurements of eggs as given by various authors range in size from 48 to 62 $\mu$ long by 29 to 37 $\mu$ wide.

**Life Cycle.**    With the adult females embedded in the liver, the eggs appear in clumps around the worms. They are uncleaved when laid in the liver and do not develop beyond the eight-cell stage while still in it, possibly because of insufficient oxygen. Development starts again only after they are free from the liver by death and decomposition of the host or when it is eaten and digested by some animal. In the latter case, the freed eggs are passed undeveloped in the feces where they develop and become infective. Some eggs from females in the liver regularly enter damaged bile ducts, are carried to the intestine, and are passed in the feces. This usually happens early in infections, especially in heavy ones. Later, fibrosis and calcification of the lesions seals off the worms and clumps of eggs, thus preventing their escape in the feces.

The eggs are resistant to adverse conditions in the environment. They withstand 1 to 2 weeks of near freezing temperature and desiccation above 50 percent relative humidity for an

## Explanation of Plate 137

A, Anterior end of adult female. B, *En face* view of adult female. C, Vulvar region. D, Ventral view of posterior end of male, with spicule withdrawn. E, Lateral view of posterior end of male with spicule and spicular sheath extended. F, Early first stage larva squeezed from egg. G, Late first stage larva from liver of mouse 2 days after infection. H, Anterior end of first stage larva. I, Stichosome of molting first stage larva from liver of mouse 3 days after infection. J, Second stage larva from liver of mouse 3 days after infection. K, Detail of stichosome of second stage larva. L, Third stage larva from liver of mouse 7 days after infection. M, Detail of stichosome of third stage larva. N, Stichosome of fourth stage larva from mouse 9 days after infection. O, Mouse definitive host. P, Carnivore in whose alimentary canal the definitive host is digested, the eggs released from the liver, and voided with the feces where development to the infective stage takes place. Q, Dead mouse host decomposes, releasing eggs from liver with subsequent development.

1, Muscular portion of esophagus; 2, stichosome, or glandular portion of esophagus; 3, intestine; 4, vulva; 5, evaginated portion of vagina; 6, vagina; 7, unembryonated egg; 8, mouth; 9, amphid; 10, outer circle of cephalic papillae; 11, inner circle of papillae; 12, anus of male; 13, caudal papilla; 14, spicular pouch; 15, spicule; 16, spicular sheath; 17, early stage of esophagus, with nerve ring; 18, early stage of intestine; 19, undifferentiated esophagus; 20, beginning of buccal capsule; 21, differentiating stichosome; 22, undifferentiated muscular portion of esophagus of second stage larvae; 23, differentiating stichosomal portion of esophagus; 24, celomocytes; 25, ventral cell; 26, binucleate genital primordium; 27, gonad; 28, nerve ring.

a, Adult worm embedded in liver parenchyma of mouse; b, mass of undeveloped eggs in liver parenchyma; c, definitive host about to be eaten by carnivore; d, definitive host digested; e, eggs released from liver in alimentary tract; f, eggs travel through digestive tract without further development; g, eggs voided with feces; h, embryonated egg containing first stage larva; i, infection takes place when embryonated eggs are swallowed by mouse or other definitive host; j, mucoid plugs released from egg by digestive juices; k, larvae escape from eggs; l, first stage larvae penetrate wall of intestine or cecum, entering hepatic portal vein; m, some larvae enter celom; n, larvae escape from capillaries of liver to enter hepatic tissue; o, molt of first stage larvae; p, second stage larvae; q, molt of second stage larvae; r, third stage larvae which molt and continue development, finally reaching adult stage after fourth molt (not shown).

Figure A adapted from various sources; B, F–N redrawn from Wright, 1961, Can. J. Zool. 38:167; C, from Pavlov in Skrjabin *et al.*, 1954, Opredelitel Paraziticheskikh Nematod, Vol. 4, Pl. 143, Fig. B; D, E, from Baylis, 1931, Parasitology 23:533.

---

equal time. They are able to develop in rat pellets and remain alive in them for 7 months. Survival up to 714 days occurs in rat livers kept unfrozen in a refrigerator. Winter temperature under the snow and in nests of mice does not damage the eggs.

Development of the eggs outside the host depends on a favorable combination of temperature, moisture, and the presence of sufficient oxygen. Under these conditions at room temperature, the embryos are fully developed and infective in 4 to 5 weeks. There is no molt or further development in the egg.

Infection of the host is by swallowing fully developed eggs. Hatching takes place in the cecum within 24 hours and first stage larvae, 140 to 190 $\mu$ long by 7 to 11 $\mu$ in diameter, are burrowing into the cecal wall within 6 hours. They enter the hepatic portal vein and are in the liver within 52 hours after infection. There is no increase in size during this time. The first molt in the liver takes place during days 3 and 4 postinfection.

Second stage larvae retain the shed cuticle as a sheath. Differentiation in this stage is primarily growth in the esophageal region. The anterior muscular and posterior stichosomal parts of the esophagus appear. Larvae attain an average length of 160 to 220 $\mu$ and diameter of 10 to 14 $\mu$. Those in the most advanced stage of development begin the second molt by day 5 after infection.

In the third stage larvae, most of the growth is in the stichosome and reproductive organs. The longitudinal bacillary bands on the body appear. Larvae are 270 to 670 $\mu$ long by 16 to 33 $\mu$ in diameter. The third molt takes place about day 9. Fourth stage larvae are ensheathed in the third stage cuticle. The length increases to 1.1 to 3.6 mm and the diameter to 0.023 to 0.038 mm. The spicule is distinguishable by day 13 and fully developed by day 16. The ensheathing cuticle is lost by day 18 by the males and day 20 by the females. Gravid females appear at 21 days after infection. The life span of males is around 40 days and that of females 59 days.

**Plate 137**    *Capillaria hepatica*                                                      505

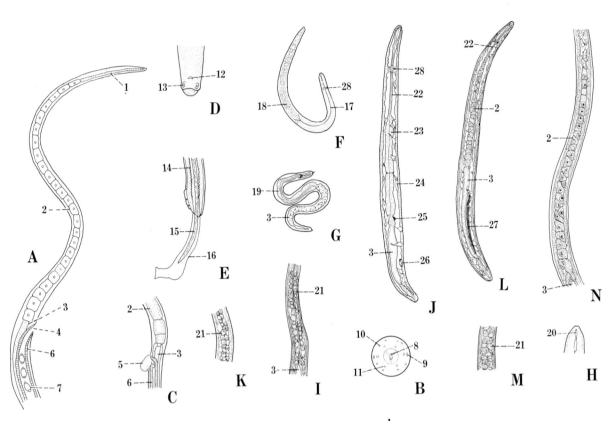

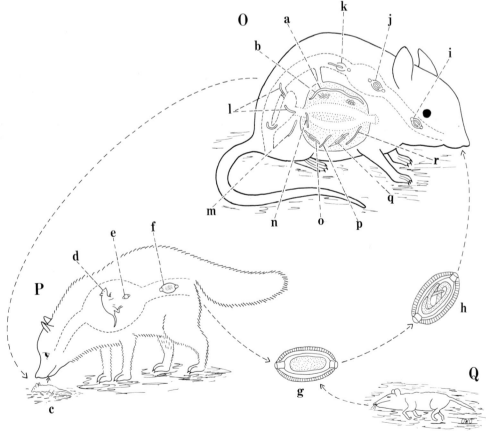

About 19 days after infection, the non–egg-bearing lesions around worms in the liver begin cellular infiltration and fibrosis, followed by calcification. Egg-bearing lesions are slower to calcify. Dead worms become calcified.

Since some eggs are passed regularly in the feces during the early stages of infection, they are an important source of infection, especially when in the nest of the hosts. A higher incidence of infection occurs among old animals in the population. There seems to be little resistance to reinfection. In Florida, infections are restricted to scrub areas.

Rodents constitute a reservoir of infection from which humans may be exposed. The passage of eggs in the feces and the close contact of some species with man's habitat and stores of food provide an epizootiological relationship favorable to infection.

### EXERCISE ON LIFE CYCLE

When infected livers are fed to rats or mice, the eggs are freed and passed undeveloped in the feces, from which they may be retrieved by standard methods. Artificial digestion of the liver would also free them, but it is not known whether eggs freed by this method would develop and hatch. Infective larvae appear after about 6 weeks of incubation under warm, moist conditions.

Feed fully embryonated eggs to a series of mice and examine some of them at close intervals for a period of 60 days. Determine the date of hatching, migration through the wall of the cecum, entry into the liver, time and number of molts, length of the prepatent period, and longevity of the males and females in the liver.

### SELECTED REFERENCES

Baylis, H. A. (1931). On the structure and relationships of the nematode *Capillaria* (*Hepaticola*) *hepatica* Bancroft. Parasitology 23:533–534.

Freeman, R. S., and K. A. Wright. (1960). Factors concerned with the epizootiology of *Capillaria hepatica* (Bancroft, 1893) (Nematoda) in a population of *Peromyscus maniculatus* in Algonquin Park, Canada. J. Parasitol. 46:373–382.

Layne, J. N. (1968). Host and ecological relationships of the parasitic helminth *Capillaria hepatica* in Florida mammals. Zoologica 53:107–123.

Lubinsky, G. (1956). On the probable presence of parasitic liver cirrhosis in Canada. Can. J. Comp. Med. 20:457–465.

Luttermoser, G. (1938). Factors influencing the development and viability of the eggs of *Capillaria hepatica*. Amer. J. Hyg. 27:275–289.

Luttermoser, G. (1938). An experimental study of *Capillaria hepatica* in the rat and mouse. Amer. J. Hyg. 27:321–340.

Otto, G. F., and Th. von Brand. (1941). Natural and experimental calcification in Capillaria infection. Amer. J. Hyg. 34(D):13–17.

Rausch, R. (1961). Notes on occurrence of *Capillaria hepatica* (Bancroft, 1893). Proc. Helminthol. Soc. Wash. 28:17–18.

Shorb, D. A. (1931). Experimental infestation of white rats with *Hepaticola hepatica*. J. Parasitol. 17:151–154.

Wright, K. A. (1961). Observations on the life cycle of *Capillaria hepatica* (Bancroft, 1893) with a description of the adult. Can. J. Zool. 39:167–182.

### Capillaria plica
(Rudolphi, 1819) (Plate 138)

This species is a common parasite of the urinary bladder of dogs, wolves, and, in particular, foxes. Martens, badgers, and cats also are reported as hosts. Occasionally the worms occur in the pelvis of the kidney. Typical capillarid eggs in an undeveloped stage are passed in the urine.

**Description.** Males are 13 to 30 mm long, with the thin anterior part of the body only slightly shorter than the thicker posterior portion. There is a small bursa-like structure at the caudal end of the body. The spicular sheath is long and thin, with a punctate appearance due to the crossing of the transverse and longitudinal striae. The spicule is thread-like, rounded terminally, and 4.49 mm long.

Females are 30 to 60 mm long, with the slender portion of the body consisting of two-thirds of the total length. The caudal end is blunt and the anus is terminal. The vulva is at the end of a cylindrical appendage. Eggs measure 60 by 30 $\mu$ and are unembryonated when laid.

**Life Cycle.** Adult females in the urinary bladder deposit their eggs which are voided with the urine. Under favorable conditions of moisture and temperature, the first stage larvae develop in 30 to 36 days. Further growth and hatching do not occur until the embryonated eggs are swallowed by earthworms (*Lumbricus terrestris, L. rubellus, Dendrobaena subrubi-*

*cunda*). Shortly after reaching the posterior two-thirds of the intestine, the eggs hatch and the larvae soon burrow through the intestinal wall into the connective tissue, where the first stage larvae become infective to the final host in about 24 hours.

Infection of the vertebrate host takes place when earthworms harboring first stage larvae are eaten. Upon escaping from the annelids, the larvae molt again and burrow into the wall of the small intestine, where the second molt occurs within 7 days. The third stage larvae enter the hepatic portal vein sometime before day 20, going to the lungs via the liver and heart. They pass through the pulmonary capillaries, return to the heart, and enter the general circulation.

It is believed that larvae enter the kidneys by way of the renal arteries and follow the blood vessels to the glomeruli. Here they migrate into Bowman's capsule, go through the tubular portion of the nephron, and enter the pelvis of the kidney, going down the ureter to the urinary bladder. Third and fourth stage larvae are present in the bladder on or about day 33 after infection. Sexually mature worms appear 58 to 63 days after infection, as indicated by the appearance of eggs in the urine.

*Capillaria aerophila*, another capillarid parasite of dogs, occurs in the lungs. The life cycle is direct. Eggs develop in the soil to the infective stage in 5 to 7 weeks, depending on the temperature. They are cold resistant and capable of surviving a year in the soil. Upon being swallowed by dogs, the embryonated eggs hatch in the small intestine and the larvae make their way to the lungs, presumably via the portal or lymphatic systems, or both, arriving in 7 to 10 days. Sexual maturity is attained in 40 days. Eggs cease to appear in the feces after 8 to 11 months.

## EXERCISE ON LIFE CYCLE

Eggs collected from gravid females obtained from the urinary bladder or from the urine of infected dogs should be incubated in a 2 percent solution of potassium bichromate. When the first stage larvae have developed, they may be released from the eggs for purposes of study by gently applying pressure on the cover slip.

Earthworms may be infected by mixing embryonated eggs with the soil in which they are kept. Examine the intestinal contents and the connective tissue surrounding the gut for first stage larvae. This may be done effectively by digesting the gut in artificial media *in vitro*. Determine the hatching time and how soon thereafter larvae migrate into the celom. Ascertain in what part of the gut migration from it into the body cavity takes place.

If a dog is available for feeding experiments, introduce infected earthworms into its food and determine the prepatent period of the nematodes by observing the date when eggs appear in the urine. These parasites are not especially pathogenic when present in small numbers.

## SELECTED REFERENCES

Chitwood, M. B., and F. D. Enzie. (1953). The domestic cat as a new host for *Capillaria plica* in North America. Proc. Helminthol. Soc. Wash. 20: 27–28.

Christensen, R. O. (1938). Life history and epidemiological studies on the fox lungworm, *Capillaria aerophila* (Creplin, 1839). Livro Jub. Travassos, Rio de Janeiro, pp. 119–136.

Enigk, K. (1950). Die Biologie von *Capillaria plica* (Trichuroidea, Nematoda). Z. Tropenmed. Parasitol. 1:560–571.

### FAMILY TRICHINELLIDAE

This family consists of a single species. The adults occur in the lumen and mucosal lining of the small intestine of mammals. Gravid females give birth to hatched larvae which make their way via the blood and lymph vessels to the skeletal muscles where they encyst. This parasite is prevalent in the northern hemisphere and parts of South America, but rare in the tropics and other areas of the southern hemisphere. Humans are commonly infected with it.

*Trichinella spiralis*
(Owen, 1835) (Plate 139)

*Trichinella spiralis* is a common parasite of many species of carnivorous and omnivorous mammals, including humans. On the basis of recent surveys in the United States, the incidence of infection in the general population is estimated to be about 4 percent, down from 15 to 20 percent of 25 years ago. The decline is due to 1) fewer pigs being infected, as a result of the legal requirement to cook all garbage fed to pigs to prevent the spread of vesicular xanthema which incidentally kills trichinella, 2) reduced

## Explanation of Plate 138

A, Caudal end of adult male with spicule and spicular sheath extended. B, Caudal end of body of adult male showing lobes. C, Vulvar region of adult female. D, Cross section of body showing lateral bacillary bands. E, Unembryonated egg. F, First stage larva. G, Second stage larva from intestine of earthworm. H, Fox definitive host. I, Earthworm intermediate host.

1, Basal portion of spicule; 2, spicular pouch; 3, distal portion of spicule; 4, caudal lobe; 5, stichosome; 6, capillary tube of esophagus; 7, intestine; 8, bacillary band; 9, vulva; 10, vagina; 11, spear; 12, nerve ring; 13, anterior portion of esophagus; 14, rectum; 15, reproductive primordium.

a, Adult worm in urinary bladder; b, eggs laid in urinary bladder; c, eggs pass from body through urethra in urine; d, eggs undeveloped when voided with urine; e, fully developed egg; f, embryonated eggs swallowed by earthworms; g, embryonated eggs hatch in middle third of intestine; h, newly hatched larvae molt and as second stage larvae migrate through intestinal wall into celom; i, larvae encyst in loose connective tissue of celom; j, earthworms eaten by foxes; k, digestion of infected earthworms, releasing second stage larvae; (l–t not demonstrated experimentally but postulated to occur); l, second stage larvae molt to third stage; m, third stage larvae enter hepatic portal vein; n, larvae pass through capillaries of liver; o, larvae pass through right side of heart; p, larvae pass through capillaries of lungs; q, larvae pass through left side of heart; r, larvae enter dorsal aorta; s, larvae enter renal artery; t, larvae leave glomeruli in kidneys, entering Bowman's capsule and travel through renal tubules to pelvis of kidneys; u, larvae in pelvis of kidney; v, larvae travel down ureter, enter urinary bladder, and develop to maturity.

Figures A–E adapted from de Freitas and Lent, 1936, Mem. Inst. Oswaldo Cruz. 31:85; F, G, from Enigk, 1950, Z. Tropenmed. Parasitol. 1:560.

---

per capita consumption of pork, 3) extensive use of low-temperature storage, and 4) general awareness of the necessity for proper cooking of pork. The principal source of infection for humans in the United States is pork and bearmeat when eaten. In the arctic regions, polar bears, walruses, and seals constitute the primary sources. Extremely high incidences of infection occur in the wild rats on community garbage dumps.

Among domestic animals other than swine, dogs and cats commonly are infected. Wild mammals such as foxes, coyotes, bobcats, mink, opossums, raccoons, bears, walruses, hair seals, whales, squirrels, mice, muskrats, and others serve as natural hosts.

**Description.**   These are nematodes of small size with the posterior portion of the body only slightly greater in diameter than the anterior. The esophagus consists of a short anterior muscular part and a long posterior stichosome, or row of cells.

Males measure 1.4 to 1.6 mm long by 40 $\mu$ in diameter. A pair of lobes is located at the posterior extremity with one pair of papillae between them. There is no spicule or spicular sheath. Females are 3 to 4 mm long by 60 $\mu$ in diameter with the vulva near the middle of the stichosome. They are ovoviparous.

**Life Cycle.**   Transmission of the parasite from host to host is by ingesting flesh containing encysted larvae, swallowing viable cysts passed in feces resulting from meals of trichinous meat, eating carrion beetles or marine amphipods that have fed on trichinous carcasses, or prenatally by larvae entering fetuses in the blood or postnatally through the milk of infected mothers that are nursing their young during the period when larvae are migrating in the tissues.

Upon reaching the stomach, the encysted larvae are released from the muscle tissue when it is digested by the gastric juices. Most of them excyst in the duodenum and begin development there. However, some are passed in a viable and infective state in bits of undigested tissue in the feces. Animals swallowing them become infected in the same manner as when trichinous meat is eaten.

Development in the intestine includes molting and growth leading to sexual maturity. When about to molt, the larvae enter a period of quiescence during which growth and morphological reorganization of the organs occurs preparatory to the next stage in the life cycle.

The number of molts and location of the larvae at the time is a matter of controversy. It has been reported to be from one molt *in vitro* to two, three, and four *in vivo*, according to Shanta and Meerovitch, who reviewed the literature on the subject. In their carefully conducted experiments with mice, only two molts were observed, both in the intestine, which is in accord with observations of some other investigators. No molt occurs in the egg or uterus of the female

Plate 138 *Capillaria plica* 509

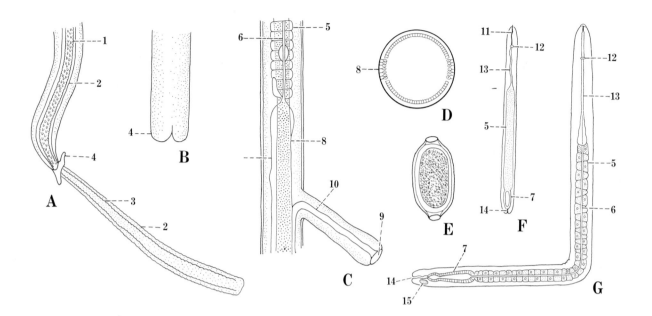

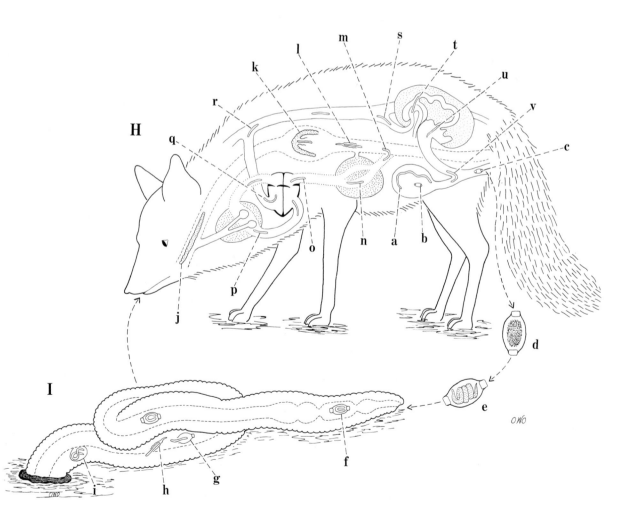

## Explanation of Plate 139

A, Adult female. B, Adult male. C, Ventral view of posterior end of male. D, Unencysted and fully developed larva. E, Unencysted young larvae in skeletal muscles. F, Encysted larvae in skeletal muscles. G, Swine host. H, Sylvatic cycle with carnivores, scavengers, and carrion as sources of infection of swine and other animals that eat them. I, Marine cycle. J, Rat cycle. K, Urban cycle with garbage containing scraps of infected pork or bodies of infected animals fed to swine.

1, Mouth; 2, anus; 3, copulatory appendages; 4, genital papillae; 5, muscular portion of esophagus; 6, glandular portion of esophagus, or stichosome; 7, intestine; 8, ovary; 9, uterus; 10, vulva; 11, testis; 12, sperm duct; 13, ejaculatory duct; 14, cloaca; 15, nerve ring; 16, skeletal muscle; 17, unencysted young larvae in skeletal muscle; 18, older larvae with cyst forming from host material; 19, cyst formed by reaction of host tissues.

a, Flesh containing first stage encysted and infective larvae; b, cyst being freed from muscle tissue in stomach; c, excystation of larvae in stomach and duodenum; d, first molt in anterior part of small intestine (stages d–i occur in duodenum and jejunum but are spread out because of lack of space); e, second stage larvae; f, second molt; g, third stage larva growing into sexually mature form; h, males and females mate; i, impregnated females enter lacteals where prelarvae are born; j, prelarva migrates through lymph vessels to and through lymph nodes; k, prelarva traveling by lymph vessels; l, prelarva entering thoracic lymph duct on way to vena cava; m, prelarva going through right side of heart; n, prelarva going through pulmonary artery; o, prelarva going through lungs; p, prelarva in pulmonary vein; q, prelarva going through left side of heart; r, prelarva in dorsal aorta being distributed to all parts of body; s, prelarva entering skeletal muscles; t, prelarva extended in muscle fibers; u, encysted first stage larva in skeletal muscles; v, prelarva migrating through peritoneal cavity toward muscles; w, excysted larva passing from intestine in feces; x, encysted first stage larvae in bits of undigested tissue leaving body in feces; aa, viable cyst passed in feces is infective to swine and other animals, course of development same as a–x.

a′ (H section), Rodents as scavengers of carrion or feces of large animals become infected; b′, predators eat parasitized rodents and become infected; c′, carcasses of predators (and rodents) become available to rodents; d′, swine eat rodents and carcasses of predators, becoming infected; offal from slaughtered swine is left for carnivores, scavengers (including swine).

a″ (I section), Polar bears; b″, whales; c″, hair seals; d″, walruses infected with trichina.

a (J section), Rats transmit infection among themselves through cannibalism; b, rat feces containing viable cysts infective to rats, swine, and any mammal that might ingest them; c, (K section), scraps of pork, or carcasses of small mammals, containing viable cysts are a source of infection to swine that are fed garbage and to mammals (rats, cats, dogs, etc.) that eat them.

Figures A, B adapted from Faust and Russell, 1957, Craig and Faust's Clinical Parasitology, 6th ed., Lea & Febiger, Philadelphia, p. 330; C, from Wu, 1955, J. Parasitol. 41:40; D, from Skrjabin, Shikobalova, Paramonov, and Sudarikov, 1954, Opredelitel Paraziticheskikh Nematod, Vol. 4, Pl. 147, Fig. B.

worms or in the muscles of the vertebrate hosts.

Upon release of larvae from the cysts, they must adjust to the transition from the isotonic environment of the cyst to the hypertonic one of the intestine. The evidence of this adaptation is the appearance of large amphids on the anterior end of the body. They are evident within 2 hours after excystment, increase in size up to hour 6, and then regress, becoming barely visible after a few days. Their principal function during this time is probably one of osmoregulation. Having served this need, the residual amphids possibly function as chemoreceptors.

The first molt occurs between hours 12 and 16, with no distinction between the sexes at this time; but shortly thereafter these second stage larvae show developing copulatory protuberances at the posterior end of the male and a small indentation in the body wall of the female, indicating the vagina.

The second molt of the males is between hours 24 and 32, when they are 1.11 to 1.15 mm or more in length, and that of females during hours 22 and 30 at 1.21 to 1.26 mm in length. In the third stage larvae, the sexual organs are well formed and the worms develop to maturity without further molts. The copulatory appendages are well formed and the seminal vesicle is filled with sperms by 30 hours postinfection. The vagina is patent at the end of the molt and some females are inseminated by hour 36. The general population of worms is sexually mature after 2 to 4 days postinfection. At the outset, there are about three times as many females (2.93 : 1) as males, but toward the end of the life

**Plate 139**   *Trichinella spiralis*                                                                          511

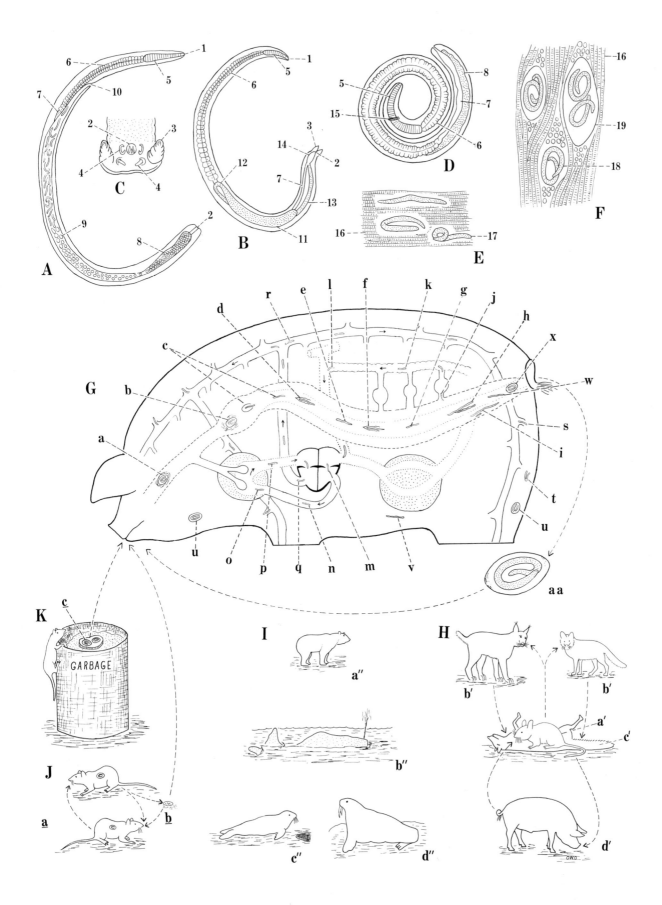

span, there are more than three times (3.4 : 1) as many males as females. Other reports state that the males die earlier than the females.

Following copulation, in which the females may participate several times, they migrate through the ducts of the glands of Leiberkühn into the intestinal mucosa and lacteals. As the embryos develop in the uterus, they elongate, stretching the eggshell until it has the appearance of a thin, hyaline sheath surrounding the larvae. When developed, the embryos hatch by escaping through a slit leaving the empty shell free in the uterus.

Birth of embryos, or prelarvae, begins 5 days after infection and continues for about 40 days. By this time, the worms begin disappearing from the intestine, when about 6 weeks of age. In the process of birth, the vaginal sphincter relaxes and the uterus contracts, forcing out 20 to 30 larvae at a time. Each act of expulsion is followed by a brief period of rest before another batch is born. Empty eggshells are passed along with the larvae. The number of larvae produced by a single female varies greatly, according to reports. Larvae recovered by Shanta and Meerovitch from the muscles of mice infected individually with a single pair of worms numbered 90 to 625. Other reports on the number produced by a single female range from 1000 to 1500, and even up to 10,000 during the lifetime.

The larvae are released in the lacteals and lymph spaces. Some are carried via the lymph vessels, lymph nodes, and thoracic lymph duct to the vena cava, and thence through the heart, pulmonary artery, lungs, pulmonary vein, heart, and into the general circulation to be distributed throughout the body during days 6 to 9. Upon leaving the blood vessels, they enter the fibers of the skeletal muscles, particularly the most active ones such as the masseter, intercostal, lingual, and diaphragmatic muscles. Other larvae enter the peritoneal cavity by day 6, reach peak numbers during days 12 and 13, and taper off until day 25, after which all have entered the muscles.

Larvae are in the diaphragmatic muscles by day 6. Upon entering the muscle fibers, the larvae begin development. The cuticle is permeable up to day 19, but not thereafter. At first, the larvae in the muscles lie more or less extended parallel with the fibers. By day 17, the larvae are 800 to 1000 $\mu$ long by 30 $\mu$ in diameter. As they coil, a thin-walled, lemon-shaped collagenous cyst of host origin 400 to 600 $\mu$ long

by 250 $\mu$ in diameter begins to form around them and is completed within 3 months. Calcification of the cysts begins after 6 months and is completed within a year, but the larvae live much longer, up to 6 years or more. Encysted larvae are infective to mammals and some birds that swallow them.

Development of the parasite constitutes three critical periods for the host. The first is the incubation period, consisting of excystation, molting, and penetration by the females into the duodenal and jejunal muscosa. Symptoms in humans begin soon after infection and include nausea, vomiting, diarrhea, intestinal pain, and sweating. These may persist for about 5 days.

The second period includes birth and migration of the larvae, with penetration into and destruction of the muscle fibers. The symptoms appear as muscular pain during respiration, speech, eye movement, and chewing, edema of the eyes and enlargement of lymph nodes. There may be intermittent fever.

The third period is that of encystation, with its physical and toxic effect. It constitutes a period of illness, edema, and dehydration. In heavy infections, blood pressure falls. There are nervous disorders, probably from damage to the central nervous system. If death does not intervene, there is slow recovery with chronic infection as the cysts calcify.

Pigs weighing 50 to 90 pounds fed 5 to 100 encysted larvae digested from muscle tissue produce an average of less than 0.5 larvae per gram of diaphragm pillars. Similar pigs fed 100 larvae in rat flesh had 13 larvae per gram of diaphragm. In general these numbers of larvae are not enough to be serious in humans.

A few larvae (1 to 50) per gram of muscle in humans are considered only as slight infection, 50 to 100 moderate, 500 to 1000 heavy to severe, and over 1000 critical.

Three ecological types of life cycles are recognized. They are 1) the urban cycle, in which rats and swine serve as reservoirs from which each may be infected, and humans are infected by eating parasitized pork that is insufficiently cooked to kill the infective larvae; 2) the sylvatic cycle, in which predators and scavengers serve as hosts; and 3) the marine cycle, in which hair seals, walruses, whales, and polar bears are the hosts. Each of these types is related to infection of humans.

Necrophagous insects commonly ingest encysted larvae of trichina which are able to survive for several days in them. Such arthropods serve as short term reservoir hosts for transferring cysts from infected carcasses to any mammals that might eat them.

The occurrence of trichina in marine mammals may be explained in a similar manner. Bodies of infected mammals entering the sea are fed upon by amphipods which ingest the cysts and harbor them in the intestine. Plankton-feeding fish eating such crustacea in turn serve as a second reservoir. Mammals such as whales feeding on the crustacea or seals on fish containing infected crustaceans become infected.

Carnivorous mammals of the arctic region constitute an important item of food of the people. Much of the meat is eaten raw or insufficiently cooked to kill the trichinella and is a constant source of infection.

## EXERCISE ON LIFE CYCLE

Rats living on municipal dumps usually provide a plentiful source of trichina larvae for experimental purposes. The presence of larvae in them may be determined by examining with the aid of a dissecting microscope bits of muscle tissue pressed between two plates of glass.

Feed infected flesh to a few white rats or mice to obtain heavily infected muscles for further experiments. The larvae become infective in 17 to 18 days, which is before encapsulation. Muscle tissue in equal amounts from a heavily infected animal should be fed to 28 white rats or mice on the same day. Kill one at intervals of 3 hours the first day and daily thereafter for a period of 20 days. Exsanguinate each animal the first day, lake the blood, and examine it for larvae. Open the body and rinse the peritoneal cavity with physiological saline to recover migrating larvae.

Examine the intestine of each rat by slitting it open and washing the contents into a dish of normal saline, and the mesenteric lymph nodes and striated muscle by pressing bits of them between glass plates. Sections of intestine and lymph nodes are useful for showing developmental stages in them. Note when and where each stage in the life cycle occurs. Make permanent mounts of each stage, using the glycerin method.

To determine the extent to which cysts and larvae are passed in the feces, feed a rat a fairly large number of cysts (1000 to 2000 by counting them) and examine the pellets passed during the first 24 hours afterward to determine the percentage of larvae that was retained. About 3 to 4 weeks later, feed it a similar number of cysts and ascertain whether more or fewer larvae appear in the feces. Test the larvae and cysts passed in the feces for infectivity by feeding them to separate uninfected rats.

## SELECTED REFERENCES

Carlin, A. F., C. Mott, D. Cash, and W. Zimmermann. (1969). Destruction of trichina larvae in cooked pork roasts. J. Food Sci. 34:210–212.

Denham, D. A. (1966). Infections with *Trichinella spiralis* passing from mother to filial mice pre- and postnatally. J. Helminthol. 40:291–296.

Fay, F. H. (1968). Experimental transmission of *Trichinella spiralis* via marine amphipods. Can. J. Zool. 46:597–599.

Gould, S. D. (1970). Trichinosis in Man and Animals. Charles C Thomas, Springfield, Ill.

Hill, C. H. (1966). Survival of encysted larvae of *Trichinella spiralis:* effect of exposure to O F, using precooled and fresh pork. Proc. Helminthol. Soc. Wash. 33:130–133.

Hill, C. H. (1968). Fecal transmission of *Trichinella spiralis* in penned hogs. Amer. J. Vet. Res. 29:1229–1234.

Hill, C. H. (1968). Light infections produced in swine dosed with 5 to 100 capsulated larvae of *Trichinella spiralis*. J. Parasitol. 54:1235–1236.

Most, H. (1965). Trichinellosis in the United States. Changing epidemiology during past 25 years. J. Amer. Med. Ass. 193:871–873.

Rausch, R. L. (1970). Trichinosis in the Arctic. In: S. E. Gould, ed. Trichinosis in Man and Animals. Charles C Thomas, Springfield, Ill., pp. 348–373.

Read, C. P., and E. L. Schiller. (1969). Infectivity of *Trichinella* from temperate and arctic zones of North America. J. Parasitol. 55:72–73.

Robinson, H. A., and O. W. Olsen. (1960). The role of rats and mice in the transmission of the porkworm, *Trichinella spiralis* (Owens, 1835) Railliet, 1895. J. Parasitol. 46:589–597.

Shanta, C. S., and E. Meerovitch. (1967). The life cycle of *Trichinella spiralis*. I. The intestinal phase of development. Can. J. Zool. 45:1255–1260.

Shanta, C. S., and E. Meerovitch. (1967). The life cycle of *Trichinella spiralis*. II. The muscle phase of development and its possible evolution. Can. J. Zool. 45:1261–1267.

Thomas, H. (1965). Beiträge zur Biologie und

## Explanation of Plate 140

A, Lateral view of anterior end of adult male. B, *En face* view of adult. C, Caudal end of adult male showing bursa and spicule. D, Egg in early stage of cleavage. E, First stage larva 227 days old. F, Penetration glands of first stage larva. G, Second stage larva penetrating tissue of annelid. H, Second stage larva encysted in annelid. I, Third stage larva migrating in tissue of bullhead. J, Third stage larva encysted in bullhead. K, Fourth stage larva migrating from intestine of ferret to kidney. L, Mink (*Mustela vison*), the common natural definitive host. M, Annelid (*Cambarincola philadelphica*) first intermediate host. N, Crayfish to which annelids habitually attach. O, Bullhead (*Ameiurus melas*) second intermediate host.

1, Papillae on anterior end of adult male; 2, esophagus; 3, anterior loop of testis; 4, outer ring of cephalic papillae; 5, inner ring of cephalic papillae; 6, mouth; 7, fleshy rayless copulatory bursa; 8, spicule; 9, terminal plug; 10, embryo in first cleavage; 11, stylet; 12, penetration glands; 13, excretory reservoir; 14, excretory cell; 15, elongated papilla; 16, pharyngeal rods; 17, anterior portion of body designated as a reservoir; 18, tissue of annelid host; 19, second stage larvae encysted in annelid; 20, cyst; 21, small intestine of ferret; 22, fourth stage larva migrating through intestinal wall toward kidney.

a, Adult male giant kidney worm in kidney of mink; b, eggs laid by females in kidneys escape by way of ureters; c, eggs in urinary bladder; d, undeveloped eggs voided with urine into water where embryonation occurs; e, embryonated eggs eaten by branchiobdellids; f, eggs hatch in intestine of annelid; g, first stage larvae migrate through intestinal wall into celom; h, molting of first stage larvae occurs in tissues; i, second stage larvae encyst in annelids; j, annelids attach to gills of crayfish, especially during the winter (newly laid coccoons appear in March, and by the end of May many young annelids can be found free in water); k, crayfish carrying infected annelids eaten by bullheads serve as a means of introducing cysts into second intermediate host; l, second stage larvae are released from cysts in intestine of bullheads; m, larvae migrate through intestinal wall into celom; n, in bullheads, larvae molt second time to form third stage larvae; o, third stage larvae encyst in viscera of bullheads and molt for third time, this time in the cyst, and become fourth stage larvae; p, cast cuticle from third molt; q, cysts containing larvae freed from fish by action of digestive juices; r, larvae escape through rupture in cyst wall into digestive tract; s, larvae pentrate wall of duodenum at point where it makes contact with right kidney in mink, thus affecting entrance into kidney at that point.

Adapted from Woodhead, 1950, Trans. Amer. Microsc. Soc. 69:21.

---

mikroskopischen Anatomie von *Trichinella spiralis* Owen, 1835. Z. Tropenmed. Parasitol. 16:148–180.

Villella, J. B. (1970). Life cycle and morphology. In: S. E. Gould, ed. Trichinosis in Man and Animals. Charles C Thomas, Springfield, Ill., pp. 19–60.

Zimmermann, W. J. (1970). Trichinosis in the United States. In: S. E. Gould, ed. Trichinosis in Man and Animals. Charles C Thomas, Springfield, Ill., pp. 378–400.

Zimmermann, W. J., and E. D. Hubbard. (1963). Wildlife reservoirs of *Trichinella spiralis*. Sci. Proc. Amer. Vet. Med. Ass. 100th Ann. Meet., pp. 194–199.

Zimmermann, W. J., E. D. Hubbard, and J. Mathews. (1959). Studies on the fecal transmission of *Trichinella spiralis*. J. Parasitol. 45:441–445.

### Superfamily Dioctophymoidea

These nematodes are medium to very large in size. The males are characterized by a muscular bell-shaped bursa without supporting rays and a single bristle-like spicule. The eggs have a thick pitted shell that is lighter in color at the poles. There are two families: Dioctophymatidae without and Soboliphymatidae with a muscular cephalic sucker. They are parasites of mammals and birds.

### FAMILY DIOCTOPHYMATIDAE

The family consists of the genera *Dioctophyma*, *Eustrongylides*, and *Hystrichis*. *Dioctophyma* occurs in the kidneys and abdominal cavity of mammals. The other two are in the proventriculus of aquatic birds.

### *Dioctophyma renale*
(Goeze, 1782) (Plate 140)

This is the giant kidney worm found in the kidneys and peritoneal cavity of mammals, particularly mink and dogs. Other common hosts include coyotes, wolves, foxes, jackals, raccoons, weasels, otters, martens, and polecats. Humans, horses, and pigs are occasionally infected. Nearly 8 percent of 388 mink from one locality in Michigan were infected. The parasite is cosmopolitan. It is fairly common in North America.

**Plate 140** *Dioctophyma renale* 515

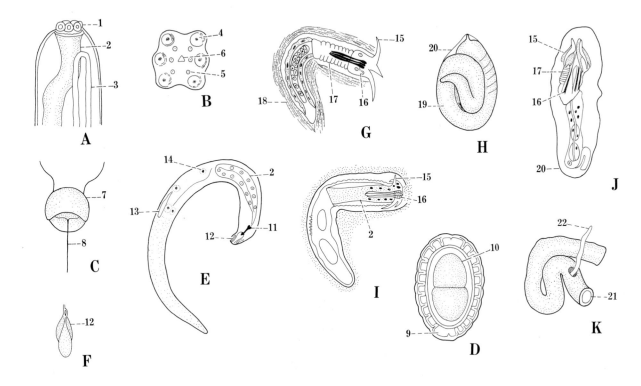

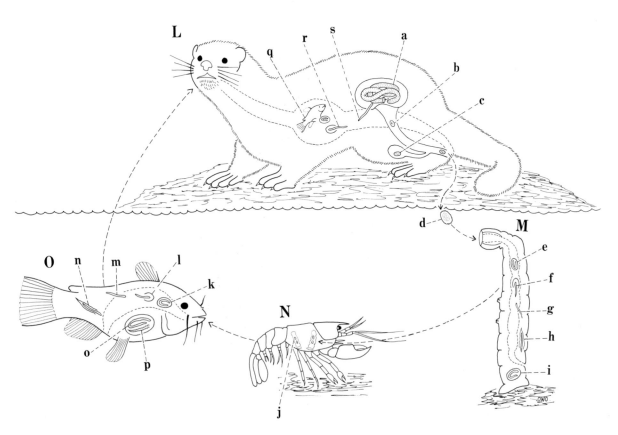

**Description.**    Living worms are red in color. Males measure 140 to 200 mm long by 4 to 6 mm in diameter. The copulatory bursa is a fleshy bell-shaped structure without supporting rays, as occur in the Strongylina. There is a single bristle-like spicule. Females attain a length of 200 to 1000 mm and a diameter of 5 to 12 mm. The vulva is located in the anterior part of the body. Eggs are in the two-cell stage when laid; the shell is thick, deeply pitted, and lighter in color at each end; the average size is 74.3 by 46.7 $\mu$.

**Life Cycle.**    Eggs laid in the kidneys pass to the urinary bladder and are voided with the urine. Motile larvae appear after 21 days of incubation at room temperature, but hatching does not occur until they are ingested by branchiobdellid annelids (*Cambarincola philadelphica*), a genus of worldwide distribution. Just when the eggs become infective to the annelids is not known but they are still infective after being kept in glass dishes at room temperature for nearly one year.

Branchiobdellids feeding on the bottom of ponds frequented by infected mink and other hosts ingest the infective eggs. Hatching occurs in the intestine in 2 days for eggs incubated 9 months and in 7 days for those incubated 7 months. The first stage larvae quickly migrate into the celom. The second stage larvae soon develop six long pointed papillae on the head. Encystment takes place after the larvae cease their migration in the celom. The cyst is formed from secretions extruded from the head. No noticeable development takes place in the cyst.

The annelids commonly attach to the gills of crayfish after a period of living free in the water. When crayfish bearing annelids on their gills are eaten by bullheads, cysts freed from the tissues by the action of the digestive juices rupture, releasing the second stage larvae. They migrate through the intestinal wall of the fish into the celom where the second molt takes place. The third stage larvae attach to the mesenteries and encyst. They continue to grow, eventually molting to the fourth stage, as evidenced by the presence of the cast cuticle in the cyst. If they remain long periods in the cyst, development continues with the formation of preadults.

Infection of the vertebrate host occurs when bullheads containing fourth stage larvae or preadults are eaten. The cysts are freed from the fish in the stomach and the larvae from the cysts in either the stomach or duodenum. Migration from the intestine is through the wall of the duodenum. In mink, this portion of the intestine is in contact with the right kidney, so the parasite continues its progress directly into that renal organ. In dogs, the kidney is separated from the duodenum a short distance which probably accounts for the high percentage of worms being in the peritoneal cavity of these hosts. Sexual maturity is attained in about 3 months or more after the worms reach the kidneys. Evidence indicates a longevity of 1 to 3 years, after which the worms die and degenerate and the remaining shell of the infected kidney shrivels. A total of about 2 years is required to complete the cycle.

Karmanova's studies in Russia showed that eggs eaten by oligochaetes hatch and the larvae enter the vascular system where they undergo three molts, becoming infective to the final host. If the larvae pass from an oligochaete to a fish, the latter serves as a reservoir only. According to these studies, fish do not appear to be a biological necessity for completion of the life cycle of *D. renale*. However, they do serve an important ecological function in bringing the parasite more directly into the food chain of the final host.

It is Karmanova's opinion that Woodhead was dealing with the larvae of one of the Gordiacea instead of *Dioctophyma*.

The life cycle of *Hystrichis tricolor* of the proventricular glands of ducks involves only earthworms as intermediate hosts. The first stage larva develops in the egg as in *D. renale*. Hatching takes place in the intestine of oligocheates and the larvae develop to the fourth stage in the blood vessels. When the earthworms are eaten by ducks, the larvae bore into the proventricular wall, molt the fourth time, and develop to maturity in a month. In both *Dioctophyme renale* and *Hystrichis tricolor*, annelids serve as intermediate hosts.

## EXERCISE ON LIFE CYCLE

Since this life cycle does not lend itself to a general study, no instructions are given for it. These are available in the papers by Woodhead and Hallberg for *D. renale* and Karmanova for *H. tricolor*.

## SELECTED REFERENCES

Crichton, V. J., and R. E. Urban. (1970). *Dioctophyme renale* (Goeze 1782) (Nematoda: Dioctophymata) in Manitoba mink. Can. J. Zool. 48:591–592.

Halberg, C. W. (1953). *Dioctophyma renale* (Goeze, 1782) a study of the migration routes to the kidneys of mammals and resultant pathology. Trans. Amer. Micros. Soc. 72:351–363.

Karmanova, E. M. (1956). [An interpretation of the biological cycle of the nematode *Hystrichis tricolor* Dujardin, 1845, a parasite of domestic and wild ducks.] Dokl. Akad. Nauk SSSR 111:83–86.

Karmanova, E. M. (1959). [Biology of *Hystrichis tricolor* Dujardin, 1845 and some information on the epizootiology of hystrichosis in ducks.] Tr. Gel'mintol. Lab. Akad. Nauk SSSR 9:113–125.

Karmanova, E. M. (1959). [On elucidation of the developmental cycle of the nematode *Dioctophyme renale* (Goeze, 1782)—parasite of kidney of carnivores and humans.] Dokl. Akad. Nauk SSSR Transl. Biol. Sci. Sec. 132:456–457.

Karmanova, E. M. (1962). [Development of the nematode *Dioctophyme renale* (Goeze, 1782) in its intermediate and definitive hosts.] Tr. Gel'mintol. Lab. 12:27–36.

Karmanova, E. M. (1963). [Interpretation of the development cycle in *D. renale.*] Med. Parazitol. Parazit. Bolez. 32:331–334.

McLeod, J. A. (1967). *Dioctophyma* [*Dioctophyme*] *renale* infections in Manitoba. Can. J. Zool. 45:505–508.

Woodhead, A. E. (1950). Life history cycle of the giant kidney worm, *Dioctophyme renale* (Nematoda), of man and many other mammals. Trans. Amer. Microsc. Soc. 69:21–46.

**Plate 141**   *Summary of Direct Life Cycles of Some Nematodes*                518

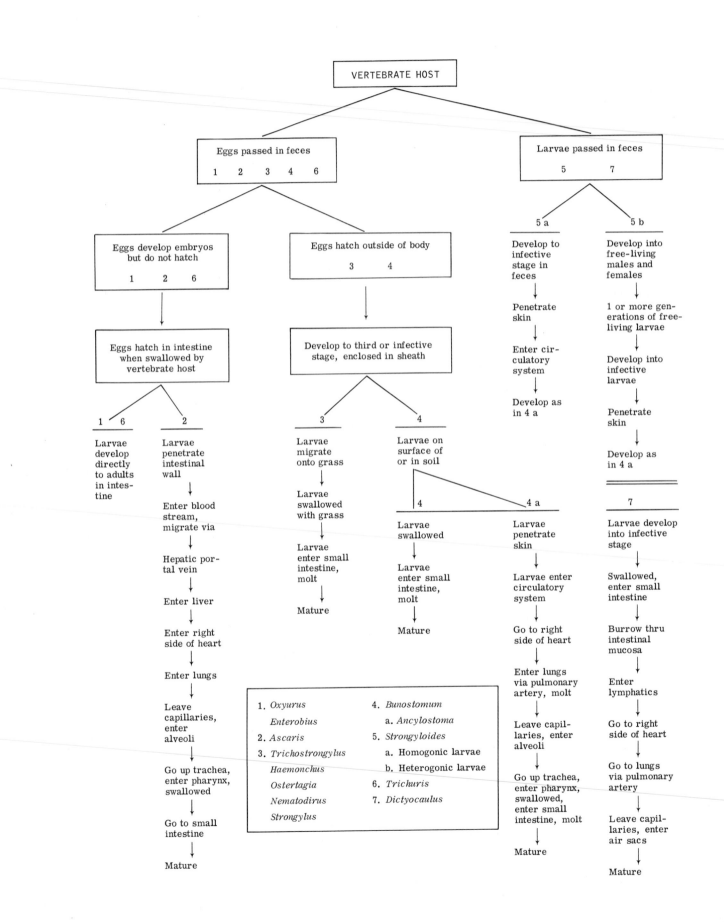

**Plate 142** *Summary of Indirect Life Cycles of Some Nematodes* 519

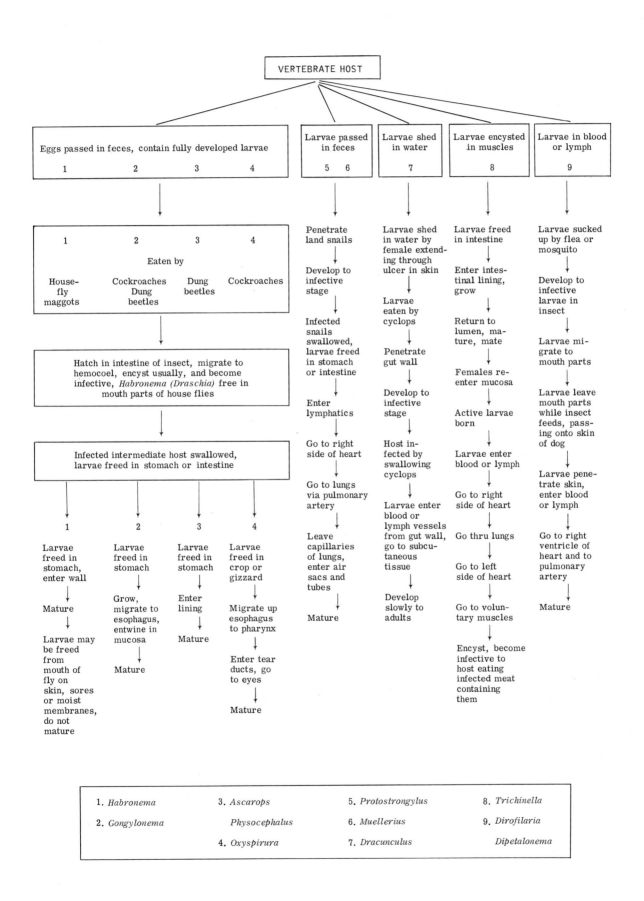

# Index